Lecture Notes in Electrical Engineering

Volume 332

About this Series

"Lecture Notes in Electrical Engineering (LNEE)" is a book series which reports the latest research and developments in Electrical Engineering, namely:

- Communication, Networks, and Information Theory
- Computer Engineering
- Signal, Image, Speech and Information Processing
- Circuits and Systems
- Bioengineering

LNEE publishes authored monographs and contributed volumes which present cutting edge research information as well as new perspectives on classical fields, while maintaining Springer's high standards of academic excellence. Also considered for publication are lecture materials, proceedings, and other related materials of exceptionally high quality and interest. The subject matter should be original and timely, reporting the latest research and developments in all areas of electrical engineering.

The audience for the books in LNEE consists of advanced level students, researchers, and industry professionals working at the forefront of their fields. Much like Springer's other Lecture Notes series, LNEE will be distributed through Springer's print and electronic publishing channels.

More information about this series at http://www.springer.com/series/7818

Tong-Cun Zhang · Motowo Nakajima
Editors

Advances in Applied Biotechnology

Proceedings of the 2nd International Conference on Applied Biotechnology (ICAB 2014)-Volume I

Springer

Editors
Tong-Cun Zhang
Tianjin University of Science and Technology
Tianjin
China

Motowo Nakajima
SBI ALApromo
Tokyo
Japan

ISSN 1876-1100 ISSN 1876-1119 (electronic)
Lecture Notes in Electrical Engineering
ISBN 978-3-662-45656-9 ISBN 978-3-662-45657-6 (eBook)
DOI 10.1007/978-3-662-45657-6

Library of Congress Control Number: 2014955624

Springer Heidelberg New York Dordrecht London

Printed on acid-free paper

Springer-Verlag GmbH Berlin Heidelberg is part of Springer Science+Business Media
(www.springer.com)

Preface

The 2014 International Conference on Applied Biotechnology (ICAB 2014), organized by Chinese Society of Biotechnology and Tianjin University of Science, was held from November 28 to 30, 2014 in Tianjin, China.

The conference served as a forum for exchange and dissemination of ideas and the latest findings in aspects of applied biotechnology. The conference was complemented by talks given by more than 30 professors and researchers.

The conference papers were submitted by more than 100 authors from different universities, institutes and companies. Numerous fields were covered, ranging from fermentation engineering, cell engineering, genetic engineering, enzyme engineering to protein engineering.

Special thanks are given to Secretary Staff of the conference for the commitment to the conference organization. We would also like to thank all the authors who contributed with their papers to the success of the conference.

This book gathers a selection of the papers presented at the conference. It contains contributions from both academic and industrial researchers focusing on the research and development of applied biotechnology from all over the world. The scientific value of the papers also helps researchers in this field to get more valuable results.

Tianjin, China Tong-Cun Zhang
Tokyo, Japan Motowo Nakajima

Preface

[illegible]

Contents

Part I
Microbial Genetics and Breeding

Chapter 1
Cloning and Bioinformatics Analysis of *spsC* Gene from *Sphingomonas sanxanigenens* NX02

Xiaoyan Li, Haidong Huang, Mingming Zhou and Peng Zhang

Abstract *Sphingomonas sanxanigenens* strain NX02 synthesizes a novel sphingan Ss, which can be used as drilling mud and thickening agent in the recovery of petroleum by water flooding. In order to research genes involved in the biosyntheses of sphingan Ss, strain NX02 was induced by transposon mini-Tn5 on suicide plasmid pUT, and a mutant strain T163, which cannot produce sphingan Ss, was screened. The *spsC* gene of NX02 was obtained by the method of Tn5 flanking PCR and LP-RAPD. The predicted amino acid sequence of the *spsC* protein possessed 493 amino acids and a calculated molecular mass of 53.5 kDa. Bioinformatics analysis revealed that *spsC* had the highest 60 % amino acid sequence identity with polysaccharide biosynthesis protein of *Novosphingobium lindaniclasticum* LE124. *spsC* protein had typical polysaccharide polymerases family transmembrane helices, located between amino acids Y13-V44 and P411-L436. The N-terminal sequence of *spsC* had high identity to chain length determinant protein of Wzz superfamily.

Keywords *Sphingomonas sanxanigenens* · Polysaccharide · Sphingan Ss · Bioinformatics

X. Li
College of Food Science and Bioengineering, Tianjin Agricultural University, Tianjin 300384, China
e-mail: lxy@tjau.edu.cn

H. Huang (✉) · M. Zhou · P. Zhang
College of Agronomy and Environmental Resources, Tianjin Agricultural University, Tianjin 300384, China
e-mail: hdhuang@tjau.edu.cn

T.-C. Zhang and M. Nakajima (eds.), *Advances in Applied Biotechnology*, Lecture Notes in Electrical Engineering 332, DOI 10.1007/978-3-662-45657-6_1

1.1 Introduction

A number of bacteria of the genus *Sphingomonas* produce polysaccharides called sphingans, including gellan, welan, S-88, rhamsan, and diutan [1–3]. Sphingans share the similar tetrasaccharide backbone structures and divergent side chains. Because of their excellent rheological characteristics, sphingans have been utilized for a wide range of biotechnological applications in the food, oilfield, and pharmaceutical industries [4–7]. In recent years, with the continuous exploration of microbial resources, some new sphingan-secreting strains have been isolated from diverse environments [8]. *Sphingomonas sanxanigenens* strain NX02 is a new species of the genus *Sphingomonas sensu stricto* that was isolated from soil [9]. Strain NX02 synthesizes a novel sphingan called sphingan Ss, with a linear tetrasaccharide repeat unit consisting of glucose, glucuronic acid, rhamnose, and mannose [10]. Although sphingan Ss has been used in the field of oil exploitation, its mechanism of synthesis is still unknown.

The complete biosynthetic pathway of gellan, S-88, and diutan are presently known. It is a multistep process that can be divided into three sequential steps: intracellular synthesis of the nucleotide-sugar precursors, assembly of the tetrasaccharide repeat units linked to the inner membrane, and translocation of the repeat units to the periplasmic space followed by their polymerization and export through the outer membrane [11–13]. Polymerase, encoded by the *spsC* gene, catalyzes the tetrasaccharide repeat units to polysaccharide. The *spsC* protein involves in sphingan polysaccharide chain length determination [14, 15].

In this paper, a mini-Tn5 transposon mutant strain of *S. sanxanigenens* NX02, which cannot produce sphingan Ss, was screened and isolated. The complete ORF sequence of *spsC* gene was obtained by TAIL PCR. The phylogenetic relation and protein characteristic was analyzed with bioinformatics method.

1.2 Materials and Methods

1.2.1 Bacterial Strains, Plasmids, and Growth Conditions

Escherichia coli strains DH5a (Transgen, Beijing, China) were used as host cells for gene cloning. *E. coli* strains S17-1(mini-Tn5) were used as donor strains for transposon mutagenesis. *S. sanxanigenens* strain NX02 was cultured on NK medium (15 g glucose l^{-1}, 5 g peptone l^{-1}, 3 g beef powder l^{-1}, 1 g yeast extract l^{-1}, and 15 g agar l^{-1}, pH 7.0) at 30 °C. The fermentation medium contained the following: 45 g glucose l^{-1}, 2.5 g $NaNO_3$ l^{-1}, 0.2 g yeast extract l^{-1}, 1.2 g K_2HPO_4 l^{-1}, 1 g $CaCO_3$ l^{-1}, 0.005 g $FeSO_4$ l^{-1}, 0.4 g NaCl l^{-1}, and 1 g $MgSO_4$ l^{-1}, pH 7.5. pEASY-Blunt (Transgen) was employed as gene cloning. When required, the culture medium was supplemented with 100 mg ampicillin l^{-1}, 30 mg chloramphenicol l^{-1}, or 10 mg kanamycin l^{-1}. Peptone, beef powder, yeast extract, agar, and other chemicals were purchased from Dingguo Limited (Tianjin, China).

1.2.2 Transposon Mutagenesis

Suicide plasmid with transposon mini-Tn5 was transferred from donor strain *E. coli* S17-1 into recipient strain *S. sanxanigenens* NX02 by mobilization with a filter mating technique [16]. *E. coli* S17-1(mini-Tn5) was incubated for 12 h at 37 °C with 10 mg kanamycin l^{-1}, and *S. sanxanigenens* NX02 was incubated for 24 h at 30 °C with 30 mg chloramphenicol l^{-1}. Filters with the mixture of donor and recipient strains in a 1:4 ratio were incubated for 8 h at 30 °C on the surface of NK medium plates. Cells were then suspended in 10 mM $MgSO_4$, and the appropriate dilutions were plated on selective medium with kanamycin and chloramphenicol. The mini-Tn5 transposon mutant strains were screened by the viscous phenotype of colony.

1.2.3 DNA Techniques

Standard procedures, including DNA isolation, restriction enzyme digestion, agarose gel electrophoresis, DNA ligation, transformation of *E. coli*, and *S. sanxanigenens*, were performed using conventional methods [17]. Genomic DNA was extracted by LiCl precipitation [18]. Plasmid DNA was purified from *E. coli* by the alkaline lysis procedure or using the AxyprepTM Plasmid Miniprep Kit [19].

1.2.4 Cloning and Sequence Determination of *spsC* Gene

The flanking sequences of mini-Tn5 transposon insertion site was obtained by the method of Tn5 external direction PCR amplification and long primer RAPD, using the following primers: Wt1 (5'- CAATAGCGTTATCAACCCGCT-3'), Wt2 (5'-CCAAACGTTGACACCCAGTT-3'), Ric1(5'-ATGTAAGCTCCTGGGGATT-CAC-3'), Ric2(5'-AAGTAAGTGACTGGGGTGAGCG-3'), Box(5'-CTACGGCA-AGGCGACGCTGACG-3'), Rep1(5'-IIIICGICGICATCIGGC-3'), Rep2 (5'-ICGIC TTATCIGGCCTAC-3'). The PCR product was sequenced, and analysis of the deduced amino acid sequence confirmed that it contained an incomplete open reading frame (ORF) and that the deduced amino acid sequence was homologous to GelC protein sequences in data banks. The complete ORF sequence of *spsC* was obtained by thermal asymmetric interlaced (TAIL) PCR.

1.2.5 Sequence Alignment and Bioinformatics Analysis

Sequence similarity searches were performed using BLAST 2.0 [20] at NCBI. Alignments to determine protein and DNA similarities were performed using the CLUSTAL method [21] and a phylogenetic tree was constructed using MEGA 4.0

[36] with the neighbor-joining method [22]. Sequence data were analyzed with DNAMAN 5.0 (Lynnon Biosoft, Quebec, Canada). The physicochemical and hydrophobic properties of protein *spsC* were obtained with program ProtParam and ProtScale, respectively. The protein secondary structure prediction was analyzed with program PSIPRED [23].

1.3 Results and Discussion

1.3.1 Tn5-Induced Sphingan Ss-Deficient Mutants of S. sanxanigenens NX02

A library of random mini-Tn5 insertions was constructed in *S. sanxanigenens* NX02 as described in the experimental section. Colonies were individually screened for sphingan Ss deficient at NK medium plate with chloramphenicol and kanamycin. The morphological character of wide-type strain NX02 was convex and viscous (Fig. 1.1a). The flat and tenuous colony of mutant T163 indicated that sphingans Ss was not secreted from the mutant (Fig. 1.1b). This result was then confirmed by shake flask fermentation experiment. The result of PCR showed that the phenotypic change of mutant T163 was caused by mini-Tn5 insertion.

1.3.2 Cloning of spsC Complete ORF Sequence

The flanking sequences of mini-Tn5 insertion site were amplified by PCR. As shown in Fig. 1.2, two electrophoretic bands of about 1,200 and 1,100 bp were obtained (lane 1–2). With DNA sequencing and TAIL PCR, the complete ORF

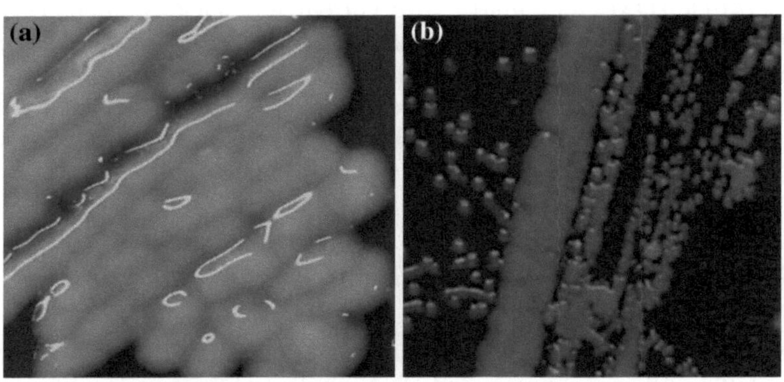

Fig. 1.1 Colony characteristics of strain NX02 and transposon mutant T163

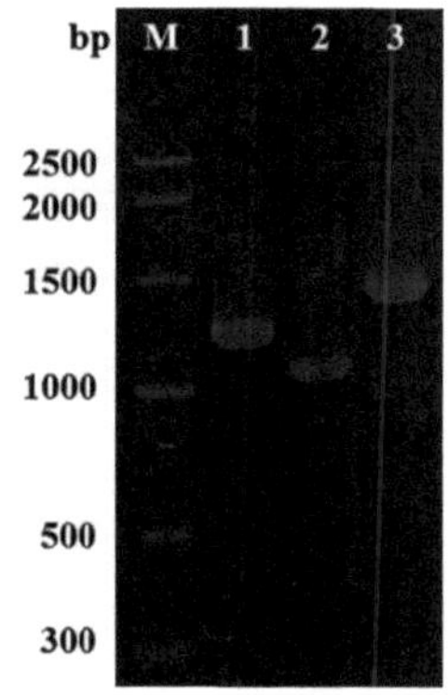

Fig. 1.2 PCR result of *spsC* gene

sequence of *spsC* was obtained as shown in lane 3. The nucleotide sequence of *spsC* gene has been deposited in the GenBank database under the accession number AGQ04616.

Nucleotide sequencing of *spsC* in *S. sanxanigenens* revealed a unique 1,482-nt ORF, starting with a putative ATG start codon. Preceding the start codon (8 nt upstream), a putative ribosome-binding site (RBS) (5'-GGGGA-3') was identified by taking into consideration previous descriptions of RBSs from *S. paucimobilis* ATCC31461 [14]. However, typical −10 and −35 regions were not identified upstream of the predicted Shine-Dalgarno (SD) sequence. The *spsC* gene has a high G +C content (68 %) and a high frequency of G or C in the third position (87 %), which is characteristic of *Sphingomonas* genes [24] and consistent with that of *S. sanxanigenens* [25].

1.3.3 Phylogenetic Analysis of spsC Amino Acid Sequence

The putative amino acid sequence encoded by the *spsC* was compared with data deposited in the GenBank database. The following high levels of identity with other proteins from a variety of organisms were detected: 60 % identity with polysaccharide biosynthesis protein of *Novosphingobium lindaniclasticum* LE124 (EQB15321) and 58 % identity with *Sphingomonas* sp. LH128 (EJU14361), followed by 55 % identity with *Novosphingobium* sp. AP12 (EJL23329), 52 and 51 % identity with protein from *Sphingomonas* sp. PR090111-T3T-6A (WP_019832308) and *Sphingobium* sp. YL23 (WP_022681617). Construction of a phylogenetic tree for the *spsC* proteins (Fig. 1.3) revealed two obviously divergent phylogenetic groups of prokaryotes. *spsC* of *S. sanxanigenens* was in the group including protein of *N. lindaniclasticumand* LE124 and *Sphingomonas* sp. LH128, but further apart from the group including protein of *Sphingopyxis baekryungensis* and *Sphingomonas wittichii* RW1. Homologous analysis showed that the most sequence of *spsC*

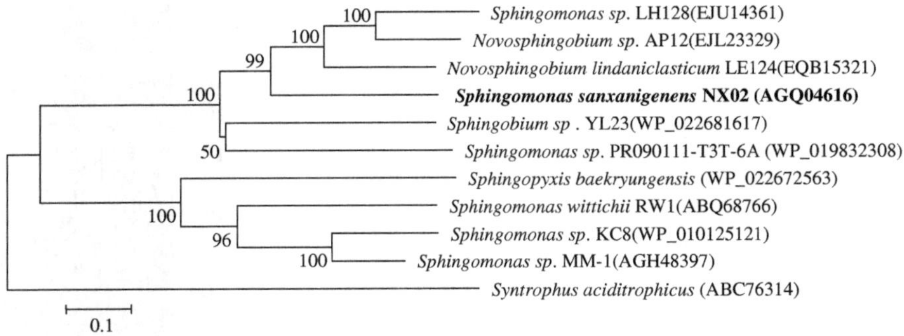

Fig. 1.3 Phylognetic tree of *spsC* amino acid sequence

had high identity with GumC protein which involved in exopolysaccharide Xanthan biosynthesis, and the N-terminal sequence of *spsC* had high identity to chain length determinant protein of Wzz superfamily.

1.3.4 Properties of Protein spsC from S. sanxanigenens

The protein *spsC* gene is composed of 493 amino acids, with a calculated molecular mass of 53.51 kDa and a predicted isoelectric point (PI) of 9.42. Analysis of the amino acid composition of *spsC* revealed a composition of 54 % polar residues and 46 % hydrophobic residues. The amount of basic and acidic residues was 66 and 55. The result of hydrophobic analysis showed that the aliphatic index of *spsC* was 0.93, and the instability index was computed to be 52.49 (Fig. 1.4). The analysis of PSIPRED showed that *spsC* have two transmembrane domains flanking a central

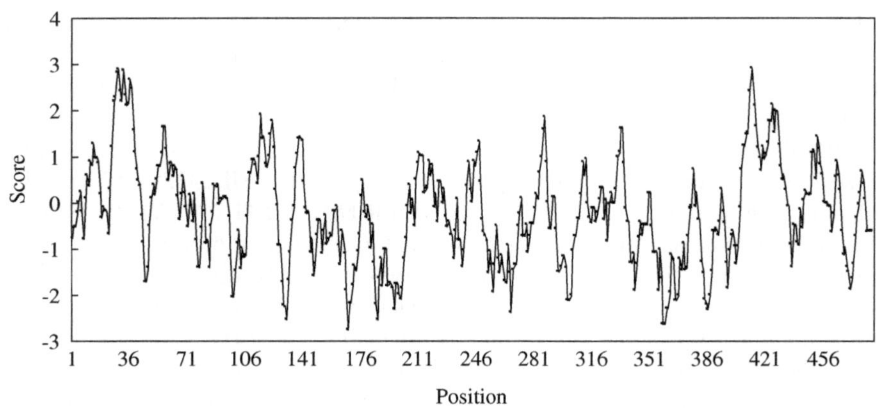

Fig. 1.4 Hydrophobicity and hydrophilicity analysis of *spsC*

extracellular segment. The determination of length distribution of the polysaccharide chains is controlled by a family of proteins termed polysaccharide polymerases (PCP). PCP enzymes involved in extracellular polysaccharides synthesis systems in Gram-negative bacteria have, in addition to the membrane/periplasmic domain, a cytoplasmic domain of protein tyrosine kinases, and the prototype of this family is Wzc from *Escherichia coli* [26]. The PCP enzyme in *S. sanxanigenens* NX02 is encoded by the gene *spsC*. The hydrophobic plot for *spsC* suggested the presence of two putative transmembrane α-helices, located between amino acids Y13-V44 and P411-L436, respectively. The protein of *spsC* shows the typical PCP N- and C-terminal transmembrane helices separated by a segment with a predicted coil region located in the periplasm.

1.4 Conclusion

Sphingomonas sanxanigenens NX02 is a new species of the genus *Sphingomonas* and has low homology with other known sphingan producing strains. The complete ORF sequence of sphingan gene cannot be obtained by standard PCR with degenerate primers. Screening deficient mutants is the necessary way to obtain the gene information about sphingan Ss. In this study, the complete ORF sequence of *spsC* gene from *S. sanxanigenens* was cloned and characterized for the first time. Bioinformatics analysis showed that the sequence of *spsC* had high identity with GumC protein and chain length determinant protein of Wzz superfamily. The protein of *spsC* showed the typical polysaccharide polymerases family transmembrane helices and periplasm coil region. This work should prove useful for further research into sphingan Ss synthesis pathways and genetic engineering with a view to control sphingan Ss production.

Acknowledgments The authors are grateful to Tianjin Research Program of Application Foundation and Advanced Technology (11JCZDJC16600) for the financial support for this work.

References

1. Pollock TJ (1993) Gellan-related polysaccharides and the genus *Sphingomonas*. J Gen Microbiol 139:1939–1945
2. Sá-Correia I, Fialho AM, Videira P, Moreira LM, Marques AR, Albano H (2002) Gellan gum biosynthesis in *Sphingomonas paucimobilis* ATCC 31461: genes, enzymes and exopolysaccharide production engineering. J Ind Microbiol Biotechnol 29:170–176
3. Yabuuchi E, Yano I, Oyaizu H, Hashimoto Y, Ezaki T, Yamamoto H (1990) Proposals of *Sphingomonas paucimobilis* gen. nov. and comb. nov., *Sphingomonas parapaucimobilis* sp. nov., *Sphingomonas yanoikuyae* sp. nov., *Sphingomonas adhaesiva* sp. nov., *Sphingomonas capsulata* comb. nov., and two genospecies of the genus *Sphingomonas*. Microbiol Immunol 34:99–119
4. Prajapati VD, Jani GK, Zala BS, Khutliwala TA (2013) An insight into the emerging exopolysaccharide gellan gum as a novel polymer. Carbohydr Polym 93:670–678

5. Smith AM, Shelton RM, Perrie Y, Harris JJ (2007) An initial evaluation of gellan gum as a material for tissue engineering applications. J Biomater Appl 22:241–254
6. Banik RM, Kanari B, Upadhyay S (2000) Exopolysaccharide of the gellan family: prospects and potential. World J Microbiol Biotechnol 16:407–414
7. Ishwar BB, Shrikant AS, Parag SS, Rekha SS (2007) Gellan gum: fermentative production, downstream processing and applications. Food Technol Biotechnol 45:341–354
8. Seo EJ, Yoo SH, Oh KW, Cha J, Lee HG, Park CS (2004) Isolation of an exopolysaccharide-producing bacterium, *Sphingomonas* sp. CS101, which forms an unusual type of sphingan. Biosci Biotechnol Biochem 68:1146–1148
9. Huang HD, Wang W, Ma T, Li GQ, Liang FL, Liu RL (2009) *Sphingomonas sanxanigenens* sp. nov., isolated from soil. Int J Syst Evol Microbiol 59:719–723
10. Huang HD, Wang W, Ma T, Li ZY, Liang FL, Liu RL (2009) Analysis of molecular compositioni and properties of a novel biopolymer. Chem J Chin Univ 30:324–327
11. Yamazaki M, Thorne L, Mikolajczak M, Armentrout RW, Pollock TJ (1996) Linkage of genes essential for synthesis of a polysaccharide capsule in *Sphingomonas* strain S88. J Bacteriol 178:2676–2687
12. Coleman RJ, Patel YN, Harding NE (2008) Identification and organization of genes for diutan polysaccharide synthesis from *Sphingomonas* sp. ATCC 53159. J Ind Microbiol Biotechnol 35:263–274
13. Li H, Xu H, Xu H, Li S, Ouyang PK (2010) Biosynthetic pathway of sugar nucleotides essential for welan gum production in *Alcaligenes* sp. CGMCC2428. Appl Microbiol Biotechnol 86:295–303
14. Fialho AM, Moreira LM, Granja AT, Popescu AO, Hoffmann K, Sá-Correia I (2008) Occurrence, production, and applications of gellan: current state and perspectives. Appl Microbiol Biotechnol 79:889–900
15. Moreira LM, Hoffmann K, Albano H, Becker A, Niehaus K, Sá-Correia I (2004) The gellan gum biosynthetic genes gelC and gelE encode two separate polypeptides homologous to the activator and the kinase domains of *tyrosine autokinases*. J Mol Microbiol Biotechnol 8:43–57
16. Rather PN, Ding X, Lancey RB, Siddiqui S (1999) Providencia stuartii genes activated by cell-to-cell signaling and identification of a gene required for production or activity of an extracellular factor. J Bacteriol 181:7185–7191
17. Sambrook J, Fritsch EF, Maniatis T (2001) Molecular cloning: a laboratory manual, 3rd edn. Cold Spring Harbor, New York
18. Cashion P, Holder-Franklin MA, McCully J, Franklin M (1977) A rapid method for the base ratio determination of bacterial DNA. Anal Biochen 81:461–466
19. Birnboim HC, Doly J (1979) A rapid alkaline extraction procedure for screening recombinant plasmid DNA. Nucleic Acids Res 7:1513–1523
20. Altschul SF, Gish W, Miller W, Myers EW, Lipman DJ (1990) Basic local alignment search tool. J Mol Biol 215:403–410
21. Higgins DG, Sharp PM (1988) CLUSTAL: a package for performing multiple sequence alignment on a microcomputer. Gene 73:237–244
22. Saitou N, Nei M (1987) The neighbor-joining method: a new method for reconstructing phylogenetic trees. Mol Biol Evol 4:406–425
23. Guffin LJ, Bryson K, David TJ (2000) The PSIPRED protein structure prediction server. Bioinform 16:404–405
24. Videira PA, Fialho AM, Geremia RA, Breton C, Sá-Correia I (2001) Biochemical characterization of the beta-1,4-glucuronosyltransferase GelK in the gellan gum-producing strain *Sphingomonas paucimobilis* ATCC 31461. Biochem J 358:457–464
25. Huang HD, Li XY, Wu MM, Wang SX, Li GQ, Ma T (2013) Cloning, expression and characterization of a phosphoglucomutase/phosphomannomutase from sphingan-producing *Sphingomonas sanxanigenens*. Biotechnol Lett 35:1265–1270
26. Whitfield C (2006) Biosynthesis and assembly of capsular polysaccharides in *Escherichia coli*. Ann Rev Biochem 75:39–68

Chapter 2
Preliminary Study on Salt Resistance Seedling Trait in Maize by SRAP Molecular Markers

Chunyang Xiang, Jin Du, Peipei Zhang, Gaoyi Cao and Dan Wang

Abstract In this study, different genotypes of maize salt tolerance inbred line and salt sensitive inbred line were used as the parent hybrid combinations to obtain F2 populations. Two salt tolerance extreme types of DNA pools were established, where BSA method was used to select polymorphic SRAP markers. The result showed that 48 pair primers can be amplified and clear and stable bands can be obtained by parental, tolerant, and sensitive gene pools. Six pair primers of M2E1, M2E7, M6E15, M7E7, M11E4, and M14E6 showed polymorphism between two parents and between tolerant and sensitive bulks. The six SRAP molecular markers closely linked to salt tolerance were determined. The best maize SRAP-PCR reaction system was established. This research will accelerate maize marker-assisted selection breeding and lay the foundation for salt-tolerant gene cloning.

Keywords Maize · Salt tolerance · SRAP molecular marker

In China, salt-affected soils are found in large areas [1]. Saline soil is one type of middle and low yield soil. It not only affects the growth of plants, but also limits the yield and quality of crops [2]. Improving and reusing salinized soil have played an important part in increasing agricultural production. Maize is an important food and feed crop in China, however, the salt tolerance of maize is relatively poor. Some researchers have proposed extreme salinity at 0.017 mol/L NaCl [3]. Therefore, cultivation of resistant maize salt has been attracting considerable attention.

Sequence-related amplified polymorphism (SRAP) is a kind of newly developed molecular marker system with advantages of stabilization, simplicity, high co-dominance, moderate throughput ratio, and easily obtainable sequence of selected bands. Especially, it can be amplified by PCR without any sequence information [4]. SRAP markers were mainly studied in domestic maize that revealed the genetic diversity and heterotic grouping of maize germplasm using SRAP markers [5]. Although studies on SRAP markers of salt tolerance in maize have been few for a

C. Xiang (✉) · J. Du · P. Zhang · G. Cao · D. Wang
College of Agriculture and Resource and Environment, Tianjin Agricultural University, Tianjin 300384, China
e-mail: xiang5918@sina.com

T.-C. Zhang and M. Nakajima (eds.), *Advances in Applied Biotechnology*,
Lecture Notes in Electrical Engineering 332, DOI 10.1007/978-3-662-45657-6_2

long time, some researchers have studied it in other plants. Seven SRAP molecular markers closely linked to salt tolerance were obtained in zoysiagrasses [6]. Its research can not only provide theoretical guidance for breeding of salt tolerant maize, but also benefit the selection and cultivation of new varieties of salt-tolerant maize. Salt resistance seedling traits of maize that were preliminarily studied by SRAP molecular markers in this research provide the scientific basis for gene mapping studies of salt tolerance in maize.

2.1 Materials and Methods

2.1.1 Experimental Materials

The F2 segregated population originating from the selfing of F1 hybrids, and F1 from a cross between maize inbred line N1 (A060233, salt tolerant) and M2 (A06148, salt sensitive).

2.1.2 Experimental Methods

2.1.2.1 DNA Extraction and DNA Pools Construction

Total DNA in F2 generations seedlings of maize was extracted with improved CTAB.

The two relative DNA pools (salt-tolerant DNA pool and susceptible DNA pool) which come from the F_2 populations were made according to the method of BSA (bulked segregant analysis).

2.1.2.2 SRAP Analysis

SRAP Primer

SRAP primer was designed with reference to the literature [4, 7, 8], and synthesized by Sangon biotech Shanghai Co. Ltd. as shown in Table 2.1.

PCR Amplification System

The total volume of the reaction system was 20 μL, and concentration of the components was according to Table 2.2.

Table 2.1 Primers used in this study

Code	Forward primers	Code	Reverse primers
Me1	TGAGTCCAAACCGGATA	Em1	GACTGCGTACGAATTAAT
Me2	TGAGTCCAAACCGGAGC	Em2	GACTGCGTACGAATTTGC
Me3	TGAGTCCAAACCGGAAT	Em3	GACTGCGTACGAATTGAC
Me4	TGAGTCCAAACCGGACC	Em4	GACTGCGTACGAATTTGA
Me5	TGAGTCCAAACCGGAAG	Em5	GACTGCGTACGAATTAAC
Me6	TGAGTCCAAACCGGTAA	Em6	GACTGCGTACGAATTGCA
Me7	TGAGTCCAAACCGGTCC	Em7	GACTGCGTACGAATTCAA
Me8	TGAGTCCAAACCGGTGC	Em8	GACTGCGTACGAATTCTG
Me9	TGAGTCCAAACCGGACG	Em9	GACTGCGTACGAATTCGA
Me10	TGAGTCCAAACCGGACT	Em10	GACTGCGTACGAATTCAG
Me11	TGAGTCCAAACCGGAGG	Em11	GACTGCGTACGAATTCCA
Me12	TTCAGGGTGGCCGGATG	Em12	GACACCGTACGAATTTGC
Me13	TGGGGACAACCCGGCTT	Em13	GACACCGTACGAATTGAC
Me14	GTAGCACAAGCCGGAAG	Em14	GACACCGTACGAATTTGA
Me15	CGAATCTTAGCCGGATA	Em15	CGCACGTCCGTAATTAAC

Table 2.2 PCR reaction optimization

Reaction components	Adding quantity (μL)	Final concentration
ddH2O	10.42	–
10 × Buffer	2	1.5 mM
dNTP	0.4	0.2 mM each
Taq E	0.18	0.04 U/μL
Forward and reverse primers (1 μM)	5	0.25 μM each
DNA(20 ng/μL)	2	3 ng/μL
Total	20	

2.1.2.3 Screening of SRAP Primers

Salt tolerance and salt sensitive DNA pool were used in this study, which was developed from a cross between N1 and M2. The material of this study included F2 generation of salt tolerance and salt sensitive, and their parental lines. A total of 225 pairs of SRAP primers were composed of 15 forward primers and 15 reverse primers, which were used to amplify the mapping population. The amplified products were checked by polyacrylamide gel electrophoresis (PAGE) electrophoresis.

2.1.2.4 SRAP Primers Authentication

SRAP primers were verified using 20 individuals in salt tolerance and sensitive DNA pools. Identification selection criteria of primers with salinity-tolerance is that differential DNA bands in initial screening primers was of good consistency with 10 individuals of salt tolerance or salt sensitive.

2.2 Results and Analysis

2.2.1 Quality Analysis of Total DNA

The genomic DNA samples of maize were isolated by improved CTAB method, which based on optimizing extraction, deposition, and dissociation. The purity and quantity of DNA was evaluated by agarose gel electrophoresis and SRAP analysis. The results are as shown in Figs. 2.1 and 2.2. The concentration range of genomic DNA was among 20–23 ng/μL. Figure 2.1 shows that DNA bands identified by SRAP-PCR were all clear, and the brightness of bands was the same during parent N1, parent M2, salt tolerant DNA pool, and susceptible DNA pool. Figure 2.2 shows that brightness of DNA bands was the same in 20 individual plants that identify SRAP markers lined in the salt tolerance gene.

2.2.2 Preliminary Screening of SRAP Markers

225 SRAP primer combinations were screened by patent parent N1 and parent M2 from salt-tolerant gene pool and susceptible gene pool. 48 pair primers could be amplified and clear and stable bands were obtained. There were six pair primers with high polymorphism fragments (Fig. 2.3).

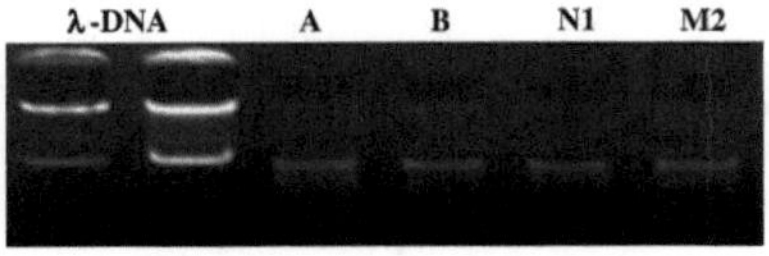

Fig. 2.1 Results of genomic DNA after diluted

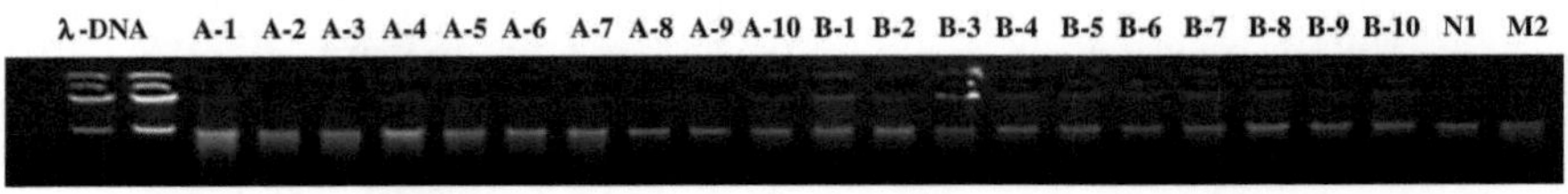

Fig. 2.2 Results of genomic DNA after diluted

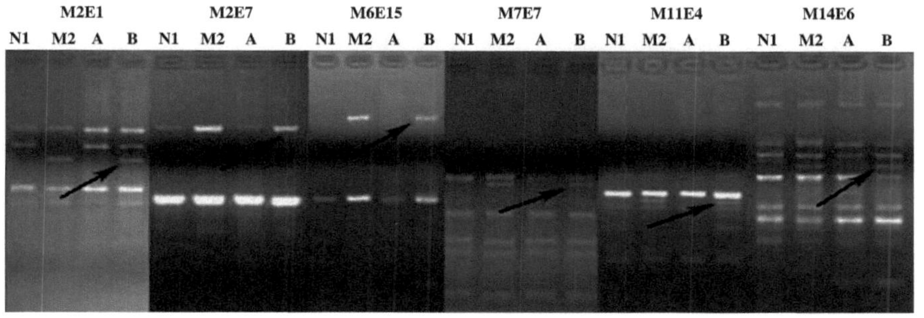

Fig. 2.3 Electrophoresis figure of SRAP primer screening in N1 × M2 populations

2.2.3 Verification of SRAP Markers

The present experiment screened six pair of SRAP primers, which could distinguish the two pools; then, these diversity pairs were used to amplify the individual of F2 plants that used built gene pools. M2E1 primer could amplify five bands clearly and stably. There is a specific band at 700 bp (Fig. 2.4). M2E7 primer could amplify two bands clearly and stably. There is a specific band at 1,500 bp (Fig. 2.5). M6E15 primer could amplify two bands clearly and stably. There is a specific band at

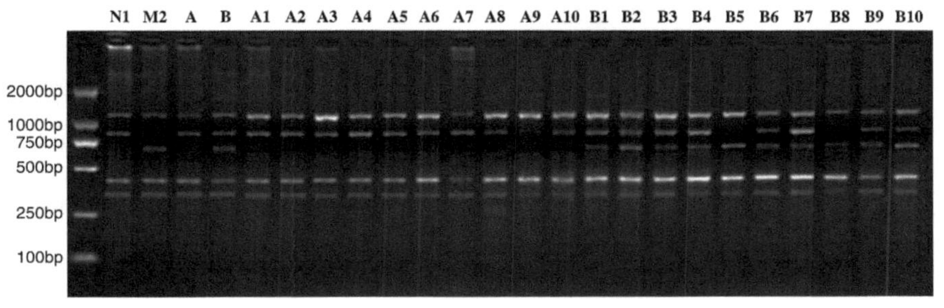

Fig. 2.4 The amplified result of M2E1primer among salt tolerance maize individuals

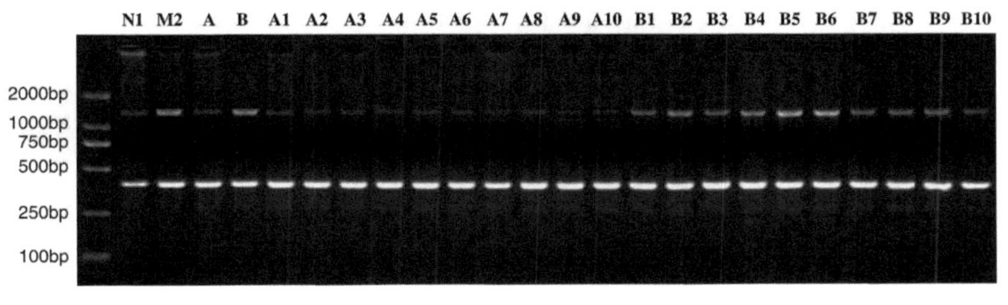

Fig. 2.5 The amplified result of M2E7 primer among salt tolerance maize individuals

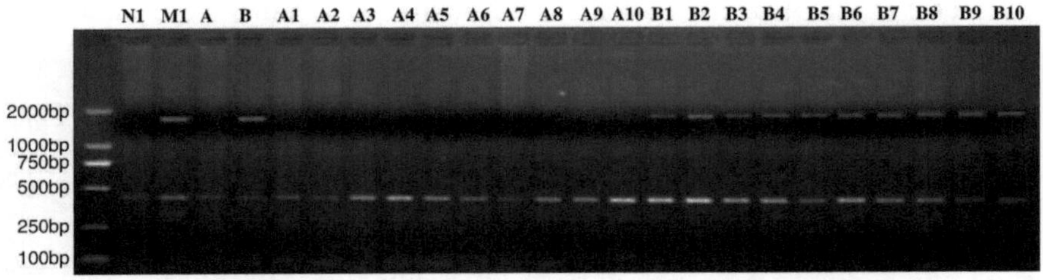

Fig. 2.6 The amplified result of M6E15 primer among salt tolerance maize individuals

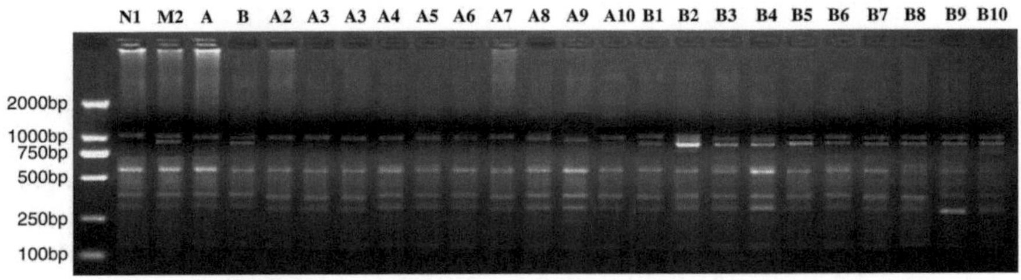

Fig. 2.7 The amplified result of M7E7 primer among salt tolerance maize individuals

1,800 bp (Fig. 2.6). M7E7 primer could amplify five bands clearly and stably. There is a specific band at 950 bp (Fig. 2.7). M11E4 primer could amplify seven bands clearly and stably. There is a specific band at 625 bp (Fig. 2.8). M14E6 primer could amplify eight bands clearly and stably. There is a specific band at 625 bp (Fig. 2.9).

Amplification bands of each pair primers are consistent among salt tolerance maize individuals in parent N1, parent M2, salt-tolerant DNA pool, and susceptible DNA pool. This indicated that six SRAP molecular markers were closely linked to salt tolerance.

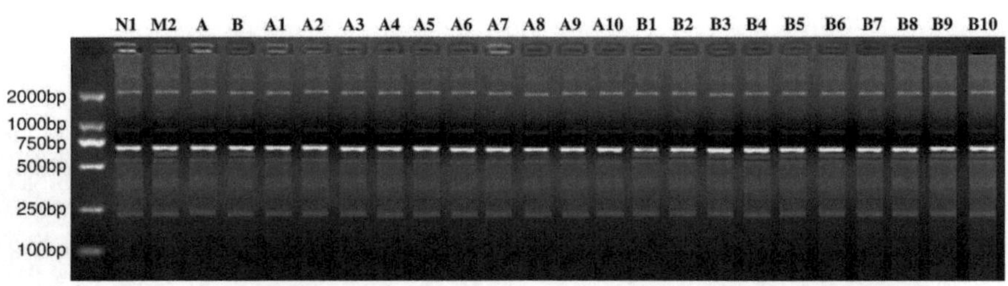

Fig. 2.8 The amplified result of M11E4 primer among salt tolerance maize individuals

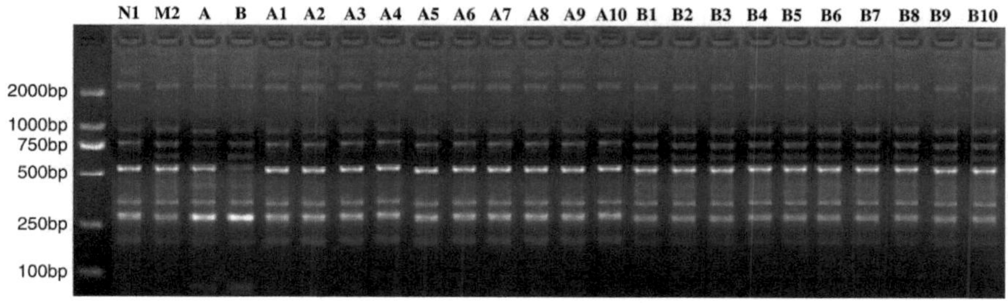

Fig. 2.9 The amplified result of M14E6 primer among salt tolerance maize individuals

2.3 Discussion

Salt-tolerant characteristics of plants that belong to complex quantitative traits are controlled by multiple genes, and are easily affected by the environment. Some researches have shown that salt tolerance can be controlled by the main gene. Exploring and screening SRAP markers associated with main salt tolerance gene have important practical significance for cultivation of salt-tolerant crops. The near isogenic lines of high and low salt tolerance maize were used as experiment materials. DNA was extracted from the seedlings to develop high and low salt tolerance DNA gene pools, respectively. SRAP and BSA were the most efficient strategies for looking for molecular markers related to salt-tolerant genes in maize.

Currently, there are four major DNA-based molecular mapping methods: restriction fragment length polymorphism (RFLP), random amplified polymorphism DNA (RAPD), amplified fragment length polymorphism (AFLP), and simple sequence repeats (SSR). These technologies are being widely applied in the aspect of plants' salt tolerance [9–13], and these methods exist with their own advantages and defects. For example, AFLP marker has high polymorphism, but high false-positive and complex operation. RAPD marker has instability. The SSR marker is of high cost. SRAP, which possesses advantages over other molecular markers, possesses extraordinary characters such as simple design of primers, low cost, stable amplification, and high polymorphism. It is a simple, economic, effective, and reliable molecular marker. In order to screen molecular markers of salt-tolerant genes, 225 pair primers, which were designed, were mined using genomic DNA of maize. The 48 pairs of primers could amplify stable and clear bands. Six pairs of primers could amplify polymorphic PCR products among parent N1, parent M2, salt tolerant DNA pool, and susceptible DNA pool. And the six primers produced clear, stable, and reproducible polymorphic patterns. Therefore, SRAP could be used for screening marker which is associated with the salt-tolerant gene. However, the little tested population, incomplete genetic population, and less primer may have some influence on the results, and which need expanded population and increased primers for further exploration and study in future experiments.

In this paper, BSA was used to identify SRAP molecular markers linked to salt tolerance in maize using two salt-tolerant, extreme types of DNA pools for the first time. Six SRAP molecular markers closely linked to salt tolerance were obtained. This suggests that the salt tolerance of maize was determined by one incomplete dominant gene. The molecular markers obtained can be applied in salt-tolerant identification of the maize germplasm resources, and markers-assisted selection in salt-tolerant breeding of maize. This research will accelerate maize marker-assisted selection breeding and lay the foundation for salt-tolerant gene cloning.

References

1. Liu W, Wang T, Hseung Y (1978) Delimitation of salt-affected soils for reclamation and utilization in china. Acta Pedol Sin 15(2):101–112
2. Yang J (2008) Development and prospect of the research on salt-affected soils in china. Acta Pedol Sin 45(5):837–845
3. Zhang Y, Bo Y (2008) Advances in study of salt-stress tolerance in maize. J Maize Sci 16 (6):83–85
4. Li G, Quiros CF (2001) Sequence-related amplified polymorphism(SRAP). A new marker system based on a simple PCR reaction: its application to mapping and gene tagging in Brassica. Theor Appl Genet 103:455–461
5. Jiang S, Ma H, Liu J et al (2007) Genetic diversity in maize inbred lines revealed by SRAP marker. Mol Plant Breed 15(3):412–416
6. Chen X, Guo H, Xue D et al (2009) Identification of SRAP molecular markers linked to salt tolerance in Zoysia grasses. Acta Prataculturae Sinica 18(2):66–75
7. Zhang Y, Zhang X, Hua W et al (2010) Analysis of genetic diversity among indigenous landraces from sesame (sesamum indicum L.) core collection in China as revealed by SRAP and SSR markers.Gene Genomics 32:207–215
8. Yves C, Jean C, Annick B et al (2010) SRAP polymorphisms associated with superior freezing tolerance in alfalfa (Medicago sativa spp. Sativa). Theor Appl Genet 12:1611–1619
9. Qian YL, Engelke MC, Foster MJV (2000) Salinity effect on Zoysiagrass cultivars and experimental lines. Crop Sci 40:488–492
10. Alshammary SF, Qian YL, Wallner SJ (2004) Growth response of four turfgrass species to salinity. Agric Water Manag 66:97–111
11. Hideki O, Masumi E (2005) Isolation of cDNA and enzymatic properties of betaine aldehyde dehydrogenase from Zoysia tenuifolia. J Plant Physiol 162:1077–1086
12. Weng YJ, Chen DM (2002) Molecular markers and its clone for salt tolerance gene in wheat. Acta Genetica Sinica 29(4):343-349
13. Yang Q, Han J (2005) Identification of molecular marker to salt tolerance gene in alfalfa. Sci Agric Sin 38(3):606–611

Chapter 3
Isolation of Differentially Expressed Genes from Groundnut Genotypes Differing in Seed Dormancy

Bo Qu, Yue Yi Tang, Xiu Zhen Wang, Qi Wu, Quan Xi Sun, Shu Yan Guan, Chuan Tang Wang and Pi Wu Wang

Abstract In situ sprouting may result in considerable reduction in quantity and quality of groundnut produce. The most feasible and convenient solution is to develop groundnut cultivars with seed dormancy. No molecular marker or gene associated with this important trait of groundnut has been documented. In the present study, 105 unique cDNA fragments were isolated from two groundnut materials with contrasting dormancy phenotypes using GeneFishing technology. Three of the 16 DEGs were related to seed germination as suggested by BLAST2GO analysis. Differential expression of a gene with functions indicated in seed germination [85-3-1 (oleosin kda-like)] was confirmed by qRT-PCR. We postulate that expression of the gene is positively related to intensity of groundnut seed dormancy, and that the gene product protects oil bodies thereby preventing lipid mobilization from providing energy for germination.

Keywords Peanut · Differentially expressed gene · Seed dormancy · Germination

3.1 Introduction

As a rich source of protein and oil for human consumption, the cultivated groundnut, *Arachis hypogaea* L., is an important cash crop widely planted across tropical, subtropical, and temperate regions in the world. In situ sprouting is considered one of the major constraints to groundnut production in China, India, Argentina, Indonesia, Zimbabwe, and Senegal, especially for Spanish-type and

S.Y. Guan · P.W. Wang (✉)
Jilin Agricultural University, Changchun 130118, China
e-mail: peiwuw@163.com

B. Qu · Y.Y. Tang · X.Z. Wang · Q. Wu · Q.X. Sun · C.T. Wang (✉)
Shandong Peanut Research Institute, Qingdao 266100, China
e-mail: chinapeanut@126.com

T.-C. Zhang and M. Nakajima (eds.), *Advances in Applied Biotechnology*,
Lecture Notes in Electrical Engineering 332, DOI 10.1007/978-3-662-45657-6_3

Valencia-type cultivars, which generally lack seed dormancy. Reportedly, yield reduction caused by untimely rainfall during harvest of these varieties varied from 10 to 40 %. Older Virginia-type cultivars generally possess seed dormancy. However, with the intensive exploitation of intersubspecific hybridization in nowadays groundnut breeding programs, some newer large-podded varietal releases, though with shorter duration and higher productivity, are found to have no seed dormancy [1]. In situ sprouting not only causes large yield losses, but also renders the seeds susceptible to *Aspergillus flavus* infection and subsequent aflatoxin production [2]. In addition, groundnut sprouts cannot be used as seed, and their value as oil stock is lowered owing to significant reduction in oil content. Theoretically, sowing dates of groundnut may be adjusted, and cultivars with shorter/longer duration may be adopted to prevent in situ sprouting, but these measures are not economically viable due to difficulty in rearrangement of preceding and succeeding crops. The most feasible and convenient means is to develop groundnut cultivars with seed dormancy [3].

Studies on groundnut seed dormancy have been concentrated on evaluation methodologies, genetics, physiochemical mechanisms, and influential factors. Several groundnut cultivars with seed dormancy have been released in Taiwan and India. However, no molecular marker or gene associated with this important trait of groundnut has been documented.

The aim of the present study was to isolate differentially expressed genes (DEGs) from imbibed seeds of two groundnut materials with contrasting seed dormancy phenotypes. These candidate genes, once confirmed by functional analysis, may provide a basis for the genetic improvement of the groundnut seed dormancy through molecular means.

3.2 Materials and Methods

3.2.1 Groundnut Materials

In the present study, we employed two groundnut genotypes, y-1 and x-166. y-1 is a Virginia-type cultivar with seed dormancy, whereas x-166 is a Spanish-type genotype lacking seed dormancy. In primary evaluation, groundnut was cultivated under polythene mulch, harvested and sundried as described by Wan [4]. Seeds thus obtained were preserved in a refrigerator (4 °C) to retain seed dormancy. In germination test, all the 30 seeds of x-166 germinated after 24 h incubation, but only 4 of the 30 seeds of y-1 germinated even after 14 days incubation. In secondary evaluation, groundnut pods were harvested and dried in shade. As expected, y-1 proved to have seed dormancy (None of the 30 seeds germinated after 14 days of incubation), while x-166 was nondormant (all the 30 seeds germinated on day three of incubation).

3.2.2 Seed Treatments

Sundried groundnut seeds were first immersed in a Carbendazim (Longdengfulian) (Jiangsu Longdeng Chemical Co. Ltd, China): double distilled water (ddw) (*W*/*V* = 1/1,000) solution at 25 °C for 3 h, and then rinsed thoroughly in running water to remove the disinfectant. Afterward, the seeds underwent the subsequent treatments (4 seeds per treatment per genotype, incubation temperature: 25 °C): germination in ddw for 0.5 h (y-1 and x-166), germination in ddw for 1.5 h (y-1 and x-166), germination in ddw with 0.05 % etherel for 0.5 h (y-1), or germination in ddw with 0.05 % etherel for 1.5 h (y-1).

3.2.3 Cloning and Sequencing of DEGs

Total RNAs were extracted from groundnut seeds with different treatments stated above and their integrity and quantity determined as per the procedures previously described by Ding et al. [5]. DEGs from groundnut seeds were identified using GeneFishing™ DEG Premix Kit (Seegene, Korea) based on manufacturer's instructions.

Three micrograms of total RNA and 1 μl of 10 μM dT-ACP1 were added to a PCR tube placed in an ice bath, and then RNase-free water was supplemented to the mixture to reach a total volume of 9.5 μl. After gently mixed, the mixture was incubated at 70 °C for 10 min, cooled on ice for 2 min and spun briefly. Four micrograms of 5 × M-MLV buffer, 0.5 μl of RNase inhibitor (40 U/μl) (Tiangen, Beijing), 2 μl of 10 mM dNTP mix, and 200 U of M-MLV reverse transcriptase (RNase H^-) (TaKaRa, Japan) were then added, and RNase-free water was supplemented to reach a total volume of 20 μl. Reverse transcription was conducted at 42 °C for 90 min, followed by incubation at 94 °C for 2 min, cooling on ice for 2 min and a brief spin.

For use in subsequent Genefishing PCR, first strand cDNA products were diluted with DNase-free water. The PCR mixture (20 μl) was composed of 52.8 ng of first strand cDNA, 0.5 μM arbitrary ACP, 0.5 μM dT-ACP2 and 10 μl of 2 × See Amp ACP Master-mix. Thermal cycling was performed on a PCR machine preheated to 94 °C, and the program was 94 °C for 5 min, 50 °C for 3 min, and 72 °C for 1 min, followed by 40 cycles of 94 °C for 40 s, 65 °C for 40 s, and 72 °C for 40 s, and a final extension of 72 °C for 10 min.

PCR products were separated on a 2 % agarose gel, and amplicons of interest were cloned into a pGM-T vector (Promega, USA) and sequenced by Genscript Inc., Nanjing, China.

Removal of poor quality reads and vector sequences and sequence assembly were performed with DNAStar (DNASTAR Inc., London, UK) package. Transcript

Table 3.1 Sequences of qRT-PCR primers

Primer ID	Forward primer (5′ → 3′)	Reverse primer (5′ → 3′)
p-108-2-1	ATGTATGTTGCTGGGTTATTTG	CTGTGAATGCTGCGACTA
p-103-25	GCCAAGAGATACAGACTAAGG	TGTGAATGCTGCGACTAC
p-85-3-1	GTTGCTTGTATTACCGTTGTT	CTGTGAATGCTGCGACTA
p-108-7	TTGAAGGGATAGAGTGGAAAGT	CTGTGAATGCTGCGACTAC
p-86-6	GTAGTGAGTAGCAGCATT	GATTAGCATCATTCATATCCAT
p-β-actin	TTG GAA TGG GTC AGA AGG ATG C	AGT GGT GCC TCA GTA AGA AGC

annotation and functional assignment were carried out using BLAST2GO (http://blast2go.org) and GenBank nr database.

3.2.4 *qRT-PCR Validation of DEGs*

qRT-PCR was conducted in a PikoREAL (Thermo Scientific, USA) PCR machine. The PCR mixture (20 μl) contained 80 ng of cDNA, 10 μl of SYBR Premix Ex TaqTM (TaKaRa, Japan), and 0.2 μ mol l-1 of each primer (Table 3.1). Thermal cycling profile and the procedure for melting curve analysis were the same as depicted by Tang et al. [6]. Reactions were performed in triplicates and resultant data averaged. Fold changes in RNA transcripts were calculated following the $2^{-\Delta\Delta Ct}$ method described by Livak and Schmittgen [7] using β-actin gene as an internal control (Table 3.1).

3.3 Results

3.3.1 *Isolation of DEGs*

A total of 30 ACP primers were used to screen for DEGs among x-166, y-1 and ethrel-treated y-1 during the two germination periods, resulting in 105 unique fragments after removal of redundant DNA sequences (Part of the results were shown in Fig. 3.1). But only 16 of them had known functions as indicated by GO annotation analysis (Table 3.2). Of these, three were directly related to seed

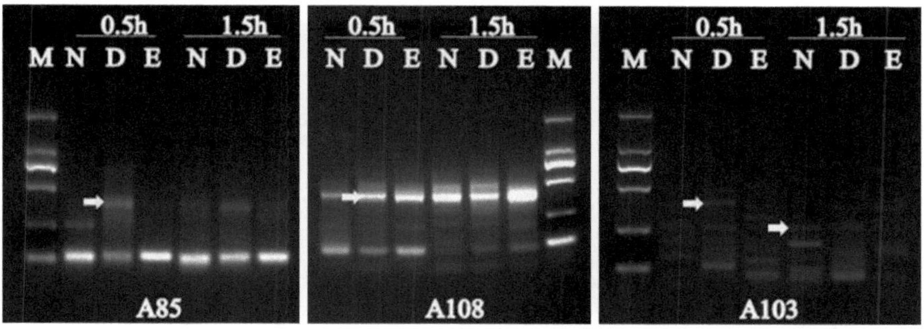

Fig. 3.1 DEGs identified by using the GeneFishing technology (only part of the results were shown), *arrows* indicating four DEGs with known functions as suggested by GO annotation analysis (85-3-1, 108-2-1, 103-21-2, and 103-25) and 0.5 and 1.5 h denoted different germination time. A85, A108, and A103 were ACP primers from the GeneFishing kit. M: D2000 DNA Marker (Tiangen Biotech, Beijing) (MW of bands from top to bottom: 2,000, 1,000, 750, 500, 250, and 100 bp), N:nondormant x-166, D:dormant y-1, E:ethrel-treated y-1

germination. They were 103-25 (oleosin 5), 108-2-1 (late embryogenesis abundant protein d-34), and 85-3-1 (oleosinkda-like) (Table 3.2).

3.3.2 Validation of DEGs

Three genes, namely 103-25, 108-2-1, and 85-3-1, with functions related to seed germination were initially chosen for further qRT-PCR validation. But, of them, only 85-3-1 gave believable results (Fig. 3.2a, b). Therefore, two additional sequences, viz., 108-7 and 86-6, randomly selected from DEGs with no known functions as suggested by BLAST2GO analysis, were included in qRT-PCR validation (Figs. 3.3a, b and 3.4a, b).

85-3-1 was isolated from y-1 (ddw, 0.5 h). Its expression was lowered as the extension of germination periods. This was true in x-166, y-1 and ethrel-treated y-1 (Fig. 3.2b). Expression of 85-3-1 in y-1 at 0.5 h or 1.5 h was consistently higher than that in x-166 or ethrel-treated y-1, at the same incubation time (Fig. 3.2a). Relative expression ranged from 2.82 to 11.31. The results suggested that the gene may play an important role in retaining groundnut seed dormancy.

108-7 and 86-6 were isolated from x-166 (ddw, 1.5 h). Their relative expression indicated that both genes had no direct relationships to groundnut seed germination/dormancy (Figs. 3.3a, b and 3.4a, b).

Table 3.2 Homology analysis of the unique transcripts and annotation using BLAST2GO (only those with known functions were shown)

Sequence name (source)	Description	Size	Similarity (%)	Gos
103-21-2(y-1, ddw, 0.5 h)	Dna damage repair toleration protein drt102	409	92.40	P:carbohydrate metabolic process; P:photoreactive repair; C:cytoplasm; P:response to UV; P:response to cold; C:nucleus; C:cytosol
103-25(x-166, ddw, 1.5 h) (103-25-1,103-25-2)	Oleosin 5	271	74.40	C:lipid particle; C:integral to membrane; C:membrane; C:monolayer-surrounded lipid storage body; P:seed germination; F:molecular_function; P:lipid storage; P:response to freezing
108-2-1(y-1,ddw,0.5)(108-6-1,108-9,108-1-1,81-8-1)	Late embryogenesis abundant protein d-34	405	83.90	P:seed germination; P:multicellular organismal development; P:cellular ion homeostasis; P:response to lithium ion; C:nucleolus; C:nucleus
101-3(y-1,ddw,0.5)(101-56,101-65,101-8,101-11,101-14,101-9)	Aba-responsive protein abr17	690	74.90	P:response to biotic stimulus; P:defense response; F:hydrolase activity
109-4(x-166, ddw, 1.5 h)	Anti-virus transcriptional factor	504	81.85	F:zinc ion binding; P:oxidation-reduction process; F:oxidoreductase activity; P:metabolic process; F:metal ion binding
111-2(y-1, ddw, 0.5 h)	Ubiquitin-associated ts-n domain-containing protein	366	82.09	C:plasma membrane
111-3-1(x-166, ddw, 1.5 h)	23 s ribosomal rna	402	71.30	F:hydrolase activity; C:ribosome; C:plastid; C:chloroplast
118-2-1-M13F.seq(y-1, ddw, 0.5 h)	Sister chromatid cohesion	210	73.75	F:metal ion binding; F:zinc ion binding
83-8-1(y-1, ethrel, 1.5 h) (83-4-2,85-1)	Cyclophilin 1	509	97.90	F:peptidyl-prolyl cis-trans isomerase activity; P:protein folding; P:protein peptidyl-prolyl isomerization; F:isomerase activity
85-3-1(y-1, ddw, 0.5 h)	Oleosin kda-like	385	90.65	C:integral to membrane; C:monolayer-surrounded lipid storage body; C:lipid particle; C:membrane; P:lipid storage; P:seed oilbody biogenesis; P:seed germination; P:response to freezing
86-2-1(y-1, ddw, 0.5 h)	cyclophilin 1	488	98.35	F:peptidyl-prolyl cis-trans isomerase activity; P:protein folding; P:protein peptidyl-prolyl isomerization; F:isomerase activity; C:apoplast; P:response to cadmium ion; C:plasma membrane; C:chloroplast

(continued)

Table 3.2 (continued)

Sequence name (source)	Description	Size	Similarity (%)	Gos
86-5-1(x-166, ddw,1.5 h) (86-1,86-5,86-7)	Eukaryotic translation initiation factor 5a	618	93.60	F:RNA binding; F:translation elongation factor activity; P:positive regulation of translational termination; P:positive regulation of translational elongation; F:ribosome binding; P:peptidyl-lysine modification to hypusine; P:translational frameshifting; P:translational initiation; F:translation initiation factor activity; P:translation; P:response to wounding; P:xylem development; C:nucleus; P:host programmed cell death induced by symbiont; P:defense response to bacterium; P:response to bacterium; P:response to cadmium ion; P: programmed cell death
113-9-1(y-1, ethrel,1.5 h)	Heat shock protein 70	745	94.35	F:ATP binding; P:response to stress; F:nucleotide binding; F: protease binding; C:vacuolar membrane; P:defense response to fungus; C:cytosol; P:defense response to bacterium; P:response to bacterium; P:response to virus; C:apoplast; C:vacuole; C:chloroplast; C:nucleus; C:membrane; C:cell wall; P:response to cadmium ion; P: response to cold; P:response to karrikin; P:response to heat; C: cytosolic ribosome; C:nucleolus; C:chloroplast envelope; C:plasma membrane
90-5-1(x-166, ddw,1.5 h) (90-2,119-6,114-2,115-6)	Rna polymerase beta subunit	293	90.50	C:plastid; F:nucleotidyltransferase activity; F:transferase activity; C: chloroplast; P:transcription, DNA-dependent; F:DNA-directed RNA polymerase activity; F:DNA binding
87-1-2(y-1, ddw, 0.5 h)	Vacuolar-processing enzyme	309	80.15	P:proteolysis; F:cysteine-type endopeptidase activity; F:hydrolase activity; F:cysteine-type peptidase activity; F:peptidase activity
87-2-1(y-1,ethrel, 0.5 h)	Vacuolar-processing enzyme	296	80.15	F:cysteine-type endopeptidase activity; P:proteolysis

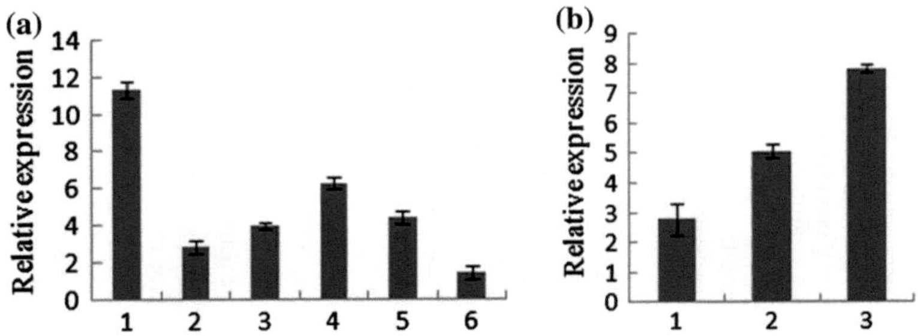

Fig. 3.2 **a** Relative expression of 85-3-1 at the same time under different treatments. *1*. (y-1, ddw, 0.5 h)/(x-166, ddw, 0.5 h) *2*. (y-1, ddw, 0.5 h)/(y-1, ethrel, 0.5 h) *3*. (y-1, ethrel, 0.5 h)/(x-166, ddw, 0.5 h) *4*. (y-1, ddw, 1.5 h)/(x-166, ddw, 1.5 h) *5*. (y-1, ddw, 1.5 h)/(y-1, ethrel, 1.5 h) *6*. (y-1, ethrel, 1.5 h)/(x-166, ddw, 1.5 h), **b** Expression of 85-3-1 at 0.5 h relative to 1.5 h under the same treatment. *1*. (x-166, ddw, 0.5 h)/(x-166, ddw, 1.5 h) *2*. (y-1, ddw, 0.5 h)/(y-1, ddw, 1.5 h) *3*. (y-1, ethrel, 0.5 h)/(y-1, ethrel, 1.5 h)

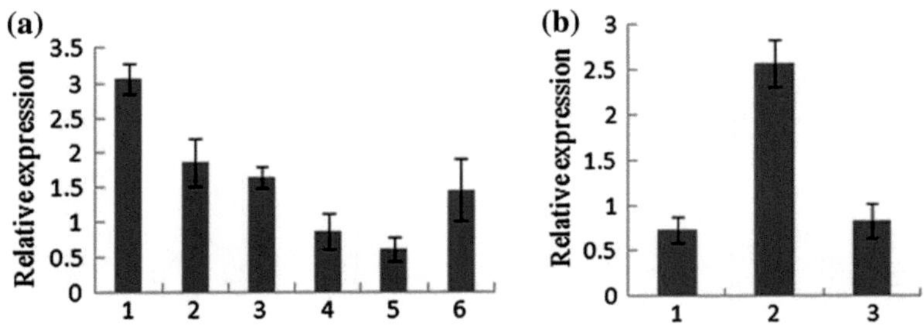

Fig. 3.3 **a** Relative expression of 108-7 at the same time under different treatments. *1*. (y-1, ddw, 0.5 h)/(x-166, ddw, 0.5 h) *2*. (y-1, ddw, 0.5 h)/(y-1, ethrel, 0.5 h) *3*. (y-1, ethrel, 0.5 h)/(x-166, ddw, 0.5 h) *4*. (y-1, ddw, 1.5 h)/(x-166, ddw, 1.5 h) *5*. (y-1, ddw, 1.5 h)/(y-1, ethrel, 1.5 h) *6*. (y-1, ethrel, 1.5 h)/(x-166, ddw, 1.5 h), **b** Expression of 108-7 at 0.5 h relative to 1.5 h under the same treatment. *1*. (x-166, ddw, 0.5 h)/(x-166, ddw, 1.5 h) *2*. (y-1, ddw, 0.5 h)/(y-1, ddw, 1.5 h) *3*. (y-1, ethrel, 0.5 h)/(y-1, ethrel, 1.5 h)

3.4 Discussion

Seed dormancy is not only a key character in plant physiology, but also a target for genetic improvement in crop breeding science. In situ sprouting may result in considerable reduction in quantity and quality of groundnut produce. To test groundnut seed dormancy, pods were generally dried in shade to avoid dormancy loss caused by hot temperature. Presently, seed dormancy in segregation populations can only be tested after harvest, which necessitates the development of tools for selection prior sowing. Previous studies have shown that seed dormancy of the groundnut crop is controlled by oligogenes or polygenes, and that the trait is related to high ABA-like inhibitor content, high phenolic content, high coumarin content,

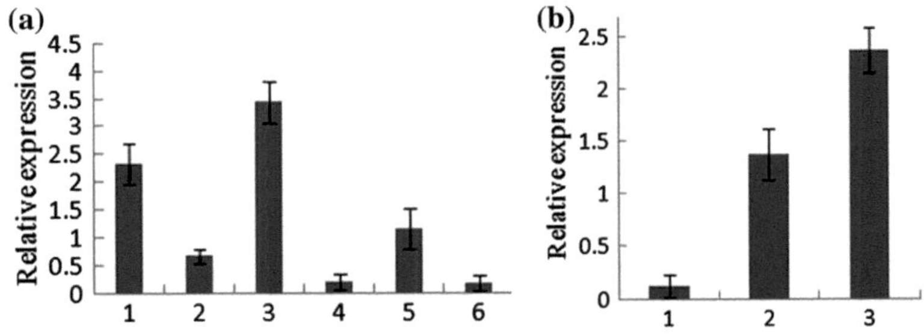

Fig. 3.4 **a** Relative expression of 86-6 at the same time under different treatments. *1.* (y-1, ddw, 0.5 h)/(x-166, ddw, 0.5 h) *2.* (y-1, ddw, 0.5 h)/(y-1, ethrel, 0.5 h) *3.* (y-1, ethrel, 0.5 h)/(x-166, ddw, 0.5 h) *4.* (y-1, ddw, 1.5 h)/(x-166, ddw, 1.5 h) *5.* (y-1, ddw, 1.5 h)/(y-1, ethrel, 1.5 h) *6.* (y-1, ethrel, 1.5 h)/(x-166, ddw, 1.5 h), **b** Expression of 86-6 at 0.5 h relative to 1.5 h under the same treatment. *1.* (x-166, ddw, 0.5 h)/(x-166, ddw, 1.5 h) *2.* (y-1, ddw, 0.5 h)/(y-1, ddw, 1.5 h) *3.* (y-1, ethrel, 0.5 h)/(y-1, ethrel, 1.5 h)

and low cytokinin content, low gibberellin content, and low ethylene-producing capacity [8]. The above-mentioned knowledge is helpful for us to understand the genetic and physiochemical mechanisms underlying seed dormancy, but unfortunately, it cannot be utilized directly as selection tools.

To isolate DEGs, Genefishing technology has several advantages inclusive of low cost, ease of use, and rapidity. In groundnut, the technology proved to be competent in identification of genes related to chilling tolerance [9] and high oil content [10]. In our research group, based on DEGs isolated using Genefishing technology, full length cDNAs of *cyp* and *arf*, two genes responsive to *Ralstonia solanacearum* [the causal pathogen of groundnut bacterial wilt (BW)] infection, have been obtained [5]. Using the groundnut transformation protocol developed at our lab [11], transgenic plants were produced. Overexpression of the genes in a susceptible groundnut cultivar enhanced BW resistance, while transfer of antisense constructs of the genes into a BW-resistant groundnut cultivar led to loss of resistance (Dr Qi Wu, unpublished data). Genefishing technology coupled with our high efficiency *in planta* transformation protocol [11] provides an ideal platform for fast isolation and functional analysis of DEGs in groundnut.

In the present study, 105 unique cDNA fragments were isolated from two groundnut materials with contrasting dormancy phenotypes using Genefishing technology. Three of the 16 DEGs with known functions were related to seed germination as suggested by BLAST2GO analysis. Differential expression of one gene with functions indicated in seed germination [85-3-1 (oleosin kda-like)] was confirmed by qRT-PCR. In addition to seed germination, other functions of the gene include lipid storage. As seed storage lipid mobilization is a common phenomenon during germination, we postulate that expression of the gene is positively related to intensity of groundnut seed dormancy, and that the gene product may protect oil bodies thereby preventing lipid mobilization from providing energy for germination. The next step is to study the expression of the remaining DEGs by qRT-PCR; based on the results, full-length cDNAs of interests including 85-3-1 will be amplified, and plant expression vectors constructed. The functions of these genes will then be analyzed using transgenesis technology, which may help to elucidate the mechanisms underlying groundnut seed dormancy and pave the way to incorporating the dormancy trait to high-yielding nondormant groundnut genotypes through marker-aided selection or transgenesis technology.

Acknowledgments Thanks are due to China Agricultural Research System (Grant No. CARS-14), Ministry of Agriculture, China, SAAS (Shandong Academy of Agricultural Sciences) Key Science and Technology Achievements Incubation Project (2014CGPY09), and Shandong Peninsula Blue Sea Economic Region Human Resource Development Program for providing financial support for this study.

References

1. Wang CT, Zhang JC, Tang YY, Guan SY, Wang XZ, Wu Q, Shan L, Zhu LG, Su JW, Yu ST (eds) (2013) Genetic improvement of peanut. Shanghai Science and Technology Press, Shanghai, China
2. Martin J, Dimanche P, Schilling R (1999) Comment lutter contre la contamination de l'arachide par les aflatoxines? Expériences conduites au Sénégal. Agriculture et Développement 23:58–67
3. Huang MD (1990) Breeding for seed dormancy in peanut. In: Xie SJ (ed) Plant breeding in Taiwan. Agricultural Press, Beijing, pp 201–215
4. Wan SB (ed) (2003) Peanut cultivation science in China. Shanghai Science and Technology Press, Shanghai,
5. Ding YF, Wang CT, Tang YY, Wang XZ, Wu Q, Hu DQ, Yu HT, Zhang JC, Cui FG, Song GS, Gao HY, Yu SL (2012) Isolation and analysis of differentially expressed genes from peanut in response to challenge with *Ralstonia solanacearum*. Electron J Biotechnol. doi:10.2225/vol15-issue5-fulltext-1
6. Tang YY, Wang CT, Yang GP, Feng T, Gao HY, Wang XZ, Chi XY, Xu YL, Wu Q, Chen DX (2011) Identification of chilling-responsive transcripts in peanut (*Arachis hypogaea* L.). Elect J Biotechnol. doi:10.2225/vol14-issue5-fulltext-5
7. Livak KJ, Schmittgen TD (2001) Analysis of relative gene expression data using realtime quantitative PCR and the $2^{-\Delta\Delta Ct}$ method. Methods 25:402–408
8. Sanders TH, Schubert AM, Pattee H (1982) Maturity methodology and postharvest physiology (Chapter 16). In: Pattee HE, Young CT (eds) Peanut science and technology. American Peanut Research and Education Society, Yoakum, pp 641–647
9. Wang XZ, Tang YY, Wu Q, Gao HY, Hu DQ, Zhang SW, Chen DX, Chi YC, Wang CT (2013) Cloning and analysis of differentially expressed genes from peanut in response to chilling stress. J Nulear Agric Sci 27(2):0152–0157
10. Tang YY, Wang XZ, Wu Q, Fang CQ, Guan SY, Yang WQ, Wang CT, Wang PW (2013) Identification of differentially expressed genes from developing seeds of a normal oil peanut cultivar and its high oil EMS mutant. Res Crops 14(2):511–516
11. Wang CT, Wang XZ, Tang YY, Wu Q, Li GJ, Song GS, Yu HT, Hu DQ, Guo BT (2013) Transforming peanut (*Arachis hypogaea* L.): a simple in planta method. Res Crops 14(3):850–854

Chapter 4
Increase of the Lycopene Production in the Recombinant Strains of *Escherichia coli* by Supplementing with Fructose

Tong-Cun Zhang, Wen Li, Xue-Gang Luo, Cui-Xia Feng, Ming-Hui Zhang, Wen Du and De-Yun Ma

Abstract Lycopene is an important carotenoid which has been widely used as functional foods and feed supplements, and pharmaceutical compounds. In the present study, the host strain and the auxiliary carbon source were optimized to enhance the lycopene yield in the recombinant *Escherichia coli*. The results showed that lycopene concentration of the cells could be significantly increased when the recombinant strain was grown on the LB medium supplemented with 6 g l^{-1} fructose, whereas glucose has little effect. Under the optimized conditions, the recombinant bacteria could exhibited a lycopene yield up to 1,050 mg l^{-1} for 24 h in a 250 ml baffled flask. This is the first report of enhancing lycopene production by supplementing with fructose.

Keywords Lycopene · *Eschericiha coli* · Fructose · Glucose

4.1 Introduction

Numerous carotenoids are known to be naturally biosynthesized by plants and microorganisms and their diversity is derived from differences in type and levels of desaturation and other modifications of the C40 backbone [1]. Currently, lycopene,

---

Wen Li and Tong-Cun Zhang contributed equally to the project.

T.-C. Zhang (✉) · W. Li · X.-G. Luo (✉) · C.-X. Feng · M.-H. Zhang · W. Du · D.-Y. Ma
Key Laboratory of Industrial Microbiology of the Ministry of Education, Tianjin Key Laboratory of Industrial Microbiology, College of Biotechnology, Tianjin University of Science and Technology, 300457 Tianjin, China
e-mail: tony@tust.edu.cn

X.-G. Luo
e-mail: luoxuegang@tust.edu.cn

T.-C. Zhang and M. Nakajima (eds.), *Advances in Applied Biotechnology*,
Lecture Notes in Electrical Engineering 332, DOI 10.1007/978-3-662-45657-6_4

an important carotenoid which has striking anti-cancer, anti-inflammatory [2], and anti-oxidative activities, has been widely used as functional food and feed supplements, and pharmaceutical compound [3].

Various strategies for the lycopene production such as natural extraction, chemical synthesis [4], and fermentation [5] have been described in prior reports. Up to now, a large number of reports on the production of lycopene in *Escherichia coli* by using metabolic engineering [6] have been published in the literature. In the previous study, the pACLYCipi plasmid containing crtE, crtB, and crtI, which plays key roles in the lycopene biosynthetic pathway, has been constructed. In this work, we further attempted to enhance the lycopene yield by optimizing of the bacteria strain and culture medium, and the results showed that supplementation with fructose could have stronger effects than glucose, glycerol and L-arabinose on the lycopene production in recombinant *E. coli*.

4.2 Materials and Methods

4.2.1 Bacterial Strains, Plasmids, and Growth Conditions

Diverse strains of *E. coli* K-12 (2T1R), DH5α, JM109, and BL21 (DE3) were used in this study. The pACLYCipi plasmid containing genes of *crtE*, *crtB*, and *crtI* from Pantoeaagglomerans, were constructed in our previous study. All the *E. coli* strains were grown in LB medium (10 g l^{-1}tryptone, 5 g l^{-1}yeast extract, and 10 g l^{-1}NaCl) supplemented or not with additional carbon source at 37 °C with shaking. The media were supplemented with the antibiotics chloramphenicol (50 mg l^{-1}) as needed.

4.2.2 Extraction and Quantification of Lycopene

The lycopene content of recombinant *E. coli* strains was quantified as previously reported [7]. The recombinant *E. coli* cells were harvested by centrifugation at 12,000 rpm for 5 min. The cell pellet was washed with water and then extracted in 1 mL of acetone at 55 °C for 15 min with intermittent vortexing. The mixture was then centrifuged at 12,000 rpm for 10 min, and the acetone supernatant was transferred to a new tube. The absorbance of the resulting extract was measured at 475 nm and converted to lycopene concentration (mg/l) using a standard curve obtained using commercial lycopene (Sigma).

4.2.3 Detection of Lycopene Yield Curve

A single colony of the recombinant *E. coli* was inoculated into a 20 ml test tube containing 5 ml of the growth medium and grown overnight at 37 °C with shaking; then the starter culture was separately diluted into LB, LB supplemented with 6 g l^{-1} glucose, and LB supplemented with 6 g l^{-1} fructose to an OD_{600} of 0.025. Chloramphenicol (50 g/ml) was added to the above medium. Every other 4 h, the absorbance of the resulting extract was measured at 475 nm and the lycopene concentration (mg/l) was calculated using commercial lycopene as a standard (Sigma). Each sample was measured in triplicate.

4.3 Result

4.3.1 Optimization of the Host Strain for Lycopene Production

To select an optimal host strain of recombinant strains for lycopene production, *E. coli* K-12, DH5α, JM109, and BL21 (DE3) were transformed with the pACLY-Cipi plasmid respectively, and then cultured in a 250 ml baffled flask in LB medium which is popular with bacteriologists because it permits fast growth and good growth yields for many species [8] (Fig. 4.1). As shown in Fig. 4.1, *E. coli* K-12 (pACLYCipi) exhibited the highest cell mass and lycopene content among the used strains. The final concentration of lycopene produced by the used strains followed the order K-12(pACLYCipi) > DH5α (pACLYCipi) > JM109 (pACLYCipi) > BL21 (pACLYCipi). Thus, *E. coli* K-12 strain was selected as the optimal host for lycopene production.

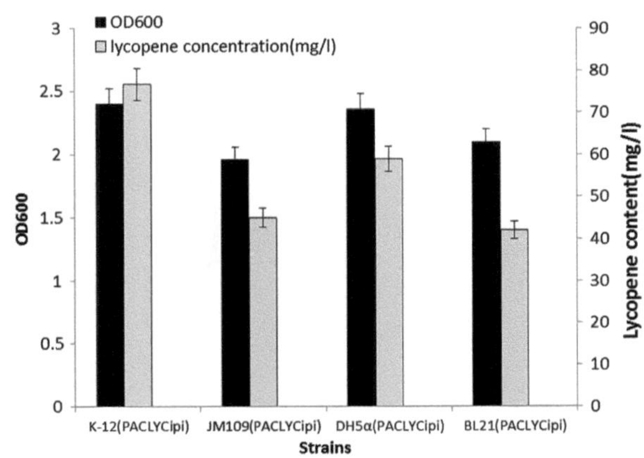

Fig. 4.1 Lycopene yield (mg/l) and cell concentration by various *E. coli* strains as hosts in a 250 ml baffled flask

4.3.2 *Optimization of the Auxiliary Carbon Source for Lycopene Production*

When LB was used as main medium for the recombinant strains, both the cell mass and the lycopene production were very low, and the lycopene yield was below 80 mg l^{-1} (Fig. 4.1). Since LB medium contains little carbohydrate, in the present study, an auxiliary carbon source, including of glucose, fructose, glycerol, arabinose, was supplemented respectively to overcome this limitation (Fig. 4.2). The auxiliary carbon source of were all supplemented at 6 g l^{-1} into the LB medium. Fructose supplementation could exhibited a lycopene yield up to 1,050 mg l^{-1} for 24 h in a 250 ml baffled flask. After 24 h, the culture by fructose supplementation was the most effective for increases of the lycopene yield, which were 14.9, 5.1, 13.9 and 11.9 fold higher than the culture by control, glycerol, glucose, arabinose respectively. As a result, supplementation with fructose exhibited the highest lycopene production and the cell mass, whereas supplementation with glucose showed the lowest cell mass and the lycopene production, suggesting that fructose was the optimal auxiliary carbon source for lycopene production.

Furthermore, the detailed influence of fructose on lycopene production was evaluated with initial concentrations of 4, 5, 6, 7, and 8 g l^{-1} (Fig. 4.3). After 24 h, the final lycopene concentrations and the cell mass of the cultures with fructose supplementation for all initial concentrations were higher than the culture with no fructose supplementation. In addition, Fig. 4.3 showed that there was a clear positive relationship between cell growth and lycopene yield in both cases. The maximal lycopene content of 1,050 mg l^{-1} was observed at an initial fructose concentration of 6 g l^{-1}. Thus, 6 g l^{-1} was considered as the optimal initial concentration of fructose for lycopene production.

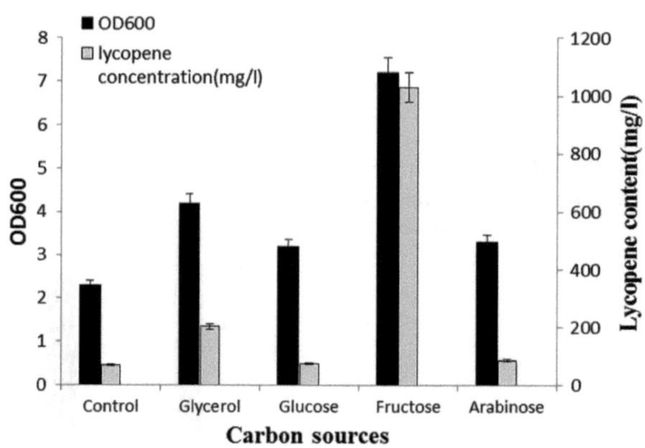

Fig. 4.2 Effects of different auxiliary carbon sources on lycopene yield (mg/l) and cell concentration (OD600) in a 250 ml baffled flask

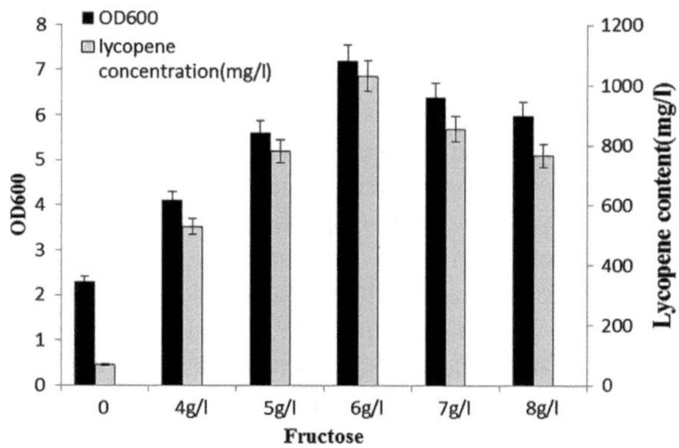

Fig. 4.3 The relationship between the concentration of fructose and the lycopene yield (mg/l) and cell concentration

4.3.3 Influence of Fructose and Glucose as an Auxiliary Carbon Source on Cell Growth and Lycopene Yield

Figure 4.2 showed that fructose and glucose supplementation respectively exhibited the highest and the lowest lycopene production and the cell mass. To further investigate the difference between the effect of fructose on lycopene production and that of glucose, the lycopene yield curve was drawn. As shown in Fig. 4.4, apparently, by contrast to the control, it caused little change on lycopene concentration to supplement glucose into LB medium. Meanwhile, there was no significant increase in lycopene production between the culture by glucose supplementation and the culture by LB medium in all growth phases. However, the culture by fructose showed an obvious index growth in lycopene production. And fructose supplementation dramatically stimulated the lycopene production especially when the recombinant *E. coli* was cultured over 12 h, suggesting that *E. coli* K-12 (pACLYCipi) prefer fructose to glucose.

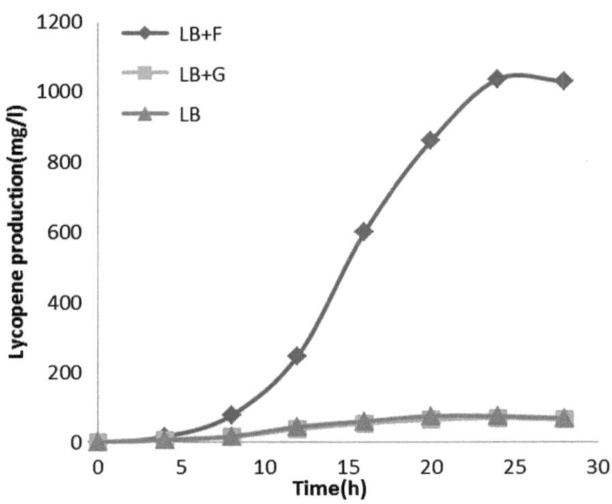

Fig. 4.4 Lycopene yield curve under different concentration of fructose and glucose

4.4 Discussion

As LB medium contains little carbohydrate, peptides and free amino acids [8], a number of carbon sources were added into LB medium to enhance lycopene production in our experiments. Based on the current reports, glycerol, glucose, and L-arabinose have been respectively shown to be the optimal carbon source for lycopene production [9, 10]. However, we have demonstrate that supplementation with fructose as an auxiliary carbon source resulted in the highest lycopene concentration and final cell mass (Fig. 4.2). And the maximum lycopene concentration was harvested when 6 g l^{-1} fructose was supplemented into LB medium (Fig. 4.3). This is the first report of enhancing lycopene production by supplementing with fructose.

Since supplementation with fructose could increase the final cell mass and lycopene concentration (Fig. 4.1), we were interested in evaluating the effect of the lycopene production by supplementation with fructose. Compared to supplementation with or without glucose, it caused a significant change on lycopene concentration by supplementation with fructose especially when the recombinant *E. coli* K-12 (pACLYCipi) was cultured for more than 12 h. By contrast to the previous report, we proposed that the fructose might be a more effective carbon source to stimulated lycopene production.

Besides, our study also found that *E. coli* K-12 showed the highest lycopene productivity among the tested strains. In our opinion, it is likely that *E. coli* K-12 contains a more effective expression of the enzyme relative to the lycopene production.

Acknowledgments This study was financially supported by the 863 (Hi-tech research and development program of China) program under contract NO. 2012AA022108 and NO. 2012AA021505, and the Program for Changjiang Scholars and Innovative Research Team in University of Ministry of Education of China (NO. IRT1166).

References

1. Umeno D, Arnold FH (2003) A C35 carotenoid biosynthetic pathway. Appl Environ Microbiol 69:3573–3579
2. Bignotto L, Rocha J, Sepodes B (2009) Anti-inflammatory effect of lycopene on carrageenan-induced paw oedema and hepatic ischaemia- reperfusion in the rat. Br J Nutr 102:126–133
3. Erdman JW Jr, Ford NA, Lindshield BL (2009) Are the health attributes of lycopene related to its antioxidant function? Arch Biochem Biophys 483:229–235
4. Michael McClain R, Bausch J (2003) Summary of safety studies conducted with synthetic lycopene. Regul Toxicol Pharmacol 37:274–285
5. Mantzouridou F, Tsimidou MZ (2008) Lycopene formation in *Blakeslea trispora*. Chemical aspects of a bioprocess. Trends Food Sci Technol 19:363–371
6. Bhataya A, Schmidt Dannert C, Lee PC (2009) Metabolic engineering of *Pichiapastoris* X-33 for lycopene production. Process Biochem 44:1095–1102

7. Alper H, Jin YS, Moxley JF, Stephanopoulos G (2005) Identifying gene targets for the metabolic engineering of lycopene biosynthesis in *Escherichia coli*. Metab Eng 7:155–164
8. GuennadiSezonov Daniele Joseleau-Petit, Dari Richard (2007) *Escherichia coli* Physiology in Luria-Bertani Broth. J Bacteriol 189:8746–8749
9. Yoon SH, Lee YM, Kim JE et al (2006) Enhanced lycopene production in *Escherichia coli* engineered to synthesize isopentenyl diphosphate and dimethylallyl diphosphate from mevalonate. Biotechnol Bioeng 94:1025–1032
10. Yoon SH, Park HM, Kim JE, Lee SH, Choi MS, Kim JY, Oh DK, Keasling JD, Kim SW (2007) Increased beta-carotene production in recombinant *Escherichia coli* harboring an engineered isoprenoid precursor pathway with mevalonate addition. Biotechnol Prog 23:599–605

Chapter 5
Isolation of Differentially Expressed Genes from Developing Seeds of a High-Protein Peanut Mutant and Its Wild Type Using Genefishing™ Technology

Shu Tao Yu, Hong Bo Yu, Guo Qing Yu, Li Ren Zhao, Hong Xi Sun, Yue Yi Tang, Xiu Zhen Wang, Qi Wu, Quan Xi Sun and Chuan Tang Wang

Abstract Peanut is a good source of dietary protein. Raising protein content in peanut will not only fill the growing need for vegetable protein, but also in most cases lower oil content, which is good news to health-conscious populations. However, no attempt has been made to isolate genes related to protein content in peanut. In the present study, a total of 40 unique differentially expressed genes in developing seeds of high-protein peanut mutant (SDPM) and its normal-protein wild type (SDPW) at 46 or 49 days after flowering were isolated using Genefishing technology. Of them, 8 sequences were undescribed previously; the rest 32 were found to be significantly similar to the sequences in GenBank nr database. Three genes potentially related to protein content in peanut, viz., P2-2-2, P2-92-2 and P1-89-1-5 with high homology to thioredoxin h, arachin ahy-4 and abc transporter, respectively, were selected for further analysis. All the 3 genes validated by qRT-PCR showed differential expression between SDPM and SDPW, with relative expression ranging from 0.41–10.60. The detailed functions of the differentially expressed genes isolated from developing seeds in the present study in conditioning peanut seed protein content are yet to be validated by transgenic experiments.

Keywords Differentially expressed gene · Genefishing technology · Mutant · Peanut · Protein content

S.T. Yu (✉) · H.B. Yu · G.Q. Yu · L.R. Zhao · H.X. Sun
Sandy Land Amelioration and Utilization Research Institute of Liaoning, 123000 Fuxin, China
e-mail: shutaoyu@163.com

Y.Y. Tang · X.Z. Wang · Q. Wu · Q.X. Sun · C.T. Wang (✉)
Shandong Peanut Research Institute, Qingdao 266100, China
e-mail: chinapeanut@126.com

T.-C. Zhang and M. Nakajima (eds.), *Advances in Applied Biotechnology*, Lecture Notes in Electrical Engineering 332, DOI 10.1007/978-3-662-45657-6_5

5.1 Introduction

The cultivated peanut, *Arachis hypogaea* L., as a major cash crop valued worldwide, is cultivated in more than eighty countries/regions. In western world, peanut is a main dietary protein source. In developing nations, even though most peanut produce is crushed for edible oil, the portion for food uses has been on a steady increase. Raising protein content in peanut seeds will not only fill the growing need for vegetable protein, but also in most cases lower oil content, which is good news to health-conscious populations. To lower oil content, defatting is generally useful; however, it cannot be applied to peanut for in-shell consumption.

In literature, Spanish- or Valencia-type peanut landraces/cultivars with high-protein content are not uncommon, and there are reports on heritability, heterosis and combinability for peanut protein content [1]. Recently, [2] analyzed the data from 2 environments, and concluded that protein content in peanut is conditioned by polygenes. Quantitative trait loci (QTLs) with lower than 15 % contribution to phenotypic variations have been reported from two independent research groups [3, 4]. But to the best of our knowledge, no attempt has been made to isolate genes related to this valuable trait.

The present study represented an effort aimed at isolation of candidate genes governing protein content in peanut through Genefishing technology using a large-podded peanut genotype and its high-protein chemical mutant.

5.2 Materials and Methods

5.2.1 Peanut Materials

SDPM, a peanut EMS (ethyl methane sulfonate) mutant with 28.67 % protein, developed at Shandong Peanut Research Institute (SPRI), Qingdao, China, and its wild type counterpart with 17.69 % protein (SDPW) were used in the present study. The high-protein mutant was first identified in near infra-red reflectance spectroscopy (NIRS) analysis of intact seeds [5] and further confirmed by nitrogen amount determination using an automated Kjeldahl Analyzer (model 2300 II, Foss, Sweden). A conversion factor of 5.46 was used to convert the amount of nitrogen to amount of protein [6].

5.2.2 Peanut Cultivation and Seed Sampling

Peanut seeds were sown in field under polythene film mulch and routine agronomic practices were followed as per the description by Wan et al. [7]. Plant population was 150,000 hills per ha, with two plants per hill. Flowers on the first and second node of the cotyledonary branches were tagged and pods were harvested 46 and 49 days after flowering (DAF). Seeds were then stored in liquid nitrogen.

5.2.3 RNA Isolation

Total RNA was extracted from developing seeds of SDPM and SDPW using RNAprep pure Plant Kit (Tiangen, Beijing, China) following manufacturer's instructions. RNA concentration and integrity were determined by spectrophotometry and relative intensity of brightness of GelRed (Biotium, CA, USA) stained bands resolved on a 1.2 % agarose gel [8].

5.2.4 Cloning and Sequence Analysis of DEGs

Differentially expressed genes (DEGs) from developing seeds of SDPM or SDPW harvested at 46 DAF and 49 DAF were identified using Genefishing™ DEG Premix Kit (Seegene, Korea). RNase-free water was added to the mixture of 3 μg of total RNA and 2 μl of 10 μM dT-ACP1 to a total volume of 9.5 μl. The mixture was incubated at 72 °C for 10 min., cooled on ice for 2 min., centrifuged briefly, and then 4 μl of 5 × RT buffer, 5 μl of 2 mM dNTP mix, and 0.5 μl of RNase inhibitor (40 U/μ1) (Tiangen, Beijing, China) along with 200 U of M-MLV reverse transcriptase (TaKaRa, Japan) were added. Reverse transcription was conducted at 42 °C for 90 min, followed by incubation at 70 °C for 10 min to terminate the reaction. First strand cDNA products were then diluted with 80 μl of DNase-free water and directly used in subsequent isolation and analysis of differentially expressed genes from peanut using Genefishing PCR. PCR mixture (20 μl) contained 50 ng of first strand cDNA, 0.5 μM arbitrary ACP, 0.5 μM dT-ACP2 and 2 × SeeAmp ACP Master-mix. PCR program was 94 °C for 5 min, 50 °C for 3 min and 72 °C for 1 min, followed by 40 cycles of 94 °C for 40 s, 65 °C for 40 s and 72 °C for 40 s, and a final extension of 72 °C for 5 min. PCR products were separated on a 2 % agarose gel, stained with Gelred and visualized under UV light. Amplicons of interest from treated samples were cloned into a pGM-T vector (Tiangen, Beijing, China), and sequenced by Genscript Inc., Nanjing, China. Resultant sequences, after trimmed to remove poor quality reads and vector sequences, were assembled with DNAStar (DNASTAR Inc., London, UK) package. Transcript annotation and functional assignment were carried out with BLAST2GO (http://blast2go.org).

Table 5.1 List of realtime PCR primers

Primer ID	Forward primer(5′–3′)	Reverse primer(5′–3′)
p-P2-2-2	ACTCATAGAATGGCAGAGG	TGTTTCGTCTTTCGTTTCC
p-P2-92-2	GGCATTCAAGACAGACTCA	CGGTGGAACGAAGAACTT
p-P1-89-1-5	AAGTTGCGGATAGGAAGAT	GGATGGAGACGAAGAAGATA
β-actin	TTGGAATGGGTCAGAAGGATGC	AGTGGTGCCTCAGTAAGAAGC

5.2.5 qRT-PCR

Realtime PCR primers were designed using Beacon Designer 7 (Premier Biosoft International, CA, USA) (Table 5.1). qRT-PCR was performed in a Lightcycler 2.0 PCR machine (Roche, USA) as per Tang et al. (2011), with 3 replications for each reaction. β-actin gene was utilized as an internal control. Fold changes in RNA transcripts were calculated using the $2^{-\Delta\Delta Ct}$ method.

5.3 Results and Discussion

Of 36 primers (ACPs) provided with the Genefishing kit, 14 produced bands that could clearly differentiate SDPM and SDPW (Fig. 5.1). Cloning, sequencing and assembly of the DEGs resulted in a total of 40 unique sequences (Table 5.2). There were 8 sequences undescribed previously; the rest 32 DEGs were found to be significantly similar to the sequences in GenBank nr database, of which 12 and 20 DEGs were from SDPW and SDPM, respectively.

Three of the unique DEGs potentially related to protein content in peanut, viz., P2-2-2, P2-92-2 and P1-89-1-5 with high homology to *thioredoxin h*, *arachin ahy-4* and *abc transporter*, respectively, were selected for further analysis. The annotation results were shown in Table 5.3. Relative expression of the 3 genes between high-protein EMS mutant (SDPM) and normal-protein wild type (SDPW) in developing

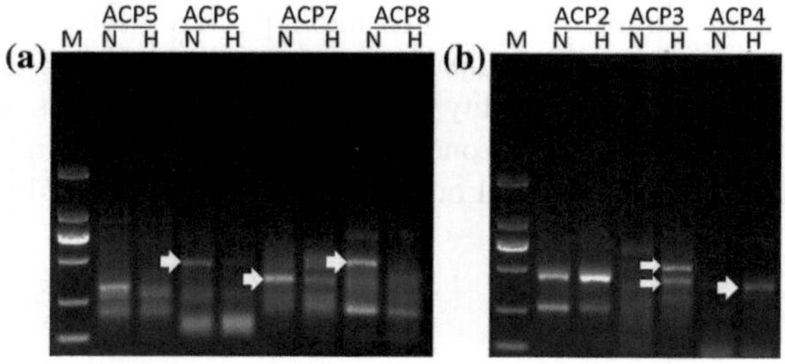

Fig. 5.1 Differentially expressed genes between SDPW (N) and SDPM (H) as indicated by *arrows* in developing seeds harvested at 46 DAF (**a**) or 49 DAF (**b**)

Table 5.2 BLAST* results of 40 unique DEGs

Seq. name	ACP primer	Seq. description	Seq. length (bp)	Min. eValue	Mean similarity (%)
C1-2-1-2	ACP2	Unnamed protein product [Vitis vinifera]	395	1.25E-20	68.50
C1-2-2-4	ACP2	High mobility group family	779	1.23E-52	85.80
C1-22-2	ACP22	—NA—	247		
C1-28-1	ACP28	—NA—	426		
C1-3-1-2	ACP3	Type 2 metallothionein	527	1.60E-39	87.55
C1-3-1-4	ACP3	Type 2 metallothionein	519	8.05E-39	87.60
C1-3-2-1	ACP2	Arachin ahy-4	427	1.24E-16	90.20
C1-3-2-5	ACP3	—NA—	477		
C1-8-1	ACP8	Sterile alpha motif domain-containing protein	488	7.38E-36	81.20
C1-83-3	ACP83	Phage tail tape measure lambda family	752	2.17E-125	98.75
C1-83-5	ACP83	Cyclophilin	379	4.16E-26	97.40
C1-85-1	ACP85	*Arachis hypogaea* seed storage protein SSP1 mRNA, partial cds	511	1.00E-141	98
C1-92-1	ACP92	*Arachis hypogaea* iso-Ara h3 mRNA, complete cds	719	0	96
C1-92-2	ACP92	Arachin ahy-4	556	9.35E-49	91.5
C2-28-1	ACP28	—NA—	375		
C2-92-4	ACP92	Arachin ahy-4	461	8.06E-39	91.95
P2-2-2	ACP2	Thioredoxin h	348	5.91E-12	87.35
P1-22-5	ACP22	DNA-directed rna polymerases and iii subunit rpabc2-like	487	1.83E-39	97.70
P1-3-2	ACP3	60 s ribosomal protein	625	6.14E-70	95.50
P1-36-1-1	ACP36	Endothelial differentiation-Related factor 1	345	2.71E-25	93.30
P1-36-2-3	ACP36	—NA—	130		
P1-36-3-1	ACP36	—NA—	130		
P1-7-1	ACP7	Gdp-mannose-dehydratase	454	1.14E-54	96.15
P1-7-2	ACP7	Sterile alpha motif domain-containing protein	477	6.63E-36	81.20
P1-7-3	ACP7	Germin-like protein	477	2.57E-71	84.50
P1-81-1	ACP81	Uncharacterized protein loc100804883	419	9.45E-09	73.50
P1-81-3	ACP81	—NA—	374		
P1-84-2	ACP84	Aminotransferase-like protein	382	2.63E-21	65.65
P1-84-3	ACP27	Protein	532	4.94E-66	93.45
P1-84-4	ACP28	Photosystem II reaction center protein m	330	1.80E-10	75.00
P1-84-5	ACP29	Aminotransferase-like protein	385	1.49E-21	65.20
p1-89-1-1	ACP89	Stf2-like protein	776	3.50E-43	78.60
P1-89-1-5	ACP89	Abc transporter	952	1.59E-149	94.95

(continued)

Table 5.2 (continued)

Seq. name	ACP primer	Seq. description	Seq. length (bp)	Min. eValue	Mean similarity (%)
P1-89-2-5	ACP89	—NA—	137		
P1-92-1	ACP92	Arachin ahy-4	555	8.78E-49	91.50
P2-3-3	ACP3	Type 2 metallothionein	471	1.35E-38	86.45
P2-3-5	ACP3	Kda class i heat shock protein	366	2.27E-16	97.20
p2-90-3	ACP90	Rar1 protein	382	1.38E-48	95.20
P2-92-2	ACP92	Arachin ahy-4	556	9.35E-49	91.5
p2-84-3	ACP84	*Arachis hypogaea* oleosin 18.5 gene, promoter region and complete cds	203	3.00E-49	96

* BLAST = basic local alignment search tool

Table 5.3 Annotation of three DEGs using BLAST2GO

Seq. name	Seq. description	#Hits	Min. e value	Mean similarity (%)	#GOs	GOs
P2-2-2	Thioredoxin h	20	6.07E-12	87.35	7	P: cell redox homeostasis; F: electron carrier activity; P: glycerol ether metabolic process; F: protein disulfide oxidoreductase activity; C: cytoplasm; P: electron transport chain; P: transport
P2-92-2	Arachin ahy-4	20	4.77E-49	90.95	1	F: nutrient reservoir activity
P1-89-1-5	Abc transporter	20	1.70E-149	94.95	10	F: ATPase activity; F: ATP binding; F: hydrolase activity; F: nucleotide binding; F: nucleoside-triphosphatase activity; F: phosphonate transmembrane-transporting ATPase activity; P: ATP catabolic process; F: transporter activity; C:mitochondrion; C: plastid

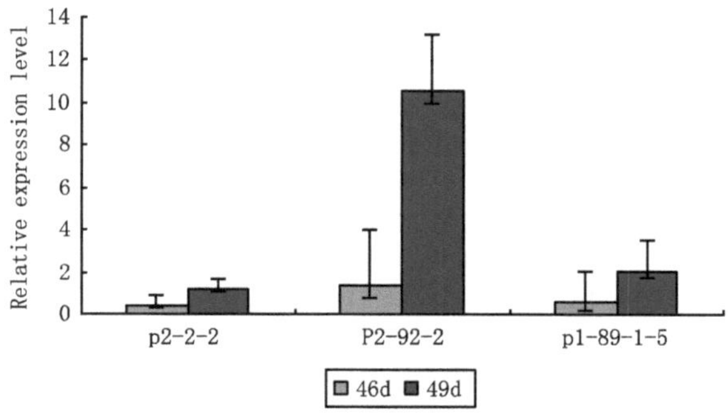

Fig. 5.2 Relative expression of 3 genes between the high-protein EMS mutant (SDPM) and the normal-protein wild type (SDPW) in developing seeds harvested at 46DAF(*46d*) or 49DAF(*49d*), respectively. Relative expression of SDPM was computed based on the corresponding gene expression of SDPW. The error bars indicate standard deviation of mean

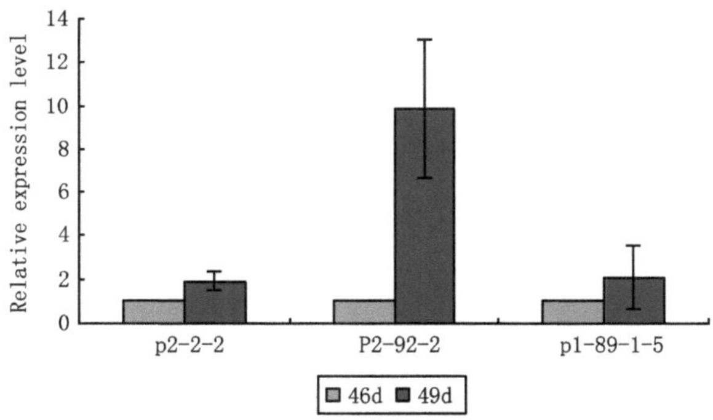

Fig. 5.3 Relative expression of the 3 genes in developing peanut seeds of SDPM harvested at 46 DAF (*46d*) and 49DAF. Error bar indicating standard deviation of mean. Relative expression at 49 DAF (*49d*) was computed based on the corresponding gene expression at 46 DAF

seeds harvested at 46 DAF or 49 DAF was illustrated in Fig. 5.2. Notably, expression of P2-92-2 (*arachin ahy-4*) at 49 DAF in SDPM was 10.60 times as high as in SDPW. Expression of the same 3 genes in seeds of SDPM harvested at 49 DAF relative to the expression at 46 DAF was shown in Fig. 5.3. P2-92-2 exhibited a marked increase in expression at 49 DAF as compared with that at 46 DAF.

Through NIRS-aided selection of mutagenized populations, SPRI scientists were able to identify peanut quality mutants, providing materials for the present study [5]. In contrast to randomly selected peanut materials with different protein content, mutant and wild types have similar genetic backgrounds, thereby precluding a large number of genes unrelated to the target trait when transcriptional profiling strategy is used.

Thus far, there have been few studies on genes related to protein content in plants [9]. Two reports have shown that *PEPC* (encoding phosphor-enolpyruvate carboxylase) and *VfAAP* (encoding *Vicia faba* amino acid permease) were genes conditioning protein content in legume seeds. As compared with untransformed control, transgenic bean seeds with the overexpression construct of *PEPC* accumulated up to 20 % more protein per gram seed dry weight [10]. Overexpression of the transporter gene *VfAAP* in pea resulted in 43 % increase in total globulins production in seeds [11].

In the present study, through comparison of gene transcription in developing seeds of a high-protein mutant and its normal-protein wild type peanut genotype at the stage of rapid seed protein accumulation (46 DAF and 49 DAF), totally 40 DEGs were isolated using Genefishing™ technology. All the 3 DEGs further validated by qRT-PCR showed differential expression between the high-protein mutant and the peanut wild type, with relative expression ranging from 0.41–10.60. As indicated by BLAST2GO analysis, the 3 genes may have protein disulfide oxidoreductase activity, nutrient reservoir activity or ATPase/transporter activity, respectively. Further studies are still needed to validate the differential expression of the rest genes by qRT-PCR and to investigate the detailed and exact functions of confirmed DEGs through transgenic experiments.

Acknowledgments We wish to express our sincere thanks to the financial support from China Agricultural Research System (CARS-14), Liaoning Natural Science Foundation Project for Fostering Talent (2014027029), Key Project of Liaoning Provincial Science and Technology Department (2011201021) and China National Science and Technology Support Project (2012BAD36B00).

References

1. Yu SL, Wang CT, Yang QL et al (2011) Peanut genetics and breeding in China. Shanghai Science and Technology Press, Shanghai, pp 18–24
2. Zhang XY, Han SY, Xu J et al (2011) Genetic analysis of protein using major gene plus polygene methods in peanut (*Arachis hypogaea* L.). Chinese J Oil Crop Sci 33(2):118–122
3. Liang X, Zhou G, Hong X et al (2009) Overview of research progress on peanut (*Arachis hypogaea* L.) host resistance to aflatoxin contamination and genomics at the guangdong academy of agricultural sciences. Peanut Sci. 36:29–34
4. Sarvamangala C, Gowda MVC, Varshney RK (2011) Identification of quantitative trait loci for protein content, oil content and oil quality for groundnut (*Arachis hypogaea* L.). Field Crops Res 122:49–59
5. Wang CT, Zhang JC, Tang YY et al (2013) Peanut genetic improvement. Shanghai Science and Technology Press, Shanghai, p 531
6. Wang CT, Tang YY, Wang XZ et al (2011) Evaluation of groundnut genotypes from China for nutritional quality. J SAT Agric. Res. vol. 9. Available at: http://ejournal.icrisat.org/Volume9/Groundnut/Evaluation.pdf. (Accessed 31 May 2013)
7. Wan SB (ed) (2003) China peanut cultivation science. Shanghai Science & Technology Press, Shanghai pp 408–430

8. Tang YY, Wang CT, Yang GP et al (2011) Identification of chilling-responsive transcripts in peanut (*Arachis hypogaea L.*). Electronic J. Biotechnol. 14(5). DOI: 10.2225/vol14-issue5-fulltext-5
9. Weber H, Sreenivasulu N, Weschke W (2010) Molecular physiology of seed maturation and seed storage protein biosynthesis. In: Pua EC, Davey MR (eds) Plant developmental biology–biotechnological perspectives, Springer, Berlin Heidelberg, pp 83–104
10. Rolletschek H, Borisjuk L, Radchuk R et al (2004) Seed-specific expression of a bacterial phosphoenolpyruvate carboxylase in Vicia narbonensis increases protein content and improves carbon economy. Plant Biotechnol J 2(3):211–219
11. Rolletschek H, Hosein F, Miranda M et al (2005) Ectopic expression of an amino acid transporter (VfAAP1) in seeds of Vicia narbonensis and pea increases storage proteins. Plant Physiol 137(4):1236–1249

Chapter 6
Identification of the Binding Domains of Nedd4 E3 Ubiquitin Ligase with Its Substrate Protein TMEPAI

Lei Jing, Xin Huo, Yufeng Li, Yuyin Li and Aipo Diao

Abstract To investigate which domains of Nedd4 are responsible for the interaction with its substrate protein TMEPAI, the plasmids encoding GST-tagged WW domains of Nedd4 and His-HA-tagged TMEPAI were constructed, and the fusion proteins were expressed in *E. coli* BL21-CodonPlus (DE3)-RIL cells and purified. GST pull-down experiment indicated that Nedd4 directly interacts with His-HA-TMEPAI, and the WW2 and WW3 domains of Nedd4 are responsible for the interaction with TMEPAI, which represents a general mechanism for the interaction between Nedd4 and its substrate protein.

Keywords Nedd4 · WW domain · TMEPAI · GST pull-down

6.1 Introduction

Ubiquitination is a common protein modification, which regulates protein degradation, as well as protein transport and a variety of metabolic pathways in the cells [1]. The process of ubiquitination involves the ubiquitin-activating enzyme E1, ubiquitin-conjugating enzyme E2, and ubiquitin-protein ligating enzymes E3. Nedd4 (neuronal precursor cell expressed developmentally down-regulated 4) belongs to the Nedd4 of HECT E3 ubiquitin ligase family including Nedd4, Nedd4-2, Smurf1, Smurf2, Itch, WWP1, WWP2, Nedl1, and Nedl2 [2, 3]. Nedd4 contains a C2 domain, four WW domains, and a HECT domain [4]. The C2 domain binds to phosphor-lipids. The WW domain is 35–40 amino acids long containing two conserved tryptophan (W) residues

L. Jing · X. Huo · Y. Li · Y. Li · A. Diao (✉)
Key Lab of Industrial Fermentation Microbiology of the Ministry of Education,
School of Biotechnology, Tianjin University of Science and Technology,
Tianjin 300457, China
e-mail: diaoaipo@tust.edu.cn

T.-C. Zhang and M. Nakajima (eds.), *Advances in Applied Biotechnology*,
Lecture Notes in Electrical Engineering 332, DOI 10.1007/978-3-662-45657-6_6

spaced 21 amino acids apart. The WW domains interact with a variety of proline-based motifs, and recognize proline-containing phosphor-serine/phosphor-threonine sequences on the substrate protein. The HECT domain is comprised of approximately 350 amino acids and is responsible for ubiquitin transfer from a conserved cysteine residue to a lysine residue in the substrate protein [5–7].

TMEPAI is a Nedd4-binding protein that is originally identified as a prostate abundant, highly androgen-induced protein, and the *TMEPA1* gene is mapped to chromosome 20q13 [8]. TMEPAI protein contains two PY motifs (PPPY and PPTY), which are necessary for binding to the WW domains of Nedd4 ubiquitin ligase [8, 9].

It has been reported that TMEPAI is a substrate of Nedd4 through the interaction with the WW domains, however, which WW domains interact with TMEPAI is still unclear. In this study, we investigate the binding affinity of the four WW domains of Nedd4 with its substrate protein TMEPAI [10].

6.2 Materials and Methods

6.2.1 Construction of Plasmid

Four DNA fragments encoding the WW domains of Nedd4 were amplified by PCR using the primers listed in Table 6.1, and the four fragments were digested with *Bam*HI and *Eco*RI and inserted into the pGEX-4T-2 (GE Healthcare) vector. The *TMEPAI* gene was amplified by PCR with the forward primer 5′-ATGAGCCACTACAAGCTGTCTG-3′ and reverse primer 5′-CTAGAGAGGGTGTCCTTTC-3′, and the PCR product was purified and digested with *Bam*HI and *Eco*RI, and inserted into the *Bam*HI and *Eco*RI sites of the pET21b vector.

Table 6.1 Primers used for PCR experiments

Target gene	Primer sequence
Nedd4-WW1 domain	
Forward	5′-CAACAACAAGAACCTTCTCCTCTAC-3’
Reverse	5′-CTTAGCCATTCTCAGCATCTGTTAGG-3’
Nedd4-WW2 domain	
Forward	5′-CTTTTGCCTACTTCATCTGGATTAC-3’
Reverse	5′-TTAGGTCAGCTGACTGGTCTCCACTG-3’
Nedd4-WW3 domain	
Forward	5′-TCTGAAATTGAGCAAGGATTC-3’
Reverse	5′-TTATGTCTTTCCTCTCAGATGGGCTG-3’
Nedd4-WW4 domain	
Forward	5′-ACTTCCAATGATCTAGGGCCTTTAC-3’
Reverse	5′-TTACACTGCTGGTCCAGTTATTGC-3’

6.2.2 Protein Expression and Purification

Plasmids encoding GST-fusion proteins of Nedd4 were transformed into *E. coli* BL21-CodonPlus (DE3)-RIL (Stratagene). Cells were induced with 0.1 mM IPTG overnight at 16 °C, and lysed in lysis buffer (20 mM Tris–Cl pH 7.4, 100 mM NaCl, 2 mM DTT) containing protease inhibitors. The recombinant protein was purified on Glutathione–Sepharose beads (GE Healthcare). After sonication, the extracts were clarified by centrifugation. The recovered supernatants were allowed to bind to Glutathione Sepharose beads for 3 h at 4 °C. The matrix was washed three times with lysis buffer, then the beads were eluted with elution buffer (100 mM Tris–Cl pH 7.4, 100 mM NaCl, 20 mM reduced glutathione). His-HA-TMEPAI was expressed in *E. coli* BL21-CodonPlus (DE3)-RIL (Stratagene) cells that transformed with pET21b-His-HA-TMEPAI plasmid, and induced with 0.5 mM IPTG overnight at 16 °C. The His-HA-tagged protein was purified on Ni-Sepharose beads after lysis of the bacteria in Na–P lysis buffer (50 mM Na–P pH 8.0, 300 mM NaCl, 5 mM β-mercaptoethanol) plus protease inhibitors. After sonication, the extracts were clarified by centrifugation. The recovered supernatants were allowed to bind to Ni-Sepharose beads for 3 h at 4 °C. The matrix was washed three times with Na–P lysis buffer containing 50 mM imidazole, the beads were eluted with elution buffer (50 mM Na–P pH 8.0, 300 mM NaCl, 5 mM β-mercaptoethanol, and 250 mM imidazole).

6.2.3 GST Pull-Down Assays and Western Blot

For the GST pull-down assays, 5 μg GST-fusion proteins were first immobilized on the Glutathione–Sepharose beads. The matrix was incubated for 2 h in rotation at 4 °C, then washed three times with cold GST pull-down buffer (20 mM Tris–Cl pH 7.4, 100 mM NaCl, 1 % Triton-100, 2 mM DTT). At the same time, 0.5 μg protein of His-HA-TMEPAI was resuspended in the GST pull-down buffer and clarified by centrifugation at 14,000 rpm for 10 min. The supernatant was added into the GST-beads and incubated for 2 h in rotation at 4 °C, then washed three times with cold GST pull-down buffer, and washed once with 1 × PBS. Finally, the beads were resuspended and boiled in 40 μL 2 × SDS loading buffer; after boiled for 5 min, the supernatants were collected and separated by 10 % SDS-PAGE, electroblotted onto a polyvinylidene difluoride membrane (Millipore) and analyzed by Western blot using anti-HA-tag antibody.

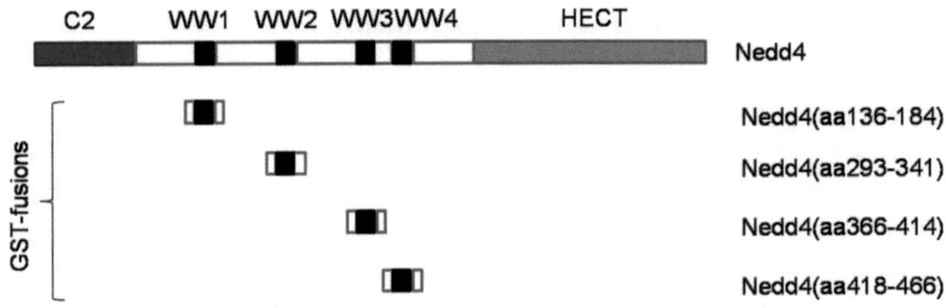

Fig. 6.1 The schematic structure of Nedd4. Amino acid positions of individual domains are denoted with corresponding numbers. C2, calcium-dependent lipid binding domain; WW, two conserved tryptophan (W) residues spaced 21 amino acids apart; HECT, homologous to E6 carboxyl terminus

6.3 Result

6.3.1 The Schematic Structure of Nedd4

Nedd4 (NP-006145) contains 900 amino acids, including a N-terminal C2 domain for membrane binding, a central region containing four WW domains for protein–protein interaction, and a C-terminal HECT domain for ubiquitin-protein ligation (Fig. 6.1).

6.3.2 Nedd4 Interacts with TMEPAI Directly

To verify that Nedd4 directly interacts with TMEPAI, GST-Nedd4, and His-HA-TMEPAI proteins were expressed in *E. coli* BL21-CodonPlus (DE3)-RIL and purified. GST pull-down experiment showed that GST-Nedd4, but not GST, pulled down purified His-HA-TMEPAI (Fig. 6.2), which indicates that GST-Nedd4 interacts with His-HA-TMEPAI directly.

6.3.3 Expression and Purification of the GST-Tagged WW Domains of Nedd4

To investigate which domains of Nedd4 are responsible for binding with TMEPAI, four plasmids encoding GST-tagged WW1, WW2, WW3, and WW4 domains of Nedd4 were constructed (Fig. 6.1). These four GST-fusion proteins were expressed and purified. The purified fusion proteins were subjected to SDS-PAGE and stained with Coomassie blue (Fig. 6.3).

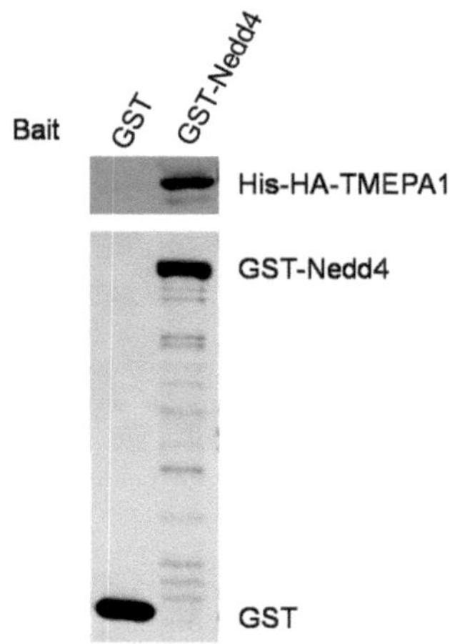

Fig. 6.2 Direct interaction between Nedd4 and TMEPAI determined by the GST pull-down assay

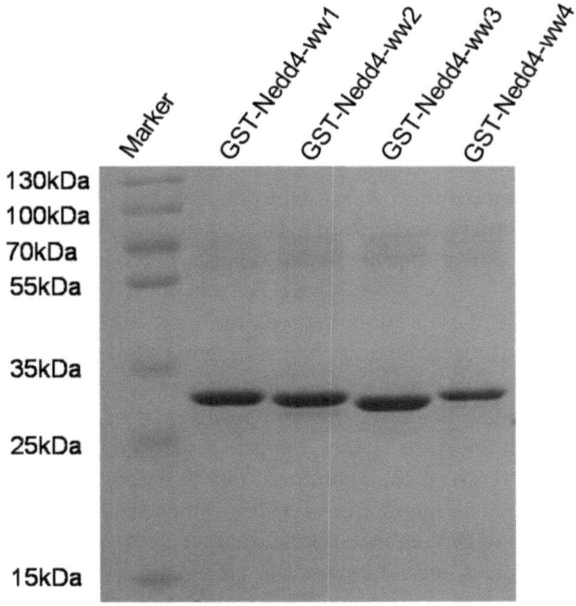

Fig. 6.3 GST-tagged WW domains of Nedd4 were expressed and purified using Glutathione–Sepharose beads. The protein marker lane indicates the protein masses

6.3.4 The WW2 and WW3 Domains of Nedd4 Interact with TMEPAI

To measure the binding affinity of the Nedd4 WW domains with TMEPAI, the GST pull-down assay was performed. 5 μg GST or GST-WW domain fusion proteins were first bound to Glutathione–Sepharose beads, then incubated with 0.5 μg

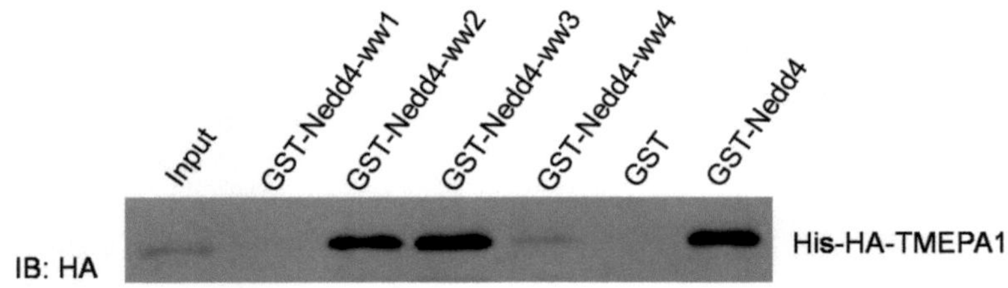

Fig. 6.4 GST pull-down assay to identify the interaction domain of Nedd4 with TMEPAI. GST or GST-WW domains fusion proteins coupled to the Glutathione–Sepharose beads were incubated with His-HA-TMEPAI for 2 h at 4 °C and analyzed by Western blot. 5 % input represents 5 % of the material used for loading

His-HA-TMEPAI protein. GST pull-down assay showed that the WW2 and WW3 domain had a higher binding affinity with His-HA-TMEPAI, with a much less binding affinity for WW4 domain, but no binding affinity for WW1 domain (Fig. 6.4). These data indicate that the WW2 and WW3 domains of Nedd4 are mainly responsible for the interaction with TMEPAI.

6.4 Discussion

Previous studies have demonstrated that TMEPAI is a substrate of Nedd4 E3 ubiquitin ligase [5, 9, 11]. The mouse Nedd4 contains three WW domains and it has been shown that the WW2 domain plays an important role in protein interaction [12, 13]. In this study, we have identified that the WW2 and WW3 domains of Nedd4 play an important role in interacting with TMEPAI. This may represent a general mechanism for the interaction between Nedd4 and its substates. However, we need to make a series of mutants of these proteins for further study.

Protein ubiquitination is a posttranslational modification in which a ubiquitin, or a multiubiquitin chain, is attached onto proteins by the formation of an isopeptide bond between the C-terminal Gly76 carboxyl group of ubiquitin and the 3-amino group of an internal Lys residue of the substrates, so as to control its stability or as a signal of protein transportation in the cells [14–17]. Given lacking of the specificity of proteasome inhibitors, the enzymatic nature, abundance, and specific substrate recognition properties of E3 ubiquitin ligase indicate that E3 might be used to as even more specific and effective therapeutic targets. Targeting individual E3 would selectively stabilize a particular or limit a number of protein substrates, thus possibly limiting untoward side effects. Hence, drugs targeting E3 promise a better therapeutic ratio than proteasome inhibitors, and the potential E3 inhibitors have a stronger chance to inhibit the active catalytic site.

Acknowledgments This research is supported by the program for Changjiang Scholars and Innovative Research Team in University (IRT1166).

References

1. Rotin D, Kumar S (2009) Physiological functions of the HECT family of ubiquitin ligases. Nat Rev Mol Cell Biol 10(6):398–409
2. Chen C, Matesic LE (2007) The Nedd4-like family of E3 ubiquitin ligases and cancer. Cancer Metastasis Rev 26(3–4):587–604
3. Bernassola F, Karin M, Ciechanover A et al (2008) The HECT family of E3 ubiquitin ligases: multiple players in cancer development. Cancer Cell 14(1):10–21
4. Lu PJ, Zhou XZ, Lu KP et al (1999) Function of WW domains as phosphoserine- or phosphothreonine-binding modules. Science 283(5406):1325–1328
5. Scheffner M, Staub O (2007) HECT E3s and human disease. BMC Biochem 8:S1–S6
6. Magnifico A, Ettenberg S, Lipkowitz S (2003) WW domain HECT E3s target Cbl RING finger E3s for proteasomal degradation. J Biol Chem 278(44):43169–43177
7. Sudol M, Hunter T (2000) NeW wrinkles for an old domain. Cell 103(7):1001–1004
8. Xu LL, Shanmugam N, Srivastava S et al (2000) A novel androgen-regulated gene, PMEPA1, located on chromosome 20q13 exhibits high level expression in prostate. Genomics 66 (3):257–263
9. Xu LL, Shi Y, Srivastava S et al (2003) PMEPA1, an androgen-regulated NEDD4-binding protein, exhibits cell growth inhibitory function and decreased expression during prostate cancer progression. Cancer Res 63(15):4299–4304
10. Macias MJ, Wiesner S, Sudol M (2002) WW and SH3 domains, two different scaffolds to recognize proline-rich ligands. FEBS Lett 513(1):30–37
11. Li H, Xu LL, Srivastava S et al (2008) A feedback loop between the androgen receptor and a NEDD4-binding protein, PMEPA1, in prostate cancer cells. J Biol Chem 283(43):28988–28995
12. Yang B, Kumar S (2010) Nedd4 and Nedd4-2: closely related ubiquitin-protein ligases with distinct physiological functions. Cell Death Differ 17(1):68–77
13. Giuseppe G, Lucia DM, Alberto G et al (2003) EGF- and cell-cycle-regulated STAG1/PMEPA1/ERG1.2 belongs to a conserved gene family and is overexpressed and amplified in breast and ovarian cancer. Mol Carcinog 38(4):188–200
14. Ciechanover A (1998) The ubiquitin-proteasome pathway: on protein death and cell life. EMBO J 17(24):7151–7160
15. Hershko A, Ciechanover A (1998) The ubiquitin system. Ann Rev Biochem 67:425–479
16. Ciechanover A (2005) Intracellular protein degradation: from a vague idea thru the lysosome and the ubiquitin–proteasome system and onto human diseases and drug targeting. Cell Death Differ 12(9):1178–1190
17. Ing B, Kotler AS, Rotin D et al (2007) Regulation of commissureless by the ubiquitin ligase DNedd4 is required for neuromuscular synaptogenesis in drosophila melanogaster. Mol Cell Biol 27(2):481–496

Chapter 7
Optimization of the Fermentation Conditions of Pep-1-Fused EGF in *Escherichia coli*

Tong-Cun Zhang, De-Yun Ma, Xue-Gang Luo and Yue Wang

Abstract Human epidermal growth factor (hEGF), a well-known polypeptide agent which has been widely used in the medicine and cosmetics, is a 6.2 kDa single-chain polypeptide consisting of 53 amino acids. Here, to enhance its trans-membrane ability, a recombinant EGF fused with Pep-1, a cell-penetrating peptides (CPP) that has been previously shown to be powerful transport vector tool for the intracellular delivery of a variety of cargos through the cell membrane, was expressed in *Escherichia coli*. Furthermore, The expression conditions was optimized, and the results showed that the Pep-1-fused EGF (P-EGF) could be successfully expressed in *E. coli* BL21-TrixB (DE3) using an expression vector, pGEX-6P-3, which contains a GST tag. The recombinant product reached the highest soluble expression when the expression strain was induced by 0.2 mmol/l IPTG and cultivated at the temperature of 20 °C with a rotation speed of 200 rpm for 8 h.

Keywords Epidermal growth factor (EGF) · P-EGF · Optimization · Fermentation conditions

T.-C. Zhang (✉) · D.-Y. Ma · X.-G. Luo · Y. Wang (✉)
Key Laboratory of Industrial Fermentation Microbiology, Ministry of Education, College of Biotechnology, Tianjin University of Science and Technology, Tianjin 300457, China
e-mail: tony@tust.edu.en

X.-G. Luo
e-mail: luoxuegang@tust.edu.cn

T.-C. Zhang · D.-Y. Ma · X.-G. Luo · Y. Wang
Tianjin Key Laboratory of Industrial Microbiology, Tianjin 300457, China

T.-C. Zhang and M. Nakajima (eds.), *Advances in Applied Biotechnology*, Lecture Notes in Electrical Engineering 332, DOI 10.1007/978-3-662-45657-6_7

7.1 Introduction

Epidermal growth factor (EGF) is a 6.2 kDa single-chain small polypeptide containing 53 amino acids secreted from various tissues [1] and it has huge potential application value and broad market prospects both as medicines and cosmetics. However, the transdermal absorption of EGF is very weak because of the intestine mucosal and stratum corneum which functions as barriers to prevent penetration of most foreign large molecules [2, 3]. To solve this problem, the EGF could be fused with Pep-1, a cell-penetrating peptide (CPP) that has been previously shown to be a powerful transport vector tool for the intracellular delivery of a variety of cargos through the cell membrane [4, 5].

In our previously study, we have successfully constructed a recombinant strain to express the protein P-EGF, a fusion of EGF and PEP-1. The gene encoding EGF and DNA fragment encoding Pep-1 were combined together using overlapping PCR technology and then inserted into the vector pGEX-6P-3 with a GST tag to construct the expression vector. In order to obtain the maximum soluble expression of the fusion protein P-EGF, the expression conditions were optimized in this study.

7.2 Materials and Methods

7.2.1 Bacterial Strain and Plasmid

Escherichia coli BL21-TrixB (DE3) was used as the expression host. pGEX-6P-3 with a GST tag was used as the expression vector. The recombination plasmid P-EGF was constructed by our laboratory.

7.2.2 Expression of the Fusion Protein

Escherichia coli BL21-TrixB (DE3) harboring recombinant plasmid was cultured in LB medium containing 50 μg/ml ampicillin and 15 μg/ml kanamycin at 37 °C to OD_{600} of 0.6. The cells were then induced with 0.2 mM isopropy-beta-D-thiogalactopyranoside (IPTG) at 20 °C for 12 h. After harvest, cells were centrifuged at 5,000 × g for 20 min, and then lysed by sonication in phosphate buffer. After centrifugation at 13,000 × g for 30 min, the supernatant was taken out as the soluble protein sample for the subsequent experiment.

7.2.3 Protein Quantification

The total protein concentration was determined by BCA assay. The protein was detected by sodium dodecyl sulfate-polyacrylamide gel electrophoresis (SDS-PAGE). The proportion of fusion protein P-EGF in total protein was analyzed using specialized software Quantity One.

7.3 Results

7.3.1 Optimization of Temperature

The optimum growth temperature for *E. coli* is 37–39 °C, but this temperature was not suitable to express P-EGF for the formation of inclusion bodies. In order to find optimal temperature for the soluble expression of protein, several different temperature gradients were tested, including 37, 30, 25, and 20 °C. P-EGF expressed under different temperature conditions showed in Fig. 7.1 and the result demonstrated that 20 °C was the most suitable temperature tested.

7.3.2 Optimization of Rotation Speed

Different speeds will result in different dissolved oxygen and then will lead to different protein expression. Therefore, four rotation speeds, including 250, 200, 150, and 100 rpm, were chosen to be analyzed. The result showed that highest expression of the soluble protein could be obtained when the rotation speed was 200 rpm (Fig. 7.2).

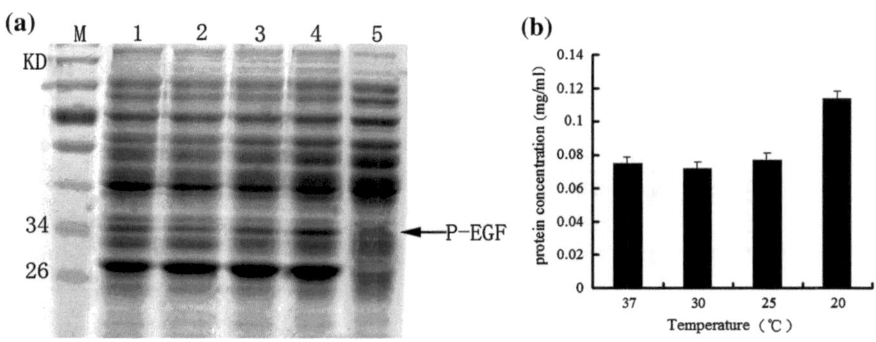

Fig. 7.1 Optimization of temperature for the expression of P-EGF **a** SDS-PAGE analysis of P-EGF expressed under different temperature conditions. *M* protein marker; *1* expression of P-EGF at 37 °C; *2* expression of P-EGF at 30 °C; *3* expression of P-EGF at 25 °C; *4* expression of P-EGF at 20 °C; *5* expression of P-EGF without induction. **b** Quantitative analysis of P-EGF expressed under different temperature conditions

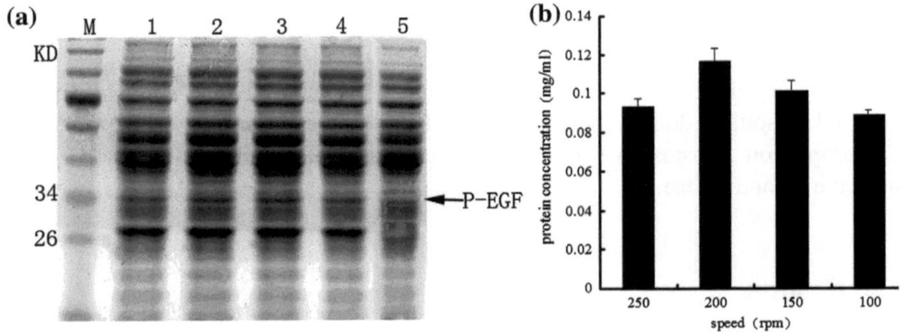

Fig. 7.2 Optimization of rotation speed for the expression of P-EGF **a** SDS-PAGE analysis of P-EGF expressed under different revolutions per minute. *M* protein marker; *1* expression of P-EGF at 250 rpm; *2* expression of P-EGF at 200 rpm; *3* expression of P-EGF at 150 rpm; *4* expression of P-EGF at 100 rpm; *5* expression of P-EGF without induction. **b** Quantitative analysis of P-EGF expressed under different rotation speeds

7.3.3 Optimization of IPTG Concentration

The inducer IPTG could reduce protein expression for its certain toxicity to cells when it was added too much. Therefore, the concentration of IPTG is also very important for the expression of fusion protein P-EGF. In this study, the effect of different concentration of IPTG, including of 0.1, 0.2, 0.4, and 0.6 mmol/l, on the yield of the recombinant protein, were detected. As shown in Fig. 7.3, the optimal IPTG concentration for the expression of the fusion protein was 0.2 mmol/l.

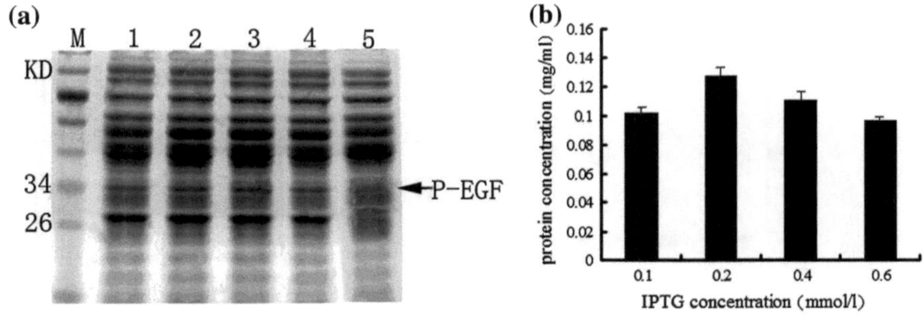

Fig. 7.3 Optimization of IPTG concentration for the expression of P-EGF **a** SDS-PAGE analysis of P-EGF expressed with different IPTG concentrations. *M* protein marker; *1* expression of P-EGF with 0.1 mmol/l IPTG; *2* expression of P-EGF with 0.2 mmol/l IPTG; *3* expression of P-EGF with 0.4 mmol/l IPTG; *4* expression of P-EGF with 0.6 mmol/l IPTG; *5* expression of P-EGF without induction. **b** Quantitative analysis of P-EGF expressed with different IPTG concentrations

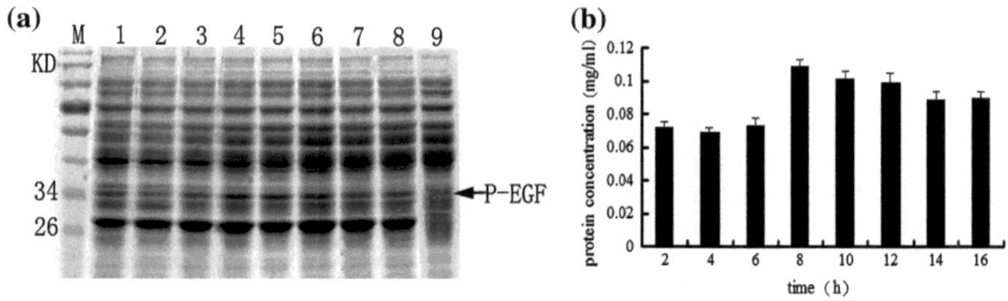

Fig. 7.4 Optimization of induction time for the expression of P-EGF **a** SDS-PAGE analysis of P-EGF expressed with different induction times. *M* protein marker; *1* expression of P-EGF for 2 h; *2* expression of P-EGF for 4 h; *3* expression of P-EGF for 6 h; *4* expression of P-EGF for 8 h; *5* expression of P-EGF for 10 h; *6* expression of P-EGF for 12 h; *7* expression of P-EGF for 14 h; *8* expression of P-EGF for 16 h; *9* expression of P-EGF without induction. **b** Quantitative analysis of P-EGF expressed for different induction times

7.3.4 Optimization of Induction Time

The IPTG was added when then the OD_{600} of the culture reached 0.6. After induction, the sample was collected every 2 h. The expression condition at different time was showed in Fig. 7.4 and the result demonstrated that the optimal induction time was 8 h.

7.4 Conclusion

Administration of macromolecule drugs nonintravenously always exhibited low bioavailability due to the penetration limitation of cell membrane barrier [6]. Increasing evidence has indicated that CPPs could facilitate proteins' transmembrane transport [7]. In our previous work, we have successfully constructed the recombinant plasmid P-EGF. And in this study, we mainly carried out the work to optimize the fermentation conditions for the soluble expression of P-EGF. The expression conditions were optimized, and the results showed that the P-EGF reached the highest soluble expression when the expression strain was induced by 0.2 mmol/l IPTG and cultivated at the temperature of 20 °C with a rotation speed of 200 rpm for 8 h.

References

1. Kim DG, Min MK, Ahn SC, Kim JK, Kong IS (2009) Expression of a fusion protein containing human epidermal growth factor and the collagen-binding domain of *Vibrio mimicus* metalloprotease. Biotechnol Lett 31(2):259–264
2. Bau mann L (2007) Skin ageing and its treatment. J Pathol 211(2):241–251

3. Brown MB, Martin GP, Jones SA, Akomeah FK (2006) Dermal and transdermal drug delivery systems: current and future prospects. Drug Deliv 13(3):175–187
4. Deshayes S, Morris MC, Divita G, Heitz F (2005) Cell-penetrating peptides: tools for intracellular delivery of therapeutics. Cell Mol Life Sci 62(16):1839–1849
5. Koren E, Torchilin VP (2012) Cell-penetrating peptides: breaking through to the other side. Trends Mol Med 18(7):385–393
6. Morris MCVP, Chaloin L, Heitz F, Divita G (1997) A new peptide vector for efficient delivery of oligonucleotides into mammalian cells. Nucleic Acids Res 25(14):2730–2736
7. Ruan RQ, Wang SS, Wang CL, Zhang L, Zhang YJ, Zhou W et al (2013) Transdermal delivery of human epidermal growth factor facilitated by a peptide chaperon. Eur J Med Chem 62:405–409

Chapter 8
Characterization of Rhamnolipid Production in a *Pseudomonas aeruginosa* Strain

Cuikun Zhang and Hongjiang Yang

Abstract Rhamnolipid is a class of glycolipid biosurfactant and has the potential to replace chemical surfactants due to its low toxicity and biodegradability. In this study, 38 *Pseudomonas* strains were screened for rhamnolipid production on cetyl trimethyl ammonium bromide (CTAB)-methylene blue agar plates. Among them, strain 487 was found to produce the relative high level of rhamnolipid. The rhamnolipid extract could reduce the surface tension of the medium from 67.2 to 32.7 mN/m and form stable emulsions with liquid paraffin. Strain 487 was identified as *Pseudomonas aeruginosa* with 16S rDNA gene sequence analysis. Rapeseed oil and glycerol were investigated for rhamnolipid production. Strain 487 produced 42.4 g/L rhamnolipid in rapeseed oil medium versus 33.4 g/L rhamnolipid using glycerol as carbon resource after 72 h fermentation. In addition, thin layer chromatography (TLC) and Fourier transform infrared spectroscopy (FTIR) were performed to analyze the structure of the rhamnolipid molecule synthesized by strain 487.

Keywords Biosurfactant · *Pseudomonas aeruginosa* · Rhamnolipid

8.1 Introduction

Biosurfactants are produced by a wide variety of microorganisms. Like their chemical counterparts, biosurfactants can reduce the surface tension. They can be used in emulsification, foaming, or dispersion. The most important advantages of biosurfactant are ecological acceptance, low toxicity, and biodegradability [1]. Biosurfactants have been widely applied in different areas such as oil industry, food industry, cosmetics, and pharmaceutical industries [2].

C. Zhang · H. Yang (✉)
Tianjin University of Science and Technology, 300457 Tianjin, China
e-mail: hongjiangyang@tust.edu.cn

T.-C. Zhang and M. Nakajima (eds.), *Advances in Applied Biotechnology*,
Lecture Notes in Electrical Engineering 332, DOI 10.1007/978-3-662-45657-6_8

Microbial surfactants could be classified into different types according to their chemical groups, such as glycolipids, lipopeptides, fatty acids, phospholipids, peptides, and neutral lipids [3]. Rhamnolipids belong to glycosides, and are composed of a glycon part and an aglycon part linked to each other via O-glycosidic linkage. The glycon part is composed of one or two rhamnose moieties, while the aglycon part is composed of mainly one or two β-hydroxy fatty acid chains [4].

Pseudomonas strains, belonging to gram-negative bacteria, have been reported to produce rhamnolipids since 1949 [5]. The *Pseudomonas* species could utilize both water-soluble substrates (such as glycerol, glucose, mannitol, and ethanol) and water-immiscible carbon sources (like nalkane and vegetable oils) for the rhamnolipid production [6]. Normally, the carbon sources immiscible in water can synthesize a higher level of rhamnolipid production [7].

This work aimed at screening biosurfactant producers from 38 collected strains. A semi-quantitative method, cetyl trimethyl ammonium bromide (CTAB) with methylene blue, was used to differentiate between the rhamnolipid producing and non-producing strains. Rhamnolipid production and characteristics of the rhamnolipid extract were investigated.

8.2 Materials and Methods

8.2.1 Media and Solutions

PIA medium was used for routinely cultivation and maintenance of bacteria, and it was composed of $MgCl_2$ 1.4 g/L, K_2SO_4 10 g/L, peptone 10 g/L. The medium was adjusted to pH 7.0. To make the solid medium, 15 g/L agar is added to PIA medium.

LB broth was used for preparation of the seed culture, and it was composed of NaCl 10 g/L, peptone 10 g/L, and yeast extracts 5 g/L.

CTAB-methylene blue agar plate was used for the selection of biosurfactant producing strains, and it was composed of KH_2PO_4 0.7 g/L, Na_2HPO_4 0.9 g/L, $NaNO_3$ 2 g/L, $MgSO_4$ 2.5 g/L, $CaCl_2$ 0.1 g/L, glycerol 20 g/L, CTAB 0.2 g/L, methylene blue 0.005 g/L, trace element solution 1 mL/L, and agar 15 g/L. The medium was adjusted to pH 6.8.

The mineral medium (pH 6.8) was composed of NaCl 1 g/L, KCl 1 g/L, $NaNO_3$ 6 g/L, KH_2PO_4 3 g/L, Na_2HPO_4 0.3 g/L, $MgSO_4$ 2.5 g/L, $CaCl_2$ 0.1 g/L, and trace element solution 1 mL/L. 30 g/L carbon resource was added to the medium for biosurfactant production.

Trace element solution was composed of $ZnSO_4$ 0.75 g/L, $CuSO_4$ 0.075 g/L, $MnSO_4$ 0.75 g/L, and H_3BO_3 0.15 g/L.

2 % anthrone-sulfuric acid solution was used for quantitative detection of rhamnolipid.

8.2.2 Isolation of Rhamnolipid Producing Strain

The anionic rhamnolipid could react with cationic bromide salt and form an insoluble complex revealed by methylene blue staining. Bacteria which had a dark blue halo around the colony were determined as rhamnolipid producing strains [8]. The collected 38 laboratory strains were incubated on CTAB-methylene blue agar plates by spotting at 37 °C for 72 h to observe whether there is a dark blue halo around the colony. The diameter of the halos and colonies were measured.

To screen the strains which could produce higher level of rhamnolipid, single colony of each strain was inoculated into 5 mL mineral medium (supplemented with 30 g/L glycerol) with a shaking speed 220 rpm at 37 °C for 72 h. At the end of cultivation, OD_{600}, emulsification activity and the production of rhamnolipid were measured, respectively.

8.2.3 Determination of Emulsification Activity

For determination of the emulsification activity, 2 mL of suitable diluted supernatant and equal volume of liquid paraffin was mixed in a tube and then vortexes for 30 s. The test tubes were maintained at room temperature, and the height of the emulsion layer was measured after 72 h. Emulsification activity was evaluated as the equation as below.

$$\text{Emulsification activity} = \frac{\text{the height of emulsification layer}}{\text{the original height of liquid paraffin}}.$$

8.2.4 Production Detection of Rhamnolipid

One of the most widely used methods for rhamnolipid quantification is the anthrone (9, 10-dihydro-9-oxoanthracene) test. The rhamnolipid groups are hydrolyzed and transformed into methyl furfural, which then reacts with the anthrone to produce a blue-green color that can be measured at 625 nm [9]. A standard curve is prepared with rhamnose, a correction factor must be applied to compensate for the extra mass of the lipidic portion of rhamnolipid, rhamnolipid (g/L) = 3.4 × rhamnose (g/L) [10]. 2.5 mL 2 % anthrone-sulfuric acid was slowly added to 1.25 mL suitable diluted supernatant, at the time the tube should gently shaken in an ice bath to remove the rapidly generated heat. The anthrone-added sample then needs to be heated in boiled water for additional 20 min to complete the reaction part of the anthrone method. To make the standard curve, 0, 20, 40, 60, 80, 100 mg/L rhamnose was reacted and detected at 625 nm, every concentration repeated 3 times and its mean was used as the result.

8.2.5 Identification of Selected Organism

Bacterial genomic DNA was extracted as described previously [11]. Specific primers 27F and 1492R were used to amplify bacterial 16S rDNA gene fragment, and their sequences were as follows: 27F: 5′-AGAGTTTGATCCTGGCTCAG-3′, 1492R: 5′-GGTTACCTTGTTACGACTT-3′. The PCR system was 50 μL and the parameters include denaturation at 94 °C for 5 min, 1 cycle; denaturation at 94 °C for 1 min, annealing at 53 °C for 45 s, and extension at 72 °C for 1.5 min, 30 cycles; 72 °C for 10 min.

PCR product was purified from gel and subjected to sequencing analysis with the same primers used in PCR. Sequence homology search was carried out with BLAST provided by NCBI to identify the strain [12].

8.2.6 The Time Course of Rhamnolipid Production

To study the rhamnolipid accumulation of different periods, overnight seed culture of the strain was grown on mineral medium (30 g/L rapeseed oil added) at 35 °C, 200 rpm in an orbital shaker. 1 mL samples of culture broth were collected at 6 or 12 h intervals for a period of 158 h. Biomass was estimated as OD_{600}, we also evaluated their production and pH. Then we could draw a time curve about biomass and production.

8.2.7 Extraction of Rhamnolipid

Production was done with 487 in mineral medium containing 3 % rapeseed oil at 35 °C, 200 rpm. After 72 h incubation, cell-free supernatant was adjusted to pH 2.0 used HCl, standing at 4 °C for 24 h. Equal volume of chloroform and methanol (v/v, 2:1) were added to biosurfactant solution and shaken for extraction, the upper phase was extracted twice. Combined the organic phase and evaporated using a rotary vacuum evaporator at 45 °C, then suspended the crude rhamnolipid with 0.05 M $NaHCO_3$. For getting solid crude rhamnolipid, the crude solution was freeze-dried.

8.2.8 Thin Layer Chromatography Assay

10 μL of crude rhamnolipid was applied on a side of silica gel-coated sheet, the eluent was composed of chloroform: methanol: acetic acid (65: 15: 2, V: V: V). The chromatography was stopped when the solvent almost reached another side.

The silica gel plate was dyed by anthrone-sulfuric acid, and baked until the color appeared. The retardation factors (Rf) of the spots were determined. This allows for straight forward separation of mono- and di-rhamnolipid according to the eluting distance. Although this method provides good structural information, it is not suitable for quantification purposes.

8.2.9 FTIR Spectra Analysis

Fourier transform infrared spectroscopy (FTIR) spectrometer is very useful for identifying types of chemical bonds [13]. The extracted biosurfactant was freeze-dried and used for FTIR spectra analysis. Briefly, 2 mg of crude biosurfactant was milled with 200 mg of KBr to form powder, and then pressed with 10,000 kg for 1 min to obtain a thin pellet. The pellet was analyzed in the wave number range of 4,000–400 cm^{-1}.

8.2.10 Determination of the Critical Micelle Concentration (CMC)

Determination of CMC was performed by measuring surface tension of serial dilution of the biosurfactant at room temperature. We could get a curve about surface tension and concentration of rhamnolipid. CMC is defined as the concentration of surfactants above which micelles form and all additional surfactants added to the system go to micelles. So the minimum concentration which could reach the lowest surface tension was defined as the CMC of the rhamnolipid under studied.

8.3 Results and Discussion

8.3.1 Screening of Biosurfactant Producing Strains

Thirty-eight different colonies were cultured on the CTAB-methylene blue agar plate, and incubated at 37 °C for 72 h. We found that four strains could not grow on the plate, another two strains did not produce the dark blue halo. There are 32 strains who have dark halos around the colonies, most of them (22 strains) the ratio of halo diameter and colony diameter are between 1 and 2, 11 strains the ratio exceed 2, among them the 487 has the biggest ratio of 2.661.

The bacteria that could grow on CTAB plate were further measured. Single colony was inoculated into mineral medium (3 % glycerol added), incubated at 37 °C for 72 h. OD_{600}, emulsification activity and the yield of rhamnolipid were measured for every strain. The emulsification activity was evaluated using fivefold dilution of supernatant. There are 10 strains whose production exceeded 10 g/L, among them the strain 487 has the highest production of 25.2 g/L.

8.3.2 Identification of Strain 487

The 16S rDNA was amplified (Fig. 8.1) and deposited in the NCBI database under the GenBank accession number KJ883617. Homology analysis of the 16S rDNA sequence showed that it was 100 % identical to the 16S rDNA sequence of *P. aeruginosa* strain PAO1 (accession number NR074828) [14]. Strain 487 was identified as *P. aeruginosa* (Fig. 8.2).

8.3.3 Rhamnolipid Production in Different Carbon Sources

Strain 487 could reach a production of 33.4 g/L using glycerol as carbon resource after 72 h of fermentation. Rhamnolipid production with rapeseed oil even showed higher levels, OD_{600} and production increased by 129.0 and 26.9 %, respectively. Therefore, rapeseed oil was better carbon source than glycerol for strain 487 to produce rhamnolipid (Fig. 8.3).

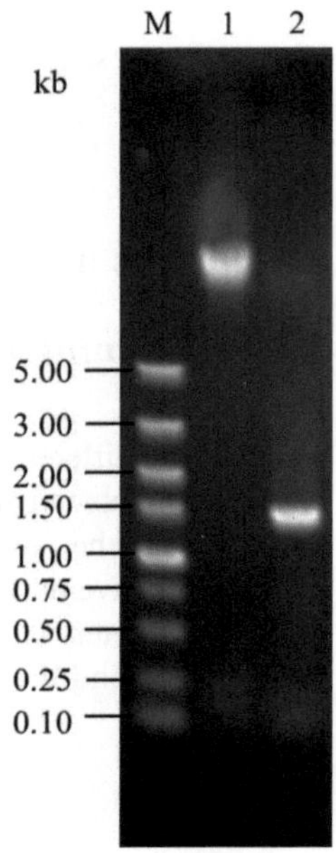

Fig. 8.1 The genomic DNA and 16S rDNA gene of strain *P. aeruginosa* 487 (M: marker; *1* genomic DNA; *2* 16S rDNA)

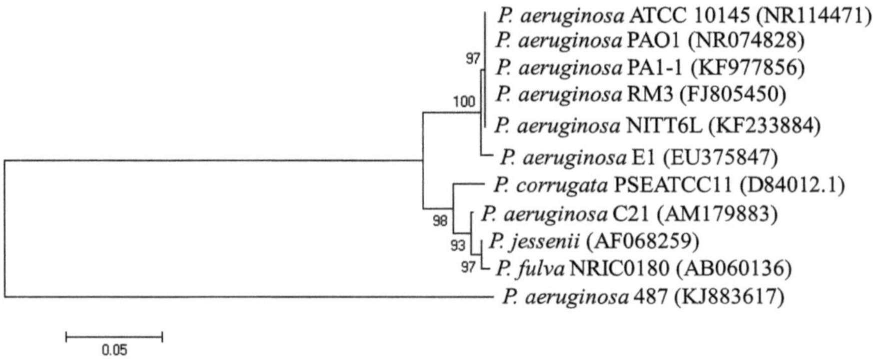

Fig. 8.2 The phylogenetic tree of *P. aeruginosa* 487

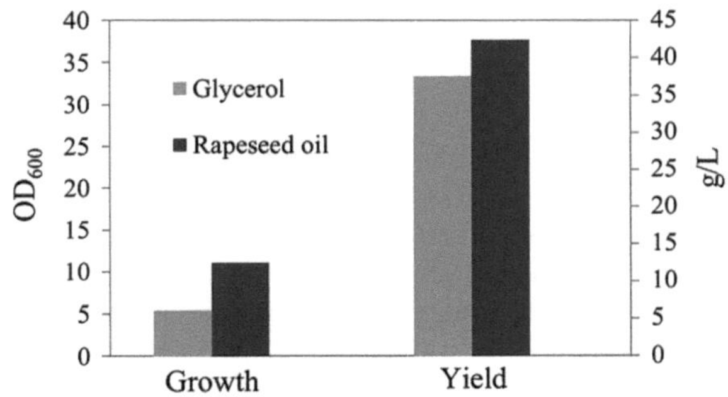

Fig. 8.3 Growth and rhamnolipid production of *P. aeruginosa* 487 in different substrates

8.3.4 Time Course of Rhamnolipid Accumulation

As described in materials and methods, overnight seed culture was added into the mineral medium supplemented with 30 g/L rapeseed oil, and incubated at 35 °C. 1 mL sample was collected at time intervals. As shown in Fig. 8.4, OD_{600} reached the maximum at 24 h and then declined steadily. The pH kept stable at 7.3–7.5 during the fermentation process, but began to rise after 120 h. Rhamnolipid began to accumulate at the late logarithmic phase or early stationary phase, and the production increased gradually to 54.9 g/L at the end of the fermentation. *Pseudomonas aeruginosa* LBI produced 15.8 g/L rhamnolipid using soapstock after 86 h [10]. Thavasi et al. found that the maximum biosurfactant production (8.6 g/L) occurred with peanut oil cake at 132 h [13].

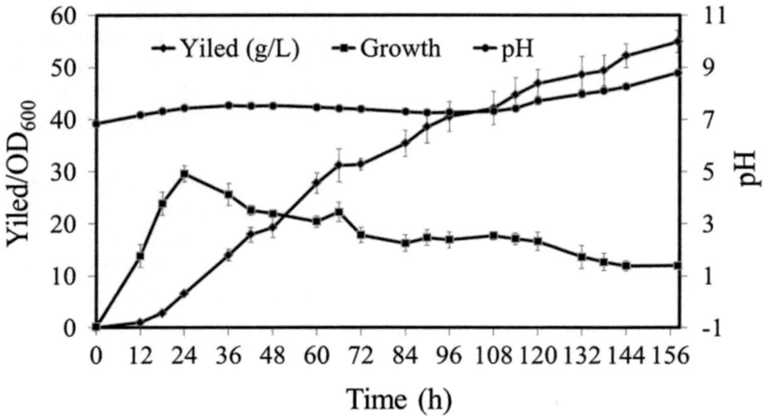

Fig. 8.4 Time course of the growth and rhamnolipid production by *P. aeruginosa* 487 using rapeseed oil as carbon source

8.3.5 TLC Analysis

We conducted thin layer chromosphere to identify the type of biosurfactant produced by *P. aeruginosa* 487. Glycolipid biosurfactant may show brown spots which are verified on the silica gel plate as shown in Fig. 8.5. So we demonstrated the biosurfactant produced by 487 is glycolipid. We got two spots whose mobilities are 0.87 and 0.6 respectively representing mono- and di-rhamnolipid, because of mono-rhamnolipid's higher solubility in the eluent.

8.3.6 Surface Tension and CMC Analysis

The crude extract was diluted into different concentrations to evaluate surface tensions of each dilution. The relation of surface tension with rhamnolipid concentrations was calculated. Surface tension decreased rapidly from 72 mN/m to 38 mN/m with the rhamnolipid concentration at 143 mg/L. Further increase in the rhamnolipid concentrations only reduced the surface tension from 38 to 32.7 mN/m.

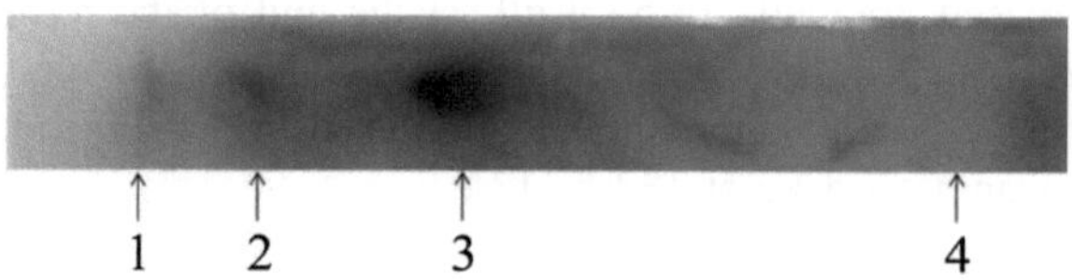

Fig. 8.5 TLC analysis of the rhamnolipid of *P. aeruginosa* 487, *1* solvent; *2* mono-rhamnolipid; *3* di-rhamnolipid; *4* the starting line of solvent

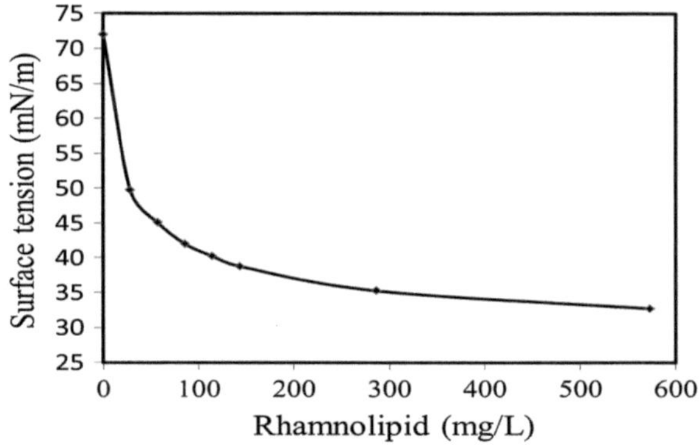

Fig. 8.6 CMC analysis of the rhamnolipid

The results indicated that low rhamnolipid concentrations had a strong effect on reducing the surface tension. As shown in Fig. 8.6, the CMC of rhamnolipid obtained from *P. aeruginosa* 487 was between 100 and 300 mN/m.

8.3.7 FTIR Analysis of Function Groups

The characteristics bonds of different groups of the rhamnolipid molecule were analyzed in Fig. 8.7. N–H bonds were found at the wavelength of 3,428 cm^{-1}, which indicated the presence of amine groups. The peaks at 2,930 and 2,857 cm^{-1}

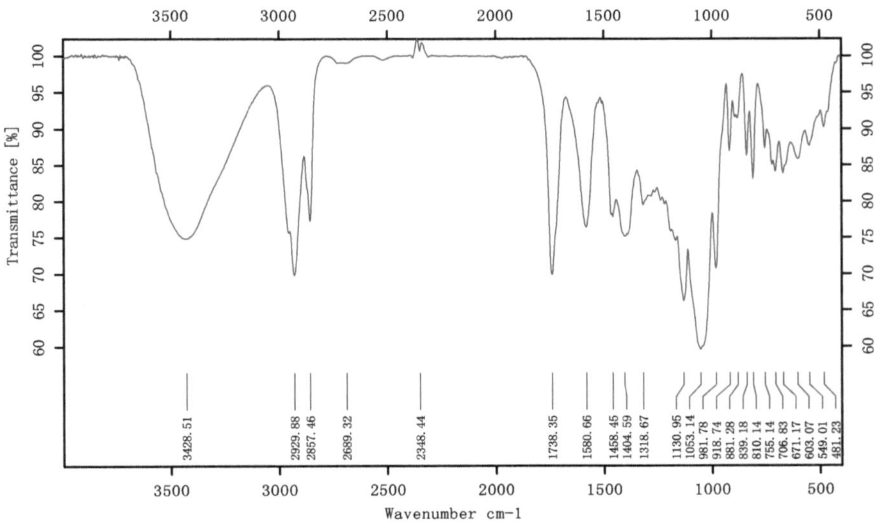

Fig. 8.7 Fourier transform infrared spectrum of the rhamnolipid of *P. aeruginosa* 487

represent the stretching bonds of CH_2 and CH_3 groups, indicating the presence of alkanes. The peak at 1,731 cm^{-1} represented carbonyl stretching bond, which was the characteristic bond for ester compounds. C–O stretching bond was also observed at the wavelength of 1,053 cm^{-1}. C–H and O–H deformations of carbohydrates were confirmed by the peaks at 1458, 1405, and 1319 cm^{-1}, respectively. The obtained data confirmed that the nature of biosurfactant was glycolipid.

8.4 Conclusion

P. aeruginosa 487 was able to produce biosurfactant rhamnolipid using water-soluble carbon source glycerol or water-insoluble carbon source rapeseed oil. The emulsification activity also very high, fivefold-diluted supernatant could 100 % emulsification of equal volume liquid paraffin. Thin layer chromatography analysis reveals that the product of strain 487 is composed of two types of rhamnolipid. Fourier transform infrared spectroscopy analysis reveals that the product is rhamnolipid. The growth and rhamnolipid accumulation of *P. aeruginosa* 487 using rapeseed oil as carbon resource were also investigated. The rhamnolipid production could reach 54.9 g/L after 158 h incubation.

Acknowledgments This work was partly supported by The National Natural Science Foundation of China (Grant No. 31370205).

References

1. Makkar RS, Cameotra SS (2002) An update on the use of unconventional substrates for biosurfactant production and their new applications. Appl Microbiol Biotechnol 58:428–434
2. Banat IM, Franzetti A, Gandolfi I et al (2010) Microbial biosurfactants production, applications and future potential. Appl Microbiol Biotechnol 87:427−444
3. Gautam KK, Tyagi VK (2006) Microbial surfactant: a review. J Oleo Sci 55:155–166
4. Sánchez M, Aranda FJ, Espuny MJ et al (2007) Aggregation behaviour of a dirhamnolipid biosurfactant secreted by *Pseudomonas aeruginosa* in aqueous media. J Colloid Interface Sci 307:246–253
5. Jarvis FG, Johnson MJ (1949) A glyco-lipid produced by *Pseudomonas aeruginosa*. J Am Chem Soc 71:4124–4126
6. Wei YH, Chou CL, Chang JS (2005) Rhamnolipid production by indigeneous *Pseudomonas aeruginosa* J4 originating from petrochemical wastewater. Biochem Eng J 27:146–154
7. Syldatk C, Lang S, Matulovic U (1985) Production of four interfacial active rhamnolipids from n-alkane or glycerol by resting cells of *Pseudomonas* species DSM 2874. Z Naturforsch 40:61–67
8. Pinzon N, Ju LK (2009) Improved detection of rhamnolipid production using agar plates containing methylene blue and cetyl trimethyl ammonium bromide. Biotechnol Lett 31:1583–1588
9. Helbert JR, Brown KD (1957) Color reaction of anthrone with monosaccharide mixtures and oligo- and polysaccharides containing hexuronic acids. Anal Chem 29:1464–1466

10. Benincasa M, Abalos A, Oliveira A et al (2004) Chemical structure, surface properties and biological activities of the biosurfactant produced by *Pseudomonas aeruginosa* LBI from soapstock. Antonie Van Leeuwenhoek 85:1–8
11. Sambrook J, Russell DW (2001) Molecular cloning: a laboratory manual (3rd ed). New York: Cold Spring Harbor Press 1.31−1.38
12. Zhang J, Madden TL (1997) Power BLAST: a new network BLAST application for interactive or automated sequence analysis and annotation. Genome Res 7(6):649–656
13. Thavasi R, Subramanyam Nambaru VRM (2011) Biosurfactant production by *Pseudomonas aeruginosa* from renewable resources. Indian J Microbiol 51(1):30–36
14. Winsor GL, Van Rossum T, Lo R et al (2009) *Pseudomonas* genome database: facilitating user-friendly, comprehensive comparisons of microbial genomes. Nucleic Acids Res 37:483–488

Chapter 9
High-Quality Protein-Encoding Gene Design and Protein Analysis

Guo-qing Huang, Lei Wang, Dong-kai Wang, Qiong Wu, Yao Li, Jin-hai Zhao and Di-fei Cao

Abstract Rapid growth of global population should be demanded more high-quality food and this might be alleviated by improving protein nutrition levels in usual foodstuff. The expense associated with food additives used for animal production have led to the study of intracellular production of high-quality protein (HQP) by transgenic crops and other organisms, as a means of obtaining efficient and less costly sources of essential amino acids. One important approach is to extract high essential amino acids from bacteria or others. We designed a HQP gene encoding about 100 amino acids which had more than 80 % essential amino acids. GOR/Gibrat and Swiss-Pdb Viewer analysis were used to predict the secondary structure of HQP-1 in pI value, hydrophilic and hydrophobic properties. Then the HQP gene was transferred into *E. coli* BL21 (DE3), and the protein was extracted and identified by SDS-PAGE.

Keywords Protein design · HQP-1

9.1 Introduction

With the rapid growth of population in developing countries and the increase in global ageing, more high-quality food should be needed. We need to pay more attention to food nutrition and this should be alleviated by improving nutrition levels in foodstuff. The amino acids regarded as essential for humans are at least 8 kinds that mainly recharge through the animal protein absorption and decomposition. Improving high essential amino acid content in food additives would have a very important economic significance.

G. Huang (✉) · L. Wang · D. Wang · Q. Wu · Y. Li · J. Zhao · D. Cao
Institute of Advanced Technology, Heilongjiang Academy of Sciences, Dao wai District, Nan ma Road, No.135, Harbin 150020, China
e-mail: huangguoqing0921@163.com

T.-C. Zhang and M. Nakajima (eds.), *Advances in Applied Biotechnology*, Lecture Notes in Electrical Engineering 332, DOI 10.1007/978-3-662-45657-6_9

Three main approaches for improving protein quality in a given organism are being explored [1]. The first approach involves modifying the genes of an organism so that more selected essential amino acids are inserted in the proteins. Attempts of changing amino acid content, however, often destabilise the protein, or prevent its recovery, prevent it from folding or still [1, 2]. The second approach involves transferring a gene coding for a HQP, from one organism to another that is more suitable for heterologous expression. In this approach, the amino acid composition of a natural protein is predetermined and may not conform to the desired essential amino acids ratio. The third approach, chosen here, involves creating a new protein with a biased composition of selected essential amino acids [3]. In theory, this strategy allows for a full control of the amino acid composition of the protein and thus is an advantage over the previously mentioned options.

The efforts before that what to express synthetic proteins with high essential amino acids content in bacteria [4] have not produced high quantities of protein, among other reasons, which is possibly attributable to the lack of stability of the final peptide structural. In order to improve the properties and quantities of such new proteins, we have tried to identify dominant folding principles in selected natural proteins and then to encode them into the new one. The basic assumption of this strategy is that by encoding information that confers stability, structure and compactness, a protein stable expressive in vivo could be designed that regardless of its biased compositions.

First of all, the HQP we hope to design would have a strong biased composition, with methionine, lysine and other essential amino acids being dominant; second, it must be expressed in vivo as a stable, soluble and full-length protein; third, the protein should be easily to digested and absorbed by the human body. Design of HQP was a challenge to be met yet since most reported advances in protein creation had not involved soluble and expressed in vivo full-length proteins [5]. The design criteria such as charged residue distribution, secondary structure propensities and other global factors found determinant for folding were used to refine the initial design [3, 6, 7]. Characterisation of HQP indicated that the design process used resulted in the stable expression of a new, largely helical protein enriched in desired amino acids. Its expression and behaviour levels in vivo were found to be far superior to that obtained from earlier attempts of high essential amino acids polypeptide design [8].

We designed the HQP gene that encoding about 100 amino acids which had more than 80 % essential amino acids. GOR/Gibrat and Swiss-Pdb viewer analysis were used to predict the secondary structure of HQP-1 in pI value, hydrophilic and hydrophobic properties. HQP-1 gene was inserted into pET32a and transferred into *Escherichia coli* BL21(DE3). The protein was extracted and then identified by SDS-PAGE. Using microbial method with low cost and high efficiency to product of high essential amino acids will also have very important significance in economic and healthy.

9.2 Materials and Method

9.2.1 The Design of HQP-1 Gene

According to *E. coli* BL21 (DE3) codon preference, we designed a HQP gene which considering of pI, trypsin digestion characteristic and its effect on the space forming stability.

9.2.2 Prediction of Physicochemical Properties of HQP-1

The Gibrat and GOR method were used to forecast the secondary structure characteristics of HQP-1 and determine whether it could express stably. The pI, hydrophilic and hydrophobic of HQP-1 were calculated and analysed by ANTHEPROT5.0. Based on these data, the instability coefficient (the instability index, II) of HQP-1 could be calculated.

9.2.3 DNA Sequencing

The gene was synthesised by SANGON Company as form as pUC57-HQP recombinant plasmid in DH5α and amplified by PCR of universal primers M13±. The PCR products were purified from the agarose gel according to Star Prep Gel Extraction Kit (Genstar) and DNA sequenced by SANGON Company.

9.2.4 Protein Expression

The HQP-1 gene was cloned into pET-32a plasmid and then transformed the recombinant plasmid pET-HQP into BL21(DE3). Bacteria carrying the recombinant vectors were grown at 37 °C, 300 rpm in 50 ml of LB Miller medium to an O.D. of 0.4. Transcription was induced using 0.5, 1 and 2 mM, isopropylthio-D-galactoside (IPTG) for 3 h. The cells were then harvested by centrifugation at 3,000 g, 5 min.

9.2.5 Protein Detection

The protein was visualised by sodium dodecyl sulphate polyacrylamide gel electrophoresis (SDS-PAGE).

9.3 Results

9.3.1 HQP-1 Gene Design

HQP-1 gene should have the following characteristics: 1. The HQP-1 gene must have the BL21(DE3) codon preference; 2. It should encode the human essential amino acid higher than 85 %; 3. Both sides of HQP-1 gene were introduced restriction sites to facilitate subsequent cloning and expression; 4. The sequence was introduced a plurality of trypsin enzyme cutting sites that could be more easily digested and absorbed by the body. The gene sequence and translated amino acid residues are shown in Fig. 9.1.

```
EcoR I      Met Gly Asp Arg Lys Lys Trp Met Asp Arg Met Leu Pro Phe His Lys Lys Trp
 5'-AATTC ATG GGG GAT CGT AAG AAA TGG ATG GAT CGT ATG CTT CCA TTC CAC AAG AAG TGG
      3'-G TAC CCC CTA GCA TTC TTT ACC TAC CTA GCA TAC GAA GGT AAG GTG TTC TTC ACC
           Leu Pro Ile Thr Leu Phe Pro His Ile Thr His Lys Trp Glu Val Leu Leu Pro

His Pro Phe Leu Thr Ile His Pro Phe Leu Lys Lys Trp Met Lys Lys Trp Met
CAT CCA TTT CTT ACG ATG CAT CCA TTT CTT AAG AAA TGG ATG AAG AAA TGG ATG
GTA GGT AAA GAA TGC TAC GTA GGT AAA GAA TTC TTT ACC TAC TTC TTT ACC TAC
Met Trp Lys Lys Arg Asp Met Trp Lys Lys Leu Phe Pro His Leu Phe Pro His

Lys Lys Trp Met Thr Ile His Pro Lys Lys Trp Phe Met Leu Lys Leu Trp Met
AAG AAA TGG ATG ACG ATC CAT CCA AAG AAG TGG TTC ATG CTT AAG CTT TGG ATG
TCC TTT ACC TAC TGC TAC GTA GGT TTC TTC ACC AAG TAC GAA TTC GAA  ACC TAG
Leu Phe Pro His Arg Asp Met Trp Leu Leu Pro Glu His Lys Leu Met Pro His

Lys Lys Trp Met Asp Arg Met Leu Pro Phe His Lys Lys Trp His Pro Phe Leu
AAG AAA TGG ATG GAT CGT ATG CTT CCA TTG CAC AAG AAG TGG CAT CCA TTT CTT
TTC TTT ACC TAC CTA GCA TAC GAA GGT AAC GTG TTC TTC ACC GTA GGT AAA GAA
Leu Phe Pro His Ile Thr His Lys Trp Glu Val Leu Leu Pro Met Trp Met Met

Thr Ile His Pro Phe Leu Lys Lys Trp Met Lys Lys Trp Met Thr Ile His Pro
ACG ATC CAT CCA TTT CTT AAG AAA TGG ATG AAG AAA TGG ATG ACG ATC CAT CCA
TGC TAG GTA GGT AAA GAA TTC TTT ACC TAC TTC TTT ACC TAC TGC TAG GTA CGT
Arg Asp Met Trp Lys Lys Leu Phe Pro His Leu Phe Pro His Arg Asp Met Trp

Phe Leu Thr Ile Lys Lys Trp Met Thr Ile Pro
TTT CTT ACG ATG AAG AAA TGG ATG ACG ATC CCC CAT GAATTCG-3'
AAA GAA TGC TAC TTC TTT ACC TAC TGC TAG GGG GTA CTTAAGCCTAG-5'
Lys Lys Arg Asp Leu Phe Pro His Leu Asp Gly Met      BamH I
```

Fig. 9.1 HQP-1 gene and amino acid residues which was translated by it

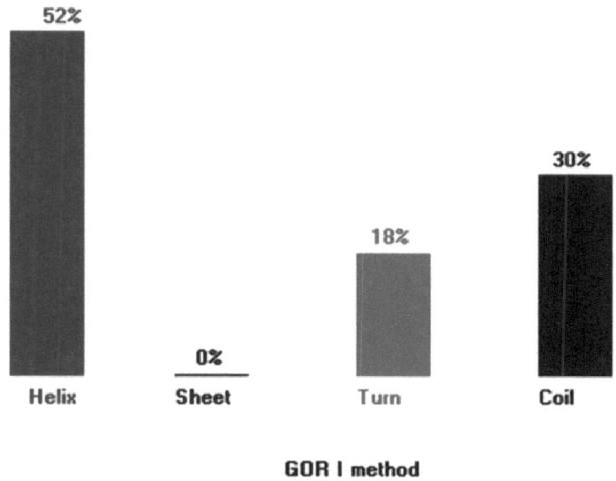

Fig. 9.2 The prediction of protein secondary structure by GOR method

9.3.2 The Secondary Structure Prediction

GOR and Gibrat protein structure prediction method were used to speculate on the secondary structure of HQP-1 protein and determine its ability to form the corresponding spatial structure and cannot collapse. As shown in Figs. 9.2 and 9.3.

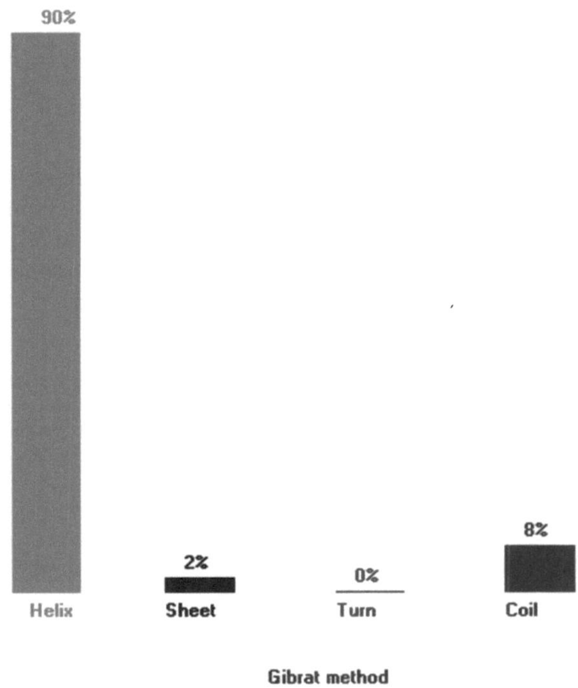

Fig. 9.3 The prediction of protein secondary structure by Gibrat method

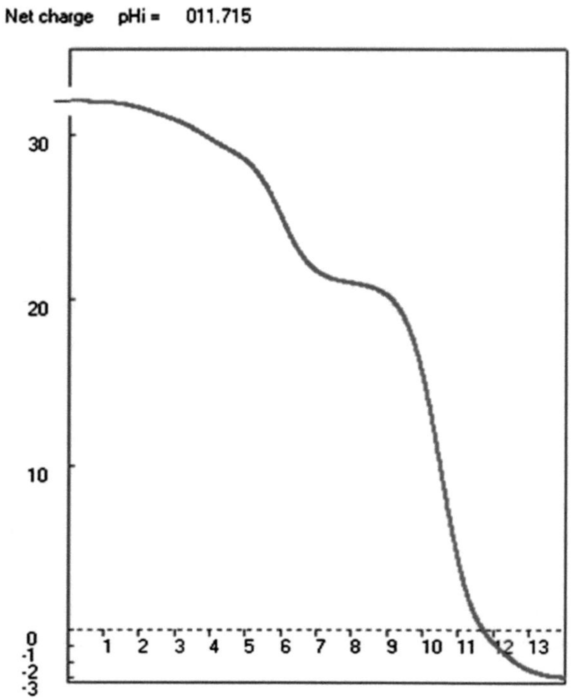

Fig. 9.4 The Isoelectric point of HQP-1 protein

9.3.3 The Physicochemical Properties of HEAAE-1 Protein

The ANTHEPROT5.0 protein analysis software was used to calculate HQP-1 protein isoelectric point what was 11.715 (Fig. 9.4). Figure 9.5 shows that hydrophilic, hydrophobic curve and the corresponding data of HQP-1 protein, concluding these, we should know that the protein has strong hydrophilic, relatively easy to dissolved in the water, easy to be digested and absorbed.

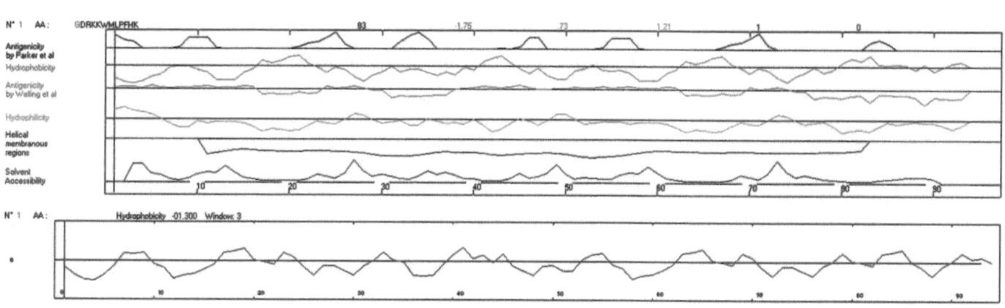

Fig. 9.5 Hydrophilic–hydrophobic of HQP-1 protein (*The red line for hydrophilic curve, the blue line for hydrophobic curve*)

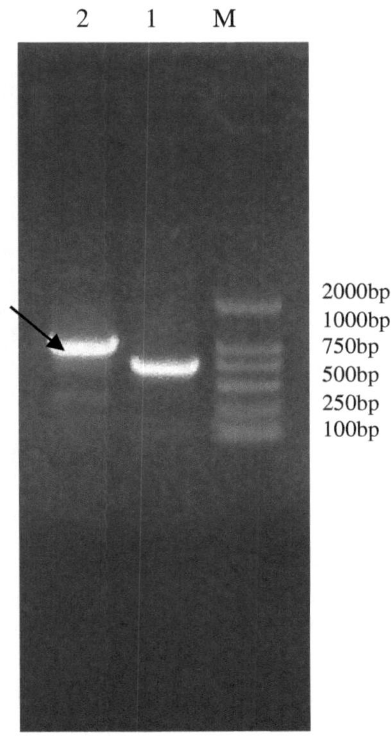

Fig. 9.6 The results of BL21 colony PCR *M* DL2000 DNA Marker

We also could know that, the instability index (II) was computed to be 19.00, less than 40, this classifies the protein as stable; its aliphatic index is 31.69; its grand average of hydropathicity (GRAVY) is 0.761; the estimated half-life is more than 10 h within *E. coli* in vivo.

9.3.4 Positive Transformants Identification

pET/HQP-1 fusion vector was transformed to BL21(DE3), we took the empty vector as control, through the colony PCR method for positive identification. The result shown in Fig. 9.6, the arrow to the positive transformants in Lane 2.

9.3.5 Expression of Recombinant Protein

The protein was visualised by sodium dodecyl sulphate polyacrylamide gel electrophoresis (SDS-PAGE) using 12 % polyacrylamide–tricine gels, followed by

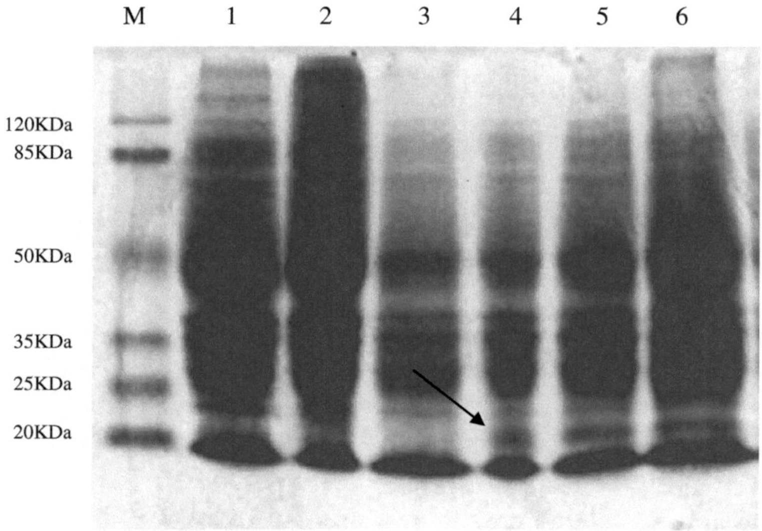

Fig. 9.7 Analysis of HQP-1 fusion protein by SDS-PAGE, *M* Standard protein 120KDa Marker; *1*, *2* and *3* Induced expression of fusion protein before inducted; *4*, *5* and *6* Induced expression of fusion protein after inducted by 0.5, 1.0, 2.0 mM IPTG

coomassie brilliant blue staining. It is approximately 20 KDa in size, Lane recombinant strain-specific bands emerged, and there are no obvious differences at 0.5, 1 and 2 mM inducers, see Fig. 9.7. Therefore, we determined that HQP-1 protein was soluble expression, about 20 KDa, concentration of inducer for 0.5 mM.

9.4 Discussion

In organisms that were engineered to produce high-quality proteins, the amount of protein stocked depends on several factors. Among these was the resistance to proteolysis, a major requirement for effective storage of such proteins [9, 10]. Therefore, the genetic engineering techniques combined with bioinformatics methods could be used for constructed synthetic the rich of high essential amino acids through prokaryotic expression system. Compared with the existing products on the market, using microorganism to produce the protein like HQP-1 protein that have high ratio of essential amino acid, which was have a obviously advantage in nutrition and price. Using microbial method with low cost and high efficiency to product of high essential amino acids will also have very important significance in economic and healthy. Whether it is stable or has nontoxic effects still need to be validated.

References

1. DeLumen BO, Krenz DC, Revilleza J (1997) Molecular strategies to improve to protein quality of legumes. Food Technol 51:67–70
2. Dyer JM, Nelson JW, Murai N (1993) Strategies for selecting mutation sites for methionine enhancement in the bean seed storage phaseolin. J Protein Chem 12:545–560
3. Beauregard M, Teather RM, Hefford MA (1995) Design, expression and initial characterization of MB-1, a de novo protein enriched in essential amino acids. Bio Technol 13:974–981
4. Beauregard M, Hefford MA, Teather RM (1994) Use of a monoclonal antibody to β-galactosidase for the detection of α-peptide fusions. Biotechniques 16:831–838
5. Kohn WD, Hodges RS (1998) De novo design of α-helical coiled coils and bundles: models for the development of protein-design principles. Trends Biotechnol 16:379–389
6. Hill CP, Anderson DH, Wesson L et al (1990) Crystal structure of α1: implications for protein design. Science 249:543–546
7. Beauregard M, Goraj K, Goffin V et al (1991) Spectroscopic investigation of structure in octarellin: a de novo protein designed to adopt the (α/β-barrel packing). Protein Eng 4:745–749
8. MacCallum J, Hefford MA, Omar S et al (1997) Prediction of folding stability and degradability of the de novo designed protein MB-1 in cow rumen. Appl Biochem Biotechnol 66:83–93
9. Hoffman LM, Donaldson DD, Herman EM (1988) A modified storage protein is synthesized, processed, and degraded in the seeds of transgenic plants. Plant Mol Biol 11:717–729
10. Williams M1, Gagnon MC, Doucet A et al (2002) Design of high essential aminoacid proteins: two design strategies for improving protease resistance of thenutritious MB-1 protein. J Biotechnol 94:245–254

Chapter 10
Isolation and Characterization of a Highly Siderophore Producing *Bacillus subtilis* Strain

Huiming Zhu and Hongjiang Yang

Abstract Bacteria, fungi, and graminaceous plants synthesize siderophore for scavenging iron from environment to inhibit the growth of pathogens by depriving of iron. At this study, 19 siderophore producing bacterial strains were isolated from heat treated samples by O-CAS agar screening and liquid CAS assay. Among them, strain MB8 produced a relatively high level of siderophore (70.38 % SU, ++++). Comparative sequence analysis of 16S rRNA gene identified MB8 as a *Bacillus subtilis* strain. With Arnow analysis, the siderophore of MB8 was confirmed to be one type of catecholate. Furthermore, fermentation parameters affecting the siderophore production by MB8, including carbon sources and amino acids, were investigated at individual levels. Antimicrobial activity assay showed that the siderophore exhibited significant antimicrobial activity against *Proteus vulgaris, Staphylococcus aureus, Micrococcus luteus, Escherichia coli, Vibrio parahaemolyticus*, and *Mucor*. Our results indicated that *B. subtilis* MB8 had the potential to be used as a probiotic organism.

Keywords Siderophore · SU · *Bacillus subtilis* · Antimicrobial activity · Catecholate

10.1 Introduction

Iron is essential for almost all organism by virtue of its unique chemical properties, the ability to coordinate and activate oxygen, and the possession of an ideal redox chemistry ($Fe^{II} \rightarrow Fe^{III} \rightarrow Fe^{IV}$) for involvement in electron transport and metabolic processes. Higher organisms acquire iron nutrient from diet, while microorganisms

H. Zhu · H. Yang (✉)
Key Laboratory of Industrial Microbiology, Ministry of Education, Tianjin Key Laboratory of Industrial Microbiology, College of Biotechnology, Tianjin University of Science and Technology, Tianjin 300457, China
e-mail: hongjiangyang@tust.edu.cn

T.-C. Zhang and M. Nakajima (eds.), *Advances in Applied Biotechnology*,
Lecture Notes in Electrical Engineering 332, DOI 10.1007/978-3-662-45657-6_10

obtain iron element only from environment. Despite the abundance of iron in the earth's crust, iron is not readily bio-available in aerobic environments due to the low solubility of iron (III). These environmental restrictions and biological imperatives have prompted microorganisms to synthesize siderophores can sequester and solubilize the nonsoluble surrounding iron in order to make it biologically available. Siderophores are relatively low molecular weight (500−1,500 dalton) iron-chelating ligands, and are synthesized under conditions of iron starvation by non-ribosomal peptide synthetases (NRPS) or NRPS independent pathways [1].

Until now, there are about 500 different siderophores discovered and of which 270 siderophores have been structurally characterized [2]. Based on their structural features, siderophores comprise three basic groups, the hydroxamate, the catecholate, and hydroxycarboxylate. The probiotics strain *Escherichia coli* Nissle 1917 produces the catecholate enterobactin and salmochelin, the hydroxamate aerobactin, and the mixed-type siderophore yersiniabactin [3]. And *Pseudomonas aeruginosa* synthesizes pyochelin and pyoverdine. Pyoverdine, which contained hydroxamates and catecholates group, is a yellow-green, fluorescent, water-soluble pigment [4].

Siderophore-mediated iron acquisition is essential for the survival of some pathogenic bacteria. The siderophores secreted by nonpathogenic strains was compete iron with pathogenic bacteria to inhibit their growth. The siderophore provide iron not only for itself, but also for others organism. The organic Fe chelators can not only enhance plant yield, but also protect the environment and water by avoiding the use of pesticides [5].

In this study, a bacillus strain secreting siderophores was isolated and identified. The siderophore secreted was catecholate and its antimicrobial activity was characterized. Moreover, the fermentation parameters for siderophore production were investigated.

10.2 Materials and Methods

10.2.1 Samples Collection

The chicken fecal samples were collected from five henneries located at Beijing Huairou district and Henan province. The samples were stored in sterile sealed bags at 4 °C refrigerator ready for use.

10.2.2 Media and Culture Condition

The medium used in this work included MKB medium [6]; Arginine medium [7] (AM); Luria-Bertani (LB) medium [8]; Czapek Dox medium [8]; CAS [9] solution; AM was modified by replacing glucose of arginine medium with 20 g/L glycerol along with the reduction of arginine to 1 g/L.

10.2.3 O-CAS Assay and CAS Liquid Assay

2 g samples were dissolved with 10 mL PBS buffer and heated at 80 °C for 20 min. When cooled to room temperature, 0.2 mL suspension was spreaded on AM agar plate and incubated at 37 °C for 24 h. Then 15 mL CAS agar medium were applied over the colonies [10].

The strains which showed positive responses in the O-CAS assay were recovered and purified by streaking. Single colonies were inoculated into AM, and incubated at 37 °C with a shaking speed 200 rpm for 24 h. Supernatant was harvested by centrifugation at 10,000 rpm for 5 min, and mixed with equal volume CAS solution. Siderophores, if present, will remove iron from the dye complex, resulting in a reduction in blue color of the solution. Measure the absorbance (OD_{680} [11]) to test the loss of blue color after 1 h. For OD_{680} measured, the AM medium were used as blanks, and the mixture of AM plus CAS assay solution was used as a reference (Ar). The samples (As) should have a lower reading than the reference. Siderophore units are defined as $SU = [(Ar - As)/Ar] \times 100\ \%$ [12]. And As/Ar stands for the siderophore activity level, ranging from 0 to 1. In general the more siderophore is produced, the lower of As/Ar value was [13]. All these experiments were conducted at least three times with three replicates for each one.

10.2.4 Siderophore Typing by Chemical Methods

The Arnow assay was used to detect catecholate type siderophore [14]; the $FeCl_3$ assay was used to detect hydroxamate type siderophore [15], and the Shenker assay was used to detect carboxylate type siderophore [16]. All the assays were conducted at least three times with three replicates for each one.

10.2.5 Identification by 16S rRNA Sequencing

Genomic DNA of MB8 was extracted as described previously [17]. Then MB8 strain was identified by 16S rRNA gene sequence analysis. Universal primers 27F (5′-AGAGTTTGATCCTGGCTCAG-3′) and 1492R (5′-GGTTACCTTGTTACG-ACTT-3′) were used to amplify 16S rRNA gene fragment PCR products were sequenced directly. Sequence homology search was carried out with BLAST provided by NCBI [18].

10.2.6 Growth and Siderophore Production

The isolate was inoculated in 100 mL AM by transferring 1 % seed cultures and incubated for 48 h at 37 °C with a shaking 200 rpm. Cell density was tested at an interval of 2 h up to 48 h. 1 mL culture supernatant was mixed with 1 mL of CAS solution to measure the siderophore production. A reference was prepared by mixing AM medium and CAS solution. Both the test and reference were read at 680 nm.

10.2.7 The Effect of Fermentation Parameters

In order to verify whether carbon source [7] and amino acid [7] had an influence on siderophore unit and growth of MB8, xylose, maltose, lactose, soluble starch, sucrose, fructose, and glycerol were used as an alternative carbon source to glucose in AM. Moreover, arginine was replaced with alanine, cysteine, aspartic acid, histidine, glutamate, glutamine, and glycine, respectively, to investigate the amino acid effect on siderophore production. All above experiments were performed for 36 h at 37 °C with a shaking speed 200 rpm.

10.2.8 Antimicrobial Activity Analysis

Overnight cultures of *E. coli, Staphylococcus aureus, Proteus vulgaris, Micrococcus luteus*, and *Vibrio parahaemolyticus* were inoculated to 30 mL MKB medium at 1 % inoculum, respectively. In the same time, the siderophore extract was added to the cultures with different ratios. 1 g/L 2, 3-DHBA and PBS buffer was used as negative control. All cultures were incubated at 37 °C for 16 h. Cell density were tested by measurement of OD_{600}, and the inhibition ratio was calculated as the following:

$$\text{Inhibition ration} = \left(OD_{600\,\text{without siderophore extract}} - OD_{600\,\text{with siderophore extract}}\right) / OD_{600\,\text{without siderophore extract}}\ [5].$$

Added 10 mL of PBS buffer to PDA slant of mucor, the suspension was transfer to 30 mL Czapek Dox medium with 3 % inoculums. In the same time, the siderophore extract was added to the cultures with different ratios. 1 g/L 2, 3-DHBA and PBS buffer was used as negative control. All cultures were incubated at 37 °C for 48 h. The cultures were filtered and the biomass was dried completely before measurement of mycelium weight. The inhibition ratio = (dry weight of mycelium of without siderophore extract − dry weight of mycelium with siderophore extract)/ dry weight of mycelium without siderophore extract [5].

10.3 Results and Discussion

10.3.1 Isolation and Identification of Siderophore Producing Bacillus Strains

According to the method described, a total of 19 strains showed a positive response to O-CAS assay. Among them, MB8 strain changed the color of CAS rapidly within 15 min, then turned into nacarat in 30 min. Therefore, strain MB8 is tentatively considered as a potentially strong siderophore producer.

The 19 isolated strains were subjected to CAS liquid assay, and they have different siderophore yields. The siderophore units of strain MB8 in liquid AM was 64.53 % (As/Ar = 0.35) and siderophore activity level was ++++. The obtained result was similar to *Bacillus spp.* ST13 isolated from human stool, and siderophore units of this strain was approximately 70 % [19]. The results suggested that strain MB8 was a highly siderophore producing strain.

The genomic DNA of MB8 was extracted and used as templates to amplify 16S rRNA gene fragments by PCR method. Homology alignments were performed via BLAST provided by NCBI. The taxonomy classification was carried out by Classifier of RDPII [18]. Both analyses gave similar results that strain MB8 was homologous to *Bacillus subtilis* with 100 % similarity [22].

10.3.2 Siderophore Typing and Siderophore Production

The culture of strain MB8 was spin and the supernatant was collected. With Arnow method, the supernatant was analyzed. The colorless supernatant turned into dark red rapidly (OD_{510} = 0.762), equivalent to 44 mg/mL 2, 3-DHBA. Therefore, the siderophore synthesized by strain MB8 was catecholate type siderophore. The result was consistent with previous findings [20, 21].

As shown in Fig. 10.1, MB8 grew at a relatively slow speed. The culture entered the stationary phase at 30 h. The production of siderophore was consistent with the growth curve of the MB8, and the siderophore unit of MB8 achieved the maximum (SU = 70.38 %) at 36 h. It took about 12 h for MB8 to enter log phase. The long log phase phenomenon was probably due to the nutrient deficiency of synthetic medium [23].

10.3.3 Investigation of the Fermentation Parameters Affecting Siderophore Production

The siderophore production by MB8 was affected by the carbon source used. As shown in Fig. 10.2, the growth on glycerol stimulated the very high siderophore

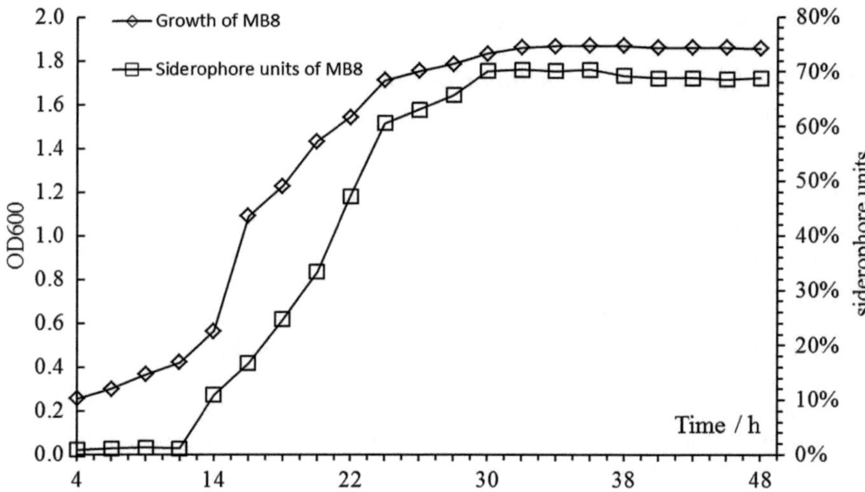

Fig. 10.1 Growth curve and siderophore production by MB8

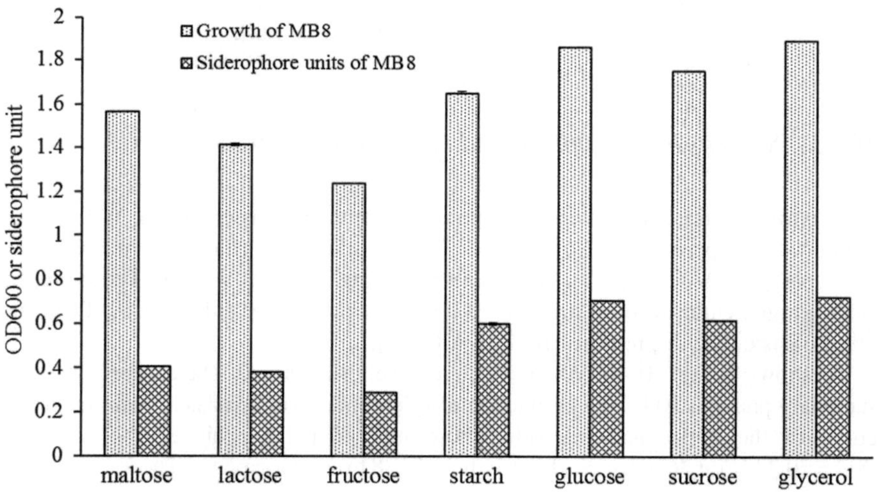

Fig. 10.2 Effect of carbon sources

production (SU = 72.13 %, OD_{600} = 1.888), while the strain produced a relatively less amount of siderophore (SU = 70.49 %, OD_{600} = 1.86) with glucose as the sole carbon source. The siderophore unit of MB8 using maltose, lactose, fructose, starch, or sucrose as carbon source were 40.54, 38, 28.76, 60.51, and 61.85 %, respectively. Glycerol was chosen as the optimal carbon source.

MB8 couldn't grow in the medium with alanine or cysteine. As shown in Fig. 10.3, in the medium with aspartic acid, glutamine, arginine, glutamic acid,

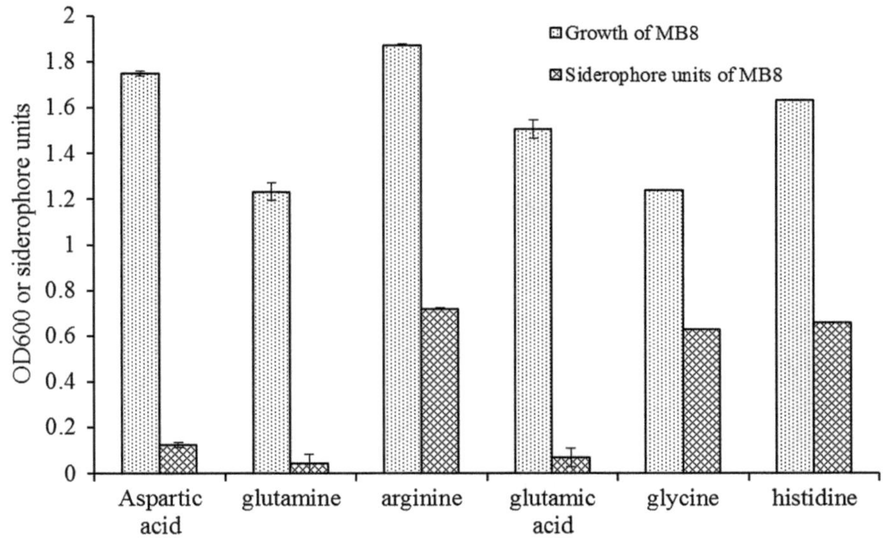

Fig. 10.3 Effect of amino acids

glycine or histidine, the siderophore units of strain MB8 was 12.44, 4.31, 72.06, 6.94, 63.06, or 65.79 %, respectively. The results demonstrated that arginine was chosen as the optimal growth amino acid, consistent with results obtained in *Bacillus megabacterium* [7].

10.3.4 Effect of Arginine Concentrations

No data were available about the relationship between the arginine concentration and the siderophore production. In order to evaluate the positive effect of arginine on the siderophore production by strain MB8, the medium was supplemented with arginine in various concentrations ranging from 0 to 4 g/L. As shown in Fig. 10.4, the increase of arginine concentration did not always irritate the siderophore production of MB8, both the siderophore production and the cell density of strain MB8 were restrained by high concentrations of arginine. In the absence of arginine, MB8 didn't grow at all, indicating MB8 couldnot synthesize arginine. In the presence of 1 g/L arginine, the siderophore unit of strain MB8 reached at the maximum level (SU = 74.89 %). The SU and cell density gradually decreased along with the incease of arginine concentration. In the presence of 4 g/L argine in the medium, SU of strain MB8 was almost zero. The results showed that the optimal concentration of arginine was 1 g/L along with 20 mL/L glycerol.

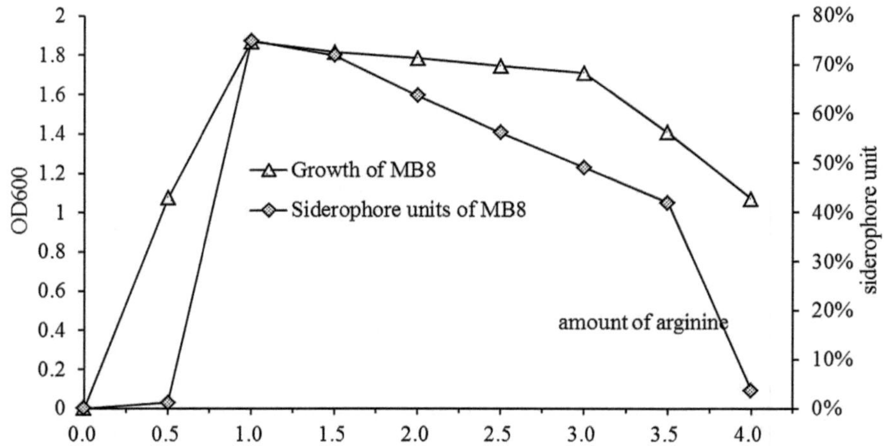

Fig. 10.4 Effect of arginine concentrations

10.3.5 Antimicrobial Activity Analysis

To determine whether the antagonistic substance of MB8 was catecholate siderophore, the siderophore of strain MB8 was preliminarily purified as described [23]. The effects of the siderophore extract on microbial growth were measured. In the presence of low concentrations of siderophore, an obvious antagonism effect on the growth of *P. vulgaris, S. aureus,* and *M. luteus* was observed, while the relatively low inhibition effect was observed in the growth of *V. parahaemolyticus* and *E. coli* (Table 10.1). In the presence of high concentrations of siderophore, all tested microorganisms grew poorly (Table 10.1).

The siderophore extract also showed inhibition effect on the growth of Mucor. In the presence of 20 % siderophore, the inhibition rate was 88.7 %. Until now, all fungal siderophores were hydroxamate-type with the exception of the polycarboxylaterhizoferrin [24]. Fungi acquired iron element by transporting siderophores into fungal cells through an energy-dependent process. In the transporting process,

Tab. 10.1 Growth inhibition by the siderophore

Strains	Siderophore (v/v)				2, 3-DHBA (%)
	5 %	10 %	15 %	20 %	
Mucor	11.4	31.1	82.5	88.7	3.1
S. aureus	89.2	97.4	100	100	2.1
V. parahaemolyticus	57.0	60.3	62.4	69.7	1.1
M. luteus	88.4	100	100	100	2.4
E. coli	43.2	56.7	69.5	80.6	2.8
P. vulgaris	91.4	100	100	100	1.09

(The final concentration of 2, 3-DHBA was 0.2 g/L)

the fungal siderophores bound with the specific receptors on the membrane. The siderophore synthesized by strain MB8 was catecholate type and could be recognized by fungal receptors. This explained why the siderophore of MB8 had an obvious inhibition effect on the growth of mucor. Usually, *Bacillus* species didn't have such extensive antagonism spectrum as reported in our study [7, 19]. Our results suggested that for the first time, the siderophore extract from a Bacillus strain exhibited the antagonism activity against microorganisms.

10.4 Conclusions

In this study, a high siderophore producing Bacillus strain MB8 was isolated. With 16S rRNA sequence analysis, MB8 was identified as *B. subtilis*. After optimization of the fermentation parameters, MB8 produced the siderophore unit up to 74.89 %. The siderophore extract had an obvious antagonistic effect on the growth of *P. vulgaris, S. aureus, M. luteus, V. parahaemolyticus, E. coli*, and *Mucor*. It was suggested that *B. subtilis* MB8 had the potential to be used as a probiotic due to its ability synthesizing the catecholate type siderophore with the broad antagonism activity against bacteria and fungus.

Acknowledgments This work was partly supported by the national natural science foundation of China (grant no. 31370205) and the science and technology program from Beichen District of Tianjin, China (grant no. BCNYKJ-18).

References

1. Oves-Costales D et al (2009) The long-overlooked enzymology of a nonribosomal peptide synthetase-independent pathway for virulence-conferring siderophore biosynthesis. Chem Commun 43:6530−41
2. Hider RC, Kong X (2010) Chemistry and biology of siderophores. | Nat Prod Rep 27:637–657
3. Valdebenitoa M et al (2006) Environmental factors influence the production of enterobactin, salmochelin, aerobactin, and yersiniabactin in Escherichia coli strain Nissle 1917. Int J Med Microbiol 296(8):513–520
4. Vandenende C (2004) Functional characterization of an aminotransferase required for pyoverdine siderophore biosynthesis in Pseudomonas aeruginosa PAO1. J Bacteriol 186 (17):5596–5602
5. Radzki W et al (2013) Bacterial siderophores efficiently provide iron to iron-starved tomato plants in hydroponics culture. Antonie Van Leeuwenhoek 104(3):321–330
6. Pi H et al (2012) Role of catecholate siderophores in gram-negative bacterial colonization of the mouse gut. PLoS ONE 7(11):1–8
7. Santos S et al (2014) Siderophore production by Bacillus megaterium: effect of growth phase and cultural conditions. Appl Biochem Biotechnol 172(1):549–560
8. Du L, Lu F (2006) The experiment technology of microbiology. China Light industry Press, Beijing

9. Schwyn B, Neilands JB (1987) Universal chemical assay for the detection and determination of siderophores. Anal Biochem 160(1):47–56
10. Pérez-Miranda S et al (2007) O-CAS, a fast and universal method for siderophore detection. J Microbiol Methods 70(1):127–131
11. Zhao X et al (2006) Improvement and analysis in over-siderophores production bacteria filtrating and detecting. J Microbiol Biotechnol 33:95–98
12. Persmark M et al (1989) Isolation, characterization, and synthesis of chrysobactin, a compound with siderophore activity from Erwinia chrysanthemi. J Biol Chem 264:3187–3193
13. Manjanatha MG et al (1992) Tn5 mutagenesis of Chinese *Rhizobium fredii* for siderophore overproduction. Soil Biol Biochem 24(2):151–155
14. Arnow LE (1937) Colorimetric determination of the components of 3, 4 - dihydroxyphenylal-aninetyrosine mixtures. J Biol Chem 118:531–537
15. Lee J et al (2012) Siderophore production by actinomycetes isolates from two soil sites in Western Australia. Biometals 25(2):285–296
16. Shenker M et al (1992) Utilization by tomatoes of iron mediated by a siderophore produced by Rhizopus arrhizus. J Plant Nutr 15(10):2173–2182
17. Sambrook JF, Russell DW (2001) Molecular cloning: a laboratory manual, 3rd edn. Cold Spring Harbor Press, New York
18. Zhang J, Madden T (1997) PowerBLAST: a new network BLAST application for interactive or automated sequence analysis and annotation. Genome Res 7(6):649–656
19. Ahire J et al (2011) Bacillus spp. of human origin: a potential siderophoregenic probiotic bacteria. Appl Biochem Biotechnol 164:386–400
20. Hu X, Boyer GL (1995) Isolation and characterization of the siderophore N-deoxyschizokinen from Bacillus megaterium ATCC 19213. Biometals 8(4):357–364
21. Kh L et al (2001) Partial characterization of polyfermenticin SCD, a newly identified bacteriocin of Bacillus polyfermenticus. Appl Microbiol 32(3):146–151
22. Lin Y et al (2014) Potential biocontrol Bacillus sp. strains isolated by an improved method from vinegar waste compost exhibit antibiosis against fungal pathogens and promote growth of cucumbers. Biol Control 71:7–15
23. Ak P et al (2009) Production, purification and chemical characterization of the catecholate siderophore from potent probiotic strains of Bacillus spp. Bioresour Technol 100:368–373
24. Renshaw JC et al (2002) Fungal siderophores: structures, functions and applications. Mycol Res 106(10):1123–1142

Chapter 11
Isolation and Identification of an Inulinase-Producing Strain and the Optimization of Its Fermentation Condition

Yang Zhang, Hongyang Zhu, Jinhai Wang, Xiuling Zhou, Wei Xu and Haiying Shi

Abstract Inulinases are classified among hydrolases, target the β-(2,1)-linkage of inulin and can be used to produce high-fructose corn syrup, fructo-oligosaccharides and alcohol, conferring upon inulinases significant potential for application in the food industry. A total of 28 strains that are capable of producing inulinase were isolated from the rhizosphere soil of *Jerusalem artichoke* (*Helianthus tuberosus* L.). Of these strains, the bacterial strain ni-3 was isolated through further screening. Based on its morphological, physiological and biochemical characteristics, ni-3 was identified by 16S rDNA sequences and systematic analysis. The 16S rDNA sequences of this strain had 99.5 % similarity with the partial sequence of *Brevibacillus centrosporus*, denominated as *B. centrosporus* ZF-9. The optimum conditions of fermentation were investigated. The optimum conditions for inulinase were as follows: inulin 45 g/L, yeast extract 11 g/L, and Na_2HPO_4 20 g/L. Under these conditions, the inulinase activity of *B. centrosporus* ZF-9 reached 6.41 U/mL.

Keywords Inulin · Inulinase · *Brevibacillus centrosporus* · Screening and identification · Optimization of conditions

Yang Zhang and Hongyang Zhu had equal contributions to this work.

Y. Zhang · X. Zhou · W. Xu · H. Shi
College of Life Science, Liaocheng University, Liaocheng 252059, Shandong, China
e-mail: zy886024@163.com

H. Zhu (✉) · J. Wang
Department of Pharmacy, Fujian Health College, Fuzhou 350101, Fujian, China
e-mail: xiangyu238@163.com

T.-C. Zhang and M. Nakajima (eds.), *Advances in Applied Biotechnology*,
Lecture Notes in Electrical Engineering 332, DOI 10.1007/978-3-662-45657-6_11

11.1 Introduction

Polymerized from the dehydration of furan-type D-fructose molecules via β-2,1-glycosidic bonds, inulin, which was first isolated from plants of the Compositae family, has a linear-chain structural polysaccharide (GFn) of glucose residues (G) that are linked to the termini of fructose residues (F) [1]. Inulin occurs as a natural carbohydrate reserve mainly in the roots and tubers of plants such as Jerusalem artichoke, chicory, and dahlia belonging to Liliaceae or Compositae [2, 3] and is a naturally occurring, nonreducing fructan that is slightly soluble in cold water and easily soluble in hot water. Inulin can lower blood lipid levels, promote the absorption of metal ions, and improve the intestinal environment, among other activities [4]. As a hydrolytic enzyme for β-2,1-D-fructan fructosidic bonds, inulinase, which is also known as β-2,1-D-fructan fructanohydrolase (EC.3.2.1.7) [5], is a member of the glycosyl hydrolase family 32 [6]. Inulinase exists as both exo-inulinase and endo-inulinase based on its actions on substrates. Targeting the glycosidic bonds at the nonreducing end of the inulin chain, the former mainly yields fructose by gradual hydrolysis. In contrast, the latter mainly produces oligofructose by randomly breaking intrachain glycosidic bonds [7]. Therefore, inulinase functions in the preparation of high fructose corn syrup and oligofructose [8].

Although fructose can be easily prepared from inulin with a high yield via the catalysis of inulinase, most related studies have been performed using yeast [9] and fungi (e.g., *Penicillium* and *Aspergillus*) [10, 11] and those using bacteria are rare. Inulinase has rarely been applied in the production of fructose due to the lack of highly active bacterial strains that are suitable for the industrial production of enzymatic agents. Therefore, an inulinase-producing, naturally occurring bacterial strain was screened and identified, and the conditions for fermentation were optimized. The results of this study are valuable for future studies on inulinase.

11.2 Materials and Methods

11.2.1 Sample Source for Bacterial Strains

Soils were sampled from the rhizospheres of wild *Helianthus tuberosus* in Forest Park, Drum Mountain, and Flag Mountain in Fuzhou City (Fujian Province, China).

11.2.2 Main Culture Media

Culture medium for primary screening: Fresh *H. tuberosus* (100 g) was chopped, boiled for 20 min, filtered through gauze, and calibrated to 500 mL. Then, 5 g of peptone, 5 g of Na_2HPO_4, 0.25 g of $MgSO_4$ and 10 g of agar were added, and the culture medium was sterilized for 20 min at 121 °C.

Culture medium for rescreening: Inulin (20 g), peptone (5 g), Na_2HPO_4 (10 g) and $MgSO_4$ (0.5 g) were added, and the culture medium was sterilized for 20 min at 121 °C.

Seeding broth (g/L): Inulin (20 g), peptone (5 g), yeast extract (5 g), Na_2HPO_4 (10 g) ,and $MgSO_4$ (0.5 g) were added, and the seeding broth was sterilized for 20 min at 121 °C.

Basic fermentation culture medium (g/L): Inulin (20 g), peptone (10 g), Na_2HPO_4 (10 g), and $MgSO_4$ (0.5 g) were added, and the culture medium was sterilized for 20 min at 121 °C.

Culture medium for preservation in agar slants: Nutrient agar was used.

11.2.3 Methods

11.2.3.1 Screening of Bacterial Strains

Primary screening: The soil suspension was spread onto the plates with culture medium for primary screening. After 24 h of culture at 30 °C, the bacterial strains were selected, isolated by streaking, purified, and stored.

Rescreening: The purified bacterial strains were inoculated in a 250 mL conical flask containing 50 mL of fermentation culture medium and shaken at 30 °C and 200 rpm for 3 days, after which the enzyme activity of the fermentation broth was determined to screen for highly active strains.

11.2.3.2 Determination of Inulinase Activity

Determination of enzyme activity: Diluted fermentation broth (0.5 mL) was reacted with 4 mL of 1.5 % inulin (prepared with 0.1 mol/L acetic acid buffer pH 4.6) at 55 °C for 20 min, and the reaction was terminated by deactivation at 100 °C for 5 min. The amount of reducing sugar was then determined via the 3,5-dinitrosalicylic acid (DNS) colorimetric method.

Definition of an enzyme unit: One U was defined as the amount of inulinase that produced 1 μmol of reducing sugar per minute.

Determination of the reducing sugar amount (*DNS method*): The reaction solution (0.5 mL) was placed in a 25-mL stoppered colorimetric tube, to which 2 mL of DNS was added. After 5 min of colorimetric reaction in boiling water, the tube was rapidly cooled under running water. Then, deionized water was added to the calibration mark and mixed. The absorbance was measured against a blank at 540 nm, and the amount of reducing sugar was calculated according to a standard curve.

11.2.4 Identification of Bacterial Strains

11.2.4.1 Detection of Physiological and Biochemical Characteristics

Physiological, structural, and biochemical characteristics used for classification were based on Bergey's Manual of Determinative Bacteriology [12].

11.2.4.2 16S rDNA Analysis

The genomic DNA of the bacterial strain ZF-9 was extracted as a template for PCR amplification using universal bacterial primers. The resulting products were sequenced, and the obtained sequences were thereafter subjected to a BLAST search in NCBI. Multiple DNA sequences alignment, distance matrice calculation (PAM Day-hoff) and phylogenetic tree construction (neighbor-joining algorithm) were performed with MEGA 4.0.

11.2.5 Optimization of Fermentation Conditions

11.2.5.1 Univariate Analysis

Effects of the carbon source on enzyme production: Soluble starch, glucose, maltose, fructose, and sucrose (20 g/L each) were used to replace inulin, with other ingredients resembling the basic culture medium. The optimum carbon source was selected by determining the enzyme activity after 3 days of culture at 30 °C and 200 rpm. Meanwhile, the effects of the concentration of the carbon source (10, 20, 30, 40, and 50 g/L) on enzyme production were evaluated.

Effects of the nitrogen source on enzyme production: Using the selected carbon source at the optimum concentration, the optimum nitrogen source was selected from yeast extract, peptone, corn syrup, beef extract, urea, $(NH_4)_2SO_4$, NH_4Cl ,and NH_4NO_3 (10 g/L each) by determining the enzyme activity under the abovementioned conditions. The effects of the nitrogen source concentration (5, 10, 15, 20, and 25 g/L) on enzyme production were evaluated.

Effects of Na_2HPO_4 on enzyme production: The effects of different concentrations of Na_2HPO_4 on enzyme production were studied under the abovementioned conditions, and the optimum concentration was determined by measuring the enzyme activity.

11.2.5.2 Response Surface Test

The factors that dominantly affected inulinase-producing bacteria and the concentrations were optimized by response surface method to determine the optimum culture medium.

11.3 Results and Analysis

11.3.1 Isolation of Inulinase-Producing Bacteria

The soil samples from the rhizospheres of wild *H. tuberosus* were placed into a conical flask containing sterile water and glass beads and shaken at 30 °C and 200 rpm for 60 min for dispersion. Then, the soil suspension was subjected to gradient dilution with sterile water, from which 0.2 ml was spread onto the culture medium for primary screening using inulin as the sole carbon source and cultured for 3–5 days. Bacterial strains with various morphologies were streaked, from which those growing vigorously were selected, purified, and preserved in agar slants. Of the 28 screened strains that managed to utilize inulin, preliminary determination that 13 were molds, seven were yeasts,five were bacteria, and three were actinomycetes by morphological observation. After rescreening by shaking, ni-3 was used as the study object denominated as ZF-9 because of its high enzyme activity (3.24 U/mL) (Table 11.1).

11.3.2 Identification of Bacterial Strain ni-3

11.3.2.1 Characteristics of Bacterial Colonies

The bacterial colonies were 2–5 mm in diameter, oval-shaped, raised with neat edges, and milky or yellowish in color. Under a microscope, the bacteria were uniformly stained, short, rod-shaped single cells that allowed endogenous spore formation (Fig. 11.1).

11.3.2.2 Physiological and Biochemical Characteristics

Culture conditions: The culture temperature ranged from 25 to 35 °C, with the optimum temperature occurring at 30 °C. The bacteria were capable of growing from pH 2 to 8, with no pigment production, and were capable of growing in culture medium containing 5 % NaCl.

11.3.2.3 16S rDNA Identification

The genomic DNA of the bacterial strain ZF-9 was used as a template, and a single band of 16S rDNA was PCR-amplified using the universal bacteria primers 27F (5′-AGAGTTTGATCCTGGCTCAG-3′) and 1492R (5′-GGTTACCTTGTTACG-ACTT-3′) with a length of approximately 1,400 bp (Fig. 11.2).

Table 11.1 Inulinase activity of the 28 strains

	Strains	Enzyme activity (U/mL)
Yeasts	ci-1	0.52
	di-4	1.24
	ji-5	2.53
	ki-9	0.26
	ni-11	0.76
	si-14	1.57
	si-15	2.11
Molds	di-1	0.31
	di-3	0.89
	di-5	0.48
	ji-7	1.29
	ki-1	1.09
	ki-6	3.12
	ki-15	1.51
	ki-18	2.75
	ni-4	0.93
	ni-27	0.71
	si-21	1.62
	si-32	0.59
	si-35	0.99
Bacteria	ji-1	1.76
	ki-4	0.73
	ni-3	3.24
	ni-13	1.55
	si-17	2.78
Actinomycetes	ki-2	2.45
	si-5	0.88
	si-8	1.37

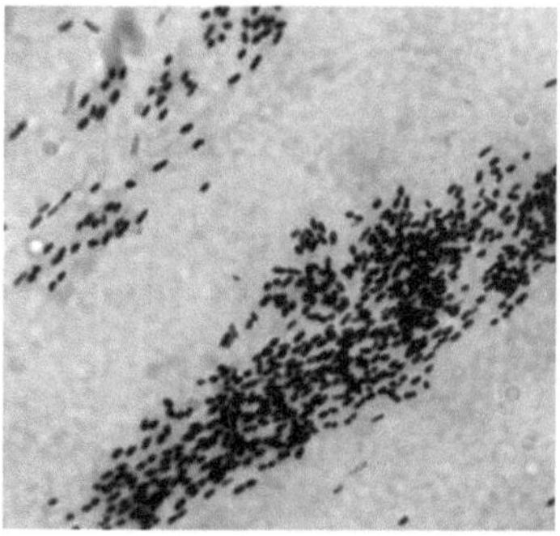

Fig. 11.1 Cell shape of inulinase producing strain ZF-9 (1,000×)

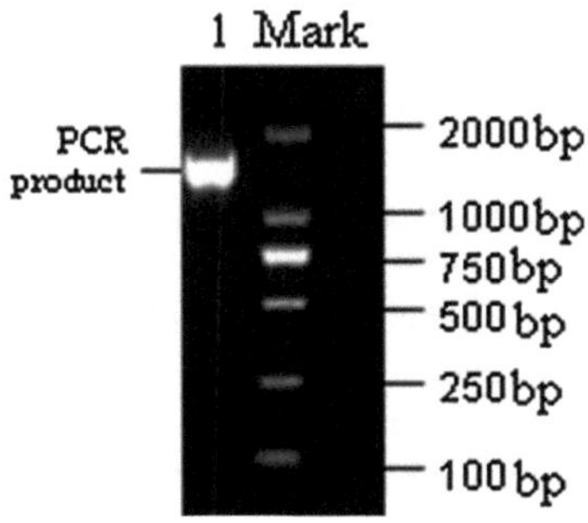

Fig. 11.2 16S rDNA PCR result of ZF-9

As verified by Sangon Biotech (Shanghai) Co., Ltd., the 16S rDNA sequence of ZF-9 was 1,440 bp in length. Then, the sequences were subjected to a BLAST search in NCBI, and a phylogenetic tree was constructed based on the complete sequence. The bacterial strain ni-3 shared the highest degree of homology (99.7 %) with *Brevibacillus centrosporus* (*B. centrosporus*) NBRC 15540 (GenBank ID: AB680893). In addition, as suggested by the morphological, physiological, and biochemical characteristics and by the 16S rDNA complete sequence analysis-based phylogenetic studies, ZF-9 was classified as *B. centrosporus* (i.e., *B. centrosporus* ZF-9) (Fig. 11.3).

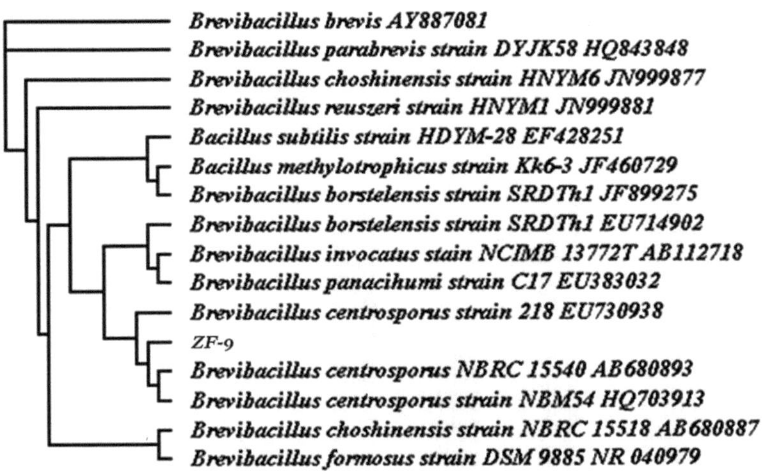

Fig. 11.3 Phylogenetic relationship of ZF-9 and related species based on their 16S rDNA sequence

11.3.3 Optimization of Fermentation Conditions for B. centrosporus ZF-9

11.3.3.1 Effects of Carbon Source on Inulinase Production by ZF-9

The carbon source, which is a crucial constituent of bacteria, mainly energizes the intracellular storage material and microbial growth. In addition, the carbon source predominantly affects the enzyme production of microorganisms. Besides being the main source of carbon skeletons in synthases, the carbon source can also induce or inhibit the production of enzymes. In this study, the influences of inulin, glucose, sucrose, fructose, maltose, and soluble starch on the inulinase-producing activity of *B. centrosporus* ZF-9 were evaluated while maintaining the other conditions (Fig. 11.4).

Inulin as the carbon source was the most conducive to bacterial growth, followed sequentially by glucose and fructose. Moreover, the enzyme activity of the fermentation broth peaked when inulin was used as the carbon source. Glucose produced to the second highest activity, while the other sources were far less effective. The bacteria grew well when fructose was used as the carbon source, albeit with the lowest enzyme activity. These results can mainly be attributed to exoinulinase, which produced copious amounts of fructose in the fermentation broth, thereby inhibiting the activity of inulinase. Therefore, inulin was finally selected as the carbon source, and the influences of its concentrations on bacterial growth and enzyme production were studied (Fig. 11.5).

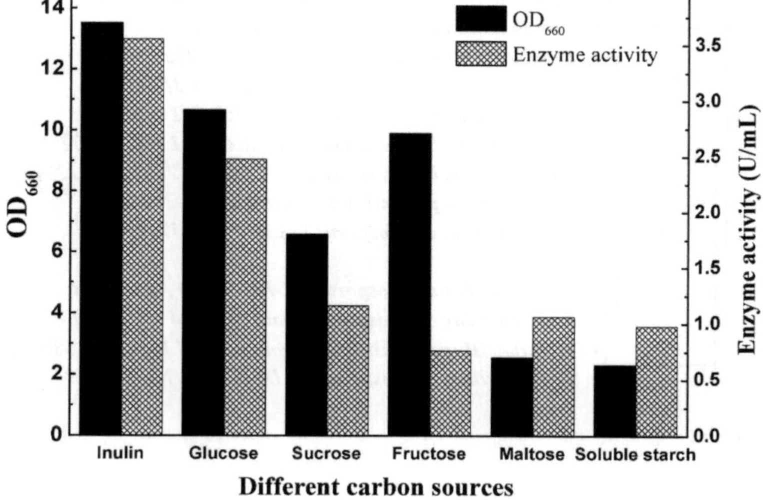

Fig. 11.4 Effect of different carbon sources on ZF-9 enzyme production and growth

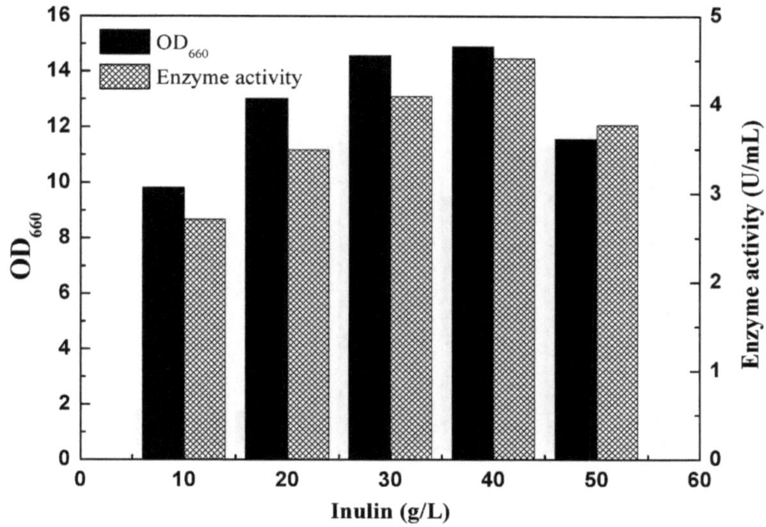

Fig. 11.5 Effect of different initial concentration of inulin on ZF-9 enzyme production and growth

The bacterial growth and enzyme activity, which were promoted by increasing concentrations of inulin, peaking at 40 g/L, were then inhibited by 50 g/L inulin. It is likely that the normal growth of bacteria was inhibited by high concentrations of inulin. Therefore, 40 g/L was selected as the optimal carbon source concentration.

11.3.3.2 Effects of Nitrogen Source on Inulinase Production by ZF-9

As an essential component of nucleic acids and proteins in microbial cells, nitrogen affects the growth and metabolites of microorganisms. In this study, the effects of yeast extract, peptone, beef extract, corn syrup, $(NH_4)_2SO_4$ and NH_4Cl on the growth, and enzyme production of *B. centrosporus* ZF-9 were investigated (Fig. 11.6).

When used as a nitrogen source, yeast extract, corn syrup, and peptone promoted bacterial growth and enzyme activity in descending order. In general, the bacteria grew better and produced enzymes more effectively when using organic rather than inorganic nitrogen sources. Therefore, yeast extract was used as the nitrogen source to evaluate the effects of its concentration on bacterial growth and enzyme production (Fig. 11.7).

Increasing the concentration of yeast extract, which is a superb source of microbial nutrition, promoted bacterial growth. However, the activity of inulinase increased and then decreased. Because this activity differed only slightly in the presence of 10 and 15 g/L yeast extract, 10 g/L was selected as the optimum concentration to consider the fermentation cost.

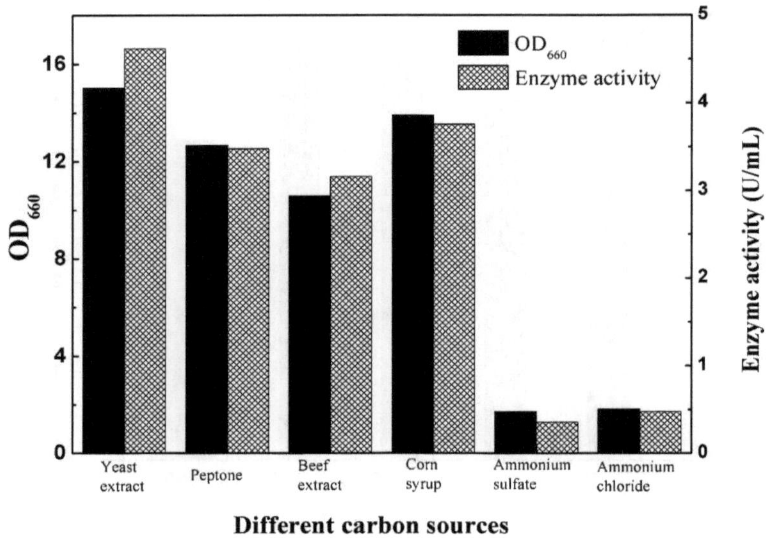

Fig. 11.6 Effect of different nitrogen sources on ZF-9 enzyme production and growth

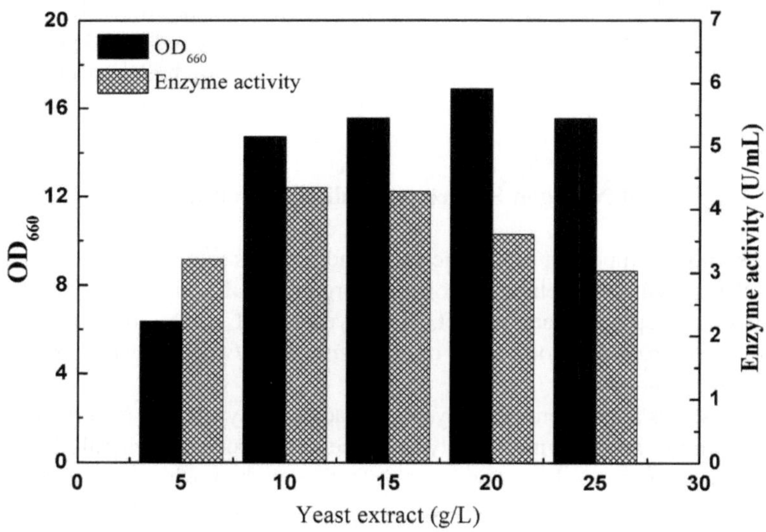

Fig. 11.7 Effect of different initial concentration of yeast extract on enzyme production and growth

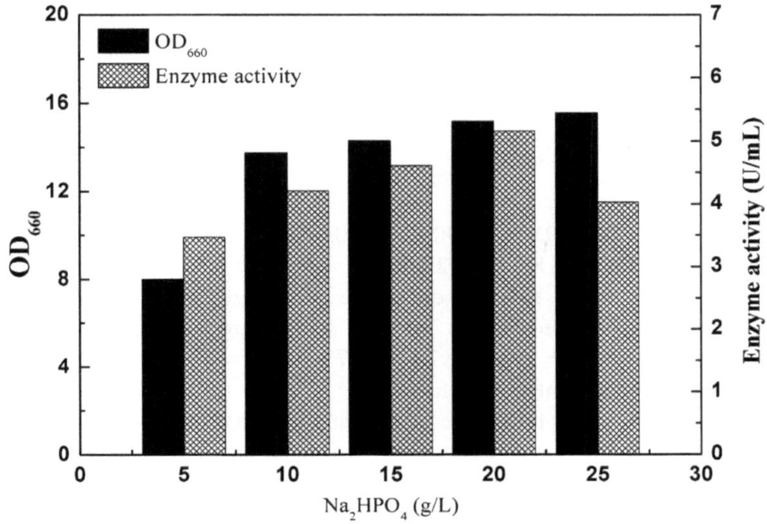

Fig. 11.8 Effect of different initial concentration of Na_2HPO_4 on ZF-9 enzyme production and growth

11.3.3.3 Effects of Na_2HPO_4 on Inulinase Production by ZF-9

Phosphates not only buffer the environment for bacterial fermentation but are also the predominant components of nucleic acids, ATP, and metabolic intermediates. The concentration of phosphates affects bacterial growth and enzyme activity. Thus, the effects of various concentrations of Na_2HPO_4 on the growth and inulinase production of ZF-9 were studied (Fig. 11.8).

Increasing the Na_2HPO_4 concentration promoted bacterial growth and enzyme activity. Given that the activity of inulinase peaked at 20 g/L Na_2HPO_4, 20 g/L was chosen as the optimum concentration.

11.3.3.4 Response Surface Optimization for Composition of Fermentation Culture Medium

Inulin, yeast extract and Na_2HPO_4, as suggested by a univariate analysis and nitrogen source composition assay, significantly affected the inulinase-producing bacteria. Accordingly, a central composite design was employed using the three factors as the independent variables and the growth and enzyme activity of ZF-9 as the response values. The experimental level and coded value of each factor are listed in Table 11.2, and the experimental design and results are summarized in Table 11.3.

A multiple regression equation was thus derived by fitting the data in Table 11.3 to a quadratic polynomial regression using STATISTICA 6.0 software as follows:

Table 11.2 Different levels of different factors in trails

Dependent variable		Independent variable				
		−1.68	−1	0	+1	+1.68
X1	Inulin (g/L)	23.2	30	40	50	56.8
X2	Yeast extract (g/L)	5.8	7.5	10	12.5	14.2
X3	Na_2HPO_4 (g/L)	15.8	17.5	20	22.5	24.2

$Y = 1.1634X1 - 0.9139X1^2 + 0.9067X2 - 0.6659X2^2 + 0.2132X3 - 0.9919X3^2 - 0.4250X1X2 - 0.0050X1X3 - 0.2250X2X3 + 6.1122$. $R^2 = 0.975$, indicating that this model was suitable to represent the experimental results. Meanwhile, the maximum Y was calculated as 6.38 U/mL by substituting the extreme values of the fitting equation, i.e., X1 = 45.18, X2 = 11.27 and X3 = 20.12, into the regression equation, aiming to optimize each factor.

The interactions of the three components and their optimal level in the inulinase production were further analyzed by the response surface methodology. The three-dimension response surface curves and respective contour plots were presented in Fig. 11.9. It was shown that the response surface was convex, which suggested that the optimal conditions were well defined and there existed a maximum for each variable. Moreover, Fig. 11.9a demonstrated a better ellipse whereas both Fig. 11.9b, c almost displayed circularity. The results suggested that the interaction between inulin and Yeast extract was more significant but the interactions between inulin and Na_2HPO_4, Yeast extract and Na_2HPO_4 were almost insignificant. This result was in good agreement with that obtained by the P-test.

Table 11.3 Factors-coding, levels, and result of central composite design

Run	Inulin (g/L)	Yeast extract (g/L)	Na_2HPO_4 (g/L)	OD_{660}	Enzyme activity (U/mL)
1	30	7.5	17.5	10.55	3.12
2	30	7.5	22.5	10.5	3.71
3	30	12.5	17.5	9.12	4.96
4	30	12.5	22.5	10.23	4.83
5	50	7.5	17.5	8.23	5.01
6	50	7.5	22.5	10.11	5.32
7	50	12.5	17.5	12.28	5.73
8	50	12.5	22.5	13.03	5.86
9	23.2	10	20	11.95	4.05
10	56.8	10	20	9.31	5.62
11	40	5.8	20	8.85	4.6
12	40	14.2	20	13.57	5.77
13	40	10	15.8	12.45	4.56
14	40	10	24.2	12.73	4.89
15	40	10	20	13.95	6.03
16	40	10	20	13.72	6.19

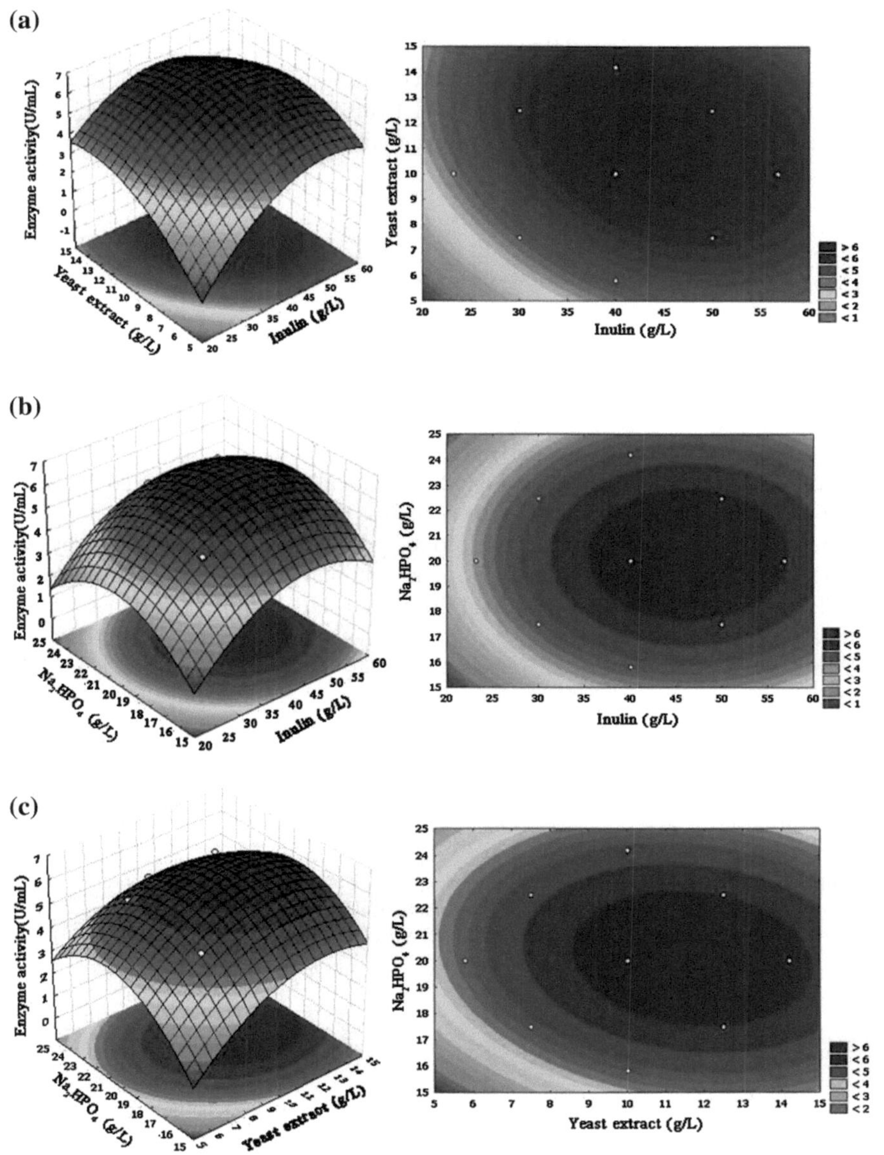

Fig. 11.9 The response surface plot and the corresponding contour plot. **a** The response surface plot and the corresponding contour plot showing the effects of inulin and Yeast extract on inulinase production by *B. centrosporus* ZF-9; **b** The response surface plot and the corresponding contour plot showing the effects of inulin and Na_2HPO_4 on inulinase production by *B. centrosporus* ZF-9; **c** The response surface plot and the corresponding contour plot showing the effects of Na_2HPO_4 and yeast extract on inulinase production by *B. centrosporus* ZF-9

Because the solutions of the regression equation were similar, the concentrations of inulin, yeast extract, and Na_2HPO_4 were eventually selected as 45, 11, and 20 g/L, respectively, for convenience. Under these conditions, the activity of inulinase was calculated via triplicate experiments as 6.41 U/mL. This study, which successfully predicted a culture medium for increasing inulinase activity, is of great practical significance.

11.4 Conclusion

Twenty-eight bacterial strains that utilized inulin as the sole carbon source were isolated from soils that were sampled from the rhizospheres of wild *H. tuberosus*, from which the strain ni-3 was screened. This strain was identified as *B. centrosporus* by observing the microscopic morphology and cultural characteristics, detecting the physiological and biochemical characteristics, and analyzing the 16S rDNA and is referred to as *B. centrosporus* ZF-9.

The conditions under which ZF-9 produced inulinase were optimized by univariate and response surface analyses. The optimum culture medium contained 45 g/L inulin, 11 g/L yeast extract and 20 g/L Na_2HPO_4, in which the activity of inulinase peaked at 6.41 U/mL.

Brevibacillus centrosporus ZF-9, as a naturally occurring wild strain, failed to meet the requirements of industrial production for enzyme activity even after the optimization of the fermentation conditions. Therefore, additional industrially applicable strains should be screened based on the mutation breeding of ZF-9. Furthermore, the enzymatic properties and purification techniques of inulinase require in-depth studies.

Acknowledgments This work was supported by the Natural Science Foundation of Fujian Province (No. 2012J05015), Educational Commission of Fujian Province (JA12430), and Doctoral Research Startup Foundation of Liaocheng University (No. 31805).

References

1. Chi Z, Zhang T, Cao T et al (2011) Biotechnological potential of inulin for bioprocesses. Bioresour Technol 102:4295–4303
2. Lu W, Li A, Guo Q (2014) Production of novel alkalitolerant and thermostable inulinase from marine actinomycete *Nocardiopsis* sp. DN-K15 and inulin hydrolysis by the enzyme. Ann Microbiol 64:441–449
3. Kango N, Jain S (2011) Production and properties of microbial inulinases: recent advances. Food Biotechnol 25:165–212
4. Tan X, Dong Q, Ding H (2007) Research and development prospects of functional health food inulin. Food Nutr 1:22–24
5. Abeer A, Hala R, Faten A (2014) Optimization of inulinase production from low cost substrates using Plackett-Burman and Taguchi methods. Carbohydr Polym 102:261–268

6. Mahsan K, Isthier C, Cheng J et al (2014) Immobilization of endo-inulinase on non-porous amino functionalized silica nanoparticles. J Mol Catal B Enzym 104:48–55
7. Singh R, Rupinder P (2010) Production of fructooligosaccharides from inulin by endoinulinases and their prebiotic potential. Food Technol Biotechnol 48:435–450
8. Porna V, Kulkani P (1995) A study of inulinase production in *Aspergillus niger* using fractional factorial design. Bioresour Technol 54:315–320
9. Possoa A, Vitolo M (1999) Inulinase from *Kluyveromyces marxianus*: culture medium composition and enzyme extraction. Braz J Chem Eng 16:237–245
10. Tatyaso Y, Rekha S, Alankar A (2013) Immobilization of inulinase from *Aspergillus niger* NCIM 945 on chitosan and its application in continuous inulin hydrolysis. Biocatal Agric Biotechnol 2:96–101
11. Porna V, Kulkani P (1995) A study of inulinase production in *Aspergillus niger* using fractional factorial design. Bioresour Technol 54:315–320
12. Holt J, Krieg N, Sneath P et al (1994) Bergey's manual of determinative bacteriology, 9th edn. Williams and Wilkins Company, Baltimore

Chapter 12
Isolation and Identification of a Cellulose-Producing Bacterial Strain from the Genus *Bacillus*

Hongyang Zhu, Yang Zhang, Jinhai Wang, Yongning Li and Weiling Lin

Abstract Strain ZF-7 was isolated from defective fruits in this study, and its product was determined as bacterial cellulose in view of the results of cellulose special reaction, FTIR spectroscopy analysis, and cellulose hydrolysis experiments. Based on morphological, physiological, and biochemical characteristics, ZF-7 was identified by 16S rDNA sequences and systematic analysis. The results indicated that 16S rDNA sequence of ZF-7 had 99.5 % similarity to the partial sequence of *Bacillus amyloliquefaciens*, suggesting that ZF-7 was categorized as a species of *B. amyloliquefaciens*, and denominated as *B. amyloliquefaciens* ZF-7. The fermentation performance of ZF-7 under shaking and static flasks was inspected preliminary, and the BC production was 6.6 and 6.2 g/L, respectively.

Keywords Bacterial cellulose · *Bacillus amyloliquefaciens* · Isolation · Identification

12.1 Introduction

Bacterial cellulose (BC), as an unbranched polymer with nanofibrils, made up of glucose units via (1 → 4) β-glycosidic bonds, has many excellent properties such as highly biocompatible, biodegradable, water holding capacity, and high tensile strength [1, 2]. Compared with plant-derived cellulose, BC has better specific chemical and physical properties, which offer applications such as in food,

Hongyang Zhu and Yang Zhang had equal contributions to this work.

H. Zhu (✉) · J. Wang · Y. Li · W. Lin
Department of Pharmacy, Fujian Health College, Fuzhou, 350101 Fujian, China
e-mail: xiangyu238@163.com

Y. Zhang
College of Life Science, Liaocheng University, Liaocheng 252059, China
e-mail: zy886024@163.com

T.-C. Zhang and M. Nakajima (eds.), *Advances in Applied Biotechnology*,
Lecture Notes in Electrical Engineering 332, DOI 10.1007/978-3-662-45657-6_12

cosmetics, paper-making, acoustics, and optics [3, 4]. Due to these good properties, BC has been used in the medical industry such as skin tissue, cartilage, bone soft tissue blood vessel, etc., and used in food industry such as food packing material, food thickener, stabilizer, texture modifier, etc. [5–8]. At present, bacterial strains such as *Agrobacterium*, *Rhizobium*, *Acetobacter*, *Aerobacter*, *Gluconacetobacter* have been reported for BC production under optimum bioprocess conditions [9–12]. *Acetobacter xylinum* is one of the most extensively studied sources of BC [13]. With the synthesis processes and physiochemical properties studied as well as fermentation conditions optimized, BC has now become potentially commercially applicable [14–16]. However, the production cost of BC should be substantially decreased before large-scale manufacture. Therefore, the goal of this study was to screen a bacterial strain with highly productive from local soil samples, fruit and vegetable resources by using glucose as the carbon source. The results herein provide valuable basis for future studies on BC.

12.2 Materials and Methods

12.2.1 Sample Source for Bacterial Strains

Soils and defective fruits in the Gushan Mountain and the Qishan Mountain (Fuzhou City, Fujian Province, China), as well as retail distillers' grains and pickled vegetables in Minhou County (Fuzhou City, Fujian Province, China) were collected as samples.

12.2.2 Main Culture Media

All the chemicals used were of analytical grade and commercially available, unless specified. Cell growth and BC production were studied in three different media. The three different media used were designated enrichment medium M1 (4 % glucose, 0.5 % yeast extract, 0.5 % peptone, 0.2 % NaAc, and 1 % $CaCO_3$, pH 5.0), solid medium M2 (2 % glucose, 0.5 % peptone, 0.5 % yeast extract, 0.1 % citric acid, 0.5 % $Na_2HPO_4 \cdot 12H_2O$, 1 % $CaCO_3$, and 2 % agar), and fermentation media M3 (5 % glucose, 0.5 % peptone, 0.5 % yeast extract, 0.1 % citric acid, 0.5 % $Na_2HPO_4 \cdot 12H_2O$, and 1.4 % ethanol were added, and pH was adjusted to 6.5). Fermentation studies were performed in 50 or 100 ml culture media inoculated with 1 % (v/v) cells suspension for 24 h at 30 °C.

12.2.3 Methods

12.2.3.1 Enrichment of Bacterial Strains

The samples were smashed and diluted by sterile water (1:10, V:V). The diluent (1 ml) was inoculated into a test tube already containing 10 ml of sterilized M1 for 3 days at 30 °C. The sample with gel membrane on the liquid surface was inoculated on the solid medium M2, and cultured for another 5 days at 30 °C. Subsequently, the strains with transparent circles around were picked and inoculated in the enrichment culture medium to observe the production of membrane. The enrichment and isolation operations were repeated twice.

12.2.3.2 Screening and Purification of Bacterial Strains

After 5 days of plate culture, the bacterial strains which produced membrane rapidly and massively were picked and preserved on M2.

12.2.3.3 Membrane Production by Fermentation of Bacterial Strains

A circle of activated slant seeds was inoculated in the fermentation medium M3 at 30 °C, 100 r/min for 24 h. Then they were inoculated in a 250-ml conical flask containing 50 ml M3 at 30 °C for another 5 days.

12.2.4 Extraction and Treatment of BC

The floating BC membrane that had been cultured for 5 days at constant temperature was taken out and washed many times by distilled water to remove the residual culture medium on the membrane surface. Then the membrane was immersed in 1 % NaOH solution at 80 °C for 30 min, then immersed again in 0.5 % acetic acid solution for 30 min, and washed to neutral by distilled water.

12.2.5 Yield and Moisture Content of BC

The treated BC gel was centrifuged at 4,000 r/min for 30 min, and weighed as the wet weight. The dry weight of BC was obtained after the gel was dried at 80 °C until constant weight. The mass of BC was measured by a precision electronic balance. The yield of BC was expressed as its mass in 1 L of culture medium.

$$\text{Moisture content of BC gel}(\%) = (\text{wet weight} - \text{dry weight})/\text{wet weight} \times 100\,\%$$

12.2.6 *Identification of Bacterial Strains*

12.2.6.1 Physiological and Biochemical Characteristics [17]

The strains were identified according to Bergey's Manual of Systematic Bacteriology.

12.2.6.2 16S rDNA Analysis

By using genomic DNA of bacterial strain ZF-7 as the template, PCR amplification was performed with universal primers. The sequences of the PCR products were then subjected to BLAST search in NCBI. Meanwhile, the phylogenetic tree was constructed by NT Vector, finally identifying the genus and species of the BC-producing bacterium ZF-7.

12.2.7 *Product Analysis*

12.2.7.1 Qualitative Analysis by Staining [18]

The gel membrane on the surface of fermentation broth that had been cultured for 1–2 days was picked, placed in a culture dish, and naturally unfolded on a glass slide that was thereafter taken out and air-dried. Afterward, 66.5 % H_2SO_4 was dropped onto the dry membrane, on which KI solution was then dropped to allow color development that was observed by the naked eye.

12.2.7.2 FTIR Analysis

The gel membrane was characterized using FTIR spectroscopy. The cellulose membrane was freeze-dried under vacuum, and ground into powders. Sample pellets were prepared by mixing equal amount of cellulose samples spectroscopic grade KBr (1:100) and subsequently palletizing the mixture at 10 Torr. The absorbance mode FTIR data were recorded with 2 cm^{-1} resolution in the wave number region of 4,000–400 cm^{-1} with a Nicolet 360 Smart infrared spectrometer.

12.2.7.3 Cellulase Hydrolysis Assay

Dry gel membrane (1 g) was thoroughly crushed and added with a cellulase solution prepared with pH 5.0 PBS, hydrolyzed at 50 °C until the solution became clarified, and filtrated through a filter paper which has dried to constant weight at 80 °C. The content of glucose in the filtrate was detected by an SBA-40C sensor

(Shandong Academy of Sciences, Jinan, China) with glucose dehydrogenase, and the hydrolyzed residue was washed twice and dried to constant weight together with the filter paper. Then the weight of cellulose was derived by subtracting that of the filter paper, from which the content of cellulose was converted. This method provides useful evidence for the qualitative detection of cellulose owing to the specific hydrolytic capacity of cellulase against cellulose and the specific detection of SBA-40C biosensor.

12.3 Results and Analysis

12.3.1 Isolation of BC-Producing Bacterial Strains and Product Analysis

Of the 50 soil, defective fruit and vegetable samples, only three from fruits produced gel membranes. Since the membranes did not dissolve in 0.1 M NaOH and the specific blue-black reaction of H_2SO_4-KI staining (Fig. 12.1) were conducted, the results indicated that the membranes were mainly comprised of cellulose [19]. Particularly, a strain that produced maximum BC (i.e., ZF-7), as observed by the naked eye, was selected as the study object.

The fermentation products of ZF-7 were extracted, freeze-dried under vacuum, and characterized by FTIR (Fig. 12.2).

The absorption peak at 1,164 cm^{-1} can be assigned to the stretching vibration of C–O, as the characteristic of cellulose molecule, while that at 3,435 cm^{-1} corresponds to the stretching vibration of O–H. The absorption peak at 1,644 cm^{-1} represents hemi-acetal at the 4'-end of cellulose, and that at 1,428 cm^{-1} represents the stretching vibration of C–H, with the depth associated with the degree of crystallinity. Moreover, the peaks at 2,896 and 900 cm^{-1} are induced by the stretching of CH_2–CH and the characteristic glycosidic bond respectively [19]. The results indicated that the fermentation products of ZF-7 resembled BC.

Dry membrane (1.011 g) was freeze-dried under vacuum and hydrolyzed by 50 ml of cellulase solution for 24 h until the solution became clear. Subsequently, the solution was filtered by an oven-dried filter paper (0.752 g), and 0.99 g glucose was detected in the filtrate by SBA-40C biosensor. Given that the weight of

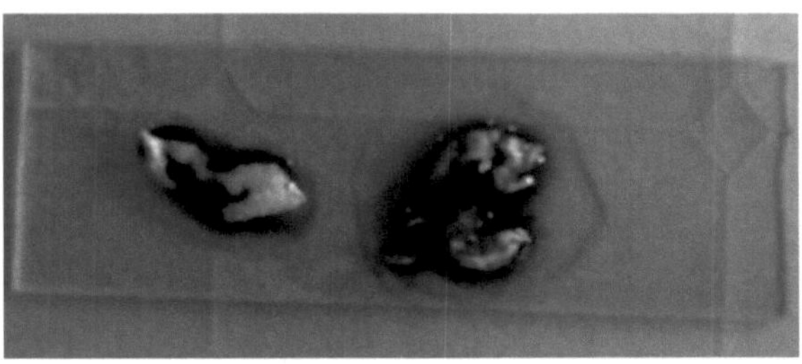

Fig. 12.1 H_2SO_4-KI reaction

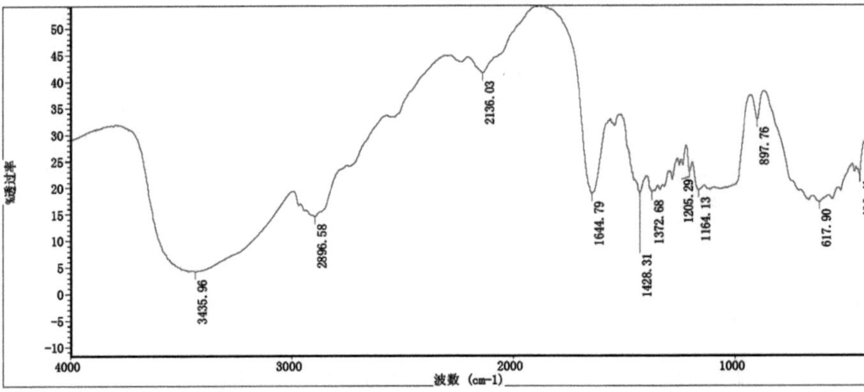

Fig. 12.2 FTIR of bacterial cellulose

Table 12.1 The result of cellulose hydrolysis

Cellulose membrane (g)	Filter paper weight (g)	Filter paper and residue weight (g)	Residue weight (g)	Cellulose content (%)	Glucose content (g)
1.011	0.752	0.845	0.093	90.8	0.99

residue was 0.093 g (i.e., weight of combined filter paper and residue subtracted by that of filter paper), the content of cellulose was calculated by the formula of $[(1.011 - 0.093)/1.011] \times 100\ \%$. As evidenced by the outcomes of cellulase hydrolysis and glucose detection by SBA-40C biosensor, the membrane produced by ZF-7 was cellulose, with the content of 90.8 % (Table 12.1).

12.3.2 Morphology, and Physiological and Biochemical Characteristics of ZF-7

The bacterial colonies were 2–5 mm in diameter, circle-shaped, yellowish, and raised with neat edges and smooth, wet surface, forming a leather-like membrane. The colonic morphology in the solid culture medium is shown in Fig. 12.3, and the scanning electron microscopic morphology is exhibited in Fig. 12.4.

In addition, the physiological and biochemical characteristics of ZF-7 are listed in Table 12.2.

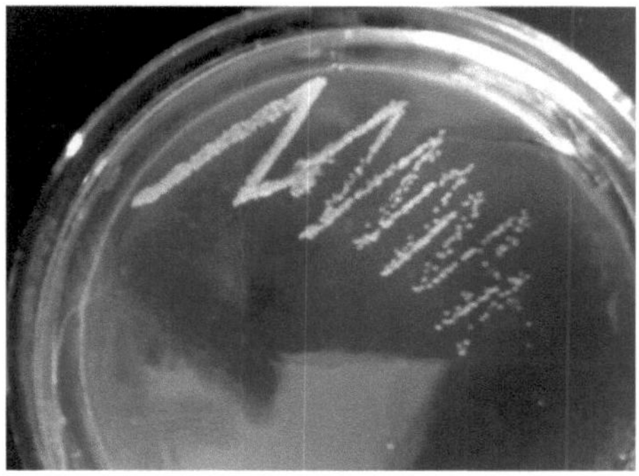

Fig. 12.3 Colony morphology of bacterial strain ZF-7

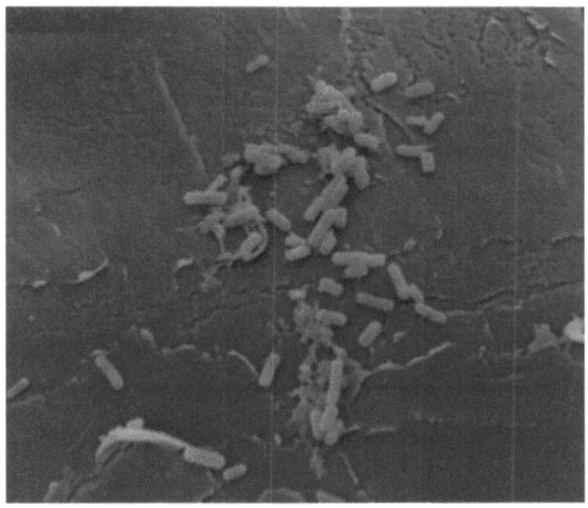

Fig. 12.4 The SEM morphology of bacterial strain ZF-7

12.3.3 16S rDNA Identification

Genomic DNA of ZF-7 was used as the template, and a single band of 16S rDNA was PCR-amplified with universal bacteria primers 27F (5'-AGAGTTTGATCCTGGCTC AG-3') and 1492R (5'-GGTTACCTTGTTACGACTT-3'), with the length of approximately 1,500 bp (Fig. 12.5).

Sequencing of the PCR products, which was completed by Sangon Biotech (Shanghai) Co., Ltd., showed that the 16S rDNA sequence of ZF-7 was 1,460 bp in length. Then the sequences were subjected to BLAST search in NCBI, and the phylogenetic tree was constructed on the basis of the complete 16S rDNA sequence (Fig. 12.6). ZF-7 shared the highest degree of homology (99.5 %) with *Bacillus amyloliquefaciens* subsp. (GenBank entry ID: JF899275). Hence, ZF-7 was, as suggested by the results of morphological, physiological, and biochemical characteristics, and 16S rDNA complete sequence analysis-based phylogenetic studies, classified as *B. amyloliquefaciens* (*B. amyloliquefaciens* ZF-7).

Table 12.2 The main physiology and biochemistry characters of ZF-7

Physiology and biochemistry projects		Result	Physiology and biochemistry projects		Result
Growth performance	3 % NaCl	+[a]	Carbon source	Acetates	+
	5 % NaCl	−		Citrate	−
	0.001 % lysozyme	+		Fumarate	−
	Catalase	+		Maleate	−
Hydrolysis performance	Casein	−		Succinate	−
	Tween 80	−	Fermentation	D-Sorbitol	+
	Urea	−		L-Rhamnose	+
	Starch	+	Cellulose-producing		+
	Gelatin	−	Relation to O_2		+
Ketogenic activity		+	Ethanol is oxidized to acetic acid		+
With ammonium as the sole nitrogen source		−	Lactate oxidation		+
Quinones		Q_{10}	Spore test		+
Gram staining		Positive			

[a]+ positive; − negative

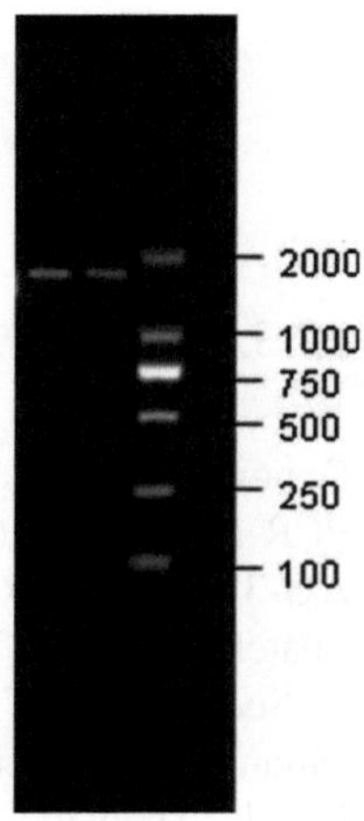

Fig. 12.5 16S rDNA PCR result of ZF-7

12.3.4 Primary Fermentation Assay for BC Production by ZF-7

Static fermentation and shaking fermentation were performed with the fermentation medium M3 respectively. During the former assay, membrane that formed on the surface of fermentation broth gradually thickened with prolonged culture, and filled

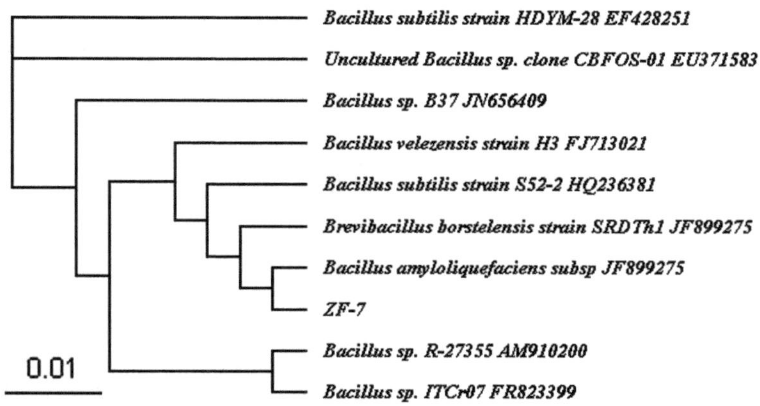

Fig. 12.6 Phylogenetic relationship of ZF-7 and related species based on their 16S rDNA sequence

the culture medium eventually (Fig. 12.7). In the latter assay, however, BC membrane, instead of remaining intact, was dispersed into smooth-surfaced microballs suspending on the surface of fermentation broth owing to shear force (Fig. 12.8). The experimental results with the two fermentation methods are summarized in Table 12.3. After 5 days of shaking fermentation, the wet and dry weights of BC membrane were 425.2 and 6.6 g/L respectively, and the moisture content was 98.45 %. Under the conditions of static fermentation, the wet and dry weights of BC membrane were 502.6 and 6.2 g/L respectively, and the moisture content was 98.77 % after 5 days. Therefore, the bacterial strain produced more BC during shaking fermentation, with lower water content. Given the high yield of ZF-7, it is highly industrially applicable as the initial strain, when assisted with industrial mutation breeding techniques, for the screening of excellent strains with high BC production capacity.

Fig. 12.7 Appearance in shape of the bacterial cellulose after 3 days of stationary culture

Fig. 12.8 Appearance in shape of the bacterial cellulose after 3 days of flask agitated culture

Table 12.3 The fermentation results of ZF-7 under different culture conditions

Fermentation	Bacterial cellulose wet weight (g/L)	Bacterial cellulose dry weight (g/L)	Moisture content (%)
Shake cultivation	425.2	6.6	98.45
Static culture	502.6	6.2	98.77

12.4 Conclusion

A naturally occurring bacterial strain ZF-7 was isolated, the fermentation products of which were determined as BC by specific staining for cellulose and FTIR analysis. ZF-7 was ultimately identified as *B. amyloliquefaciens* (referred to as *B. amyloliquefaciens* ZF-7) by observing the microscopic morphological and culture characteristics, by detecting physiological and biochemical characteristics, and by performing 16S rDNA identification. Moreover, the strain exhibited high BC production capacities under the conditions of both static fermentation and shaking fermentation. After 5 days of oscillating fermentation, the wet and dry weights of BC membrane were 425.2 and 6.6 g/L respectively, and the moisture content was 98.45 %. Under the conditions of static fermentation, the wet and dry weights of BC membrane were 502.6 and 6.2 g/L respectively, and the moisture content was 98.77 % after 5 days. In summary, ZF-7 is highly applicable to industrial fermentation.

Acknowledgements This work was supported by the Natural Science Foundation of Fujian Province of China (No. 2012J05015), Educational Commission of Fujian Province of China (No. JA12430, JA13414), and Doctoral Research Startup Foundation of Liaocheng University of China (No. 31805).

References

1. Yu B, Zhou HL (2007) Progress in research of bacterial cellulose. Biotechnol Bull 2:87–90
2. Brown RM (2004) Cellulose structure and biosynthesis: what is in store for the 21st century? J Polym Sci Part A: Polym Chem 42:487–495
3. Gullo M, Giudici P (2008) Acetic acid bacteria in traditional balsamic vinegar: phenotypic traits relevant for starter cultures selection. Int J Food Microbiol 125:46–53
4. Jonas R, Farah LF (1998) Production and application of microbial cellulose. Polym Degrad Stab 59(1):101–106
5. Jeong SI, Lee SE, Yang H et al (2010) Toxicologic evaluation of bacterial synthesized cellulose in endothelial cells and animals. Mol and Cell Toxicol 6(4):370–377
6. Azuma C, Yasuda K, Tanabe Y et al (2006) Biodegradation of high-toughness double network hydrogels as potential material for artificial cartilage. J Biomed Res Part A 81(2):373–380
7. Mori R, Nakai T, Enomoto K et al (2011) Increased antibiotic release from a bone cement containing bacterial cellulose. Clin Orthop Relat Res 469(2):600–606
8. Wang J, Gao C, Zhang Y et al (2010) Preparation and in vitro characterization of BC/PVA hydrogel composite for its potential use as artificial cornea biomaterial. Mater Sci Eng: C 30 (1):214–218
9. Shi ZJ, Zhang Y, Philips GO et al (2014) Utilization of bacterial cellulose in food. Food Hydrocolloids 35:539–545
10. Rani MU, Udayasankar K, Appaiah KA (2011) Properties of bacterial cellulose produced in grape medium by native isolate *Gluconacetobacter* sp. J Appl Polym Sci 120(5):2835–2841
11. Wu RQ, Li ZX, Yang JP et al (2010) Mutagenesis induced by high drostatic pressure treatment: a useful method to improve the bacterial cellulose yield of a *Gluconoacetobacter xylinus* strain. Cellulose 17(2):399–405
12. Son HJ, Kim HG, Kim KK et al (2003) Increased production of bacterial cellulose by *Acetobacter* sp. V6 in synthetic media under shaking culture conditions. Bioresour Technol 86 (3):215–219
13. Cheng KC, Catchmark JM, Demirci A (2009) Effect of different additives on bacterial cellulose production by *Acetobacter xylinum* and analysis of material property. Cellulose 16:1033–1045
14. Bodin A, Backdahl H, Fink H (2007) Influence of cultivation conditions on mechanical and morphological properties of bacterial cellulose tubes. Biotechnol Bioeng 97:425–434
15. Jipa IM, Stoica-Guzun A, Stroescu M (2012) Controlled released of sorbic acid from bacterial cellulose based mono and multilayer antimicrobial films. LWT-Food Sci and Technol 47 (2):400–406
16. Sureshkumar M, Siswanto DY, Lee C (2010) Magnetic antimicrobial nanocomposite based on bacterial cellulose and silver nanoparticles. J Mater Chem 20(33):6948–6955
17. George MG, Julia AB, Timothy GL (1984) Bergey's manual of systematic bacteriology (In: Chinese translation). Science Press, Beijing
18. Shi QS, Feng J, Feng J et al (2009) Isolation and identification of a cellulose-producing bacterial strain. Fajiao Keji Tongxun 38:11–15
19. Liu JL (2002) Instrumental Analysis. Science Press, Beijing

Chapter 13
Improved Lactose Utilization by Overexpression β-Galactosidase and Lactose Permease in *Klebsiella pneumoniae*

Xuewu Guo, Yazhou Wang, Xiangyu Guan, Yefu Chen, Cuiying Zhang and Dongguang Xiao

Abstract As an important bio-based chemical product, 2,3-butanediol has a wide range of applications in many fields, such as chemical, fuel, food, and aerospace. Cheese whey powder (CWP), an inexpensive, available, and abundant material, is considered to be an ideal substrate for 2,3-BD fermentation. To improve 2,3-butanediol production, the previous studies mainly focus on the metabolic pathway from pyruvate to 2,3-butanediol or the metabolic pathway of by-products, but studies about improving lactose utilization rate are rarely reported. In the present study, adding exogenous β-galactosidase was proved to favor the lactose utilization and lactose utilization might be the limiting step of lactose fermentation to 2,3-butanediol. *ElacY* (encoding lactose permease of *Escherichia coli*) and *bgaB* (encoding β-galactosidase of *K. pneumonia*) were overexpressed in *K. pneumonia* CICC10781. Of the two genes, only overexpression of *ElacY* promoted lactose utilization of CICC10781, and meanwhile the 2,3-butanediol generation capacity was not affected.

Keywords *Klebsiella pneumoniae* · *bgaB* gene · *Lacy* gene · 2,3-butanediol · Overexpression

13.1 Introduction

2,3-Butanediol (2.3-BD), an important chemical, also known as 2,3-butylene glycol or dimethylene glycol, is a four-carbon diol synthesized from mixed acid and alcoholic fermentation [1, 2]. As an important bio-based chemical product,

X. Guo · Y. Wang · X. Guan · Y. Chen · C. Zhang · D. Xiao (✉)
Key Laboratory of Industrial Fermentation Microbiology Ministry of Education, Tianjin Industrial Microbiology Key Laboratory, College of Biotechnology, Tianjin University of Science and Technology, Tianjin 300457, People's Republic of China
e-mail: xiao99@tust.edu.cn

T.-C. Zhang and M. Nakajima (eds.), *Advances in Applied Biotechnology*,
Lecture Notes in Electrical Engineering 332, DOI 10.1007/978-3-662-45657-6_13

2,3-butanediol has a wide range of applications in many fields, such as chemical, fuel, food, and aerospace. Microbial fermentation production of 2,3-butanediol has attracted more attentions [3, 4].

For the production of 2,3-butanediol, the major cost in most biomass conversion processes appears to be the substrate cost [5]. So finding an inexpensive raw materials is critical for 2,3-butanediol production in a large scale. In recent years, many studies about different materials for 2,3-butanediol have been reported. The possible materials for 2,3-butanediol production include cheese whey [6, 7], molasses [8], glycerol [9, 10], Jerusalem artichoke [11, 12], wood fibers [10], and corn stover [13].

Cheese whey is a major by-product of the dairy industry and the world's annual production of cheese whey is more than one hundred million tons [14]. Cheese whey has abundant nutrients, which retain 55 % nutrients of milk. Cheese whey contains 3–5 % lactose, 1 % protein, 0.45–0.5 % fat, inorganic salts, and other minerals [15]. Thus, the efficient use of cheese whey has been one of the most serious problems to be solved in dairy industry.

A number of species are capable of fermenting CWP to 2,3-BD, such as *Bacillus polymyxa, Enterobacter cloacae, Klebsiella oxytoca*, and *K. pneumonia* [6, 16, 17]. A variety of enzymes of the above strains involve in the metabolic pathway from lactose to 2,3-butanediol. To improve 2,3-butanediol production,the previous studies mainly focus on the overexpression of enzymes in the metabolic pathway from pyruvate to 2,3-butanediol or knockout of the key enzymes of the metabolic pathway of by-products [3, 18], but studies about improving lactose utilization rate are rarely reported. The lactose enters cells through the lactose permease, and it is hydrolyzed into glucose and galactose moieties by β-galactosidase. The previous study show that the production of 2,3-BD significantly increased if the lactose hydrolysed enzymatically prior. And the lactose utilization is the limiting step in the 2,3-BD production process from cheese whey [6].

In the present study, to improve the lactose consumption rate and shorten the fermentation period, an appropriate expression plasmid was selected. *ElacY* (encoding lactose permease of *E. coli*) and *bgaB* (encoding β-galactosidase of *K. pneumonia*) were overexpressed in the wild strain CICC1078.

13.2 Materials and Methods

13.2.1 Strains and Plasmids

All of the strains and plasmids as well as their relevant genotypes used in this study are listed in Table 13.1. The parental strain *K. pneumonia* CICC10781 is stored at China Center of Industrial Culture Collection (CICC). The plasmid pUC18 K were provided by Yu Zheng, associate professor of college of Biotechnology, Tianjin University of Science and Technology.

Table 13.1 Characteristics of strains and plasmids used in this study

Strains and plasmid	Relevant characteristic	Reference
Klebsiella pneumonia		
CICC10781	Wild- type	CICC, China
10781-ElacY	CICC10781 containing plasmid pUC-ElacY	This study
10781-pUC18 K	CICC10781 containing plasmid pUC18 K	This study
Escherichia coli		
DH5α	Φ80 *bgaB*ΔM15 Δ*lacU169 recA1 endA1 hsdR17 supE44 thi-1 gyrA relA1*	This lab
Plasmid		
pUC18 K	Kan^r, pUC ori, P_{lac}, MCS	This study
pUC-ElacY	*ElacY* gene cloned into pUC18 K	This study

13.2.2 Media and Growth Conditions

LB medium (composed of 1 % (m/v) NaCl, 1 % (m/v) tryptone, and 0.5 % (m/v) yeast extract) were used as a rich medium for the routine growth of *K. pneumonia* and *E. coli* DH5α. The LB medium supplemented with 30 μg/mL kanamycin sulfate (Kan) was used to select positive DH5α transformants. The LB plates supplemented 25 μg/mL Kan were used to selected the positive *K. pneumomia* positive transformants harboring the Kan-resistant selective marker. All of the solid media used in this study contained 2 % (m/v) agar (Difco, USA).

Medium used for cultivation of inoculum culture consisted of 20 g/L lactose, 5 g/L yeast extract, 6 g/L KH_2PO_4, 14 g/L K_2HPO_4, and 2 g/L $(NH_4)_2SO_4$ with the initial pH 6.0–7.0. The lactose fermentation medium consisted of 60 g/L lactose, 10 g/L yeast extract, 2 g/L $(NH_4)_2SO_4$, 6 g/L KH_2PO_4, 14 g/L K_2HPO_4, and 4 g/L sodium acetate with the initial pH 6.0. For seed preparation, a full loop of strains from a fresh slant tube was inoculated in a 250 mL flask containing 50 mL fresh seed medium in a rotary shaker at 200 rpm and 37 °C for 12 h. Seed culture (10 %, v/v) was then inoculated into the fermentation medium. Flask experiments were carried out in 500 mL flasks loaded with 100 mL sample solution at 150 rpm and 35 °C.

13.2.3 Construction of Recombination Plasmids and Strains

pUC18 K, which was constructed by substituting the ampicillin-resistant marker of pUC18 with kan-resistant marker was selected as the expression vector. The *bgaB* gene encoding β-galactosidase was cloned from *K. pneumonia* CICC10781 genome by primers of bgaB-U(5′CCGGAATTCATGAATAAATTTGCACCTTTACATCCG)/bgaB-D (5′CCGGAATTCTTACGCGTGACGACGGAGAAT). The *ElacY* gene encoding lactose permease were cloned from *E. coli* MG 1655 by the primers of ElacY-U (5′CCGGAATTCATGTACTATTTAAAAAACACAAAC)/ElacY-D

(5′CCCAAGCTTTTAAGCGACTTCATTCACC). The underlined parts in these primers indicated restriction enzyme sites (*Eco*R I, *Eco*R I, *Eco*R I, and Hind III, respectively). The gene fragments were inserted into pUC18 K plasmid to form pUC-bgaB and pUC-ElacY recombinant plasmids, following restriction endonuclease and ligase reactions. Then, the pUC18 K plasmid and recombinant plasmids were separately transferred into *K. pneumonia* CICC10781 to generate recombinant strains (10781-pUC18 K, 10781-bgaB, and 10781-ElacY).

13.2.4 Recombinant Protein Overexpression Analysis

The recombination protein β-galactosidase was detected by polyacrylamide gel electrophoresis (SDS-PAGE) according to Bai et al. [19]. The strains CICC10781 and 10781-bgaB were cultivated in lactose fermentation medium and the samples were collected for SDS-PAGE analysis at 12 h. Protein samples for SDS-PAGE were prepared using the following steps: (1) harvesting cells by centrifugation (6,000 rpm, 10 min); (2) washing and resuspending the precipitates in phosphate-buffered saline (PBS); (3) breaking the cells of the suspension by sonication and getting the supernatant as the protein samples for SDS-PAGE by centrifugation.

13.2.5 Activity Analysis of β-Galactosidase and Lactose Permease

The activity of β-galactosidase was determined by our previous reported [20]. The β-galactosidase activity was quantified by the amount of o-nitrophenol (ONP) evolved from o-nitrophenyl-β-D-galactopyranoside (ONPG), which could be hydrolyzed to galactose and ONP by β-galactosidase. The crude enzyme of β-galactosidase were prepared as described in preparation of SDS-PAGE.

The lactose permease activity was assayed according to a modified method of Beney and Marechal [21]. 1 mL of the fermentation broth was centrifuged to collect the cells. The cells were washed twice with 0.1 M phosphate buffer (pH 6.5) and finally the cells were resuspended with 1 mL of buffer. At 35 °C, 1 mL of lactose standard solution was added to the cell suspension. At regular intervals, 50 μL reaction mixture was drawn and added to 4 mL precooled distilled water to terminate the reaction tube. After centrifugation, the supernatant was used to determine the lactose concentration by liquid chromatography.

13.2.6 Analysis of Biomass, Substrate and Products

The biomass concentration was measured as absorbance at 600 nm. The concentration of lactose, 2,3-butanediol and by-products including acetoin, acetic acid, lactic acid, succinic acid, and ethanol were measured by high-performance liquid chromatography (HPLC, Agilent 1,100 series, Hewlett-Packard), equipped with a Bio-Rad Aminex HPX-87H column (300 × 7.8 mm) and a refractive index detector. The temperature of column and injector were set at 75 and 45 °C, respectively. The column was isocratically eluted with 0.005 M H_2SO_4 at a flow rate of 0.6 mL per minute.

13.2.7 Determination of Maximum Specific Lactose Consumption Rate (MSLCR)

At two points of stationary phase, by measuring the variation of lactose concentration and cell dry weight biomass, MSLCR was determined.

$$\mathrm{MSLCR} = \frac{\text{glucose concentration variation(g)}}{\text{cell dry weight varaition(g)}} * \mu;\ \mu = \frac{\ln(X_2/X_1)}{T_2 - T_1}$$

μ was the specific growth rate; X1 and X2 were the cell dry weight at time T1 and T2 of stationary phase.

13.3 Results and Discussions

13.3.1 Effect of Exogenously Adding β-Galactosidase

To improve lactose utilization, β-galactosidase was added. In lactose fermentation experiments by CICC10781, 10, 50, and 100 μL of β-galactosidase (lactozym pure 2,600 L, purchased from Novozymes) were added to the 100 mL fermentation medium at the beginning, respectively. Effects of different amounts of β-galactosidase on lactose utilization were compared.

As shown in Fig. 13.1, adding exogenous β-galactosidase could accelerate the lactose consumption rate. Lactose was broken down into glucose and galactose which would be used quickly then. More β-galactosidase was added, more the lactose consumption rate was improved. When the initial lactose was 56 g/L, the fermentation was accomplished in 30 h by CICC10781 without adding exogenous β-galactosidase and when 10, 50, and 100 μL of β-galactosidase were added, the fermentation time was shorted to be 15, 9 and 6 h. Glucose and galactose were undetected in the fermentation. adding exogenous β-galactosidase had almost no influence on the concentration of desired product of 2,3-butanediol, and other by-product (data not shown).

The results above indicated that adding exogenous β-galactosidase favored the lactose utilization and lactose utilization might be the limiting step of lactose fermentation to 2,3-butanediol [6]. To improve the lactose utilization, genes encoding β-galactosidase and lactose permease were overexpressed in CICC10781.

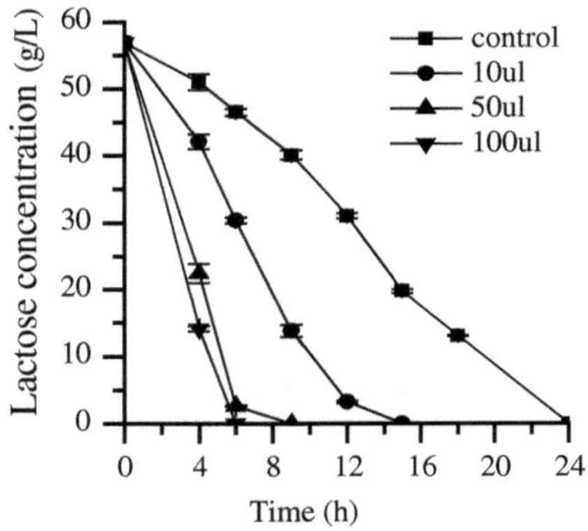

Fig. 13.1 Change of lactose in lactose fermentation by CICC10781

13.3.2 Overexpression of bgaB in CICC10781

After cultivation of CICC10781 in the lactose fermentation medium for a regular period, protein expression of β-galactosidase was assayed with SDS-PAGE. According to the nucleotide and amino acid composition, the molecular weight (Mw) of β-galactosidase was predicted to be about 78 kDa. The specific protein band of about 78 kDa in SDS-PAGE (Fig. 13.2) indicated that *bgaB* was successfully expressed in the recombinant strains.

The samples were collected at regular intervals and the β-galactosidase activity was measured (Fig. 13.3). During the whole fermentation period, the β-galactosidase activity of 10781-bgaB was much higher than CICC10781. At 18 h, the activity of 10781-bgaB was 8.639 IU/mg cell, which was 33.08 times of CICC10781. These results showed that the recombinant strain exhibited higher β-galactosidase activity than the wild-type strain and the β-galactosidase overexpression in 10781-bgaB was achieved.

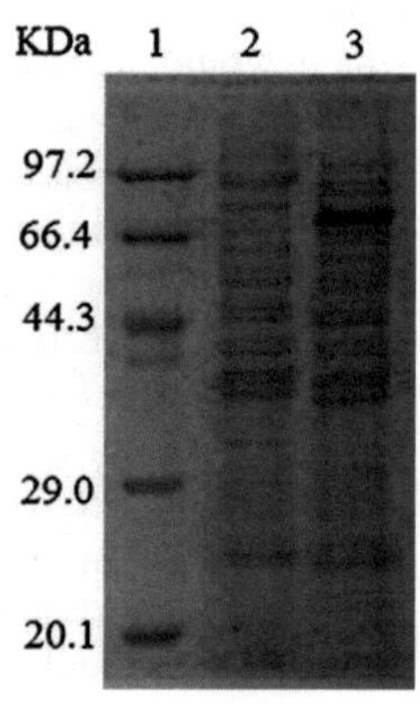

Fig. 13.2 SDS-PAGE results of CICC10781 and 10781-bgaB. *1.* Marker; *2.* CICC10781 (pUC18 K); *3.* 10781-bgaB

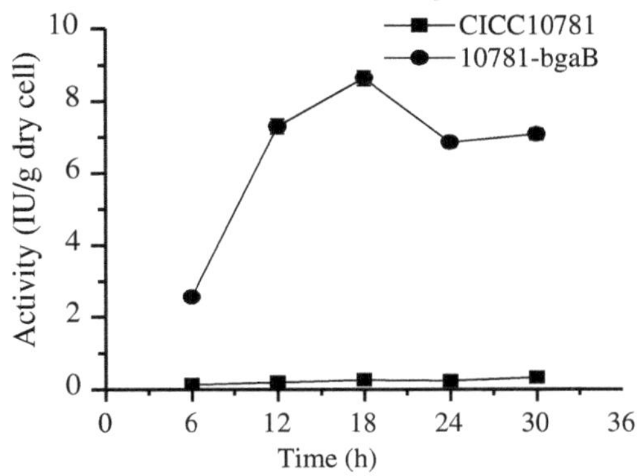

Fig. 13.3 β-galactosidase activity in lactose fermentation

13.3.3 Overexpression of ElacY in CICC10781

The strains CICC10781 and 10781-ElacY were cultured in lactose fermentation culture medium. The samples were collected at 12 h and the lactose permease activity was measured to confirm the proper and successful expression. The lactose permease activities of recombinant strains 10781-ElacY were increased by 101 % compared to the wild strains CICC10781. The results demonstrated that the *lacY* gene from *E. coli* MG 1655 was properly and successfully expressed.

13.3.4 Comparison of Lactose Consumption by Engineered Strains

After incubation at 37 °C and 200 rpm for 12 h, the seed inoculum was transferred to lactose fermentation medium. Samples were collected at regular intervals and the concentration of lactose and absorbance at 600 nm were measured. The comparison of different strains was shown in Fig. 13.4. Furthermore, MSLCR was determined (Fig. 13.5).

The recombinant strain 10781-pUC18 K, grow more slowly than the wild-type strain since they carry an extra plasmid [22]. Furthermore, the recombinant strains, 10781-bgaB and 10781-ElacY grow more slowly than 10781-pUC18 K (data not shown). These results indicated that the synthesis of large quantities of foreign proteins bring a heavy metabolic burden to the recombinant strain, reducing the growth rate of recombinant strains [22, 23].

The recombinant strains, which contained empty plasmid pUC18 K, exhibited a decrease of lactose consumption rate. This phenomenon might be ascribed to the inhibition of cell growth. Compared with the starting strain CICC10781, the lactose consumption rate of the recombinant strain 10781-bgaB was lower. After fermentation for 30 h by 10781-bgaB, the residual lactose concentration was still 41.03 g/L. the maximum specific lactose consumption rate (MSLCR) was decreased to 0.364 g

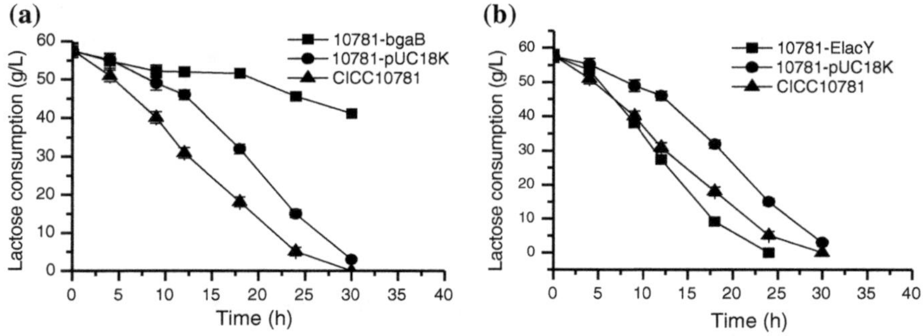

Fig. 13.4 Profiles of lactose consumption in lactose fermentation by 10781-bgaB (**a**) and 10781-ElacY (**b**)

Fig. 13.5 MSLCR of wild and recombinant strains

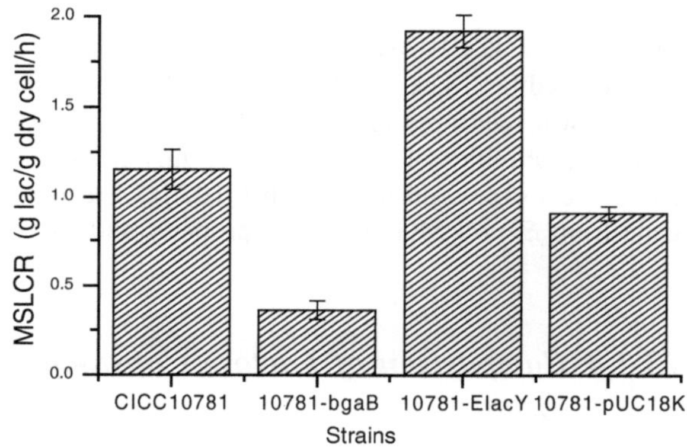

lactose/g dry cell/h. The results above showed that overexpression of *bgaB* gene was not able to improve the lactose consumption rate of CICC10781.

The recombinant strain 10781-ElacY exhibited a higher lactose consumption rate than the wild strain. The fermentation periods were shortened by 6 h. The MSLCR was increased to 1.918 lactose/g dry cell/h, representing an increase of 66.2 %, compared with the wild strain. The results above showed that overexpression of ElacY gene improved the lactose consumption rate of CICC10781.

13.3.5 Comparison of 2,3-Butanediol Production

The final concentration of 2,3-butanediol and other by-products were depicted in Table 13.2. Because of the difference of lactose consumption rate, the fermentations were completed at different time. When the residual lactose concentration was below 1 g/L, the fermentations were considered completed.

Table 13.2 Final concentration of metabolites by CICC10781, 10781-bgaB, and 10781-ElacY

Strains	Residual lactose (g/L)	Fermentation time (h)	2,3-butanediol (g/L)	Cell weight (g/L)	Acetoin (g/L)	Lactic acid (g/L)	Acetic acid (g/L)	Ethanol (g/L)
CICC10781	–	30	18.83	9.23	1.54	5.93	0.77	4.97
10781-bgaB	41.03	30	2.50	6.19	0.78	0.72	0.34	1.24
10781-ElacY		24	18.92	7.46	1.31	4.71	1.41	3.92

The lactose consumption rate of 10781-bgaB decreased. At the end of fermentation, the residual lactose concentration of KG-bgaB was still 41.03 g/L and a small amount of 2,3-butanediol was produced. Compared to wild strain CICC10781, the fermentation period of 10781-ElacY was reduced by 6 h. The final 2,3-butanediol concentration and other by-products of 10781-ElacY changed very little. Therefore, of the two genes *bgaB* and *ElacY*, only overexpression of *ElacY* could promote lactose utilization of CICC10781, and meanwhile the 2,3-butanediol generation capacity was not affected.

The lactose enters cells through the lactose permease, and it is hydrolyzed into glucose and galactose moieties by β-galactosidase. But only overexpression of *ElacY* could promote lactose utilization of CICC10781. So the lactose permease may be the rate-limiting enzyme of lactose utilization, which was in accordance with the conclusion of Fekete et al. [24]. Overexpression of β-galactosidase not only inhibited cell growth but also repressed lactose utilization, which might be because of the large metabolic burden from β-galactosidase overexpression.

13.4 Conclusion

In the present study, adding exogenous β-galactosidase was proved to favor the lactose utilization of CICC10781 and CICC10011 and lactose utilization might be the limiting step of lactose fermentation to 2,3-butanediol. *bgaB* and *ElacY* were all properly and successfully overexpressed in CICC10781 by constructing the expression plasmids pUC-bgaB and pUC-ElacY. Of the two genes, only overexpression of *ElacY* promoted lactose utilization of CICC10781 and meanwhile the 2,3-butanediol generation capacity was not affected. On the other hand, overexpression of β-galactosidase repressed lactose utilization of CICC10781.

Acknowledgments This research was financed by the Cheung Kong Scholars and Innovative Research Team Program in University of Ministry of Education, China (Grant Number IRT1166), the National High Technology Research, and the Development Program of China (863 Program) (Grant Number 2012AA022108), the National Agricultural Research Projects Funded (Grant Number 2012AA101805).

References

1. Kim DK, Rathnasingh C, Song H et al (2013) Metabolic engineering of a novel Klebsiella oxytoca strain for enhanced 2,3-butanediol production. J Biosci Bioeng 116:186–192
2. Han SH, Lee JE, Park K et al (2013) Production of 2,3-butanediol by a low-acid producing Klebsiella oxytoca NBRF4. New Biotechnol 30:166–172
3. Guo XW, Cao CH, Wang YZ et al (2014) Effect of the inactivation of lactate dehydrogenase, ethanol dehydrogenase, and phosphotransacetylase on 2,3-butanediol production in *Klebsiella pneumoniae* strain. Biotechnol Biofuels 7:44

4. Yu EKC, Levitin N, Saddler JN (1982) Production of 2,3-Butanediol by *Klebsiella pneumoniae* grown on acid hydrolyzed wood hemicelluloses. Biotechnol Lett 4:741–746
5. Song Y, Xu Y, Li Q et al (2011) Fermentation of bio-based product 2,3-butanediol. Chem ind and Eng Prog 30:1069–1077 (in Chinese)
6. Lee HK, Maddox IS (1984) Microbial production of 2,3-Butanediol from whey permeate. Biotechnol Lett 6:815–818
7. Lee HK, Maddox IS (1986) Continuous production of 2,3-butanediol from whey permeate using *Klebsiella pneumoniae* immobilized in calcium alginate. Enzyme and iol technol 8:409–411
8. Wang A, Wang Y, Jiang T et al (2010) Production of 2,3-butanediol from corncob molasses, a waste by-product in xylitol production. Appl Microbiol Biotechnol 87:965–970
9. Petrov K, Petrova P (2010) Enhanced production of 2,3-butanediol from glycerol by forced pH fluctuations. Appl Iol Biotechnol 87:943–949
10. Petrov K, Petrova P (2009) High production of 2,3-butanediol from glycerol by *Klebsiella pneumoniae* G31. Appl Iol Biotechnol 84:659–665
11. Sun LH, Wang XD, Dai JY et al (2009) Microbial production of 2,3-butanediol from Jerusalem artichoke tubers by *Klebsiella pneumonia*. Appl Microbiol Biotechnol 82:847–852
12. Fages J, Mulard D, Rouquet JJ et al (1986) 2,3-Butanediol production from Jerusalem artichoke, *Helianthus tuberosus*, by *Bacillus polymyxa* ATCC 12 321. Optim of k L a profile: Appl Microbiol and Biotechnol 25:197–202
13. Cheng K, Liu Q, Zhang JA et al (2010) Improved 2,3-butanediol production from corncob acid hydrolysate by fed-batch fermentation using *Klebsiella oxytoca*. Process Biochem 45:613–616
14. Guo XW, Wang RS, Chen YF et al (2012) Intergeneric yeast fusants with efficient ethanol production from cheese whey powder solution: construction of a Kluyveromyces marxianus and Saccharomyces cerevisiae AY-5 hybrid. Eng Life Sci 12(6):656–661
15. Guo XW, Zhou J, Xiao DG (2010) Improved ethanol production by mixed immobilized cells of Kluyveromyces marxianus and Saccharomyces cerevisiae from cheese whey powder solution fermentation. Appl Biochem Biotechnol 160(2):532–538
16. Martinez SB, Speckman RA (1988) 2,3-Butanediol production from hydrolyzed whey permeate by immobilized cells of Bacillus polymyxa. Appl Biochem Biotechnol 18:303–313
17. Speckman R, Collins E (1982) Microbial production of 2,3-butylene glycol from cheese whey. Appl Environ Microbiol 43:1216–1218
18. Guo XW, Zhang YH, Cao CH et al (2014) Enhanced production of 2,3-butanediol by overexpressing acetolactate synthase and acetoin reductase in *Klebsiella pneumoniae*. Biotechnol Appl Biochem. doi:10.1002/bab.1217
19. Bai LP, Wu XB, Jiang LJ et al (2012) Hydrogen production by over-expression of hydrogenase subunit in oxygen tolerant Klebsiella oxytoca HP_1. Int J Hydrogen Energy 37:13227
20. Zou J, Guo X, Shen T et al (2013) Construction of lactose-consuming Saccharomyces cerevisiae for lactose fermentation into ethanol fuel. J Ind Microbiol Biotechnol 40(3–4):353–363
21. Beney L, Marechal PA, Gervais P (2001) Coupling effects of osmotic pressure and temperature on the viability of Saccharomyces cerevisiae. Appl Microbiol Biotechnol 56:513–516
22. Ji XJ, Xia ZF, Fu NH et al (2013) Cofactor engineering through heterologous expression of an NADH oxidase and its impact on metabolic flux redistribution in *Klebsiella pneumonia*. Biotechnol Biofuels 6:7–15
23. Kleiner D, Paul W, Merrick MJ (1988) Construction of multicopy expression vectors for regulated over-production of proteins in *Klebsiella pneumoniae* and other enteric bacteria. J Gen Microbiol 134:1779–1784
24. Fekete E, Karaffa L, Seiboth B et al (2012) Ident of a permease gene involved in lactose utilisation in Aspergillus nidulans. Fungal Gen Biol 49:415–425

Chapter 14
Breeding High Producers of Enduracidin from *Streptomyces fungicidicus* by Combination of Various Mutation Treatments

Dong Zhang, Qingling Wang and Xinle Liang

Abstract Enduracidin is a kind of lipodepsipeptides and composed by a 17 amino acid residual conjugated with a fatty acid strain. It is used as a feed additive approved by EU because of its high safety, low toxicity, low-residue, and activity against a wide variety of Gram-positive bacteria. Here, we combined a physical mutagenesis of ^{137}Cs γ-irradiation with the atmospheric and room temperature plasma (ARTP) biological breeding system to improve enduracidin productivity in the industrial Streptomyces. The mutation rates of spores treated by ARTP for 1 min and by ^{137}Cs γ-irradiation for 500 Gy dosages were 26.51 and 19.41 %, respectively. An optimal process for breeding by the combination was proposed. Through the high throughout screening method, the mutant *Streptomyces fungicidicus* S-224 was obtained, and its enduracidin yield reached to 1.58 g/L that was 1.65-fold to the original strain. The mutation breeding combined ARTP with ^{137}Cs γ rays was the first trial and provided an alternative method for industrial strains improvement.

Keywords Enduracidin · Streptomyces · ARTP · Breeding · ^{137}Cs · Mutation

14.1 Introduction

Enduracidin is produced by a soil bacterium *Streptomyces fungicidicus*. Especially, it is added to the feed of pig and chicken widely for improving the growth and disease resistance of livestock [2]. Enduracidin shows the strong activity against

D. Zhang · Q. Wang · X. Liang (✉)
College of Food and Biotechnology, Zhejiang Gongshang University, Xuezheng Jie 18, Hangzhou 310018, People's Republic of China
e-mail: dbiot@mail.zjgsu.edu.cn

T.-C. Zhang and M. Nakajima (eds.), *Advances in Applied Biotechnology*, Lecture Notes in Electrical Engineering 332, DOI 10.1007/978-3-662-45657-6_14

Gram-positive bacteria which inhibits the activity of glucosamine-phosphate N-acetyltransferase specifically, and then reduces the absorption rate of N-acetyl-glucosamine. Because of the lack of N-acetyl-glucosamine concentration and inactivation of glucosamine-phosphate N-acetyltransferase, the cytoderm precursor accumulated heavily in cell [1]. In clinical applications, enduracidin was reported to inhibit hepatitis B virus (HBV) and hepatitis B virus e antigen (HBeAg) [17].

Physical mutagenesis has been applied to increase the antibiotics production in industrial strains. Conventional microorganism mutagenesis was applied for enduracidin high-producers in the past. Atmospheric and room temperature plasma (ARTP) is a new kind of atmospheric pressure nonequilibrium discharge plasma source [10]. Because of its low puncture voltage, more uniformity, large area glow discharges, and low temperatures, many applications of ARTP concerning microbiology effects have been reported [15]. ARTP had been used in sterilization and environmental management [8]. DNA can be destroyed when treated by ARTP, rather than by heat, UV radiation, charged particles, and intense electric field [21]. This result implies that ARTP has the potential to be an effective method for microbial mutagenesis. ARTP was used to generate mutations in *S. avermitilis* for high avermectin production, the obtained the mutants had a genetic stability [14]. In this study, we tried to combine two mutation treatments, ARTP and ^{137}Cs γ rays, for industrial strain breeding. The optimal conditions of mutation process were determined and genetic stability was verified.

Based on the concept of ribosome engineering in the industrial *Streptomyces*, drugs included in the breeding plates would direct the microbial evolution and initiate the secondary metabolites production [12, 19]. Among all of the drug-resistance mutation breeding, streptomycin is often successfully used [6, 11, 16]. Mutants with streptomycin resistance frequently carried mutations within ribosomal protein S12, resulting in activation of silent genes [5]. The streptomycin resistance screening has been used as an effective approach for the high productive strains. By streptomycin resistance screening of *Streptomyces viridochromogenes* mutants, avilamycin production was increased 36.8-fold than that of the original strain [11]. To erase the parental strains and improve the breeding yield of mutation, streptomycin resistance was introduced in this study.

14.2 Materials and Methods

14.2.1 Strains

Streptomyces fungicidicus L-69 (deposited by our own laboratory) was used as the initial strain and was preserved in 30 %(v/v) glycerol at −80 °C. It was unable to grow on the agar plates containing more than 10 μg/mL streptomycin. Its enduracidin production in shaking flask was 0.34 g/L. *Bacillus subtilis* ATCC6633 was used as the indicator bacteria. It was purchased from China industrial microbial preservation management center.

14.2.2 Media and Culture Conditions

Slant and plate solid medium contained (per liter) 20 g starch, 0.5 g K_2HPO_4, 0.5 g $MgSO_4$, 0.5 g NaCl, 0.5 g L-valine, 1 g KNO_3, 0.01 g $FeSO_4$ $7H_2O$, and 20 g agar. The pH value was adjusted to 7.0–7.2 before autoclaving. Seed medium and enduracidin production medium (EPM) were prepared as described previously [9]. The solid medium for the indicator bacteria growing was beef extract peptone medium.

For enduracidin fermentation in 500-mL Erlenmeyer flask with 100 mL EPM, a loop full of slant culture was inoculated into 50 mL of seed medium in a 250-mL Erlenmeyer flask, respectively. After incubation for 72 h, 10 mL of seed broth was transferred into 100 mL EPM in 500-mL Erlenmeyer flask and incubated at 28 °C, 220 rpm shaker speed. Other culture conditions were as previously described [22].

14.2.3 Determination of Minimal Inhibitory Concentration (MIC)

MIC determination of streptomycin was conducted as the previous description [3]. The spores were spread on the solid medium containing streptomycin (5, 10, 15, 20, and 30 mg/L). Then, they were cultivated at 28 °C for 6–7 days. The MIC was determined based on the growth. Then a concentration of streptomycin up to the MIC was adopted in the next screening plates to erase the parental strains.

14.2.4 Preparation of Single Spore Suspension

The strain was inoculated to slant solid medium at 28 °C for 3–4 days. The mature spores were scraped with appropriate volume of normal saline (0.85 %). Spore suspension was interrupted by pipette and filtered with a few of sterile absorbent cotton. So the uniform single spore suspension was obtained. The number of spores was counted under a microscope using hemacytometer. For the mutation process, the concentration of spores was adjusted to 10^7/mL.

14.2.5 Mutagenesis Method for the ARTP Biological Breeding System

The original strain was mutated by atmospheric and room temperature plasma (ARTP). The ARTP biological breeding system was supplied by Biotechnology and Pharmaceutical Engineering, Nanjing Tech University, Nanjing, China. Its system parameters were set up as follows: the working gas was pure helium, the gas flow velocity was 10 L/min, the output power was 750 VA, the distance between plasma generator slider and sample plate was 2 mm, the gas jet temperature was <40 °C [7]. The volume of single spore suspension on sample plate was 20 μl. These samples on aseptic plate were jetted by plasma generator for predetermined time. Then, the plate was put into a sterilized centrifuge tube and washed with 1 mL normal saline for getting viable spores. The treated spore solution was spread on the solid medium to obtain purposeful mutant strain.

14.2.6 Physical Mutagenesis Process by ^{137}Cs γ-Irradiation

The physical mutagenesis breeding system was supplied by Zhejiang Academy of Agricultural Sciences. ^{137}Cs γ-irradiation was used for mutation treatment as described previously. Briefly, 5 ml of spore suspension from a slant culture were transferred to an aseptic tube. After appropriately mutation treatment, the suspension of spores was spread on the drug medium. A number of colonies were selected and enduracidin production was evaluated by fermentation tests.

14.2.7 Determination of the Optimal Mutation Dosage

In the mutation process, the lethal rate (Y_{let}), mutation rate (Y_{SM}), and positive mutation rate (Y_{PM}) are often adopted to assess the mutative dosage whether it is optimal to the process. In ARTP breeding and ^{137}Cs γ-irradiation mutation process, the lethal rate (Y_{let}) under different mutation dosages was calculated based on the formula as below:

$$\text{Lethal rate}\,(Y_{let}) = (A - B)/A * 100\,\% \tag{14.1}$$

$$\text{Mutation rate}\,(Y_{SM}) = (C/B) \times 100\,\%; \tag{14.2}$$

$$\text{Positive mutation rate}\,(Y_{PM}) = (D/B) \times 100\,\% \tag{14.3}$$

A is the total colony count of strain without treatment on solid medium streptomycin. *B* is the total colony count that treated with ARTP or ^{137}Cs γ rays on solid medium without streptomycin. *C* is the total colony count that treated with ARTP or ^{137}Cs γ rays on solid medium with streptomycin. *D* is the colony count that had an improved enduracidin yield.

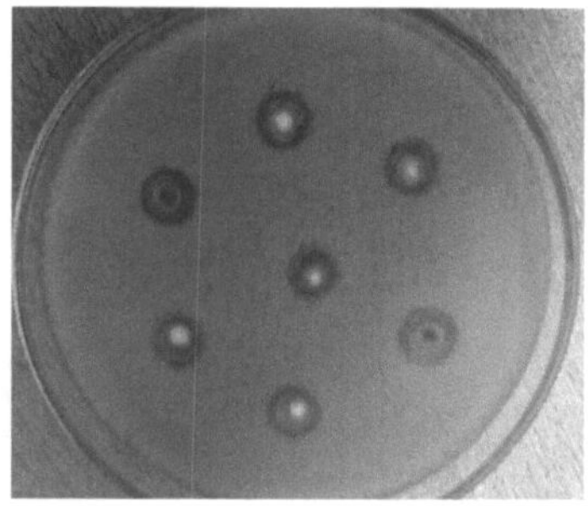

Fig. 14.1 Selective establishment of high-producing mutants by agar block methodology

14.2.8 Selective Establishment of High-Producing Mutants

In order to select the positive mutants efficiently we carried out the rapid plate selection by agar block methodology. These mutation colonies on solid medium with streptomycin were punched by puncher with 8 mm of diameter. Then, these agar lumps with the colonies were transferred into sterile empty culture dish and continued the cultivation at 28 °C for 7 days. The agar block was put onto the solid medium with moderate *Bacillus subtilis* at 37 °C for 16–18 h. After the time, the inhibition zone was observed and the diameter size was recorded and compared them with control (Fig. 14.1). Those positive mutants with bigger inhibition zone were selected and conducted fermentation test in Erlenmeyer flask. The determination of enduracidin was assayed by HPLC as described before [20].

14.3 Results and Discussion

14.3.1 The Lethal Rate Introduced by ARTP and ^{137}Cs γ Rays Treatments

For the ARTP treatment, the dosages were set up at 0.5, 1, 2, 3 and 4 min. Along with e increase of the dosage, the lethal rate was increased by 88.01, 92.31, 98.44, 98.67, and 99.24 %, respectively (Fig. 14.2). Since the treatment time was longer than 1 min, the lethal rate was high up to 92.31 %. Therefore, the treatment time by 1 min would be a rational dosage and adopted in the following tests. In *S. avermitilis* mutagenesis with ARTP treatment, the lethal rate obtained by 77.8, 89.3, and 98.2 % in according to 1, 2, and 3 min, respectively. When the sample was treated for 5 min, no spores could survive [14]. ARTP treatment with 3.5 min dosage introduced a 95.5 % lethal rate on *Oleaginous Yeast* [13]. In *Clostridium beijerinckii* mutant screening by ARTP, about 10 % of them survival rate was obtained after 3 min dosage [4]. It seemed that about 90 % lethal rate would be applicable dosage in the ARTP.

As the same time, the different dosage of ^{137}Cs γ rays were also evaluated at 300, 500, and 700 Gy (Gy is the ray radiation intensity unit). In Fig. 14.3, the lethal rate

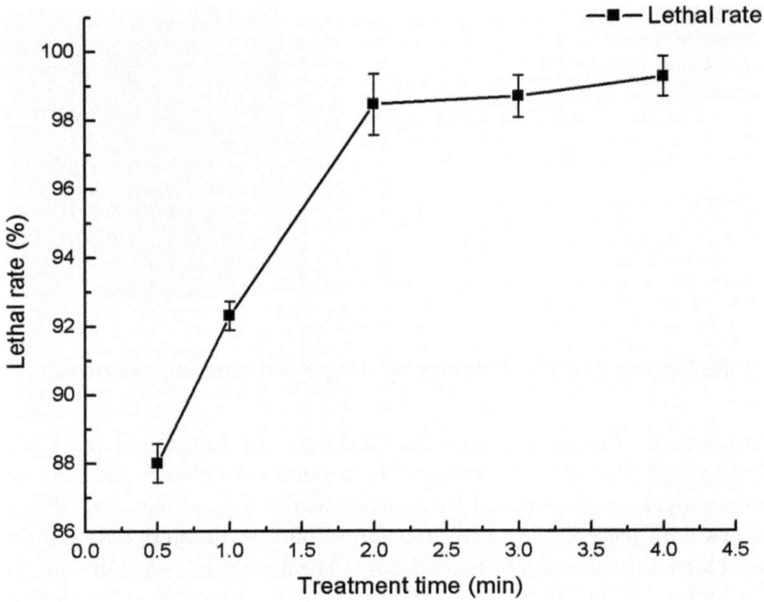

Fig. 14.2 Variation of the lethal rate by ARTP with different exposure times

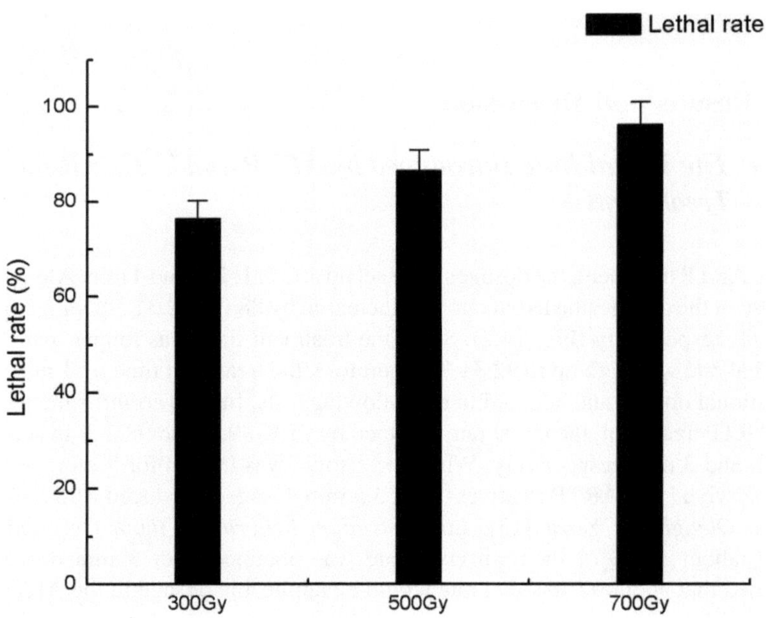

Fig. 14.3 Variation of the lethal rate by ^{137}Cs γ-rays with different dosages

of spores was increased to 76.4, 86.5, and 96.1 % respectively. So 500 Gy of radiation dose was chosen for the next test.

14.3.2 Determination of MIC of Streptomycin to **Streptomyces fungicidicus** *for Screening Plate*

Streptomycin as the screening pressure was determined and designed to screening plate. Streptomycin concentration was set up at a higher than MIC. The streptomycin in sub-high concentration could eliminate the parent strain, promote determinate evolution, and increase the positive mutation as described by ribosome engineering. In Fig. 14.4, the obtained results showed that there was no colony grown out on the plate with 15 mg/L streptomycin. And 15 mg/L concentration was regarded briefly as the minimum inhibition concentration and critical point of drug pressure. Then, the sub-high streptomycin concentration 20 mg/L was designed to apply to the following selective condition.

14.3.3 Screening High-Producers Derived from ARTP and ^{137}Cs γ Rays Treatments

After the spore solution was treated by ARTP for 1 min and cultivated on the solid medium with 20 mg/L streptomycin for 7 days at 28 °C, about 347 single colonies were obtained from plates. And then the colonies were transferred to solid medium respectively. After incubation on the plates for 5 days, 92 colonies were obtained which generated inhibition zones larger than that of the control. The mutant productivities were evaluated by fermentation tests in Erlenmeyer flasks and detected by HPLC. Strain *S. fungicidicus* A-08 showed the highest enduracidin yield 1.37 g/L, which was 1.43 times as much as that of the original strain (Fig. 14.5). In this process, the mutation rate and positive mutation rate were at 26.49 and 17.60 %,

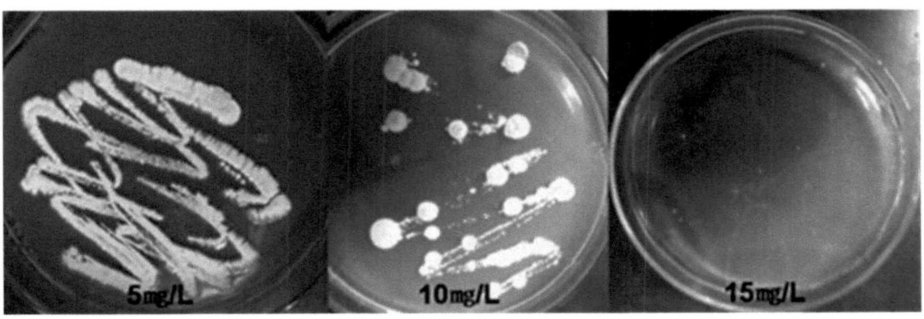

Fig. 14.4 Determination of MIC of streptomycin to *S. fungicidicus*

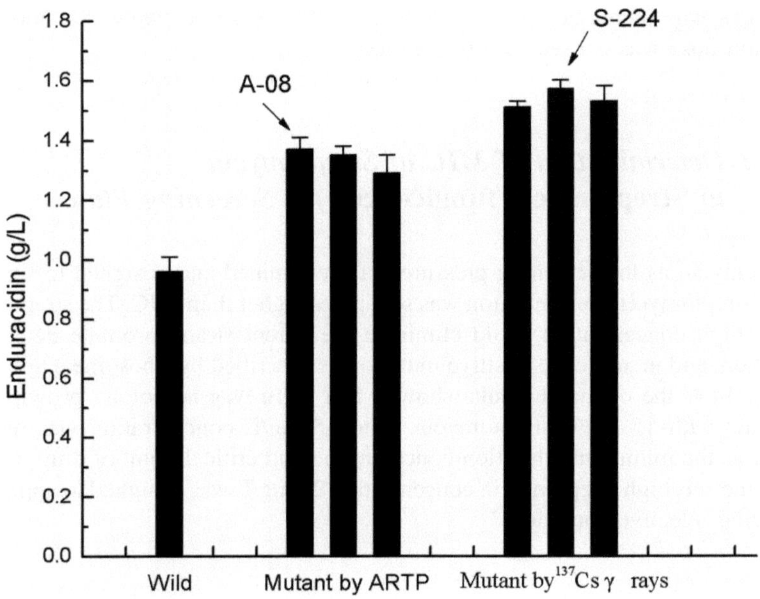

Fig. 14.5 Improvement of enduracidin productivity of mutants by ARTP and ^{137}Cs γ rays treatments

respectively (Table 14.1). It was as similar to previous reports that the *S. avermitilis* mutation rate and positive mutation rate were 29.86 and 20.93 % in *S. avermitilis* [14]. In *Enterobacter aerogenes* positive mutant rate was estimated to be 10 % which showed the improved hydrogen productivity [10]. The mutants with higher transglutaminase production were obtained by ARTP treatment, and t the positive mutant rate was at 20.6 % [18].

Following the ARTP treatment, the derived mutant *S. fungicidicus* A-08 spore solution was prepared and treated by ^{137}Cs γ rays at 500 Gy. And then the spores were cultivated on the solid medium with 50 mg/L streptomycin for 7–10 days at 28 °C. About 227 single colonies were obtained from plates. These colonies were

Table 14.1 The mutation rate and positive mutation rate by ARTP and ^{137}Cs γ rays treatment on *S. fungicidicus*

	B	C	D
ARTP treatment			
Colony amounts	1310	347	61
Y_{SM}	26.49 %		
Y_{PM}	17.60 %		
^{137}Cs γ rays treatment			
Colony amounts	1186	227	65
Y_{SM}	19.14 %		
Y_{PM}	28.63 %		

screened. Of which strain *S. fungicidicus* S-224 showed the highest enduracidin yield 1.58 g/L, which was 1.15 times as much as *S. fungicidicus* A-08 and 1.65 times as much as original strain (Fig. 14.5). In this process, the mutation rate and positive mutation rate were 19.14 and 28.63 %, respectively. Comparison mutation between ARTP and ^{137}Cs γ rays treatments (Table 14.1), it suggested that ARTP method introduce a higher mutation rate (Y_{SM}), while ^{137}Cs γ rays method do a higher positive mutation rate (Y_{PM}).

The genetic stability of *S. fungicidicus* S-224 was evaluated by five successive subcultivation tests. The range of production levels among five generations was 1.52 ± 0.03 to 1.75 ± 0.04 g/L, indicating that the hereditary character of the high enduracidin producing recombinant, *S. fungicidicus* S-224, was stable.

14.4 Conclusion

A combination of ARTP to ^{137}Cs γ rays treatment was proposed to improve industrial *Streptomyces* strain breeding. The combination method was effective significantly in the industrial enduracidin producers breeding. This technology showed a high mutation rate in *S. fungicidicus*. The enduracidin yield of *S. fungicidicus* S-224 increased 65 % compared to the original production, which would be a promise in the industrial value of enduracidin production.

Acknowledgments We thank the technical support by Biotechnology and Pharmaceutical Engineering, Nanjing Tech University, Nanjing, China, and Zhejiang Academy of Agricultural Sciences. The work was financially supported by College of Food Science and Biotechnology, Zhejiang Gongshang University, Hangzhou, China.

References

1. Fang X, Tiyanont K, Zhang Y et al (2006) The mechanism of action of ramoplanin and enduracidin. Mol BioSyst 2(1):69–76
2. Gao YJ, Zhang Y (2010) Peptides feed additive–enduracidin. Feed Rev (4):32–35
3. Gong GL, Sun X, Liu XL et al (2007) Mutation and a high-throughput screening method for improving the production of epothilones of *Sorangium*. J Ind Microbiol Biotechnol 34 (9):615–623
4. Guo T, Tang Y, Xi YL et al (2011) *Clostridium beijerinckii* mutant obtained by atmospheric pressure glow discharge producing high proportions of butanol and solvent yields. Biotechnol Lett 33(12):2379–2383
5. Hosaka T, Ohnishi-Kameyama M, Muramatsu H et al (2009) Antibacterial discovery in actinomycetes strains with mutations in RNA polymerase or ribosomal protein S12. Natl Biotechnol 27:462–464
6. Irina A, Vsevolod P (2004) Structural basis for transcription regulation by alarmone ppGpp. Cell 117(3):299–310

7. Jin L, Fang M, Zhang C et al (2011) Operating conditions for the rapid mutation of the oleaginous yeast by atmospheric and room temperature plasmas and the characteristics of the mutants. Chin J Biotechnol 27(3):461–467
8. Lerouge S, Tabrizian M, Wertheimer MR et al (2002) Safety of plasma-based sterilization: surface modifications of polymeric medical devices induced by sterrad and plazlyte processes. Biomed Mater Eng 12(1):3–13
9. Li HF, Gao XJ, Bai LJ et al (2012) Rapid Selection of high yield enramycin. China J Vet Med 46(5):40–42
10. Lu Y, Wang LY, Ma K et al (2011) Characteristics of hydrogen production of an *Enterobacter aerogenes* mutant generated by a new atmospheric and room temperature plasma (ARTP). Biochem Eng J 55(1):17–22
11. Lv XA, Jin YY, Li YD et al (2012) Genome shuffling of *Streptomyces viridochromogenes* for improved production of avilamycin. Appl Microbiol Biotechnol 97:641–648
12. Ochi K (2007) From microbial differentiation to ribosome engineering. Biosci Biotechnol Biochem 71(6):1373–1386
13. Tan YY, Jin LH, Zhang C et al (2012) Mutation of oleaginous yeast using atmospheric and room temperature plasma for xylose utiliability and characteristics of typical mutant. Food Sci 33(1):212–217
14. Wang LY, Huang ZL, Li G et al (2010) Novel mutation breeding method for *Streptomyces avermitilis* using an atmospheric pressure glow discharge plasma. J Appl Microbiol 108 (3):851–858
15. Wang Q, Feng LR, Wei L et al (2014) Mutation breeding of lycopene-producing strain *Blakeslea trispora* by a novel atmospheric and room temperature plasma (ARTP). Appl Biochem Biotechnol 174(1):452–460
16. Wang Q, Zhang D, Li Y et al (2014) Genome shuffling and ribosome engineering of *Streptomyces actuosus* for high-yield nosiheptide production. Appl Biochem Biotechnol 173 (6):1553–1563
17. Wu T, Huang H, Zhou P (1998) The Inhibitory effects of enduracidin on hepatitis B virus in vitro. Virol Sinica 13(1):46–49
18. Xia SQ, Liu L, Zhang DX et al (2010) Mutation and selection of transglutaminase producing strain by atmospheric pressure glow discharge plasma. Microbiol China 37(11):1642–1649
19. Xie SJ, Xiao J, Xu J (2009) Advance in microbial ribosome engineering. Acta Microbiologica Sinica 49(8):981–986
20. Yin XH, Zabriskie TM (2006) The enduracidin biosynthetic gene cluster from *Streptomyces fungicidicus*. Microbiol 152(10):2969–2983
21. Zhang X, Zhang XF, Li HP et al (2014) Atmospheric and room temperature plasma (ARTP) as a new powerful mutagenesis tool. Appl Microbiol Biotechnol 98(12):5387–5396
22. Zhu CH, Lu FP, He YN et al (2007) Regulation of avilamycin biosynthesis in *Streptomyces viridochromogenes*:effects of glucose, ammonium ion, and inorganic phosphate. Appl Microbiol Biotechnol 73(5):1031–1038

Chapter 15
Expression of *Stichopus japonicus* Lysozyme Gene in *Bacillus subtilis* WB600

Zhiwen Liu, Xingyu Liao, Lu Sun, Dan Zou, Dan Li and Lina Cong

Abstract In this study, a genetic engineering bacteria *Bacillus subtilis* pHT43-SjLys/WB600 was successfully constructed for the expression of the lysozyme gene from sea cucumber (*Stichopus japonicus*) by the method of recombinant DNA technique. The growth trend of engineering bacteria was consistent with the wild-type strain WB600, and the results demonstrated that insertion of foreign gene did not affect its physiological and biochemical metabolism. In the absence of selection pressure, the analysis of the stability revealed that there was no gene rearrangement and lost of the recombinant plasmid in the bacteria which showed that it has high genetic stability. The SDS-PAGE results demonstrated that pHT43-SjLys/WB600 successfully expressed soluble SjLys protein after incubated for 48 h induced by ITPG. The heterologous expression protein of pHT43-SjLys/WB600 displayed remarkable inhibitive effect on the growth of the *Vibrio parahaemolyticus*. To our knowledge, this is the first report about the *SjLys* gene authentic heterologous expression in *B. subtilis*. It should provide a robust secretion expression system for genetic engineering of *B. subtilis* and was thus proposed a potentially new way for producing recombinant SjLys protein.

Keywords Sea cucumber · Lysozyme · *Bacillus subtilis* · Genetic engineering bacteria · Antimicrobial activity

15.1 Introduction

Sea cucumber (*Stichopus japonicus*) is one of the most important and valuable holothurian species in coastal fisheries and commonly consumed echinoderms because of their good flavor and medicinal value. Aquaculture of sea cucumber has rapidly developed in many Asian countries in recent decades [7, 20]. However, in

Z. Liu · X. Liao · L. Sun · D. Zou · D. Li · L. Cong (✉)
College of Bioengineering, Dalian Polytechnic University, Dalian 116034, Liaoning, People's Republic of China
e-mail: alzw@dlpu.edu.cn

T.-C. Zhang and M. Nakajima (eds.), *Advances in Applied Biotechnology*,
Lecture Notes in Electrical Engineering 332, DOI 10.1007/978-3-662-45657-6_15

recent years, because of rapid expansion and intensification of cultivation, sea cucumber in China has suffered from frequent disease outbreaks. *Bacteria*, especially *vibrios*, often caused major diseases such as skin ulcer and bacterial ulceration syndrome in *S. japonicus* at breeding, aestivation, and outdoor cultivation stages. It poses a threat to the aquaculture industry and lead to heavy economic losses. In addition, the abuse of antibiotics in aquaculture has accelerated the development of drug-resistant bacteria and more virulent pathogens [6, 12].

Lysozyme (EC3.2.1.17), a well-known bacteriolytic enzyme, is widely distributed in the animal and plant kingdoms. Their key role is to lyse bacteria by hydrolyzing the β-1,4-glycosidic bonds between N-acetylmuramic acid (NAM) and N-acetylglucosamine (GlcNAc) of peptidoglycan in the bacterial cell walls. Thus, the major function of lysozyme is host defense, as it acts as an antimicrobial and immune-modulating agent, and furthermore, it can display digestive activity or nonenzymatic activity in some species [1, 9]. Based on the differences in their structure, biological functions, catalytic character, and original source, lysozymes are classified into six groups: chicken-type lysozyme (c-lysozyme) present in many vertebrates and insects is the most extensively studied lysozyme; goose-type lysozyme (glysozyme) identified mainly in vertebrates including mammals, birds, and fish; invertebrate-type lysozyme (i-lysozyme), plant lysozyme, bacterial lysozyme, and phagelysozyme [3].

Sea cucumber (*S. japonicus*) relies on their innate immune system to resist invasion of prokaryotic and eukaryotic pathogens, whose digestive system are simple and lack of specific immune. Therefore, the lysozyme would play an important role in the immune, digestive system, and autolysis for the invertebrates sea cucumber which signifies that it is related to both immunity and growth [21, 23]. Many comprehensive studies have been performed and showed that the *S. japonicus* lysozyme is belong to i-type lysozyme, whereas it is widely existed in every organization of sea cucumber, even in the no cell gastrovascular cavity liquids. The further research and utilization of the sea cucumber i-type lysozyme should have important value and significance for the sea cucumber farming, product development, biopharmaceutical, and other marine creatures immune [6]. The i-type lysozyme gene from *Asterias rubens* was the first reported i-type lysozyme gene. Recently, many lysozyme genes have been identified through sequencing or protein purification. Some of them were well-studied for their functions in immunity [24]. Cong et al. [6] previously reported and presented the full-length cDNA sequence encoding lysozyme from the sea cucumber (*S. japonicus*) body wall and registered in the NCBI GenBank under the accession number EF036468. They found that the recombinant *S. japonicus* lysozyme (SjLys) protein in *Escherichia coli* has broad-spectrum antimicrobial activity; the enzyme can dissolve not only the gram-positive bacteria but also the gram-negative bacterium. Especially, it has significant antibacterial effect on the serious aquatic animal diseases caused by pathogenic *vibrio* and *pseudomonas*. Therefore, the recombination SjLys protein should play important role on the prevention and control of the aquatic animal disease and substitute for antibiotic medicines applying in green aquaculture, and it has aroused attention of the domestic and overseas experts and scholars.

The traditional methods to produce lysozyme with high cost and complex production process, which cannot meet the increasing market demand. At present, the biological engineering methods to produce lysozyme are the important trend and means. The recombinant lysozyme producing genetic engineering strains of *E. coli* have been successfully constructed in many studies; however, its expressions in *E. coli* have been demonstrated that there were many problems and difficulties, such as low production and unstable enzyme activity, even more the lysozyme protein existed in inclusion body so that the subsequently operation of the protein denaturation and renaturation were very complicated [8, 25]. Although the transformation of gram-negative bacteria such as *E.coli* is routine for most molecular biologists, many laboratories struggle with incorporating foreign DNA into gram-positive bacteria such as *Bacillus* [13, 14]. The *Bacillus subtilis* has been considered as an attractive and ideal host for expression and secretion of heterologous proteins in gene engineering operation and a kind of beneficial probiotics that is widely dispersed in soil, lakes, and oceans. It is nonpathogenic and does not produce any endotoxins and is generally regarded as safe organism (GRAS) [15, 19]. The other advantages of *B. subtilis* include direct secretion of the functional proteins into the medium, no significant bias in codon usage, its well-known genetics, simple fermentation processes, and the development of simplified downstream processing procedures [16, 22]. These further make the *B. subtilis* a potential *bacillus* in genetic engineering and industrial application [4, 11].

To our knowledge, so far, there is no report about the *S. japonicus* lysozyme gene (*Sjlys*) heterologous expression in *B. subtilis*. In this study, the objective was to isolate and clone the *SjLys*, and then it would be inserted into the *B. subtilis* expression plasmid pHT43. At last, a genetics engineering bacteria *B. subtilis* pHT43-SjLys/WB600 for recombinant expression of *Sjlys* gene would be successfully constructed. This study can provide data on the application of i-type lysozymes from sea cucumber.

15.2 Materials and Methods

15.2.1 Materials

Expression vector pHT43 and six protease defect type *B. subtilis* strain WB600 (his Δ nprB Δ nprE18 Δ aprE Δ epr Δ bpf Δ mpr) was selected as the *Sjlys* gene expression host in this study were purchased from Shanghai Genemy Biological co., Ltd.

Escherichia coli DH5$_{\alpha}$, pMD18-T Simple Vector, *Taq* DNA polymerase, gel extraction and purification Kit, plasmid extraction kit, and restriction enzymes *Bam*HI and *Sma*I were purchased from Takara Biotechnology (Dalian) co., Ltd.

Recombinant plasmid pMD18T-SjLys contained the *Sjlys* gene was constructed by our lab, and the indicator *bacteria Staphylococcus aureu*, *Micrococcus lysodeikticus*, *Vibrio parahaemolyticus*, and *Pseudomonas aeruginosa* are preserved in our lab.

15.2.2 Methods

15.2.2.1 Primers Design and Amplification of the Target *SjLys* Gene

Based on the published cDNA sequence (EF036468) from *S. japonicus* lysozyme gene, a pair of specific primers was designed by using the Primer Premier 5.0 software. The upstream primer and downstream primer are designed, respectively, as following, HS-Q-P11: 5'-GCC GGATCCATGCAAGTTCCTTCTG-3', HS-Q-P12: 5'-GCCCGGGAATTCTCAGTTGTTGCTC-3'. The PCR product was approximately 400 bp. The restriction sites of *Bam*HI and *Sma*I were introduced into the amplified fragments by the primers (underline), and the box indicated the initial or stop codon. They were synthesized by Beijing genomics institute (BGI).

The PCR assays were carried out as the following reagent concentrations: 100 ng pMD18T-SjLys plasmid template DNA; 5.0 μl 10 × Taq PCR buffer; 2.0 mM of $MgCl_2$; 125 μM of each dNTP; 1.0 U of Taq DNA Polymerase; 1.5 μM of each upstream and downstream primer pair. The final reaction volume was adjusted to 50 μl with the sterilized ultrapure $dddH_2O$.

Amplifications were done using a PTC 225 Peltier Thermal Cycler (MJ Research Inc. USA). The amplification profile was performed as following: 5 min at 94 °C; 30 cycles of 30 s at 94 °C, 30 s at 59 °C, 1 min at 72 °C; and 10 min at 72 °C for final extension, 4 °C for holding. 5.0 μl PCR products were resolved using 1.2 % agarose gel electrophoresis for verification of the *SjLys* fragment. The remaining was used for extraction.

15.2.2.2 Construction of the Expression Vector pHT43-SjLys

The target *SjLys* PCR amplified fragment was extracted and purified with the agarose gel DNA fragment recovery Kit Ver.2.0 (Takara), and then cleaved and ligated into vector pMD18T-Simple between the sites of *Sam*I and *Bam*HI to yield positive recombinant plasmid pMD18T-Simple-SjLys.

The products of the plasmid pMD18T-Simple-SjLys and pHT43 digested with the same restriction enzymes *Bam*HI and *Sma*I, were ligated by the isolation buffer to construct the recombinant a express vector in *B. subtilis* and designated as pHT43-SjLys which is confirmed by restriction enzyme analysis and sequencing, the nucleotide sequence was compared to GenBank data using the Blastx (http://www.ncbi.nlm.nih.gov/blast) algorithm and the alignments were done using ClustalW2 (http://www.ebi.ac.uk).

15.2.2.3 Transformation of *Bacillus subtilis*

The pHT43-SjLys plasmid with a chloramphenicol resistance marker was used for integration. It was transformed into *B. subtilis* WB600 competent cells using electrotransformation. Transformants were screened on LB agar plates containing 10 μg/mL chloramphenicol. Finally the recombinant engineering strains named pHT43-SjLys/WB600 were identified and confirmed by PCR, restriction enzyme analysis, and sequencing.

15.2.2.4 The Growth Curve and Genetic Stability of the pHT43-SjLys/WB600

The growth curve of the strain was measured using a method described by Shoham et al. [17] with minor modifications. The WB600 and pHT43-SjLys/WB600 (with10 μg/mL chloramphenicol) were activated in Luria-Bertani medium (LB). The cells were routinely grown at 30 °C in 50 ml Erlenmeyer flasks with a culture volume of 10 ml in a rotary shaker at 200 rpm for overnight. And then the absorbance value at a wavelength of 600 nm (OD_{600}) at regular intervals of the two bacteria was determined and measured by ultraviolet spectrophotometer. All assays were performed in triplicates and the results are the means of three independent experiments.

The plasmid genetics stability of the recombinant pHT43-SjLys/WB600 for continuous passage culture with chloramphenicol was determined through the method of Avsaroglu et al. [2] and Leen et al. [10]. For the strain, stability experiments were performed in triplicate too.

15.2.2.5 SDS-Polyacrylamide Gel Assay

The recombinant *bacterium* pHT43-SjLys/WB600 was cultured with 10 μg/mL chloramphenicol and induced by 1 mmol/L isopropylthiogalactoside (IPTG) for 12, 24, 36, and 48 h, respectively. The fermentation broth was centrifuged at 12,000 rpm and 4 °C for 10 min to remove cells. Culture supernatant was collected for sodium dodecyl sulfate (SDS)-polyacrylamide gel electrophoresis (PAGE) analysis of recombinant protein. Proteins were visualized with Coomassie Brilliant Blue.

15.2.2.6 Antimicrobial Activity Assay

Antimicrobial activity of the recombinant SjLys protein was assessed with the radial diffusion assay as described by Cong et al. [5]. Four bacterial strains were used as substrates, including Gram-negative bacteria *P. aeruginosa* and *V. parahaemolyticus*, Gram-positive bacteria *Micrococcus lysodeikticus* and *S. aureus*. The bacterial concentration in the liquid broth was adjusted to an absorbance of 0.1 at 600 nm and poured onto 90 mm LB agar plates. Wells (diameter 5 mm) were cut into the freshly poured plates after the agar solidification. For radial diffusion assays, 10 μl of the recombinant engineering strain was pipetted into individual wells and the agar plates were incubated at 37 °C. After 16 h incubation, the radius of the clearing zone was measured. The original strain WB600 was used as negative control.

15.3 Results

15.3.1 The Fragment of the Stichopus japonicus Lysozyme Gene

The sea cucumber (*S. japonicus*) lysozyme (*SjLys)* gene was reamplified with the specific PCR primers HS-Q-P11and HS-Q-P12 and the plasmid pMD18T-SjLys as the template. The amplification results are shown in Fig. 15.1. In the Fig. 15.1, there was a gene fragment with approximately 400 bp, and it was in accordance with the anticipated objective strap size 400 bp. The sequencing result revealed that the nucleotide sequence of the reamplified fragment was absolutely identical to the reported cDNA of *S. japonicus* body wall in NCBI (EF036468). These results showed that the *SjLys* gene from *S. japonicus* have been successfully reamplified and obtained. Additionally, the analysis of DNA sequence further showed that the inserted lysozyme gene *SjLys* would contain 375 bp length DNA fragment that encodes a mature polypeptide of 125 amino acid (Fig. 15.2).

15.3.2 Construction and Identification of the pHT43-SjLys

The target *SjLys* fragment was extracted and purified with the recovery Kit. The product was then digested with *Sam*I and *Bam*HI, and finally subcloned to vector pMD18T-Simple plasmid to yield a positive recombinant plasmid pMD18T-Simple-SjLys with 3,077 bp length that was identified and verified by PCR and restriction enzyme digestion (Fig. 15.3).

After that, the pMD18T-Simple-SjLys and vector pHT43 were digested with the *Bam*HI and *Sma*I for 4 h at 37 °C. And their products were thus ligated to construct a recombinant expression vector pHT43-SjLys with 8,432 bp length (Fig. 15.4). It was further confirmed and verified by PCR and restriction enzyme digestion.

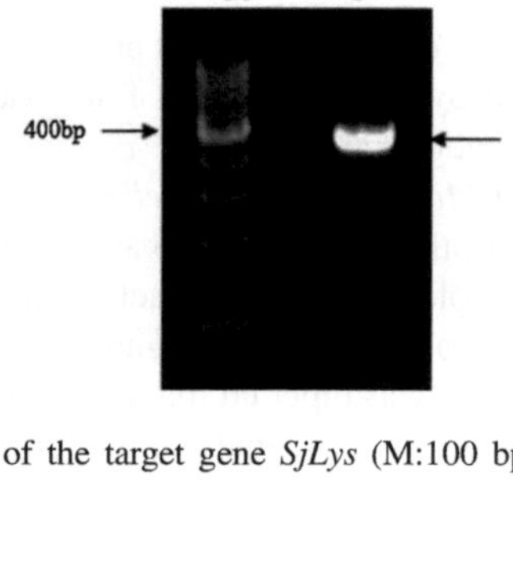

Fig. 15.1 PCR amplification of the target gene *SjLys* (M:100 bp DNA ladder marker, *lane 1*: target gene *SjLys*)

>*SjLys* sequences

GCCGGATCCATGCAAGTTCCTTCTGATTGCCTAAAGTGCATCTGTTTTGTAGAGTCCACTTGCACTATACCTTCCCCAT
TGTGTCATATGGATGTAGGATCACTGTCATGTGGTCCTTACCAAATCAAACTAGGCTACTGGCAGGATGCTAGGCTGA
AGGGAGGTAGTCTGGATGGAGATTGGCAGAAATGTTCAGCAACCTTTGACTGCAGTGAACGGGCTGTACAGGGTTAT
ATGGCACGGTACGCAACCTATGCCCGTCTAGAGCATAATCCTACCTGTGAGGATTTTGCGCGGATACACAACGGCGG
ACCAAATGGGTTCAAGAATCCAGCAACTGAAAAATATTGGTTGAGAGTGAAGAAATGTCTTGACATGAGCAACAAC
TGA GAATCCCGGGC

Fig. 15.2 Sequence of the reamplified *SjLys* fragment. (The *boxes* indicate the initial or stop codon, the *gray* regions represent the primers sequences and the *underlines* indicate the restrict enzyme sites)

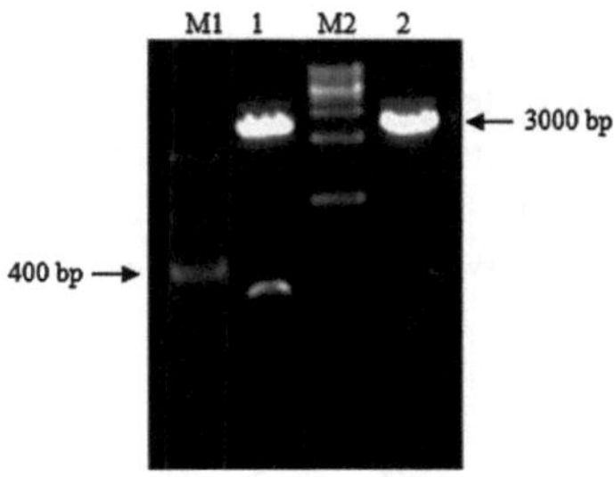

Fig. 15.3 Recombinant plasmid pMD18T-simple-SjLys double digestion verification. (M1 and M2: 100 and 1,000 bp DNA ladder marker, lane 1: double enzyme digestion, lane 2: pMD18T-Simple-SjLys)

15.3.3 Transformation of B. subtilis

The recombinant expression plasmid pHT43-SjLys was selected and transformed to *B. subtilis* WB600. The *B. subtilis* transformants were first selected by antibiotic screening at a concentration of 10 μg/ml chloroamphenicol. As a result, a genetics

Fig. 15.4 The map of the recombinant expression vector pHT43-SjLys

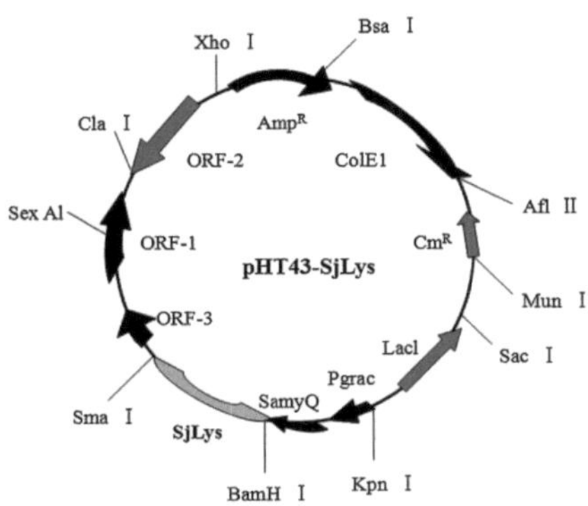

engineering strain designed pHT43-SjLys/WB600 was successfully obtained by further verification with PCR amplification, double digestion (Fig. 15.5), and sequencing of the *SjLys* gene of its recombinant plasmid extracted from the selected genetics engineering strain.

15.3.4 Growth Curve of the pHT43-SjLys/WB600

The growth curves of the pHT43-SjLys/WB600 and WB600 are presented in Fig. 15.6. In the Fig. 15.6, it can be found that the engineering strain and its original strain growth tendency are mostly identical. WB600 entered the logarithmic growth phase after inoculating 1.5 h and ended after 6 h, while the engineering strain were correspondingly 1 and 5 h. After that, the growth rate of these strains was consistent. The growth curve results demonstrated that the heterologous *SjLys* gene insertion did not affect the physiological metabolism and the growth of the engineering bacteria pHT43-SjLys/WB600.

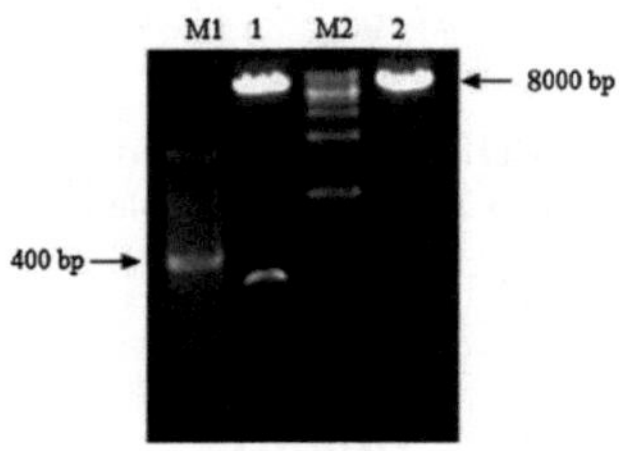

Fig. 15.5 The double digestion verification from the pHT43-SjLys/WB600 plasmid. (M1 and M2: 100 and 1,000 bp DNA ladder marker, *lane1*: double enzyme digestion, *lane 2*: pHT43-SjLys)

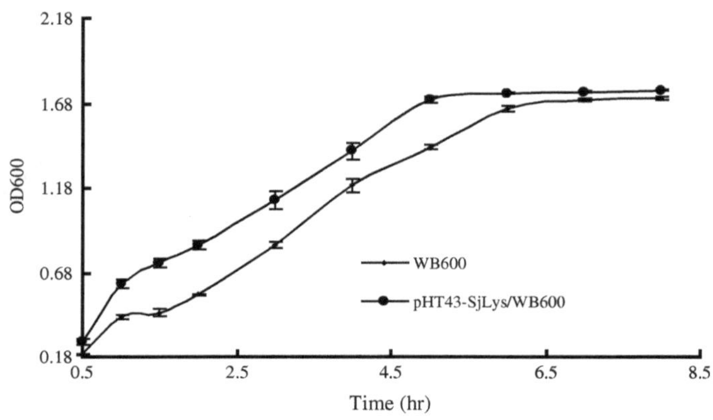

Fig. 15.6 The growth *curves* of *B. subtilis* WB600 and pHT43-SjLys/WB600

15.3.5 The Stability Analysis of the pHT43-SjLys/WB600

To investigate whether the bacterial host affects the stability of the plasmid, the pHT43-SjLys/WB600 strain was continuously cultured in the absence of antibiotics LB medium and the bacterium fluid are measured at each time. Figure 15.7 showed that recombinant plasmid has excellent genetic stability in the absence of selection pressure with approximately 94 % after 5 times (100 generation) continuous cultured while the biology yield were steady.

Plasmid DNA of the pHT43-SjLys/WB600 with 3 and 5 times continuous culture were extracted and digested, respectively. The result demonstrated that the strains grew on the chloramphenicol plates were the same as its initial strain, however, there were no plasmid extraction for the strains that cannot grow on the chloramphenicol plates. There was no rearrangement or lost phenomenon found in the engineered bacteria. These results together suggest that the engineered bacteria have a good genetic stability.

15.3.6 SDS-PAGE Analysis

The recombinant construct pHT43-SjLys/WB600 containing the coding region for the mature SjLys protein, was selected for heterologous expression. After IPTG induction for 12, 24, 36 and 48 h respectively, the recombinant SjLys protein was purified and analyzed by SDS-PAGE on a 12 % gel and autoradiography.

The SDS-PAGE analysis of the rSjlys protein demonstrated that there was a faint band with a molecular mass of approximately 14.0 kDa when it was induced for 36 and 48 h (Fig. 15.8). Furthermore, its molecular mass was accordance with the theoretical size of the fusion protein. These confirmed that the rSjlys protein were successfully expressed but with relatively low level.

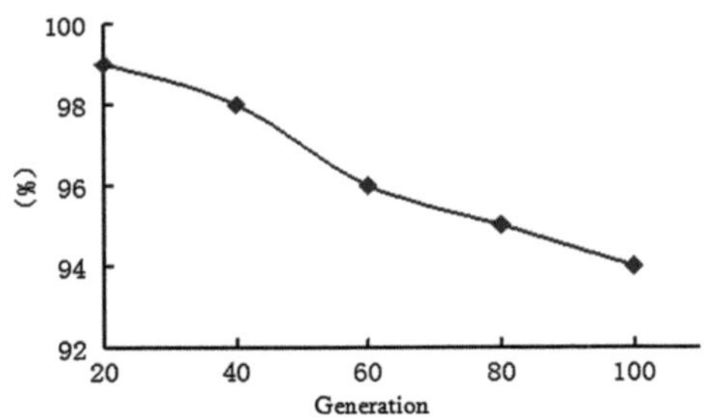

Fig. 15.7 The stability of engineered bacteria pHT43-SjLys/WB600

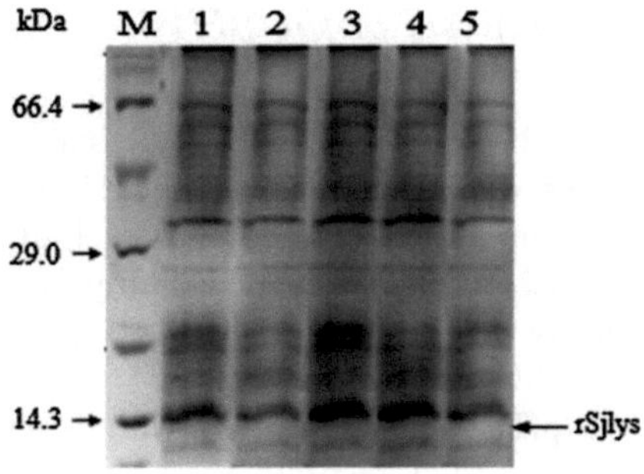

Fig. 15.8 SDS-PAGE analysis of the recombinant SjLys protein secreted by pHT43-SjLys/WB600. (M: Molecular mass markers; *lanes 1, 2, 3* and *4*: Total cellular protein induced by IPTG for 12, 24, 36, and 48 h, respectively; *lane 5*: the control WB600)

15.3.7 Antimicrobial Activities of the Recombinant Proteins

The radial diffusion assay was employed to evaluate the antimicrobial spectrum of the recombinant SjLys proteins against several gram-positive and gram-negative bacteria, including pathogenic bacteria *Vibro parahaemolyticus*, which are commonly present in coastal marine environments and a common finding and the leading agent causing disease of *Stichopus japonicas*. The original *B. subtilis* WB600 was used as control.

Based on the radius of the antimicrobial zone, the recombinant SjLys protein from the pHT43-SjLys/WB600 displayed a remarkable inhibitory effect on the growth of *Vibro parahaemolyticus* (Fig. 15.9). It revealed that rSjLys effectively inhibited the growth of *V. parahaemolyticus*, however, it did not obviously affect the growth of the other strains and inhibit the remaining 3 indicator bacteria. In summary, combined with the results of the 2.6, it is suggested that the heterologous expression of *Sjlys* was successfully realized in *B. subtilis*.

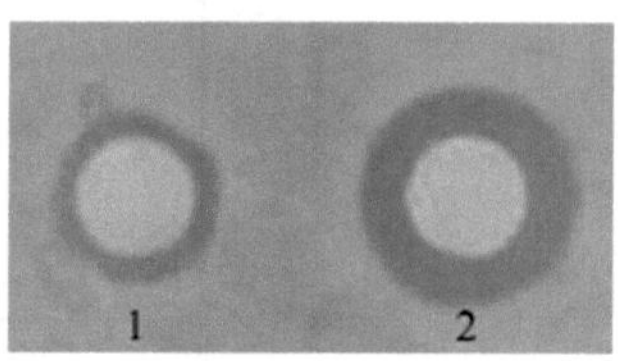

Fig. 15.9 Assay of antimicrobial activity of the recombinant SjLys protein to *V. parahaemolyticus*. (*1* WB600; *2* pHT43-Sjlys/WB600)

15.4 Discussion

Sea cucumber (*S. japonicus*) is one of the economically important farmed echinoderm species in Northern China. However, infectious diseases are becoming a severe problem with increasing culturing. Disease caused by *Vibrio* is most widespread in sea cucumber farming. Antibiotics and chemotherapeutics used to control these diseases can result in the development of drug-resistant bacteria, environmental pollution and unwanted residues in aquaculture. One of the most promising methods for controlling sea cucumber diseases in aquaculture is addition of Lysozyme in feed [18].

To our knowledge, this is the first study about the *S. japonicus* lysozyme gene expression in *B. subtilis*, which provide a new way for *SjLys* gene expression and efficient utilization. Undoubtedly, this knowledge should accelerate the biotechnological application of *B. subtilis* in industry field, and be helpful for the prevention and treatment of the sea cucumber disease [8].

Unfortunately, the protein concentration of genetic engineering strain pHT43-Sjlys/WB600 is not high and can't well meet production's need, which meant that the production of rSjlys protein in *B. subtilis* pHT43-SjLys/WB600 could still be improved. Therefore, it is necessary to improve the genetic engineering strain for effective expression. We are focusing on increasing its expression to achieve more products and taking measures to gradually resolve them, such as optimization of the fermentation conditions and separation and purification technology. Moreover, we are trying to insert the *Sjlys* gene in a new and excellent *B. subtilis* HS-38A for further genetic modification, which it was isolated from the intestine of wild sea cucumber in Dalian sea area [5]. Meanwhile, what the foreign *Sjlys* gene introduced and integrated into *B. subtilis* chromosome for integration expression was ongoing.

Acknowledgments This research was supported by the Natural Science Foundation of China (Grant no. 31072224) and Natural Science Foundation of Liaoning Province of China (Grant no. 20102009).

References

1. Alma BP, Adriana TM, Aldo AA et al (2012) Shrimp invertebrate lysozyme i-lyz: Gene structure, molecular model and response of c and i lysozymes to lipopolysaccharide (LPS). Fish Shellfish Immun 32:230–236
2. Avsaroglu MD, Buzrul S, Sanlibaba P et al (2007) A kinetic study on the plasmid stability of three Lactococcus lactis strains. J Ind Microbiol Biotechnol 34(11):729–737
3. Bathige SD, Umasuthan N, Kasthuri SR et al (2013) A bifunctional invertebrate-type lysozyme from the disk abalone, Haliotis discus: Genome organization, transcriptional profiling and biological activities of recombinant protein. Dev Comp Immunol 41:282–294
4. Chittibabu G, Ma C, Netter HJ et al (2014) Production, characterization, and immunogenicity of a secreted form of Plasmodium falciparum merozoite surface protein 4 produced in *Bacillus subtilis*. Appl Microbiol Biotechnol 98:3669–3678

5. Cong LN, Ruan QK, Liu ZW et al (2013) Isolation of an antimicrobial marine strain HS-A38 and purification of its bioactive substances. J Pure Appl Microbio 7(1):379–385
6. Cong LN, Yang XJ, Wang XX et al (2009) Characterization of an i-type lysozyme gene from the sea cucumber *Stichopus japonicus*, and enzymatic and nonenzymatic antimicrobial activities of its recombinant protein. J Biosci Bioeng 107(6):583–588
7. Dong XP, Zhu BW, Sun LM et al (2011) Changes of collagen in sea cucumber (*Stichopus japonicas*) during cooking. Food Sci. Biotechnol. 20(4):1137–1141
8. Fatma MM, Carlos DD, Marc B (2012) High transformation efficiency of *Bacillus subtilis* with integrative DNA using glycine betaine as osmoprotectant. Anal Biochem 424:127–129
9. Gao FY, Qu L, Yu SG et al (2012) Identification and expression analysis of three c-type lysozymes in *Oreochromis aureus*. Fish Shellfish Immun 32:779–788
10. Leen DG, Jos MP, Paul J et al (2007) Stability of a promiscuous plasmid in different hosts: no guarantee for a long-term relationship. Microbiology 153:452–463
11. Li SS, Wen JP, Jia XQ (2011) Engineering *Bacillus subtilis* for isobutanol production by heterologous Ehrlich pathway construction and the biosynthetic 2-ketoisovalerate precursor pathway overexpression. Appl Microbiol Biotechnol 91:577–589
12. Liu HZ, Zheng FR, Sun XQ et al (2012) Construction of cDNA library from intestine, Mesentery and Coelomocyte of *Apostichopus japonicus* selenka infected with *Vibrio* sp. and a preliminary analysis of immunity-related genes. J. Ocean Univ. China 11(2):187–196
13. Liu YH, Lin S, Zhang XQ et al (2014) A novel approach for improving the yield of *Bacillus subtilis* transglutaminase in heterologous strains. J Ind Microbiol Biotechnol 41:1227–1235
14. Mu L, Wen JP (2013) Engineered *Bacillus subtilis* 168 produces L-malate by heterologous biosynthesis pathway construction and lactate dehydrogenase deletion. World J Microbiol Biotechnol 29:33–41
15. Ranjita B, Masaru Y, Hideki N et al (2012) Enhanced production of 2,3-butanediol by engineered *Bacillus subtilis*. Appl Microbiol Biotechnol 94:651–658
16. Reza P, Ebrahim VF, Seyed AS et al (2014) Induction of *Bacillus subtilis* expression system using environmental stresses and glucose starvation. Ann Microbiol 164:879−882
17. Shoham Y, Israeli E, Sonensheim AL et al (1991) Inhibition of growth of *Bacillus subtilis* by recombinant plasmid pCED3. Arch Microbiol 156:204–212
18. Sun LM, Zhu BW, Wu HT et al (2011) Purification and characterization of *Cathepsin B* from the Gut of the Sea Cucumber (*Stichopus japonicas*). Food Sci Biotechnol 20(4):919–925
19. Wang XY, Chen WJ, Tian YL et al (2014) Surface display of *Clonorchis sinensis* enolase on *Bacillus subtilis* spores potentializes an oral vaccine candidate. Vaccine 32:1338–1345
20. Wu HT, Li DM, Zhu BW et al (2013) Characterization of acetylcholinesterase from the gut of sea cucumber *Stichopus japonicus*. Fish Sci 79:303–311
21. Yang AF, Zhou ZC, Dong Y et al (2010) Expression of immune-related genes in embryos and larvae of sea cucumber *Apostichopus japonicus*. Fish Shellfish Immunol 29:839–845
22. Yuan Y, Feng F, Chen L et al (2014) Surface display of *Acetobacter pasteurianus AdhA* on *Bacillus subtilis* spores to enhance ethanol tolerance for liquor industrial potential. Eur Food Res Technol 238:285–293
23. Yue X, Wang HX, Huang XH et al (2012) Single nucleotide polymorphisms in i-type lysozyme gene and their correlation with vibrio-resistance and growth of clam *Meretrix meretrix* based on the selected resistance stocks. Fish Shellfish Immunol 33:559–568
24. Zhang HW, Sun C, Sun SS et al (2010) Functional analysis of two invertebrate-type lysozymes from red swamp crayfish, *Procambarus clarkia*. Fish Shellfish Immunol 29:1066–1072
25. Zou MY, Li XZ, Shi WJ et al (2013) Improved production of alkaline polygalacturonate lyase by homologous overexpression *pelA* in *Bacillus subtilis*. Process Biochem 48:1143–1150

Chapter 16
Mega-Genome DNA Extraction from Pit Mud

Huimin Xie, Yali Dai and Lin Yuan

Abstract Molecular analyses for the study of microbial community often depend on the direct extraction of the mega-genome DNA from pit mud. The present work compared six different methods to extract the mega-genome DNA from pit mud, and investigated the relationship between the quantity and the diversity of the mega-genome DNA. Using the six different extraction methods, the DNA ranged from 3.18 to 20.17 ng DNA/μL. Both the quantity and the diversity of the mega-genome DNA, the "repeated freezing and thawing" method is superior to other five methods, so it is the best choice for the extraction of mega-genome DNA from pit mud.

Keywords Mega-genome DNA · Extraction · Diversity · Pit Mud

16.1 Introduction

During Chinese liquor production, the pit mud of cellars plays an important role, and a sufficient quantity of beneficial microbes is important in the fermentation of the liquor-making raw materials [1]. The pit mud is regarded as the carrier for the propagation and growth of the functional bacteria, while the functional bacteria could produce flavor substance in liquor [2].

The methods of mega-genome DNA extraction is divided into direct extraction and indirect extraction. The direct in situ lysis extraction method has been widely used, which are summarized in two critical steps: the disruption of the microbial cell wall and the extraction of nucleic acids. The most commonly used disruption methods are freezing–thawing or freezing–boiling [3–5] and bead-mill homogenization [6–9]. Though these methods generally provide the highest DNA yields within acceptable processing times, each method has its own disadvantages. The

H. Xie · Y. Dai · L. Yuan (✉)
College of Life Science, Inner Mongolia University, Huhhot, China
e-mail: yuan0079@163.com

T.-C. Zhang and M. Nakajima (eds.), *Advances in Applied Biotechnology*,
Lecture Notes in Electrical Engineering 332, DOI 10.1007/978-3-662-45657-6_16

lysis efficiency in any nucleic acid extraction procedure is critical in determining its success, so that an accurate representation of the microbial community can be achieved. Combinations and modifications of different protocols might be needed at some conditions.

PCR combined with single strand conformation polymorphism (SSCP) has also been applied to the analysis of microbial community [10] and microbial community composition [11]. James [12] investigation also highlights the utility of SSCP in molecular microbial ecology.

In the present study, we compared and evaluated the six different methods for extraction of mega-genome DNA from pit mud. The yields, purity, and diversity of the mega-genome DNA were determined to compare the efficiency of the six methods.

16.2 Materials and Methods

16.2.1 Sample Collection

The pit mud samples were collected from the liquor industry in west Inner Mongolia.

16.2.2 Extraction of Pit Mud DNA

In the experiment, we compared six methods: control method; repeated freezing and thawing; PBS buffer cleaning; PBS buffer cleaning + repeated freezing and thawing; Method-B [13]; Method-C [13]. The direct extraction method is summarized in two critical steps, the disruption of the microbial cell wall and the separation of the nucleic acids. Because the disruption of the microbial cell wall is the key factor, so the main difference of the six methods is in this step.

For the control method, pit mud DNA was extracted by the protocol of Yan liang zhen et al. [5] with a little modification. 1 mL of the extraction buffer (100 mM Tris-HCl, 100 mM EDTA, 100 mM Na_3PO_4, 1.5 M NaCl and 1 % CTAB, pH8.0) and 2 μL lysozyme was added to 0.25 g pit mud samples. The mixture was incubated at 37 °C for 90 min. 2 μL helicase, 200 μL 10 % SDS and 1 μL Proteinase-K were added. The mixture was incubated at 65 °C for 60 min. And then centrifuged at 5,000 r/min for 10 min. The supernatant was transferred into a new tube. 400 μL of the DNA extraction buffer and 100 μL 10 % SDS were added to the sediment. The mixture was incubated at 65 °C for 20 min. And then centrifuged at 5,000 r/min for 5 min. Supernatants obtained from all of the two steps were combined and added 0.6 % (v/v) isopropanol, then freeze at −20 °C for 30 min. The mixture was centrifuged at 13,000 r/min for 10 min. The sediment was dissolved by

400 μL 1 × TE, 80 μL CTAB/NaCl and 100 μL NaCl, incubated at 65 °C for 15 min. Mixed with an equal volume of Phenol chloroform isoamyl alcohol (vol/vol vol/, 25:24:1), and then centrifuging at 13,000 r/min for 10 min. The supernatant was transferred into a new tube and mixed with an equal volume of chloroform isoamyl alcohol (vol/vol, 24:1), then centrifuged at 13,000 r/min for 10 min. The supernatant was transferred into a new tube, and added 0.6 % (v/v) isopropanol, then freeze at −20 °C for 30 min. The mixture was centrifuged at 13,000 r/min for 10 min. The pellet was collected. After drying, the pellet was suspended in 30 μL 0.1 × TE (10 mM Tris, 1 mM EDTA, pH 8.0), and added to 2 μL RNase. The mixture was incubated at 37 °C for 2 h and stored at −20 °C. For the repeated freezing and thawing, 0.25 g samples were mixed with 1 mL DNA extraction buffer, then three cycles of freezing at −20 °C and thawing at 65 °C was conducted to release DNA from the microbial cells. For the buffer cleaning, 0.25 g samples were mixed with 1 mL PBS, vortexed for 10 s. For the PBS buffer cleaning + repeated freezing and thawing, 0.25 g samples were mixed with 1 mL PBS, vortexed for 10 s, and then centrifuged at 5,000 r/min for 10 min, the sediment were mixed with 1 mL DNA extraction buffer, three cycles of freezing in −20 °C and thawing at 65 °C was conducted to release DNA from the microbial cells. The remaining steps of the three method are the same as the control method. The purity and yield of the extracted mega-genome DNA was determined by NanoDrop 1,000.

16.2.3 PCR-SSCP Analysis

The diversity of the mega-genome DNA was determined by PCR-SSCP. The V3 region of 16S rDNA was amplified to analysis the diversity of prokaryotic microorganisms, while the V8-9 region of 18S rDNA was amplified to analysis the diversity of eukaryotic microorganisms. Amplification of the V3 region of 16S rDNA was performed with the universal primers Com1 (5′-ACT CCT ACG GGA GGC AGC AG-3′) and Com2 (5′-TTA CCG CGG CTG CTG GCA C-3′). The primers Come3 (5′-GAG GCA ATA ACA GGT CTG TGA TGC -3′) and Come4 (5′-TCC GCA GGT TCA CCT ACG GA -3′) were used to amplify the V8-9 region of 18S rDNA. The PCR reaction system containing: 20 ng of mega-genome DNA, 2.5 mM Mg^{2+}, 2.5 mM dNTPs, 20 μM of each of the primers, 2.5U Easy Taq polymerase. The PCR reaction was performed with the following conditions: 94 °C for 2 min; 35 cycles of denaturation at 94 °C for 30 s, annealing at 57 °C for 30 s, and prolonging at 72 °C for 20 s; Final elongation at 72 °C for 5 min.

In SSCP electrophoresis analysis, the 16S rDNA PCR products were separated by 10 % polyacrylamide (29:1) at 200 V for 16 h at 4 °C, while the 18S rDNA PCR products were separated by 16 % polyacrylamide (29:1) at 300 V for 24 h at 4 °C. Gels were stained by silver according to the standard procedure [14].

16.2.4 Statistical Analysis

The data was analyzed using the software Quantity One (from Bio-Rad). The band abundance (S) is the number of the bands in the SSCP map. The Shannon−Wiener diversity (H) is the index of the diversity of the micro-ecosystem. H is the value for Shannon−Wiener diversity; Pi is the ratio of the DNA quantity of the i band to the DNA quantity of all the bands of the sample.

$$H = -\sum_{i=1}^{S} p_i \ln p_i \qquad (16.1)$$

16.3 Results

16.3.1 Purity and Yield of the Meta-Genome DNA

The purity of the mega-genome DNA extracted by the six methods was determined by NanoDrop 1,000 (Table 16.1). UV absorption ratio of $A_{260/A280}$ ranged from 1.45 to 1.64. UV absorption ratio of $A_{260/A230}$ ranged from 1.05 to 1.26. Among then, the DNA extracted by the repeated freezing and thawing method was relatively free from humic acids, RNA and protein. The values of the PBS buffer cleaning method were comparatively lower than the other methods.

Total extracted yield of DNA from pit mud samples were determined by NanoDrop 1,000 (Table 16.1). The extracted yields of DNA were varied from 159.60 to 956.65 ng/μL, respectively. Also, the yield of the repeated freezing and thawing method is higher than the other methods.

Considering the purity as well as the yield of the mega-genome DNA, the repeated freezing and thawing method is the best choice among the six methods.

Table 16.1 Comparison of the yield and purity of the pit mud DNA extracted by the six methods

Method	ng/μL	A260/280	A260/230
1. Control methods	791.69	1.56	1.12
2. Repeated freezing and thawing	956.65	1.62	1.26
3. Buffer cleaning + repeated freezing and thawing	363.19	1.50	1.11
4. PBS buffer cleaning	704.27	1.45	1.05
5. Method-B	159.60	1.64	1.15
6. Method-C	477.17	1.52	1.10

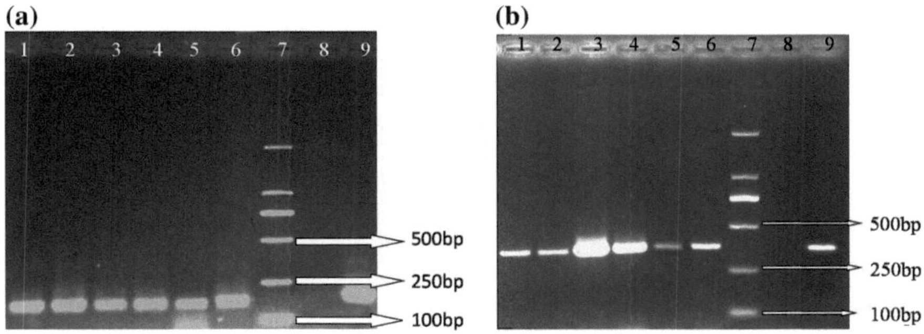

Fig. 16.1 Agarose gel electrophoresis results of the PCR amplification of 16S rDNA and 18S rDNA gene fragment. **a** 16S rDNA; **b** 18S rDNA. *Lane1* control methods; *lane2* repeated freezing and thawing; *lane3* PBS buffer cleaning + repeated freezing and thawing; *lane4* PBS buffer cleaning; *lane5* Method-B; *lane6* Method-C; *lane7* molecular weight marker (DL2,000);*lane8* negative control; *lane9* positive control

16.3.2 PCR-SSCP Analysis

To investigate the diversity, the mega-genome DNA of the pit mud was determined by PCR-SSCP. The V3 region of 16S rDNA was amplified to analysis the diversity of prokaryotic microorganisms, while the V8-9 region of 18S rDNA was amplified to analysis the diversity of eukaryotic microorganisms. The amplified fragments of 16S rDNA and 18S rDNA were about 200 and 400 bp, respectively (Fig. 16.1a, b). The PCR products of 16S rDNA and 18S rDNA were separated by SSCP (Fig. 16.2a, b).

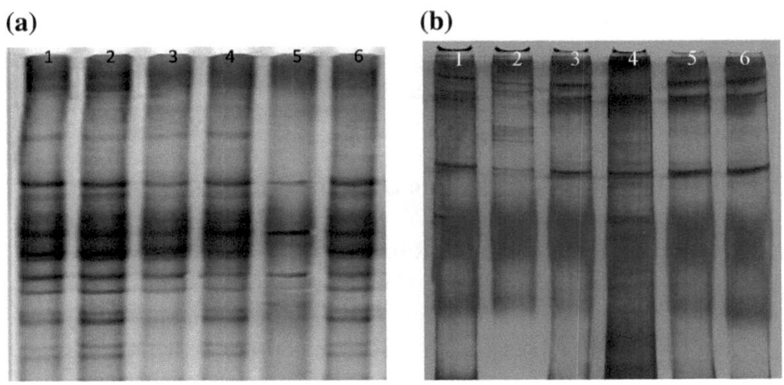

Fig. 16.2 SSCP map of the PCR products of 16S rDNA and 18S rDNA. **a** 16S rDNA; **b** 18S rDNA. *Lane1* control methods; *lane2* repeated freezing and thawing; *lane3* PBS buffer cleaning + repeated freezing and thawing; *lane4* PBS buffer cleaning; *lane5* Method-B; *lane6* Method-C

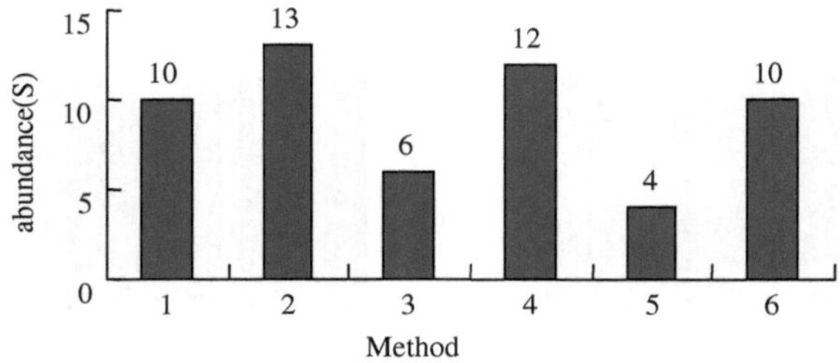

Fig. 16.3 The band abundance of 16S rDNA in PCR-SSCP map. *Lane1* control methods; *lane2* repeated freezing and thawing; *lane3* PBS buffer cleaning + repeated freezing and thawing; *lane4* PBS buffer cleaning; *lane5* Method-B; *lane6* Method-C

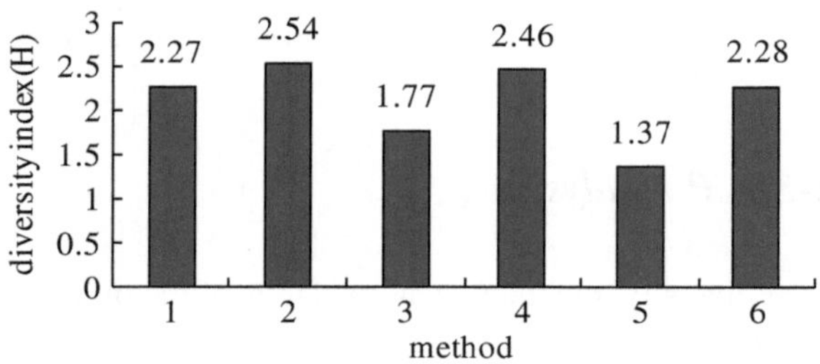

Fig. 16.4 Diversity index of 16S rDNA in PCR-SSCP map. *Lane1* control methods; *lane2* repeated freezing and thawing; *lane3* PBS buffer cleaning + repeated freezing and thawing; *lane4* PBS buffer cleaning; *lane5* Method-B; *lane6* Method-C

16.3.2.1 Prokaryotic Microorganisms Diversity Analysis

In Fig. 16.2a, all of the 16S rDNA PCR products were separated well by SSCP. The band abundance was between 4 and 13 (Fig. 16.3). The calculation revealed that the H values ranged from 1.37 to 2.54 (Fig. 16.4). Both the band abundance and the H values of the repeated freezing and thawing method are higher.

16.3.2.2 Eukaryotic Microorganisms Diversity Analysis

In Fig. 16.2b, all of the 18S rDNA PCR products were separated well by SSCP. The band abundance was between 4 and 13 (Fig. 16.5). The calculation revealed that the H values ranged from 1.37 to 2.56 (Fig. 16.6). Both the band abundance and the H values of the repeated freezing and thawing method are higher.

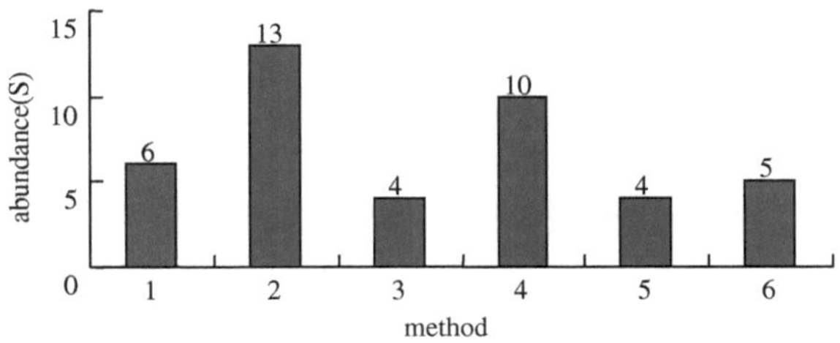

Fig. 16.5 The band abundance of 18S rDNA in PCR-SSCP map. *Lane1* control methods; *lane2* repeated freezing and thawing; *lane3* PBS buffer cleaning + repeated freezing and thawing; *lane4* PBS buffer cleaning; *lane5* Method-B; *lane6* Method-C

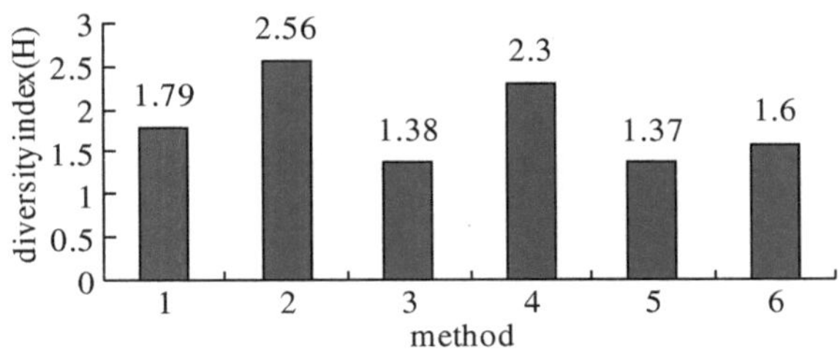

Fig. 16.6 Diversity index of 16S rDNA in PCR-SSCP map. *Lane1* control methods; *lane2* repeated freezing and thawing; *lane3* PBS buffer cleaning + repeated freezing and thawing; *lane4* PBS buffer cleaning; *lane5* Method-B; *lane6* Method-C

16.4 Discussion

It is known that about 99 % microorganisms in natural environment couldn't be obtained their pure culture by traditional culture methods, but for extremeophiles of extreme conditions is poorly understood. With the emergence of metagenomic technology, people could investigate the immense and unknown world of microorganisms in many ways [15]. Total DNA was directly extracted from the environment without the need of the microorganism cultures [16]. The sensitivity and efficiency of the DNA extraction method are very important for research of microbial diversity of pit mud and its activity. Only by efficiently extracting can the typical and practical be achieved. It is, however, important to recognize that no single method of DNA extraction will be able to extract all DNA, as there are multiple factors that may affect the performance of an extraction method. Combinations and modifications of different protocols might be needed for some conditions.

The aim of this study was to compare extraction methods for mega-genome DNA from pit mud. The purity, yield and diversity of the total DNA were determined to evaluate the efficiency of the six different methods. The results of this comprehensive evaluation of the DNA extraction methods suggest that the repeated freezing and thawing method was superior to other methods, and was the best choice to extract total DNA to analysis the diversity from pit mud samples.

But in our study, it can be seen that the ratios of $A_{260/A230}$ and $A_{260/A280}$ of the total DNA extract from pit mud were all above 1.00 (Table 16.1), suggesting that DNA extracted by the six methods contained low humic acid and protein impurities [17, 18].

PCR-SSCP has the advantages of being accurate, sensitive and convenient in analyzing [19] the microbial communities and in probing the changes in the complicated microbial communities in the pit mud, which is helpful in investigating the fermentation mechanisms. Nevertheless, the PCR-SSCP methodology still requires improvement because of the following disadvantage. Such as that in the analyses of SSCP, the smaller the fragments are, the higher the detection rate is. So you had better choose a fragment which is shorter than 300 bp [20]. In order to analyses well the PCR products in PCR-SSCP, we could digest the long fragments (>400 bp) using restrict enzyme [21].

Acknowledgments This work was funded by the doctoral program fund from Chinese Ministry of Education (20121501120001).

References

1. Deng Bo, Shen Cai-hong, Shan Xiao-hu, Ao Zong-hua, Zhao Jin-song, Shen Xiao-juan, Huang Zhi-guo (2012) PCR-DGGE analysis on microbial communities in pit mud of cellars used for different periods of time. Inst Brew Distilling 118:120–126
2. Huang YG, Huang P (2004) Research on Extraction and purification of total DNA microbes in pit mud. Liquor-Making Science and Technology 123(3): 41−42
3. Degrange V, Bardin R (1995) Detection and counting of nitrobacter populations in soil by PCR. Appl Environ Microbiol 61:2093–2098
4. More MI, Herrick JB, Silva MC, Ghiorse WC, Madsen EL (1994) Quantitative cell lysis of indigenous microorganisms and rapid extraction of microbial DNA from sediment. Appl Environ Microbiol 60:1572–1580
5. Tsai YL, Park MJ, Olson BH (1991) Rapid method for direct extraction of mRNA from seeded soils. Appl Environ Microbiol 57:765–768
6. Maarit Niemi R, Heiskanen I, Wallenius K, Lindstrom K (2001) Extraction and purification of DNA in rhizosphere soil samples for PCR-DGGE analysis of bacterial consortia. J Microbiol Methods 45:155–165
7. Miller DN (2001) Evaluation of gel filtration resins for the removal of PCR-inhibitory substances from soils and sediments. J Microbiol Methods 44:49–58
8. Miller DN, Bryant JE, Madsen EL, Ghiorse WC (1999) Evaluation and optimization of DNA extraction and purification procedures for soil and sediment samples. Appl Environ Microbiol 65:4715–4724
9. Steffan RJ, Goksoyr J, Bej AK, Atlas RM (1988) Recovery of DNA from soils and sediments. Appl Environ Microbiol 54:2908–2915
10. Stach James EM, Bathe Stephan, Clapp Justin P, Burns Richard G (2001) PCR-SSCP comparison of 16S rDNA the diversity in soil DNA obtained using different isolation and purication methods. FEMS Microbiol Ecol 36:139–151

11. Orita M, Iwahana H (1989) Detection Of Polymorphisms Of Human DNA By Gel Electrophoresis As Single-Strand Conformation Polymorphisms. Proc Natl Acad Sci USA 86:2766–2770
12. Stach James EM, Bathe Stephan, Clapp Justin P, Burns Richard G (2001) PCR-SSCP comparison of 16S rDNA the diversity in soil DNA obtained using di¡erent isolation and puri¢ cation methods. FEMS Microbiol Ecol 36:139–151
13. Islam, MR, Sultana, T, Joe, MM, Cho, J-C, Sa T (2012) Comparisons of direct extraction methods of microbial DNA from different paddy soils. Saudi J Biol Sci 19: 337−342
14. Liang Hong Wei, Wang Chang Zhong, Li Zhong et al (2008) Improvement of the silver-stained technique of polyacrylamide gel electrophoresis. Hereditas 30(10):1379–1382
15. Zhu Yun hua, Li Jian, Fang Jun, Tian Yun, Lu Xiang yang (2011) Application of Metagenomic Technique in the Exploring of Uncultured Microbial in Extreme Environments [CH]. Biotech Bull 9:52–58
16. Fei MENG, Chun-Na YU, Qiu-Yan WANG, Tian XIE (2010) Metagenome and Metagenomics. Chin J Biochem Mol Biol 26(2):116–120
17. Yeates C, Gillings MR, Davison AD, Altavilla N, Veal DA (1998) Methods for microbial DNA extraction from soil for PCR amplification. Biol. Proc. Online 1:40–47
18. Roh C, Villatte F, Kim BG, Schmid RD (2006) Comparative study of methods for extraction and purification of environmental DNA from soil and sludge samples. Appl Biochem Biotechnol 134:97–112
19. GUI F, ZHANG Z (2005) PCR-SSCP technology and its application in microbiological study. J microbiol[CH] 25(1): 89−93
20. Zh OU Ming-tao, Hong GAO, Peng XIAO, Hui ZHOU (2009) Advance in PCR-SSCP and Its Application. Prog Vet Med 30(6):90–93
21. Qiang Chen, Yuangui Chen, Genlin Wang (2008) Optimization of the conditions affecting PCR-SSCP experiment of HSP70 gene. Anim Husbandry Vet Med 140(12):66–67

Chapter 17
Evidence for a Link of SDPR and Cytoskeleton

Baoxia Zhang, Jun Zhu, Liqiao Ma, Yuyin Li, Aipo Diao and Yinchuan Li

Abstract Serum deprivation response (SDPR) is a plasma membrane binding protein and a substrate of protein kinase C (PKC) phosphorylation, which recruits polymerase I and transcript release factor (PTRF) to caveolae to stabilize and define their morphology. But how they were transported to membrane remain unclear. In order to clarify the link of SDPR and cytoskeleton, we observed the localized relationship between SDPR and cytoskeleton proteins by Immunofluorescence. Here, we discovered that SDPR colocalizes with cortactin (CTTN) in A7r5. Interestingly, SDPR was found to be highly concentrated in podosomes and dorsal ruffles induced by Phorbol 12, 13-dibutyrate (PDBu), and platelet-derived growth factor (PDGF), respectively, which suggest a possible role of SDPR in cell motility and cell invasion.

Keywords SDPR · Cytoskeleton · Podosome · Dorsal ruffle

17.1 Introduction

Caveolae are flask-shaped plasma membrane invaginations that are extremely abundant in endothelial, epithelial, adipocytes, and smooth muscle cells [1, 2]. Many reports show that caveolae are involved in cell signaling transduction, clathrin-independent endocytosis [3] and lipid homeostasis. The characterized

B. Zhang · J. Zhu · L. Ma · Y. Li · A. Diao (✉) · Y. Li (✉)
Key Lab of Industrial Fermentation Microbiology of the Ministry of Education, College of Bioengineering, Tianjin University of Science and Technology, Tianjin 300457, China
e-mail: diaoaipo@tust.edu.cn

Y. Li
e-mail: ycli@tust.edu.cn

T.-C. Zhang and M. Nakajima (eds.), *Advances in Applied Biotechnology*, Lecture Notes in Electrical Engineering 332, DOI 10.1007/978-3-662-45657-6_17

components of caveolae are cholesterol, sphingolipid and two key structural components, caveolins [1, 4] and cavins [1]. Caveolins are a kind of multiply acylated integral membrane proteins. As the major marker protein of caveolae, caveolin1 has significant roles in adjusting many signaling pathways associated with diseases. Mice lacking caveolin1 have multiple phenotypes including hyperglycemia, lipidosis, and changes in endothelial permeability [5]. Cavins are essential for the formation and dynamics of caveolae [6–8]. As two members of cavins family, cavin-1/PTRF and cavin-2/SDPR can interact with each other forming soluble and unique complexes that binds to caveolin scaffolds in the plasma membrane for stabilizing caveolae [8, 9]. Like caveolins, cavins regulate various cellular functions by affecting caveolae distribution, composition, and other unknown consequences.

Several studies have suggested that caveolins and cavins are associated with actin microfilaments [1, 10]. Interactions between components of the cytoskeleton and caveolae are likely to control dynamics of caveolae [11]. However, the localization and distribution relationships between cavins and cytoskeleton are still unclear. Our data provides an evidence that cavin-2/SDPRs concentrate in podosomes and forming dorsal ruffles, which implies the involvement of SDPR in cell cytoskeleton reorganization and cell motility.

17.2 Materials and Methods

17.2.1 Construction of Plasmids

cDNAs for mouse PTRF (polymerase I and transcript release factor), SDPR (serum deprivation response), CNN1 (Calponin1) and CTTN (Cortactin) were reversely transcribed from mRNAs derived from NIH3T3 cells. The cDNA encoding PTRF was PCR amplified with primers: forward 5′ GCAGAATTCGAGGATGTCACGCTCCA 3′ and reverse primer 5′ GAGTCTAGAGTCGCTGTCGCTCTTGTC 3′, and was inserted into the *E*coRI and *X*baI sites of pCDNA-HA-C1. Primers for amplification of SDPR were: forward primer 5′ CAGCCATGGGAGAGGACGCTGCACAGG 3′ and reverse primer 5′ CTGAAGCTTCAGGCAGTCTGATCCACATGCAG 3′. The SDPR PCR amplified products were digested with *B*amHI and *N*otI and inserted into the pCDNA-HA-C1 and pEGFP-C1 vectors respectively. The CTTN gene was amplified by PCR with forward primer 5′ AAGCTCGAGATGTGGAAAGCCTCTG 3′ and reverse primer 5′ GCAAAGCTTCTGCCGCAGCTCCA 3′. The product was purified and digested with *X*hoI and *H*indIII, and inserted into the pEGFP-C1 vector.

17.2.2 Cell Culture and Transfection

A7r5 (vascular smooth muscle cells) from ATCC were grown in Dubelcco's modified Eagle's medium (DMEM; Gibco) supplemented with 10 % fetal bovine serum (FBS), 2 mM L-Glutamine, 1,000 U/ml penicillin, and 0.1 mg/mL streptomycin and maintained at 37 °C and 5 % CO_2. Cells were transfected using Turbofect (Thermo Scientific) according to manufacturer's recommendations.

17.2.3 Antibodies and Other Reagents

Monoclonal anti-HA antibody and anti-MyosinII antibodies were obtained from Sigma-Aldrich. Rabbit pAb anti-Caveolin1 was from Santa Cruz (N20, sc-894). Phalloidin-Alexa Fluor 555 for staining F-actin was from Invitrogen. Secondary antibodies Alexa Fluor 555 donkey anti-Rabbit IgG and Alexa Fluor 555 Donkey anti-mouse IgG were also from Invitrogen. Recombinant human platelet-derived growth factor (PDGF-BB) reagent was obtained from ACROBiosystems. Phorbol 12, 13-dibutyrate (PDBu), and Hoechst 33342 were purchased from Sigma-Aldrich.

17.2.4 Immunofluorescence Imaging and Analysis

Cells plated in 12-well plates were fixed with 4 % paraformaldehyde in phosphate buffered saline (PBS), and were permeabilized using 0.5 % Triton X-100. After blocking of nonspecific antigens with 5 % BSA in PBS, hybridization with the primary antibody at an appropriate dilution ratio was performed at RT for 1 h. Cells were then stained with secondary antibody and Hoechst 33342 for 40 min at RT. Images were captured via an Olympus FluoView confocal laser scanning system FV1000, and images were disposed with Olympus FV10-ASW 3.0 Viewer and Image J softwares.

17.2.5 The Induction of Dorsal Ruffles and Podosomes

24 h after transfection with plasmids, cells were serum starved and then treated with 20 ng/mL PDGF for 6 min to induce the formation of dorsal rings. Cells without serum starvation were treated with 1 μM PDBu for 30 min to induce podosome formation.

17.3 Results

17.3.1 SDPR Is Associated with Cytoskeleton Proteins in A7r5

Recently, it was revealed that PTRF modulates cholesterol dynamics and impairs recruitment of actin cytoskeleton to detergent-resistant membrane [12]. To test whether SDPR takes roles in cytoskeleton, we then overexpressed SDPR-HA in A7r5 cells to monitor its subcellular distribution and its colocalization with several cytoskeleton proteins. We found that Cellular ATPase motor protein -Myosin II and the thin filament-associated protein CNN1 (Calponin1) rarely colocalized with SDPR (Fig. 17.1a, b). Interestingly, when GFP tagged CTTN (Cortactin) and SDPR-HA co-expressed, they are highly colocalized with each other which also can be demonstrated in their localization figures (Fig. 17.1c).

CTTN is a monomeric protein locates in the cytoplasm that can recruit Arp2/3 complex proteins to existing actin microfilaments and facilitating and stabilizing nucleation sites for actin branching when it was activated by external stimuli to promote polymerization and rearrangement of the actin cytoskeleton, especially the actin cortex around the cellular periphery [13, 14]. CTTN is crucial in promoting lamellipodia formation, invadopodia formation, and cell migration. CTTN together

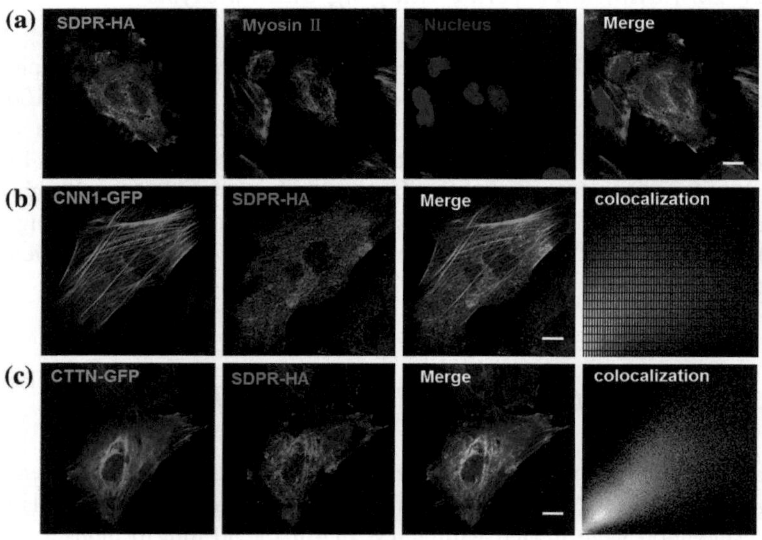

Fig. 17.1 The relationship between SDPR and cytoskeleton. **a** A7r5 cells expressing SDPR-HA were immunostained with anti-HA antibodies and endogenous Myosin II and analyzed by confocal microscopy. **b, c** Confocal images of A7r5 cells were co-transfected SDPR-HA with CNN1-GFP, CTTN-GFP, respectively. Cells were fixed and stained for anti-HA antibody. All Bars, 10 μm. The colocalization signals were plotted via the plugin (colocalization indices) of Image J (**b**, **c**)

with cofilin are usually involved in F-actin severing and barbed end formation in invasive protrusions [15, 16]. So SDPR, as an organizer of certain subset of caveolae [17], may have tight spatial correlation with the nucleation sites and severing of F-actin.

17.3.2 SDPR Is Present at Podosome Forming Sites but Have Low Affinity with F-Actin

The high colocalization between SDPR and CTTN suggests that SDPR may participate in cellular functions involving in F-actin organization and remodeling. To further verify this hypothesis, we treated A7r5 cells with PKC activator PDBu to induce podosome formation. Podosomes are cylindrical, actin-rich structures found on the outer surface of the plasma membrane of animal cells, which display a polarized pattern of distribution in migrating cells [18, 19]. PDBu not only facilitates actin polymerization and podosome formation, but also induces a general alteration in the adhesion system of smooth muscle cells, which triggers the conversion of focal adhesions into podosomes in A7r5 smooth muscle cells [20]. We found that SDPR was seldomly colocalized with long fibrous actin. However, SDPR highly accumulated at podosome induced by PDBu (Fig. 17.2).

17.3.3 SDPR Localizes to the Circular Dorsal Ruffles (CDRs) Induced by PDGF

Another evidence showing SDPR involving in F-actin severing and reorganization process was its enrichment on CDRs induced by growth factors. CDRs are highly dynamic and ring-shaped structures rich in filamentous actin (F-actin), which are

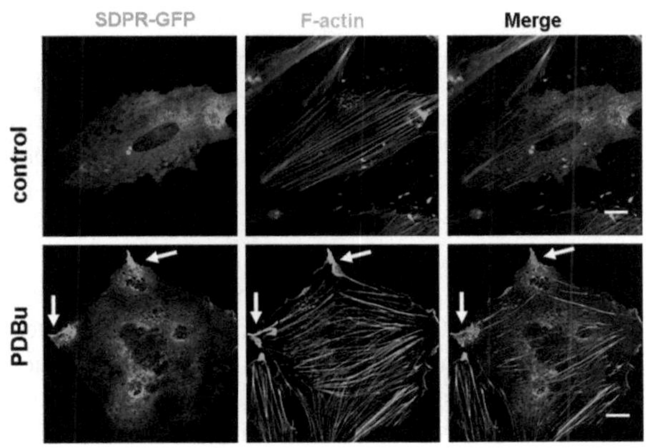

Fig. 17.2 SDPR is present at podosome sites. A7r5 cells expressing SDPR-GFP were stimulated with 1 μM PDBu for 30 min, fixed, and stained with phalloidin-Alexa Fluor 555. The *arrows* highlight the podosomes. Bars, 10 μm

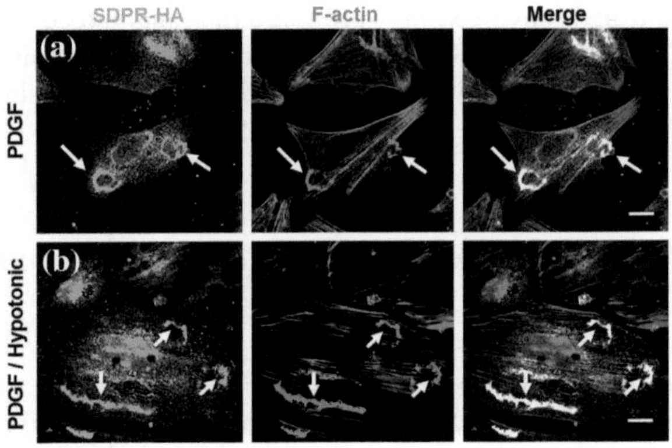

Fig. 17.3 SDPR localizes to the CDRs. **a** A7r5 cells expressing SDPR-HA were serum starved and then stimulated with 20 ng/ml PDGF for 6 min, fixed, and stained with anti-HA and phalloidin-Alexa Fluor 555. **b** The forming CDRs 6 min after treatment with PDGF were treated with 120 mOsm/L hypotonic PBS buffer. *Arrows* indicates CDRs. Bars, 10 μm

induced at the dorsal membrane surface within minutes after stimulation with platelet-derived growth factor (PDGF) [21]. Under serum-starvation, SDPR accumulated at CDRs with severed and reorganizing F-actin structures (Fig. 17.3a). Upon acute hypotonic treatment of the forming CDRs, SDPR still localizes to the interceptive dorsal ring (Fig. 17.3b). This implies that SDPR-located membrane ruffles have relatively strong physical contact with the severed F-actin filament. CDRs had been implicated have specific mode in cell locomotion [22, 23]. So SDPR may coordinately or synergistically participates in cytoskeleton organization.

17.4 Discussion

Both podosome and CDRs require cytoskeleton rearrangements accompanying membrane curvature formation. SDPR has been reported to play a role in membrane curvature formation and PS binding capability [7, 24]. In this context, SDPR may take roles in the interplay between plasma membrane and cytoskeleton, similar to other phosphoinositide-binding proteins with actin binding/modulating functions, for example, I-, N-, F-BAR domain-containing proteins. But it needs to further monitor its direct binding partner of cytoskeletons and its exact role as a curvature scaffolding protein. We should test whether SDPR binds to CTTN directly. For a membrane curvature formation, in addition to a driving force exerted by cytoskeletons, it also needs interplay between complexes that contain activators of actin polymerization and membrane-bound curved proteins of both types of curvature [25, 26]. For example, TUBA is a concave curved activator of N-WASP; the I-BAR domain-containing IRSp53 is a convex curved activator that recruits WAVE1 and mediate Rac-WAVE binding during the formation of dorsal ruffles [26, 27]. Since both podosome and CDRs are two similar key structures for cell invasion, we propose SDPR may also regulate tumor cell metastasis process. But how SDPR, plasma membrane, and cytoskeleton are coordinated in cell protrusion formation, remains to be investigated.

Acknowledgments This study was supported by National Natural Science Foundation of China (NO. 31340001, 31271448) and research start-up fund of Tianjin University of Science and Technology.

References

1. Hansen CG, Nichols BJ (2010) Exploring the caves: cavins, caveolins and caveolae. Trends Cell Biol 20:177–186
2. Pilch PF, Liu L (2011) Fat caves: caveolae, lipid trafficking and lipid metabolism in adipocytes. Trends Endocrinol Metab 22:318–324
3. Chaudhary N, Gomez GA, Howes MT et al (2014) Endocytic crosstalk: cavins, caveolins, and caveolae regulate clathrin-independent endocytosis. PLoS Biol 12:e1001832
4. Rothberg KG, Heuser JE, Donzell WC et al (1992) Caveolin, a protein component of caveolae membrane coats. Cell 68:673–682
5. Le Lay S, Kurzchalia TV (2005) Getting rid of caveolins: phenotypes of caveolin-deficient animals. Biochim Biophys Acta 1746:322–333
6. McMahon KA, Zajicek H, Li WP et al (2009) SRBC/cavin-3 is a caveolin adapter protein that regulates caveolae function. EMBO J 28:1001–1015
7. Hansen CG, Bright NA, Howard G, Nichols BJ (2009) SDPR induces membrane curvature and functions in the formation of caveolae. Nat Cell Biol 11:807–814
8. Hill MM, Bastiani M, Luetterforst R et al (2008) PTRF-Cavin, a conserved cytoplasmic protein required for caveola formation and function. Cell 132:113–124
9. Liu L, Pilch PF (2008) A critical role of cavin (polymerase I and transcript release factor) in caveolae formation and organization. J Biol Chem 283:4314–4322
10. Richter T, Floetenmeyer M, Ferguson C et al (2008) High-resolution 3D quantitative analysis of caveolar ultrastructure and caveola-cytoskeleton interactions. Traffic 9:893–909
11. Mundy DI, Machleidt T, Ying YS et al (2002) Dual control of caveolar membrane traffic by microtubules and the actin cytoskeleton. J Cell Sci 115:4327–4339
12. Inder KL, Zheng YZ, Davis MJ et al (2012) Expression of PTRF in PC-3 cells modulates cholesterol dynamics and the actin cytoskeleton impacting secretion pathways. Mol Cell Proteomics 11(2):M111.012245. doi:10.1074/mcp.M111.012245
13. Weed SA, Parsons JT (2001) Cortactin: coupling membrane dynamics to cortical actin assembly. Oncogene 20:6418–6434
14. Buday L, Downward J (2007) Roles of cortactin in tumor pathogenesis. Biochim Biophys Acta 1775:263–273
15. Bravo-Cordero JJ, Magalhaes MA, Eddy RJ et al (2013) Functions of cofilin in cell locomotion and invasion. Nat Rev Mol Cell Biol 14:405–415
16. Oser M, Yamaguchi H, Mader CC et al (2009) Cortactin regulates cofilin and N-WASp activities to control the stages of invadopodium assembly and maturation. J Cell Biol 186:571–587
17. Aboulaich N, Vainonen JP, Stralfors P, Vener AV (2004) Vectorial proteomics reveal targeting, phosphorylation and specific fragmentation of polymerase I and transcript release factor (PTRF) at the surface of caveolae in human adipocytes. Biochem J 383:237–248
18. Calle Y, Burns S, Thrasher AJ, Jones GE (2006) The leukocyte podosome. Eur J Cell Biol 85:151–157

19. Rottiers P, Saltel F, Daubon T et al (2009) TGFbeta-induced endothelial podosomes mediate basement membrane collagen degradation in arterial vessels. J Cell Sci 122:4311–4318
20. Kaverina I, Stradal TE, Gimona M (2003) Podosome formation in cultured A7r5 vascular smooth muscle cells requires Arp2/3-dependent de-novo actin polymerization at discrete microdomains. J Cell Sci 116:4915–4924
21. Mellstrom K, Heldin CH, Westermark B (1988) Induction of circular membrane ruffling on human fibroblasts by platelet-derived growth factor. Exp Cell Res 177:347–359
22. Suetsugu S, Yamazaki D, Kurisu S, Takenawa T (2003) Differential roles of WAVE1 and WAVE2 in dorsal and peripheral ruffle formation for fibroblast cell migration. Dev Cell 5:595–609
23. Sero JE, Thodeti CK, Mammoto A et al (2011) Paxillin mediates sensing of physical cues and regulates directional cell motility by controlling lamellipodia positioning. PLoS ONE 6: e28303
24. Burgener R, Wolf M, Ganz T, Baggiolini M (1990) Purification and characterization of a major phosphatidylserine-binding phosphoprotein from human platelets. Biochem J 269:729–734
25. McMahon HT, Gallop JL (2005) Membrane curvature and mechanisms of dynamic cell membrane remodelling. Nature 438:590–596
26. Eden S, Rohatgi R, Podtelejnikov AV et al (2002) Mechanism of regulation of WAVE1-induced actin nucleation by Rac1 and Nck. Nature 418:790–793
27. Peleg B, Disanza A, Scita G, Gov N (2011) Propagating cell-membrane waves driven by curved activators of actin polymerization. PLoS ONE 6:e18635

Chapter 18
CREB Regulated Transcription Coactivator 1 (CRTC1) Interacts with Microtubules

Liqiao Ma, Yu Sun, Baoxia Zhang, Yuyin Li, Aipo Diao and Yinchuan Li

Abstract CRTCs are found to be coactivators of Creb1 that constitutively regulate cell energy balance and longevity by regulating glucose metabolism. To accomplish this, dephosphorylated CRTCs translocate to nuclear to enhance the activation of Creb1 to activate gluconeogenesis. Recently, CRTC1 was reported to be located in hippocampal neuron synapses and dendrites with long distance transport along neurites. In this report, we found that CRTC1 was distributed mostly in the cytosol along microtubules and was also able to target membrane ruffles, which suggests CRTC1 may relay signals far from cell periphery region and possibly directly from plasma membrane to cell nuclei. However, its mechanism is still unknown. Our result shows that CRTC1 binds to polymerized tubulins, which implies that signal transduction of CRTC1 complex may be mediated via microtubules. This finding helps to further understand the detailed role of CRTC1 in cell cytosol.

Keywords CRTC1 · Tubulin · Microtubules

18.1 Introduction

The CRTC family has three members: CRTC1, CRTC2, and CRTC3. CRTC1, also named as MECT1 or TORC1, initially found from a mucoepidermoid carcinoma translocation where exon 1 of CRTC1 fuses with exons 2–5 of MAML2 to form a novel oncogene as MECT1–MAML2 [1]. Lately, CRTC1 is found to enhance Creb1 activity by binding to its bZIP DNA binding/dimerization domain to facilitate the interaction of Creb with TFIID on target gene promoters [2, 3].

L. Ma · Y. Sun · B. Zhang · Y. Li · A. Diao (✉) · Y. Li (✉)
Key Lab of Industrial Fermentation Microbiology of the Ministry of Education, College of Bioengineering, Tianjin University of Science and Technology, Tianjin 300457, China
e-mail: diaoaipo@tust.edu.cn

Y. Li
e-mail: ycli@tust.edu.cn

T.-C. Zhang and M. Nakajima (eds.), *Advances in Applied Biotechnology*, Lecture Notes in Electrical Engineering 332, DOI 10.1007/978-3-662-45657-6_18

As coactivator of Creb1, CRTCs shuttle between cytoplasm and nucleus controlled by hormone, calcium, or synaptic stimulation [3–8]. CRTCs usually augment calcium, cAMP signals, and also enhance insulin resistance [9]. The role of CRTCs in nuclei has been in focus for a long time, but its role in cytosol is still unknown. Recently, CRTC1 has been reported to reach synapses of neurons and mediate the long distance signal transduction from synapses to nuclei [5, 7, 10]. But how CRTC1 signal complexes are transported is still unknown. In this report, we obtained evidence indicating that CRTC1 may deliver signals via microtubules.

18.2 Materials and Methods

18.2.1 Cell Culture, Antibodies and Reagents

Hela cells and 293T cell line were cultured in Dulbecco's modified Eagle's medium, supplemented with 10 % fetal bovine serum and 1 % penicillin and streptomycin at 37 °C in a CO_2 incubator. TurboFect transfection reagent was purchased from Thermo Scientific. The mouse monoclonal antibodies against β-Tubulin and HA tag, and rabbit polyclonal antibodies against GST tag were from Sigma Aldrich. HRP conjugated goat anti-mouse IgG and HRP conjugated goat anti-rabbit IgG were purchased from ABGENT. Alexa Fluor 555 donkey anti-mouse IgG and Alexa Fluor 488 donkey anti-rabbit IgG were from Invitrogen. Glutathione Sepharose 4B beads were from GE Health. All other biological or chemical reagents were from Thermo Scientific, Shanghai Sangon, Roche, or Sigma Aldrich.

18.2.2 Construction of Plasmids

Expression plasmid for HA-tagged tubulin (pHA-β-Tubulin) was constructed by amplifying β-Tubulin using the primers listed in Table 18.1 via RT-PCR. The fragments were digested with *Bam*HI and *Not*I and then inserted into the pcDNA3.1+ vector with an HA tag on its N-terminus. Human CRTC1 was RT-PCR amplified from 293T cells with primers listed in Table 18.1 and was inserted into

Table 18.1 Primers used for PCR experiments

Target gene	Primer sequence
Tubulin	
Forward	5'—CGGATCCCGTGAGTGCATCTCCATC-3'
Reverse	5'—CGGCGGCCGCGTATTCCTCTCCTTCTTC-3'
CRTC1	
Forward	5'—TTAGAATTCGGATCCGGCTCAGGTTCTGCGACTTCGAAC AATCCGCGGAAATTCAGC-3'
Reverse	5'—AGCTCTAGAGCGGCCGCCAGGCGGTCCATCCGGA-3'

pcDNA3.1+ vector in frame with GST tag on its C-terminus to form pCRTC1-GST. The same fragment was inserted into the same vector with HA tag on its N-terminus to obtain pHA-CRTC1 construct.

18.2.3 Immunofluorescence Staining Procedure

For immunofluorescence staining, Hela cells grown on gelatin coated coverslips were transfected with the indicated plasmids for 24 h, and fixed for 5 min with ice-cold methanol. Cover glasses with cells were blocked with 3 % BSA in PBS for 30 min. Cells were incubated with HA tag and β-Tubulin antibodies, washed and then incubated for 30 min with appropriate secondary antibodies conjugated with Alexa Fluor fluorochrome plus Hoechst 33342 to visualize nuclei. An Olympus FV-1000 confocal microscope was used to capture images. The images were then disposed with FV10-ASW 3.0 Viewer and Image J.

18.2.4 GST Pull-Down Assays and Western Blotting

HEK293 cells were transfected with plasmids encoding HA-tubulin, CRTC1-GST, or pEBG empty vector encoding GST with combinations as indicated in the below results. After 24 h, cells were lysed in lysis buffer (20 mM Tris–HCl, pH 7.4, 100 mM NaCl, 0.2 % Triton X-100, 5 % glycerol, 1 mM $MgCl_2$ and 1 mM DTT) and supplemented with protease inhibitor cocktails plus PMSF without EDTA. After being centrifuged at 16,000 g for 15 min, the supernatant was added to Glutathione Sepharose 4B beads (GST beads), incubating all samples gently for 4 h at 4 °C. Then the supernatant was discarded and the GST beads were washed thrice with ice cold lysis buffer and washed once with ice cold 1 × PBS. Finally, GST fusion protein CRTC1 was eluted by reduced glutathione. Samples were then subjected to western blot with anti-Tubulin antibody and the corresponding HRP conjugated secondary antibodies.

18.3 Results

18.3.1 Subcellular Localization of CRTC1

CRTCs have been found to shuttle between cytosol and nuclei. HA-CRTC1s were mainly located in cell cytoplasm with some located in membrane ruffles (arrows), but relatively few CRTC1 were found in nuclei. Microtubules were found to be partially colocalized with HA-CRTC1s (red) in cytosol (Fig. 18.1).

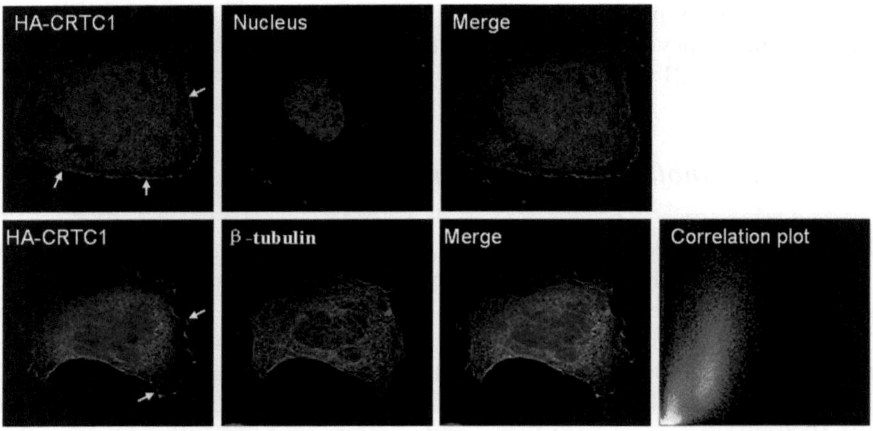

Fig. 18.1 More CRTC1 is distributed in cell cytoplasm and less in nuclear with partial colocalization with β-tubulins. The plasmid encoding HA-CRTC1 was transfected into Hela cells for 24 h before being fixed. Cells were stained with anti-HA antibody (*upper panel*) and antibody against endogenous β-tubulins (*lower panel*). Representative cell images were picked. Nuclei were stained with Hoechst 33342. Colocalization correlation plot was generated by Image J. *Arrows* show cell plasma membrane and membrane ruffles

18.3.2 CRTC1 Interacts with Tubulin

The plasmid encoding CRTC1-GST and plasmid encoding HA-β-Tubulin were transfected into HEK293 cells to perform in vivo GST pull-down assay of the interaction of CRTC1 with Tubulins. pEBG vector encoding GST and pHA-β-Tubulin were co-transfected as negative control. Result revealed that CRTC1-GST but not GST interacted with HA-Tubulin (Fig. 18.2a). We then checked the interaction of CRTC1-GST with endogenous tubulin. CRTC1-GST was able to pull down endogenous tubulin but GST control did not (Fig. 18.2b). We then determined whether the interaction was mediated by polymerized tubulins or by free tubulin subunits by treated cells with 100 nM taxol for 3 h to stabilize microtubules at 37 °C before extracting cell total proteins. In cells pretreated with taxol, tubulin showed better binding capability to CRTC1-GST than in untreated sample (Fig. 18.2c). This result implies that CRTC1 binds to microtubules rather than free tubulin proteins, but we still do not know whether the binding is direct or not. So it still needs further detailed investigation.

18.4 Discussion

The binding of CRTC1 to polymerized tubulins implies two possibilities. One is that CRTC1 might be a structural component of microtubules; the other possibility is that CRTC1 is a kind of cargo being transported by microtubules. The former is

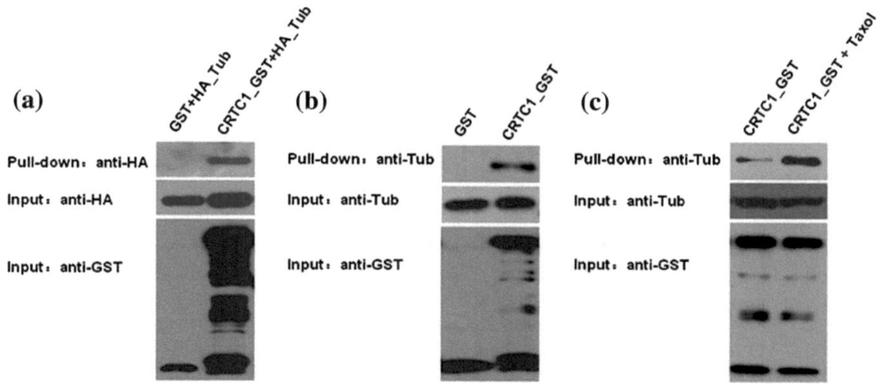

Fig. 18.2 In vivo GST pull-down assay showed CRTC1-GST bound to β-Tubulins. **a** pEBG empty vector with HA-β-Tubulin (negative control) and plasmids encoding CRTC1-GST with plasmids encoding HA-β-Tubulin were co-transfected respectively into 293T cells to monitor the binding of CRTC1 to HA-β-Tubulin by GST pull-down assay. Pull-down products were blotted with anti-HA tag antibody to stain HA-β-Tubulin. **b** pEBG empty vector alone and plasmids encoding CRTC1-GST were transfected respectively to pull down the endogenous β-Tubulin. Pull-down products were blotted with anti-β-Tubulin antibody. **c** CRTC1-GST plasmids were transfected into 293T cells, and 3 h before harvesting cells, one sample was treated with 100 nM taxol to stabilize microtubules. Pull-down products were blotted with anti-β-Tubulin antibody as in **b**. anti-Tub, anti-β-Tubulin antibody. HA-Tub, plasmids encoding HA-β-Tubulin

less likely since no evidence supports this point. The possibility of CRTC1 as part of an unknown signalosome transported along microtubules is very likely, especially in neurite [7, 10, 11]. Several motor proteins can be used for cell cargo transport and dyneins and kinesins are the two main cargo transporting motors along microtubule [12]. Different motor proteins have distinct cargoes with unique transporting directions [13]. So next we should focus more on which kind of motor proteins and which direction is used by CRTC1 transport along microtubules.

Too much focus has been made on the nuclei translocation of CRTCs for transcription activation in previous reports [4]. Our results show, in most circumstances, CRTC1 is distributed throughout the cytosols and also on cell membrane ruffles. This implies the distribution of CRTC1 in the juxta-plasma membrane region may also take roles. An express way for CRTC1 signal may exist from plasma membrane to nuclei. But how this signal is relayed is still unknown, though it is found this process needs the dephosphorylation of CRTCs and dissociation from 14-3-3 [13]. A preliminary try to link calcium signal relay and CRTC1 pathway was made by direct linking of CRTC1 with IP3 receptor in ER [8]. However, the knowledge of ER calcium releasing via IP3R regulated by CRTC1 still has many limitations in explaining the role of CRTCs in cytosol to relay signals from synapse to nuclei. Our finding of the CRTC1 binding to microtubules may provide some clues to this signal transduction process.

Acknowledgments This work was supported by the National Natural Science Foundation of China (No. 31271448) and by Research Start-up fund of Tianjin University of Science and Technology.

References

1. Tonon G, Modi S, Wu L, Kubo A, Coxon AB, Komiya T, O'Neil K, Stover K, El-Naggar A, Griffin JD et al (2003) t(11;19)(q21;p13) translocation in mucoepidermoid carcinoma creates a novel fusion product that disrupts a Notch signaling pathway. Nat Genet 33:208–213
2. Conkright MD, Canettieri G, Screaton R, Guzman E, Miraglia L, Hogenesch JB, Montminy M (2003) TORCs: transducers of regulated CREB activity. Mol Cell 12:413–423
3. Jansson D, Ng AC, Fu A, Depatie C, Al Azzabi M, Screaton RA (2008) Glucose controls CREB activity in islet cells via regulated phosphorylation of TORC2. Proc Natl Acad Sci USA 105:10161–10166
4. Altarejos JY, Montminy M (2011) CREB and the CRTC co-activators: sensors for hormonal and metabolic signals. Nat Rev Mol Cell Biol 12:141–151
5. Kovacs KA, Steullet P, Steinmann M, Do KQ, Magistretti PJ, Halfon O, Cardinaux JR (2007) TORC1 is a calcium- and cAMP-sensitive coincidence detector involved in hippocampal long-term synaptic plasticity. Proc Natl Acad Sci USA 104:4700–4705
6. Screaton RA, Conkright MD, Katoh Y, Best JL, Canettieri G, Jeffries S, Guzman E, Niessen S, Yates JR 3rd, Takemori H et al (2004) The CREB coactivator TORC2 functions as a calcium- and cAMP-sensitive coincidence detector. Cell 119:61–74
7. Riera CE, Huising MO, Follett P, Leblanc M, Halloran J, Van Andel R, de Magalhaes Filho CD, Merkwirth C, Dillin A (2014) TRPV1 pain receptors regulate longevity and metabolism by neuropeptide signaling. Cell 157:1023–1036
8. Wang Y, Li G, Goode J, Paz JC, Ouyang K, Screaton R, Fischer WH, Chen J, Tabas I, Montminy M (2012) Inositol-1,4,5-trisphosphate receptor regulates hepatic gluconeogenesis in fasting and diabetes. Nature 485:128–132
9. Wang Y, Inoue H, Ravnskjaer K, Viste K, Miller N, Liu Y, Hedrick S, Vera L, Montminy M (2010) Targeted disruption of the CREB coactivator Crtc2 increases insulin sensitivity. Proc Natl Acad Sci USA 107:3087–3092
10. Ch'ng TH, Uzgil B, Lin P, Avliyakulov NK, O'Dell TJ, Martin KC (2012) Activity-dependent transport of the transcriptional coactivator CRTC1 from synapse to nucleus. Cell 150:207–221
11. Jagannath A, Butler R, Godinho SI, Couch Y, Brown LA, Vasudevan SR, Flanagan KC, Anthony D, Churchill GC, Wood MJ et al (2013) The CRTC1-SIK1 pathway regulates entrainment of the circadian clock. Cell 154:1100–1111
12. Vale RD (2003) The molecular motor toolbox for intracellular transport. Cell 112:467–480
13. Hook P, Vallee RB (2006) The dynein family at a glance. J Cell Sci 119:4369–4371

Chapter 19
The Biological Effects of Carbon Nanotubes in Plasma Membranes Damage, DNA Damage, and Mitochondrial Dysfunction

Zhuo Zhao, Zhi-Peng Liu, Hua Wang, Feng-Juan Liu, Hui Zhang, Cong-Hui Zhang, Chen-Guang Wang and Xiao-Chuan Jia

Abstract In the last ten years, accompanied by deep development of nanotechnology, a large number of nanomaterials have come into people's horizons. Carbon nanotubes (CNTs), one of the important nanometer materials, have been widely applied to various biomedical applications such as cancer photothermal therapy, specific drugs delivery, and so on due to their unique physicochemical property. Along with applications of CNT, the relationship between the biotoxicity of CNT and health was widely concerned. This review provides a general overview of the biological damage of overexposure of single-walled carbon nanotubes (SWCNTs) and mutiwalled carbon nanotubes (MWCNTs) and discusses some of the challenges associated with CNTs toxicity.

Keywords CNT · Biosafety · Nanomaterials · Nanotechnology

19.1 Introduction

As the units of length, 1 nm is equal to 1.0×10^{-9} m. In the fields of nanotechnology, the means of nanoscale is equal to or less than 100 nm. This nanoscale is close or less than physical characteristics length of electron [1, 2]. As is well

Z. Zhao and Z.-P. Liu had contributed equally to this work.

Z. Zhao · Z.-P. Liu (✉) · H. Wang · F.-J. Liu · H. Zhang · C.-H. Zhang
C.-G. Wang · X.-C. Jia
Technical Center for Safety of Industrial Products, Tianjin Entry-Exit Inspection Quarantine Bureau, No. 2 Dong Wu Road Airport Economics Area, Tianjin 300308, China
e-mail: andelu824777@163.com

T.-C. Zhang and M. Nakajima (eds.), *Advances in Applied Biotechnology*,
Lecture Notes in Electrical Engineering 332, DOI 10.1007/978-3-662-45657-6_19

known, compared with micromaterial, the substance within nanoscale shows completely different electrical, mechanical, optical, chemical, and biological characteristics such as black nanosized golden particles [3, 4]. Actually, many natural biological macromolecules such as protein, DNA, and some plasma membranes all belong to natural nanomaterials [5, 6].

Carbon nanotube, a new typed carbon crystal, is firstly discovered by Iijima in (1991) in the process of synthesizing carbon fiber by electric arc method. Hexagonal-grid fake graphite layer is curled up into a seamless nanopipe and the paired ends are closed by pentagon or heptagon, that's CNT. According to composed amount of flake graphite layers, CNT is divided into single-walled CNT and mutiwalled CNT (Fig. 19.1). The diameter of SWCNT is between 0.4 and 2.5 nm, but the diameter of SWCNT is various and is mainly determined by manufacture conditions (maximum diameter is up to 100 nm). The length of CNT is in a few to a few hundred nanometers. The unique structures of CNT make it have many excellent physical and chemical properties and furthermore make it have a wide range of industrial and medical purposes. In present, the research field of CNT almost focuses on its biomedical applications, but is CNT completely safe? Now, some researches have been reported that overexposure of CNT could induce much biological damage due to its significant physicochemical property.

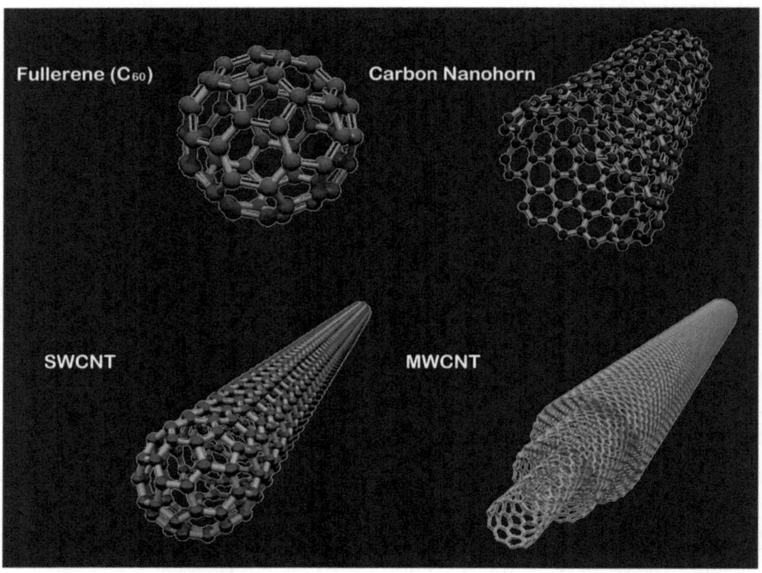

Fig. 19.1 Structures of various important carbon nanomaterials: fullerene (C_{60}), single-walled carbon nanotube (*SWCNT*), multiwalled carbon nanotube (*MWCNT*), carbon nanohorn, graphene, few layer graphene and graphene oxide. Adapted from [7]

In this paper, we will systematically review the biological toxicity of overexposure of CNT in vivo and in vitro such as plasma membranes damage, DNA damage, mitochondrial dysfunction, and so on.

19.2 Plasma Membranes Damage

It is important to understand the interactions of CNTs with cells and their components, and especially with cell membranes. As is well known, human cells can ingest both functionalized [8] and natural CNTs [9]. Therefore, more understanding of the fundamental interactions between CNTs and cell components is required to potential risks to health.

With regard to toxicity of CNT, several studies exist dealing with plasma membranes damage in eukaryotes, but there are limited data concerning human genetics. It has been reported that the overexposure to a high concentration of CNT (30 mg/m^3) could cause subpleural fibrosis in mice, while a low concentration (1 mg/m^3) could not [10]. Ma-Hock's and his collaborates thought the inhalation biotoxicity caused by CNT is due to the burden on the lung [11]. Additionally, some researchers reported that inhalation of CNTs induces chronic inflammation of the lungs and pleura, resulting the development of pulmonary fibrosis and pleural mesothelioma [12, 13]. What's more, Poland and his collaborator found that the pathogenicity of CNT is very similar to that of asbestos-induced mesothelioma [14]. Takagi had also showed that intraperitoneal administration of MWCNTs to mice (p53±) could develop mesothelioma and the effects of MWCNTs are similar to the effects of crocidolite [15]. So far, the mechanism of CNT-induced mesothelioma development is still unclear although some one reported that the upregulated expression of inflammatory cytokines is stimulated by CNTs [16, 17]. Together, these finds indicate that CNTs could affect macrophage function and this effect may a switch of the development of fibrosis and mesothelioma stimulated by CNTs.

In the present studies, Kosuke and his cooperators demonstrated that exposure of MWCNTs could cause biomembrane damage of macrophage, and this find reveal a possible mechanism of CNT-induced fibrosis and mesothelioma [18]. According to the experiment of LDH release assay, they demonstrated that overexposure of MWCNTs (100 μg/ml) could acute death of macrophages and they considered the reason is due largely to physically dysfunction of plasma membrane through the interaction between MWCNTs and lipid bilayer of membrane. In particular, they also demonstrated that MWCNTs could directly interact with lipid bilayer by using liposomes as a biomembrane and these results are consistent with Mickaël Lelimousin's study [19]. The present studies indicated that direct biomembrane perturbation stimulated by CNTs is a route of inducing cell death [18].

Compared with fullerene C_{60}, CNTs could cause the drastic perturbation of the liposomal membrane despite fullerene C_{60} could directly interact with lipid bilayer too. The different effects between fullerene C_{60} and CNTs may be caused by various shapes of nanomaterials. Mickaël Lelimousin concluded that after insertions

CNTs could remain within the membrane for the duration of the simulations and do not spontaneously exit to either face of the membrane indicating that active endocytosis is easier rather than passive physical penetration in the process of CNTs entering into mammalian cell [20, 21]. In addition, Hirano demonstrated that CNTs induce dysfunction of the plasma membrane through reacting with collagenous structure [22]. But to date, it is still unclear whether the perturbation of biomembranes by CNT directly caused the cell death of the macrophages, because some reports suggest that CNT exposure induces the production of reactive oxygen species (ROS) and further acute cell death [23, 24].

Together, some researchers found that CNTs could directly cause the liposomal membranes damage and suggested that the interaction of CNTs on biomembrane is a potential trigger of causing cell death, inflammation, pulmonary and pleural mesothelioma.

19.3 DNA Damage

The biotoxicity of CNTs in vitro has also been reported in several types of cells such as macrophages, epithelial cells, mesothelial cells, and keratinocytes [16, 22, 25–30]. From these reports, we can obviously concluded CNTs are cytotoxic under overexposure. In present, many of investigations indicated that the genotoxicity of carbon nanotubes is mediated by DNA damage through generating reactive oxygen species (ROS) and reactive nitrogen species (RNS) that interact with DNA bases to form oxidative and nitrative DNA lesions [31, 32].

According to the comet assay and micronucleus test, explore CNT can induce DNA damage [33–36]. Meanwhile, Zhu et al. have been reported that the expression of DNA repair enzymes (p53, OGG1, Rad51) was upregulated by CNTs [37]. It is well known that 8-Nitroguanine, a mutagenic nitrative DNA lesion formed during chronic inflammation, is formed from the interaction between guanine and peroxynitrite. Recently, Guo et al. reported that MWCNT could induce the formation of 8-nitroguanine in a lung epithelial cell line through caveolae- and clathrin-mediated endocytosis. Furthermore, they found that MWCNT particles could increase iNOS expression and RNS generation in inflammatory reactions indicating that MWCNT play a potential role in DNA damage and carcinogenesis [38].

Recently, MWCNTs have been shown to generate reactive oxygen species (ROS), induce oxidative stress, and exert toxic effects [26, 39–41]. As we know, ROS such as hydroxyl radical and oxygen radical are high-risk factors to DNA damage and can cause the oxidation of DNA, DNA strand breaks, 8-hydroxyguanine formation, and lipid peroxidation-mediated DNA adducts [42]. Now, the generation of ROS is considered to be a major mechanism in the induction of genotoxicity by CNTs although other mechanisms may also be relevant to interference with DNA, chromatin structures, or the mitotic apparatus. Additionally, Ye et al. reported that MWCNTs could increase the nucleus translocation and accumulation of p65 in A549 cells through degrading IκBα [26]. Based on their research, we can conclude that MWCNTs can activate NF-κB signal pathway and promote NF-κB binding activity. Especially, the consistent conclusions are obtained in human keratinocyte HaCaT cells [43, 44]. In the meantime, Ye et al. also found that after activation of NF-κB signal transduction stimulated by MWCNTs, the release and expression of IL-8 were enhanced suggesting that MWCNT could promote cytokine production through activating NF-κB signal transduction. In addition, it is noteworthy that many of cytokines such as tumor necrosis factor alpha (TNF-α), IL-1β, IL-6, IL-8, iNOS, and so on are regulated by NF-κB signal transduction in the process of inflammatory response. In vivo studies, abundant experimental evidences demonstrate that CNTs increase cytokines release and inducing inflammatory reactions through NF-κB signal pathway suggesting that CNTs may contribute to the carcinogenesis [26, 34, 45].

The long thin tubular-shaped carbon nanotubes are very similarity to cellular microtubules. In other words, they have a striking similarity on the function suggesting that there is a potential interaction between CNT and cellular biomolecules, such as mitotic spindle and motor proteins which play a crucial role in the chromosomes separating and cell division [46]. The disruption of centrosomes and mitotic spindles could result in monopolar, tripolar, and quadrupolar divisions of chromosomes and aneuploidy. That has been considered to be a key mechanism in the potential carcinogenesis.

In summary, MWCNTs are able to induce DNA damage such as DNA strand breakage, 8-hydroxyguanine and 8-Nitroguanine adduct formation, upregulating DNA response gene and altered chromosomes separating and cell division through inducing ROS and RNS release. Furthermore, oxidative stress mediated by CNTs induces inflammation response and in this progress cytokines production induced by NF-κB activation seem to play an important role. Altogether, CNTs play a potential role in the proinflammatory and carcinogenesis through CNTs-induced DNA damage (Fig. 19.2).

19.4 Mitochondrial Dysfunction

It is known that CNTs could be able to enter the organelles due to their unique tiny structures. In present, the research of mitochondrial dysfunctions caused by CNTs gradually becomes very hot. In previous studies, the interaction between biological macromolecules and carbon nanomaterials has been reported to cause structural and functional changes of protein [47–52]. Additionally, the report showed that the enzymatic activity of α-chymotrypsin was directly inhibited by functionalized CNTs under the condition that CNTs recognize and bind to the catalytic sites [51]. In a word, many of experimental evidence demonstrated that CNTs could modify the function of cell factors through direct and/or indirect interactions.

Mitochondria, a vital and lethal organelle in cells, have such interesting roles, and after the dysregulation of mitochondria cells will implicate in pathological conditions such as cancer, neurodegenerative and neuromuscular diseases, diabetes,

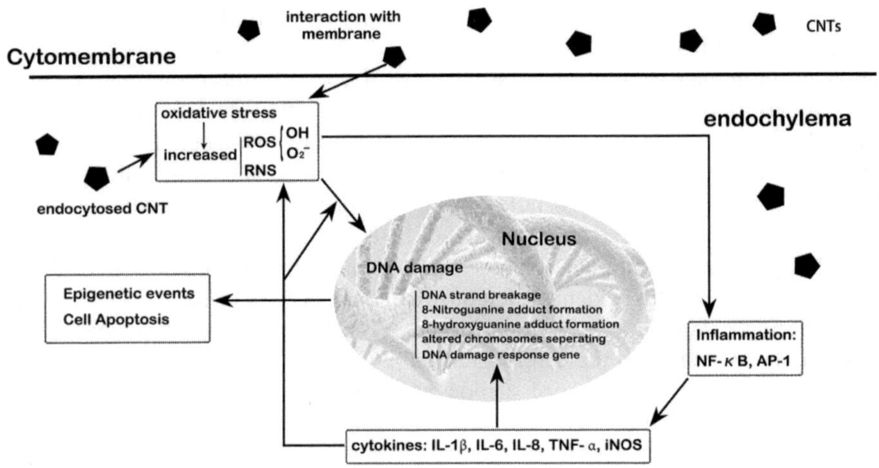

Fig. 19.2 The schematic diagram of CNTs-induced DNA damage

and so on [53–55]. Zhou et al. firstly demonstrated that SWNT-PL-PEG could localize in the mitochondria of cells and the translocation resulted in the modification of mitochondrial transmembrane potential [56]. Along with the further studies, many of reports have also shown that CNT including SWCNT and MWCNT could localize exclusively in mitochondria of both tumor and normal cells and induce cytochrome c release and furthermore induce the mitochondrial dysfunctions [57–62]. In mitochondrial dysfunctions mediated by CNT, the releasing of cytochrome c is the key linkage. As we know, cytochrome c is a crucial component of mitochondrial respiratory chain which exists in cytosol between the inner and outer membranes of mitochondria. In fact, cytochrome c plays an important role in the biological respiratory chain and is essential for mitochondrial function as well. Furthermore, the biological activity of cytochrome c is associated with many biological progresses such as cellular proliferation, differentiation, apoptosis, and cell cycle [63]. Meanwhile, some studies in present reveal that mitochondrial pathway induced by CNT, especially functionalized MWCNT, can change the cellular redox state and leads to the opening of mitochondrial permeability transition pore (PTP) [64, 65]. Generally, the activation of mitochondrial pathway mediated by CNT is usually considered to be the result of that oxidative stress is triggered and induce the release of cytochrome c from mitochondria to cytosol [66]. After releasing, together with Apaf-1 and pro-caspase-9, cytochrome c can form the apoptosome that activates caspase-9 and thus promotes subsequent activation of a cascade of caspases including caspase-3, leading to cell death [67].

All in all, the fact of translocation of CNTs into mitochondria is beyond dispute and CNT can induce mitochondrial dysfunction mainly through decreasing mitochondrial transmembrane potential and release cytochrome c and therefore influence multiple biological progresses.

19.5 Discussion

Carbon nanotubes have a wide impact on our daily life by providing various useful products and solutions to global problems. Still, the evaluation system between human health and natural environment and CNTs is not completely established. Therefore, it is vital to investigate the biosafety of CNTs for using in our daily life, especially for using of drug delivery. In accordance with recently published studies, there are still many gaps in available experimental data devoted to risk assessment of CNTs, even it is controversy about CNTs biotoxicity.

In the present studies, the researches about potential hazards of CNTs were mainly focused on the inhalation and epidermal exposure. However, according to the latest reports, CNTs are able to penetrate beyond the bronchus deep into the alveolar region of the lung after inhalation, and then translocate into the blood circulatory system and distributed in the entire body [68]. Functionalized CNT is a kind of potential drug delivery carriers, and thus might be exposed to human body through intravenous injection and accumulated in the liver [69, 70] Therefore, it is pivotally important to investigate the effect of CNT on the liver cell, which has been scarcely addressed thus far.

Together, in this paper, we reviewed the biotoxicity of CNTs in plasma membranes damage, DNA damage, and mitochondrial dysfunction. We proposed these finds might provide better basis for the human and environmental risk researches induced by CNTs.

References

1. Lucovsky G, Phillips JC (2010) Nano-regime length scales extracted from the first sharp diffraction peak in non-crystalline SiO(2) and related materials: device applications. Nanoscale Res Lett 5:550–558
2. Huang T, Nancy XX (2010) Synthesis and characterization of tunable rainbow colored colloidal silver nanoparticles using single-nanoparticle plasmonic microscopy and spectroscopy. J Mater Chem 20:9867–9876
3. Baughman RH, Zakhidov AA, de Heer WA (2002) Carbon nanotubes–the route toward applications. Science 297:787–792
4. Stoner BR, Brown B, Glass JT (2014) Selected topics on the synthesis, properties and applications of multiwalled carbon nanotubes. Diam Relat Mater 42:49–57
5. Ye P, Zhang W, Yang T et al (2014) Folate receptor-targeted liposomes enhanced the antitumor potency of imatinib through the combination of active targeting and molecular targeting. Int J Nanomed 9:2167–2178
6. Nagy L, Magyar M, Szabo T et al (2014) Photosynthetic machineries in nano-systems. Curr Protein Pept Sci 15:363–373
7. Uo M, Akasaka T, Watari F et al (2011) Toxicity evaluations of various carbon nanomaterials. Dent Mater J 30:245–263
8. Kam NW, Dai H (2005) Carbon nanotubes as intracellular protein transporters: generality and biological functionality. J Am Chem Soc 127:6021–6026
9. Porter AE, Gass M, Muller K et al (2007) Direct imaging of single-walled carbon nanotubes in cells. Nat Nanotechnol 2:713–717

10. Ryman-Rasmussen JP, Cesta MF, Brody AR et al (2009) Inhaled carbon nanotubes reach the subpleural tissue in mice. Nat Nanotechnol 4:747–751
11. Ma-Hock L, Treumann S, Strauss V et al (2009) Inhalation toxicity of multiwall carbon nanotubes in rats exposed for 3 months. Toxicol Sci 112:468–481
12. Mitchell LA, Gao J, Wal RV et al (2007) Pulmonary and systemic immune response to inhaled multiwalled carbon nanotubes. Toxicol Sci 100:203–214
13. Chou CC, Hsiao HY, Hong QS et al (2008) Single-walled carbon nanotubes can induce pulmonary injury in mouse model. Nano Lett 8:437–445
14. Poland CA, Duffin R, Kinloch I et al (2008) Carbon nanotubes introduced into the abdominal cavity of mice show asbestos-like pathogenicity in a pilot study. Nat Nanotechnol 3:423–428
15. Takagi A, Hirose A, Nishimura T et al (2008) Induction of mesothelioma in p53± mouse by intraperitoneal application of multi-wall carbon nanotube. J Toxicol Sci 33:105–116
16. Jia G, Wang H, Yan L et al (2005) Cytotoxicity of carbon nanomaterials: single-wall nanotube, multi-wall nanotube, and fullerene. Environ Sci Technol 39:1378–1383
17. Montes-Fonseca SL, Orrantia-Borunda E, Aguilar-Elguezabal A et al (2012) Cytotoxicity of functionalized carbon nanotubes in J774A macrophages. Nanomedicine 8:853–859
18. Shimizu K, Uchiyama A, Yamashita M et al (2013) Biomembrane damage caused by exposure to multi-walled carbon nanotubes. J Toxicol Sci 38:7–12
19. Lelimousin M, Sansom MS (2013) Membrane perturbation by carbon nanotube insertion: pathways to internalization. Small 9:3639–3646
20. Firme CR, Bandaru PR (2010) Toxicity issues in the application of carbon nanotubes to biological systems. Nanomedicine 6:245–256
21. Yaron PN, Holt BD, Short PA et al (2011) Single wall carbon nanotubes enter cells by endocytosis and not membrane penetration. J Nanobiotechnol 9:45
22. Hirano S, Kanno S, Furuyama A (2008) Multi-walled carbon nanotubes injure the plasma membrane of macrophages. Toxicol Appl Pharmacol 232:244–251
23. Herzog E, Byrne HJ, Davoren M et al (2009) Dispersion medium modulates oxidative stress response of human lung epithelial cells upon exposure to carbon nanomaterial samples. Toxicol Appl Pharmacol 236:276–281
24. Pulskamp K, Diabate S, Krug HF (2007) Carbon nanotubes show no sign of acute toxicity but induce intracellular reactive oxygen species in dependence on contaminants. Toxicol Lett 168:58–74
25. Kagan VE, Tyurina YY, Tyurin VA et al (2006) Direct and indirect effects of single walled carbon nanotubes on RAW 264.7 macrophages: role of iron. Toxicol Lett 165:88–100
26. Ye SF, Wu YH, Hou ZQ et al (2009) ROS and NF-kappaB are involved in upregulation of IL-8 in A549 cells exposed to multi-walled carbon nanotubes. Biochem Biophys Res Commun 379:643–648
27. Wick P, Manser P, Limbach LK et al (2007) The degree and kind of agglomeration affect carbon nanotube cytotoxicity. Toxicol Lett 168:121–131
28. Pacurari M, Yin XJ, Zhao J et al (2008) Raw single-wall carbon nanotubes induce oxidative stress and activate MAPKs, AP-1, NF-kappaB, and Akt in normal and malignant human mesothelial cells. Environ Health Perspect 116:1211–1217
29. Shvedova AA, Castranova V, Kisin ER et al (2003) Exposure to carbon nanotube material: assessment of nanotube cytotoxicity using human keratinocyte cells. J Toxicol Environ Health A 66:1909–1926
30. Monteiro-Riviere NA, Nemanich RJ, Inman AO et al (2005) Multi-walled carbon nanotube interactions with human epidermal keratinocytes. Toxicol Lett 155:377–384
31. Hussain SP, Hofseth LJ, Harris CC (2003) Radical causes of cancer. Nat Rev Cancer 3:276–285
32. Ohshima H, Tatemichi M, Sawa T (2003) Chemical basis of inflammation-induced carcinogenesis. Arch Biochem Biophys 417:3–11
33. Karlsson HL, Cronholm P, Gustafsson J et al (2008) Copper oxide nanoparticles are highly toxic: a comparison between metal oxide nanoparticles and carbon nanotubes. Chem Res Toxicol 21:1726–1732

34. Kisin ER, Murray AR, Sargent L et al (2011) Genotoxicity of carbon nanofibers: are they potentially more or less dangerous than carbon nanotubes or asbestos? Toxicol Appl Pharmacol 252:1–10
35. Lindberg HK, Falck GC, Suhonen S et al (2009) Genotoxicity of nanomaterials: DNA damage and micronuclei induced by carbon nanotubes and graphite nanofibres in human bronchial epithelial cells in vitro. Toxicol Lett 186:166–173
36. Yamashita K, Yoshioka Y, Higashisaka K et al (2010) Carbon nanotubes elicit DNA damage and inflammatory response relative to their size and shape. Inflammation 33:276–280
37. Zhu L, Chang DW, Dai L et al (2007) DNA damage induced by multiwalled carbon nanotubes in mouse embryonic stem cells. Nano Lett 7:3592–3597
38. Guo F, Ma N, Horibe Y et al (2012) Nitrative DNA damage induced by multi-walled carbon nanotube via endocytosis in human lung epithelial cells. Toxicol Appl Pharmacol 260:183–192
39. Di Giorgio ML, Di Bucchianico S, Ragnelli AM et al (2011) Effects of single and multi walled carbon nanotubes on macrophages: cyto and genotoxicity and electron microscopy. Mutat Res 722:20–31
40. Guo YY, Zhang J, Zheng YF et al (2011) Cytotoxic and genotoxic effects of multi-wall carbon nanotubes on human umbilical vein endothelial cells in vitro. Mutat Res 721:184–191
41. Patlolla AK, Berry A, Tchounwou PB (2011) Study of hepatotoxicity and oxidative stress in male Swiss-Webster mice exposed to functionalized multi-walled carbon nanotubes. Mol Cell Biochem 358:189–199
42. Migliore L, Saracino D, Bonelli A et al (2010) Carbon nanotubes induce oxidative DNA damage in RAW 264.7 cells. Environ Mol Mutagen 51:294–303
43. Manna SK, Sarkar S, Barr J et al (2005) Single-walled carbon nanotube induces oxidative stress and activates nuclear transcription factor-kappaB in human keratinocytes. Nano Lett 5:1676–1684
44. Sharma CS, Sarkar S, Periyakaruppan A et al (2007) Single-walled carbon nanotubes induces oxidative stress in rat lung epithelial cells. J Nanosci Nanotechnol 7:2466–2472
45. Davoren M, Herzog E, Casey A et al (2007) In vitro toxicity evaluation of single walled carbon nanotubes on human A549 lung cells. Toxicol In Vitro 21:438–448
46. Liu Y, Zhao Y, Sun B et al (2013) Understanding the toxicity of carbon nanotubes. Acc Chem Res 46:702–713
47. Prato M, Kostarelos K, Bianco A (2008) Functionalized carbon nanotubes in drug design and discovery. Acc Chem Res 41:60–68
48. Bayraktar H, Ghosh PS, Rotello VM et al (2006) Disruption of protein-protein interactions using nanoparticles: inhibition of cytochrome c peroxidase. Chem Commun (Camb) 7:1390-2
49. Karajanagi SS, Vertegel AA, Kane RS et al (2004) Structure and function of enzymes adsorbed onto single-walled carbon nanotubes. Langmuir 20:11594–11599
50. Yi C, Fong CC, Zhang Q et al (2008) The structure and function of ribonuclease a upon interacting with carbon nanotubes. Nanotechnology 19:095102
51. Zhang B, Xing Y, Li Z et al (2009) Functionalized carbon nanotubes specifically bind to alpha-chymotrypsin's catalytic site and regulate its enzymatic function. Nano Lett 9:2280–2284
52. Sandanaraj BS, Bayraktar H, Krishnamoorthy K et al (2007) Recognition and modulation of cytochrome c's redox properties using an amphiphilic homopolymer. Langmuir 23:3891–3897
53. Armstrong JS (2007) Mitochondrial medicine: pharmacological targeting of mitochondria in disease. Br J Pharmacol 151:1154–1165
54. Maloyan A, Sanbe A, Osinska H et al (2005) Mitochondrial dysfunction and apoptosis underlie the pathogenic process in alpha-B-crystallin desmin-related cardiomyopathy. Circulation 112:3451–3461
55. Irwin WA, Bergamin N, Sabatelli P et al (2003) Mitochondrial dysfunction and apoptosis in myopathic mice with collagen VI deficiency. Nat Genet 35:367–371
56. Zhou F, Xing D, Wu B et al (2010) New insights of transmembranal mechanism and subcellular localization of noncovalently modified single-walled carbon nanotubes. Nano Lett 10:1677–1681

57. Ma X, Zhang LH, Wang LR et al (2012) Single-walled carbon nanotubes alter cytochrome c electron transfer and modulate mitochondrial function. ACS Nano 6:10486–10496
58. Ye S, Jiang Y, Zhang H et al (2012) Multi-walled carbon nanotubes induce apoptosis in RAW 264.7 cell-derived osteoclasts through mitochondria-mediated death pathway. J Nanosci Nanotechnol 12:2101–2112
59. Wang X, Guo J, Chen T et al (2012) Multi-walled carbon nanotubes induce apoptosis via mitochondrial pathway and scavenger receptor. Toxicol In Vitro 26:799–806
60. Wang LR, Xue X, Hu XM et al (2014) Structure-dependent mitochondrial dysfunction and hypoxia induced with single-walled carbon nanotubes. Small 10:2859–2869
61. Cheng WW, Lin ZQ, Wei BF et al (2011) Single-walled carbon nanotube induction of rat aortic endothelial cell apoptosis: Reactive oxygen species are involved in the mitochondrial pathway. Int J Biochem Cell Biol 43:564–572
62. Zhou F, Wu S, Wu B et al (2011) Mitochondria-targeting single-walled carbon nanotubes for cancer photothermal therapy. Small 7:2727–2735
63. Li K, Li Y, Shelton JM et al (2000) Cytochrome c deficiency causes embryonic lethality and attenuates stress-induced apoptosis. Cell 101:389–399
64. Cai J, Yang J, Jones DP (1998) Mitochondrial control of apoptosis: the role of cytochrome c. Biochim Biophys Acta 1366:139–149
65. Debatin KM, Poncet D, Kroemer G (2002) Chemotherapy: targeting the mitochondrial cell death pathway. Oncogene 21:8786–8803
66. Budihardjo I, Oliver H, Lutter M et al (1999) Biochemical pathways of caspase activation during apoptosis. Annu Rev Cell Dev Biol 15:269–290
67. Newmeyer DD, Ferguson-Miller S (2003) Mitochondria: releasing power for life and unleashing the machineries of death. Cell 112:481–490
68. Ding L, Stilwell J, Zhang T et al (2005) Molecular characterization of the cytotoxic mechanism of multiwall carbon nanotubes and nano-onions on human skin fibroblast. Nano Lett 5:2448–2464
69. Yang S, Guo W, Lin Yi et al (2007) Biodistribution of pristine single-walled carbon nanotubes in vivo. Phys Chem C 111:17761–17764
70. Wang J, Deng X, Yang S (2008) Rapid translocation and pharmacokinetics of hydroxylated single-walled carbon nanotubes in mice. Nanotoxicology 2:28–32

Chapter 20
Evidence of the Interplay of Menin, CRTC1 and THOC5 Triangles

Lichang Wu, Qiwen Zhang, Liqiao Ma, Yu Sun, Baoxia Zhang, Caicai Kang, Aipo Diao and Yinchuan Li

Abstract Multiple endocrine neoplasia type 1 (MEN1) is an endocrine cancer syndrome occurred on multiple endocrine tissues caused by *MEN1* gene. Its protein product is 610 amino acids in length, named Menin. Creb-regulated transcriptional coactivator 1 (*CRTC1*) activates transcription through enhancing the interaction of CREB1 with TAF4. THOC5 was the fifth member of a subcomplex of the transcription/export complex (THO), which is involved in the nuclear export of a subset of mRNAs. In this report, we found an interesting regulation pattern between these three genes. CRTC1 promotes THOC5 nuclear export and Menin reverses this process. Both Menin and THOC5 inhibit CRTC1 activation on G6Pase promoter in a synergistic manner, suggesting the interplay of these three genes regulates gene expression.

Keywords Menin · CRTC1 · THOC5

20.1 Introduction

Multiple endocrine neoplasia type 1 (MEN1) syndrome includes the concurrence of tumors of the parathyroid glands (95 % of cases), the anterior pituitary (15–90 % of cases), and the pancreatic islet cells (30–80 % of cases). Other endocrine and nonendocrine neoplasms including adrenocortical and thyroid tumors, visceral and cutaneous lipomas, meningiomas, facial angiofibromas and collagenomas, thymic, gastric, and bronchial carcinoids are also found in some patients [1, 2]. The large

L. Wu · Q. Zhang · L. Ma · Y. Sun · B. Zhang · C. Kang · A. Diao (✉) · Y. Li (✉)
Key Lab of Industrial Fermentation Microbiology of the Ministry of Education, College of Bioengineering, Tianjin University of Science and Technology, Tianjin 300457, China
e-mail: diaoaipo@tust.edu.cn

Y. Li
e-mail: ycli@tust.edu.cn

T.-C. Zhang and M. Nakajima (eds.), *Advances in Applied Biotechnology*, Lecture Notes in Electrical Engineering 332, DOI 10.1007/978-3-662-45657-6_20

majority of patients with MEN1 have mutations in the *MEN1* gene [3]. *MEN1* gene has 10 exons spanning about 10 kb. According to a recent *MEN1* mutation report, *MEN1* mutations are located along the full 1830-bp coding region and flanking splicing sites of the *MEN1* gene [2]. *MEN1* mutations approximately include frameshift insertions and deletions (41 %), nonsense mutations (23 %), missense mutations (20 %), in-frame deletions or insertions (6 %), splice site defects (9 %), and whole or partial gene gross deletions (1 %) [2]. More than two-thirds of all identified mutations (frameshifts, nonsense, and some splicing site mutations) predict a loss-of-function of Menin, and therefore support the hypothesis that *MEN1* is a tumor suppressor gene.

CRTC1 activates transcription through enhancing the interaction of CREB1 with TAF4 on cAMP response element (CRE) sites in promoter [4]. Diseases associated with *CRTC1* include clear cell hidradenoma, and mucoepidermoid salivary gland carcinoma via fusion of exon 1 of CRTC1 with exons 2–5 of MAML2 to form a novel oncogene as *CRTC1-MAML2* [5]. CRTC1 is a nuclear/cytoplasm shuttling protein and usually acts as a transcriptional coactivator to enhance glucose metabolism [4].

THOC5/FMIP is a member of the THO complex which is a subcomplex of the transcription/export complex (TREX) [6, 7]. TREX is recruited to spliced mRNAs by a transcription-independent mechanism, binds to mRNA upstream of the exon junction complex (EJC) and is recruited in a splicing- and cap-dependent manner to a region near the 5′ end of the mRNA where it is involved in mRNA export to the cytoplasm via the TAP/NFX1 pathway [8]. THO complex components are not essential for bulk poly (A)+ RNA export in higher eukaryotes, but for the nuclear export of a subset of mRNAs [9]. THOC5 is a nuclear/cytoplasm shuttling protein and substrate for several kinases such as several tyrosine kinases [7], protein kinase C [10], and the downstream kinase from insulin stimulus [11]. Its exact roles in THO complex are still unknown.

For the above mentioned MEN1, CRTC1, and THOC5, there are no reports describing any correlation among them. In this report, we found a unique interplay among CRTC1, Men1, and THOC5 in gene expression regulation.

20.2 Materials and Methods

20.2.1 Plasmids

cDNAs for human MEN1, THOC5, and CRTC1 were reversely transcribed from mRNAs derived from 293t cells. The PCR products were purified and digested with enzymes recognizing the restriction sites on both primers, and then were inserted into the pcDNA3.1+ vector with HA, Flag, or T7 tags on its upstream. The promoter of Glucose-6-Phosphatase, Catalytic Subunit (G6Pase) with a consensus CRE element (−700 to +30) was PCR amplified with 293T genomic DNA and inserted into pGL3-Basic vehicle plasmids (Promega) to form pG6P-LUC reporter vector. All truncated mutations of human Menin were PCR amplified and cloned into pcDNA3.1+ vector.

20.2.2 Cell Culture and Transfection

Hela cells or 293t cells were grown in Dulbecco's modified Eagle's medium (DMEM; Gibco) supplemented with 10 % fetal bovine serum (FBS), 2 mM L-Glutamine, 1,000 U/ml penicillin, and 0.1 mg/mL streptomycin and maintained at 37 °C and 5 % CO_2. Turbofect transfection reagent was from Thermo Scientific.

20.2.3 Antibodies and Other Reagents

Monoclonal anti-HA and anti-Flag M2 antibody was obtained from Sigma-Aldrich. Rabbit polyclonal antibody against HA was from Abgent. Rabbit monoclonal antibody against Menin was from Proteintech. Secondary antibodies of Alexa Fluor 555 donkey anti-Rabbit IgG and Alexa Fluor 488 donkey anti-mouse IgG were from Invitrogen. Luciferase assay kits were from Promega. All other biological or chemical reagents were from Thermo Scientific, Shanghai Sangon, Roche, or Sigma-Aldrich.

20.2.4 Immunofluorescence Imaging and Analysis

Cells plated on coverslips coated with 0.01 % gelatin were fixed with 4 % paraformaldehyde in phosphate-buffered saline (PBS), and were permeabilized using 0.25 % Triton X-100. Incubating with the primary antibody at an appropriate dilution ratio was performed at RT for 1 h. Cells were then stained with secondary antibody and nuclei were counterstained with Hoechst 33342. Images were captured via an Olympus FluoView confocal laser scanning FV1000, and images were disposed with Olympus FV10-ASW 3.0 Viewer.

20.2.5 Luciferase Illumination Assay

Twenty four hours after transfection with indicated plasmids, cells were lysed and were immediately subjected to illumination assay on a Synergy™ 4 (Bioteck).

20.3 Results

20.3.1 Menin Inhibits the Regulation of CRTC1 on THOC5 Nuclei–Cytosol Translocation

THOC5 is normally accumulated in nucleoplasm with a similar distribution pattern with Menin (Fig. 20.1a). Overexpression of Menin did not alter the distribution of THOC5 in nucleoplasm (Fig. 20.1b). While overexpression of CRTC1, however, enormous THOC5 was found to be translocated from nuclei into cytosol, forming very large ring structures, which were possibly membranous structures (Fig. 20.1c). No available previous reports have addressed such phenomenon to THOC5. The result implies that CRTC1 promotes the nuclei export of THOC5. THOC5 acts as a component of THO complex to couple a subset of mRNA transcription, processing, and nuclear export [7]. THOC5 was previously found to be able to exert effects on

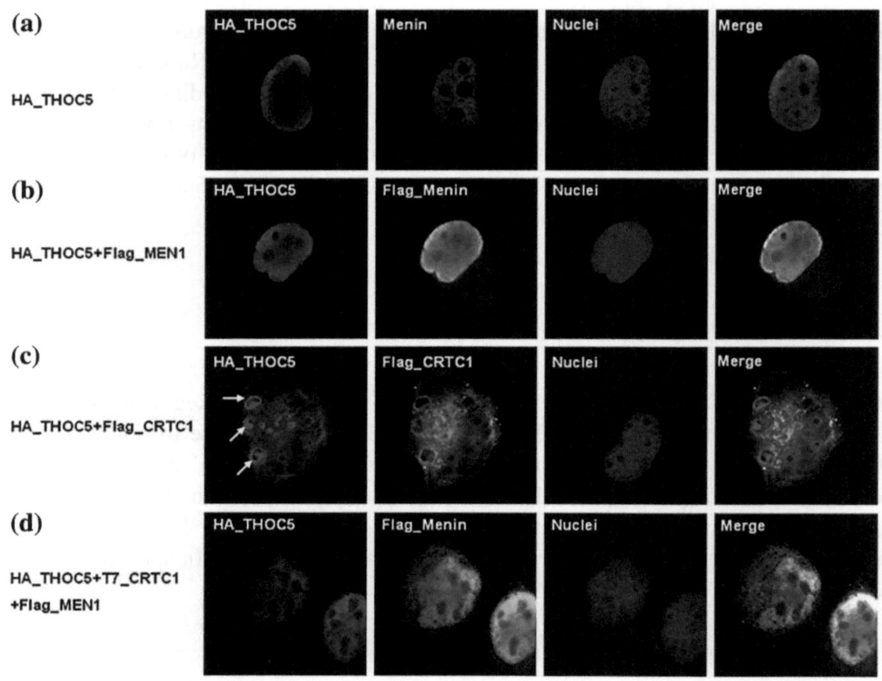

Fig. 20.1 Overexpression of CRTC1 expels THOC5 from nuclei, while Menin reverses this process in Hela cells. **a** Representative immunofluorescence image of HA-THOC5 and endogenous Menin. **b** Representative immunofluorescence image of overexpression of both HA-THOC5 and Flag-Menin. **c** Upon overexpression of both HA-THOC5 and Flag-CRTC1, predominant HA-THOC5s were found to be exported into cytosol, forming ring-like structures (*arrows*). **d** Triple overexpression of Flag-Menin together with HA-THOC5 and T7-CRTC1 restored the HA-THOC5 from cytosol back into nuclei

metabolism, for example, increasing PIP3 levels and affecting the level of HDL cholesterol [12, 13]. The composition of these rings needs further investigation. Overexpression of Menin reverses these processes, which suggests that Menin inhibits the regulation of CRTC1 (Fig. 20.1d).

20.3.2 Menin and THOC5 Synergistically Inhibit CRTC1 in Activating G6Pase Promoter Activity

Either Menin or THOC5 inhibit the activation of CRTC1 on G6P-Luc reporter (Fig. 20.2). Both further decrease the activation of CRTC1 on G6P-Luc, implying that Menin and THOC5 are on two different knots exerting effect on CRTC1 (Fig. 20.2). G6Pase hydrolyzes glucose-6-phosphate to glucose in the endoplasmic reticulum to control glucose homeostasis [14]. A consensus CRE element was found in its promoter which is strictly regulated by creb1 and PKA signaling pathway [15]. Interestingly, for Menin, the strongest inhibition was derived from both of its two terminus, especially its N-terminus (aa1-154). Its C-terminus (aa479-end) also displayed strong inhibition. This result may suggest that mutations of Menin resulting in C-terminal deletion may be different to the effect of MEN1

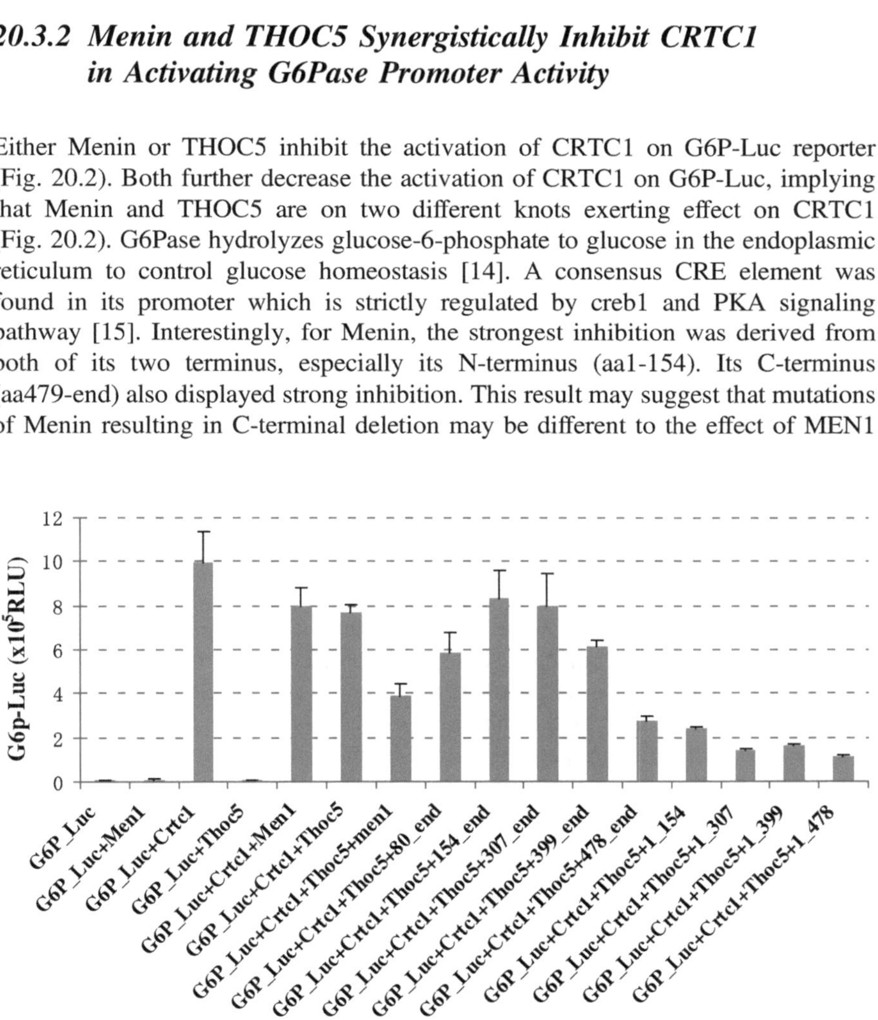

Fig. 20.2 Full-length wild type Menin, especially truncated Menin (the N-terminus or C-terminus) and THOC5 synergistically inhibit the activation of CRTC1 on G6Pase promoter. Different combinations of plasmids as indicated were cotransfected into 293T cells. Each treatment was performed in triplicates. 24 h after transfection, cell lysates were subjected to illumination assay according to manufacturer's instruction. Data were expressed as means ±SD

knocking down, with an effect of dominant negative blocking. So any try to mimic the MEN1 mutation via RNAi should evaluate how much is equivalent for both. This needs further verification.

20.4 Discussion

Menin may take roles in regulation of DNA transcription, in DNA replication, in maintenance of genome integrity, or in cell cycle, displaying a role of tumor inhibitor. In this report, we found Menin was able to downregulate G6pase via inhibiting CRTC1-mediated metabolic activity. THOC5 has been reported to regulate nuclear export of a subset of mRNAs, some of which are key genes in metabolism. For example, THOC5 knockout led to the downregulation of Slc2a3 (GLUT3) for fourfold to sevenfold change [9]. Defects of THO complex upregulate PI3K/AKT signaling [16]. CRTC1 increases the export of THOC5 implying CRTC1 increasing metabolism was partially mediated via THOC5 pathway and the inhibiting activity of Menin was also partially, if not all, via this pathway. This raises a possibility that the regulation of gene expression by Menin and CRTC1 is at posttranscriptional level, which reveals a previously unrecognized pathway for CRTC1 and MEN1 in posttranscriptional control. Next, we should elucidate how Menin exerts its influence on CRTC1 and THOC5. The negative results of yeast two hybridization of Menin with CRTC1 or Menin with THOC5 in our lab have ruled out the direct interaction of these proteins (data not shown). Therefore, revealing the exact converging knot of these three proteins in signaling transduction is really important.

Many mutations of MEN1 were found to lose its C-terminus due to frameshift mutations. According to our data, these mutations, however, increased the inhibiting activity of Menin. So we postulate that Menin versus MEN1 syndrome is not simply attributed to lose of the inhibiting activity of Menin. The role of its C-terminus and its interacting partners should be paid more attention in the future.

Acknowledgments This study was supported by National Natural Science Foundation of China (No. 31271448).

References

1. Gibril F, Schumann M, Pace A, Jensen RT (2004) Multiple endocrine neoplasia type 1 and Zollinger-Ellison syndrome: a prospective study of 107 cases and comparison with 1009 cases from the literature. Medicine 83:43–83
2. Lemos MC, Thakker RV (2008) Multiple endocrine neoplasia type 1 (MEN1): analysis of 1336 mutations reported in the first decade following identification of the gene. Hum Mutat 29:22–32
3. Lakhani VT, You YN, Wells SA (2007) The multiple endocrine neoplasia syndromes. Annu Rev Med 58:253–265

4. Altarejos JY, Montminy M (2011) CREB and the CRTC co-activators: sensors for hormonal and metabolic signals. Nat Rev Mol Cell Biol 12:141–151
5. Tonon G, Modi S, Wu L et al (2003) t(11;19) (q21;p13) translocation in mucoepidermoid carcinoma creates a novel fusion product that disrupts a Notch signaling pathway. Nat Genet 33:208–213
6. Katahira J, Inoue H, Hurt E, Yoneda Y (2009) Adaptor Aly and co-adaptor Thoc5 function in the Tap-p15-mediated nuclear export of HSP70 mRNA. EMBO J 28:556–567
7. Mancini A, Niemann-Seyde SC, Pankow R et al (2010) THOC5/FMIP, an mRNA export TREX complex protein, is essential for hematopoietic primitive cell survival in vivo. BMC Biol 8:1
8. Katahira J (2012) mRNA export and the TREX complex. Biochim Biophys Acta 1819:507–513
9. Guria A, Tran DD, Ramachandran S (2011) Identification of mRNAs that are spliced but not exported to the cytoplasm in the absence of THOC5 in mouse embryo fibroblasts. RNA 17:1048–1056
10. Mancini A, Koch A, Whetton AD, Tamura T (2004) The M-CSF receptor substrate and interacting protein FMIP is governed in its subcellular localization by protein kinase C-mediated phosphorylation, and thereby potentiates M-CSF-mediated differentiation. Oncogene 23:6581–6589
11. Gridley S, Lane WS, Garner CW, Lienhard GE (2005) Novel insulin-elicited phosphoproteins in adipocytes. Cell Signal 17:59–66
12. Pierce A, Carney L, Hamza HG et al (2008) THOC5 spliceosome protein: a target for leukaemogenic tyrosine kinases that affects inositol lipid turnover. Br J Haematol 141:641–650
13. Keller M, Schleinitz D, Forster J et al (2013) THOC5: a novel gene involved in HDL-cholesterol metabolism. J Lipid Res 54:3170–3176
14. Sukalski KA, Nordlie RC (1989) Glucose-6-phosphatase: two concepts of membrane-function relationship. Adv Enzymol Relat Areas Mol Biol 62:93–117
15. Schmoll D, Wasner C, Hinds CJ et al (1999) Identification of a cAMP response element within the glucose-6-phosphatase hydrolytic subunit gene promoter which is involved in the transcriptional regulation by cAMP and glucocorticoids in H4IIE hepatoma cells. Biochem J 338(Pt 2):457–463
16. Moon S, Chung YD (2013) p53 and PI3K/AKT signalings are up-regulated in flies with defects in the THO complex. Mol Cells 35:261–268

Part II
Optimization and Control of Biological Process

Chapter 21
Classification of Lymphoma Cell Image Based on Improved SVM

Ting Yan, Quan Liu, Qin Wei, Fen Chen and Ting Deng

Abstract Due to the diversity of lymphoma, its classification must rely on experienced pathologist in clinic pathologic analysis. In order to improve the accuracy of lymphoma classification, lots of image processing technologies and recognition methods were presented. The support vector machine (SVM) has been widely applied in medical image classification as an effective classification method. However, the application of SVM is blocked by the limitation that each classifier must adopt the same feature vector. In this paper, an improved SVM is proposed to overcome this limitation. Through the analysis of features of different classes, different feature vectors are obtained for each class of objects respectively. And then the improved SVM based on "one-against-one" strategy is applied to classify each class one by one. According to the results of classifying seven different lymphoma images, our classification method is effective to acquire the higher precision than conventional SVM and PSO-SVM model in lymphoma classification.

Keywords Support vector machine · Lymphoma cell · Feature selection

21.1 Introduction

Lymphoma is a kind of cancer affecting lymph nodes and breaks the immunity of body quickly. Automatic classification of lymphoma images is able to help for more accurate and less labor-consuming in diagnosis of this disease. In particular, the

T. Yan (✉) · Q. Liu · Q. Wei · F. Chen · T. Deng
Key Laboratory of Fiber Optic Sensing Technology and Information Processing, Ministry of Education, Wuhan University of Technology, 430070 Wuhan, China
e-mail: 504491872@qq.com

Q. Wei
School of Mechanical and Electronic Engineering, Wuhan University of Technology, 430070 Wuhan, China

T.-C. Zhang and M. Nakajima (eds.), *Advances in Applied Biotechnology*, Lecture Notes in Electrical Engineering 332, DOI 10.1007/978-3-662-45657-6_21

support vector machine (SVM) has become an effective tool in automatic classification of medical image. Varol et al. [1] presented that an ensemble of linear support machine classifiers was employed for classification of structural magnetic resonance images of the brain after ranking individual image feature through Welch's *t*-test. Schwamborn et al. [2] applied a five-dimensional genetic algorithm and an SVM algorithm to distinguish classical Hodgkin lymphoma (HL) from lymphadenitis. Lu et al. [3] proposed a novel method for liquor classification based on SVM after considering the deficiency of the conventional liquor classification method. Moreover, SVM was referred to as a predominant role for classification of breast cancer [4, 5] and brain states [6].

SVM often adopts the same feature vector for each class which restricts its application and affects the final results of classification in practice. With regard to this limitation, an improved SVM algorithm is proposed in this paper. It not only includes a feature selection method to construct one effective and adaptive feature vector for each class, but also presents a two-class classifier consisting of two parts: feature combination and a "one-against-one" SVM. The rest of the paper is organized as follows. Section 21.2 describes the improved SVM algorithm. Section 21.3 introduces the application of the proposed algorithm to the case of lymphoma images and experimental results. Section 21.4 demonstrates other classical classification algorithm for comparison purposes. Section 21.5 gives conclusions and points out our future work in lymphoma image processing.

21.2 Improved Support Vector Machine

21.2.1 Feature Selection

In classification, better discrimination ability of classifier is accompanied with the greater differences in feature vectors of various classes. There are many efficient and robust feature selection methods which are desired to extract meaningful features and eliminate noisy ones. However, these methods extract the same features for all classes, but neglect the correlation of different features. The remaining classes may be classified better with some features after eliminating some classes. For example, there are three classes with three features. In Fig. 21.1, classes 3 and 1 are capable of being divided from mixed classes according to the different distribution of features 1 and 2 (Fig. 21.1a) and features 1 and 3 (Fig. 21.1b), respectively.

A feature selection approach is proposed, in order to extract different features for different classes. The feature vector for each class is searched from original features using Euclidean distance as the criterion of identification. At first, according to the known classes and kinds of features, one class is selected randomly as analyzing object (named as C_1). The residual classes are integrated into one as another class (named as C_2). In order to get effective features for the analyzing object, the distance between two classes (C_1 and C_2) is calculated for each feature. Let $\{V_{1\max}, V_{2\max}\}$

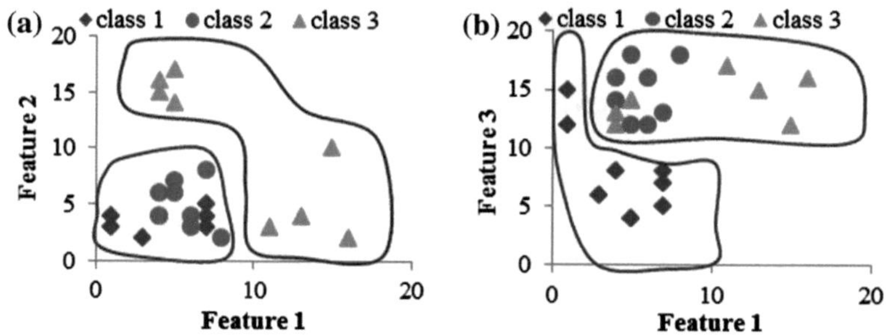

Fig. 21.1 The distribution of different feature vectors. **a** Distribution of feature 1 and feature 2. **b** Distribution of feature 1 and feature 3

and $\{V_{1\min}, V_{2\min}\}$ represent the maximum and minimum values of the two classes in one specified feature (such as F_1), respectively. The distance is calculated as follows:

$$D = \begin{cases} V_{1\min} - V_{2\max}, & V_{1\min} \geq V_{2\max} \text{ or } V_{2\max} \geq V_{1\min} \\ V_{2\min} - V_{1\max}, & V_{2\min} \geq V_{1\max} \text{ or } V_{1\max} \geq V_{2\min} \end{cases} \quad (21.1)$$

The specified feature is regarded as an effective feature of the analyzing object, when the distance is larger than zero (as shown in Fig. 21.2a). It means that this feature has better performance of identification. The amount of effective features of each class is capable of being counted by calculating the distance for all features one by one. The selected class is determined by the counted number (N) of effective features and sum (S) of the calculated distance for the effective features. Then the corresponding class is selected when the maximum N is unique. Otherwise, the class with the maximum S is the one. Furthermore, any classes have no effective features if the maximum N is zero. Under this condition, the class with the maximum distance for one feature is the selected class, and this feature is the effective feature for this class. After eliminating selected class from original classes, the rest classes reiterate this process until effective features and order of all classes can be confirmed.

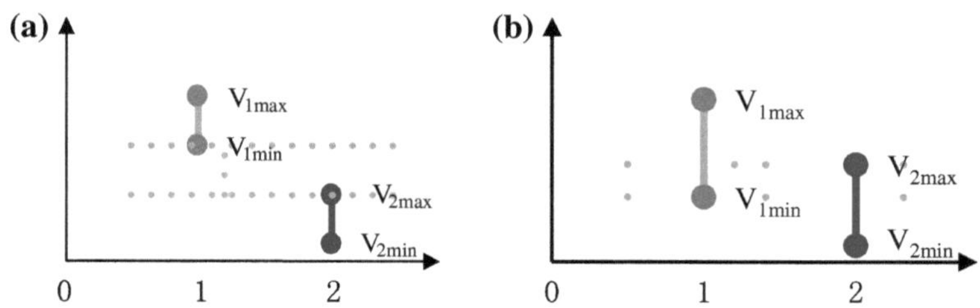

Fig. 21.2 The distribution of a specified feature. **a** The distance between two classes is larger than zero. **b** The distance between two classes is less than zero

21.2.2 Improved SVM Classifier

21.2.2.1 Support Vector Machine

The main idea of SVM is to find the optical classification plane of two classes that maximized the margin between the two. Image training sample is set as (x_i, y_i), $i = 1, 2, \ldots, l$, $x \in R^n$, $y \in \{+1, -1\}$, l is the number of samples and n is the input dimension, also expresses the number of characteristic value in an image. The following is the equation of finding best optical classification plane:

$$\begin{aligned} &\frac{\min\|w\|^2}{2} + C\sum_{i=1}^{l} \varepsilon_i \\ &\text{s.t. } y_i(w^* x_i + b) \geq 1 - \varepsilon_i \\ &\varepsilon_i \geq 0, \quad i = 1, 2, \ldots, l \end{aligned} \tag{21.2}$$

where $\varepsilon_i \geq 0$, $i = 1, 2, \ldots, l$ is non-negative slack variable, C is the penalty parameter, bigger C said bigger punishment for fault classification. By introducing Lagrange multipliers for the problem of optimization, the optimal decision function is shown as follows:

$$f(x) = \operatorname{sgn}\left[\sum_{i=1}^{l} y_i \alpha_i (x * x_i) + b\right] \tag{21.3}$$

in which α is the Lagrange coefficient.

21.2.3 Design of SVM Classifier

Standard SVM only divides two types of objects in classification, thus how to construct multiclass support vector machine (MSVM) continues to be an unsolved problem. There are two widely used methods to extend binary SVM to multiclass problems. One is the one-against-one (OAO) method and the other is one-against-all (OAA) method. Studies indicate that OAO method has better performance than OAA method [7]. Hence, OAO policy is adopted for multiclass classification.

Same feature vector adopted for all classes restricts conventional OAO-MSVM application in practice. An improved classification algorithm for solving this drawback is designed in this paper. The improved SVM consists of many two-class classifiers, as shown in Fig. 21.3. As mentioned above, effective features and division order of each class can be gained by using the feature selection method. A two-class classifier is constructed for each class, which consists of feature combination and OAO-MSVM. The process of feature combination is to reconstruct feature vector according to the efficient features of corresponding class. Then new

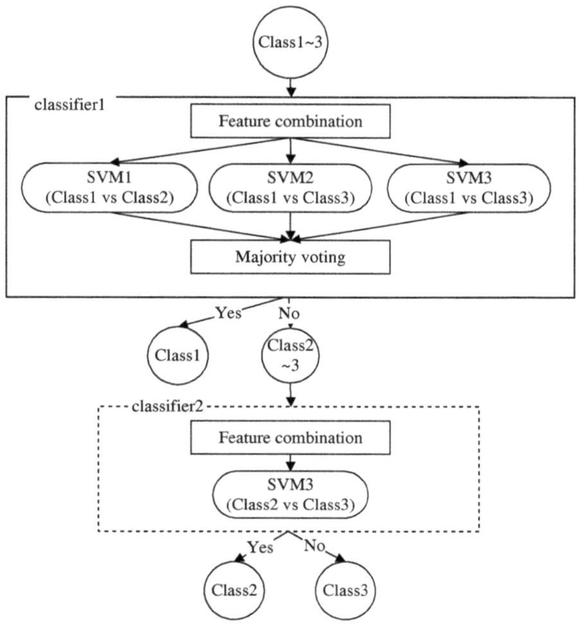

Fig. 21.3 The flow chart of proposed classification algorithm

feature vector is the input data of OAO-MSVM, which uses voting model to classify all samples into two classes which are corresponding class and the other classes, respectively. The other classes are regarded as the input data of the next classifier.

21.3 Experiment

21.3.1 Experimental Object

Malignancies originated from the lymphocyte are known as malignant lymphoma, which are divided into Hodgkin's diseases (HD) and Non-Hodgkin's lymphoma (NHL). Hodgkin lymphoma and mixed cellularity Hodgkin lymphoma (MCHL) belong to HD. But diffuse large B-cell lymphoma (DLBCL), Follicular lymphoma (FL), Burkitt lymphoma (BL), mantle cell lymphoma (MCL), and reactive lymphoid hyperplasia (RLH) belong to NHL. Each kind of lymphoma has its own unique feature, for example, as for as distribution, those with diffuse distribution of large cells are DLBCL, those with diffuse distribution of medium sized cells are BL [8]. Seven kinds of lymphoma images are displayed in Fig. 21.4.

Different kinds of cells have difference in shape, arrangement, and distribution. In order to classify different kinds of lymphoma cells, texture analysis is used to extract the feature of cell image. Twelve common statistic values of texture analysis

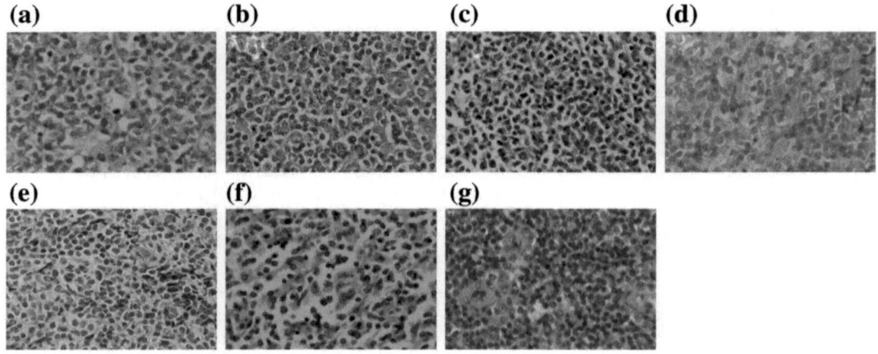

Fig. 21.4 Selection diagrams of various types of lymphoma cells. **a** BL, **b** DLBCL, **c** FL, **d** HL, **e** MCL, **f** MCHL, **g** RLH

are extracted, such as mean, standard deviation, smoothness, third moment, consistent, entropy based on gray-level histogram and the maximum probability, contrast, correlation, energy, homogeneity, relative entropy based on gray-level co-occurrence matrix [9], respectively labeled F_1–F_{12}.

21.3.2 Experimental Results

The scatterplot of twelve features is presented in Fig. 21.5. As is seen, the patterns related to the different classes are located close to each other and are relatively well separated from the other classes within the feature space.

Efficient features and division order of classes are obtained by the proposed feature selection approach for those lymphoma images, and results are presented in Table 21.1. It is obvious that F_5, F_6, F_7, F_{10} are useless features because these features do not belong to effective features of any class.

Six classifiers are constructed for seven different lymphoma cells. From Table 21.1, it can be known that the input feature vector of the first classifier consists of four features F_8, F_9, F_{11}, F_{12} with aim of distinguishing the DLBCL from other lymphoma cells. The input feature vector of second classifier consists of two features F_2, F_3 with the aim of distinguishing FL from other rest lymphoma cells, and so on.

In this paper, the improved SVM adopts the popular Gaussian kernel. When the penalty factor C is 200 and the kernel parameter σ is 3, the classification accuracies in training and testing groups are approximately equal to 99.48 and 97.08 %, respectively.

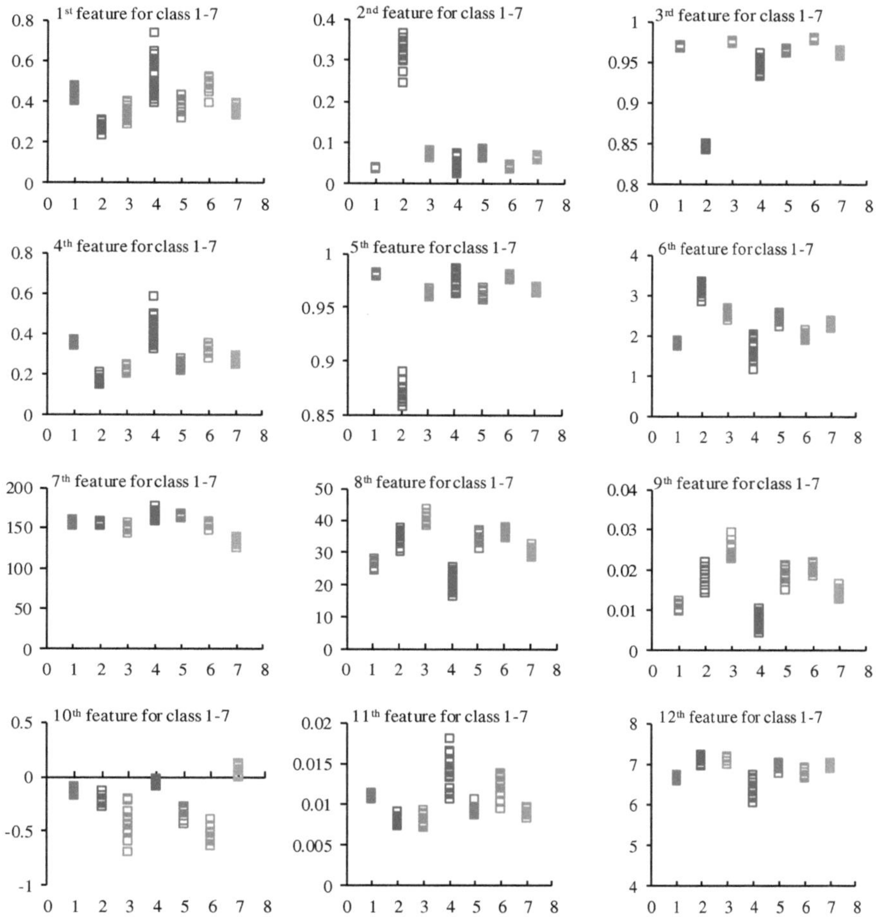

Fig. 21.5 The scatterplot of 12 features for different classes (1 = BL, 2 = DLBCL, 3 = FL, 4 = HD, 5 = MCL, 6 = MCHL, 7 = RLH)

Table 21.1 The result of feature selection

All lymphomas	Feature combination	Division order
DLBCL	F_8, F_9, F_{11}, F_{12}	1
FL	F_2, F_3	2
RLH	F_1, F_4	3
MCL	F_9	4
MCHL	F_2, F_3, F_4, F_9,	5
BL	F_9	6
HL	F_9	7

21.4 Performance Analysis and Comparison

The performances of different SVM classifiers are evaluated by using a threefold cross-validation. In total, there are 84 samples (12 samples per class), which are divided into three equal groups. Classifiers are trained by using two-thirds of the samples, and then are tested by the remaining one-third. This procedure is repeated three times, the one-third of all samples is randomly selected as the test group. To obtain a more reliable estimation, the threefold cross-validation process is repeated five times using different partitions of the samples. The average of train/test accuracies is used as the estimator for the corresponding algorithm.

The largest problems encountered in establishing the SVM model are how to select the kernel function and its parameters. All MSVM classifiers described in this paper used Gaussian kernel due to its popularity in practical use. So the variation of parameters is taken into consideration to compared with the different algorithms. And the OAO-MSVM classifier utilized directly all features.

The performances of the OAO and the proposed (Pro) methods using the SVM classifier with different penalty factor are compared, shown in Fig. 21.6a. In OAO-Train and Pro-Train two curves, it denotes recognition accuracy of training set of two classifier, when the value of C is 0.5, the classification accuracy of two classifiers are approximately equal to 100 %, but while the value of penalty factor C ranges from 0.1 to 0.5, these results achieved by our classifier are better than those achieved by OAO method. According to OAO-test and Pro-test results, it indicates recognition accuracy of testing set of two classifiers, the percentage recognition accuracy of the proposed method is about 92 %, which is obviously higher than OAO method (about 75 %). The experimental results prove that the effectiveness of the proposed algorithm in comparison with OAO classifier. According to Fig. 21.6b, the same conclusion can be obtained in comparison with similar analytical method.

In [10], authors applied particle swarm optimization algorithm to choose the suitable parameter combination for an SVM model, called PSO-MSVM model. This paper also applies PSO-MSVM model in classification of lymphoma cells, and makes a comparison with the proposed algorithm. The input feature and cross-validation process of PSO-MSVM are the same with OAO-MSVM. Table 21.2

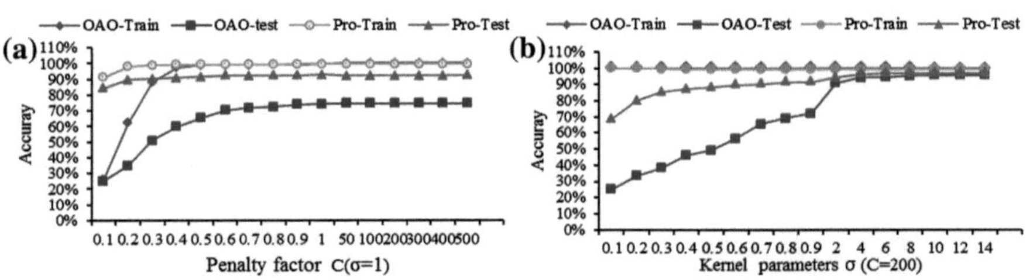

Fig. 21.6 Recognition accuracy of training set and testing set. **a** Penalty factor tuned in the range [0.1, 600]. **b** Kernal parameter tuned in the range [0.1, 8]

Table 21.2 Performance comparison of OAO-MSVM and PSO-MSVM

All classifiers	Train accuracy (%)	Test accuracy (%)
OAO-MSVM	99	95
PSO-MSVM	98	90
Pro-MSVM	99	97

shows the best performance of three classifiers. The results imply that the proposed algorithm is more effective than OAO-MSVM and PSO-MSVM for lymphoma classification.

21.5 Conclusions

The application of conventional SVM is restricted in the same feature vector adopted for each class. In order to overcome the limitation in SVM, an improved classification algorithm is presented to classify the different kinds of lymphoma cell images. The improvement is concentrated on a novel feature selection method and multiclass SVM adopted OAO strategy. A comparison among our method, PSO-MSVM and common OAO-MSVM indicate that our improved method is able to correctly select the input features and also achieve high classification accuracy. Furthermore, it also has the high efficiency in classification, which is achieved by the variable feature vectors for each classifier. The experiments on classification of seven kinds of lymphoma images show that the effectiveness and efficiency of our method.

Like most issues in multiclass classification algorithm, there is an error accumulation in the proposed method, especially the identification error of the root node. Therefore, how to solve this problem needs further study.

Acknowledgments The research work was supported by National Natural Science Foundation of China under Grant Nos. 50935005 and 51175389, and the Key Grant Project of Chinese Ministry of Education No. 313042.

References

1. Varol E, Gaonkar B, Erus G et al (2012) Feature ranking based nested support vector machine ensemble for medical image classification. In: 9th IEEE international symposium on biomedical imaging (ISBI), 2012. IEEE, pp 146–149
2. Schwamborn K, Krieg RC, Jirak P et al (2010) Application of MALDI imaging for the diagnosis of classical Hodgkin lymphoma. J Cancer Res Clin Oncol 136(11):1651–1655
3. Lu J, Du L, Ding H, et al (2014) Application of support vector machine in base liquor classification. In: Proceedings of the 2012 international conference on applied biotechnology (ICAB 2012). Springer, Berlin, pp 1051–1056

4. George G, Raj VC (2011) Review on feature selection techniques and the impact of svm for cancer classification using gene expression profile. arXiv:1109.1062, doi:10.5121/ijcses.2011.2302
5. Huang CL, Liao HC, Chen MC (2008) Prediction model building and feature selection with support vector machines in breast cancer diagnosis. Expert Syst Appl 34(1):578–587
6. Sitaram R, Lee S, Ruiz S et al (2011) Real-time support vector classification and feedback of multiple emotional brain states. Neuroimage 56(2):753–765
7. Hsu CW, Lin CJ (2002) A comparison of methods for multiclass support vector machines. IEEE Trans Neural Netw 13(2):415–425
8. Vardiman JW (2010) The World Health Organization (WHO) classification of tumors of the hematopoietic and lymphoid tissues: an overview with emphasis on the myeloid neoplasms. Chem Biol Interact 184(1):16–20
9. Gonzalez RC, Woods RE, Eddins SL (2009) Digital image processing using MATLAB. Gatesmark Publishing, Knoxville
10. Meng Q, Ma X, Zhou Y (2012) Application of the PSO-SVM model for coal mine safety assessment. In: Eighth international conference on natural computation (ICNC), 2012. IEEE pp 393–397

Chapter 22
Foam Control in Epothilones Fermentation of *Sorangium cellulosum*

Yue Liu, Lin Zhao, Hongrui Zhang, Fuming He and Xinli Liu

Abstract Eight different antifoam agents were added in shake flasks or in a 50 L fermenter, and the foam control abilities and effects on cultures were evaluated in this paper. Results showed that THIX-298 and SXP107 antifoam agents could eliminate foam rapidly and suppress foam during the fermentation process in the shake flasks. The biomasses and yields of epothilones were affected by different antifoam agents' addition when 0.06 % (v/v) THIX-298 antifoam agent was added in the shake flasks. The yields of epothilone A and B had a significant increase of about 48.12 and 48.55 % compared to the control, respectively, representing the largest effects of the antifoam agents tested. Reducing sugar and amino nitrogen were also utilized more rapidly than the control. Additionally, efficiency parameters of the different antifoam agents were compared in the fermentor. Finally, THIX-298 antifoam agent was chosen as the optimal one.

Keywords Antifoam agents · Epothilones fermentation · Foam destruction · Foam suppression

22.1 Introduction

Epothilones are cytotoxic macrolides that are naturally produced by the *myxobacterium Sorangium cellulosum* [1, 2]. They are regarded as potential anticancer drugs and the mechanism of action is similar to the anticancer drug paclitaxel [3]. However, cultivation of *S. cellulosum* is seriously limited by several difficulties in fermentation [4]. Formation of excessive foam is one of the key limiting factors during fermentation

These authors contributed equally to this work.

Y. Liu · L. Zhao · H. Zhang · F. He · X. Liu (✉)
Shandong Provincial Key Laboratory of Microbial Engineering, Qi Lu University of Technology, Jinan 250353, People's Republic of China
e-mail: vip.lxl@163.com

T.-C. Zhang and M. Nakajima (eds.), *Advances in Applied Biotechnology*,
Lecture Notes in Electrical Engineering 332, DOI 10.1007/978-3-662-45657-6_22

processes because of the requirement for intense agitation and aeration combined with the presence of surface-active species in the culture medium [5]. Furthermore, foams significantly accelerated when the resins are added in the fermentation medium. It can lead to a series of problems, such as decrease of the working volume of the reactor, removal of microbial cells and metabolites, increased risk of contamination, and decrease of the oxygen transfer rate [5, 6].

Currently, a number of methods are used for foam control, such as mechanical foam breakers, ultrasound, and the addition of chemical antifoam agents [7]. Chemical antifoam agents are most employed for being effective and economical. There are several classes of antifoam agents, including treated silica, synthetic or natural waxes, poly-glycol, and silicones or silicone derivatives. Diverse control mechanisms of antifoam agents are also found, such as bridging de-wetting, spreading fluid entrainment, and bridging-stretching [5, 6, 8, 9].

Although antifoam agents are important in the fermentation process, little research has done in epothilones fermentation, since selecting an appropriate antifoam agent for a specific fermentation system must be with caution and understanding [10]. In this paper, eight different antifoam agents were compared to select an appropriate one for epothilones fermentation.

22.2 Materials and Methods

22.2.1 Microorganisms and Culture Conditions

Sorangium cellulosum strain So2161 was preserved in our lab. It was inoculated in CNST agar medium (KNO_3 0.05 %, Na_2HPO_4 0.025 %, $MgSO_4 \cdot 7H_2O$ 0.1 %, agar 1.5 %, EDTA-Fe^{3+} 0.1 %, TE solution 0.1 %) for activation. The strain was then spread on a plate at 30 °C for 3 days. Single colony was transferred into M26 medium (potato starch 0.8 %, soy peptone 0.2 %, glucose 0.2 %, yeast extract 0.2 %, $MgSO_4 \cdot 7H_2O$ 0.1 %, $CaCl_2$ 0.1 %, TE solution 0.1 %) for propagation at 30 °C for 4 days. Then the culture was cultivated in fermentation medium at 30 °C for 7 days.

22.2.2 Antifoam Agents

The antifoam agents tested in this study were SB2121 (a polyalkylene glycol), J673A (an alkoxylated fatty acid ester on a vegetable base), P2000 (a polypropylene glycol), TBP (tributyl phosphate), SXP101 (a silicone polymer), SXP107 (a silicone mixed with polyether polymer), HR (a polyether), and THIX-298 (a polyether modified polysiloxane). All the antifoam agents were autoclaved prior to use and each shake flask experiment was performed in triplicate, with the undiluted antifoam added directly to the medium.

22.2.3 Analysis of the Antifoam Abilities

Antifoam abilities include foam destruction and foam suppression. Sparge test was carried out to compare the antifoam abilities of antifoam agents [5]. In foam destruction test, five 500 mL graduated measured cylinders were filled with 100 mL fermentation medium and airflow was supplied at the bottom of the cylinders at a rate of 3 L/min. When the foams reached at 400 mL scale, 0.01 % (v/v) antifoam agents were added to the cylinder. Meanwhile, the times until foam faded away were recorded and regarded as defoam time. In the foam suppression test, the same amounts of antifoam agents were added in graduated measured cylinders before aeration. Then airflow was supplied at a rate of 3 L/min. Recording the time until foaming, the time value represents the suppressibility of antifoam agents.

22.2.4 Determination of Biomasses and the Yields of Epothilones

The strains were cultured in 300 mL Erlenmeyer flasks containing 50 mL medium and 1 mL XAD-16 resin. After incubation at 30 °C for 7 days, culture broth was filtrated with 80 mesh strainer to get rid of the resins and cells which were harvested by centrifugation at 10,000 rpm for 10 min. The cells were dried overnight at 40 °C and then weighed. Then the sample was extracted with 1 mL of methanol. HPLC analysis was used to determine epothilones yields.

22.2.5 Determination of the Reducing Sugar and Amino Nitrogen

Two milliliters of cultivation broth were centrifuged at 10,000 ×g for 10 min. The supernatant was used to determine the reducing sugar with Fully Automatic reducing Sugar Tester (SGD- IV, Shandong Academy of Sciences, China). Amino nitrogen was determined by formaldehyde titration.

22.2.6 Fermenter Studies

A 50 L fermenter with two impellers was used to examine the efficiency and effects of antifoam agents on the yields of epothilones. The fermenter with dissolved oxygen and pH probes was filled with 30 L of fermentation medium and sterilized in situ at 120 °C for 30 min. Culture temperature was controlled at 30 °C during the process. The airflow rate was fixed at 10 L/min and stirring speed was 120 rpm, respectively. 0.02 % (v/v) of antifoam agents was added in the bioreactor before sterilization. During the fermentation process, antifoam agents were added in the fermenter with peristaltic pump automatically when needed.

22.2.7 Comparison of Different Antifoam Agents

In bioreactor experiments, several efficiency parameters were used to compare the antifoam agents [11]. Volumetric antifoam fraction (V_f), foam suppression coefficient (FSc), cost coefficient (Cc), and efficiency coefficient (Ec) are defined by Eqs. (22.1)–(22.4):

$$V_f = V_x / V_l \quad (22.1)$$

$$\mathrm{FSc} = h_f - h_{f\min} / h_f \quad (22.2)$$

$$\mathrm{Cc} = C_x / C_{\mathrm{THIX-298}} \quad (22.3)$$

$$\mathrm{Ec} = \mathrm{FSc} / \left(V_f \cdot \mathrm{Cc}\right) \quad (22.4)$$

where V_x = total volume of an antifoam agent used to control foam; V_l = total medium volume in the bioreactor; h_f = average foam height before addition of antifoam agents; h_{fmin} = residual foam height after addition of antifoam agents; C_x = unit cost of antifoam agents(¥/1); $C_{THIX\text{-}298}$ = unit cost of THIX-298.

22.3 Result and Discussion

22.3.1 Comparison of Antifoam Abilities of Different Antifoam Agents

Figure 22.1 shows that persistent foams were found in fermentation medium, foam height reduced slowly and did not reach zero in the 30 min testing time. The defoam times and antifoam times of different antifoam agents are shown in Table 22.1. The most effective agent for foam destruction was THIX-298, although all the antifoam agents could quickly destruct the initial foams (<10 s). Silicone-based antifoam agents (SXP101, SXP107) also showed excellent defoam abilities. For foam suppression, THIX-298, HR, J673A, SXP107 antifoam agents showed excellent antifoam abilities. According to the results in Table 22.1, THIX-298 and SXP107 antifoam agents were effective in foam destruction and suppression.

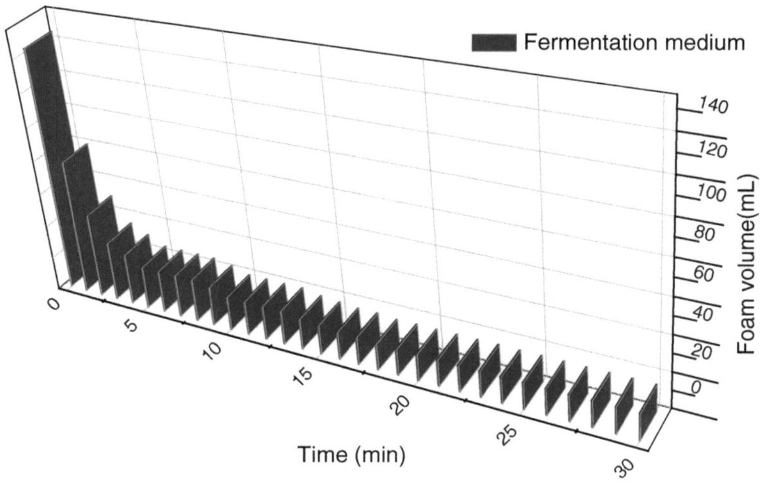

Fig. 22.1 The change of foam volume with respect to time duration in fermentation medium without antifoam agents

Table 22.1 Comparison of antifoam abilities of different antifoam agents

Antifoam agents	Defoam time (s)	Antifoam time (s)
TBP	1.58	845
P2000	3.67	760
SB2121	2.49	850
J673A	4.23	>900
SXP101	1.35	660
SXP107	1.43	>900
HR	6.65	>900
THIX-298	1.13	>900

These experiments were repeated five times independently
All results are given as the mean

22.3.2 *The Effects of Antifoam Agents on Biomasses and Epothilones*

In order to compare the physiological effects of different antifoam agents in *S. cellulosum* fermentation, 0.02, 0.06 and 0.10 % volume of them were added in the 300 mL shake flasks containing 50 mL fermentation medium.

The shake flask data showed that the addition of different antifoam agents affected the biomasses and yields of epothilones in the 300 mL shake flasks after 7 days. TBP decreased the biomasses slightly, but increases in the yields of epothilones were observed even at a lower volume fraction of TBP in the fermentation medium (Fig. 22.2a). Especially, Epothilones (A, B) increased 31.41 and 18.05 %

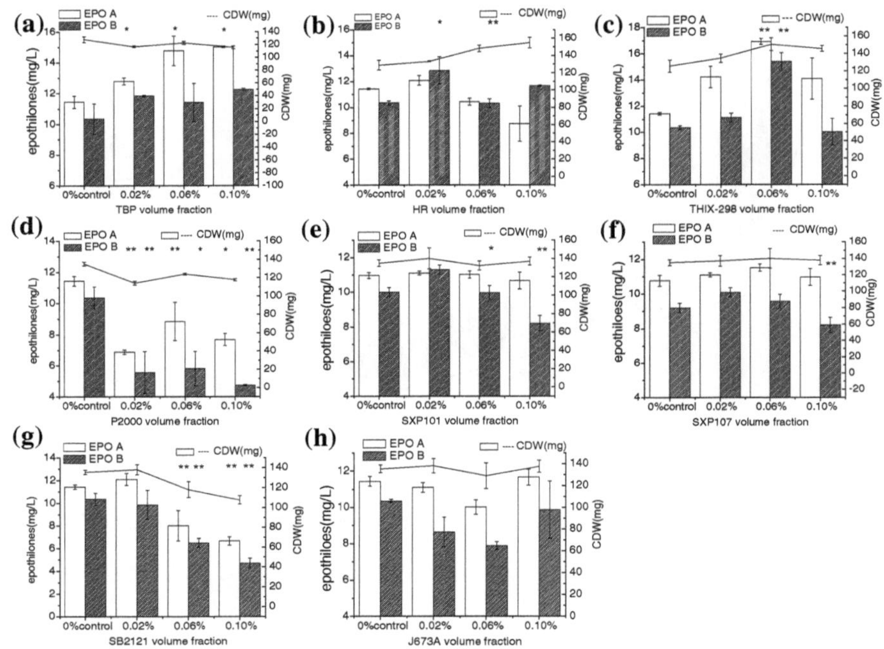

Fig. 22.2 Biomasses and yields of epothilones A and B with different antifoam agents addition (All the experiments were conducted thrice, where * = $P \leq 0.05$ and ** = $P \leq 0.01$.). **a** TBT volume fraction, **b** HR volume fraction, **c** THIX-298 volume fraction, **d** P2000 volume fraction, **e** SXP 101 volume fraction, **f** SXP 107 volume fraction, **g** SB2121 volume fraction, **h** J673A volume fraction

at 0.10 % volume fraction compared to the control ($P < 0.05$). This suggested that TBP might change the cell membrane permeability and increase the secretion of epothilones. Addition of 0.1 % HR increased the total biomasses from 125 to 155 mg compared to the control, representing the most significant increase of all the antifoam agents evaluated. However, the yields of epothilones were not improved at high volume fraction of HR. Addition of P2000 and SB2121 decreased the biomasses and yields of epothilones (Fig. 22.2d, g); it indicated that polypropylene glycol may inhibit the growth of *S. cellulosum.* While SXP101, SXP107 and J673A showed slight effects on the biomasses and the yields of epothilones. The highest yields of epothilones (A, B) were obtained when 0.06 % (v/v) THIX-298 was added, and increased about 48.12 and 48.55 % compared to the control, respectively, representing the largest effects of the antifoam agents tested. It also resulted in increased total biomasses in the range of 0.02–0.10 % (v/v). Overall, the different antifoam agents tested showed different effects on the cultures at 0.02–0.10 % (v/v) concentrations. These data showed more than one effect of antifoam action: one possibly changing the culture growth and a second due to increased cellular production levels of epothilones.

22.3.3 The Effects of THIX-298 on Culture Growth in Shake Flasks

From the results above, THIX-298 antifoam agents showed excellent abilities on suppression of foam. In addition, both the biomasses and yields of epothilones obtained the maximum at a concentration of 0.06 % volume fraction. So we subsequently observed the change in biomasses, epothilones, amino nitrogen, reducing sugar, and pH with 0.06 % THIX-298 addition during the fermentation process. Results are shown in Fig. 22.3; pH, amino nitrogen, and reducing sugar were affected compared to the control. The pH variation of the control medium was higher than that of antifoam-containing media. It was concluded that the addition of antifoam agent to the medium maintained the pH at a suitable level, which could provide a better growing condition for *S. cellulosum.* Reducing sugar and amino nitrogen were utilized more rapidly when the antifoam agent was added. The results obtained in Fig. 22.3b show that more biomasses were obtained in antifoam-containing media, which explained the rapid consumption of reducing sugar and amino nitrogen. Meanwhile, there was a correlation between cultures growth and yields of epothilones.

22.3.4 Fermenter Studies

Because of intense agitation and aeration in fermenter, there are different situations in the small vessels and fermenter. Based on the performance of foam control and the effects on the cultures, THIX-298 and TBP could improve the yields of epothilones. SXP107 showed excellent foam-control abilities and slight effects on cultures, so THIX-298, TBP, and SXP107 were selected to test in 50 L fermenter.

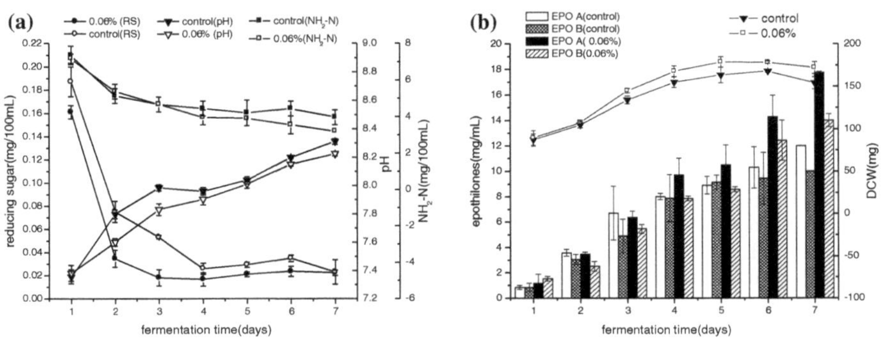

Fig. 22.3 Time course profile of reducing sugar, pH and NH_2–N in fermentation medium with 0.06 % v/v THIX-298 (Fig. 22.3a); the change of biomasses and epothilones with respect to time duration in fermentation medium (Fig. 22.3b)

Table 22.2 Foam suppression coefficient of antifoam agents

Antifoam agent	Foam height, h_f (cm)	Minimum foam height, $h_{f\min}$(cm)	Foam suppression coefficient, FSc
THIX-298	20.25	1.45	0.928
SXP107	19.85	6.85	0.848
TBP	21.25	10.65	0.499

Table 22.3 Comparison of antifoam agents in the fermenter

Antifoam agents	Final volume fraction, V_f (%)	Yields of epothilones, (A/B mg/L)	Cost coefficient, Cc	Efficiency coefficient, Ec
THIX-298	0.020	15.21/13.49	1	4640
SXP107	0.035	12.34/8.59	0.95	2544
TBP	0.260	10.62/7.58	1.6	119

As shown in Table 22.2, THIX-298 showed the best foam suppression followed by SXP107; the results obtained were similar to that observed in shake flasks. Compared to the other antifoam agents, THIX-298 showed better foam suppression with less volumetric fraction (Table 22.3). The time required to suppress the foam was also shorter. This allowed the foam to rise for a longer time before being destroyed by chemical antifoam agents. When choosing an antifoam agent, there are three factors to be considered: the volume required for foam control; not affecting the yields of products; fast destruction of foam. Therefore, several coefficients were used to compare these antifoam agents. Addition of THIX-298 led to the highest yields of epothilones among the antifoam used (Table 22.3). Also, the efficiency coefficient was higher than other antifoam agents. Thus, 0.02 % (v/v) THIX-298 was chosen to control the foam in epothilones fermentation.

Acknowledgments This work received financial support from Provincial Post-Doctoral Innovation Foundation (201203006) and Research projects of Science and Technology Division, Shandong (2012GSF12107).

References

1. Reichenbach H, Höfle G (2008) Discovery and development of the epothilones: a novel class of antineoplastic drugs. Drugs R.D 9:1–10
2. Gerth K, Bedorf N, Höfle G, Irschik H, Reichenbach H (1996) Epothilons A and B: antifungal and cytotoxic compounds from *Sorangium cellulosum* (*Myxobacteria*)-productions, physicochemical and biological properties. J Antibiot 49:560–564
3. Bollag DM, McQueney PA, Zhu J, Hensens O, Koupal L, Liesch J, Goetz M, Lazarides E, Woods CM (1995) Epothilones, a new class of microtubule-stabilizing agents with a taxol-like mechanism of action. Cancer Res 55:2325–2333

4. Gong G, Sun X, Liu X, Hu W, Cao W, Liu H, Liu W, Li Y-Z (2007) Mutation and a high-throughput screening method for improving the productions of Epothilones of *Sorangium*. J Ind Microbiol Biotechnol 34:615–623
5. Routlegde SJ, Hewitt CJ, Bora N, Bill RM (2011) Antifoam addition to shake flask cultures of recombinant *Pichia pastoris* increases yield. Microb Cell Fact 10:17
6. Joshi K, Jeelani S, Blickenstorfer C, Naegeli I, Windhab E (2005) Influence of fatty alcohol antifoam suspensions on foam stability. Colloids Surf A 263:239–249
7. Varley J, Brown A, Boyd R, Dodd P, Gallagher S (2004) Dynamic multipoint measurement of foam behavior for a continuous fermentation over a range of key process variables. Bio chem Eng J 20:61–72
8. Denkov ND, Krastanka M, Christova C, Hadjiiski A, Cooper P (2000) Mechanisms of action of mixed solid-liquid antifoams: 3. Exhaustion and reactivation. Langmuir 21:8163–8619
9. Ju LK, Sundararajan A (1995) The effects of cells on oxygen transfer in bioreactors. Bioprocess Eng 13:271–278
10. Duitschaever CL, Buteau C, Kamel BS (1988) An investigation on the efficiency of antifoaming agents in aerobic fermentation. Process Biochem 23(6):163–165
11. Vidyarthi AS, Desrosiers M, Tyagi RD (2000) Foam control in biopesticide productions from sewage sludge. J Ind Microbiol Biotechnol 25:86–92

Chapter 23
Acute Toxicity by Four Kinds of Oil Dispersants in *Cynoglossus semilaevis*

Jinwei Gao, Wenli Zhou and Ruinan Chen

Abstract This study evaluated the in vivo toxicity of four kinds of oil dispersants to *Cynoglossus semilaevis*, in acute exposures of 24, 48, 72, and 96 h through mortality. The 24 h LC_{50} of RS-1, HH, GM-1, and FK were 17.0793, 4.6957, 3.2758, and 4.7510 mg/L, respectively. The 48 h LC_{50} were 16.0173, 3.7251, 2.2135, and 2.5568 mg/L. The 72 h LC_{50} were 14.6422, 2.9266, 1.9983, and 1.8877 mg/L. The 96 h LC_{50} were 13.7879, 2.3060, 1.4391, and 1.5962 mg/L. The decreasing order of toxicity was GM-1 > HH > FK > RS-1. Compared with the grading table of toxicity of chemical dispersants, four kinds of oil dispersants were highly toxic.

Keywords *Cynoglossus semilaevis* · Oil dispersant · Toxicity · Median lethal concentration (LC_{50})

23.1 Introduction

Cynoglossus semilaevis are the warm-temperated demersal fishes, which belong to Pleuronectiformes, Cynoglossidae, and *Cynoglossus Buchanan-Hamiltou*. They distribute on the coastal area of our country, which are the large, precious, and economical fishes located in the bottom of the offshore in our country [1]. They have become ideal culture and proliferation target, which have wide developing prospect [2–4].

With crude oil demand increase, oil spill pollution is increasing constantly which threaten seriously Marine Ecological Environment [5]. According to statistics, the oil and their products into ocean can reach about 10–15 million tons every year [6, 7]. Oil pollution has become one of the three main marine pollutants in China.

J. Gao · W. Zhou (✉) · R. Chen
Key Laboratory of Aqua-ecology and Aquaculture of Tianjin, College of Fisheries, Tianjin Agricultural University, Tianjin 300384, People's Republic of China
e-mail: wenlizhou@tjau.edu.cn

T.-C. Zhang and M. Nakajima (eds.), *Advances in Applied Biotechnology*, Lecture Notes in Electrical Engineering 332, DOI 10.1007/978-3-662-45657-6_23

Oil dispersants are one of the most important treatments to oil pollution. However, they have harm to the marine organism. There is less and exact related reports [8–11]. Because of the lack of basic research data on the pollution of oil and oil dispersants, the evaluations of the accident lost lack evidence and the progress and result of environmental remediation lack theoretical guidance. This paper aims to analyze the acute toxic effect of the four kinds of oil dispersants on the *C. semilaevis* and provide theoretical basis and technical support for screening of oil dispersant, determining applicable condition and optimal using dosage.

23.2 Materials and Methods

Carebara semilaevis was provided by Tianjin Lida Seawater Resources Limited Company, weight 2.71 ± 0.43 g and length 7.93 ± 0.51 cm. Fish were kept for 1 week of acclimation before the start of the experiment. During the toxicity test, the fish did not receive any food. The variables of the water (salty = 31.08 ± 0.18, pH = 8.05 ± 0.03, temperature = 20.5 ± 0.4 °C, dissolved oxygen >5 mg/L) were kept within the comfort range for these fish, so that these parameters would not influence the results.

Heng Hai green oil dispersant RS-1 type (RS-1), Han Hai oil dispersant (HH), Guang Ming oil dispersant GM-1 type (GM-1), and Fu Ken oil dispersant of Guangzhou (FK) were provided by CNOOC (Tianjin) Environmental Protection Service Co. Ltd. Four kinds of oil dispersants were diluted with aseptic seawater (salinity 30 ± 1) into 10 g/L and preserved in brown volumetric flask by mixing for 1 h with continuous gradient dilution. The treatment was according to the standard of GB/18188.1-2000.

From preliminary range-finding toxicity tests, the lowest concentration that fish died all in 24 h and the highest concentration that fish survived all in 96 h were determined. Thus, seven concentrations of oil dispersants were used in these trials (Table 23.1). In each group, 10 fish were randomly distributed in aquaria. The experimental conditions are kept consistent with acclimation and, respectively, record the number of dead fish in 24, 48, 72, and 96 h.

Table 23.1 Concentration gradient of four kinds of oil dispersants

Kinds of oil dispersants	Concentration gradient (mg/L)						
RS-1	0	10	11.7	13.7	16	18.8	22
HH	0	1	1.6	2.7	4.4	7.3	12
GM-1	0	1	1.5	2.2	3.2	4.7	7
FK	0	1	1.6	2.5	4	6.3	10

Table 23.2 Toxicity classification of chemical dispersants

Toxicity classification	Concentration (mg/L)
High toxicity	1–100
Medium toxicity	100–1,000
Low toxicity	1,000–10,000

For this trial, a fully randomized design was used, with seven treatments and three repetitions. The experiment was performed using average. The median lethal concentration (LC_{50}) of oil dispersants in 24, 48, 72, and 96 h were obtained by Bliss method and determined the toxicity compared with the grading table of toxicity of chemical dispersants (Table 23.2).

23.3 Results and Discussion

According to the experimental results, LC_{50} of four kinds of oil dispersants in 24, 48, 72, and 96 h were obtained (Table 23.3).

Compared with the grading table of toxicity of chemical dispersants, four kinds of oil dispersants were highly toxic on *C. semilaevis*. The decreasing order of toxicity was GM-1 > HH > FK > RS-1. Compared with the effects of environmental factors which had been researched [2–4, 12, 13], oil dispersants had significant acute toxic effects on *C. semilaevis*. So in order to get more accurate experimental results, observation indexes should be enriched and experimental time should be prolonged.

With crude oil demand increase, oil spill pollution is increasing constantly which threaten seriously Marine Ecological Environment [14]. Oil dispersant is composed of surfactant, solvent, and some additives [15–17]. In the present study, the acute toxicity in *C. semilaevis* increased with the concentration of oil dispersants increasing, but decreased with time. These results were not in agreement with those found by several authors who proved that the toxic effects of many kinds of oil dispersants on *Venerupis philippinarum* and *Aboma lactipes* increased with the increase in time and the concentration [18]. Consequently, different species of marine organism had different degrees of sensitivity to oil dispersants.

Table 23.3 LC_{50} of four kinds of oil dispersants in different time

Kinds of oil dispersants	LC_{50} (mg/L)			
	24 h	48 h	72 h	96 h
RS-1	17.0793	16.0173	14.6422	13.7879
HH	4.6957	3.7251	2.9266	2.3060
GM-1	3.2758	2.2135	1.9983	1.4391
FK	4.7510	2.5568	1.8877	1.5962

Acknowledgments This work was supported by Key Projects of Applied Basic and Frontier technology Research of Tianjin (13JCZDJC29300); Scientific research plan program of Tianjin Agricultural University (2009D005).

References

1. Ma AJ, Wang XA, Zhuang ZM et al (2007) Structure of retina and visual characteristics of the half-smooth tongue-sole *Cynoglossus semilaevis* Günter. Acta Zool Sin 53(2):354–363
2. Wan RJ, Jiang YW, Zhuang ZM (2004) Morphological and developmental characters at the early stages of the tonguefish *Cynoglossus semilaevis*. Acta Zool Sinica 50(1):91–102
3. Sun ZZ, Liu XZ, Xu RJ et al (2007) Technique and technology of productive scale seedling rearing of *Cynoglossus semilaevis* Günter. J Fish Sci China 14(2):244–248
4. Ji XS, Chen SL, Ma HY et al (2010) Natural sex reversal of female *Cynoglossus semilaevis* in rearing populations. J Fish Sci China 34(2):322–327
5. Li PF, Chen LL (2002) Discussion on chemical dispersants. Environ Prot Transp 23(3):30–32
6. Huang YJ, Chen QZ, Zeng JN et al (2010) Acute toxicity of crude oil and dispersant to marine copepods. Chin J Appl Environ Biol 16(4):566–571
7. Yang QX, Gao GZ, Su J et al (1997) Effect of chemical dispersants on the growth of *Ulua pertusa*. Acta Oceanol Sinica 19(3):45–49
8. Chen HJ, Fang JM, Yin CS et al (2012) Acute toxicity of chemical enhanced water accommodated fraction of dispersed oil to *Apostichopus japonicas* Juveniles. Mar Environ Sci 31(3):414–417
9. Lv FR, Xiong DQ (2010) Effect of dispersant and No. 0 diesel oil pollution on sea urchin, *Hem icentrotus pulcherrimus*. Mar Environ Sci 29(3):328–331
10. Bao TM, Guan LJ, Ma AQ et al (2012) Study on influencing factors of chemical oil dispersant emulsification. Period Ocean Univ China 42(9):053–058
11. Zhao YH, Zhang YC, Sun PY (2006) Application and further development of dispersants in pollution control of oil spill. Mar Environ Sci 25:97–100
12. Zhou XZ, Xu YJ, Liu NZ et al (2009) Study on histological and morphometric characters of gonad development of *Cynoglossus semilaevis* Günter. Prog Fish Sci 30(6):25–35
13. Zhou LQ, Yang AG, Liu XZ et al (2005) The karyotype of the tonguefish *Cynoglossus semilaevis*. J Fish China 29(3):417–419
14. Ma H, Chen G, Ma L, Hu Q (2011) A first set of gene-derived polymorphic microsatellite markers in half-smooth tongue sole (Cynoglossus semilaevis). Environ Boil Fish 91:501–504
15. Brookfield JFY (1996) A simple new method for estimating null allele frequency from heterozygote deficiency. Mol Ecol 5:453–455
16. Li Q, Wan JM (2005) SSRHUNTER: development of local searching software for SSR sites. Hereditas 27:808–810
17. Liao XL, Shao CW, Tian YS et al (2007) Polymorphic dinucleotide microsatellites in tongue sole (Cynoglossus semilaevis). Mol Ecol Resour 7:1147–1149
18. Yang B, Guan M, Xu HG et al (1991) Toxic effect of oil dispersants on marine biology. Mar Environ Sci 10(4):14–20

Chapter 24
Imprinted Cross-Linked Enzyme Aggregate (iCLEA) of Phenylalanine Ammonia Lyase: A New Stable Biocatalyst

Jian Dong Cui, Rong Lin Liu and Lian Lian Li

Abstract The industrial use of phenylalanine ammonia lyase (PAL), an interesting biocatalyst for manufacture of L-phenylalanine (L-Phe) by reversing the enzyme reaction with high concentration of trans-cinnamic acids (t-CA) and ammonia, has been hampered by a lack of long-term stability and low activity toward substrates. In this study, it is shown that the PAL activity of such a CLEA can be improved by molecular imprinting with a suitable substrate. PAL was imprinted with t-CA and subsequently cross-linked with glutaraldehyde (iCLEAs). Compared to free PAL, PAL stability in the iCLEAs against substrate inhibition was significantly improved, furthermore, the iCLEAs exhibited good reusability. These results indicated that the procedure might be used as a feasible and efficient solution for improving properties of immobilized enzyme in industrial application.

Keywords Phenylalanine ammonia lyase · Immobilization · Imprinting · CLEA · L-phenylalanine

24.1 Introduction

Phenylalanine ammonia-lyase (E.C.4.3.1.24-PAL) belongs to the family of lyases [1–3]. It has been used chiefly in the manufacture of L-phenylalanine (L-Phe) by reversing the enzyme reaction with high concentration of trans-cinnamic acids

J.D. Cui (✉) · R.L. Liu · L.L. Li
College of Bioscience and Bioengineering, Hebei University of Science and Technology, 050000 Shijiazhang, People's Republic of China
e-mail: cjd007cn@163.com

J.D. Cui
Key Laboratory of Industry Fermentation Microbiology, Ministry of Education, Tianjin University of Science and Technology, Tianjin 300457, People's Republic of China

T.-C. Zhang and M. Nakajima (eds.), *Advances in Applied Biotechnology*,
Lecture Notes in Electrical Engineering 332, DOI 10.1007/978-3-662-45657-6_24

(t-CA) and ammonia at an elevated pH. L-Phe is an important amino acid and widely used in food and pharmaceutical industries. Its production has been stimulated by the great demand on the diet sweetener aspartame, a peptide made up of L-Phe and L-aspartic acid. The enzyme is therefore of particular interests to researchers in the biotechnological industry. The industrial-scale production of PAL mainly utilizes the genus *Rhodotorula* [4, 5]. Unfortunately, the use of PAL as industrial biocatalyst has been limited by a lack of long-term stability and low activity during the bioconversion [2, 4, 6]. Especially, PAL activity is restrained by high concentration of substrate [6, 7]. These drawbacks have promoted the search for strategies dedicated to enhance the stability of the enzyme such as enzyme immobilization (with or without support), enzyme modification, genetic modification and medium engineering [6, 8–10]. Although these stabilization methods enable to improve enzyme stability, lower specific activity and enzyme activity decrease still hamper the development of enzyme based bioprocesses. Recently, cross-linked enzyme aggregates (CLEA), a promising carrier free immobilization strategy, has been proposed as an efficient immobilization method to stabilize enzyme activity [11–13]. Successful preparation of CLEAs from a broad range of enzymes, including laccase, lipases, and amidase, has been reported [11, 14, 15]. In addition, the bioimprinting technique is also a versatile tool to improve enzymes in catalytic performance, enantioselectivity, operational stability, and recycling in industry. The substrate affinity or enantioselectivity of an enzyme may be rationally adapted by this technique with substrate-like ligands [7, 16–18]. Consequently, the formation of a new generation of biocatalysts based on a combination of CLEAs and bioimprinting techniques is a way to enhance efficiency of biocatalyst. The first objective of the present work was to synthesize an imprinted CLEAs of PAL from *Rhodotorula glutinis* (*R. glutinis*) by the combination of bio-imprinting and CLEAs technology. The second objective was to assess the feasibility of utilizing imprinted PAL-CLEAs (iCLEAs) as the biocatalyst for synthesis of L-Phe.

24.2 Materials and Methods

24.2.1 Materials

Glutaraldehyde was obtained from Sigma-Aldrich Inc. (St. Louis, MO, U.S.A.). All other reagents were of analytical grade.

24.2.2 Production and Partially Purified of PAL in R. glutinis

The strain of *R. glutinis* (CICC 32 917) used throughout this study was obtained from China Center of Industrial Culture Collection (CICC, Beijing, China). The strain was maintained on potato-glucose agar slant at 4 °C. The medium for seed culture consisted of (g/L) glucose 5, peptone 10, yeast extract 10, NaCl 5 and KH_2PO_4 1, pH 5.5. The PAL production medium (g/L) included glucose 1, peptone 35, NaCl 5, KH_2PO_4 0.25 and $(NH_4)_2HPO_3$ 1.5, pH 5. For seed preparation, the cells on the slant stock medium were cultivated for 3 days at 30 °C, and were inoculated into 50 ml seed medium in a 250 Erlenmeyer flask and cultivated on a 150 rpm reciprocal shaker for 20 h at 30 °C. Then, 2 ml seed culture was inoculated into a 250 ml Erlenmeyer flask with 48 ml PAL induction medium and cultivated on a reciprocal shaker for 72 h at 150 rpm and 30 °C. The culture broth was harvested by centrifugation at 6000 × g for 10 min. After washing with 0.9 % sterilized saline and water, the cells were resuspended in a potassium phosphate buffer (10 mM, pH 7.5). Then, the cells were disrupted by intermittent ultrasonication, the cell extracts were centrifuged at 10 000 × g for 15 min. Solid $(NH_4)_2SO_4$ was added to bring the supernatant to 55 % saturation and centrifuged (10 000 × g) after 30 min. The precipitate was dissolved in 25 mM-Tris-HCI buffer, pH8.8, and centrifuged. The clear supernatant (10 ml) was loaded on a column (15 cm × 1 cm) of ion exchange chromatography in DEAE-Sephacel pre-equilibrated with the same buffer. The enzyme was eluted with a linear gradient of 75 ml of 25 mM-Tris-HCI buffer, pH8.8, and 75 ml of the same buffer containing 300 mM-NaCl. The active fractions from the DEAE-Sephacel column (Beijing DingGuo ChangSheng Biotechnology Co., Ltd, China) were pooled, and a partially purified PAL was obtained and utilized for the preparation of CLEAs and iCLEAs.

24.2.3 Preparation of CLEAs and iCLEAs

For the preparation of CLEAs, aggregates were made using 60 % ammonium sulfate saturation to a solution of partially purified PAL in 25 mM Tris-HCI buffer at pH 8.0 under gentle stirring at 4 °C for 1 h. The glutaraldehyde (10 % v/v in water) was slowly added to the mixture. The mixture was kept at 4 °C for 2 h with constant shaking at 200 rpm. Then the suspension was centrifuged at 10 000 × g for 10 min at 4 °C. The pellet was washed for three times by 25 mM Tris-HCl buffer pH 8.0 and finally stored in 25 mM-Tris-HCI buffer pH 8.8 at 4 °C.

For iCLEAs, the imprinting molecules was added to a solution of partially purified PAL in 25 mM Tris-HCI buffer at pH 8.0, and allowed to react with PAL for 30 min at 30 °C. Next, 60 % ammonium sulfate saturation was added to the imprinted PAL solution under agitation. After 1 h, a certain amount of glutaraldehyde (10 % v/v in water) were slowly added to the mixture and stirred at 4 °C for 2 h. Finally, the iCLEA was harvested by centrifugation (10 000 × g, 4 °C, 10 min) and washed five times with 25 mM-Tris-HCI buffer at pH 8.8.

24.2.4 Bioconversion Experiments

For the preparation of the substrate solution, a certain amount of t-CA was dissolved in 28 % ammonium hydroxide. The resulting solution was adjusted to pH 10.0 with H_2SO_4 and diluted with distilled water in a final volume of 100 ml. t-CA was converted to L-Phe as follows: A reaction mixture of 1.0 ml enzyme samples and 4.0 ml substrate solution was incubated at 30 °C for a specified time. The reaction mixture was centrifuged, and L-Phe concentration in supernatant was analyzed by high performance liquid chromatography (HPLC). HPLC analysis was performed on a reverse phase C18 column (4.6 × 250 nm, ODS-100S, Shimadzu, Japan). The highly purified L-Phe (>99 %) was dissolved in HPLC-grade methanol to make a 5 g/L L-Phe stock solution. A standard curve was established for L-phenylalanine using 0.5, 1.0, 2.0, 3.0, 4.0 g/L L-Phe versus the corresponding area number of spectra. Samples were eluted isocratically with methanol: water (20/80, V/V) at a flow rate of 1.0 ml/min and detected at 259 nm using an UVD 170U detector (Dionex, P680, USA). Reusability of iCLEAs was evaluated in batch operation mode in the biotransformation of L-phenylalanine reaction. The reaction using 50 mg iCLEAs and 10 g/L t-CA in 28 % ammonium hydroxide (pH 10.0) was carried out at 30 °C for 2 h. After each cycle, iCLEAs were separated using centrifugation separation, washed with distilled water and then suspended again in a fresh reaction mixture. L-Phe concentration in supernatant was analyzed by HPLC.

24.3 Results and Discussion

As shown in Fig. 24.1, the steps in the preparation of the iCLEA consist of imprinting, followed by precipitation. Figure 24.2 showed the effects of different imprinting molecules for iCLEA on L-Phe production. The results showed that L-Phe production was affected significantly by different imprinting molecules. The highest L-Phe production was obtained when t-CA was used as imprinting molecules. L-Phe concentration was improved 63 % as compared with that of a non-imprinted CLEA (control). However, imprinting with nitrocinnamic acid, fluorocinnamic acid or coumaric acid was found to almost have no positive influence on L-Phe production. It was indicated that imprinting with t-CA almost improved the specific activity on t-CA. The previous reports showed that all PAL preparation were inhibited by t-CA, the concentrations of t-CA greater than 135 mM were enough cause substrate inhibition in the reversed reaction, the strong substrate inhibition leads to low reaction rates [4]. To further assess iCLEA's applicability in L-Phe conversions, the effects of high concentration of substrate on the iCLEA preparation was evaluated.

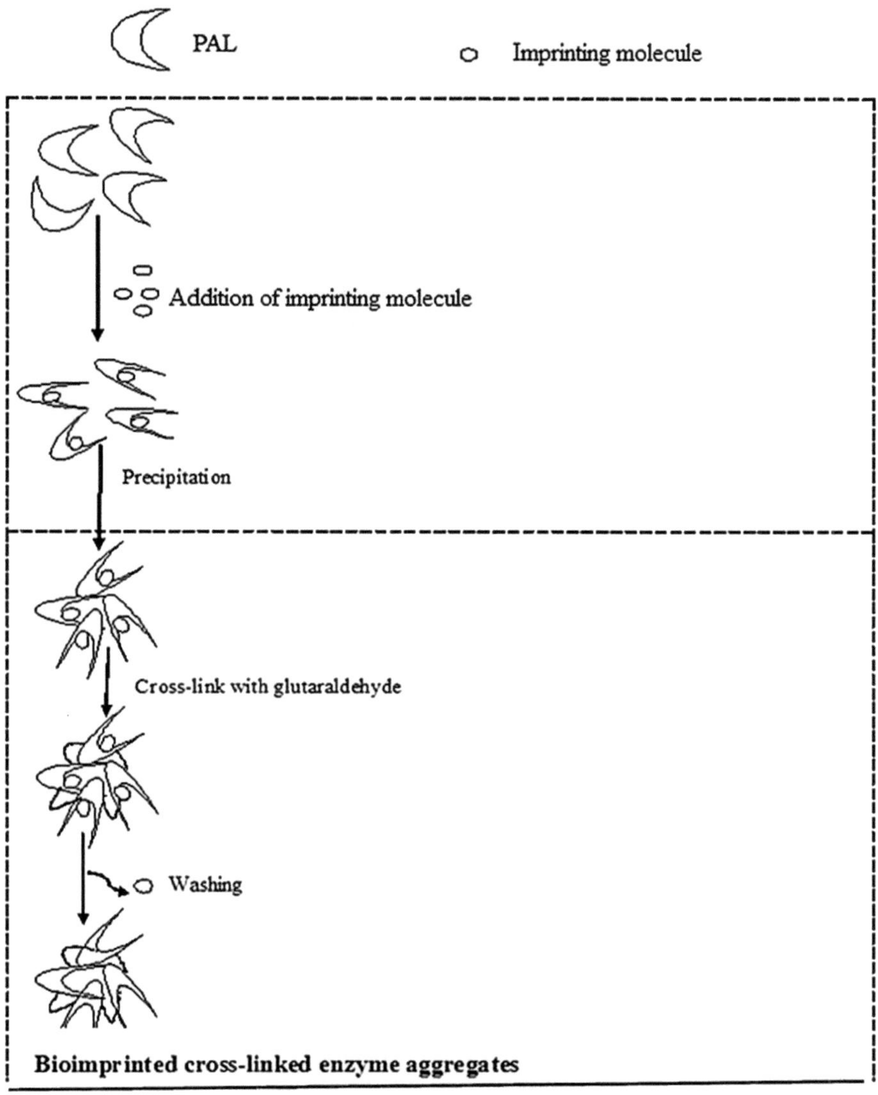

Fig. 24.1 General scheme for the production of iCLEAs

As shown in Fig. 24.3, free PAL activity was inhibited significantly by 20 g/L t-CA, however, for CLEAs and iCLEAs, this inhibitory effect on PAL by t-CA were decreased obviously. Especially, compared to non-imprinted CLEAs, all of iCLEAs exhibited more tolerant against high concentration substrate inhibition.

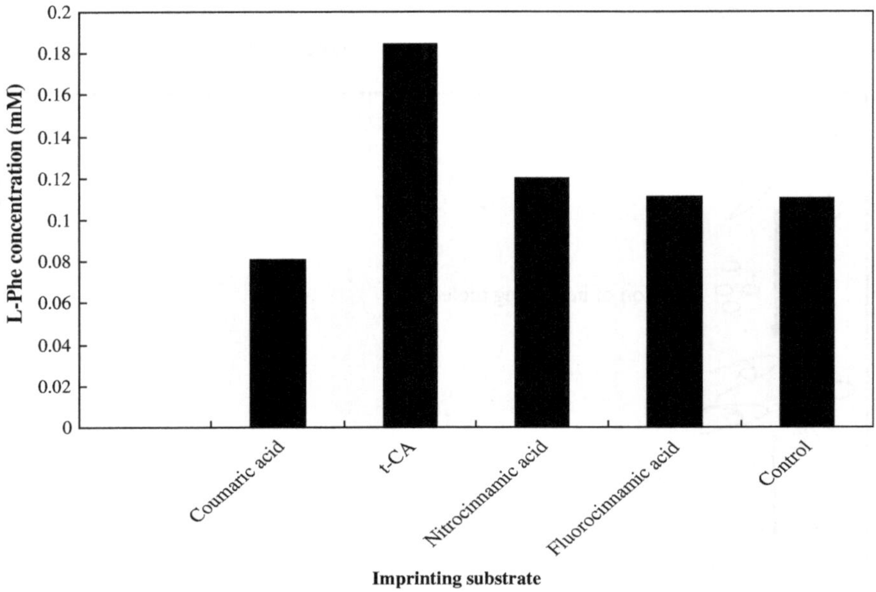

Fig. 24.2 The effects of different imprinting molecules for iCLEA on L-Phe production

Moreover, among iCLEAs, imprinting with t-CA (t-CA-CLEAs) showed the higher resistance against high concentration of the substrate than that of other imprinting molecules. Generally, PAL activity is known to decline rapidly during the bio-conversion process, thus improving stabilization of PAL is necessary to enhance L-Phe yield. The effects of conversion time on L-Phe production in high concentration of substrate (20 g/L t-CA) were evaluated. The results were showed in Fig. 24.4. The highest L-Phe yield was increased observed at 3 h when free PAL was used as a biocatalyst. However, after 3 h conversion, L-Phe yield was not increased further. In contrast, for CLEAs and iCLEAs (t-CA-CLEAs), L-Phe yield was enhanced with increasing conversion time, after 3 h conversion, L-Phe yield was still increased. Furthermore, the highest L-Phe yield was obtained when the t-CA-CLEAs was used as a biocatalyst, which indicated that CLEAs imprinted with t-CA were stability in high concentration of substrate during continuous conversion process.

In addition, Recyclabilities of t-CA-CLEAs in batch operation mode was assessed. Figure 24.5 showed the L-Phe concentration in each cycle. In Fig. 24.5, although the L-Phe concentration decreases to 0.23 g/L after the first cycle and it drops to 0.1 g/L after 5 cycles. However, we also found that t-CA-CLEAs still produced 0.03 g/L of L-Phe after 9 cycle of conversion. From these results, we can conclude that the iCLEAs is reusable.

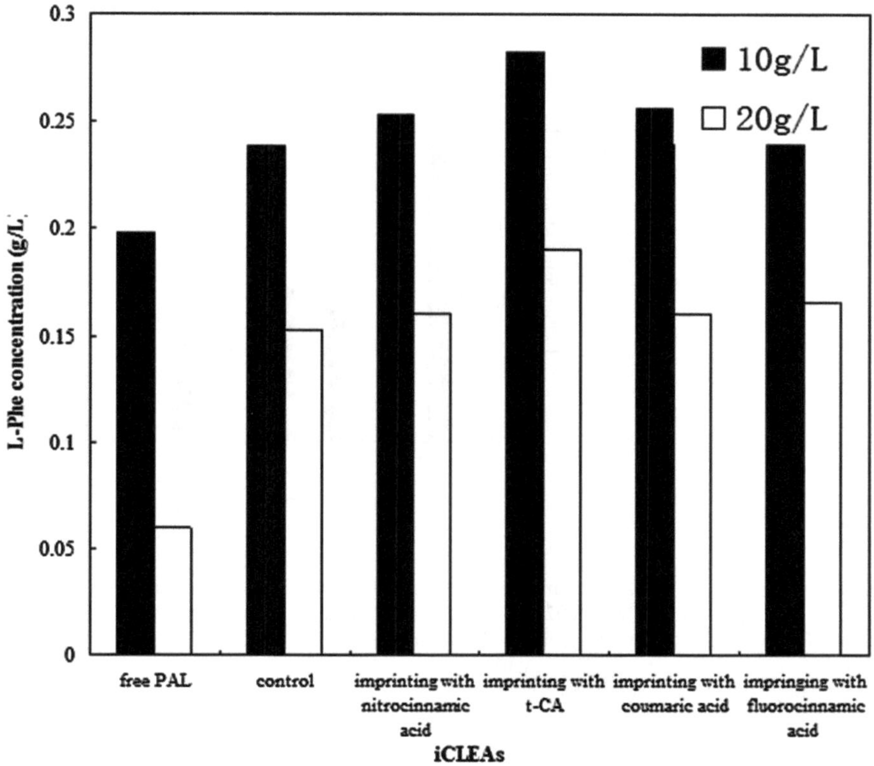

Fig. 24.3 The effects of tran-cinnima acids concentration on activity of iCLEAs

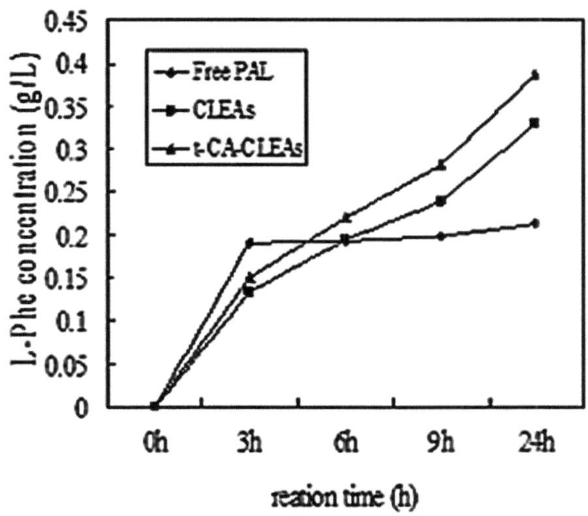

Fig. 24.4 The effects of conversion time on L-Phe production

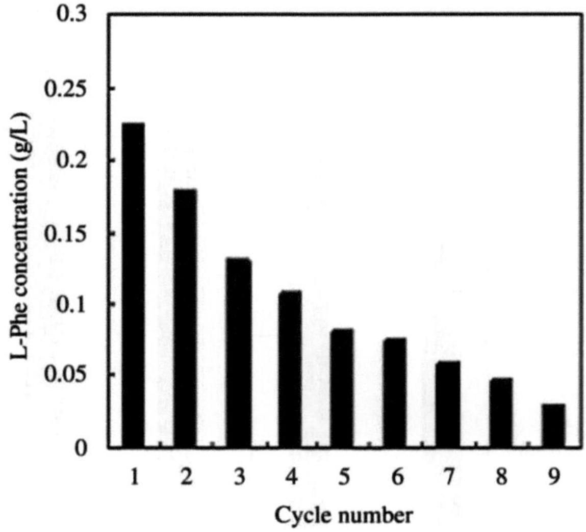

Fig. 24.5 Recyclabilities of t-CA-CLEAs in high concentration of substrate

Acknowledgments The Project Partially Supported by the National Natural Science Foundation of China (NSFC, Project No. 21072041), the Natural Science Foundation of Hebei Province, China (Project No. B2014208054), and the Foundation (No. 2012IM004) of Ministry of Education Key Laboratory of Industrial Fermentation Microbiology (Tianjin University of Science and Technology), People's Republic of China,

References

1. Kalghatgi KK, Subba RPV (1975) Microbial L-phenylalanine ammonia-lyase: purification, subunit structure and kinetic properties of the enzyme from *Rhizoctonia solani*. Biochem J 149:65–70
2. D'Cunha GB, Satyanarayan V, Nair PM (1996) Purification of phenylalanine ammonia lyase from *Rhodotorula glutinis*. Phytochemistry 42:17–22
3. Moffitt MC, Louie GV, Bowman ME et al (2007) Discovery of two cyanobacterial phenylalanine ammonia lyases: kinetic and structural characterization. Biochemistry 46:1004–1012
4. Yamada S, Nabe K, Izuo N et al (1981) Production of L-phenylalanine from trans-cinnamic acid with R. glutinis containing L-phenylalanine ammonia lyase activity. Appl Environ Microbiol 42:773–778
5. El-Batal AI (2002) Continuous production of L-phenylalanine by *R. glutinis* immobilized cells using a column reactor. Acta Microbiol Pol 51:153–158
6. Zhang BZh, Cui JD, Zhao GX et al (2010) Modeling and optimization of phenylalanine ammonia Lyase stabilization in recombinant *Escherichia coli* for the continuous synthesis of L-phenylalanine on the statistical based experimental designs. J Agric Food Chem 58:2795–2800
7. El-Batal AI (2002) Optimization of reaction conditions and stabilization of phenylalanine ammonia lyase-containing *Rhodotorula glutinis* cells during bioconversion of *trans*-cinnamic acid to L-phenylalanine. Acta Microbiol Pol 51:139–152

8. D' Cunha GB, Satyanarayan V, Nair PM (1996) Stabilization of phenylalanine ammonia lyase containing *Rhodotorula glutinis* cells for the continuous synthesis of L-phenylalanine methyl ester. Enzyme Microb Technol 19:421–427
9. Jia ShR, Cui JD, Li Y (2008) Production of L-phenylalanine from *trans*-cinnamic acids by high-level expression of phenylalanine ammonia lyase gene from *Rhodosporidium toruloides* in *Escherichia coli*. Biochem Eng J 42:193–197
10. Quinn AJ, Pickup MJ, D'Cunha GB (2011) Enzyme activity evaluation of organic solvent-treated phenylalanine ammonia lyase. Biotechnol Prog 27:1554–1560
11. Wilson L, Fernández-Lorente G, Fernández-Lafuente R et al (2006) CLEAs of lipases and poly-ionic polymers: A simple way of preparing stable biocatalysts with improved properties. Enzyme Microb Technol 39:750–755
12. Dong T, Zhao L, Huang Y et al (2010) Preparation of cross-linked aggregates of aminoacylase from *Aspergillus melleus* by using bovine serum albumin as an inert additive. Bioresource. Technol. 101:6569–6584
13. Cui JD, Zhang S, Sun LM (2012) Cross-linked enzyme aggregates of phenylalanine ammonia lyase: novel biocatalysts for synthesis of L-phenylalanine. Appl Biochem Biotech 167:835–844
14. Joo PH, Uhm KN, Kim HK (2010) Biotransformation of amides to acids using a co-cross-linked enzyme aggregate of *Rhodococcus erythropolis* amidase. J Microbiol Biotechnol 20:325–330
15. Vinoth Kumar V, Prem Kumar MP, Thiruvenkadaravi KV et al (2012) Preparation and characterization of porous cross linked laccase aggregates for the decolorization of triphenyl methane and reactive dyes. Bioresource Technol. 119:28–34
16. Stahl M, Jeppssonwistrand U, Mansson MO et al (1991) Induced stereo- and substrate selectivity of bioimprinted alpha-chymotrypsin in anhydrous organic media. J Am Chem Soc 113:9366–9368
17. Fishman A, Cogan U (2003) Bio-imprinting of lipases with fatty acids. J. Mol. Catal. B-Enzym. 22:193–202
18. Nie GJ, Zheng ZM, Gong GH et al (2012) Characterization of bioimprinted tannase and its kinetic and thermodynamics properties in synthesis of propyl gallate by transesterification in anhydrous medium. Appl Biochem Biotechnol 167:2305–2317

Chapter 25
Effects of Calcium on the Morphology of *Rhizopus oryzae* and L-lactic Acid Production

Yong-Qian Fu, Long-Fei Yin, Ru Jiang, Hua-Yue Zhu and Qing-Cheng Ruan

Abstract The effects of exogenous calcium on fungal pellet morphology during preculture and L-lactic acid production were studied. The results showed that addition of exogenous calcium could induce pellet formation. The diameter of the pellet increased with increasing concentration of exogenous calcium, including $CaCl_2$ and $CaCO_3$. The smaller pellet precultured with low concentration of soluble calcium ($CaCl_2$) was beneficial for L-lactic acid production because the pellet was dense and the large inner part of the pellet was inactive. By contrast, the larger pellet precultured with high concentration of insoluble calcium ($CaCO_3$), except 8.0 g/L $CaCO_3$, was beneficial for L-lactic acid production. Supported by the $CaCO_3$ powder, the entire biomass layer was fully active, and the highest L-lactic acid productivities of 1.22 g/L h and 58.6 g/L L-lactic acid were reached using the 1.5 mm pellet.

Keywords *Rhizopus oryzae* · Morphology · Exogenous calcium · L-lactic acid · Pellet

25.1 Introduction

Submerged cultures of filamentous fungi are widely used to provide important biotechnological products, such as enzymes, organic acid, and antibiotics, which have a lot of applications in food, medical, pharmaceutical, chemical, and textile industries [1]. However, the filamentous fungi growth characteristic brings a number of process engineering problems attributed to the morphological change accounted during the fermentation process in large scale [2]. Three extreme morphologies of

Y.-Q. Fu (✉) · L.-F. Yin · R. Jiang · H.-Y. Zhu · Q.-C. Ruan
Institute of Biomass Resources, Taizhou University, Jiaojiang 318000, Zhejiang, People's Republic of China
e-mail: fuyq@tzc.edu.cn

T.-C. Zhang and M. Nakajima (eds.), *Advances in Applied Biotechnology*,
Lecture Notes in Electrical Engineering 332, DOI 10.1007/978-3-662-45657-6_25

filamentous fungi have been reported, namely suspended mycelial, pellet, and clump morphology [3]. The morphology of filamentous fungi in submerged cultivation has been a subject of considerable interest for many years. The fungal growth in pellet form is a favorable alternative [4, 5] because it makes repeated-batch fungal fermentation possible. This form is also favored because it significantly improves culture rheology, which results in better mass and oxygen transfer into the biomass and lower energy consumption for aeration and agitation. Numerous studies have been carried out to control fungal morphology in pellet form [6–8]. However, most early studies mainly focused on environmental factors, such as medium composition, inoculum, pH, medium shear, additives (polymers, surfactants, and chelators), culture temperature, and medium viscosity. For individual strains, each factor has a different importance to the growth morphologies; some strains (e.g., *Rhizopus* sp.) need strong agitation to form pellets, whereas other strains (e.g., *Penicillium chrysogenum*) require high pH [4, 9]. Thus, most studies on fungal pellet formation are limited to the level of the individual strain because of the shortage of mechanisms of morphogenesis.

Environmental conditions maybe markedly influence the growth pattern of filamentous fungi, which can range from a dispersed filamentous form to pellet. However, Braun and Vecht-Lifshitz [10] reported in their study that the pellet morphology of a filamentous microorganism developing in any fermentation system may be represented as a final result of the competing influences, which is the equilibrium between the forces of cohesion and disintegration. Shear forces may be unambiguously assigned the function of disintegrating factors. The hyphal extension and branching rate thereby affected the mycelium cohesion. The morphology of a mycelium and final fungal morphology are mainly determined by the mechanisms that regulate the polarity and direction of hyphal growth, as well as the frequency with which they branch [11, 12]. A typical fungal hypha grows out of a single cell-spore as a multinucleate tube containing cytoplasm, which moves within a hypha toward the hyphal tip, where it grows. During normal tip growth, a delicate balance must exist among the deposition of new material, synthetic activity, lytic activity, and turgor pressure, which provides the force for elongation [2, 3, 11]. Phosphoinositides, calcium, calmodulin, and cyclic nucleotides, especially Ca^{2+}, are involved in the mechanisms that regulate hyphal extension and branching, and ultimately affect fungal morphology [13, 14]. Several studies have investigated the effects of calcium on the hyphal extension and branching. For example, Robson et al. [15] studied the effects of Ca^{2+} on the regulation of hyphal extension and branching in *Fusarium graminearum* A 3/5, and illustrated that low Ca^{2+} concentrations increase the mean hyphal extension rate and hyphal growth unit length. Their results showed that the treated mycelia become more sparsely branched in *F. graminearum*, which was similar to the results of Robson et al. [16]. Jackson and Heath [17] also found in their research that increasing the external Ca^{2+} concentration generally resulted in an increased rate of hyphal extension and in a decreased frequency of branching. Very high concentration of external Ca^{2+} (>50 mM) will inhibit tip extension. A few studies also researched the effects of Ca^{2+} on the morphology of fungi in the macro-morphology. For example, Žnidaršič et al. [13] found that the addition of 1×10^{-3} M Ca^{2+} to basal medium resulted in the formation of smooth large pellets and clump. The average diameter of pellets was 3.84 ± 0.84 mm. When the medium was supplemented with Ca^{2+} in concentration above 1×10^{-2} M, the whole mycelium of *Rhizopus nigricans* was aggregated in clumps.

In conclusion, most studies mainly focused the effects of Ca^{2+} on the morphology of filamentous fungi in micro-morphology, including hyphal tip growth, morphology, extension, and branching. No systematic reports have been published to discuss the effect of exogenous calcium on the macro-morphology of filamentous fungi, especially *Rhizopus oryzae*. The present study addresses the challenging task of investigating the influence of exogenous calcium on the macro-morphology of *R. oryzae*, and controlling the morphology of *R. oryzae* in pellet form to produce L-lactic acid efficiently. In this work, the effects of exogenous calcium, including soluble and insoluble calcium, on the pellet form, growth characteristics, and L-lactic acid production in a mutant strain of *R. oryzae* is discussed.

25.2 Materials and Methods

25.2.1 Microorganism and Growth

Rhizopus oryzae TZ-45, the mutant of *R. oryzae* NRRL 395, was used in this study. The fungus was grown on a potato dextrose agar (PDA) plate at 30 °C for 7 d. For the experiments, fungal spores were collected by shaving the PDA surface with a sterile loop and extracting spores with sterile water. Fungal spores were then stored at 4 °C.

25.2.2 Preculture Conditions

The preculture medium contained the following components (per liter): 20.0 g of glucose; 2.0 g of peptone; 0.2 g of KH_2PO_4, 0.2 g of $MgSO_4 \cdot 7H_2O$; 0, 2.0, 4.0, and 6.0 g of $CaCl_2$ or 0, 2.0, 4.0, 6.0, and 8.0 g of $CaCO_3$ (the initial concentration in the preculture medium); and natural pH. Approximately 50 mL of medium without $CaCl_2/CaCO_3$ was loaded into a 250 mL Erlenmeyer flask and heat sterilized (121 °C for 20 min).

Before preculture, 15.0 mL of 100.0 g/L $CaCl_2$ solution and 0.1–0.4 g of $CaCO_3$ powders were separately placed in a 20 mL vial. All vials were then tightly closed with screw caps to avoid moisture adsorption from the outside, and autoclaved at 115 °C for 30 min. The sterilized $CaCl_2$ solution with definite volume and $CaCO_3$ powders were then added to each sterilized 250 mL Erlenmeyer flask to ensure that $CaCl_2/CaCO_3$ in the medium reached the initial concentration. The spore solution was inoculated in the Erlenmeyer flask with a spore concentration of 1×10^6 spores/mL, and cultured in a rotary shaker (150 rpm) at 30 °C for 18 h. All values presented in this study are averages of at least three independent trials.

25.2.3 *Fermentation*

The production medium contained the following components (per liter): 80.0 g of glucose, 3.0 g of $(NH_4)_2SO_4$, 0.25 g of $MgSO_4{\cdot}7H_2O$, 0.04 g of $ZnSO_4{\cdot}7H_2O$, 0.2 g of KH_2PO_4, and 40 g of $CaCO_3$. The media without $CaCO_3$ were autoclaved at 121 °C for 20 min. The calcium carbonate powder was sterilized separately (115 °C at 30 min). L-lactic acid production was performed in a 3.0 L stirred tank (New Brunswick Scientific, USA) with a 2.0 L working volume, which was inoculated with 300 mL of preculture. The cultivation temperature in the stirred tank was maintained at 30 °C throughout the experiments. The aeration rate and agitation speed were set at 0.5 vvm and 300 rpm, respectively. Sterile $CaCO_3$ was used as a neutralizer, which was added to the tank before fermentation to maintain a pH of approximately 6.0 during culture. The cultivation time in the experiments ranged from 48 to 68 h. Samples were periodically obtained for high-performance liquid chromatography (HPLC) analysis.

25.2.4 *Analytical Methods*

To determine glucose and ethanol concentrations, samples were centrifuged, and the resulting supernatants were used. To determine lactic acid concentration, samples were diluted by the addition of distilled water and hydrochloric acid, heated at 80 °C until the broth was clear, and centrifuged. The resulting supernatants were then used for analysis.

Glucose, ethanol, and lactic acid concentrations were measured by HPLC (Summit P 680 HPLC, Dionex, USA; Shodex RI-101 Refractive Index Detector, Showa Denko, Japan; Aminex HPX-87 H Ion Exclusion Column 300 mm × 7.8 mm, Bio-Rad, USA) under the following conditions: sample volume, 20 μL; mobile phase, 0.005 M H_2SO_4; flow rate, 0.8 mL/min; and column temperature, 60 °C [18]. Biomass was determined by weighing the mycelial mass after drying at 60 °C overnight. Seed morphology was determined using an Olympus microphotograph (Tokyo, Japan).

25.3 Results and Discussion

25.3.1 *Effects of Exogenous Calcium on the Growth of* R. oryzae

All morphologies observed during the cultivation of *R. oryzae* with different exogenous calcium and different initial concentrations are summarized in Table 25.1. The diameter and other characteristics of the obtained pellets are also included in Table 25.1.

Table 25.1 Overview of the morphology of *R. oryzae* after 18 h of growth with different calcium compounds and concentrations

Additive	Concentration (g/L)	Morphology	pH		Pellet diameter (mm)		Pellet characteristics	Dry weight (g/L)
			Initial	Final	Average	Size range		
$CaCl_2$	6.0	Clumps	5.3	4.8	–	–	–	6.01 ± 0.30
	4.0	Pellets	5.2	3.7	1.2	1.0–1.5	Radial, slightly hairy	6.42 ± 0.32
	2.0	Pellets	5.2	3.6	0.8	0.5–1.0	Smooth	6.21 ± 0.31
	0	Filaments	5.0	4.5	–	–	–	5.90 ± 0.30
$CaCO_3$	2.0	Pellets and filaments	5.5	4.6	1.0	0.5–1.5	Small, hollow, sticking together	6.30 ± 0.32
	4.0	Pellets	5.8	5.1	1.2	1.0–1.8	Smooth	6.86 ± 0.34
	6.0	Pellets	6.0	5.8	1.5	1.0–2.0	Radial, fluffy	7.02 ± 0.35
	8.0	Pellets	5.9	6.1	2.3	1.5–3.0	Large, smooth, mixed with $CaCO_3$	8.14 ± 0.41

The morphology and final cell dry weight clearly varied with different exogenous calcium. Table 25.1 showed that the final cell dry weight initially increased, and then decreased when the $CaCl_2$ concentration changed from 0 to 6.0 g/L. The maximum cell dry weight of 6.42 g/L was obtained at 4.0 g/L $CaCl_2$, which differed by 8.8 % from the minimum weight (5.9 g/L). The final cell dry weight increased with the $CaCO_3$ increased from 0 to 8.0 g/L. The maximum cell dry weight reached 8.14 g/L, which differed by 38 % from the minimum weight (6.3 g/L), possibly because the pellet was mixed with excess $CaCO_3$ powder. The changes in pH also proved this phenomenon.

25.3.2 *Effects of Exogenous Calcium on the Morphology of* R. oryzae

Figure 25.1 illustrates the representative morphological forms generated in the precultures with different calcium types. It was found that exogenous calcium significantly influenced the morphology and pellet size of *R. oryzae*. The experiments performed in shake flasks using the precultures with exogenous calcium mostly resulted in pellets, except 6.0 g/L $CaCl_2$. Long and entangled filaments (Fig. 25.1d) were formed in the preculture without exogenous calcium, whereas large mycelial clumps were observed in the preculture with 6.0 g/L $CaCl_2$ (Fig. 25.1a). Fluffy pellets mixed with filaments were observed in the preculture with 2.0 g/L $CaCO_3$. Small and smooth pellets with an average pellet diameter of 1.0 and 1.2 mm (Fig. 25.1b, f) were found in the precultures with 2.0 g/L $CaCl_2$ and 4.0 g/L $CaCO_3$, respectively. Larger radial and hairy pellets (Fig. 25.1c, g) were formed in the precultures with 4.0 g/L $CaCl_2$ and 6.0 g/L $CaCO_3$ as the

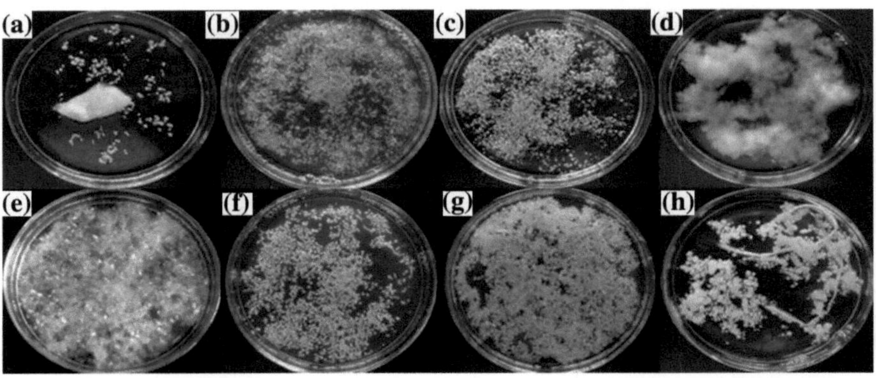

Fig. 25.1 Influence of exogenous calcium on the morphology of *R. oryzae* NRRL 395 in submerged cultures in shaken flasks. $CaCl_2$ and $CaCO_3$ were added at various concentrations: **a** 6.0 g/L $CaCl_2$; **b** 4.0 g/L $CaCl_2$; **c** 2.0 g/L $CaCl_2$; **d** 0 g/L $CaCl_2$; **e** 2.0 g/L $CaCO_3$; **f** 4.0 g/L $CaCO_3$; **g** 6.0 g/L $CaCO_3$; and **h** 8.0 g/L $CaCO_3$

concentrations of $CaCl_2$ and $CaCO_3$ increased. The average pellet diameters were 1.2 and 1.5 mm. Large and smooth pellets mixed with $CaCO_3$ powder were formed in a higher $CaCO_3$ concentration (8.0 g/L). The average pellet diameters were 2.3 mm. Notably, higher soluble calcium ($CaCl_2$) was not beneficial for pellet formation, but the opposite was observed in insoluble calcium ($CaCO_3$).

This study showed that an appropriate concentration of calcium (different solubilities with different concentrations) could promote pellet formation. Very low or very high concentrations were not beneficial for pellet formation. Similar results were reported by Pera and Callieri [14], Jackson and Heath [12], and Robson et al. [15]. These results indicated that in low or deficient exogenous calcium, fungi lack a Ca^{2+}-CTC membrane-associated gradient, grow slowly, display hyper-branching, and have abnormal swollen hyphae. The increase in the external Ca^{2+} concentration resulted in an increased rate of hyphal extension [16, 17, 19] and a decrease rate in the branching frequency [17]. Very high concentrations of external Ca^{2+} can inhibit tip extension [16]. However, whether this inhibition is due to the direct Ca^{2+} interactions with the cell wall (i.e., Ca^{2+} induces rigidity of the apical cell wall [16]) or a general toxic response to high cytosolic Ca^{2+} concentration remains unclear. Žnidaršič and Pavko [11] reported that the pellet morphology of a filamentous microorganism developing in any fermentation system may be represented as a final result of the competing influences, which is the equilibrium between the forces of cohesion and disintegration. Shear forces may be unambiguously assigned the function of disintegrating factors. Ca^{2+} may affect the hyphal extension and branching rate, thereby affecting mycelium cohesion. This result possibly explains why the pellet cannot form without exogenous calcium (Fig. 25.1d) or high exogenous calcium (Fig. 25.1a). Meanwhile, pellets obtained from a low concentration of exogenous calcium showed a smooth surface (Fig. 25.1c, f), whereas pellets obtained from a high calcium concentration showed a rough surface (Fig. 25.1b, g).

25.3.3 Effect of Exogenous Calcium on L-lactic Acid Production

Table 25.2 summarizes the L-lactic acid production, residual glucose concentration, lactic acid yield, lactic acid productivity, and by-product concentrations as measurable indicators of different *R. oryzae* morphologies, which resulting from different precultures. It was found that the residual glucose concentrations reached 21.4 and 32.2 g/L when the precultures with 0 g/L $CaCl_2$ and 8.0 g/L $CaCO_3$, respectively. The final L-lactic acid production increased from 30.3 to 57.2 g/L in the precultures with $CaCl_2$. The highest L-lactic acid production of 57.2 g/L was obtained in the fermentation with 2.0 g/L $CaCl_2$. Meanwhile, the final L-lactic acid production increased from 39.2 to 58.6 g/L in the precultures with $CaCO_3$. The highest L-lactic acid production of 58.6 g/L was obtained when using 6.0 g/L $CaCO_3$. Although the highest L-lactic acid production using precultures with $CaCl_2$

Table 25.2 Experimental data from *R. oryzae* fermentation carried out using precultures with different exogenous calcium

Calcium factor	$CaCl_2$			$CaCO_3$ powder			
	0	2.0	4.0	2.0	4.0	6.0	8.0
Initial glucose concentration (g/L)	82	81	83	82	84	83	82
Residual glucose concentration (g/L)	21.4	2	3	9	2	3	32.2
Final L-lactic acid production (g/L)	30.3	57.2	54.1	39.2	53.2	58.6	25.1
Average L-lactic acid yield on glucose (g/g)	0.50	0.72	0.68	0.54	0.65	0.73	0.504
Average L-lactic acid productivity ($g/L^{-1}\ h^{-1}$)	0.446	1.02	0.902	0.576	1.02	1.22	0.369
Final ethanol (g/L)	7.8	5.4	5.8	6.1	4.2	2.7	9.4
Fermentation time (h)	68	56	60	68	52	48	68

and $CaCO_3$ had little difference, the average L-lactic acid productivity had a difference of 20 % (1.02 $g/L^{-1}\ h^{-1}$ for 2.0 g/L $CaCl_2$ and 1.22 $g/L^{-1}\ h^{-1}$ for 6.0 g/L $CaCO_3$). Ethanol production was also higher for 2.0 g/L $CaCl_2$ than that for 6.0 g/L $CaCO_3$.

Pellets have been reported with desired morphology for the production of lactic acid [20], itaconic acid [21], citric acid [22], or penicillin [23]. Mass transfer within a typically dense pellet is regarded as a severe disadvantage, and the smaller pellet is much more beneficial for fermentation [11, 24]. However, a large pellet was more beneficial for fermentation in our study.

25.3.4 Microscopic Analysis for the Morphology of Mycelial Pellet

Microscopic analysis (Fig. 25.2) revealed that the application of pellets was limited by the low mass transfer inside the fungal aggregate [25, 26].

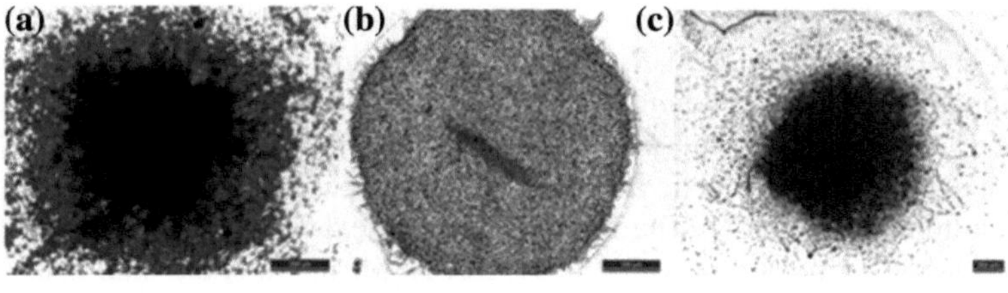

Fig. 25.2 Morphology of mycelial pellet generated by growing *R. oryzae* with different exogenous calcium (**a** pellet with 2.0 g/L $CaCl_2$; **b** pellet with 6.0 g/L $CaCO_3$; and **c** pellet with 8.0 g/L $CaCO_3$). The images were captured using a Leica microscope (dm2500p)

Pellets are fully supplied with nutrients or oxygen only up to a critical diameter. For *Aspergillus niger* pellets, this diameter is approximately 0.4 mm. Recent fluorescence analysis of GFP reporter strains of *A. niger* confirmed that only a thin layer at the pellet surface contributes to protein production, whereas the large inner part of the pellet is inactive [27]. The pellets of approximately 0.8 mm in size and precultured with 2.0 g/L $CaCl_2$ consisted of a dense outer layer of biomass, and exhibited an unfilled center (Fig. 25.2a). The thickness of the outer layer was approximately 0.2 mm. The supply of oxygen and other nutrients to the cell in the interior layer was limited. The addition of $CaCO_3$ in the preculture, especially a certain amount of $CaCO_3$, significantly changed the morphology (Fig. 25.2b). Most strikingly, the insoluble $CaCO_3$ powder was associated with the biomass and occurred inside the pellets, which created a loose interior structure with a better biomass filling in the pellet core. Thus, the $CaCO_3$ powder was randomly distributed within the aggregate [28, 29]. At higher levels, core shell aggregates were formed (Fig. 25.2b). Thus, the larger pellets were mainly composed of $CaCO_3$ powder. Supported by the $CaCO_3$ powder, the entire biomass layer was fully active. The created inner pellet structure was rather loose, which enabled higher mass transfer than the pellets precultured with $CaCl_2$. When the $CaCO_3$ concentration increased, the excess $CaCO_3$ enveloped the pellet (Fig. 25.2c) and decreased L-lactic acid productivity.

25.4 Conclusion

Exogenous calcium regulates the polarity and direction of hyphal growth, as well as the frequency with which they branch, ultimately determining the mycelial morphology. The results show that the addition of exogenous calcium could induce pellet formation. The diameter of the pellet increased with the concentration of exogenous calcium increased, including $CaCl_2$ and $CaCO_3$. The pellet precultured with soluble $CaCl_2$ (Fig. 25.2a) formed a dense interior structure, and the large inner part of the pellet was inactive. Thus, the smaller pellet precultured with low concentration of soluble calcium ($CaCl_2$), the better for the benefit of L-lactic acid production. By contrast, the larger pellet precultured with high concentration of insoluble calcium ($CaCO_3$), except 8.0 g/L $CaCO_3$, was beneficial for L-lactic acid production. Microscopic analysis revealed that the addition of $CaCO_3$ in the preculture, especially a certain amount of $CaCO_3$ (6 g/L $CaCO_3$ in this study), resulted in a remarkable change in the morphology, and a loose interior structure with a better biomass filling in the pellet core was created (Fig. 25.2b). Supported by the $CaCO_3$ powder, the entire biomass layer was fully active and the highest L-lactic acid production rates of 1.22 $g/L^{-1}\ h^{-1}$ and 58.6 g/L L-lactic acid were obtained using the 1.5 mm pellet. The excess $CaCO_3$ enveloped the pellet and decreased L-lactic acid productivity when precultured with 8.0 g/L $CaCO_3$.

Acknowledgments This work was financially supported by the National Natural Science Foundation of China (Grant No. 21106091), Zhejiang Provincial Natural Science Foundation of China (LQ12B06004).

References

1. Meussen BJ, de Graaff LH, Sanders JPM, Weusthuis RA (2012) Metabolic engineering of *Rhizopus oryzae* for the production of platform chemicals. Appl Microbiol Biotechnol 94:875–886
2. Grimm LH, Kelly S, Krull R, Hempel DC (2005) Morphology and productivity of filamentous fungi. Appl Microbiol Biotechnol 69:375–384
3. Kossen NWF (2000) The morphology of filamentous fungi. Adv Biochem Eng Biotechnol 70:1–32
4. Liao W, Liu Y, Chen SL (2007) Studying pellet formation of a filamentous fungus *Rhizopus oryzae* to enhance organic acid production. Appl Biochem Biotechnol 136–140:689–701
5. Metz B, Kossen NWF (1977) The growth of molds in the form of pellets: literature review. Biotechnol Bioeng 19:781–799
6. Engel CAR, Gulik WM, Marang L, Wielen LAM, Straathof AJJ (2011) Development of a low pH fermentation strategy for fumaric acid production by *Rhizopus oryzae*. Enzyme Microb Tech 48:39–47
7. Zhou Y, Du JX, Tsao GT (2000) Mycelial pellet formation by *Rhizopus oryzae* ATCC 20344. Appl Biochem Biotechnol 84–86:779–789
8. Posch AE, Spadiut O, Herwig C (2012) A novel method for fast and statistically verified morphological characterization of filamentous fungi. Fungal Genet Biol 49:499–510
9. Jüsten P, Paul GC, Nienow AW, Thomas CR (1998) Dependence of *Penicillium chrysogenum* growth, morphology, vacuolation, and productivity in fed-batch fermentations on impeller type and agitation intensity. Biotechnol Bioeng 59(6):762–775
10. Braun S, Vecht-Lifshitz SE (1991) Mycelial morphology and metabolite production. Tibtech 9:63–68
11. Žnidaršič P, Pavko A (2001) The morphology of filamentous fungi in submerged cultivations as a bioprocess parameter. Food Technol Biotechnol 39(3):237–252
12. Jackson SL, Heath IB (1993) Roles of calcium ions in hyphal tip growth. Microbiol Rev 57 (2):367–382
13. Žnidaršič P, Komel R, Pavko A (2000) Influence of some environmental factors on *Rhizopus nigricans* submerged growth in the form of pellets. World J Microb Biot 16:589–593
14. Pera LM, Callieri DA (1997) Influence of calcium on fungal growth, hyphal morphology and citric acid production in *Aspergillus niger*. Folia Microbiol 42(6):551–556
15. Robson GD, Wiebe MG, Trinci APJ (1991) Involvement of Ca^{2+} in the regulation of hyphal extension branching in *Fusarium graminearum* A 3/5. Exp Mycol 15:263–272
16. Robson GD, Wiebe MG, Trinci APJ (1991) Low calcium concentrations induce increased branching in *Fusarium graminearum*. Mycol Res 95(5):561–565
17. Jackson SL, Heath IB (1989) Effects of exogenous calcium ions on tip growth, intracellurar Ca^{2+} concentration, and actin arrays in hyphae of the Fungus *Saprolegnia ferax*. Exp Mycol 13:1–12
18. Fu YQ, Li S, Chen Y, Xu Q, Huang H, Sheng XY (2010) Enhancement of fumaric acid production by *Rhizopus oryzae* using a two-stage dissolved oxygen control strategy. Appl Biochem Biotechnol 162:1031–1038
19. Saavedra-Mobina A, Uribe S, Devlin TM (1990) Control of mitochondrial matrix calcium: studies using Fluo-3 as a fluorescent calcium indicator. Biochem Biopys Res Commun 167:148–153

20. Liu Y, Liao W, Liu CB, Chen SL (2006) Optimization of L-(+)-lactic acid production using pelletized filamentous *Rhizopus oryzae* NRRL 395. Appl Biochem Biotechnol 129–132:844–853
21. Metz B, Kossen NWF (1977) The growth of molds in the form of pellets: a literature review. Biotechnol Bioeng 19:781–800
22. Papagianni M (2004) Fungal morphology and metabolite production in submerged mycelial processes. Biotechnol Adv 22:189–259
23. Makagiansar HY, Shamalou PA, Thomas CR, Lilly MD (1993) The influence of mechanical forces on the morphology and penicillin production of *Penicillium chrysogenum*. Bioprocess Eng 9:83–90
24. Villena GK, Fujikawa T, Tsuyumu S, Gutiérrez-Correa M (2010) Structural analysis of biofilms and pellets of *Aspergillus niger* by confocal laser scanning microscopy and cryo scanning electron microscopy. Bioresour Technol 101:1920–1926
25. Driouch H, Hänsch R, Wucherpfennig T, Krull R, Wittmann C (2012) Improved enzyme production by bio-pellets of *Aspergillus niger*: targeted morphology engineering using titanate microparticles. Biotechnol Bioeng 109(2):462–471
26. Walisko R, Krull R, Schrader J, Wittmann C (2012) Microparticle based morphology engineering of filamentous microorganisms for industrial bio-production. Biotechnol Lett 34 (11):1975–1982
27. Hille A, Neu TR, Hempel DC, Horn H (2005) Oxygen profiles and biomass distribution in biopellets of *Aspergillus niger*. Biotechnol Bioeng 92:614–623
28. Driouch H, Roth A, Dersch P (2010) Filamentous fungi in good shape: microparticles for tailor-made fungal morphology and enhanced enzyme production. Bioeng Bugs 2:1–5
29. Kaup BA, Ehrich K, Pescheck M, Schrader J (2007) Microparticle-enhanced cultivation of filamentous microorganisms: increased chloroperoxidase formation by Caldariomyces fumago as an example. Biotechnol Bioeng 99:491–498

Chapter 26
Estimation of Dietary Copper (Cu) Requirement of *Cynoglossus semilaevis* Günther

Qingkui Wang, Yang Zhang, Dongqing Bai, Chengxun Chen, Yongjun Guo and Kezhi Xing

Abstract To evaluate the dietary copper (Cu) requirement of *Cynoglossus semilaevis*, inorganic Cu (0, 2, 6, 17, and 50 mg kg^{-1}) was added to the basal diet, providing actual dietary Cu (5.91, 8.23, 11.78, 24.16, and 56.34 mg kg^{-1}). Each diet was fed to *C. semilaevis* Günther (initial body weight, 68 g) in triplicate groups for 8 weeks in a flow-through system. The results showed that weight gain rate (WGR), specific growth rate (SGR), feed conversion rate (FCR), and protein efficiency ratio (PER) in fish supplemented with 11.78 mg kg^{-1} were significantly higher than those in fish fed with the basal diet ($P < 0.05$), and no significant difference was observed in fish fed dietary Cu ranging from 11.78 to 56.34 mg kg^{-1} ($P > 0.05$). The activities of protease, amylase, lipase, copper–zinc superoxide dismutase (Cu–Zn SOD), and lysozyme first increased and then decreased with the increasing dietary Cu concentrations, and 11.78 mg kg^{-1} diet provided maximum activities. The optimum requirement of *C. semilaevis* for dietary Cu was estimated to be about 11–12 mg kg^{-1} diet using broken-line regression analysis, based on the growth and enzyme activities.

Keywords Dietary copper · Requirement · *Cynoglossus semilaevis* · Growth · Enzyme activity

26.1 Introduction

Copper (Cu) is an essential trace mineral for fish. It functions in the electron transport chain, iron metabolism, antioxidant system, and hematopoiesis [17]. Cu deficiency can result in biochemical, structural, and functional pathologies [22].

Q. Wang · Y. Zhang · D. Bai · C. Chen · Y. Guo · K. Xing (✉)
Tianjin Key Laboratory of Aqua-Ecology and Aquaculture, Fisheries College, Tianjin Agricultural University, Tianjin 300384, People's Republic of China
e-mail: kzxing6668@126.com

T.-C. Zhang and M. Nakajima (eds.), *Advances in Applied Biotechnology*, Lecture Notes in Electrical Engineering 332, DOI 10.1007/978-3-662-45657-6_26

Fish can absorb Cu from either water or their diets. However, copper in the rearing water alone cannot meet the requirement, and dietary Cu supplementation was considered to be the major source for fish [15]. For the growth-stimulatory effects and antimicrobial property of Cu, its supplementation in fish diet often exceed requirement levels and may be on the borderline of toxic concentrations [21]. Dietary Cu supplemented at elevated levels always led to reduced food safety and toxic effects on fish [1, 7, 13, 24].

Copper needs to be supplemented in diet within narrow ranges to maintain the balance between fulfilling their nutritional function and avoiding their toxicity, and dietary Cu requirement have been quantified in many fish species. Cu is required in the diet of common carp and rainbow trout at 3 mg kg^{-1} diet [22], hybrid tilapia at 4 mg kg^{-1} diet [27], channel catfish at 5 mg kg^{-1} diet [11], Atlantic salmon parr at 5–10 mg kg^{-1} diet [17], and grouper [15, 16] ranging from 2 to 6 mg kg^{-1} diet. Rapidly growing Atlantic salmon fry have a higher Cu requirement of 35 mg kg^{-1} diet [2]. *C. semilaevis* with high filet yield and desirable taste has been a potentially important cultured fish species in China since artificial breeding was successfully achieved in 2002 [20]. However, there are few studies conducted in *Cynoglossus semilaevis* to evaluate the effects of dietary Cu, and there is a lack of research on dietary Cu requirement of *Cynoglossus semilaevis*.

In this study, *C. semilaevis* was used as an experimental animal. The effects of dietary Cu on growth, oxidative stress, digestive enzyme, and lysozyme activities of it were investigated, respectively. Moreover, the dietary Cu requirement for *C. semilaevis* was evaluated using the broken-line regression model.

26.2 Materials and Methods

26.2.1 Fish and Treatments

Fish were obtained from Haifa Aquaculture Co., Ltd, Tianjin, China. Prior to the start of experiment, fish were held in tanks (85 × 60 × 45 cm) filled with gentle flowing water for 2 weeks to be acclimated to the experimental conditions.

After acclimation, 300 fish weighing 68 g were randomly divided into five treatment groups with three replicates (20 fish per replicate). Fish were fed to visual satiety with diets (0, 2, 6, 17, and 50 mg Cu per kilogram dry weight) twice daily at 800 and 1,600 h, and unfed feed were removed 1 h later and weighed. During the experiment, fish were reared with the following water quality: water temperature 19.3–23.8 °C, pH 7.2–7.6, salinity 23–30 ppt, dissolved oxygen 5.6–7.2 mg L^{-1}, NH_4^+ 0.075–0.167 mg L^{-1}, NO_2^- 0.006–0.026 mg L^{-1}. The experiment lasted for 8-week period, and the mortality and feeding of fish were monitored every day and any dead fish were removed and not replaced.

To produce Cu-supplemented diet, copper sulfate ($CuSO_4 \cdot 5H_2O$; Analytical Reagent, Beifang Tianyi Chemical Co., Tianjin, China) was mixed with ingredients of the basal diet through gradually enlarging mixture scale, and pelleted (2 mm).

Table 26.1 Formulation and proximate composition of the basal diet

Ingredients	g/g %	Proximate composition	g/g%
Fish meal	64	Crude protein	55.36
Soybean meal	12	Crude lipid	8.25
Corn starch	9	Ash	13.79
Yeast	3	–	–
Fish oil	2	–	–
Corn oil	2	–	–
Dried tiny shrimp	2	–	–
Bonding agent	1	–	–
Mineral premix[a]	3	–	–
Vitamin premix[b]	2	–	–

[a] Mineral premix (mg g^{-1} premix): $Ca(H_2PO_4) \cdot 2H_2O$, 489.11; $ZnSO_4 \cdot 7H_2O$, 2.20; $FeSO_4 \cdot 7H_2O$, 3.65; KI, 1.09; $CoCl_2 \cdot 6H_2O$, 0.007; Na_2SeO_3, 0.011; $MnSO_4 \cdot H_2O$, 0.82

[b] Vitamin premix (IU or mg g^{-1} premix): vitamin A, 120 IU; vitamin D3, 330 IU; nicotinic acid, 11 mg; inositol, 50 mg; pantothenic acid, 3 mg; vitamin B2, 2.75 mg; vitamin B6, 0.75 mg; vitamin B1, 0.2 mg, vitamin K3, 0.15 mg; folic acid, 1.25 mg, tocopheryl acetate, 33.75 mg; ascorbic acid, 135 mg; vitamin B12, 0.0135 mg; biotin, 0.675 mg; choline chloride, 250 mg

The diets were stored at −20 °C until used. The actual dietary Cu concentrations were 5.91(0), 8.23 (2), 11.78 (6), 24.16 (17), 56.34 (50) mg kg^{-1} diet, respectively. Diet formulation and proximate composition are given in Table 26.1.

26.2.2 Growth Performance

At the end of the feeding trial, fish starved for 24 h were anesthetized and weighted. Weight gain rate (WGR), specific growth rate (SGR), feed conversion rate (FCR), and protein efficiency ratio (PER) were calculated using the following formulae:

$$\text{WGR}\ (\text{g/g\%}) = 100 \times (\text{average final body weight} - \text{average initial bodyweight})/\text{average initial body weight} \quad (26.1)$$

$$\text{SGR}\ (\text{g/g\%}) = 100 \times (\ln \text{final weight} - \ln \text{initial weight})/\text{rearing period (days)} \quad (26.2)$$

$$\text{FCR}\ (\text{g} \cdot \text{g}^{-1}) = \text{food consumed/body weight gain} \quad (26.3)$$

$$\text{PER}\ (\text{g/g\%}) = 100 \times (\text{body weight gain/protein intake}) \quad (26.4)$$

26.2.3 Collection of Tissue and Serum Samples

After fish were anesthetized, blood samples were collected from caudal vein and left at room temperature for 1 h, followed by 4 h at 4 °C and then serum was harvested by centrifuging at 4,000 r min^{-1} at 4 °C for 12 min. The serum obtained were stored at 4 °C until the oxidative stress parameters and lysozyme activity were measured within 12 h. Liver and kidney were removed quickly and stored at 4 °C for the lysozyme activity analysis within 12 h. Intestine were removed quickly and washed with cold physiological saline before being frozen in liquid nitrogen and stored at −80 °C for subsequent analysis.

26.2.4 Digestive Enzyme Activity Assay

Approximately 1 g wet weight intestine samples refrigerated were homogenized in 9-mL ice-cold physiological saline, and homogenates were centrifuged at 3,000 r min^{-1} for 30 min at 4 °C; the final supernatants were divided into subsamples for each enzyme assay and stored at 4 °C until analysis was done within 24 h.

Protease activity was determined by casein digestion method as follows: the reaction mixture consisted of 2 mL preheated 0.5 % casein solution and 1 mL homogenate which was kept at 30 °C for 3 min; 1.0 mL deionized water used as the control. The reaction was stopped by adding 10 % trichloroacetic acid (TCA) after being incubated in water bath (30 °C) for 15 min. Reactants were centrifuged at 3,000 r min^{-1} for 15 min and 1 mL supernatants were mixed with 5 mL of 0.55 mol mg L^{-1} Na_2CO_3 solution and 1 mL Folin phenol reagent. The mixtures were incubated in water bath (30 °C) for 15 min, and then the absorbance was recorded at 680 nm. One unit of protease activity was expressed as 1 μg tyrosine released from casein hydrolysis per min at 30 °C.

Amylase and lipase activities were determined using assay kits (Jiancheng Biotech Co., Ltd., Nanjing, China) following the instructions of the manufacturer. One amylase unit is the activity hydrolyzing 10 mg of starch in 30 min at 37 °C. One unit of lipase activity was defined as the hydrolysis of 1.0 microequivalent of fatty acids from triacylglycerols per minute per gram of protein at 30 °C. Total protein content of supernatant was assayed according to Bradford [3] using bovine albumin as a standard.

26.2.5 Oxidative Stress Assay

Serum malondialdehyde (MDA) level and copper–zinc superoxide dismutase (Cu–Zn SOD) activity were measured using assay kits (Jiancheng Biotech Co., Ltd., Nanjing, China). MDA level was quantified by the thiobarbituric acid (TBA) method [8]. SOD activity was measured spectrophotochemically by the ferricytochrome c method using xanthine/xanthine oxidase as the source of superoxide radicals [30]. One unit of SOD activity was defined as the amount of per mg protein that inhibited the rate of nitrite reduction by 50 %.

26.2.6 Monitoring Lysozyme Activity

Serum lysozyme activity was measured using modified turbidimetric method according to Ellis [9] with kit (Jiancheng Biotech Co., Ltd., Nanjing, China). Briefly, *Micrococcus lysodeikticus* suspension at a concentration of 0.1 mg mL^{-1} was added to serum at 10:1 ratio, and the decrease on absorbance was read at 0.5 and 2.5 min intervals at 530 nm in a spectrophotometer (UV-752N, Shanghai Precision Instrument Co., Ltd., China). One unit of lysozyme activity was defined as the amount of enzyme that caused a decrease in absorbance of 0.001 per min.

Liver and kidney samples were homogenized in ice-cold physiological saline at a ratio of 1:9, and homogenates were centrifuged at 10,000 r min^{-1} for 11 min at 4 °C. The final supernatants were used for lysozyme activity analysis with kit (Jiancheng Biotech Co., Ltd., Nanjing, China). Total protein content of supernatant was assayed according to Bradford [3].

26.2.7 Statistical Analysis

The resulting data were expressed as mean ± SE and were subjected to one-way analysis of variance (SPSS version 13.0) to determine significant differences among groups. Significant differences ($P < 0.05$) were reanalyzed by Duncan's multiple range test.

26.3 Results

As shown in Fig. 26.1 and Table 26.2, WGR and SGR of fish fed dietary Cu (≥11.78 mg kg^{-1}) were significantly higher than those of fish fed 5.91 and 8.23 mg kg^{-1} dietary Cu ($P < 0.05$). FCR and PER of fish fed with the basal diet (5.91 mg kg^{-1}) were significantly lower than those of fish fed dietary Cu more than or equal to that with 11.78 mg kg^{-1} ($P < 0.05$). No significant differences in WGR,

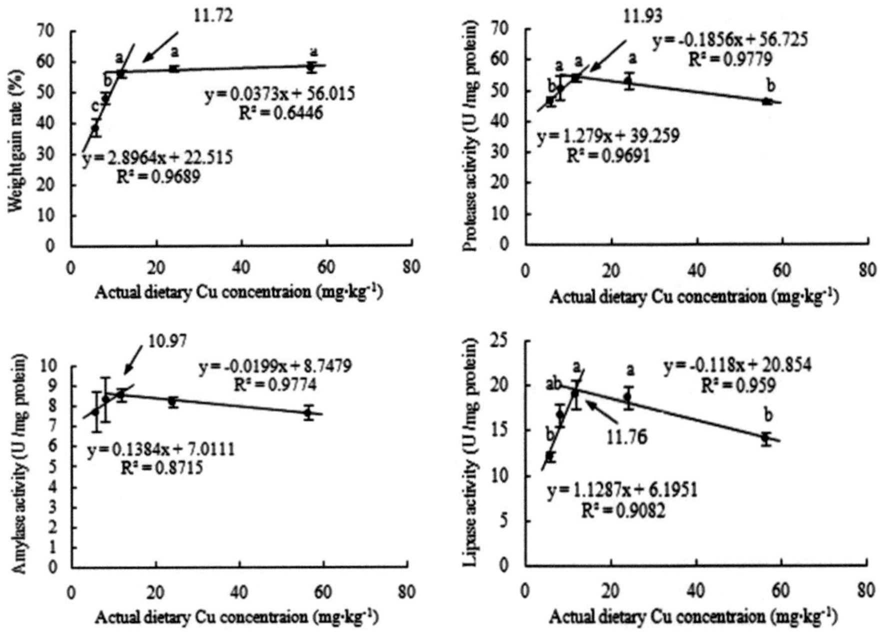

Fig. 26.1 Effects of dietary Cu on weight gain rate (WGR), intestine digestive enzyme activities of *C. semilaevis*. Values are represented as mean ± SD; values sharing the same letters do not significantly differ ($P > 0.05$), whereas those with different letters significantly differ ($P < 0.05$). Dietary Cu requirement derived by the broken-line method for WGR, protease, amylase, and lipase are 11.72, 11.93, 10.97, and 11.76 mg kg^{-1} diet, respectively

Table 26.2 Effects of dietary Cu on the growth performance of *C. semilaevis* Günther

Actual dietary Cu content	Initial weight (g)	Final weight (g)	SGR (g/g%)	FCR (g·g^{-1})	PER (g/g%)
5.91	68.97 ± 0.24	95.57 ± 2.22^{b}	0.63 ± 0.04^{c}	1.69 ± 0.09^{a}	1.08 ± 0.10^{c}
8.23	69.03 ± 1.56	102.26 ± 3.60$^{a,\ b}$	0.76 ± 0.02^{b}	1.49 ± 0.37$^{a,\ b}$	1.22 ± 0..37$^{b,\ c}$
11.78	69.40 ± 0.09	106.83 ± 0.71^{a}	0.83 ± 0.02^{a}	1.08 ± 0.08^{b}	1.68 ± 0.09$^{a,\ b}$
24.16	68.83 ± 0.05	108.50 ± 0.71^{a}	0.88 ± 0.01^{a}	1.01 ± 0.04^{b}	1.80 ± 0.05^{a}
56.34	68.47 ± 0.28	108.12 ± 0.64^{a}	0.88 ± 0.02^{a}	1.15 ± 0.11^{b}	1.58 ± 0.12$^{a,\ b}$

Values are represented as mean ± SE; Values within a column sharing the same letters or no letters do not significantly differ ($P > 0.05$), whereas those within a column with different letters significantly differ ($P < 0.05$)

SGR, FCR, and PER were observed when fish were fed with dietary Cu ranging from 11.78 to 56.34 mg kg^{-1} ($P > 0.05$).

Fish fed with 11.78 mg kg^{-1} dietary Cu have the maximal protease and lipase activities, which are significantly higher than those of fish fed with 5.91 and 8.23 mg kg^{-1} dietary Cu ($P < 0.05$) (Fig. 26.1). Highest amylase activity was observed in fish fed with 11.78 mg kg^{-1} dietary Cu, and there is no significant difference in all the treatments. With the increasing dietary Cu concentrations, Cu–Zn SOD activity was upregulated first and then downregulated, and highest value was observed in 11.78 mg kg^{-1} (Fig. 26.2a). MDA showed an inverse trend with the performance of Cu–Zn SOD activity (Fig. 26.2b). Lysozyme activities in serum, liver, and kidney were all first increased and then decreased with the increasing

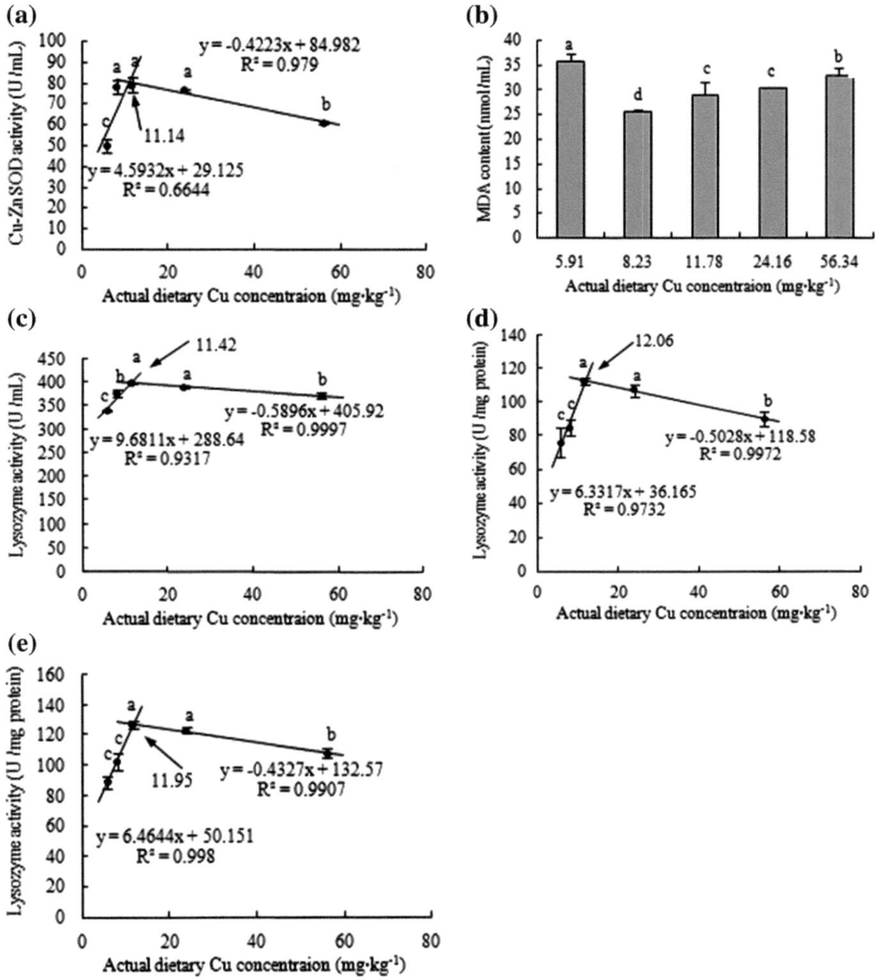

Fig. 26.2 Effects of dietary Cu on serum Cu–Zn SOD activity (**a**), serum MDA (**b**), serum lysozyme (**c**), liver lysozyme (**d**), kidney lysozyme (**e**) of *C. semilaevis*. *Values* are represented as mean ± SD; Values sharing the same *letters* do not significantly differ ($P > 0.05$), whereas those with different *letters* significantly differ ($P < 0.05$). Dietary Cu requirements derived by the *broken-line* method for Cu–Zn SOD, lysozyme in serum, liver, and kidney are 11.14, 11.42, 12.06, and 11.95 mg kg^{-1} diet, respectively

dietary Cu concentrations, and 11.78 mg kg^{-1} diet provided maximum activities (Fig. 26.2c–e). Lysozyme activities were significantly increased with dietary Cu in excess (56.34 mg kg^{-1}) as compared to that supplemented with basal diet (5.91 mg kg^{-1}).

Analysis by broken-line regression of WGR and enzyme activities of the fish indicated that the adequate requirement of dietary copper were 11–12 mg kg^{-1} (Figs. 26.1 and 26.2).

26.4 Discussion

Cowey [6] suggested that besides growth, other physiological parameters such as tissue enzyme activity could be appropriate to quantify nutrient requirement in fish. In this study, the results of enzyme activities agreed well with that obtained from growth performance of *C. semilaevis*. Dietary Cu requirement fed with inorganic Cu ($CuSO_4$) was 11–12 mg kg^{-1} diet, which was higher than that of most fish requirements ranging from 2 to 6 mg kg^{-1} diet. Dietary Cu requirements always differ among species and even within different life stages of a single species, and other factors including, supplemented Cu compound (e.g., inorganic Cu or organic Cu), duration of Cu exposure, diet type (i.e., practical, purified, or live diets), or water quality (flow rates, temperature, hardness, pH, alkalinity) will affect dietary Cu requirements of fish even within same life stages of a single species.

Maximal tolerable levels of dietary copper, characterized by a negative growth response, varied widely with fish species. Reduced growth was observed in Atlantic salmon (*Salmo salar*) exposed to 500 mg kg^{-1} diet for 3 months [18]. Nile tilapia (*Oreochromis niloticus*) fed to satiation with a Cu-loaded diet of 2,000 mg kg^{-1} for 42 days showed significant decreased growth rate [26], while African walking catfish (*Clarias gariepinus*) supplemented with excess dietary Cu of 1,500 mg kg^{-1} showed only marginal reductions in body weight [12]. No growth retardation in *Carassius auratus* gibelio fed with 9 mg kg^{-1} of dietary Cu was observed as compared to fish fed with the basal diet [25]. Reduced growth was shown in juvenile yellow catfish (*Pelteobagrus fulvidraco*) given more than 20 mg kg^{-1} Cu-supplemented diets [29]. Weight gain in blunt snout bream (*Megalobrama amblycephala*) fed diets supplemented with 150 mg kg^{-1} was significantly lower than those fed diets added with 3 mg kg^{-1} [24]. The present experiment showed no growth retardation in response to high diet Cu concentration (56.34 mg kg^{-1}). As compared to fishfed with the basal diet, elevated Cu in diet stimulated the growth of *C. semilaevis*. The mechanisms increased growth with high Cu supplementation in animals have been investigated broadly. High Cu supplemented in diet could increase daily feed intake and apparent digestibility of crude lipid weanling pigs, which in turn stimulated the growth [5]. It was hypothesized that Cu stimulated growth hormone and growth factor synthesis, and thus promoted the ability of pigs to grow [32].

Cu might excise inhibitory, stimulatory, or no effects on the digestive enzyme activities depending on exposure concentration and exposure time in vivo or in vitro [4, 19, 28]. We found that digestive enzyme activities increased with Cu concentrations at low supplementations and decreased at high supplementations. High Cu supplementation had inhibitory effects on digestive enzyme activities after saturation of the Cu-binding capacity of strong binding sites involved in catalytic function, similar to earlier studies [4].

Cu–Zn SOD, as the copper-dependent antioxidant enzyme, plays an important role in the protection of cells from free radical damages by catalyzing the dismutation of two superoxide radicals to hydrogen peroxide and oxygen [10]. It has been shown to be an excellent indicator of copper nutrition in fish [16], as well as in shellfish [31]. Insufficient dietary Cu supplementation would lead to deficiency in Cu functioning as a cofactor required for catalytic properties of Cu–Zn SOD. This might be the reason why fish fed with the basal diet exhibited the lowest Cu–Zn SOD activity. MDA, as indicator of oxidative stress by detecting lipid peroxidation, showed an inverse trend with the performance of Cu–Zn SOD activity. The high MDA values in fish fed with the basal diet might be due to low activity of Cu–Zn SOD [16]. Depressed Cu–Zn SOD activity in fish was observed when the dietary Cu was in excess (56.34 mg kg^{-1} diet) because subjecting fish to excessive dietary Cu might lead to oxidative stress and increased SOD consumption.

Experimentally induced nutritional deficiency or excesses of copper have been reported to exert inhibitory effects on immune responses and increase the severity of infection among fish and crustaceans [14, 23]. Lysozyme is released by leukocytes and plays an important role in antimicrobial activity. In the present study, depressed lysozyme activity in fish was observed when the dietary Cu was in deficiency ($\leq$8.23 mg kg^{-1} diet) and excess (56.34 mg kg^{-1} diet). Inconsistent with our result, the lysozyme activity in abalone serum, however, maintained relatively constant regardless of dietary Cu treatments (1.08–109.41 mg kg^{-1} diet) [31].

In conclusion, the present results indicated that the optimum Cu requirement of *C. semilaevis* was 11–12 mg kg^{-1}. Growth retardation was only observed in fish fed with the basal dietdeficient in Cu (5.91 mg kg^{-1}), but not in fish fed excessive dietary Cu (56.34 mg kg^{-1}). Depressed enzyme activities were found in fish when the dietary Cu was in deficiency (5.91 mg kg^{-1}) and in excess (56.34 mg kg^{-1}).

Acknowledgments The financial support provided by National "Twelfth Five-year" Science and Technology Support Project (Grant No. 2011BAD13B07), National Spark Plan Project (2012GA610003), and Science and Technology Planning Project of Tianjin (12ZCDZNC05900, 11ZCKFNC00400, 20120625, TD12-5018) is gratefully acknowledged.

References

1. Berntssen MHG, Hylland K, Wendelaar BSE et al (1999) Toxic levels of dietary copper in Atlantic salmon (*Salmo salar* L.) parr. Aquat Toxicol 46:87–99
2. Berntssen MHG, Lundebye AK, Maage A (1999) Effects of elevated dietary copper concentrations on growth, feed utilisation and nutritional status of Atlantic salmon (*Salmo salar* L.) fry. Aquaculture 174:167–181

3. Bradford MM (1976) A rapid and sensitive method for the quantitation of microgram quantities of protein utilizing the principle of protein-dye binding. Anal Biochem 72:248–254
4. Chen Z, Mayer LM, Weston DP et al (2002) Inhibition of digestive enzyme activities by copper in the guts of various marine benthic invertebrates. Environ Toxicol Chem 21(6): 1243–1248
5. Cheng ZG, Xu ZR, Lin YC (2004) Effect of high copper on growth and approach to mechanism in weanling pigs. Si Chuan Nong Ye Da Xue Xue Bao 25(6):62–64
6. Cowey CB (1976) Use of synthetic diets and biochemical criteria in the assessment of nutrient requirement of fish. J Fish Res Board Can 33:1040–1045
7. De Schamphelaere KA, Forrez I, Dierckens K et al (2007) Chronic toxicity of dietary copper to Daphnia magna. Aquat Toxicol 81(4):409–418
8. Dogru MI, Dogru AK, Gul M et al (2008) The effect of adrenomedullin on rats exposed to lead. J Appl Toxicol 28:140–146
9. Ellis AE (1999) Immunity to bacteria in fish. Fish Shellfish Immunol 9:291–308
10. Fattman CL, Schaefer LM, Oury TD (2003) Extracellular superoxide dismutase in biology and medicine. Free Radic Biol Med 35:236–256
11. Gatlin DM III, Wilson RP (1986) Dietary copper requirement of fingerling channel catfish. Aquaculture 54:277–285
12. Hoyle I, Shaw BJ, Handy RD (2007) Dietary copper exposure in the African walking catfish, *Clarias gariepinus*: Transient osmoregulatory disturbances and oxidative stress. Aquat Toxicol 83(1):62–72
13. Kim SG, Kang JC (2004) Effect of dietary copper exposure on accumulation, growth and hematological parameters of the juvenile rockfish, *Sebastes schlegeli*. Marine Environ Res 58 (1):65–82
14. Lee MH, Shiau SY (2002) Dietary copper requirement of juvenile grass shrimp, *Penaeus monodon*, and effects on non-specific immune responses. Fish Shellfish Immunol 13(4):259–270
15. Lin YH, Shie YY, Shiau SY (2008) Dietary copper requirements of juvenile grouper, *Epinephelus malabaricus*. Aquaculture 274:161–165
16. Lin YH, Shih CC, Kent M et al (2010) Dietary copper requirement reevaluation for juvenile grouper, *Epinephelus malabaricus*, with an organic copper source. Aquaculture 310:173–177
17. Lorentzen M, Maage A, Julshamn K (1998) Supplementing copper to a fish meal based diet fed to Atlantic salmon parr affects liver copper and selenium concentrations. Aquac Nutr 4 (1):67–72
18. Lundebye AK, Berntssen MHG, Bonga SEW et al (1999) Biochemical and physiological responses in Atlantic salmon (*Salmo salar*) following dietary exposure. Marine Pollut Bull 39 (1–12):137–144
19. Luo XG, Dove CR (1996) Effect of dietary copper and fat on nutrient utilization, digestive enzyme activities, and tissue mineral levels in weanling pigs. J Anim Sci 74:1888–1896
20. Lv JF (2002) Success in artificial reproduction of *Cynoglossus semilaevis* Günther. Nong Ye Zhi Shi 23:14
21. Minghetti M, Leaver MJ, Carpenè E et al (2008) Copper transporter 1, metallothionein and glutathione reductase genes are differentially expressed in tissues of sea bream (*Sparus aurata*) after exposure to dietary or waterborne copper. Comp Biochem Physiol C: Toxicol Pharmacol 147(4):450–459
22. Ogino C, Yang GY (1980) Requirements of carp and rainbow trout for dietary manganese and copper. Bull Jpn Soc Sci Fish 46:455–458
23. Rougier F, Troutaud D, Ndoye A et al (1994) Non-specific immune response of Zebra fish, *Brachydanio rerio* (Hamilton-Buchanan) follow copper and zinc exposure. Fish Shellfish Immunol 4:115–127
24. Shao XP, Liu WB, Lu KL et al (2012) Effects of tribasic copper chloride on growth, copper status, antioxidant activities, immune responses and intestinal microflora of blunt snout bream (*Megalobrama amblycephala*) fed practical diets. Aquaculture 338–341:154–159

25. Shao XP, Liu WB, Xu WN et al (2010) Effects of dietary copper sources and levels on performance, copper status, plasma antioxidant activities and relative copper bioavailability in *Carassius auratus* gibelio. Aquaculture 308(1–2):60–65
26. Shaw BJ, Handy RD (2006) Dietary copper exposure and recovery in Nile tilapia, *Oreochromis niloticus*. Aquat Toxicol 76(2):111–121
27. Shiau SY, Ning YC (2003) Estimaion of dietary copper requirements for juvenile tilapia, *Oreochromis niloticus* × *O. aureus*. Anim Sci 77(2):287–292
28. Steinhart H, Wieninger-Rustemeyer R, Kirchgessner M (1981) Effect of Cu++ ions on the activity of trypsin on natural substance. Arch Tierernahr 31(2):119–125
29. Tan XY, Luo Z, Liu X et al (2011) Dietary copper requirement of juvenile yellow catfish, *Pelteobagrus fulvidraco*. Aquac Nutr 17(2):170–176
30. Trenzado C, Carmen HM, García-Gallego M et al (2006) Antioxidant enzymes and lipid peroxidation in sturgeon *Acipenser naccarii* and trout *Oncorhynchus mykiss*. A comparative study. Aquaculture 254:758–767
31. Wang WF, Mai KS, Zhang WB et al (2009) Effects of dietary copper on survival, growth and immune response of juvenile abalone, *Haliotis discus hannai* Ino. Aquaculture 297(1–4): 122–127
32. Zhou W, Kornegay ET, van Laar H, Swinkels JW, Wong EA, Lindemann MD (1994) The role of feed consumption and feed efficiency in copper-stimulated growth. J Anim Sci 72(9): 2385–2394

Chapter 27
Influence of Different Substrates on the Production of Pigments and Citrinin by Monascus FJ46

Hongxia Mu, Liubin Huang, Xuemei Ding and Shuxin Zhao

Abstract The aim of this work is to investigate the influence of various carbon sources, including cereals, tuber crops, and argo-industrial residues as substrates on the production of pigments and citrinin by Monascus FJ46. Compared with control, all the substrates can reduce the yield of citrinin except for glutinous rice flour and potato as substrates. The relative yield (citrinin concentration/red pigment valeur) with naked oats flour was 0.0012 μg/U, followed 0.0015 μg/U with millet flour, 0.0043 μg/U with sorghum flour, 0.0080 μg/U with corn flour, 0.0088 μg/U with cordyceps sinensis residues, and 0.0091 μg/U with sweet potato. Therefore, naked oats flour, millet flour, sorghum flour, corn flour, cordyceps sinensis residues, and sweet potato can be promising substrates for the production of pigments. Additionally, this is the first report on pigments production using cordyceps sinensis residues as substrate.

Keywords Monascus · Pigments · Citrinin

27.1 Introduction

Recently, due to increasing concern in the use of natural food colorants and various synthetic coloring agents, they have been banned, as they have a potential for carcinogenicity or teratogenicity. This situation has inevitably increased demands for safety, and using naturally occurring edible coloring agents, one of which is Monascus pigment [1]. By fermentation, it is possible to develop processes for production of environmentally friendly coloring and dye intermediates that can be

---

H. Mu · L. Huang · X. Ding · S. Zhao (✉)
Tianjin University of Science and Technology, Tianjin 300457, China
e-mail: zsx999@tust.edu.cn

T.-C. Zhang and M. Nakajima (eds.), *Advances in Applied Biotechnology*,
Lecture Notes in Electrical Engineering 332, DOI 10.1007/978-3-662-45657-6_27

used as food constituents [2]. In fact, for centuries Monascus have been used as a source of pigment for coloring of fish, cookies, Chinese cheese, red rice wine, and sausages in East Asian countries [3, 4]. Monascus pigments are stable at pH 6.0–8.0 and 30–60 °C. Some Monascus pigments are stable even at higher temperatures and extreme pH values [5]. These pigments rapidly react with amino groups present in proteins, amino acids, and nucleic acids, forming hydro-soluble complexes, keeping its color stable for several months [6]. These advantages make Monascus pigments an important alternative to synthetic dyestuff.

However, another secondary toxic metabolite, known as citrinin ($C_{13}H_{14}O_5$), is also co-produced by Monascus fermentation [4, 7]. Citrinin is a secondary fungal metabolite that has been known since 1931, when it was first isolated from *Penicillium citrinum* and later from several species of *Penicillium* and *Aspergillus* [8]. Citrinin possesses antibiotic, bacteriostatic, antifungal, and antiprotozoal properties. It also has been shown to be nephrotoxic, hepatotoxic, and carcinogenic to humans and animals [9, 10]. Citrinin is a low molecular weight (250.25 g/mol) compound with a melting point of 175 °C. The LD50 of citrinin has been reported to be about 35–58 mg/kg in oral administration in mouse, 57 mg/kg in duck, 95 mg/kg in chicken, 50 mg/kg in rat, and 134 mg/kg in rabbit. Therefore, the presence of citrinin in Monascus pigments and its related products has become a threat to human health [11–13]. Because of great concern for citrinin contamination, Japan has issued an advisory limit of 200 ppb; in USA, the Food and Drug Administration (FDA) action level of citrinin is 20 ppb in agricultural products for sale, and the European Union has a recommended limit of 100 ppb [13]. However, there is no specific legislation for citrinin worldwide. It is difficult to establish widely acceptable limits for citrinin. The main reason is the lack of suitable analytical routine methods, and/or the instability of citrinin in foodstuffs [14].

Many methods have been developed to reduce the citrinin content in Monascus pigments. Medium optimization is an important process for obtaining maximum pigments and minimum citrinin yield. Several factors, such as pH, aeration, and agitation, source of carbon and nitrogen, can strongly influence Monascus growth and secondary metabolites production [15–17]. The aim of this work is to study the effect of various carious carbon sources, including cereals, tuber crops, and argo-industrial residues as substrates on the production of pigments and citrinin by Monascus FJ46.

27.2 Materials and Methods

27.2.1 Microorganism

Monascus strain: FJ46. It was maintained on Wort agar medium at 4 °C and sub-cultured once every 4 weeks.

27.2.2 Inoculum Preparation

Spore suspensions of Monascus were prepared from actively growing test-tube slants. Spore suspension (1 mL) of FJ46 was inoculated to 250 mL flasks containing 50 mL seed cultivation medium(g/L): sucrose 10, peptone 3, K_2HPO_4 0.5, $MgSO_4$ 0.5, pH 5.7, then incubated at 35 °C for 24 h in a rotary shaker incubator at 160 rpm.

27.2.3 Monascus Fermentation

Different cereal flours: rice flour, millet flour, rye flour, glutinous rice flour, mung bean flour, black soybean meal, sorghum flour, buckwheat flour, naked oats flour, corn flour, different tuber crops: potato, sweet potato, purple potato, cassava, and argo-industrial residues cordyceps sinensis residues, and molasses were used as substrates. Cultivated seed were transferred to the fermentation medium (g/L): substrate 20, glucose 20, peptone 1, pH 5.8, 50 mL in 250 mL Erlenmeyer flasks, then kept in rotary shaker at 160 rpm 35 °C for 120 h.

27.2.4 Biomass Estimation

The fermentation culture was centrifuged at 4000 × g for 10 min to separate supernatant from biomass. The precipitate was dried at 60 °C in hot air oven to constant weight and measured as biomass.

27.2.5 Pigment Estimation

Finely homogeneity samples in 2–5 mL of absolute ethyl alcohol was placed in a tube with plug, water bath (60 °C) for 1 h, and then the extract strained by a filter paper. The analysis of pigment production was done by measuring absorbance maxima (505, 465 and 410 nm) of pigment extract using a spectrophotometer and multiplied with the dilution factor.

Pigment production was expressed as a formula:

$$\text{Valeur} = \text{OD value} \times \text{dilution ratio} \tag{27.1}$$

27.2.6 Determination of Citrinin

To make citrinin extraction, homogeneity sample 2 mL mixed with 4 mL absolute ethyl alcohol was placed in a water bath (60 °C) for 1 h, and then the fungal extracts filtered through a 0.22 μm organic fiber membrane filters and analyzed for citrinin content by HPLC(Agilent 1,200 serious). A C18 column, 4.6 × 250 mm, ID. 5 μm, was used as the analytical column (COSMOSIL 5C18-AR-II). The mobile phase contained acetonitrile 35 %, water 65 % (pH 2.5) and the flow ratio was kept at 1.35 ml/min, and citrinin was detected using a FLD detector set at excitation wavelength of 331 nm and emission wavelength 500 nm. Citrinin standard solutions were used to obtain a calibration curve with six points in the range from 0.1 to 10 mg/L (r = 0.9997).

$$\text{Citrinin production was expressed as a formula: } C = C_0 \times D \tag{27.2}$$

C—citrinin concentration, mg/L
D—dilution.
C_0—citrinin concentration detect by HPLC, mg/L.

Relative yield was expressed as a formula:

$$\text{Relative yield} = \text{Citrinin concentration/Valeur} \tag{27.3}$$

27.3 Results and Discussion

27.3.1 Effect of Cereal Flours as Substrates on Fermentation of FJ46

Traditionally, rice is the most popular substrate used for Monascus fermentation. In this study we analyze the feasibility of other grains as substrates. The influence of different cereal flours as substrates on fermentation of FJ46 are shown in Table 27.1.

As shown in Table 27.1, compared with rice flour substrate, all cereal substrates can reduce the yield of citrinin except for glutinous rice flour. Although having more pigments production, glutinous rice flour is not suitable for Monascus fermentation. When mung bean flour and black soybean meal were used as substrates, the production of pigments was low. This phenomenon may be due to the fact that beans are rich in protein, so it is in favor of cell growth but is not conducive to metabolites production. Among the three pigments, the orange pigment was the least with rye flour, mung bean flour, black soybean meal, and buckwheat flour as substrates.

The influence of different cereal flours as substrates on relative yield are shown in Fig. 27.1. The relative yields were the lowest when FJ46 fermented millet flour,

Table 27.1 Effect of cereal flours on fermentation of FJ46

Substrates	Citrinin (mg/L)	Valeur (U/mL)		
		Red	Orange	Yellow
RiF	1.37 ± 0.05	58.58 ± 3.54	59.50 ± 3.09	53.34 ± 3.10
MiF	0.05 ± 0.01	33.82 ± 1.24	39.65 ± 1.97	35.29 ± 1.91
RyF	0.77 ± 0.06	67.60 ± 0.35	67.52 ± 3.94	67.80 ± 0.63
GlF	2.46 ± 0.07	63.11 ± 0.02	67.98 ± 1.42	57.30 ± 0.86
MuF	0.30 ± 0.07	23.91 ± 2.36	15.04 ± 0.02	25.66 ± 0.08
BlF	0.02 ± 0.00	16.03 ± 1.10	10.57 ± 0.38	15.60 ± 0.28
SoF	0.25 ± 0.01	59.27 ± 2.92	70.30 ± 1.78	49.85 ± 1.14
BuF	0.73 ± 0.09	58.66 ± 0.04	56.16 ± 3.32	62.21 ± 3.35
NoF	0.08 ± 0.02	62.35 ± 1.78	65.00 ± 2.36	57.38 ± 1.35
CoF	0.55 ± 0.04	67.73 ± 3.22	68.51 ± 2.60	63.37 ± 1.46

RiF rice flour, *MiF* millet flour, *RyF* rye flour, *GlF* glutinous rice flour, *MuF* mung bean flour, *BlF* black soybean meal, *SoF* sorghum flour, *BuF* buckwheat flour, *NoF* naked oats flour, *CoF* corn flour

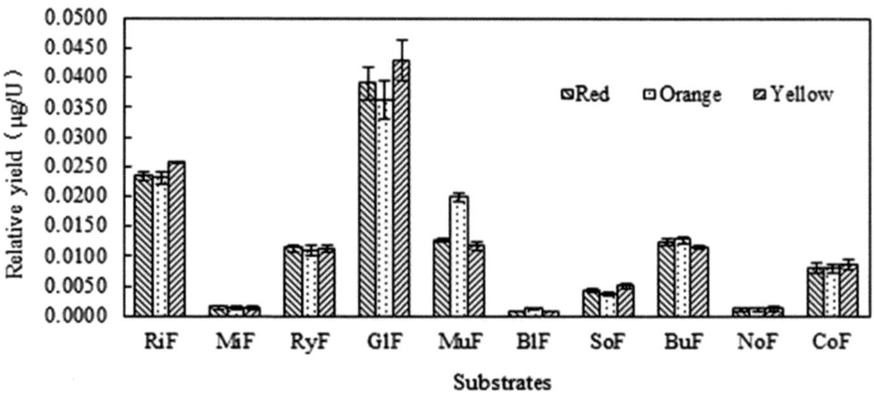

Fig. 27.1 Effect of cereal flours on relative yield of citrinin and pigments

black soybean meal, and naked oats flour. The relative yields also were very low when sorghum flour and corn flour were used as substrates. In order to obtain higher yield of pigments and lower yields of citrinin, naked oats flour was the best substrate followed by millet flour, sorghum flour and corn flour in this study.

27.3.2 *Effect of Tuber Crops as Substrates on Fermentation of FJ46*

Tuber crops are rich in starches that have been widely used as substrates in fermentation processes because of their low cost.

The influence of different tuber crop substrates on the yield of red pigments and citrinin are shown in Table 27.2 Compared with rice flour substrate, fermentation yielded lower citrinin concentration using tuber crops except for potato substrate. The yields of pigments were much lower when FJ46 fermented the tuber crop substrates. The purple potato resulted in the highest pigments production and this result may be due to purple potato itself having color. The potato and sweet potato substrates resulted in the highest yellow pigment production.

As shown in Fig. 27.2, Compared with rice flour substrates, higher relative yields were achieved when FJ46 was cultivated on potato substrate. Potato, therefore, is not a suitable substrate for FJ46 fermentation. FJ46 fermented all the other tuber crops substrates that can reduce the relative yield. Sweet potato was the most suitable substrate, next were cassava and purple potato.

Table 27.2 Effect of tuber crops on fermentation of FJ46

Substrates	Citrinin (mg/L)	Valeur (U/mL)		
		Red	Orange	Yellow
RiF	0.92 ± 0.03	52.29 ± 3.73	59.12 ± 4.80	48.84 ± 3.05
Po	1.39 ± 0.10	47.24 ± 0.11	34.53 ± 0.36	50.08 ± 1.59
SPo	0.23 ± 0.02	33.95 ± 1.48	31.22 ± 1.15	35.45 ± 1.96
RiF	11.83 ± 0.27	85.56 ± 1.35	96.09 ± 0.59	78.27 ± 0.38
PPo	7.55 ± 0.69	74.88 ± 3.16	78.72 ± 3.11	69.41 ± 2.15
RiF	9.51 ± 0.98	67.47 ± 2.95	75.83 ± 4.33	64.91 ± 5.13
Ca	4.75 ± 0.58	53.30 ± 2.10	55.86 ± 1.05	48.23 ± 1.33

RiF rice flour, *Po* potato, *SPo* sweet potato, *PPo* purple potato, *Ca* cassava

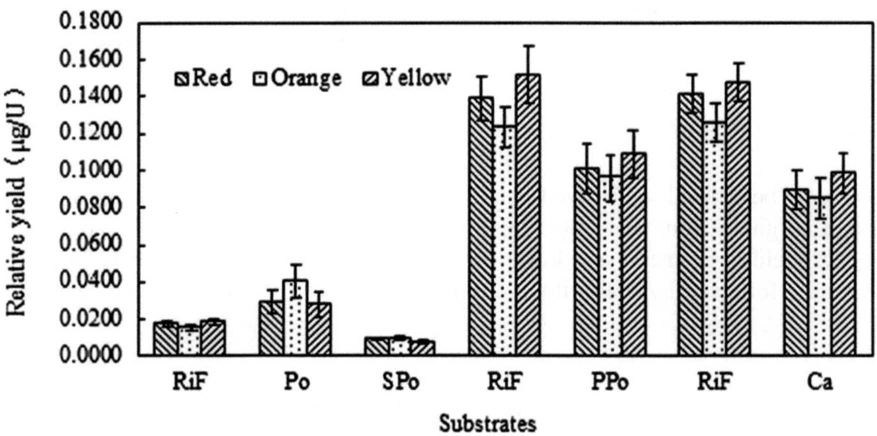

Fig. 27.2 Effect of tuber crops on relative yield of citrinin and pigments

Table 27.3 Effect of argo-industrial residues on fermentation of FJ46

Substrates	Citrinin (mg/L)	Valeur (U/mL)		
		Red	Orange	Yellow
RiF	7.89 ± 0.61	34.53 ± 0.11	38.75 ± 1.18	36.54 ± 0.36
CoS	0.56 ± 0.09	64.09 ± 0.64	54.58 ± 0.49	66.45 ± 1.65
RiF	0.92 ± 0.03	52.29 ± 3.73	59.12 ± 4.80	48.84 ± 3.05
MoL	0.06 ± 0.02	8.03 ± 1.52	5.27 ± 1.39	8.69 ± 1.53
CaM	0	11.37 ± 0.05	7.81 ± 0.21	12.02 ± 0.14

RiF rice flour, *CoS* cordyceps sinensis residues, *MoL* rock sugar molasses, *CaM* cane molasses

27.3.3 *Effect of Argo-industrial Residues as Substrates on Fermentation of FJ46*

Till date, several argo-industrial residues such as grape waste, jackfruit seed, sugarcane bagasse, palm kernel cake, and glycerol have been studied as substrates [18–20]. The production of FJ46 pigments and citrinin using cordyceps sinensis residues, Rock sugar molasses, and cane molasses were investigated.

As shown in Table 27.3, from the result it can be concluded that cordyceps sinensis residues was unfavorable for production of citrinin. At the same time higher pigments were achieved with this substrate. Although citrinin yields were reduced, we obtain low production of pigments and biomass with Rock sugar molasses and cane molasses as substrates. In other words, molasses are unsuitable for Monascus fermentation.

27.4 Conclusions

Different substrates have a significant influence on the production of pigments and citrinin by Monascus FJ46. Compared with rice flour used as substrate, all other substrates can reduce the yield of citrinin except for glutinous rice flour and potato. Lower relative yields (citrinin concentration/red pigment valeur) were achieved when naked oats flour millet flour (0.0012 μg/U), millet flour (0.0015 μg/U), sorghum flour (0.0043 μg/U), corn flour (0.0080 μg/U), cordyceps sinensis residues (0.0088 μg/U), and sweet potato (0.0091 μg/U) were used as substrates. The results obtained in this study reveal that naked oats flour, millet flour, millet flour, sorghum flour, corn flour, cordyceps sinensis residues, and sweet potato can be promising substrates for production of pigments. In addition, to our knowledge this is the first report on pigments production using cordyceps sinensis residues as substrate. Further studies on the general culture conditions for Monascus fermented cordyceps sinensis residues will be taken into consideration in the future work.

Acknowledgments This work was supported by the Major State Basic Research Development Program of China (No. 2013CB734004).

References

1. Babitha S et al (2007) Solid-state fermentation for the production of Monascus pigments from jackfruit seed. Bioresour Technol 98:1554–1560
2. Dominguez-Espinosa RM, Webb C (2003) Submerged fermentation in wheat substrates for production of Monascus pigments. World J Microbiol Biotechnol 19:329–336
3. Cheng M-J et al (2012) Secondary metabolites produced by the fungus *Monascus pilosus* and their anti-inflammatory activity. Phytochem Lett 5:567–571
4. Pattanagul P et al (2008) Mevinolin, citrinin and pigments of adlay angkak fermented by *Monascus sp*. Int J Food Microbiol 126:20–23
5. Feng Y et al (2012) Monascus pigments. Appl Microbiol Biotechnol 96:1421–1440
6. Meinicke RM et al (2012) Potential use of glycerol as substrate for the production of red pigments by *Monascus ruber* in submerged fermentation. Biocatal Agr Biotechnol 1:238–242
7. Kang B et al (2014) Production of citrinin-free Monascus pigments by submerged culture at low pH. Enzym Microb Technol 55:50–57
8. Wang Y-Z (2005) The variability of citrinin production in Monascus type cultures. Food Microbiol 22:145–148
9. Li Y et al (2012) Microsphere-based flow cytometric immunoassay for the determination of citrinin in red yeast rice. Food Chem 134:2540–2545
10. Islam MR et al (2012) Immune modulatory effects of the foodborne contaminant citrinin in mice. Food Chem Toxicol Int J Publ Br Ind Biol Res Assoc 50:3537–3547
11. Xu B-J et al (2006) Review on the qualitative and quantitative analysis of the mycotoxin citrinin. Food Control 17:271–285
12. Li Y et al (2012) Natural occurrence of citrinin in widely consumed traditional Chinese food red yeast rice, medicinal plants and their related products. Food Chem 132:1040–1045
13. Shi YC, Pan TM (2011) Beneficial effects of *Monascus purpureus* NTU 568-fermented products: a review. Appl Microbiol Biotechnol 90:1207–1217
14. Zaied C et al (2012) Natural occurrence of citrinin in Tunisian wheat grains. Food Control 28:106–109
15. Sharmila G et al (2013) Sequential statistical optimization of red pigment production by *Monascus purpureus* (MTCC 369) using potato powder. Ind Crops Prod 44:158–164
16. Silveira ST et al (2008) Pigment production by *Monascus purpureus* in grape waste using factorial design. LWT Food Sci Technol 41:170–174
17. Yongsmith B et al (1993) Culture conditions for yellow pigment formation by *Monascus sp.* KB-10 grown on cassava mediun. World J Microbiol Biotechnol 9:85–90
18. Nimnoi P, Lumyong S (2009) Improving solid-state fermentation of *Monascus purpureus* on agricultural products for pigment production. Food Bioprocess Technol 4:1384–1390
19. Silveira ST et al (2011) Stability modeling of red pigments produced by *Monascus purpureus* in submerged cultivations with sugarcane bagasse. Food Bioprocess Technol 6:1007–1014
20. Velmurugan P et al (2011) Monascus pigment production by solid-state fermentation with corn cob substrate. J Biosci Bioeng 112:590–594

Chapter 28
FAD2B from a Peanut Mutant with High Oleic Acid Content Was Not Completely Dysfunctional

Xiu Zhen Wang, Qi Wu, Yue Yi Tang, Quan Xi Sun and Chuan Tang Wang

Abstract A peanut mutant with 72.53 % oleate content was selected from sodium azide mutagenized peanut Huayu 22, an export type cultivar with 51.65 % oleate content. The chemical mutant had a point mutation (C281T) in the coding region of *FAD2B*, which caused a T94I substitution in the FAD2B protein. Though the mutated *FAD2B* gene was incompletely dysfunctional as shown in yeast expression system, the peanut mutant has a high oleic acid phenotype. This indicated the possible involvement of other unidentified genetic factor(s) conditioning oleic acid content in peanut.

Keywords Peanut · *FAD2B* · High oleate · Yeast expression · Mutant

28.1 Introduction

Peanut is not only an important oil crop but also a good energy-dense plant. Peanut with high oleic acid content is considered more suitable for the development of biodiesel products with good low temperature fluidity [1]. At the same time high oleic acid content is a preferred seed quality trait of peanut as it proved to have positive effects on shelf life and human health. In general, the normal level of oleate content in peanut varies from 36 to 67 %, but in F435, a peanut natural mutant, the oleate content reached up to 80 % according to a report by Norden et al. [2]. So, much effort has been devoted to developing peanut cultivars with high oleic acid, and several high oleate peanut mutants have been bred through chemical mutagenesis or gamma ray irradiation [3–5].

The high oleate trait was believed to be controlled by fatty acid desaturase 2 (FAD2), which is responsible for the conversion of oleic acid (C18:1) to linoleic

X.Z. Wang · Q. Wu · Y.Y. Tang · Q.X. Sun · C.T. Wang (✉)
Shandong Peanut Research Institute, Qingdao 266100, China
e-mail: chinapeanut@126.com

T.-C. Zhang and M. Nakajima (eds.), *Advances in Applied Biotechnology*,
Lecture Notes in Electrical Engineering 332, DOI 10.1007/978-3-662-45657-6_28

acid (C18:2). In *Arabidopsis* the *FAD2* gene was cloned by a forward genetics approach [6]. The seeds of *FAD2* mutants induced by T-DNA insertion contained a significantly high level of oleic acid and reduced amount of linoleic acid. In the cultivated peanut, both *FAD2A* and *FAD2B* genes have been regarded as logical candidate genes for the high oleate phenotype as the crop is an allotetraploid. In most cases, a 1-bp substitution in *FAD2A* (G:C → A:T) at position 448 after the start codon resulting in a missense amino acid substitution from aspartic acid to asparagine (D150 N) and a 1-bp insertion (A:T) at position 442 of the mutant allele *FAD2B* causing a frameshift together contributed to high oleate phenotype [7, 8].

In this paper, we cloned *FAD2A* and *FAD2B* from a peanut chemical mutant with high oleate and analyzed the function of *FAD2B* in yeast expression system.

28.2 Materials and Methods

28.2.1 Experimental Materials

Huayu 22, a high yielding Virginia type peanut cultivar (export type) was previously treated with mutagen sodium azide (NaN_3), resulting in a peanut plant with elevated oleic acid content [4]. Seeds of the derived mutants (Mutants D, E and F, M_5) along with those of the untreated controls (Samples A, B, and C) were sown on May 10, 2011 on the SPRI Experimental farm in Laixi, Qingdao, China. Each individual plant was harvested separately on September 12, 2011.

28.2.2 Screening of M_5 Seeds for High Oleate

A near infrared reflectance spectroscopy (NIRS) machine (Matrix-I, Bruker Optics, Ettlingen, Germany) was employed to collect spectral data of bulk seeds from individual single plants and individual single seeds. Each seed was scanned three times and oleate content was predicted by NIRS. To confirm the prediction results, fatty acid profiles of single seeds was determined by gas chromatograph (GC) according to Yang et al. [9].

28.2.3 Cloning and Analysis of the Genomic Sequences of FAD2A and FAD2B

The genomic DNA was isolated from a slice of peanut cotyledonary tissue according to the protocol developed at our laboratory earlier [10]. DNA was dissolved in TE buffer (10 mM Tris, 1 mM EDTA, pH 8.0) and stored at −20 °C. Primer pairs aF19

(GATTACTGATTATTGACTT)/R1 (CTCTGACTATGCATCAG) and bF19 (CCAGAACCATTAGCTTTG)/R1 [11] were used to amplify the sequences of *FAD2A* and *FAD2B* genes, respectively. PCR reaction mixtures and thermal cycling profiles were the same as those described by Fang et al. [12]. PCR products were purified, cloned, and sequenced. Pairwise alignments of nucleotide sequences and deduced amino acid sequences were performed using the Lasergene DNAStar package (DNAStar, Madison, USA).

28.2.4 Construction of Recombinant Yeast Expression Vector and Yeast Transformation

For functional analysis of wild type *FAD2B* from Huayu 22 and mutant type *FAD2B* from a high oleate peanut mutant (Mutant F), the auxotrophic *Saccharomyces cerevisiae* strain INVSc1 (MATa his3-Δ^1 leu2 trp1-289 ura3-52) and the shuttle vector pYES2 (Invirtrogen, Carlsbad, CA,USA) were utilized, and the complete coding sequences (cds) of *FAD2B* was amplified with modified gene-specific primers (bF19 and R1) with restriction enzyme site of *Bam*H I (CCGGATC) or *Xho* I (CCCTCGAGT) included. *Bam*H I/*Xho* I digestion products of *FAD2B* and pYES2 were ligated together. The pYES-FAD2B constructs with *FAD2B* cds inserted between the inducible yeast GAL1 promoter and the yeast CYC1 terminator were transferred into the INVSc1 cells using the lithium acetate method [13]. pYES2 vector with no *FAD2B* cds insert was also used to transform yeast cells. Selection was conducted on a synthetic complete agar medium without uracil (SC-U). Single colonies were cultured separately in liquid SC-U medium at 30 °C until OD_{600} reached 0.4. Yeast vector DNA was extracted with TIAN-prep yeast Plasmid DNA Kit (Tiangen, Beijing, China) and double digested with *Bam*H I and *Xho* I to identify the right yeast transformants. RNA was extracted from the yeast cultures grown in induction medium (SC-U medium supplemented with 2 % galactose and 2 % raffinose) using Tiangen RNAprep pure Kit (Tiangen, Beijing, China). RT-PCR with gene-specific primers (bF19/R1) was performed to make sure that transcription of *FAD2B* was normal.

28.2.5 Analysis of Fatty Acids in Yeast

Yeast transformants possessed pYES-FAD2B constructs were cultured individually in a 20 mL liquid SC-U medium at 30 °C with agitation (220 rpm) overnight. Transformants possessing the empty pYES2 vector were used as control. Then 2 mL cultures were added to 50 mL new induction medium and agitated until OD_{600} reached 0.4. Cell pellets were collected by centrifugation at 1,500 g at 4 °C for 5 min. Then the collection was resuspended in 50 mL liquid inducing medium and cultured at 30 °C with agitation (220 rpm) for 48 h. Cells were again collected by centrifugation at 1,500 g 4 °C for 5 min, then vacuum-dried and ground into a fine powder in liquid nitrogen. About 20 μg of the powder samples were used to analyze fatty acids by gas chromatograph-mass spectroscopy (GC-MS) [12].

28.3 Results

28.3.1 Oleate Content Analysis of the Mutants and Huayu 22

With NIRS calibration equations for main fatty acids in peanut, we were able to select peanut mutants with high oleate content. Oleate contents of the mutants and Huayu 22 as determined by GC were shown in Table 28.1. Mean oleate content from peanut mutants was 72.53 %, as against 51.65 % in wild type peanut, Huayu 22.

28.3.2 Sequence Analysis of FAD2A and FAD2B

FAD2A gene and *FAD2B* gene were cloned from a peanut mutant (Mutant F) and from wild type Huyau 22 (Sample A), respectively. Pairwise sequence alignment of *FAD2A* (Fig. 28.1) and *FAD2B* genes (Fig. 28.2) from the peanut mutant and wild type peanut (Huayu 22) showed that the sequence of *FAD2A* from the peanut

Table 28.1 Oleate content (%) in Huayu 22 (wild type, Samples A, B and C) and its mutants (Mutants D, E and F) as determined by GC

Wild type (control)				Mutant type			
Sample A	Sample B	Sample C	Average	Mutant D	Mutant E	Mutant F	Average
41.62	55.70	57.64	51.65	72.17	72.48	72.95	72.53

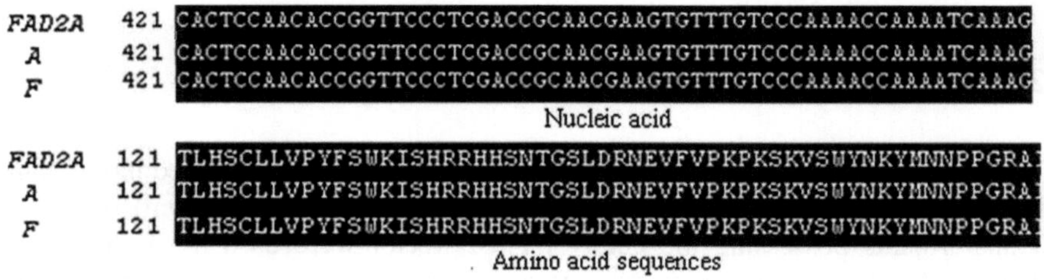

Fig. 28.1 Nucleic acid and amino acid sequences alignments of partial *FAD2A* gene from the peanut mutant with high oleate content (*F*) and wild type peanut seed (*A*). *Note FAD2A* (HM359250.1) was downloaded from NCBI

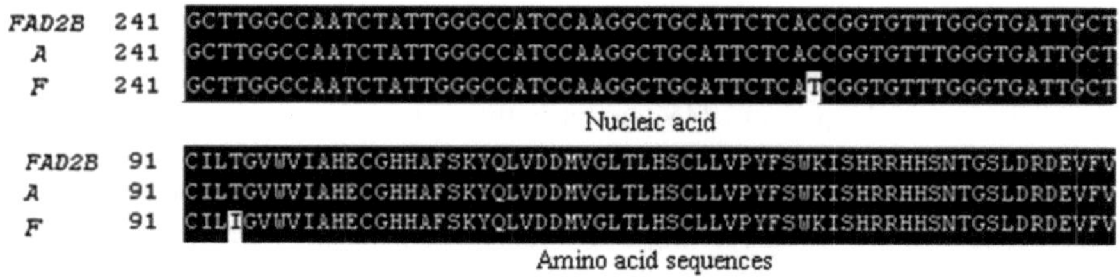

Fig. 28.2 Nucleic acid and amino acid sequences alignments of partial *FAD2B* gene from the peanut mutant with high oleate content (*F*) and wild type peanut seed (*A*). *Note FAD2B* (HM359251.1) was downloaded from NCBI

mutant was the same as that from wild type peanut. However, a C to T transition in 281 position of the coding region of *FAD2B* was found in the mutant, which caused a T94I mutation in the fatty acid desaturase. The amino acid substitution was not located in any of the known histidine boxes.

28.3.3 Functional Analysis of FAD2B Gene by Its Expression in S. cerevisiae

GC-MS analysis results showed that linoleate (C18:2$\Delta^{9,12}$) content in yeast strains with mutant *FAD2B* was only 2.14 %, which was less than half of that in yeast strains with wild type *FAD2B*, and that yeast strains harboring pYES2 did not produce any linoleate (Table 28.2). While, oleate content (C18:1Δ^{9}) was the highest in yeast harboring pYES2, followed by the yeast with mutant *FAD2B*, the yeast strain with wild type *FAD2B* contained 20.32 % oleate, ranking third. This indicated that the mutated *FAD2B* gene encoded an enzyme with reduced oleate desaturase activity.

Table 28.2 Analysis of fatty acid content in *S. cerevisiae* by GC-MS

INVSc1 yeast	Fatty acid content (%)				
	C16:0	C16:1	C18:0	C18:1	C18:2
Yeast strains harboring pYES2	25.85	38.77	6.91	23.64	0.00
Yeast strains with mutant type *FAD2B*	23.65	39.71	8.29	21.78	2.14
Yeast strain with wild type *FAD2B*	24.14	38.75	8.26	20.32	4.42

28.4 Discussion and Conclusions

Oleate content of the peanut EMS mutant in the report of Fang et al. [12] was only 66.4 %, which is significantly lower than that in the sodium azide mutant from the present study (higher than 72 %). Interestingly, even though both of the *FAD2A* and *FAD2B* genes from the EMS mutant lost their function, the oleate content in the EMS mutant was not high enough. Ironically, despite that the *FAD2B* from the present study was incompletely dysfunctional as shown by yeast expression, the sodium azide mutant was high oleic. As such, the possibility that there are mutation (s) in upstream regions (e.g., promoter) of *FAD2B* causing a completely inactive oleate desaturase cannot be fully excluded. It is likely that there exist other unidentified genetic factors conditioning oleic acid content in peanut. In addition to the report of Fang et al. [12], at least four studies supported the hypothesis. Patel et al. failed to find any differences in the *FAD2B* of 8-2122 and that of the normal, and they were unable to explain the high oleate phenotype of 8-2122 [11]. Wang et al. utilized a multiple linear regression approach to quantify the contribution of *FAD2* to fatty acid profile in peanut, and concluded that 60 % of the variation in oleic or linoleic acid content could be explained by the genotypic effect of *FAD2A* and *FAD2B* [14]. Pandey et al. found 2 marker intervals representing mutant alleles *FAD2A* and *FAD2B* contributing 23.46 and 39.3 % phenotypic variance (PVE), respectively, for oleic acid [15]. Chamberlin et al. noted normal oleic seeds in seed lots of the high oleic peanut cultivar Brantley [16].

To summarize, in the present study, a peanut high oleic mutant was identified, whose *FAD2B* gene had a C281T mutation in its coding region, causing a T94I substitution in the FAD2B protein. To our surprise, the mutated *FAD2B* gene from the high oleic acid mutant was incompletely dysfunctional as shown in yeast expression studies. The molecular basis for high oleate phenotype of the peanut mutant is still unclear, which deserves further investigation.

Acknowledgments We would like to thank China Agricultural Research System (Grant No. CARS-14) and SAAS (Shandong Academy of Agricultural Sciences) Key Science and Technology Achievements Incubation Project (2014CGPY09) for providing financial support for this study.

References

1. Liu CS, Yang M, Wang JW, Wang MX, Huang FH (2007) Effects of fatty acids on oxidation stability and low temperature fluidity of biodiesel. In: Proceedings of the 12th international rapeseed conference, pp 350–353
2. Norden AJ, Gorbet DW, Knauft DA, Young CT (1987) Variability in oil quality among peanut genotypes in Florida breeding program. Peanut Sci 14:7–11
3. Mondal S, Badigannavar AM, D'Souza SF (2011) Induced variability for fatty acid profile and molecular characterization of high mutant in cultivated groundnut (*Arachis hypogaea* L.). Plant Breed 130:242–247

4. Wang CT, Tang YY, Wang XZ, Zhang SW, Li GJ, Zhang JC, Yu SL (2011) Sodium azide mutagenesis resulted in a peanut plant with elevated oleate content. Electron J Biotechnol 14 (2). doi:10.2225/vol14-issue2-fulltext-4
5. Chen J (2011) Advances in genetics and breeding of high oleic acid peanut. Plant Genet Resour 12(2):190–196
6. Okuley J, Lightner J, Feldmann K, Yadav N, Lark E, Browse J (1994) *Arabidopsis FAD2* gene encodes the enzyme that is essential for polyunsaturated lipid synthesis. Plant Cell 6:147–158
7. Jung S, Swift D, Sengoku E, Patel M, Teule F, Powell G, Moore K, Abbott A (2000) The high oleate trait in the cultivated peanut (*Arachis hypogaea* L.). Isolation and characterization of two genes encoding microsomal oleoyl-PC desaturases. Mol Gen Genet 263:796–805
8. Wang ML, Barkley NA, Chen ZB, Pittman RN (2011) *FAD2* gene mutations significantly alter fatty acid profiles in cultivated peanuts (*Arachis hypogaea*). Biochem Genet 49 (11–12):748–759
9. Yang CD, Guan SY, Tang YY, Wang XZ, Wu Q, Gong QX, Wang CT (2012) Rapid non-destruction of fatty acids in single groundnut seeds by Gas Chromatograp. J Peanut Sci 41 (3):21–26
10. Wang CT, Wang XZ, Tang YY, Zhang JC, Chen DX, Cui FG, Yu SL, Yu ST (2012) CHN Patent ZL. 200910255786.0
11. Patel M, Jung S, Moore K, Powell G, Ainsworth C, Abbott A (2004) High-oleate peanut mutants result from a MITE insertion into the FAD2 gene. Theor Appl Genet 108 (8):1492–1502
12. Fang CQ, Wang CT, Wang PW, Tang YY, Wang XZ, Cui FG, Yu SL (2012) Identification of novel mutation in *FAD2B* from a peanut EMS mutant with elevated oleate content. J Oleo Sci 61(3):143–148
13. Gietz RD, Schiestl RH, Williems AR, Woods RA (1995) Studies on the transformation of intact yeast cell by the LiAc/SSDNA/PEG procedure. Yeast 11:355–360
14. Wang CT, Tang YY, Wang XZ, Wu Q, Yang Z, Jiao K, Song GS, Xu JZ (2013) Effect of *FAD2A/FAD2B* genes on fatty acid profiles in peanut seeds. Res Crops 14(4):1110–1113
15. Pandey M, Wang ML, Qiao L, Wang H, Feng S, Barkley NA, Wang J, Culbreath A, Holbrook C, Varshney R, Guo B (2014) Mapping *FAD2* genes on peanut (*Arachis hypogaea L.*) genome and contribution to oil quality. In: Plant and animal genome, XXII. 11–15 January, San Diego, CA. https://pag.confex.com/pag/xxii/webprogram/Paper11835.html
16. Chamberlin KD, Melouk HA, Madden R, Dillwith JW, Bannore Y, Rassi ZE, Payton M (2011) Determining the oleic/linoleic acid ratio in a single peanut seed: a comparison of two methods. Peanut Sci 38(2):78–84

Chapter 29
Optimization of Sterilization Process After Spore Activation for Cereal Beverage in Large-Scale Production

Zhe Li, Liping Zhu, Shigan Yan, Junjie Liu and Wenjuan Zhao

Abstract Cereal beverage is a novel cereal drink made of whole grain. Grain seeds carry quantities of microorganisms, including spores difficult to be killed. In order to maintain its nutrition, stability, and palatability, sterilization technology of cereal beverage in a large-scale production was systematically investigated to obtain optimal sterilization process. Sterilizing strategy to kill all bacteria followed by spore activation was used for the first time to cereal beverage. Effects of activation temperature, activation time on total numbers of bacteria, spore and enduring-heating spore, pH value, and flavor freshness of cereal beverage were investigated. Optimal sterilization process after spore activation of large-scale production of cereal beverage was finally determined. Results showed that suitable temperature for spore activation was 65–85 °C, the total number of bacteria and spores in cereal beverage both decreased. Spore activation at 75 °C for 30 min, did not change the taste of fresh cereal beverage. After spore activation at 75 °C for 30 min, all the bacteria in cereal beverage could be killed at 134 °C for 16 s, however, microorganisms could be sterilized at above 140 °C for 16 s without preactivation. Finally, optimal sterilization parameters of cereal beverage in a large-scale production was set as UHT sterilization at 134 °C for 16 s after preactivation at 75 °C for 30 min.

Keywords Cereal beverage · Spore · Activation · Sterilization · UHT · Sterilization intensity

These authors contributed equally to this work.

Z. Li · L. Zhu · S. Yan (✉)
College of Food and Bioengineering, Qilu University of Technology,
Jinan 250353, Shandong, China
e-mail: yanshigan@126.com

Z. Li · J. Liu · W. Zhao
Shandong Zhaoneng Furunda Biological Technology Co. Ltd.,
Jinan 250100, Shandong, China

T.-C. Zhang and M. Nakajima (eds.), *Advances in Applied Biotechnology*,
Lecture Notes in Electrical Engineering 332, DOI 10.1007/978-3-662-45657-6_29

29.1 Introduction

Cereals are considered one of the most important sources of dietary carbohydrates, proteins, vitamins, minerals, and fibers for people all over the world [1]. In China, cereals play an important role in the people's diet, accounting for 75 % of the heat energy and 50 % of protein of the human body.

Cereal beverage is a new type of drink made of natural whole grain, reserving dietary fiber, starch, protein, and other ingredients. Cereal beverage riches in nutrition and has a pH value of 6.8. As for a convenient-edible food, it meets the current people's roughage diet needs and adapts to the fast pace of life, so cereal beverage has a bright market prospect.

Folds and concave convex parts of grain raw materials carry a large number of microbes, including bacterial spores hard to be killed. Per gram of oats, buckwheat, sesame, peanut, corn, respectively, contain about 260, 60, 1200, and 30 bacterial spores [2]. Spores have the characteristics of heat resistance, antichemical drugs, antiradiation, antistatic water pressure, and low metabolic activity. Spores can resist bad environment, generally they can keep life ability from a few years to decades [3]. Through activation under specific conditions, spores can become bacteria body in nutritional status from the dormant spore. Once activation conditions are suitable, spores turn into vegetative cells and then cause food spoilage. The vegetative cells of bacteria, in general, can be killed at 70–80 °C for 10 min, but spores have the amazing ability to withstand high temperatures. For example, the vegetative cells of *Clostridium thermosaccharolyticum* may be killed at 50 °C for several minutes, however, its spore body can endure 132 °C for 4.4 min. Cereal beverage is nutritious, so it is a natural culture of microorganism. Therefore, if cereal beverage is not completely sterilized, it is bound to affect the quality of the products, and even endanger the people's life safety.

Cereal beverage must be sterilized. The higher sterilization strength is, the better sterilization effect is in theory. But as a grain beverage, sterilization temperature is so high or long as to make the protein denature and brown, gelatinize starch, finally affect the product's nutrition, stability, and palatability; therefore, sterilization of grain beverage must keep balance of nutrition, palatability, and security at the same time. This study is the first report on optimization of sterilization process after activation of bacterial spores in cereal beverage in large-scale production.

29.2 Materials and Methods

29.2.1 Preparation of Cereal Beverage

Oat cereal was mixed with other grains at a proper proportion, cleaned with water and dried by airing, baked, coarsely ground after adding five times the volume of water, cooked for 30 min, finely ground, filtered with 40 mesh, and finally added a proper proportion of sugar and stabilizer to prepare the cereal beverage.

29.2.2 Effect of Spore Activation Conditions on Cereal Beverage

Short-time heating strategy was developed to activate spores in cereal beverage. Single factor analysis method was taken to determine the effects of spore activation temperature and time on cereal beverage, namely fixed temperature in different time treatment or fixed time in different temperature treatment.

Cereal beverage was divided into several portions, placed at different temperature treatment to activate spores under constant temperature, antiseptically sampled at a certain time interval. Total bacteria, spore number, heat-resistant spores number, pH value, and freshness of the cereal beverage were investigated.

29.2.3 Effect of Sterilization Intensity After Spore Activation on Cereal Beverage

After activation at 75 °C for 30 min, cereal beverage was sterilized using UHT sterilization method with different intensity. Total number of bacteria, spore number, and number of heat-resistant spores in the sample were detected to judge the bactericidal effect.

29.2.4 Optimization of Sterilization Process After Pre-activation for Cereal Beverage

UHT sterilization intensity F_0 represents sterilization effect of specific combination of temperature and time [4]. F_0 value calculation formula is: $F_0 = (t/60) * 10$ $(T - 121.1)/Z$, T is sterilization temperature, t is sterilization time, Z value is in the range of 10–10.8 °C (normally is 10 °C). Effect of sterilization strength under different sterilization temperatures and time on microbial number, pH value, and freshness of cereal beverage were investigated.

29.2.5 Detection of Cereal Beverage

29.2.5.1 Numbers of Total Bacteria

Ten grams of the samples were suspended into sterile physiologic solution (NaCl, 0.9 % w/v) and homogenized with a blender at room temperature for 2 min. Total numbers of mesophilic bacteria were determined on Plate Count Agar (PCA, Oxoid) at 30 °C for 48 h.

29.2.5.2 Numbers of Total Spore

The samples were held at 80 °C for 10 min, and then diluted 10 times with sterile water and homogenized, took 0.1 ml evenly coated on Plate Count Agar (PCA, Oxoid) at 30 °C for 48 h, selected single colony on a glass, smeared, spore staining, observed with 1,000 times optical microscope [5, 6].

29.2.5.3 Numbers of Heat-Resistant Spores

The samples were held at 98 °C for 10 min, diluted 10 times with sterile water and homogenized, took 0.1 ml evenly coated on Plate Count Agar (PCA, Oxoid) at 30 °C for 48 h, selected a single colony on a glass, smeared, spore staining, observed with 1,000 times optical microscope.

29.2.5.4 pH Value

The samples were kept at 25 ± 2 °C, detected its pH value under the stirring condition [7].

29.2.5.5 Freshness

The samples were stood at room temperature for 48 h, tasted by the professional tasters to evaluate its freshness in fixed environment using the unified standard [8].

29.3 Results

29.3.1 Effect of Spore Activation Temperature on Bacterial Number in Cereal Beverage

Bacterial number in the cereal beverage before spore activation was 188 CFU/ml. After activation at 25–65 °C for 15–120 min, bacterial number in cereal beverage was more than that of no-activation, indicating a large number of bacteria growth in the product. Activation at 25 °C for 15–120 min, the proliferation of bacterium was slow, the number of bacterium was 180–245 CFU/ml; Activation at 50 °C for 15–120 min, bacteria grew rapidly, bacterial number was 226–802 CFU/ml. Activation at 65–85 °C, the number of bacteria in the sample decreased, significantly lower than that of no-treatment, and the higher the temperature was, the lower the number of bacteria was. Bacteria could not withstand temperature above

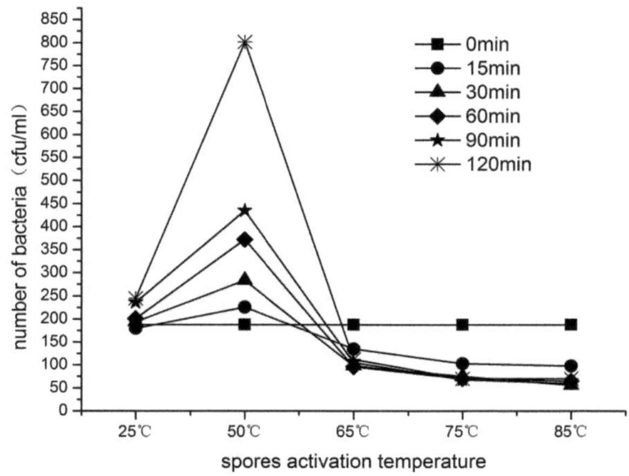

Fig. 29.1 Effect of spore activation temperature on the number of bacterium in cereal beverage

65 °C. Activation at 65, 75, and 85 °C for 15–120 min, bacterial number in the cereal beverage was 96–135, 68–103, 56–98 CFU/ml, respectively. At different activation temperature for 30 min, the number of total bacteria was the lowest. Results were shown in Fig. 29.1.

29.3.2 Effect of Activation Temperature on Spore Number in Cereal Beverage

At the temperature range of 25–85 °C for 15–120 min, spore number in the cereal beverage decreased. At 75 °C, the minimum number of spore was about 5 CFU/ml. When the temperature reached 85 °C, spore number began to increase. The results showed 75 °C was the best activation temperature, followed by 85 and 65 °C. Activation at 25 °C had no effect on bacterial death rate and spore activation. In order to reduce the total number of bacterium and spore in cereal beverage before sterilization, the sample should be activated at 65–85 °C to activate spores. Results were shown in Fig. 29.2.

At 75 °C for 15–60 min, the total number of bacterium and spore in cereal beverage was counted. The results showed that the number of total bacterium was higher in the sample activated for 15 min; however, when activated for 60 min, the number of bacterium was relatively low, but the total number of spores increased; the number of total bacteria and spores were both lower when activated for 30 min (Fig. 29.3). So activation at 75 °C for 30 min was suitable for large-scale production.

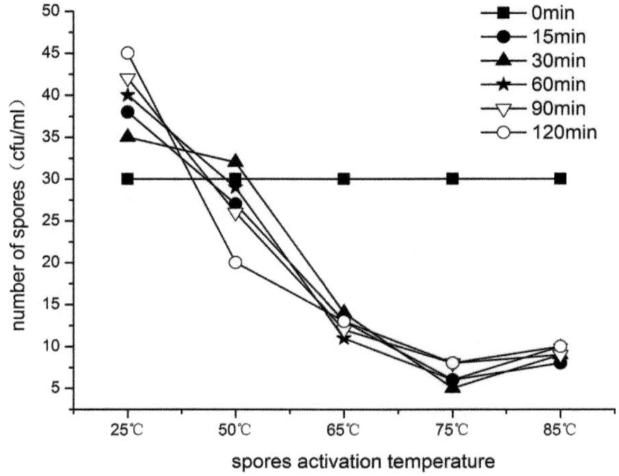

Fig. 29.2 Effect of spore activation temperature on spore number in cereal beverage

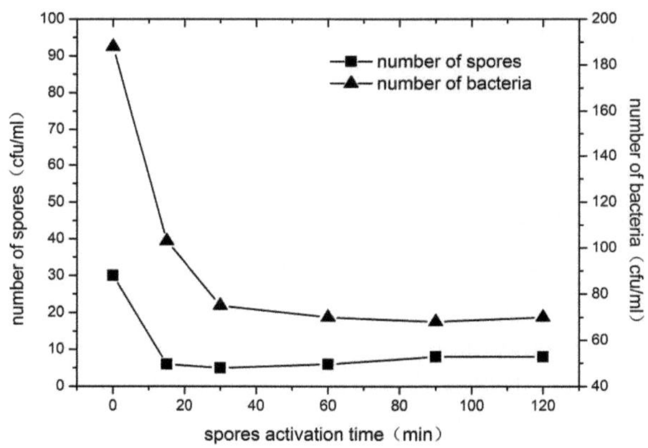

Fig. 29.3 Effect of spore activation time at 75 °C on number of bacterium and spore in cereal beverage

29.3.3 Effect of Spore Activation Conditions on pH of Cereal Beverage

During proliferation period of bacteria, acid and even trace amounts of toxins were often produced and released into the culture medium, resulting in change of the pH value of the medium; therefore the reduction of pH value of the medium was one of the indexes of the product deterioration. The results showed that spores activation at 65–85 °C, pH value of cereal beverage did not change. At 50 °C, pH value changed sharply, maybe this temperature was favorable for bacterium to grow in cereal beverage. Bacterial proliferation rate was significantly reduced at low temperature 25 °C and high temperature (Fig. 29.4). This result was consistent with bacterial count.

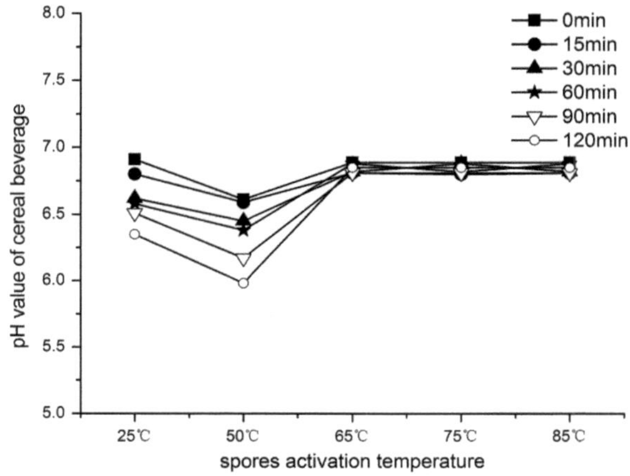

Fig. 29.4 Effect of spore activation temperature on pH value of cereal beverage

29.3.4 Effect of Activation Conditions on Freshness of Cereal Beverage

Different effects of spore activation temperature and activation time on taste freshness of cereal beverage were determined. The results showed that activation temperature and activation time to some extent affected freshness of cereal beverage. Activation at below 75 °C for 30 min, the taste freshness did not change, but at 85 °C for 30 min the freshness of the sample changed. Activation at 50 °C for 90 min, the product even got metamorphic (Table 29.1).

29.3.5 Effect of Sterilization Temperature After Spore Activation on Germicidal Efficacy and Freshness

UHT sterilization of the sample to kill all bacteria without activation needed up to 140 °C for 16 s. However, UHT sterilization temperature reached as long as 134 °C

Table 29.1 Effect of spore activation temperature and time on flavor freshness of cereal beverage

Temperature (°C)	Taste freshness					
	0 min	15 min	30 min	60 min	90 min	120 min
25	+++++	+++++	+++++	+++++	++++	++++
50	+++++	+++++	+++++	++++	Slight stale	Stale
65	+++++	+++++	+++++	+++++	++++	+++
75	+++++	+++++	+++++	++++	+++	+++
85	+++++	+++++	++++	++++	+++	++

Note "+" represents the freshness of the product

Table 29.2 Effect of sterilization temperature after spore activation on bactericidal and freshness

Group	Temperature (°C)	Taste freshness	Number of bacterium (cfu/ml)	Number of spores (cfu/ml)	Number of hot-enduring spores (cfu/ml)
Nonactivation					
	128	++++	12	8	2
	134	++++	6	5	1
	137	+++	Not	3	1
	140	++	Not	Not	Not
	143	+	Not	Not	Not
Activation					
	128	++++	Not	3	1
	134	++++	Not	Not	Not
	137	+++	Not	Not	Not
	140	++	Not	Not	Not
	143	+	Not	Not	Not

Note "+" represents the freshness of the product

for 16 s could kill all bacteria in cereal beverage after activation at 75 °C for 30 min. The results were shown in Table 29.2.

UHT sterilization temperature affected the freshness of the cereal beverage for the same time (16 s) whether or not activated. The lower the sterilization temperature was, the fresher the sample tasted. When the sterilization temperature was over 134 °C, freshness of cereal beverage became significantly worse.

29.3.6 *Optimization of Sterilization Process After Pre-activation for Large-Scale Produced Cereal Beverage*

According to the sterilization strength formula, the optimal UHT sterilization conditions of cereal beverage without spore activation was at 140 °C for 16 s, sterilization strength F_0 was 19.36; while after spores activation, the optimal UHT sterilization conditions was at 134 °C for 16 s, sterilization strength F_0 value was 5.06, which was 3.5 times less than that of the former. In the same sterilization effect, the lower the sterilization strength was, the less the nutrition and the freshness of the product lost.

UHT sterilization time was calculated according to the length of the insulating tube [9], so UHT sterilization time was relatively fixed. UHT sterilization time of current UHT device usually was 8, 16, and 32 s. Therefore, when the F_0 value was 5.06, three types of UHT sterilization parameters combination could be chosen, those were at 137 °C/8 s, or 134 °C/16 s, or 131 °C/32 s. The bactericidal experiment with the above parameters was done to determine the optimal sterilization process of

Table 29.3 Effect of UHT sterilization process on cereal beverage

Group	Number of bacterium (cfu/ml)	Number of spores (cfu/ml)	Number of hot-enduring spores (cfu/ml)	pH	Taste freshness
137 °C/8 s	No	No	No	6.88	+++
134 °C/16 s	No	No	No	6.85	+++++
131 °C/32 s	No	No	No	6.78	+++

Note "+" represents the freshness of the product

the product on basis of freshness, microorganism, and pH value. The results were shown in Table 29.3.

Under the three conditions of F_0 value as 5.06, microorganism in the products were not detected, the pH value of the product did not change, but taste fresh at 134 °C/16 s was superior to the other two groups. So the final sterilization strength of cereal beverage on a large-scale production was set as 5.06, and the optimal UHT sterilization process was at 134 °C for 16 s.

29.4 Discussion

Some bacteria turn into sporulation in cells in the later growth period, including *Bacillus, Clostridium, Sporosarcina, Sporospirillum, Sporolactobacillus, Desulfotomaculum, Coxiella, Sporomusa, and Thermoactinomyces*, etc. Bacterial spore can strongly resist to the outside environment.

Spores may be activated for bacterial vegetative state under certain conditions. Activation effect is reversible, bacterial fatrophozoite can grow in appropriate medium. Activation methods contain short-time heating, reduction pH value or reducing agent treatment. Some spores by heat treatment at 60 °C for 5 min can promote the germination of spores; some need heating in boiling water for 10 min to activate [10]. Germination by adding special chemical agents could promote spore germination of spores, such as L-alanine, Mn^{2+}, surface active agent, and glucose. Cereal beverage is a kind of food, it needs to maintain their nutrition, stability and flavor. Reduction the value of pH is obviously not suitable for spore activation of cereal beverage. Decreasing agent treatment will inevitably affect the safety of the product and taste, does not conform to the national food safety requirements. Therefore, short-time heating for spore activation is considered feasible.

The higher sterilization temperature and the longer the sterilization time is, the better the sterilization effect is. However, high intensity of sterilization has adverse consequences to the food flavor and nutrition [11]. Protein in the product will denature when met with high temperature, the flavor of the product change into smell burning [12]. Therefore, selection of sterilization time and sterilization temperature in order to achieve the best results must be taken into account the sterilization effect and the quality of the products. The UHT instantaneous sterilization technology can maintain the greatest degree of flavor and quality of food, because microorganisms are more sensitive to high temperature than the food ingredients.

This study first explores sterilization strategy after spore activation to kill bacteria in the cereal beverage, and successfully develops an optimal sterilization process for large-scale production of cereal beverage. This procedure can not only kill bacteria in the product to make the product commercial sterilized, but also reduce the intensity of sterilization in order to keep the products' taste fresh and nutritious. This chapter provides a new idea and a new paradigm for sterilization process of large-scale production of cereal beverage.

References

1. Blandino A, Al-Aseeri ME, Pandiella SS, Cantero D, Webb C (2003) Cereal-based fermented foods and beverages. Food Res Int 36:527–543
2. He GQ, Jia YM, Ding LX (2009) Food microbiology, 2nd edn. China Agricultural University Press, Beijing
3. Zhou YB, He ZX (2006) Bacterial spores: the most resistant life in nature. J Shaanxi Norm Univ (natural science edition) 34:261–263
4. Zhu HS, Yang K (2002) Quality comparison of pasteurized milk and ultra high temperature sterilized milk. Dairy Process 12:52–54
5. Li ZM (2009) Food hygiene microbiological examination. Chemical Industry Press, Beijing
6. Zhou DQ (2009) Microbiology experiment tutorial, 2nd edn. Higher Education Press, Beijing
7. Suzanne Nielsen S (2010) Food analysis, 4th edn. Springer, New York
8. Li M (2008) Juice preparation technology and examples. Chemical Industry Press, Beijing
9. Li HF, Zhang HH (2009) Production technology and equipment of the ultra high temperature sterilization (UHT) milk. Food Eng 4:18–21
10. Chen GH, Feng SL, Cao WP et al (2004) Effect of temperature pretreatment on the germination of *Bacillus thuringiensis* spores. Chin Biol Control 20(2):127–130
11. Liu ZG (2009) Food nutrition, 2nd edn. China Light Industry Press, Beijing
12. Ding NK (2008) Food flavor chemistry. China Light Industry Press, Beijing

Chapter 30
Optimization of Medium for Exopolysaccharide Production by *Agaricus brunnescens*

Li-tong Ban, Yu Wang, Liang Huang and Hongpeng Yang

Abstract The exopolysaccharide of *Agaricus brunnescens* is a newly produced biomaterial for various applications. In this paper, we optimized the culture medium for this exopolysaccharide production by *A. brunnescens* using orthogonal methods. The optimal medium for enhancing production was determined as wheat bran 120 g/L, Glucose 25 g/L and peptone 2.0 g/L for exopolysaccharide yield. This optimization strategy in shake flask culture leads to an exopolysaccharide production of 12.13 g/L, which was considerably higher than that obtained in preliminary studies.

Keywords *Agaricus brunnescens* · Exopolysaccharide · Optimization · Medium

30.1 Introduction

Agaricus brunnescens is a new mushroom, which belongs to *Agaricus bisporus*, also called brown mushroom. This mushroom is regarded in the Orient as a popular healthy food for its special flavor after cooking and as an effective medicine used to treat various diseases for its exopolysaccharide (EPS) [3, 8, 9]. *A. bisporu* has been known for many years, however, reports on this important mushroom were always in focus for their identification and preservation.

Royse and May [7] in the Pennsylvania State University used isozyme variation to identify genotypic classes of *A. brunnescens* [7]. In the same year, they also confirmed intraspecific crosses between putative homokaryotic lines of *A. brunnescens* [6] and without apparent changes in morphological or physiological characteristics if it were preserved in liquid nitrogen for 10 years [4]. Castle et al. [1] in University

L.-t. Ban · Y. Wang (✉) · L. Huang · H. Yang
Department of Biology Technology, College of Agronomy and Resources Environment, Tianjin Agriculture University, Tianjin 300384, China
e-mail: wytjac@163.com

L.-t. Ban
e-mail: banlitong@126.com

T.-C. Zhang and M. Nakajima (eds.), *Advances in Applied Biotechnology*, Lecture Notes in Electrical Engineering 332, DOI 10.1007/978-3-662-45657-6_30

of Toronto researched its restriction fragment length polymorphisms [1]. In 1983, Arthur et al. of University of Toronto estimated that the genome size of *A. brunnescens* DNA was 3.42×10^7 nucleotide pairs.

However, as we know, there are no reports on the production of exopolysaccharide from *A. brunnescens*. In this paper, we focused on investigation of the effects of the medium on exopolysaccharide formation by *A. brunnescens*.

30.2 Materials and Methods

30.2.1 Microorganism and Culture Conditions

Agaricus brunnescens (maintained at the Microbiology Lab, Tianjin Agriculture University) was used in all our experiments. The stock culture was maintained on potato dextrose agar slant (g/L): potato 200, glucose 20, agar 20, and subcultured once a month. Slants were incubated at 25 °C for 7 days and then stored at 4 °C. The medium for inoculum preparation (seed medium) contained the following components (g/L): potato 200, glucose 20, peptone 2, KH_2PO_4 2.0, $MgSO_4 \cdot 7H_2O$ 1, V_{B1} 0.01. Flask cultures were carried out in 250 mL Erlenmeyer flasks containing 100 mL medium. The fermentation medium for flask culture consisted of the following components (g/L): wheat bran 100, glucose 20, peptone 2, KH_2PO_4 1.5, $MgSO_4 \cdot 7H_2O$ 1.5 [5].

30.2.2 Inoculum Preparation, Flask Culture

The seed culture was started by transferring five pieces of approximately 0.5 cm^2 of the slant culture into a 250 mL Erlenmeyer flask containing 100 mL seed medium at 25 °C on a rotary shaker at 150 rpm for 3 days. The flask culture experiments were performed in 250 mL flasks containing 100 mL of the fermentation medium after inoculating with 10 % (v/v) the inoculum at 25 °C on a rotary shaker incubator at 150 rpm for 7 days.

30.2.3 Analytical Methods

The fermentation broth (100 ml) was centrifuged at 8,000 rpm for 10 min, and the supernatant liquor was used for detection of exopolysaccharide concentration. The crude exopolysaccharide in the filtrates was precipitated with the addition of two volumes of absolute ethanol, stirred vigorously, and left overnight at room temperature. The precipitated polysaccharide was collected by centrifugation at 8,000 rpm for 10 min and then dried at 60 °C to remove the residual ethanol [3].

30.3 Results and Discussions

30.3.1 The Orthogonal Matrix Method

To investigate the relationships between factors and optimize their concentrations for EPS production, the orthogonal matrix $L_9(3^4)$ method can be used [10]. According to preliminary experiments, we selected and varied three levels as shown in Table 30.1, with experimental results included in the last column. The fermentation conditions of temperature, agitation rate, and growth period were fixed to be 25 °C, 150 rpm, and 7 days, respectively.

30.3.2 Order of Effects of Factors

According to the orthogonal method, the effect of these factors on EPS production was calculated and the results are shown in Table 30.2. According to the magnitude order of R (Max Dif), the order of effect of all factors on EPS could be determined. The order of effects of factors on EPS was wheat bran > glucose > peptone. This result pointed out that the effect of wheat bran was more important than that of other factors.

The arrangements of column A, B, and C were decided by orthogonal design for 4 (factor) × 9(run number). $M_i^A = \sum \text{EPS at} A_i$; $m_i^A = M_i^A/3$; $R_i^A = \max\{m_i^A\} - \min\{m_i^A\}$.

30.3.3 Optimum Levels of Each Factor

The significance of each factor was judged according to *F* value. According to *F* distribution table, $F_{0.05}(2, 4) = 19.000$. It was obvious that factor A was significant while factor B and C were less significant to the results [2].

According to the *F* values, the rank of the significance was: A > B > C. Finally, as shown in Table 30.3, the optimal media for EPS production were A3B3C2,

Table 30.1 Experimental factors and their levels for orthogonal projects

Experiment nos.	Wheat bran (g/L)	Glucose (g/L)	Peptone (g/L)	Blank
	A	B	C	D
1	80	15	1.5	1
2	100	20	2.0	2
3	120	25	2.5	3

Symbols *A*, *B*, and *C* represent factors of wheat bran, glucose, and peptone. Symbols *1*, *2*, and *3* represent concentration levels of each factor

Table 30.2 Application of $L_9(3^4)$ orthogonal projects to the EPS production by *A. brunnescens*

Run	A	B	C	D	EPS (g/L)
1	1	1	1	1	6.120
2	1	2	2	2	6.032
3	1	3	3	3	6.007
4	2	1	2	3	8.349
5	2	2	3	1	5.924
6	2	3	1	2	6.692
7	3	1	3	2	9.190
8	3	2	1	3	7.886
9	3	3	2	1	11.999
M_1	18.159	23.659	20.698	24.043	
M_2	20.965	19.842	26.38	21.914	
M_3	29.075	24.698	21.121	22.242	
m_1	6.053	7.886	6.899	8.014	
m_2	6.988	6.614	8.793	7.305	
m_3	9.692	8.233	7.040	7.414	
R_f	3.639	1.619	1.894	0.710	

Table 30.3 The variance analysis of orthogonal experiment

Factor	S^a	D^b	F^c	$F_{0.05}$	S^d
A	21.423	2	24.455	19.000	*
B	4.359	2	4.976		
C	6.680	2	7.626		
e	0.88	2			

A (wheat bran), *B* (glucose), *C* (peptone)
[a] Sum of squares of deviations; [b] Degree of freedom; [c] F value; [d] Significance

which was wheat bran 120 g/L, glucose 25 g/L, and peptone 2.0 g/L; in this condition, EPS was increased by 101.67 % compared with the original media.

30.3.4 The Verification Experiments

Under optimal culture condition (wheat bran 120 g/L, glucose 25 g/L and peptone 2.0 g/L), the maximum EPS production indicated 12.13 g/L after 7 days of fermentation.

30.4 Conclusions

Agaricus brunnescens has a special flavor after cooking and may be an effective medicine for its exopolysaccharide (EPS). In order to attain higher EPS production, several media compositions were determined to influence the EPS production. Through one-factor-at-a-time and orthogonal matrix method, it was possible to determine optimal media (wheat bran 120 g/L, glucose 25 g/L, and peptone 2.0 g/L) to obtain a high EPS yield in *A. brunnescens*. This optimization strategy in shake flask culture led to EPS production of 12.13 g/L, which was considerably higher than that obtained in preliminary studies.

References

1. Castle AJ, Horgen PA et al (1987) Restriction fragment length polymorphisms in the mushrooms *Agaricus brunnescens* and *Agaricus bitorquis*. Appl Environ Microb 53:816–822
2. Chang X, He L et al (2010) Screening of optimum condition for combined modification of ultra-stable Y zeolites using multi-hydroxyl carboxylic acid and phosphate. Catal Today 158:198–204
3. Hao L, Xing X et al (2010) Optimization of effect factors for mycelial growth and exopolysaccharide production by Schizophyllum commune. Appl Biochem Biotech 160:621–631
4. Jodona MH, Roysea DJ et al (1982) Productivity of *Agaricus brunnescens* stock cultures following 5-, 7-, and 10-year storage periods in liquid nitrogen. Cryobiology 19:602–606
5. Liu C (2000) Medium test of *Agaricus bisporus* seed. Edible Fungi 22:24
6. May B, Royse DJ (1982) Confirmation of crosses between lines of *Agaricus brunnescens* by isozyme analysis. Exp Mycol 6:283–292
7. Royse DJ, May B (1982) Use of isozyme variation to identify genotypic classes of *Agaricus brunnescens*. Mycologia 74:93–102
8. Stasinopoulos JS (1990) Stimulation of exopolysaccharide production in the fungus *Acremonium persicinum* with fatty acids. Biotechnol Bioeng 36:778–782
9. Wang Y, Li Z et al (2011) The optimization of fermentation medium and conditions of *Agaricus brunnescens*. China Brew 9:138–140
10. Xu CP, Kim SW et al (2003) Optimization of submerged culture conditions for mycelial growth and exo-biopolymer production by *Paecilomyces tenuipes* C240. Process Biochem 38:1025–1030

Chapter 31
Effect of Attapulgite on Cell Activity of Steroid-Transforming *Arthrobacter simplex*

Yanbing Shen, Hengsheng Zhao, Yanhua Liu, Rui Tang and Min Wang

Abstract Attapulgite (AT) is a kind of nanostructure material. In this present study, the effect of AT on cell growth, permeability, and dehydrogenase activity of *Arthrobacter simplex* TCCC11037 (*ASP*) was investigated. The cell growth was significantly inhibited by adding AT to *ASP* medium. However, the cell growth increased when less than 0.1 % (w/w) of AT was added in the late logarithmic phase. The cell permeability increased with the increase of AT concentration, and the dehydrogenase activity of *ASP* significantly improved when the cell was treated by 0.3 % (w/w) of AT. Scanning electron microscope (SEM) analysis revealed that AT showed little effect on *ASP* cell integrity. This study provides the basic data for the practical application of AT in steroids biotransformation.

Keywords *Arthrobacter simplex* · Attapulgite · Biotransformation · Cell growth

31.1 Introduction

The low substrate aqueous solubility is regarded as the limiting step in steroid biotransformation [1]. To improve the solubility, a series of new catalytic technology has appeared, such as the organic phase bioconversion, and the nonaqueous biocatalytic medium system [2–4]. However, organic solvent inhibits the enzyme activity and stability, and limits the further increase of feed concentration. Therefore, it is particularly important to look for a reaction medium which can improves the steroids solubility and has better biocompatibility [5, 6].

Y. Shen · H. Zhao · Y. Liu · R. Tang · M. Wang (✉)
Key Laboratory of Industrial Fermentation Microbiology, Ministry of Education,
The College of Biotechnology, Tianjin University of Science and Technology,
Tianjin 300457, People's Republic of china
e-mail: minw@tust.edu.cn

T.-C. Zhang and M. Nakajima (eds.), *Advances in Applied Biotechnology*,
Lecture Notes in Electrical Engineering 332, DOI 10.1007/978-3-662-45657-6_31

Attapulgite (AT) is a kind of nanostructure material which has layered structure that is rich in hydrous magnesium silicate clay [7]. AT has large specific surface area and unique physical and chemical properties such as adsorption ability, catalytic and rheological property. It can solve the problems which restrict the catalytic efficiency of substrate dispersion and solubility in the hydrophobic compounds' biocatalytic reaction. In recent years, nanoscale AT has been more and more applied in cell culture and the probiotics immobilization, and has showed good biocompatibility [8].

Currently, there are few reports on the application of AT in biocatalytic reaction or the influence on industrial catalytic microbial growth characteristics. In this research, steroid-transforming *Arthrobacter simplex* was chosen to investigate the biocompatibility of AT by analyzing its influences on cell growth, metabolic activity, cell membrane permeability, and cell integrity of *Arthrobacter simplex.*

31.2 Materials and Methods

31.2.1 Materials

Substrate cortisone acetate (CA, 99.4 % purity) was obtained from Tianjin Pharmaceutical Company. Standard CA and prednisone acetate (PA) were purchased from Sigma-Aldrich Co. Ltd. AT (over 98 % purity) was supplied by Jiuchuan Nanomaterial Technology Co. (Jiangsu, China). 3-Aminopropyltriethoxysilane (KH550) was provided by the Institute of Petrochemistry, Heilongjiang Academy of Science (Heilongjiang, China). All chemical solvents and salts used were of analytical grade or higher.

31.2.2 Effect of AT on Cell Growth

Arthrobacter simplex TCCC11037 (*ASP*) used in this study was obtained from Tianjin University of Science and Technology Culture Collection Center (TCCC), Tianjin, China. The strain was cultivated in shake flasks in two consecutive cultivation steps as described by Zhang et al. [6]. Whole cells were grown in 250 ml shake flasks containing 30 ml culture medium with AT of 0, 0.03, 0.05, 0.1, 0.3, 0.5, and 0.7 % (w/v) on a rotary shaker (160 r/min) at 32 °C. During incubation, optical density at 600 nm was used to measure cell growth.

31.2.3 Cell Membrane Permeability Analysis

Cell membrane permeability was examined by measuring leakage of nucleic acids from the cells. *ASP* cells (2 g dry weight per liter) were suspended in 30 ml sodium phosphate buffer (50 mmol/L, pH 7.4) with AT of 0, 0.1, 0.3, 0.5, and 0.7 % (w/v) at 32 °C and 160 r/min for 5 h. Samples of 3 ml were withdrawn at various time intervals, and filtered through a 0.22 μm filter, and the nucleic acid content of the supernatant was estimated by measuring absorbance at 260 nm.

31.2.4 Determination of Cell Activity of Dehydrogenase

Cell activity of dehydrogenase was examined by TTC. The treated cells as described above were washed thrice with 50 mmol/L sodium phosphate buffer (pH 7.4). The washed cells were then resuspended in the same buffer. 2 mL of 2,3,5-triphenyltetrazolium chloride (TTC) solution (0.5 %) was added to the tube with 1 mL cell suspension, after incubating at 37 °C for 30 min. The cells were centrifuged at 6,000 × *g* for 10 min at 4 °C, and triphenyl formazan (TF) extraction was carried out using 80 % acetone. The absorbance of the supernatants was measured at 485 nm. The dehydrogenase activity was expressed by the ratio of A_{485} of treated sample with the control.

31.2.5 Preparation of SEM

The fresh cells were washed thrice with 50 mmol/L sodium phosphate buffer (pH 7.2) and fixed in 2.5 % glutaraldehyde in 50 mmol/L cacodylate buffer (pH 7.2). A pretreated sample was placed on a cover glass, air dried and examined using a SU-1510 scanning electronic microscope instrument (Hitachi, Japan).

31.3 Results and Discussion

31.3.1 Effect of AT on the Growth of Arthrobacter simplex

Effects of AT on cell growth of *Arthrobacter simplex* in various concentrations of AT were investigated as described above. The results revealed that AT obviously influenced the *ASP* growth. As shown in Fig. 31.1, the biomass of *ASP* cultivated with AT was lower than that of the control group, and the biomass decreased with the increase of AT concentration in the growth medium. Low AT concentrations slightly inhibited *ASP* growth, as evidenced by the 6.5 % growth inhibition of *ASP* cells treated with 0.03 % AT when compared with the control. However, the growth of *ASP* was significantly inhibited when AT concentration was higher than 0.05 %. The biomass of *ASP* cell was inhibited about 89.5 % in the presence of 0.3 % AT.

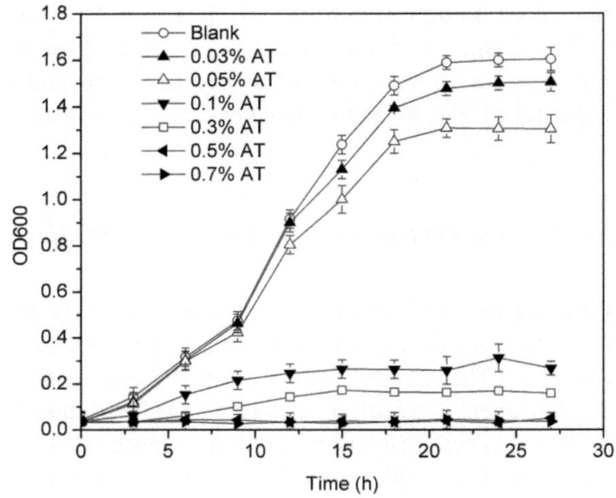

Fig. 31.1 Growth curves of *ASP* in culture medium supplemented with AT

At concentrations of 0.5 % and more, no growth was observed, even after 27 h incubation.

The direct addition of AT to the fermentation medium had obviously inhibitory effect on the *ASP*. However, when AT was added in exponential growth phase (12 h), it affected the cell growth differently. As shown in Fig. 31.2, at concentrations lower than 0.05 %, AT moderately promoted the growth of *ASP*. The biomass increased by approximately 12.5 % at AT concentration of 0.03 % than that of control. With the further increase of AT concentration, the growth of *ASP* was inhibited.

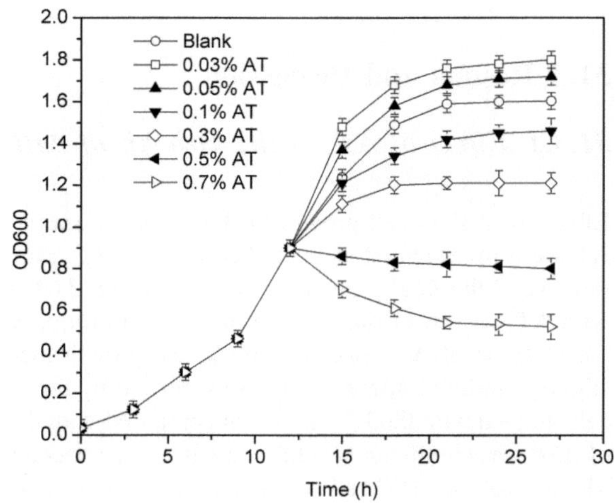

Fig. 31.2 Effect of AT on growth of *ASP* (AT was added in exponential growth phase)

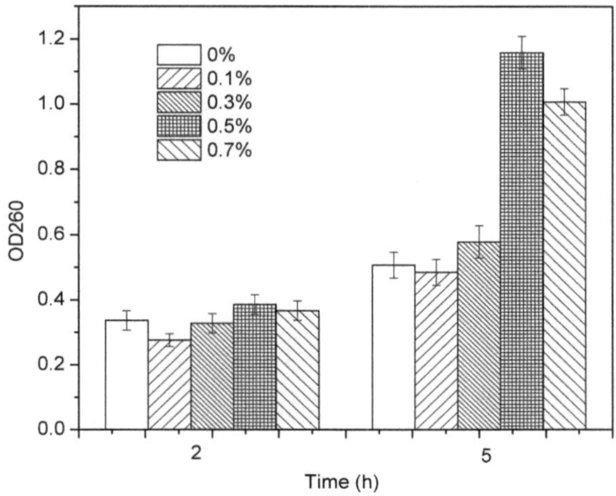

Fig. 31.3 Leakage of Nucleic acid from *ASP* cells in the presence of AT

31.3.2 Effect of AT on Cell Permeability

Cell permeability was examined by measuring leakage of nucleic acids. *ASP* cell incubated in a phosphate buffer with or without AT showed a gradual leakage of nucleic acid-like material into the medium, as estimated by absorbance measurements at 260 nm (Fig. 31.3). As shown in Fig. 31.3, the concentration and treated time significantly alter the cell permeability. When the treated time was 2 h, there were no obvious differences. However, nucleic acid leakage of that treated by 0.5 % AT was 2.1 times than that of the blank group when the treated time was 5 h.

31.3.3 Effect of AT on Cell Activity of Dehydrogenase

The effect of AT on dehydrogenase activity was presented in Fig. 31.4. The dehydrogenase activity increased with the increase of AT. The maximal increase of the dehydrogenase activity was about 26.4 % at the concentration of AT was 0.3 %, indicating that the introduction of AT had a certain influence on the enzyme activity of *ASP*.

Generally, the enzyme activity depends on temperature, pH, activators and inhibitors, and so on. Because AT contains K^+, Ca^{2+}, Mg^{2+}, and other metal ions, which might affect the activity of enzymes, the change of dehydrogenase activity of *ASP* was probably related with the concentration of AT.

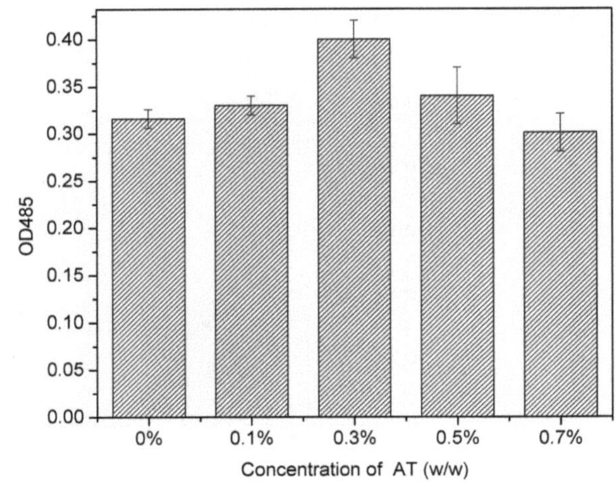

Fig. 31.4 Effect of AT on the dehydrogenation activity of *ASP*

31.3.4 Effect of AT on Arthrobacter simplex Integrity

The effects of AT on *ASP* cells surface were analyzed by SEM. Typical SEM images of *ASP* cells in the absence and presence of AT were shown in Fig. 31.5. It was revealed that the control cell was rod-like and had a nearly smooth cell body. The integrity of *ASP* cell membrane remained intact in AT-containing medium, and no destructive alterations in cell size were observed when the concentration of AT was 0.3 % (w/w), which indicated that AT was a kind of mild material for *ASP*.

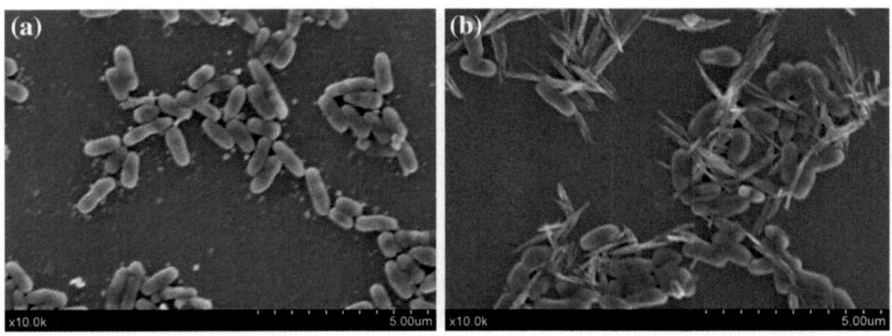

Fig. 31.5 Observation of cell morphology change by scanning electron microscope: **a** control; **b** cells treated by 0.3 % (w/w) AT

31.4 Conclusion

The low concentration of AT had little influence on the cell growth, dehydrogenase activity, and cell integrity of *Arthrobacter simplex*. The results showed that low concentration of AT had a better biocompatibility. In addition, the timing of addition of AT and its final concentrations played important roles on cell growth and cell permeability of *ASP*, which needed further investigation so as to use further in biotransformation.

Acknowledgments This work was supported by the National Natural Science Foundation of China (Nos. 21076158 and 21276196), Natural Science Foundation of Tianjin (13JCQNJC04900), and Research Fund of Tianjin University of Science and Technology (20120115).

References

1. Donova MV, Egorova OV (2012) Microbial steroid transformations: current state and prospects. Appl Microbiol Biotechnol 94:1423–1447
2. Cruz A, Angelova B, Fernandes P et al (2004) Study of key operational parameters for the side-chain cleavage of sitosterol by free mycobacterial cells in Bis-(2-ethylhexyl) phthalate. Biocatal Biotransform 22:189–194
3. Bie ST, Lu FP, Du LX et al (2008) Effect of phase composition on the bioconversion of methyltestosterone in a biphasic system. J Mol Catal B-Enzym 55:1–5
4. Heipieper HJ, Neumann G, Cornelissen S (2007) Solvent-tolerant bacteria for biotransformations in two-phase fermentation systems. Appl Microbiol Biotechnol 74:961–973
5. Fernandes P, Cruz A, Angelova B et al (2003) Microbial conversion of steroid compounds: recent developments. Enzyme Microb Technol 32:688–705
6. Zhang LT, Wang M, Shen YB et al (2009) Improvement of steroid biotransformation with hydroxypropyl-β-cyclodextrin induced complexation. Appl Biochem Biotechnol 3(159):642–654
7. Yang HM, Tang AD, Ouyang J et al (2010) From natural attapulgite to mesoporous materials: methodology, characterization and structural evolution. J Phys Chem 114:2390–2398
8. He L, Ding KJ, Wu H et al (2013) Evaluation of cytotoxicity of natural nano-attapulgite and its enhancement of vero cell productivity. Dig J Nanomater Bios 8(2):551–560

Chapter 32
Establishment of a Method to Measure the Interaction Between Nedd4 and UbCH5c for Drug Screening

Kunyuan Kou, Jianli Dang, Baoxia Zhang, Guanrong Wu, Yuyin Li and Aipo Diao

Abstract Protein ubiquitination is closely associated with tumorigenesis. E3 ubiquitin ligase Nedd4 is overexpressed in prostate, bladder, and colorectal cancer. Nedd4 interacts with UbCH5c ubiquitin transferase through its HECT domain. Two GST-tagged Nedd4 proteins and His-HA-UbCH5c protein are expressed and purified. Using these recombinant proteins, a method is established to measure the interaction between Nedd4 and UbCH5c. This established method can be used to screen the small molecule inhibitors to block the interaction between Nedd4 and UbCH5c, which inhibit the activity of E3 ubiquitin ligase Nedd4.

Keywords Nedd4 · UbCH5c · Protein purification · Protein interaction

32.1 Introduction

Ubiquitination is a common protein modification, involving in the transfer of the ubiquitin of an evolutionarily conserved 76 amino acid protein to target proteins. K48-linked polyubiquitin chains are well known to function in protein degradation by the 26S proteasome. K63-linked polyubiquitin chains have been proposed to function in DNA repair, IκB kinase activation, translational regulation, and endocytosis [1, 2]. Regulation of protein stability through ubiquitin-dependent proteolysis plays important roles in controlling the function of many proteins, including critical regulators of cell proliferation and apoptosis [3, 4]. This proteolytic

K. Kou · J. Dang · B. Zhang · G. Wu · Y. Li (✉) · A. Diao (✉)
Key Lab of Industrial Fermentation Microbiology of the Ministry of Education,
School of Biotechnology, Tianjin University of Science and Technology,
Tianjin 300457, China
e-mail: liyuyin@tust.edu.cn

A. Diao
e-mail: diaoaipo@tust.edu.cn

T.-C. Zhang and M. Nakajima (eds.), *Advances in Applied Biotechnology*,
Lecture Notes in Electrical Engineering 332, DOI 10.1007/978-3-662-45657-6_32

system depends on conjugation of ubiquitin onto primary amino groups on substrate proteins, with appropriately linked polyubiquitin chains serving as targeting signals for proteasomal degradation. Protein ubiquitination is generally catalyzed by the sequential activity of three enzymes: ubiquitin activation enzyme (E1), ubiquitin conjugation enzymes (E2s), and ubiquitin ligases (E3s). Ubiquitylation occurs through a multiple-enzyme cascade, where recognition is specified by the ubiquitin ligase (E3) [5, 6]. E3 ubiquitin ligases are important in cellular regulation because E3s specifically recognize a substrate for modification temporally and spatially. In mammalian cells, there are more than 500 E3s. They contain either a RING (the really interesting new gene) finger domain E3s or a HECT (homologous to E6-AP COOH terminus) domain E3s [2].

The Nedd4 family belongs to the HECT type ubiquitin ligases, which regulate ubiquitin-mediated trafficking, lysosomal or proteasomal degradation, and nuclear translocation of multiple proteins, they modulate important signaling pathways involved in tumorigenesis, such as TGF-β, EGF, IGF, VEGF, SDF-1, and TNF-α signaling pathways. Additionally, the Nedd4 family directly regulates various cancer-related transcription factors such as Smad, p53, KLF, RUNX, and Jun families. Furthermore, Nedd4 family themselves are frequently regulated by phosphorylation, ubiquitination, translocation, and transcription in cancer cells [2]. Nedd4 is one number of the Nedd4 family, and is overexpressed in prostate, bladder, colorectal cancer, and promotes colonic cell growth [7, 8]. Nedd4 contains several functional motifs. In the N terminus, there are a proline-rich domain and a polyserine motif followed by a C2 domain, containing four WW domains in the middle and the HECT domain in the C terminus [9]. The HECT domains consists of an approximately 250 amino-acid-elongated amino-terminal lobe (N-lobe) connected by a short flexible linker to an approximately 100 amino acid globular C-terminal lobe [10]. The E2-binding site is located on the N-lobe and the active site cysteine is within the C-lobe. The structure of an E2 ubiquitin thioester complex with the HECT domain of Nedd4 showed the C-lobe of the HECT domain oriented toward the E2 ubiquitin complex and interacting with ubiquitin [11], providing insight into how ubiquitin is transferred from the E2 to the E3 [12].

In this report, we expressed and purified two GST-tagged Nedd4 proteins, GST-Nedd4(aa196-aa900) and GST-Nedd4(aa476-aa900), and the His-HA tagged UbCH5c E2 protein, these purified proteins were used to establish a method to measure the interaction between Nedd4 and UbCH5c for drug screening.

32.2 Materials and Methods

32.2.1 Construction of Plasmids

Two DNA fragments encoding the WW domains and HECT domain of Nedd4 were amplified by PCR using the primers listed in Table 32.1, and the two fragments were digested with BamHI and EcoRI and inserted into the pGEX-4T-2 (GE

Table 32.1 Primers used for PCR experiments

Target gene	Primer sequence
Nedd4(aa196-900)	
Forward	5′-CGGGATTCCAACAACAAGAACCTTCTCC-3′
Reverse	5′-CGGAATTCCTAATCAACTCCATCAAAGC-3′
Nedd4(aa476-900)	
Forward	5′-CGGGATTCACTTCCAATGATCTAGGGCC-3′
Reverse	5′-CGGAATTCCTAATCAACTCCATCAAAGC-3′
UbCH5c	
Forward	5′-CGGGATCCATGGCGCTGAAACGGATTAATAAG-3′
Reverse	5′-CGGAATTCTCACATGGCATACTTCTGAGTCC-3′

Healthcare) vector. The *UbCH5c* gene was amplified by PCR. The PCR product was purified and digested with BamHI and EcoRI, and inserted into the BamHI and EcoRI sites of the pET21b vector.

32.2.2 Protein Expression and Purification

Plasmids encoding GST-fusion proteins of Nedd4 were transformed into *Escherichia coli* BL21-CodonPlus (DE3)-RIL (Stratagene). Cells were induced with 0.1 mM IPTG overnight at 16 °C, and lysed in lysis buffer (20 mM Tris-Cl pH 7.4, 100 mM NaCl, 2 mM DTT) containing protease inhibitors. The recombinant protein was purified on Glutathione–Sepharose beads (GE Healthcare). After sonication, the extracts were clarified by centrifugation. The recovered supernatants were allowed to bind to Glutathione–Sepharose beads for 3 h at 4 °C. The matrix was washed three times with lysis buffer, then the beads were eluted with elution buffer (100 mM Tris-Cl pH 7.4, 100 mM NaCl, 20 mM reduced glutathione). His-HA-UbCH5c was expressed in *E. coli* BL21-CodonPlus (DE3)-RIL (Stratagene) cells that transformed with pET21b-His-HA-UbCH5c plasmid, and induced with 0.5 mM IPTG overnight at 16 °C. The His-HA-tagged protein was purified on Ni–Sepharose beads (GE Healthcare) after lysis of the bacteria in Na–P lysis buffer (50 mM Na_3PO_4 pH 8.0, 300 mM NaCl, 5 mM β-mercaptoethanol) plus protease inhibitors. After sonication, the extracts were clarified by centrifugation. The recovered supernatants were allowed to bind to Ni–Sepharose beads through the AKTA purifier, and gradient elution of 50 mM imidazole, 100 mM imidazole, 150 mM imidazole with Na–P lysis buffer. The beads were eluted with elution buffer (50 mM Na_3PO_4 pH 8.0, 300 mM NaCl, 5 mM β-mercaptoethanol, 200 mM imidazole)

32.2.3 Measurement of the Interaction Between Nedd4 and UbCH5c

The GST-fusion proteins are coated to the bottom of the plate with the buffer (15 mM Na_2CO_3, 35 mM $NaHCO_3$), for 2 h at 37 °C then the plates are washed 5 times with the washing buffer (1.48 mM KH_2PO_4, 8.10 mM $Na_2HPO_4 \cdot 12H_2O$, 137 mM NaCl, 2.68 mM KCl, 0.05 % Tween). The plates are blocked in 5 % BSA for 2 h at 3 °C, and then washed 5 times. His-HA-UbCH5c proteins are added into the plate, and incubated for 2 h at 4 °C, and then washed 5 times. The HA-tag antibody (Mouse) is added and incubated overnight at 4 °C, and then washed 5 times. At the end, the HRP-conjugated goat anti-mouse IgG is added and incubated for 30 min at 37 °C, after washed five times, the TMB substrate is added and observe the color of the solution, when color becomes blue, 2M H_2SO_4 is added to stop the reaction, when the color becomes yellow, measure the absorbance at 450 nm by Microplate Reader.

32.3 Results and Discussion

32.3.1 GST-Nedd4(aa196-aa900) and GST-Nedd4 (aa476-aa900) Protein Expression and Purification

The plasmid pGEX-Nedd4(196-900) was constructed and verified by DNA sequencing, the GST-Nedd4(196-900) protein was expressed in *E. coli* BL21 and purified with glutathione–agarose beads, the molecular weight of the expressed protein is 103 kDa (Fig. 32.1).

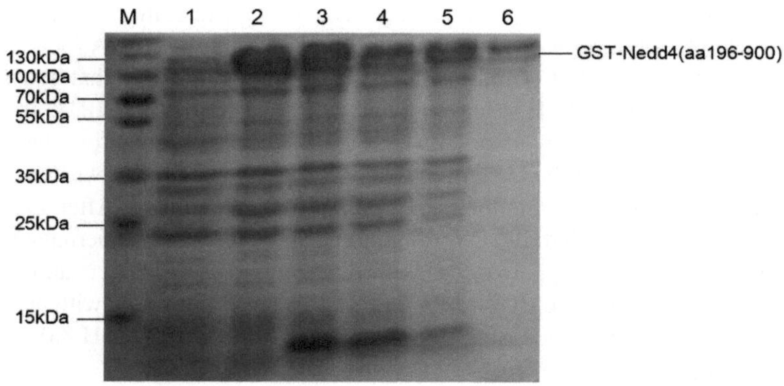

Fig. 32.1 The expression and purification of GST-Nedd4(196-900). *Line 1*, no induction; *line 3–5*, induced with 0.1 mM IPTG for overnight at 16 °C, the total extracts, soluble fraction, and insoluble fraction, respectively; *line 6*, the protein purified protein using the Glutathione–Sepharose beads

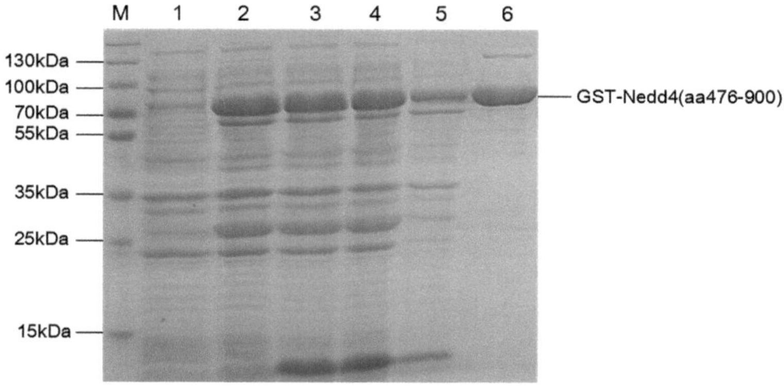

Fig. 32.2 The expression and purification of GST-Nedd4(476-900). *Line 1*, no induction; *line 3–5*, induced with 0.1 mM IPTG for overnight at 16 °C, the total extracts, soluble fraction, and insoluble fraction, respectively; *line 6*, the protein purified protein using the Glutathione–Sepharose beads

The plasmid pGEX-Nedd4(476-900) was constructed and verified by DNA sequencing, the GST-Nedd4(476-900) protein was expressed in *E. coli* BL21 and purified with glutathione–agarose beads, the molecular weight of the expressed protein is 73 kDa (Fig. 32.2).

32.3.2 His-HA-UbCH5c Expression and Purification

The plasmid pET21b-His-HA-UbCH5c was constructed and verified by DNA sequencing, the His-HA-UbCH5c protein was expressed in *E. coli* BL21 and purified with Ni–Sepharose beads, the molecular weight of the expressed protein is 16 kDa (Fig. 32.3).

32.3.3 Establish the Method to Detect the Interaction Between Nedd4 and UbCH5c

The HECT domain of Nedd4 interacts with E2 ubiquitin-conjugating enzymes UbCH5c, in order to establish a method to measure the interaction between Nedd4 and UbCH5c easily in vitro, a method that is similar to the Enzyme-Linked Immuno-sorbent Assay (ELISA) was established. The GST-Nedd4(aa196-900), GST-Nedd4 (aa476-900), or the negative control GST proteins were coated to the plates, then incubated with His-HA-UbCH5c, the interaction between these proteins was further measured using the primary antibody against HA-tag and HRP-conjugated goat anti-mouse IgG. As shown in Fig. 32.4, compared to the control, both of GST-Nedd4

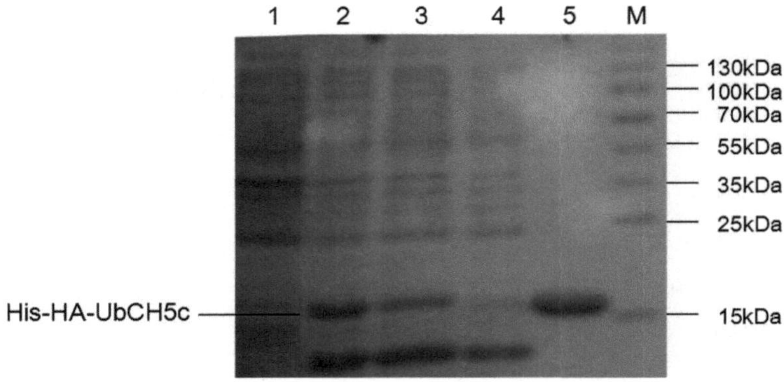

Fig. 32.3 The expression and purification of His-HA-UbCH5c. *Line 1*, no induction; *line 3–5*, induced with 0.1 mM IPTG for overnight at 16 °C, the total extracts, soluble fraction, and insoluble fraction, respectively; *line 6*, the protein purified protein using the Ni–Sepharose beads

Fig. 32.4 Establish the method to measure the interaction between Nedd4 and UbCH5c. GST/UbCH5c serves as the negative control. GST/UbCH5c

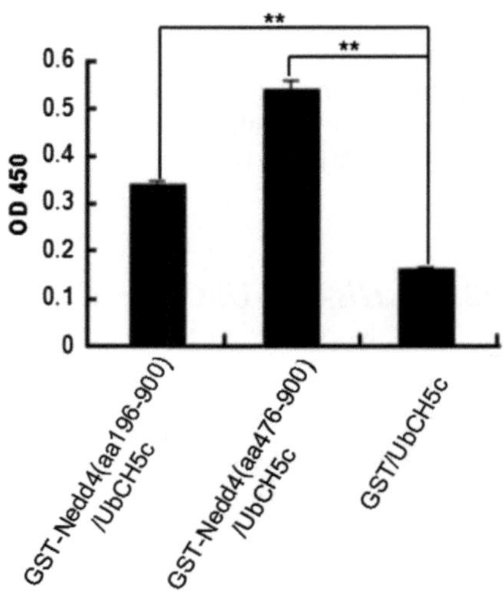

(aa196-900) and GST-Nedd4(aa476-900) interact with His-HA-UbCH5c, and the interaction is stronger for GST-Nedd4(aa476-900).

Previous studies indicate that Nedd4 is an E3 ubiquitin ligase for E2 ubiquitin-conjugating enzymes UbCH5c [13]. The Nedd4 contains a C2 domain, four WW domains, and a HECT domain. And the HECT domain plays an important role in binding with E2 ubiquitin-conjugating enzymes UbCH5c [14]. In this study, we have purified three recombinant proteins and established a method to measure the interaction between Nedd4 and UbCH5c. This established method can be used to screen the small molecule inhibitors to block the interaction between Nedd4 and UbCH5c, which inhibit the activity of E3 ubiquitin ligase Nedd4.

In conclusion, we have expressed and purified two GST-tagged Nedd4 proteins and His-HA tagged UbCH5c proteins, and established a method to detect the interaction between them. This method is more convenient than traditional GST-pulldown assay to measure the protein interaction. So this method can be widely used for high-throughput drug screening.

Acknowledgments This research is supported by the program for Changjiang Scholars and Innovative Research Team in University (IRT1166) and scientific research fund project of Tianjin University of Science and Technology (20130106).

References

1. Pickart CM, Fushman D (2004) Polyubiquitin chains: polymeric protein signals. Curr Opin Chem Biol 8:610–616
2. Ceshi C, Lydia EM (2007) The Nedd4-like family of E3 ubiquitin ligases and cancer. Cancer Metast Rev 26:587–604
3. Clarke DJ (2002) Proteolysis and the cell cycle. Cell Cycle 1:233–234
4. Yang Y, Yu X (2003) Regulation of apoptosis: the ubiquitous way. FASEB J 17:790–799
5. Fang S, Weissman AM (2004) A field guide to ubiquitylation. Cell Mol Life Sci 61:1546–1561
6. Yang Y, Robert LL et al (2005) Small molecule inhibitors of HDM2 ubiquitin ligase activity stabilize and activate p53 in cells. Cancer Cell 7:547–559
7. Chen C, Matesic LE (2007) The Nedd4-like family of E3 ubiquitin ligases and cancer. Cancer Metast Rev 26:587–604
8. Eide PW, Cekaite L, Danielsen SA et al (2013) NEDD4 is overexpressed in colorectal cancer and promotes colonic cell growth independently of the PI3K/PTEN/AKT pathway. Cell Signal 25:12–18
9. Wang X, Lloyd CT, Koppie T et al (2007) NEDD4-1 is a proto-oncogenic ubiquitin ligase for PTEN. Cell 128:129–139
10. Huang L, Kinnucan E, Wang G et al (1999) Structure of an E6AP-UbcH7 complex: insights into ubiquitination by the E2-E3 enzyme cascade. Science 286:1321–1326
11. Kamadurai HB, Souphron J, Scott DC et al (2009) Insights into ubiquitin transfer cascades from a structure of a UbcH5B approximately ubiquitin-HECT(NEDD4L) complex. Mol Cell 36:1095–1102
12. Kim HC, Alanna MS et al (2011) Structure and function of a HECT domain ubiquitin-binding site. EMBO Rep 12:334–341
13. David HS, Oberoi P et al (2004) Natural product inhibitors of specific E3 ubiquitin ligases. Meso Scale Discov 1:1–13
14. Rotin D, Kumar S (2009) Physiological functions of the HECT family of ubiquitin ligases. Nat Rev 10:398–409

Chapter 33
Determination of Phthalate Esters in Tea by Gas Chromatography–Mass Spectrometry

Yan Lu, Liping Du, Yang Qiao, Tianlu Wang and Dongguang Xiao

Abstract A method was developed for the determination of five phthalate esters (PAEs) in tea. The phthalate esters were extracted from tea samples with an optimized extraction method and quantification was achieved by gas chromatography–mass spectrometry (GC–MS). The extraction parameters (i.e., extraction type, extraction solvent, extraction time, and amount of tea) and the conditions of detection were investigated and selected. Under the optimized conditions, the linearity of five PAEs were good in their detection range, the correlation coefficient of them were above 0.99. The limits of detection (LODs) were 0.24–3.72 μg/L. The spiked recovery varies from 87.63 to 98.08 %, and the relative standard deviations (RSD) were less than 8 % ($n = 3$). The proposed method was successfully applied to determine the five phthalate esters in commercial available tea samples, which suggested that the method was simple, less interference, and good repeatability.

Keywords Simultaneous distillation extraction (SDE) · Gas chromatography–Mass spectrometry (GC–MS) · Tea · Phthalate esters (PAEs)

33.1 Introduction

Phthalate Esters (PAEs) are a group of organic compounds that are used every day as plasticizers in different materials due to its high chemical and mechanical endurance [1, 2]. In the recent years, the production and use of phthalate esters have increased significantly. The widespread presence of PAEs has attracted considerable attention since their potential risks for human health and environment [3]. The US Environmental

Y. Lu · L. Du (✉) · Y. Qiao · T. Wang · D. Xiao
Key Laboratory of Industrial Fermentation Microbiology Ministry of Education, Tianjin Industrial Microbiology Key Laboratory, College of Biotechnology, Tianjin University of Science and Technology, Tianjin 300457, People's Republic of China
e-mail: dlp123@tust.edu.cn

T.-C. Zhang and M. Nakajima (eds.), *Advances in Applied Biotechnology*,
Lecture Notes in Electrical Engineering 332, DOI 10.1007/978-3-662-45657-6_33

Protection Agency (EPA) and several other countries had made relevant use limit and management of specific PAEs [4, 5].

However, PAEs have been isolated from some organisms including plants [6], marine algae [7], fungal, and bacterial culture broths [8], and some of them belong to the natural products, produced by their biological metabolism, not contaminated during the separation process. Tea is geared to dicotyledon, a most widely consumed beverage throughout the world [9]. The quality and safety have attracted an ever-increasing attention. In the related analysis of chemical composition [10–13], we found that PAEs exist in the different teas, mainly including dimethyl phthalate (DMP), diethyl phthalate (DEP), diisobutyl phthalate (DIBP), dibutyl phthalate (DBP), and diethylexyl phthalate (DEHP). Whereas it is difficult to determine whether PAEs were produced by these organisms, contaminated during the separation process, or accumulated in the organisms. So the origin of PAEs of tea needed further research.

In order to systematically study the PAEs in tea, we need to analyze and determine the content of related PAEs first. The determination of PAEs includes an extraction step, which is followed by the instrumental analysis of them using gas chromatography (GC), high performance liquid chromatography (HPLC), gas chromatography–mass spectrometry (GC–MS), or liquid chromatography–mass spectrometry (LC–MS) [14, 15]. The conventional extraction methods are traditional liquid–liquid extraction, dispersive liquid–liquid microextraction [16], solid–phase extraction [17] and headspace solid-phase microextraction [1]. Moreover, simultaneous distillation extraction (SDE) is a technique which combined the advantages of traditional liquid–liquid extraction with steam distillation extraction, and has a better extraction ability for the high boiling point compositions, the trace constituents can be separated from the matrix.

In this chapter, we aimed to develop a method for the direct determination of five phthalate esters compounds (DMP, DEP, DIBP, DBP, and DEHP) in tea, based on an optimal extraction type, combined with GC–MS technology.

33.2 Materials and Methods

33.2.1 Sample and Reagents

Green tea, black tea, and dark tea samples were purchased in local supermarket.

Acetone, *n*-Pentane, *n*-hexane, dichloromethane, and ether were of HPLC grade, purchased from Sinopharm Chemical Reagent Co. Ltd (Tianjin, China). Dimethyl phthalate (DMP, 99.7 %), diethyl phthalate (DEP, 99.5 %), and dibutyl phthalate (DBP, 99.5 %) were purchased from Aladdin (shanghai, China), diisobutyl phthalate (DIBP, 98 %), diethylexyl phthalate (DEHP, 98 %) were purchased from TCI (shanghai, China). Anhydrous sodium sulfate was of analytical grade, purchased from BaiShi chemical Industry Co. Ltd (Tianjin, China). Pure water was obtained from a Milli-Q purification system (Millipore, Bedford, USA). To avoid PAEs contamination, all reagents should be used to detect the presence of PAEs before use.

33.2.2 Instruments and Analytical Conditions

The analysis was performed on Agilent 7890A Gas chromatograph (Agilent, USA) coupled to an Agilent 5975C MSD (Agilent, USA), equipped with HP-5MS capillary column (60 m × 0.32 mm i.d., 0.25 μm film thickness). The operating conditions of GC were: Helium (percentage purity >99.999 %) was used as carrier gas, and the flow rate was 1 mL/min, 1 μL concentration was injected in splitless mode, injector temperature 300 °C. Oven temperature was programmed from 60 °C, increased to 120 °C (held for 2 min) at 5 °C/min, then heated to 180 °C (held for 5 min) at 2 °C/min, and finally raised to 250 °C (held for 12 min) at 6 °C/min. The mass spectrum parameters were: EI 70 eV; interface temperature 280 °C; ion source temperature 230 °C; and quadrupole temperature 150 °C. Solvent delay 25 min.

A Likens–Nickerson apparatus (Tianchang huabo experiment instrument, Anhui, China) was used for simultaneous distillation extraction of tea samples and a termovap sample concentrator (Orgnomation, USA) was used to concentrate the extracts.

Because the widespread existence of PAEs, especially DBP and DEHP. The experiment cannot contact with any plastic materials, all glassware used in the study were soaked in acetone for at least 30 min, then washed with acetone, rinsed with hexane, and dried at 145 °C for at least 3 h. Moreover, the contamination level determined from the extraction procedure was also checked daily.

33.2.3 Extraction Procedure

33.2.3.1 Simultaneous Distillation Extraction

10.0 g of grounded tea was placed in a flask, and 150 mL of ultrapure water was added in the flask. 30 mL of *n*-hexane was applied as organic solvent in another flask. Both flasks were installed in a Likens–Nickerson apparatus and heated up to their boiling points. Then the extraction was refluxed for 3 h to collect the analytes. After the extraction, the organic solvent was collected and dried with an hydrous sodium sulfate over night. The extracts were then concentrated approximately to 0.4 mL with a stream of high purity nitrogen at 50 °C and volumed to 1 mL with *n*-hexane. The concentrated extract was stored at −20 °C before analysis. Each SDE process was performed in triplicate.

33.2.3.2 Liquid–Liquid Extraction

10.0 g of grounded tea was placed in a separatory funnel, and 150 mL of ultrapure water and 30 mL of *n*-hexane were added. The separatory funnel was closed and shaken manually for 3 min to homogenize, then static layering. The organic phase was removed and dried over an hydrous sodium sulfate over night. Then the extract was concentrated approximately to 0.4 mL with high purity nitrogen at 50 °C and volumed to 1 mL with *n*-hexane. The concentrated extract was stored at −20 °C before analysis. Each LLE process was performed in triplicate.

33.2.4 Optimization of the Extraction Conditions

In order to improve the extraction efficiency of five phthalate esters in tea samples, four experimental parameters including types of extraction (SDE and LLE), organic solvents (*n*-Pentane, *n*-hexane, ether, dichloromethane), amount of tea (5.0, 10.0, 15.0 and 20.0 g) and extraction time (2, 3, 4 and 5 h) were investigated.

33.3 Result and Discussion

33.3.1 Optimization of Extraction Conditions

33.3.1.1 Selection of Type of Extraction

In our experiment, the liquid–liquid extraction (LLE) and simultaneous distillation extraction (SDE) methods were applied for the determination of five PAEs in tea, the results of their extraction efficiency was compared and shown in Fig. 33.1. Compared

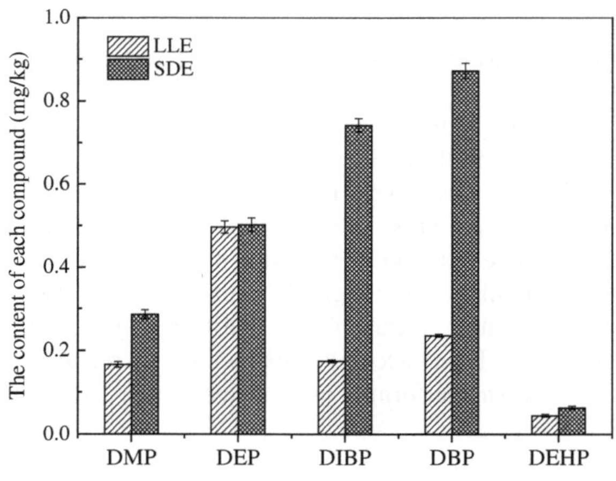

Fig. 33.1 Effect of different extraction methods on the extraction efficiency of 5 PAEs in tea samples

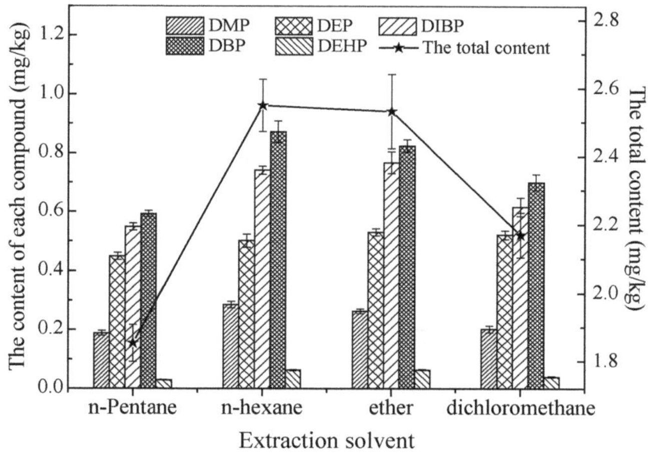

Fig. 33.2 Effect of extraction solvents on the extraction efficiency of 5 PAEs in tea samples

to traditional extraction method (LLE), SDE method showed higher extraction efficiency for five PAEs, especially for the extraction of DBP and DIBP. Thus, SDE method was used as the pretreatment method to detect five PAEs in tea samples.

33.3.1.2 Extraction Solvent

To select the extraction solvent of SDE, the impact of different solvents (*n*-pentane, *n*-hexane, ether and dichloromethane) on the extraction of five PAEs in tea was investigated. The results were shown in Fig. 33.2. The four kinds of the organic solvent had different extraction efficiency for each component. *n*-hexane had a higher extraction amount for DMP and DBP, while ether showed a higher extraction efficiency for the DEP and DIBP. Additionally, while there is a little difference between the total extraction amount of *n*-hexane and diethyl ether for 5 kinds of PAEs, because of the toxicity of *n*-hexane is lower than diethyl ether, *n*-hexane was selected for subsequent analysis.

33.3.1.3 Amount of Tea

With the certain volume of water, different amount of tea is actually the change of the ratio of tea to water. In this study, we prepared the tea sample with 150 mL ultra-water mixed in the following amount of tea: 5.0, 10.0, 15.0, and 20.0 g. As shown in Fig. 33.3, we can infer that the amount of tea played a crucial role in the extraction efficiency of 5 PAEs in tea. Except for DEHP, the content of each analyte and the total compounds reached to the maximum when using 10.0 g grounded tea. The result indicated that low amount of tea could cause higher loss during the

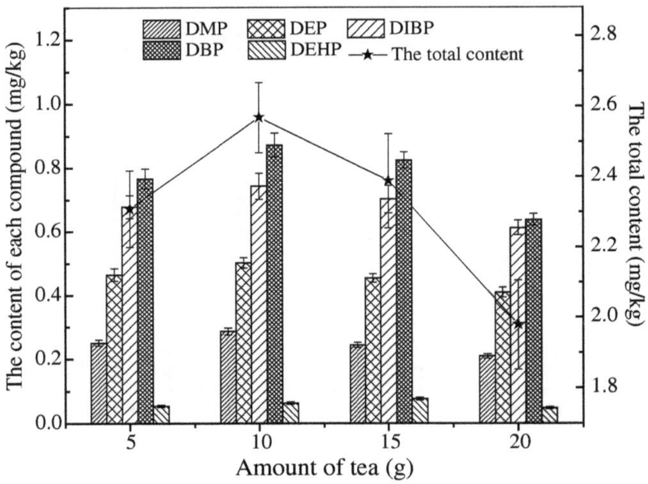

Fig. 33.3 Effect of the amount of tea on the extraction efficiency of 5 PAEs in tea samples

process, high amount of tea could not be fully immersed, and the PAEs could not be completely extracted. Therefore, a total of 10.0 g tea was considered adequate to achieve the best extraction.

33.3.1.4 Extraction Time

The extraction time was investigated by increasing the reflux time from 2 to 5 h. As can be seen from Fig. 33.4, the content of each analyte and the total compounds

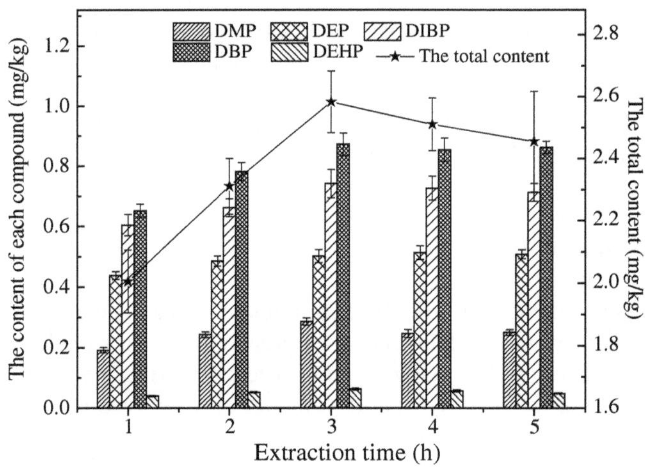

Fig. 33.4 Effect of extraction time on the extraction efficiency of 5 PAEs in tea samples

significantly increased with increasing extraction time until 3 h. When extraction time exceeded 3 h, no obvious change for the extraction efficiencies of each analyte, but the content of 5 PAEs show a decrease trend. The reason is that the extraction time of the higher boiling point of PAEs is too short to be completely extracted, but if the tea samples were extracted with a long time of a high temperature, PAEs may occur pyrolysis or volatilization losses during the distillation process. So 3 h was selected for the extraction time.

33.3.2 Establishment of Gas Chromatography–Mass Spectrometry

33.3.2.1 Selection of Chromatographic Conditions

Phthalate ester (PAEs) is a group of chemical compounds of weak polarity. According to the principle of similar compatible, we chose nonpolar or semipolar chromatographic capillary column as stationary phase. Considering the high boiling point of PAEs (generally above 200 °C), the column need to resist high temperature. Generally, the highest temperature limit of polar chromatographic column is high. As the tea is a complex matrix, in order to avoid interference with other volatile components, so HP-5MS capillary chromatographic column and the programmed ascension of oven temperature were performed for separation.

Injection port temperature was tested between 200 and 300 °C in the interval of 25 °C degrees. Figure 33.5 shows response areas of each compound obtained from the standard mixture. With the increasing of the temperature, except for DEHP, the

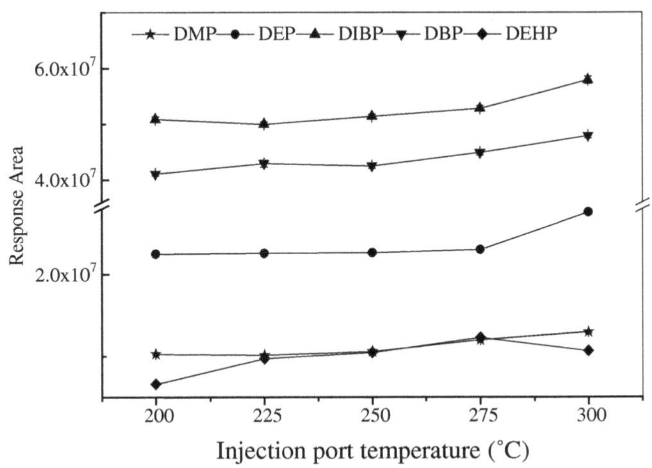

Fig. 33.5 Effect of injection port temperature on the responses of five PAEs

Table 33.1 Retention times, scan time range, selected ions, and quantitative ion of the compounds studied by GC–MS

Compound	Retention time (min)	Scan time range (min)	Select ions (m/z)	Dwell time (ms)	Quantitative ion (m/z)
DMP	28.59	25–34	163, 77, 194, 133	50	163
DEP	35.79	34–48	149, 177, 105, 121	50	149
DIBP	50.47	48–53	149, 223, 104, 167	50	149
DBP	54.44	53–65	149, 223, 205, 104	50	149
DEHP	69.15	65–73	149, 167, 279, 113	50	149

response area of the analytes increased. The best results were obtained at 300 °C, with good direct liquid injection response for the other four PAEs studied. Since the highest temperature of HP-5MS capillary column is 325 °C, and the column temperature is generally 20 °C lower than the maximum operating temperature. As a compromise, 300 °C was eventually chosen for further experiments.

33.3.2.2 Mass Spectrometry Scanning Method

There are two scan mode for MS operations, full scan and selective ion monitor (SIM), were used alternately and consecutively. Each scan was subdivided into two events. The first one was performed in full scan mode in the 35–450 amu scan range to set the retention time of the target compounds, and to divide the chromatogram into five segments. The second one in MS mode with the precursor ion selected, and four qualitative ions for each analyte, the dwell time of each ion was 50 ms. Table 33.1 lists the retention times, scan time range, selected masses, qualitative, and quantitative ion for each compound studied by GC–MS. Then the five PAEs were analyzed and quantified by GC–MS using the SIM method, and the total ion current chromatograms of the five PAEs standard solution and a real tea sample are shown in Fig. 33.6.

33.3.3 Method Validation

Under the optimal conditions, the linearity of the proposed method was evaluated by a series of standard solutions of five analytes at eight levels. And the calibration curves of each PAEs was developed by plotting the peak areas (y) of five PAEs versus corresponding concentration of the analytes (x). The LODs of five PAEs,

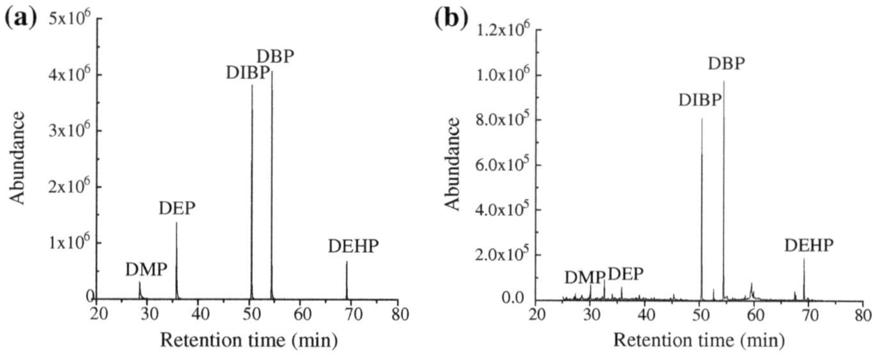

Fig. 33.6 Chromatograms of five PAEs standard solution (**a**) and a real tea sample (**b**)

calculated on the basis of signal-to-noise ratio of 3 (S/N = 3) and LOQs based on S/N = 10. The recovery was obtained from the spiked tea sample to evaluate the accuracy of the optimized method, known quantities of the five phthalate eaters were added to the real tea sample. The repeatability of the method was determined by the relative standard deviation (RSD).

The results (Table 33.2) showed that the linearity of five PAEs was satisfactory in their detection range, with R^2 values ranging from 0.9967 (DMP) to 0.9990 (DBP). LODs for them were in the range of 0.24–3.72 μg/L and the LOQs ranged from 0.80 to 12.40 μg/L. The spiked recoveries ranged from 87.63 to 98.08 %. The values of the RSD were less than 8 % for all analytes, which is considered satisfactory for the analysis of five phthalate esters in tea.

Table 33.2 Linear equations, linear ranges (n = 8), regression coefficients (R^2), limits of detection (LODs) and quantitation (LOQs), recovery and relative standard deviations (RSD) of five phthalate esters

Compound	Linear equations	R^2	Linear range (mg/L)	LODs (μg/L)	LOQs (μg/L)	Spiked value (μg)	Recovery (%)	RSD (%)
DMP	Y = 4.02X −5.65	0.9967	1.19–16.66	3.72	12.40	2.38	97.06	4.72
DEP	Y = 4.61X −4.29	0.9982	0.89–31.08	1.06	3.53	4.44	93.92	3.08
DIBP	Y = 7.20X −4.13	0.9984	0.61–42.84	0.35	1.17	6.12	94.61	6.27
DBP	Y = 5.65X −5.16	0.9990	0.62–13.68	1.08	3.60	6.24	98.08	3.98
DEHP	Y = 3.94X −1.12	0.9977	0.39–13.58	0.24	0.80	0.39	87.63	7.92

Table 33.3 The content of five PAEs in real tea samples (mg/kg)

Tea sample	DMP	DEP	DIBP	DBP	DEHP
Green tea 1#	0.12	0.11	0.49	0.58	0.055
Green tea 2#	0.29	0.50	0.74	0.87	0.063
Green tea 3#	0.20	0.33	0.16	0.59	0.055
Green tea 4#	0.34	0.21	0.23	0.73	0.051
Black tea 1#	0.15	0.34	0.58	0.51	0.044
Black tea 2#	0.17	0.63	0.66	0.42	0.064
Black tea 3#	0.29	0.37	0.35	0.77	0.038
Dark tea 1#	0.17	0.15	0.26	0.65	0.085
Dark tea 2#	0.22	0.42	0.70	0.67	0.20
Dark tea 3#	0.11	0.38	0.19	0.23	0.18

33.3.4 Sample Analysis

In order to validate the accuracy and precision of the proposed method under the selected conditions, ten commercial tea samples were tested by GC–MS. The results are shown in Table 33.3. Five PAEs were all detected out in the selected real tea samples. Among them, the content of DMP were between 0.11 and 0.34 mg/kg, DEP were 0.11–0.63 mg/kg, DIBP were 0.16–0.74 mg/kg, DBP were 0.23–0.87 mg/kg, DEHP were 0.038–0.20 mg/kg, and the total content of five PAEs were 1.09–2.46 mg/kg.

33.4 Conclusion

In this study, a successful development and application of the SDE technique, combined with GC–MS was investigated for the qualitative and quantitative analysis of five phthalate esters in tea . Under the optimization condition, the proposed method was precise, reproducible, and linear over a wide range with sufficient selectivity (using the MS detector at the SIM mode) and high sensitivity. The linear regression coefficients were between 0.9967 and 0.9990, detection limits varied from 0.24 to 3.72 μg/L and the spiked recovery was ranged from 87.63 to 98.08 %. The relative standard deviations were below 8 %. Therefore, the method could be used for the determination of five phthalate esters in tea.

Acknowledgments This work was financially supported by the program of National High Technology Research and Development Program of China (863 Program) (Grant No. SS2012AA023408), the Cheung Kong Scholars, and Innovative Research Team Program in University of Ministry of Education, China (Grant No. IRT1166).

References

1. Carrillo JD, Salazar C, Moreta C et al (2007) Determination of phthalates in wine by headspace solid-phase microextraction followed by gas chromatography–mass spectrometry: fibre comparison and selection. J Chromatogr A 1164:248–261
2. Li X, Zeng Z, Chen Y et al (2004) Determination of phthalate acid esters plasticizers in plastic by ultrasonic solvent extraction combined with solid-phase microextraction using calix [4] arene fiber. Talanta 63:1013–1019
3. Martino-Andrade AJ, Chahoud I (2010) Reproductive toxicity of phthalate esters. Mol Nutr Food Res 54:148–157
4. EPA, US (1991) National primary drinking water regulations: federal register, Part 12, 40 CFR Part 141, US Environmental Protection Agency: 395
5. Gomez HA, Aguilar CM (2003) Social and economic interest in the control of phthalic acid esters. Trends Anal Chem 22:847–857
6. Lee KH, Kim JH, Lim DS et al (2000) Anti-leukemic and anti-mutagenic effects of di(2-ethylhexyl) phthalate isolated from Aloe vera Linne. J Pharm Pharmacol 52:593–598
7. Namikoshi M, Fujiwara T, Nishikawa T et al (2006) Natural abundance ^{14}C content of dibutyl phthalate (DBP) from three marine algae. Mar Drugs 4:290–297
8. Al-Bari MA, Bhuiyan MS, Flores ME et al (2005) *Streptomyces bangladeshensis* sp. nov., isolated from soil, which produces bis-(2-ethylhexyl) phthalate. Int J Syst Evol Microbiol 55:1973–1977
9. Horanni R, Engelhardt UH (2013) Determination of amino acids in white, green, black, oolong, pu-erh teas and tea products. J Food Compos Anal 31:94–100
10. Kim ES, Liang YR, Jin J et al (2007) Impact of heating on chemical compositions of green tea liquor. Food Chem 103:1263–1267
11. Du LP, Li JX, Li W et al (2014) Characterization of volatile compounds of pu-erh tea using solid–phase microextraction and simultaneous distillation–extraction coupled with gas chromatography–mass spectrometry. Food Res Int 57:61–70
12. Gong J, Tang C, Peng C (2012) Characterization of the chemical differences between solvent extracts from Pu-erh tea and Dian Hong black tea by CP–Py–GC/MS. J Anal Appl Pyrol 95:189–197
13. Wu CL, Chen JM, Huang YH (2012) Analysis of aroma constituents of high-fragrant green tea. Mod Food Sci Technol 28:579–582
14. Rios JJ, Morales A, Márquez RG (2010) Headspace solid-phase Microextraction of oil matrices heated at high temperature and phthalate esters determination by gas chromatography multistage Mass Spectrometry. Talanta 80:2076–2082
15. López-Jiménez FJ, Rubio S, Pérez-Bendito D (2005) Determination of phthalate esters in sewage by hemimicelles-based solid-phase extraction and liquid chromatography–mass spectrometry. Anal Chim Acta 551:142–149
16. Zhang H, Chen X, Jiang X (2011) Determination of phthalate esters in water samples by ionic liquid cold-induced aggregation dispersive liquid–liquid microextraction coupled with high-performance liquid chromatography. Anal Chim Acta 689:137–142
17. Del Carlo M, Pepe A, Sacchetti G et al (2008) Determination of phthalate esters in wine using solid-phase extraction and gas chromatography–mass spectrometry. Food Chem 111:771–777

References

Chapter 34
Antibacterial Mechanism of 10-HDA Against *Bacillus subtilis*

Xiaohui Yang, Junlin Li and Ruiming Wang

Abstract 10-HDA (10-hydroxy-2-decenoic acid) is an unsaturated medium-chain fatty acid, which is the main active component of royal jelly. It has been shown to possess biological activity in antibacterial, immunoregulation, and antitumour. However, the underlying antibacterial mechanism of 10-HDA is unclear. In this study, it is shown that 10-HDA is a broad-spectrum antimicrobial agent and has obvious inhibition effects on multiple pathogenic bacteria with a concentration-dependent mode of inhibitory effect. The minimum inhibitory concentration (MIC) of *Bacillus subtilis* is 0.62 mg/mL. Furthermore, the acting mechanism of 10-HDA on *B. subtilis* is investigated by analyzing the DNA binding ability of 10-HDA with gel retardation assay and atomic force microscope. The results indicate that 10-HDA can combine with the bacterial DNA strongly and inhibit migration of DNA on Sepharose gel. By detecting the effect of 10-HDA on DNA content of bacteria, it was shown that 10-HDA had inhibitory effect on DNA synthesis of bacteria. Our results suggest that 10-HDA can combine with bacterial DNA, which further inhibits the DNA synthesis, and thus suppress the growth and activity of bacteria.

Keywords 10-HDA · *Bacillus subtilis* · Antibacterial mechanism · DNA synthesis

34.1 Introduction

Royal jelly (RJ), the exclusive food of the queen honeybee larva, is secreted from the hypopharyngeal and mandibular glands of worker honeybees. RJ has been demonstrated to possess several pharmacological activities in experimental animals, including antihypertensive activities [1], anti-hypercholesterolemic activity [2], life-span elongating action [3], antitumor activity [4, 5], and anti-inflammatory activity [6].

X. Yang · J. Li · R. Wang (✉)
Key Laboratory of Shandong Microbial Engineering, Qilu University of Technology, Jinan 250353, China
e-mail: ruiming3k@163.com

T.-C. Zhang and M. Nakajima (eds.), *Advances in Applied Biotechnology*,
Lecture Notes in Electrical Engineering 332, DOI 10.1007/978-3-662-45657-6_34

Chemical composition analysis has shown that RJ consists mainly of proteins, sugars, lipids, vitamins, and free amino acids. 10-hydroxy-2-decenoic acid (10-HDA), a major fatty acid component of RJ, has many pharmacological activities, such as antitumor activity [7, 8], immunomodulatory [9] and antioxidative activity [10, 11], collagen production promoting activity [12, 13] and antibiotic activity [14]. It is shown that the antibacterial activity of 10-HDA against *Bacillus subtilis*, *Staphylococcus aureus*, *Escherichia coli* is stronger than other C_{10} fatty acids [15].

Recently, it has been reported many acting ways of bacteriostatic agents. They may act by attacking membrane [16, 17], inhibiting cell respiration, blocking cell division, and intracellular protein synthesis. However, the underlying antibacterial mechanism of 10-HDA is unclear. In this study, the antibacterial activities of 10-HDA against *Pesudomonas pyocyaneum*, *B. subtilis*, *S. aureus*, *E. coli*, *Saccharomyces cerivisiae* and *Candida albicans* were detected. Meanwhile, the acting mechanism of 10-HDA on *B. subtilis* was investigated by analyzing the DNA binding ability of 10-HDA with gel retardation assay and atomic force microscope. For the first time, we offer evidence to indicate that 10-HDA can combine with bacterial genomic DNA, which further inhibits the DNA synthesis, and thus suppress the growth and activity of bacteria. It plays an important role for 10-HDA application in medicine and healthcare fields, and provides a new way for development and utilization of RJ.

34.2 Materials and Methods

34.2.1 Reagents

10-HDA was purchased from Shanghai Gongshuo Biotechnology Company limited (Shanghai, China). Its purity is higher than 99 %. 10-HDA was dissolved in Dimethyl sulfoxide (DMSO) and kept frozen at −20 °C. The stock solutions of 10-HDA were diluted to the desired concentration.

34.2.2 Cultivation of the Microorganism

The strains of *E. coli*, *S. aureus*, *B. subtilis*, *S. cerevisiae* were conserved in our laboratory. Luria–Bertani (LB) medium (pH 7.0 ± 0.2) contained the following: 10 g tryptone, 5.0 g yeast extract and 10 g sodium chloride. YEPD medium (pH 6.0 ± 0.2) contained the following: 10 g yeast extract, 20 g peptone and 20 g glucose. The solid medium was supplemented with 2 % agar in LB medium and YEPD medium.

34.2.3 Analysis of Antimicrobial Activity of 10-HDA

Antimicrobial activity of 10-HDA was determined by Oxford cup diffusion method using 1 mL of suspension containing 1×10^7 colony forming units (CFU)/mL of bacteria spread on nutrient agar (NA) medium. Oxford cups (6 mm in diameter) were placed on the inoculated agar, and then 100 μL of different concentrations of 10-HDA was added with a micropipette. The diameter of inhibition zone (DIZ) was measured after 24 h of incubation at 37 °C, and DMSO was used as a negative control. Tests were performed in triplicate.

34.2.4 Assay for Minimum Inhibitory Concentration of 10-HDA

The minimum inhibitory concentration (MIC) of 10-HDA on *B. subtilis* was determined by tube dilution method. Twofold serial dilutions of 10-HDA were prepared in sterile NB medium ranging from 0.16 to 5.0 mg/mL. To each tube, 50 mL of the inoculum containing approximately 1×10^7 CFU/mL microorganisms was added. A control test containing inoculated broth supplemented with only DMSO was also performed. The tubes were then incubated at 37 °C for 18 h. Then, the bacterial suspensions were spread on NA medium plate. The MIC was read as the lowest antimicrobial concentration of 10-HDA that showed no visible growth in the culture for incubating at 37 °C for 24 h. The experiments were repeated in triplicate.

34.2.5 Gel Retardation Assay of DNA Binding Ability of 10-HDA

The cultures of *B. subtilis* in a logarithmic growth phase were collected and the genomic DNA was extracted by column type extraction kit. DNA purity was detected by ultraviolet spectrophotometer. Gel retardation experiment was used to test the DNA binding ability of 10-HDA according to Park et al. [18]. One microgram of genomic DNA and 5 μL of 10-HDA with different concentrations were mixed and incubated at room temperature for 30 min. The control group was treated with DMSO solution. Subsequently, the mixture was applied to 1 % agarose gel electrophoresis.

34.2.6 Observation of 10-HDA Binding with DNA by Atomic Force Microscope

10-HDA was added into the genomic DNA of *B. subtilis* to make its final concentration of 0.62 mg/mL, and at the same time the control group with DMSO was set up. After incubation for 10 min at room temperature, 10-HDA and DNA compounds were added to fresh mica surface and incubated at 37 °C for 10 min. After drying, the samples prepared were observed by atomic force microscope scanning.

34.2.7 Effect of 10-HDA on the DNA Content of Bacteria

The cultures of *B. subtilis* in a logarithmic growth phase were prepared and 10-HDA was added to make its final concentration of 0.62 mg/mL, and at the same time the control group with DMSO was set up. After incubation at different times at 37 °C, the cultures were collected and centrifuged at 4,000 rpm for 10 min. The precipitates were suspended in sterile water and added three times volume of DAPI dye. The DNA content of bacteria was detected by fluorospectrophotometer (Hitachi, Japan) at 364 nm. The experiments were repeated in triplicate.

34.3 Results

34.3.1 Antimicrobial Activity of 10-HDA

Antimicrobial activity of 10-HDA is shown in Table 34.1. Results show that 10-HDA had significant inhibitory effects against *P. pyocyaneum*, *B. subtilis*, *S. aureus* and *E. coli*, but had no inhibition against fungi such as *S. cerevisiae* and *C. albicans*. This may be associated with the difference in the cell structure between bacteria and fungi.

34.3.2 Determination of the Minimum Inhibitory Concentration of 10-HDA

Tube dilution method was used to detect the MIC of 10-HDA against *B. subtilis*. As shown in Table 34.2, 10-HDA had good inhibitory effect on the growth of *B. subtilis*. When the concentration of 10-HDA was 0.62 mg/mL, no growth of bacteria could be detected. And when the concentration of 10-HDA was 0.31 mg/mL, the bacterial colony could be detected, indicating the MIC of 10-HDA against *B. subtilis* was 0.62 mg/ml.

Table 34.1 The inhibitory effect of 10-HDA by Oxford-cup method

Strains	10-HDA concentration (mg mL^{-1})				
	0	2.5	5.0	10.0	20.0
Pesudomonas pyocyaneum	–	+++	+++	++++	++++
Bacillus subtilis	–	+++	++++	++++	++++
Staphylococcus aureus, Micrococcus luteus	–	+++	++++	++++	++++
Escherichia coli	–	++	++	+++	++++
Saccharomyces cerevisiae	–	–	–	–	–
Candida albicans	–	–	–	–	–

Notes ++++ Inhibitory zone diameter ≥ 20 mm is significantly sensitive; +++ 15 mm ≤ inhibitory zone diameter <20 mm is highly sensitive; ++ 10 mm ≤ inhibitory zone diameter <15 mm is moderate sensitive; + 7.8 mm ≤ inhibitory zone diameter <10 mm is low sensitive; – inhibitory zone diameter <7.8 mm is invalid

Table 34.2 MIC of 10-HDA to *Bacillus subtilis*

10-HDA concentration (mg mL^{-1})	5.0	2.5	1.25	0.62	0.31	0.16
Growth of *Bacillus subtilis*	–	–	–	–	+	++

Notes ++ show many bacteria grow; + show fewer bacteria grow; – show no bacteria grow

34.3.3 DNA Binding Ability of 10-HDA

The DNA binding ability of 10-HDA was examined by analyzing the effect of 10-HDA on the electrophoretic mobility of bacterial genomic DNA on agarose gel (Fig. 34.1). As shown in Fig. 34.1, the brightness of band weakened gradually as the concentration of 10-HDA increased, and the amounts of DNA retained in the sample loading hole increased gradually, showing the aggravation of the interactions between 10-HDA and DNA. When the concentration achieved 10.0 mg/L, 10-HDA completely inhibited the migration of bacterial DNA (Fig. 34.1, lane 6). Inhibition of 10-HDA on electrophoretic mobility of bacterial genomic DNA provided direct evidence for interaction of 10-HDA with DNA.

34.3.4 Observation of 10-HDA Binding with DNA by Atomic Force Microscope

We used atomic force microscopy (AFM) to further observe and analyze the interaction of 10-HDA with DNA, as shown in Fig. 34.2. The control group treated with DMSO showed that DNA molecule was white long chain with expanding distribution. In the experiment group, which was treated 10-HDA with the final concentration of 0.62 mg/mL, the bacterial DNA were curled up with small bright spots for 10-HDA particles, which formed many beaded complex. Due to the adsorption of DNA chain with many 10-HDA particles, the image showed the

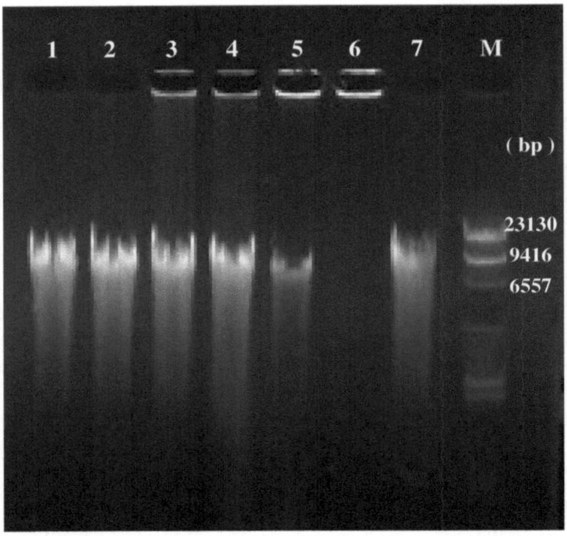

Fig. 34.1 Gel retardation assay of DNA binding ability of 10-HDA. *Lane 1–6* the concentration of 10-HDA was 0, 0.62, 1.25, 2.5, 5.0, 10.0 mg/mL, respectively; *lane 7* was control group treated with DMSO

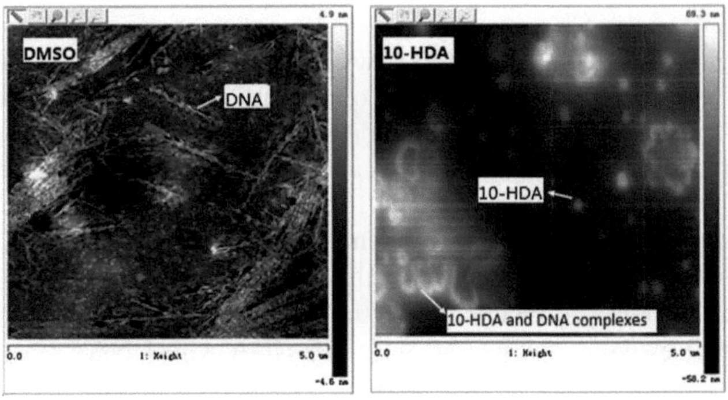

Fig. 34.2 AFM image of 10-HDA binding with *Bacillus subtilis* genomic DNA

formation of high protuberant ridges (white protrusion chain). This result indicates that 10-HDA can combine with the bacterial DNA strongly.

34.3.5 Effect of 10-HDA on the DNA Content of Bacteria

To analyze the effect of 10-HDA on the DNA synthesis on *B. subtilis*, DAPI fluorescent staining method was applied to detect the DNA content of bacteria. As shown in Fig. 34.3, the DNA content of *B. subtilis* was significantly decreased after

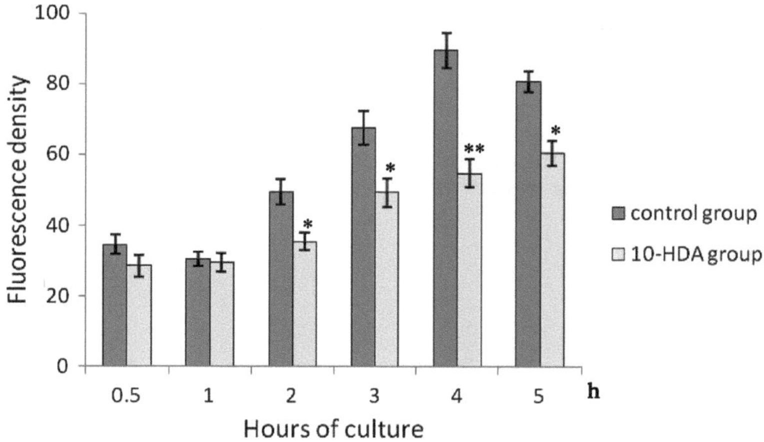

Fig. 34.3 Effect of 10-HDA on DNA synthesis on *Bacillus subtilis*. * Means that difference is significant compared with control ($p < 0.05$); ** Means that difference is most significant compared with control ($p < 0.01$)

10-HDA treatment from 2 s to 5 h. And after 10-HDA treatment in 4 h, the DNA content of bacteria decreased by 38.87 % compared to the DMSO control group, where the difference was most significant. This result indicates that 10-HDA has inhibitory effect on DNA synthesis of bacteria.

34.4 Conclusions

In our study, we first detected the antibacterial activity of 10-HDA against *P. pyocyaneum*, *B. subtilis*, *S. aureus*, *E. coli*, *S. cerivisiae* and *C. albicans* by Oxford cup method. The results showed that 10-HDA had significant inhibitory effects against bacteria, but no inhibition against fungi, such as *S. cerevisiae* and *C. albicans*. This may be associated with the difference in the cell structure between bacteria and fungi.

To understand the antibacterial mechanism of 10-HDA to bacteria, we selected *B. subtilis* as model to study the effect of 10-HDA. Results showed that 10-HDA had good inhibitory effect on the growth of *B. subtilis*, and the MIC of 10-HDA against *B. subtilis* was 0.62 mg/mL. Furthermore, the acting mechanism of 10-HDA on *B. subtilis* was investigated by analyzing the DNA binding ability of 10-HDA by gel retardation assay and atomic force microscope. The results indicate that 10-HDA can combine with the bacterial DNA strongly and inhibit the electrophoretic migration of bacterial DNA. By detecting the effect of 10-HDA on DNA content of bacteria, it was shown that as incubation time prolonged, the amounts of bacteria DNA reduced gradually, which indicates that 10-HDA had inhibitory effect on bacterial DNA synthesis. Our study results suggest that 10-HDA can combine with bacterial genomic DNA, which further inhibits the DNA synthesis and thus suppress the growth and activity of bacteria.

Acknowledgments This work was supported by grants from the National Natural Science Foundation of China (No: 31201281 and 31171727) and the Research award fund for outstanding young scientists of Shandong province (No: BS2013SW029).

References

1. Lichtenthaler R, Marx F (2005) Total oxidant scavenging capacities of common European fruit and vegetable juices. J Agric Food Chem 53:103–110
2. Vittek J (1995) Effect of royal jelly on serum lipids in experimental animals and humans with atherosclerosis. Experientia 51:927–935
3. Inoue S, Koya MS, Ushio S et al (2003) Royal jelly prolongs the life span of C3H/HeJ mice: correlation with reduced DNA damage. Exp Gerontol 38:965–969
4. Townsend GF, Morgan JF, Hazlett B (1959) Activity of 10-hydroxydecenoic acid from royal jelly against experimental leukemia and ascetic tumors. Nature 183:1270–1271
5. Townsend GF, Morgan JF, Tolnai S et al (1960) Studies on the in vitro antitumor activity of fatty acids I. 10-hydroxy-2-decenoic acid from royal jelly. Cancer Res 20:503–510
6. Kohno K, Okamoto I, Sano O et al (2004) Royal jelly inhibits the production of proinflammatory cytokines by activated macrophages. Biosci Biotechnol Biochem 68:138–145
7. Townsend GF, Morgan JF, Toinai S et al (1960) Studies on the in vitro antitumor activity of fatty acids 10-Hydroxy-2-decenoic acid from royal jelly. Cancer Res 20:503–510
8. Nakaya M, Onda H, Sasaki K et al (2007) Effect of royal jelly on bisphenol A-induced proliferation of human breast cancer cells. Biosci Biotechnol Biochem 71:253–255
9. Vucevica D, Mellioub E, Vasilijica S et al (2007) Fatty acids isolated from royal jelly modulate dendritic cell-mediated immune response in vitro. Int Immunopharmacol 7:1211–1220
10. Nagai T, Sakai M, Inoue R et al (2001) Antioxidative activities of some commercially honeys, royal jelly and propolis. Food Chem 75:237–240
11. Nagai T, Inoue R, Suzuki N et al (2006) Antioxidant properties of enzymatic hydrolysates from royal jelly. J Med Food 9:363–367
12. Satomi KM, Iwao O, Shimpei U et al (2004) Identification of a collagen production-promoting factor from an extract of royal jelly and its possible mechanism. Biosci Biotechnol Biochem 68:767–773
13. Park HM, Wang EH, Lee KG et al (2011) Royal jelly protects against ultraviolet β–induced photoaging in human skin fibroblasts via enhancing collagen production. J Med Food 14:899–906
14. Melliou E, Chinou I (2005) Chemistry and bioactivity of royal jelly from Greece. J Agric Food Chem 53:8987–8992
15. Yatsunami K, Echigo T (1985) Antibacterial action of royal jelly. Bull Fac Agric Tamagawa Univ 25:13–22
16. Xie YP, He YP, Peter LI et al (2011) Antibacterial activity and mechanism of action of zinc oxide nanoparticles against Campylobacter jejuni. Appl Environ Microbiol 77:2325–2331
17. Mariachiara C, Francesco A, Laura F et al (2007) Antimicrobial activity of various cationic molecules on foodborne. J Microbiol Biotechnol 23:1679–1683
18. Park CB, Kim HS, Kim SC (1998) Mechanism of action of the antimicrobial peptide buforin II: buforin II kills microorganisms by penetrating the cell membrane and inhibiting cellular functions. Biochem Biophys Res Commun 244:253–257

Chapter 35
Monitoring Glutamate and Glucose Concentration During the Temperature Triggered Glutamate Fermentation by Near-Infrared Spectroscopy

Yongli Gui, Jingbo Liang, Chenglin Zhang, Xixian Xie, Qingyang Xu, Ning Chen and Lei Ma

Abstract In this study, the calibration models for monitoring concentrations of glutamate and glucose in the temperature-triggered glutamate fermentation process were developed by near infrared (NIR) spectroscopy. The NIR measurements of samples were analyzed by partial least-squares (PLS) regression with selecting spectral pre-processing methods and different wavelengths. The root-mean square errors of cross-validation (RMSECV) of glutamate and glucose were 2.73 and 1.92 g/L, respectively. The determination coefficients (R^2) were 0.996 and 0.982, respectively. The residual predictive deviation (RPD) was 17.8 and 8.37, respectively. These results showed that all models had good predictive ability. New batch fermentation as an external validation was used to check the models. Compared with concentrations of predict value and measured value, the determination coefficient was 0.992 and 0.951, respectively. The average relative errors were 5.79 and 7.38 %, respectively. These results showed that prediction model could predict and monitor the temperature-triggered glutamate fermentation process accurately and quickly, and thus theoretical basis for the real-time control and optimization in the temperature-triggered glutamate fermentation process was provided.

Keywords Near infrared · Glutamate · Glucose · Model

Y. Gui · J. Liang · C. Zhang · X. Xie · Q. Xu · N. Chen
College of Biotechnology, Tianjin University of Science and Technology, Tianjin 300457, China

Y. Gui · J. Liang · C. Zhang · X. Xie · Q. Xu · N. Chen
National and Local United Engineering Lab of Metabolic Control Fermentation Technology, Tianjin 300457, China

L. Ma (✉)
College of Electronic Information and Automation, Tianjin University of Science and Technology, Tianjin 300457, China
e-mail: ningch@tust.edu.cn

T.-C. Zhang and M. Nakajima (eds.), *Advances in Applied Biotechnology*,
Lecture Notes in Electrical Engineering 332, DOI 10.1007/978-3-662-45657-6_35

35.1 Introduction

The worldwide market for glutamate is continuously increasing year by year, hence the glutamate enterprises make efforts to improve production strains and optimize fermentation processes [9, 19]. Since glutamate fermentation is a complex dynamic process, the real-time understanding of main material concentration is very important for the process control and optimization. Regrettably, the traditional analysis method is time-consuming and unsuitable for real-time analysis and optimization of the fermentation process [5, 10].

Near-infrared (NIR) spectroscopy as an emerging analysis technology has the following advantages: detection procedure is inexpensive and time-saving; easy sample preparation procedures; it has no need for skilled operators; it is not accompanied by environmentally unfriendly processes [8, 15]. With above advantages, the NIR spectroscopy has been widely applied in many fields such as food [13, 18], pharmaceutical [1, 6], wine [12], petrochemical [11], mining [14] and biological analyses for simultaneous determination of multiple constituents and real-time bioprocess monitoring.

In this study, NIR spectroscopy and the partial least squares (PLS) regression were used for developing multivariate models to quantify glutamate and glucose concentrations in the temperature-triggered process of glutamate fermentation. What is more, new batch fermentation as an external validation was used to check the models' prediction capacity of glutamate and glucose concentrations in glutamate fermentation. It is critical significance for real-time monitoring and controlling the fermentation processes and establishing suitable glucose supplementation strategies.

35.2 Materials and Methods

35.2.1 Microorganism and Medium

The temperature triggered glutamate-producing strain used throughout this study was *Corynebacterium glutamicum* TCCC11822, which was obtained from earlier work in our laboratory and stored at the Culture Collection of Tianjin University of Science and Technology.

Corynebacterium glutamicum TCCC11822 was grown in the seed medium containing 25 g/L of glucose, 20 ml/l of corn steep liquor, 10 ml/l of protein hydrolysate, 0.6 mg/L of vitamin B_1, 0.6 mg/L of vitamin H, 3 g/L of KH_2PO_4, 2 g/L of $MgSO_4$, 5 mg/L of $MnSO_4$, 5 mg/L of $FeSO_4$, 0.05 mg/L of methionine and 5 g/L of urea.

Corynebacterium glutamicum TCCC11822 was grown in the medium for glutamate fermentation containing 50 g/L of glucose, 40 g/L of beet molasses, 40 ml/l of corn steep liquor, 20 ml/l of soybean protein hydrolysate, 0.6 mg/L of vitamin B_1, 0.6 mg/L of vitamin H, 4.5 g/L of K_2HPO_4, 2 g/L of $MgSO_4$, 30 mg/L of $MnSO_4$ and 30 mg/L of $FeSO_4$.

The pH of both media was adjusted to 7.0 with 4 mol/L NaOH.

35.2.2 Process of Fermentation

A single colony of *C. glutamicum* TCCC11822 was inoculated into a 1 L baffled flask containing 0.1 L of seed medium and cultivated at 33 °C, 200 rpm for 14 h. A 0.3 L inoculum of this culture was transferred aseptically to a 5 L seed fermenter (Biotech-2002 Bioprocess controller, Baoxing, Shanghai, China), which contained 3 L seed medium. The seed was cultivated at 33 °C for 14 h. The pH was adjusted to 7.0 with 25 % (v/v) ammonia during the whole cultivation period. The dissolved oxygen (DO) level was maintained at approximately 20 % saturation by adjusting the agitation and aeration rate.

Fed-batch glutamate fermentation was performed in a 30 L jar fermenter (Biotech-2002 Bioprocess controller, Baoxing, Shanghai, China) containing 15 L production medium with 20 % (v/v) inoculum size. In the process of glutamate fermentation, pH was maintained at 7.0 with 25 % (v/v) ammonia. The temperature of growth phase was kept at 33 °C and at 39 °C during the overproduction phase to provoke glutamate production. The DO level of growth phase and production phase was respectively maintained at approximately 20 and 10 % saturation by adjusting the agitation and aeration rate. When the initial glucose was depleted, an appropriate amount of glucose solution (900 g/L) was fed into the fermenter to meet specific experimental requirement.

35.2.3 Sampling and Analytical Methods

A total of 121 glutamate fermentation samples used for the experiments were obtained from every hour during the process of glutamate fermentation. All samples were centrifuged at 13000 × g for 2 min and the supernatants were collected to determine glutamate and glucose concentrations by SBA-40E biosensor instrument (Biology Institute of Shandong Academy of Sciences, Shandong, China).

35.2.4 NIR Spectra Acquisition and Pre-processing

The NIR spectra from 833 to 2,500 nm were obtained with a resolution of 8.0 cm^{-1} and an accumulation of 32 scans using a Bruker NIR Analyzer Tensor 37 spectrophotometer (Bruker Optics, Ettlingen, Germany).

The raw spectra were mathematically transformed and visualized using the software OPUS 7.0 (Bruker Optics, Ettlingen, Germany) with the aim to reduce the spectral variability associated with the physical characteristics of samples and the instrumental variability. In order to achieve the best prediction performance, several spectral pre-processing and different wavelength ranges were investigated. The pre-processing methods included first derivative (1st), second derivative (2nd), standard normal variate (SNV), min-max normalization, straight line subtraction, multivariate scatter correction (MSC), 1st + SNV and 1st + MSC [3, 16]. The smoothing point was 13.

35.2.5 Multivariate Calibration Model

Multivariate calibration model was developed by PLS regression with cross-validation test. The fermentation samples were divided into calibration set (80 %) and validation set (20 %) randomly and equably by OPUS 7.0 (Bruker Optics, Ettlingen, Germany). The root-mean square error of cross-validation (RMSECV) and root-mean square error of prediction (RMSEP) reminded the predictive error obtained in calibration set and validation set. The selection of the number of latent variables (LVs) was due to the minimization of RMSECV of the calibration set. PLS model was developed by OPUS 7.0. The RMSECV and RMSEP were calculated as Eqs. (35.1) and (35.2):

$$\mathrm{RMSECV} = \sqrt{\frac{\sum_{i=1}^{n_C} \left(X_{Ci} - \overline{X}_{Ci}\right)^2}{n_C}} \tag{35.1}$$

in which n_C is the number of observations in the calibration set and X_{Ci} as well as $\bar{X}_{Ci}$ is the measured and predicted values, respectively.

$$\mathrm{RMSEP} = \sqrt{\frac{\sum_{i=1}^{n_p} \left(X_{pi} - \overline{X}_{pi}\right)^2}{n_p}} \tag{35.2}$$

in which n_p is the number of observations in the validation set and X_{pi} as well as $\bar{X}_{pi}$ is the measured and predicted values, respectively.

35.3 Results and Discussion

35.3.1 Chemical Analysis and NIR Spectra of Sample

In this work, all 121 fermented product samples were scanned by Bruker NIR Analyzer Tensor 37 spectrophotometer. The raw spectrum of 0, 10 and 20 h samples were shown in Fig. 35.1. The statistical parameter of glutamate and glucose concentrations was shown in Table 35.1. The range of raw spectrum of samples in 1,000–1,800 nm and 2,100–2,400 nm were relatively smooth and stable, so the differences of spectrum between the samples could fully embody and was beneficial

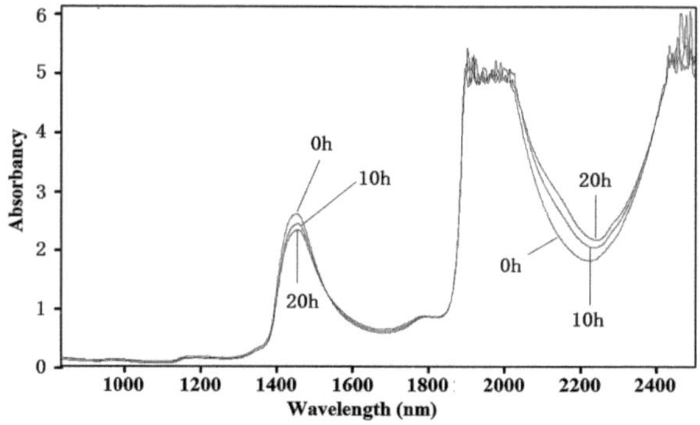

Fig. 35.1 NIR spectra of samples taken from glutamate fermentation

Table 35.1 Concentration of glutamate and glucose collected from the fermentation broth

Constituent	Glutamate		Glucose	
Subsets	Calibration set	Validation set	Calibration set	Validation set
Number of samples	97	24	97	24
Range (g/L)	9–174	10–172	1–51	1–50
Mean (g/L)	92.30	89.75	14.46	13.07
Standard deviation (g/L)	52.81	48.19	13.19	14.33

to develop multivariate calibration model. Particularly, the characteristic peaks of O–H bond are absorption at 1,450 nm, which are the main recordable phenomena in the NIR region [2]. Therefore the range of 1,000–1,800 nm was preferred to develop multivariate calibration model, and the range of 2,100–2,400 nm was used to supplement and optimize the model.

Table 35.1 shows the mean, standard deviation and range of glutamate and glucose concentration of calibration set and validation set. The range of glutamate and glucose concentration of calibration set was over the range of validation set, which were 9–174, 10–172 g/L and 1–51, 1–50 g/L, respectively. The mean of glutamate and glucose concentration of calibration set and validation set were 92.30, 89.75 g/L and 14.46, 13.07 g/L, respectively. The standard deviation of glutamate and glucose concentration of calibration set and validation set were 52.81, 48.19 g/L and 13.19, 14.33 g/L, respectively. These results showed that the category of samples was approximate and suitable to develop multivariate calibration model, respectively in the two components [4].

35.3.2 Development of a Multivariate Calibration Model

In order to establish multivariate calibration model, different spectral pre-processing methods and different wavelength ranges were tested before the development of the multivariate calibration model by OPUS 7.0. The optimal pre-processing method and wavelength range were adopted by lowest RMSECV in calibration set. The results of calibration model were showed in Table 35.2.

For all the multivariate calibration model of glutamate concentration (Table 35.2), minimization of RMSECV was 2.73 g/L. Hence SNV was identified as optimal pre-processing method and 1,000–1,800 nm was selected as optimal wavelength range. Among the parameters of the models of glutamate concentration (Table 35.3 and Fig. 35.2), the LVs was nine, the true-prediction fitting equation and coefficient of determination (R^2) of calibration set were $y = 0.998x + 0.137$ and 0.996 between experimental value and predictive value. The RPD was 17.8, which was greater than five and had good predictive ability [7, 17]. What is more, the true-prediction fitting equation and R^2 of validation set were $y = 0.984x + 0.362$ and 0.994 (Fig. 35.3).

Table 35.2 also showed that the lowest RMSECV was 1.92 g/L among the multivariate calibration model of glucose concentration with different method. Hence SNV was identified as optimal and 1,000–1,800 nm was selected as optimal wavelength range. Among the parameters of the models of glutamate concentration (Table 35.3 and Fig. 35.2), the pre-processing method was straight line subtraction,

Table 35.2 The optimal model of glucose concentration under different spectral pretreatment and different wavelength range

Spectral pre-processing method	Wavelength range (nm)	RMSECV (g/L)	
		Glutamate	Glucose
None	1,000–1,800	3.08	2.11
	1,000–1,800 + 2,100–2,400	3.54	2.37
Straight line subtraction	1,000–1,800	3.11	1.92
	1,000–1,800 + 2,100–2,400	3.51	2.14
SNV	1,000–1,800	2.73	1.98
	1,000–1,800 + 2,100–2,400	3.35	2.26
Min-max normalization	1,000–1,800	3.06	2.01
	1,000–1,800 + 2,100–2,400	3.26	2.58
MSC	1,000–1,800	2.84	1.97
	1,000–1,800 + 2,100–2,400	2.91	2.16
1st	1,000–1,800	3.46	3.02
	1,000–1,800 + 2,100–2,400	3.79	3.32
2nd	1,000–1,800	8.63	10.2
	1,000–1,800 + 2,100–2,400	8.83	16.8
1st + straight line subtraction	1,000–1,800	3.25	3.88
	1,000–1,800 + 2,100–2,400	3.52	2.99
1st + SNV	1,000–1,800	3.07	2.95
	1,000–1,800 + 2,100–2,400	3.53	3.19
1st + MSC	1,000–1,800	3.25	2.80
	1,000–1,800 + 2,100–2,400	3.31	3.30

Table 35.3 Parameters of the models of glutamate and glucose concentration

Constituent	Glutamate		Glucose	
	Calibration set	Validation set	Calibration set	Validation set
Latent variables	9		7	
Wavelength range (nm)	1,000–1,800		1,000–1,800	
Spectra pre-processing	SNV		Straight line subtraction	
RMSEC (g/L)	2.69	–	1.78	–
RMSECV (g/L)	2.73	–	1.92	–
RMSEP (g/L)	–	2.86	–	2.02
True-prediction fitting equation	$y = 0.998x + 0.137$	$y = 0.984x + 0.362$	$y = 0.981x + 0.265$	$y = 0.960x - 0.797$
Coefficient of determination	0.996	0.994	0.982	0.969
Offset	0.602	0.726	0.121	0.205
RPD	17.8	16.9	8.37	8.16

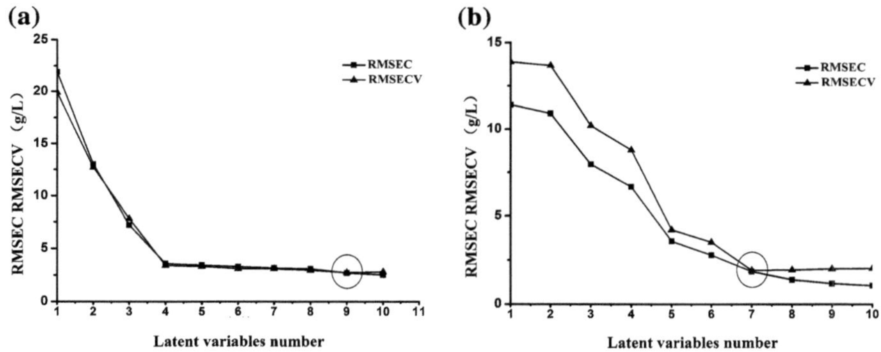

Fig. 35.2 RMSEC and RMSECV versus the numbers of LVs in the model of glutamate (**a**) and glucose (**b**) concentration

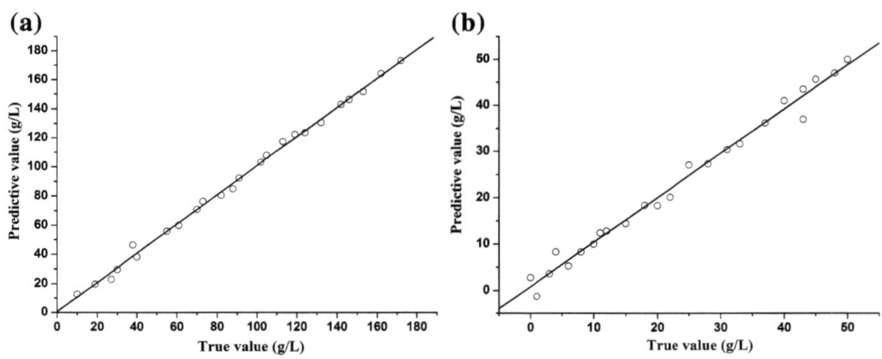

Fig. 35.3 The linear relationship for predicted values and experimental values based on glutamate (**a**) and glucose (**b**) of concentration validation set

the wavelength range and LVs was 1,000–1,800 nm and seven, respectively. The true-prediction fitting equation of calibration set and validation set was $y = 0.981x + 0.265$ and $y = 0.960x - 0.797$, respectively (Fig. 35.3). The coefficient of determination (R^2) of calibration set and validation set was 0.982 and 0.969, respectively. The RPD was 8.37 and 8.16.

All these results above showed that the models had a well predictive capacity for glutamate and glucose concentration.

35.3.3 Monitoring Glutamate and Glucose Concentration in Glutamate Fermentation

New batch glutamate fermentation was used to examine the glutamate and glucose concentration calibration models. Figure 35.4 showed that the concentration curves of glutamate and glucose, which were determined respectively by calibration model and traditional method. The predicted values and experimental values of two constituent had roughly the same variation trend. The predictive ability of calibration models was analyzed in Table 35.4. The coefficient of determination (R^2) for predictive value and true value of glutamate and glucose was 0.992 and 0.951,

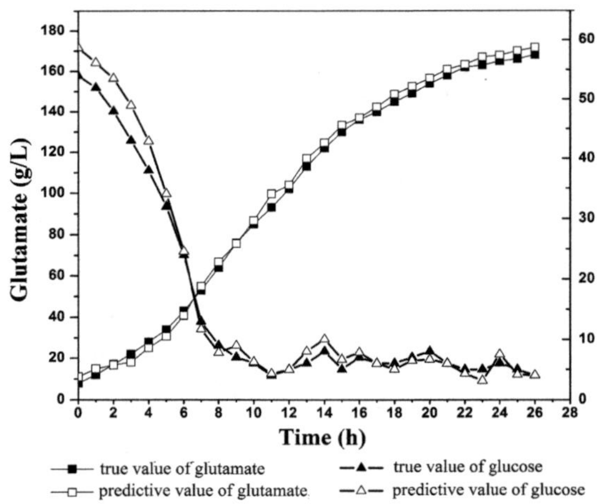

Fig. 35.4 The concentration curves of glutamate and glucose in temperature triggered glutamate fermentation

Table 35.4 The predictive ability of calibration models in temperature triggered glutamate fermentation

Mode	Predictive range (g/L)	Average relative error (%)	Coefficient of determination	RMSEP (g/L)	Variable coefficient (%)
Glutamate	9–174	5.79	0.992	3.196	5.82
Glucose	1–51	7.38	0.951	2.863	17.24

respectively. The average relative error of glutamate and glucose was 5.79 and 7.38 %, respectively. The variable coefficient of glutamate and glucose was 5.82 and 17.24 %, respectively. All the results suggested that the two models had a well predictive capacity for glutamate and glucose concentration.

35.4 Conclusion

The monitoring of the concentrations of products and substrates within a bioreactor plays an important role in bioprocess control. The results of this work demonstrate a good ability of near infrared spectroscopy and PLS to investigate glutamate and glucose concentration involved in temperature-triggered glutamate fermentation. The method was a rapid, non-destructive way for determining various compounds simultaneously with good correlations of determination and low predictive errors in fermentation industry. It provided theoretical basis for the real-time control and optimization in the temperature-triggered glutamate fermentation process.

Acknowledgments This work was supported by National High Technology Research and Development Program (2013AA102106), by Tianjin Research Program of science and technology (14JCYBJC23500), by Program for Changjiang Scholars and Innovative Research Team in University (IRT 1166) and by the Tianjin Municipal Education Commission (grant no. 20120630).

References

1. Alcalà M, Blanco M, Moyano D et al (2014) Qualitative and quantitative pharmaceutical analysis with a novel handheld miniature near-infrared spectrometer. J Near Infrared Spec 21 (6):445–457
2. Ali J, Wang WB, Zevallos M et al (2004) Near infrared spectroscopy and imaging to probe differences in water content in normal and cancer human prostate tissues. Technol Cancer Res T 3(5):491–497
3. Balabin RM, Safieva R, Lomakina EI (2007) Comparison of linear and nonlinear calibration models based on near infrared (NIR) spectroscopy data for gasoline properties prediction. Chemometr Intell Lab 88(2):183–188
4. Balabin RM, Safieva RZ, Lomakina EI (2010) Gasoline classification using near infrared (NIR) spectroscopy data: comparison of multivariate techniques. Anal Chim Acta 671(1):27–35
5. Chen Z, Lovett D, Morris J (2011) Process analytical technologies and real time process control a review of some spectroscopic issues and challenges. J Process Contr 21(10):1467–1482
6. Chirkot T, Halsey S, Swanborough A (2014) Monitoring the output of pharmaceutical hot melt extruders with near infrared spectroscopy. NIR News 25(2):15–18
7. Cozzolino D, Kwiatkowski MJ, Parker M et al (2004) Prediction of phenolic compounds in red wine fermentations by visible and near infrared spectroscopy. Anal Chim Acta 513(1):73–80
8. Ferrari M, Muthalib M, Quaresima V (2011) The use of near-infrared spectroscopy in understanding skeletal muscle physiology: recent developments. Phil Trans Roy Soc A 369 (1955):4577–4590

9. Ikeda M, Takeno S (2013) Amino acid production by *Corynebacterium glutamicum*. Springer, Berlin, pp 107–147
10. Kim J, Lim J, Lee C (2013) Quantitative real-time PCR approaches for microbial community studies in wastewater treatment systems: applications and considerations. Biotechnol Adv 31 (8):1358–1373
11. Macho S, Larrechi M (2002) Near-infrared spectroscopy and multivariate calibration for the quantitative determination of certain properties in the petrochemical industry. TrAC Trend Anal Chem 21(12):799–806
12. Negara C, Vieth KU, Lafontaine M et al (2014) Automatic fruit sorting by non-destructive determination of quality parameters using visible-near infrared to improve wine quality: II. Regression analysis. NIR News 25(1):4–6
13. Osborne BG, Fearn T (1986) Near infrared spectroscopy in food analysis. Longman, Hongkong
14. Pietrzykowski M, Chodak M (2014) Near infrared spectroscopy—a tool for chemical properties and organic matter assessment of afforested mine soils. Ecol Eng 62:115–122
15. Qu JH, Liu D, Cheng JH et al (2014) Applications of near infrared spectroscopy in food safety evaluation and control: a review of recent research advances. Crit Rev Food Sci. doi:10.1080/10408398.2013.871693
16. Shenk J, Westerhaus M (1991) Population definition, sample selection, and calibration procedures for near infrared reflectance spectroscopy. Crop Sci 31(2):469–474
17. Smyth HE, Cozzolino D, Cynkar W et al (2008) Near infrared spectroscopy as a rapid tool to measure volatile aroma compounds in Riesling wine: possibilities and limits. Anal Bioanal Chem 390(7):1911–1916
18. Williams P, Norris K (1987) Near-infrared technology in the agricultural and food industries. American Association of Cereal Chemists, Inc., St. Paul
19. Yasueda H (2014) Overproduction of l-Glutamate in *Corynebacterium glutamicum*. Microbial Production. Springer, Berlin, pp 165–176

Chapter 36
Effect of Sodium Citrate on L-tryptophan Fermentation by *Escherichia coli*

Qing-yang Xu, Li-kun Cheng, Xi-xian Xie, Cheng-lin Zhang, Yan-jun Li and Chen Ning

Abstract L-tryptophan is an essential amino acid, and the work for L-tryptophan production by microbial fermentation has important practical significance with the development of L-tryptophan market. In this study, according to the principle of metabolic engineering, the medium of L-tryptophan fermentation by the strain *Escherichia coli* TRTH was optimized. The accumulation of by-products was decreased and production of desired product was increased by adding sodium citrate to medium. The effect of sodium citrate on L-tryptophan fermentation was investigated in a 30 L fermentor. The results indicated that when L-tryptophan fermentation with the medium containing 2 g/L sodium citrate, as compared with the medium no containing sodium citrate, the biomass, L-tryptophan production and glucose conversion rate were increased by 2.57, 5.32, 4.21 %, and the accumulation of acetate, pyruvate and lactate were decreased by 5.15, 4.89, 5.23 %. With the medium containing 2 g/L sodium citrate for L-tryptophan fermentation, the biomass, L-tryptophan production and glucose conversion rate were 42.7, 35.7 and 18.2 % respectively.

Keywords *Escherichia coli* · L-tryptophan · Sodium citrate · By-products

36.1 Introduction

The L-tryptophan is an essential amino acid for humans and other animals, and is therefore widely used in food, animal feed, and medicine [1, 2]. In present, L-tryptophan is mainly produced by microbiology fermentation with *Escherichia coli* [3].

Q.-y. Xu · L.-k. Cheng · X.-x. Xie · C.-l. Zhang · Y.-j. Li · C. Ning (✉)
College of Biotechnology, Tianjin University of Science and Technology, Tianjin 300457, China
e-mail: ningch@tust.edu.cn

Q.-y. Xu · L.-k. Cheng · X.-x. Xie · C.-l. Zhang · Y.-j. Li · C. Ning
National and Local United Engineering Lab of Metabolic Control Fermentation Technology, Tianjin 300457, China

T.-C. Zhang and M. Nakajima (eds.), *Advances in Applied Biotechnology*,
Lecture Notes in Electrical Engineering 332, DOI 10.1007/978-3-662-45657-6_36

The production of L-tryptophan was increased by optimization of medium and cultivation condition [4, 5]. The study of the effect of nitrogen source on L-tryptophan fermentation indicated that the production of L-tryptophan was improved by using the medium containing 1.0 g/L yeast extract and 10.0 g/L $(NH_4)_2SO_4$ [6]. In L-tryptophan fermentation, the accumulation of acetate was decreased by controlling concentrations of initial glucose and residual glucose and specific growth rate at suitable level, leading to increasing the production of L-tryptophan [7]. Acetate is a primary inhibitory metabolite in *E. coli* cultures and is detrimental to bacterial growth and formation of desired products [8]. In L-tryptophan fermentation, the production of L-tryptophan was increased with the *pta* mutation because of decreasing the accumulation of acetate [9].

In cultivation process of *E. coli*, acetate formation is strongly affected by the composition of the culture medium [8]. Supplementing the medium with yeast extract, methionine or glycine can similarly reduce acetate formation and enhance protein production [10]. In inosine production by *Bacillus subtilis*, the EMP pathway was decreased by using the medium containing glucose and sodium citrate, which led to reducing the excretion of by-products [11]. The addition of 2.0 g/L sodium citrate in L-leucine production by *Brevibacterium flavum* could change the metabolic flux distributions of the key nodes (pyruvate, acetyl-CoA) and keep the metabolic flux balance between EMP and TCA, which finally leads to the decrease of by-products formation [12]. The optimal addition concentration of sodium citrate was 2.0 g/L in L-tryptophan fermentation.

In this study, the medium of L-tryptophan fermentation by the strain *E. coli* TRTH was optimized, 2 g/L sodium citrate was added to the production medium of L-tryptophan, and the effect of sodium citrate on biomass, production of L-tryptophan, accumulation of by-products, and consumption rate of glucose was investigated, which was to optimize the production medium of L-tryptophan.

36.2 Materials and Methods

36.2.1 Microorganism and Medium

The L-tryptophan-producing strain *E. coli* TRTH used in this study was obtained from earlier work in our laboratory and stored at the Culture Collection of Tianjin University of Science and Technology.

The seed medium contained the components as follows: 20 g/L glucose, 15 g/L yeast extract, 10 g/L $(NH_4)_2SO_4$, 0.5 g/L sodium citrate, 5 g/L $MgSO_4 \cdot 7H_2O$, 1.5 g/L KH_2PO_4, 0.015 g/L $FeSO_4 \cdot 7H_2O$, and 0.1 g/L vitamin B1. The fermentation medium for producing L-tryptophan contained the components as follows: 20 g/L glucose, 1 g/L yeast extract, 10 g/L $(NH_4)_2SO_4$, 2 g/L sodium citrate, 5 g/L $MgSO_4 \cdot 7H_2O$, 2 g/L KH_2PO_4, and 0.1 g/L $FeSO_4 \cdot 7H_2O$. The pH of both seed and fermentation media was adjusted to 7.0 with 4 mol/L NaOH.

36.2.2 Culture Conditions

A 500 mL baffled flask containing 30 mL seed medium was inoculated with a single colony of *E. coli* TRTH and cultivated at 36 °C with shaking at 200 r/min for 12 h. A 30 mL inoculum of this culture was added aseptically to a 5 L seed fermenter (Biotech-2002 Bioprocess controller, Baoxing, Shanghai, China) containing 3 L seed medium and cultivated at 36 °C for 16 h. The culture grown in the seed fermenter was inoculated aseptically into 18 L of production medium in a 30 L fermenter. The temperature and level of dissolved oxygen were maintained at 36 °C and 20 %, respectively, and the pH was adjusted to 7.0 with 25 % ammonium hydroxide (w/w) during the course of the cultivation period. When the initial glucose was decreased to 1.0 g/L, glucose solution (80 %, w/v) was added to the fermenter to maintain the concentration of residual glucose at 1.0 g/L.

36.2.3 Analysis of Fermentation Products

The biomass and concentrations of L-tryptophan in the fermentation broth were determined as described previously [2]. The concentrations of glucose and lactate were monitored using an SBA-40C biosensor analyzer (Biology Institute of Shandong Academy of Sciences, Jinan, China). Concentrations of acetate and pyruvate were measured using a Bioprofile 300A biochemical analyzer (Nova Biomedical, Waltham, MA, USA). The dynamics model of cell growth was constructed according to the study of Alam et al. [13].

36.3 Results and Discussion

36.3.1 Effect of Sodium Citrate on Biomass

In L-tryptophan fermentation, the biomass was determined every 2 h, and the growth dynamics model of *E. coli* TRTH was constructed.

The biomass and growth dynamics model of *E. coli* TRTH are presented in Fig. 36.1a, b respectively. The results indicated that the biomass was increased by adding sodium citrate to the medium. The biomass with the medium containing sodium citrate was 42.7 g/L, which was increased by 2.57 % as compared by the medium no containing sodium citrate. There was no significant difference in cell growth rate with the medium no containing or containing sodium citrate.

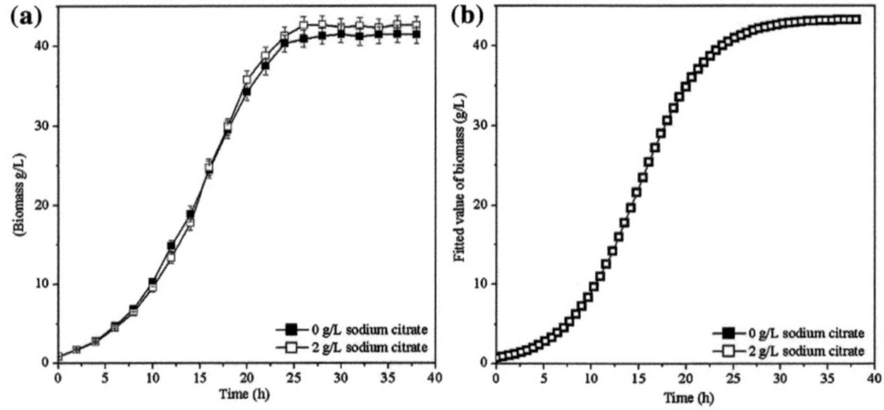

Fig. 36.1 Effect of sodium citrate on biomass

36.3.2 Effect of Sodium Citrate on Production of L-Tryptophan

The concentration of L-tryptophan is displayed in Fig. 36.2a, along with the production rate of L-tryptophan in Fig. 36.2b. The production of L-tryptophan with the medium no containing or containing sodium citrate were 33.8 and 35.7 g/L, and the results indicated that addition of sodium citrate could enhance the production of L-tryptophan. In the early fermentation period, the production rate of L-tryptophan with addition of sodium citrate was lower than that with no addition of sodium

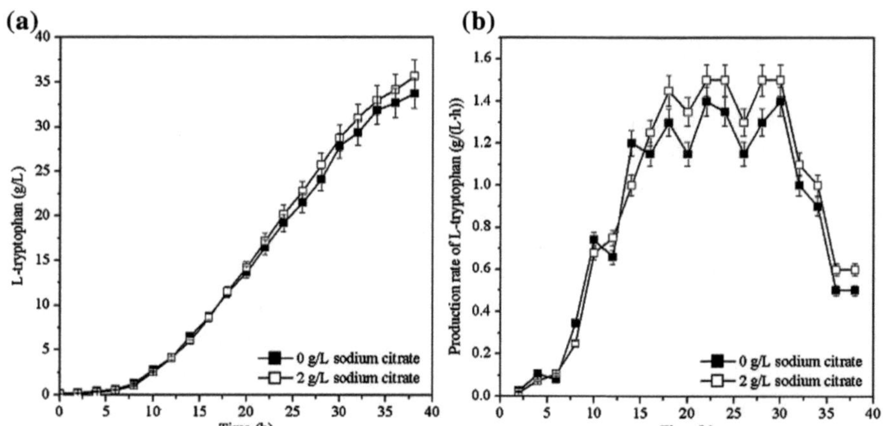

Fig. 36.2 Effect of sodium citrate on production and production rate of L-tryptophan

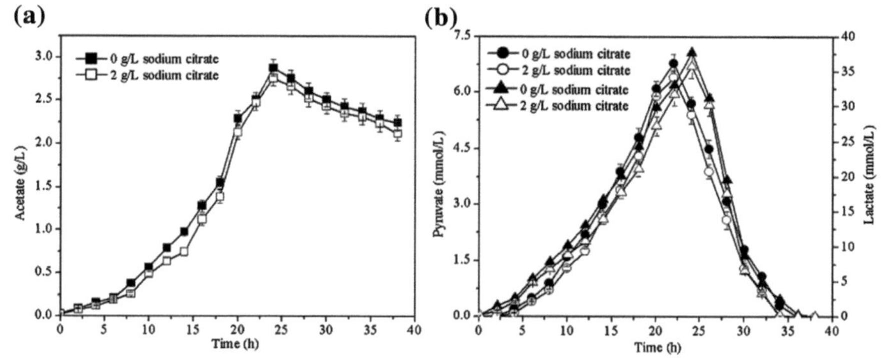

Fig. 36.3 Effect of sodium citrate on accumulation of by-products. *Note* In **b**, *solid icons* represent concentration of pyruvate, *hollow icons* represent concentration of lactate

citrate, but the production rate of L-tryptophan with addition of sodium citrate was higher than that with no addition of sodium citrate in the late fermentation period.

36.3.3 Effect of Sodium Citrate on Accumulation of By-products

The by-products including acetate, pyruvate, and lactate were excreted in L-tryptophan fermentation, and the concentration of acetate is presented in Fig. 36.3a, the concentrations of pyruvate and lactate are presented in Fig. 36.3b. The concentrations of acetate, pyruvate, and lactate increased during growth and decreased when growth ceased. The accumulation of acetate with no addition or addition of sodium citrate was 2.24 and 2.12 g/L, respectively. With the medium no containing sodium citrate, the maximum concentration of pyruvate and lactate were 6.8 and 37.8 mmol/L, which were 6.4 and 35.8 mmol/L with the medium containing sodium citrate. Pyruvate and lactate were totally consumed at the end of L-tryptophan fermentation.

36.3.4 Effect of Sodium Citrate on Consumption Rate of Glucose

Glucose was used as the sole source of carbon and energy. The consumption rate of glucose is plotted in Fig. 36.4. The consumption rate of glucose with addition of sodium citrate was lower than that with no addition of sodium citrate in the early fermentation period, and the consumption rate of glucose with addition of sodium

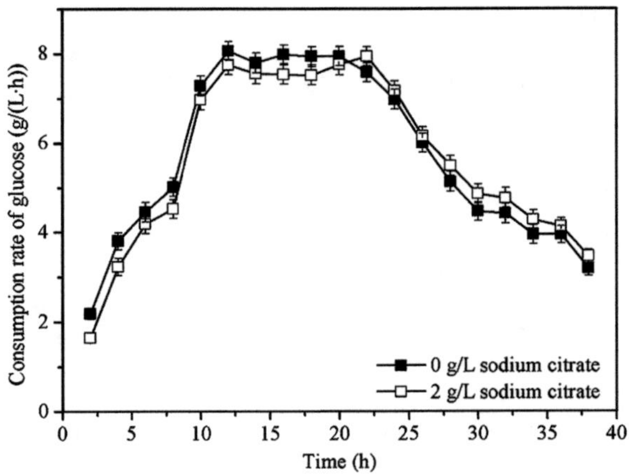

Fig. 36.4 Effect of sodium citrate on consumption rate of glucose

citrate was higher than that with no addition of sodium citrate in the late fermentation period. From the production of L-tryptophan and consumption of glucose, the glucose conversion rate with no addition or addition of sodium citrate were 17.5 and 18.2 %, and the glucose conversion rate with the medium containing sodium citrate was increased by 4.21 % as compared by the medium no containing sodium citrate.

36.4 Discussion

Because the concentration of Ca^{2+} was increased with adding sodium citrate to the glucose medium and the pyruvate kinase was inhibited by Ca^{2+}, the activity of pyruvate kinase was decrease by addition of sodium citrate [14]. The metabolic flux of EMP pathway was decreased because of the reduction activity of phosphofructokinase and pyruvate kinase [11], leading to decreasing the accumulation of by-products [15]. In addition, the accumulation of lactate was reduced with decreasing accumulation of pyruvate [16]. The intracellular levels of ATP and NADH were increased by addition of sodium citrate, and high level of ATP was provided the energy for formation of desired product [17]. But the metabolic flux of acetate was increased by the decreasing metabolic flux of TCA cycle that resulted from increasing concentration of pyruvate and the ratio of NADH and NAD [18]. The activity of citrate synthase was inhibited by addition of sodium citrate, which led to reducing the metabolic flux of acetyl-CoA to TCA cycle [19]. In L-tryptophan fermentation, the accumulations of acetate, pyruvate, and lactate were decreased. During the later fermentation period, pyruvate was reincorporated into central metabolism, and thereby decreasing the accumulation of lactate, and pyruvate and lactate were consumed by *E. coli* in the later fermentation phase [16].

At the end of L-tryptophan fermentation, pyruvate and lactate were totally consumed with presence of glucose [2, 20].

In L-tryptophan fermentation by *E. coli*, the production of L-tryptophan was increased with the reduction of acetate accumulation [7]. The metabolic flux of HMP pathway was increased because of increasing activity of glucose 6-phosphate dehydrogenase by addition of sodium citrate [11]. The reduction of metabolic flux of EMP pathway and enhancement of metabolic flux of HMP pathway was benefit for formation of L-tryptophan [9]. The oxidative branch of the HMP pathway was higher that resulted in a higher production of NADPH. As the formation of L-tryptophan requires NADPH, it might have a positive impaction improving the biosynthesis of L-tryptophan [21]. The supplication of PEP was increased by the reduction of activity of pyruvate kinase [11]. Thus, the production of L-tryptophan with addition of sodium citrate was increased by 5.32 %.

The accumulations of by-products were decreased by addition of sodium citrate, which reduced the inhibition of cell growth [8], and the biomass with addition of sodium citrate was increased by 2.57 % in L-tryptophan fermentation. Due to the reduction of accumulations of by-products and improvement of production of L-tryptophan with addition of sodium citrate, the glucose conversion rate was increased by 4.21 % as compared by that with the medium no containing sodium citrate. In L-tryptophan fermentation, the excretion of by-products was decreased and the biomass, production of L-tryptophan and glucose conversion rates were increased by the addition of sodium citrate.

Acknowledgments This work was supported by the National High Technology Research and Development Program of China (Grant No. 2012AA02A703), the Key Technologies R&D Program of Tianjin (Grant No. 14ZCZDSY00015), and the Program for Changjiang Scholars and Innovative Research Team in University (Grant No. IRT 1166).

References

1. Ikeda M, Nakagawa S (2003) The *Corynebacterium glutamicum* genome: features and impacts on biotechnological processes. Appl Microbiol Biotechnol 62:99–109
2. Wang J, Cheng LK, Wang J et al (2013) Genetic engineering of *Escherichia coli* to enhance production of L-tryptophan. Appl Microbiol Biotechnol 97:7587–7596
3. Ikeda M (2006) Towards bacterial strains overproducing L-tryptophan and other aromatics by metabolic engineering. Appl Microbiol Biotechnol 69:615–626
4. Luo W, Huang J, Zhu XC et al (2013) Enhanced production of L-tryptophan with glucose feeding and surfactant addition and related metabolic flux redistribution in the recombinant *Escherichia coli*. Food Sci Biotechnol 22(1):207–214
5. Cheng LK, Wang J, Xu QY et al (2013) Strategy for pH control and pH feedback-controlled substrate feeding for high-level production of L-tryptophan by *Escherichia coli*. World J Microbiol Biotechnol 29:883–890
6. Huang J, Shi JM, Huo WT et al (2011) The effects of nitrogen sources on the fermentation of L-tryptophan. Food Ferment Ind 37(5):21–25
7. Cheng LK, Huang J, Qing YF et al (2010) Effect of byproduct-acetic acid on L-tryptophan fermentation. Microbiol China 37(2):166–173

8. Eiteman MA, Altman E (2006) Overcoming acetate in *Escherichia coli* recombinant protein fermentations. Trends Biotechnol 24(11):530–536
9. Huang J, Shi JM, Liu Q et al (2011) Effect of gene *pta* disruption on L-tryptophan fermentation. Acta Microbiol Sinica 51(4):480–487
10. Han K, Hong J, Lim HC (1993) Relieving effects of glycine and methionine from acetic acid inhibition in *Escherichia coli* fermentations. Biotechnol Bioeng 41:316–324
11. Liu XX, Chen SX, Chu J et al (2004) Effect of sodium citrate on the growth metabolism and inosine accumulation by *Bacillus subtilis*. Acta Microbiol Sinica 44(5):627–630
12. Chen N, Liu H (2008) Effects of sodium citrate on metabolic flux distributions of L-leucine production by *Brevibacterium flavum* TK0303. J Chem Eng Chin Univ 22(3):478–483
13. Alam SI, Dube S, Reddy GSN et al (2005) Purification and characterisation of extracellular protease produced by *Clostridium sp.* from Schirmacher oasis, Antarctica. Enzyme Microb Technol 36:824–831
14. Akshay G, Jinwoon L, Michale M et al (1999) Metabolic fluxes, pool and enzyme measurements suggest a tighter coupling of energetics and biosynthetic reactions ass ociated with reduced pyruvate kinase flux. Biotechnol Bioeng 64(2):129–1341
15. Ma L, Cheng LK, Xu QY et al (2010) Effects of sodium citrate on metabolic flux distributions of L-isoleucine fermentation. J Tianjin Univ Sci Technol 25(3):14–18
16. Castaño-Cerezo S, Pastor JM, Renilla S et al (2009) An insight into the role of phosphotransacetylase (*pta*) and the acetate/acetyl-CoA node in *Escherichia coli*. Microb Cell Fact. doi:10.1186/1475-2859-8-54
17. Wang YL, Zhu J, Wei GY et al (2013) Increased co-production of S-adenosylmethionine and glutathione by sodium citrate addition. China Biotechnol 33(8):49–54
18. Vemuri GN (2006) Overflow metabolism in *Escherichia coli* during steady-state growth: transcriptional regulation and effect of the redox potential. Appl Environ Microbiol 72:3653–3661
19. Molgat GF, Donald LJ, Duckworth HW (1992) Chimeric allosteric citrate synthases: construction and properties of citrate synthases containing domains from two different enzymes. Arch Biochem Biophys 298:238–246
20. Shen T, Liu Q, Xie XX et al (2012) Improved production of tryptophan in genetically engineered *Escherichia coli* with TktA and PpsA overexpression. J Biomed Biotechnol. doi:10.1155/2012/605219
21. Báez-Viveros JL, Flores N, Juárez K et al (2007) Metabolic transcription analysis of engineered *Escherichia coli* strains that overproduce L-phenylalanine. Microb Cell Fact. doi:10.1186/1475-2859-6-30

Chapter 37
Reduction Reaction of Methyl Condensation Compound by *Saccharomyces cerevisiae*

Lu Yu, Shuhong Mao, Shaoxian Ji, Xiaoguang Liu and Fuping Lu

Abstract Microbial transformation of 13β-alkyl-3-methoxy-8, 14, secogona-1, 3, 5 (10), 9-tetraen-14, 17-dion with *Saccharomyces cerevisiae* resulted in two metabolites: product 1 (P1) and product 2 (P2). Among these, P1 was identified to be 13β alkyl-3-methoxy-8, 14-seco-1, 3, 5(10), 9(11)-estratetraene-17β-ol, 14-one, which has been studied extensively. P2 was purified, crystallized, and identified as 13β-alkyl-3-methoxy-8, 14, secogona-1, 3, 5(10), 9-tetraen-14, 17-diol by X-ray single crystal diffraction method. It was first reported in fermentation medium and the dihydroxyl reduction is a new reaction in biotransformation of steroid.

Keywords *Saccharomyces cerevisiae* · Reduction reaction · Biotransformation

37.1 Introduction

Biotransformation of steroid drugs plays an important role in commercial production [1, 2]. Compared with chemical synthesis, microbial transformation of steroid drugs has many advantages, such as the low cost, the single product and the low environmental pollution, etc.

Recently, microbial transformation of steroid drugs, first conducted by Peterson and Murray using *Rhizopus nigricans*, was employed extensively in the steroid industry [3]. And application of biocatalysis gradually replaced chemical approach

L. Yu · S. Mao · S. Ji · X. Liu · F. Lu (✉)
Key Laboratory of Industrial Fermentation Microbiology,
Ministry of Education, Tianjin University of Science and Technology,
Tianjin 300457, China
e-mail: lfp@tust.edu.cn

L. Yu · S. Mao · S. Ji · X. Liu · F. Lu
Tianjin Key Laboratory of Industrial Microbiology,
College of Biotechology, Tianjin University of Science and Technology,
Tianjin 300457, China

T.-C. Zhang and M. Nakajima (eds.), *Advances in Applied Biotechnology*,
Lecture Notes in Electrical Engineering 332, DOI 10.1007/978-3-662-45657-6_37

to complete the vital steps in the production of steroid drugs and hormones [4]. Furthermore, to obtain the commercial efficiency in the industrial production of steroid drugs, multifunctional microbial system was applied. And some of them have been developed into industrial scale. However, the successful example for steroid production is still limited [5]. Therefore, isolation and screening of the strain with the novel catalytic activity and high efficiency transformation ability are highly demanded in research of the steroid chemistry community and development of steroidal industrial fields. There is scope to improve the existing steroidal production and to find new pharmacologically active derivatives of steroids.

C-17 reduction of methyl condensates is an important intermediate for the synthesis of contraceptives. And it has been obtained from the asymmetric reduction of methyl condensates with *Saccharomyces cerevisiae* [6, 7]. In this experiment, the single hydroxyl reduction at C-17 (P1) was obtained by the incubation of methyl condensates with *S. cerevisiae*. More interestingly, dihydroxyl reduction at C-14 and C-17 of methyl condensates, which has been synthesized by chemical reduction method [8, 9], first occurred in the fermentation conversion.

37.2 Materials and Methods

37.2.1 Strains, Medium, and Transformation

The strain *S. cerevisiae*, obtained from Tianjin University of Science and Technology, Tianjin, China, was routinely maintained in a slant medium containing glucose 20 g/L, yeast extract 10 g/L, peptone 20 g/L, agar 20 g/L. After cultivation in slant medium at 30 °C for 2 days, the cell suspensions were inoculated into 50 ml sterilized fermentation medium in a 250 mL flask. The fermentation medium was composed of glucose 20 g/L, yeast extract 10 g/L, peptone 20 g/L. With continuous cultivation for 24 h at 30 °C on a rotary shaker (200 rpm), 1 g substrate was dissolved into 20 mL absolute ethanol, and 1 mL of the ethanol solution was added to each flask. Transformation of the substrate was continued under the same conditions.

37.2.2 Extraction and Purification

At the end of the conversion, the fermentation media were extracted twice with ethyl acetate. The combined organic layer was evaporated under reduced pressure. The crude product was subjected to column chromatography over silica gel, and eluted with petroleum ether/ethyl acetate (v/v = 3:1). The products were crystallized in appropriate solvents by slow evaporation.

37.2.3 Analytical Methods

500 μL of fermentation medium were extracted two times with equal volume of ethyl acetate at different incubation time. 20 μL supernatant samples were dried and diluted by 50-fold methanol of chromatographic grade and analyzed by HPLC (Agilent) with a UV detector at 240 nm and eluted with methanol/water (95:5, v/v). The column used was 4.6 mm × 250 mm × 5 μm Hypersil Symmetry C18 (Waters, Ireland). The mobile phase flow rate was 0.7 mL/min. 10 μL samples were injected with a total elution time of 10 min.

37.3 Results and Discussion

37.3.1 Identification of 13β-Alkyl-3-Methoxy-8, 14, Secogona-1, 3, 5(10), 9-Tetraen-14, 17-Diol

After the transformation, fermentation media was combined and extracted by ethyl acetate. The organic layer was collected and evaporated under reduced pressure. The crude residue was purified by silica gel column chromatography and recrystallized to give the dihydroxyl methyl carboxy compound (P2). The product was determined as 13β-alkyl-3-methoxy-8, 14, secogona-1, 3, 5(10), 9-tetraen-14, 17-diol by X-ray single crystal diffraction method. The crystal structure, which was shown in Fig. 37.1, was deposited at the Cambridge Crystallographic Data Centre (CCDC, http://www.ccdc.cam.ac.uk) and the deposition number obtained was CCDC 993502.

Crystal information: $C_{19}H_{26}O_3$, Mr = 302.40, Crystal size = 0.18 × 0.17 × 0.15 mm, Monoclinic, space group P2(1), a = 8.9498(15) Å, b = 6.8547(12) Å, c = 12.863(2) Å, α = 90°, β = 97.226(3)°, γ = 90°, V = 782.9(2) $Å^3$, Z = 2, D_{calcd} = 1.283 mg/m^3, Absorption coefficient = 0.085 mm^{-1}, Temperature 173(2)K, Reflections collected 4528, GOF = 1.05, Final R indices R_1 = 0.0365, wR_2 = 0.0884, R indices (all data) R_1 = 0.0426, wR_2 = 0.0921.

Fig. 37.1 X-ray crystal structure of P2

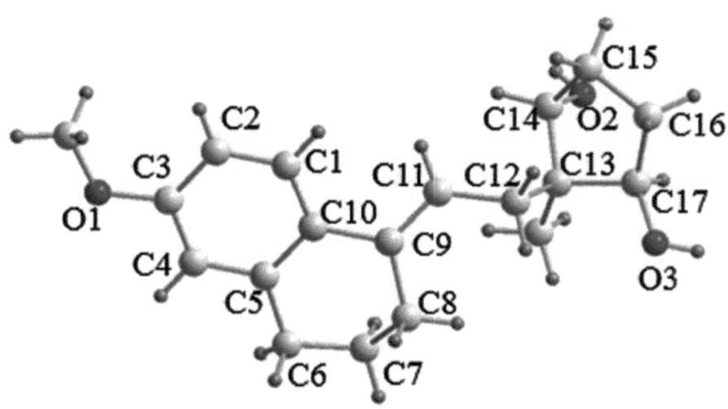

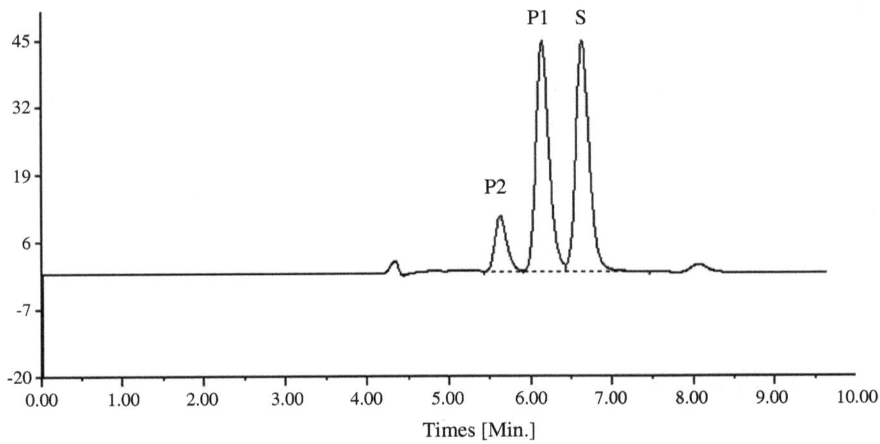

Fig. 37.2 HPLC of substrate(S), P1 and P2, whose retention times were 6.64, 6.15, and 5.64 min, respectively

37.3.2 Analyses of Conversion Process

Screening experiment displayed that *S. cerevisiae* can transform the substrate to its hydroxyl derivatives. Therefore, the transformation products at different times were subjected to HPLC (Fig. 37.2). Table 37.1 showed the yield of P1 and P2 at different incubation time. After initial 2 h transformation, yield of P1 was found to be 3.68 %. After 12 h reaction, the dihydroxyl P2 occurred. Then along with the extension of reaction time, the yield of P1 and P2 increased. And after 72 h, the substrate was transformed completely to P1 and P2, and the contents of P1 and P2 were 71.91 and 25.19 %, respectively.

To further explore the mutual transformation between P1 and P2, as substrates, P1 and P2, were transferred, respectively, into 250 mL Erlenmeyer flask containing 50 ml fermentation medium. Then the product yields of transformation, at different time, were monitored by HPLC. In case of P1, after the incubation of 72 h, the yield of P2 was 17.91 %. However, at the same conditions, no biotransformation product was detected by HPLC when the substrate P2 was incubated with *S. cerevisia*. These results indicated that P1 was first formed after the incubation of methyl

Table 37.1 Composition of crude mixtures obtained in transformation by *S. cerevisiae*

Substrate	Product	Yield (%)				
		2 h	12 h	24 h	48 h	72 h
Methyl condensation	P1	3.68	8.2	32.87	71.05	71.91
	P2	0	3.06	9.39	18.8	25.19
Product 1	P2		2.37			17.91
Product 2	–		0			0

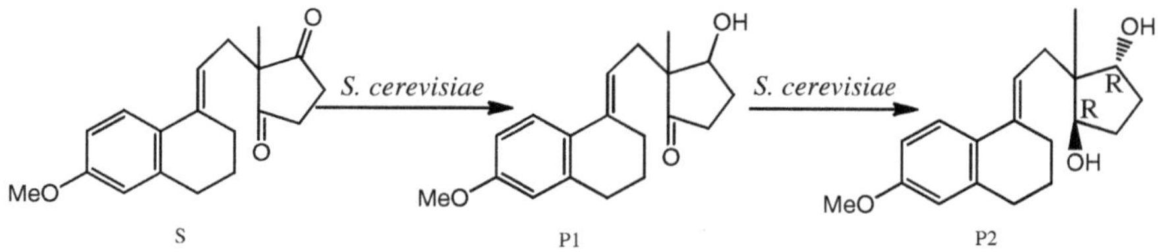

Fig. 37.3 Reduction of 13β-alkyl-3-methoxy-8, 14, secogona-1, 3, 5(10), 9-tetraen-14, 17-dion by *S. cerevisiae*

condensation by *S. cerevisiae*. Further incubation of P1 with *S. cerevisiae* led to the formation of P2. Proposed biotransformation pathway was illustrated in Fig. 37.3.

To date, few reports about the biotransformation of steroid with *S. cerevisiae were* used. Fa et al. [10] reported the 17β-ketonefrom asymmetric reduction of 13β-ethyl-3-methoxy-8, 14-secogona-1, 3, 5(10), 9-tetraen-14, 17-dion with *S. cerevisiae*. Kosmol et al. [11] studied the 14α-, 14β-, 17β-hydroxyl reduction of 13β-alkyl-3-methoxy-8, 14, secogona-1, 3, 5(10), 9-tetraen-14, 17-dioncatalyzed by *S. cerevisiae*. In this manuscript, the dihydroxyl reduction of C-14 and C-17 of steroid compound, simultaneously, was first reported. The present results suggested that there are much more work on the dihydroxyl reduction of steroid, and the corresponding results will be reported in due course.

37.4 Conclusions

Hydroxyl reduction plays a vital role in the microbial transformation of steroid. In this study, 13β-alkyl-3-methoxy-8, 14, secogona-1, 3, 5(10), 9-tetraen-14, 17-diol, was first obtained by the C-14 and C-17 dihydroxyl reduction of methyl condensation with *S. cerevisiae*. The occurrence of the biotransformation pathway provided new approach for the synthesis of the important steroid intermediate.

Acknowledgments This work was supported financially by Natural Science Foundation of China (No. 21206127) and Applied Basic Research Programs of Science and Technology Commission Foundation of Tianjin (No.12JCQNJC06400).

References

1. Chen DJ, Zhu BQ (2006) Application of microbial transformation in modern pharmaceutical industry. Chin J Antibiot 31(2):112–118
2. Wang X, Xu W, You S (2006) Application of microbialtransformation in pharmacology. J Shenyang Pharm Univ 7(23):477–482
3. Murray HC, Peterson DH (1952) Eoxygenation of steriod by muoorales fungi. US Patent 2602729

4. Manosroi J, Abe M, Manosroi A (1999) Biotransformation of steroidal drugs using microorganisms screened from various sites in Chiang Mai, Thailand. Bioresour Technol 69 (1):67–73
5. Venkataraman S, Roy RK, Chadha A (2013) Asymmetric reduction of alkyl-3-oxobutanoates by *Candida parapsilosis* ATCC 7330: insights into solvent and substrate optimisation of the biocatalytic reaction. Appl Biochem Biotechnol 171(3):756–770
6. Bhuniya R, Nanda S (2011) Asymmetric synthesis of the active form of loxoprofen and its analogue. Tetrahedron Asymmetry 22(10):1125–1132
7. Bihari V, Joshi AK, Khan AW (1985) Biotransformation of 3-methoxy-8, 14-seco-1, 3, 5(10), 9(11) estratetraen-14, 17-dione (secodione) to its 17β-hydroxy derivative (secol) using immobilized yeast cells (*Pichia farinosa* Y-118). Biotechnol Bioeng 27(9):1347–1352
8. Nara M, Terashima S, Yamada S (1980) Stereochemical studies-LVIII: synthesis of optically active compounds by the novel use of *meso*-compound-2.1 efficient synthesis of an optically pure steroid intermediate. Tetrahedron 36(22):3171–3175
9. Tomorkeny E, Makk N, Toth G et al (1976) Some 13β-alkyl-3-Methoxy-8, 14-secoestra-1, 3, 5(10), 9-gonatetraene derivatives. I. Synthesis of 14, 17-dihydroxy derivatives. Acta Chimica Academiae Scientiarum Hungaricae 91(1):81–90
10. Fa YH, Xu SW, Ma SH (1978) Studies on the microbiologucal preparation of 13β ethyl-3-methoxy-8, 14-seco-1, 3, 5(10), 9(11)-estratetraene-17β-ol, 14-one. Acta Microbiol 18 (3):232–238
11. Kosmol H, Kieslich K, Vössing R et al (1967) Totalsynthese optisch aktiver Steroide, I. Mikrobiologische stereospezifische Reduktion von 3-Methoxy-8, 14-seco-1, 3, 5(10), 9-östratetraen-14, 17-dion. Liebigs Ann Chem 701(1):198–205

Chapter 38
Study on Ultrasonic-Assisted Extraction of Essential Oil from Cinnamon Bark and Preliminary Investigation of Its Antibacterial Activity

Ping Li, Lin Tian and Tao Li

Abstract Ultrasonic-assisted extraction of essential oil from cinnamon bark was explored in this study. Effects of different parameters such as solvent, extraction time, powder granularity, ultrasonic power, solid–liquid ratio, and extraction temperature on the yield of essential oil were also investigated through single factor and orthogonal tests. The antibacterial activity of cinnamon oil obtained by ultrasonic-assisted extraction was preliminary determined. Results show that ultrasonic power has the largest effect on the oily yield, followed by powder granularity, time, and solid–liquid ratio. The optimized extraction conditions are as follows: petroleum ether (b.p. 60–90 °C) as extraction solvent, 165 W, 80–100 mesh cinnamon power, 40 min, and 6:60 (W/V) solid–liquid ratio, respectively. And under this condition, cinnamon oil yield can reach 14.8034 %, in which the content of trans-cinnamaldehyde is 82.62 % by HPLC analysis. Ultrasonic-assisted extraction of essential oil from cinnamon bark is feasible, and it can be used as a good alternative compared with other traditional methods. In addition, antibacterial activity assay demonstrates that cinnamon oil shows a strong inhibitory effect against *Escherichia coli* and *Bacillus subtilis*, and it can be applied as a natural antibacterial agent in many areas.

Keywords Cinnamon oil · Ultrasonic-assisted extraction · Single factor test · Orthogonal test · Antibacterial activity

P. Li (✉) · L. Tian · T. Li
College of Basic Science, Tianjin Agricultural University,
Tianjin 300384, People's Republic of China
e-mail: liping790520@126.com

T.-C. Zhang and M. Nakajima (eds.), *Advances in Applied Biotechnology*,
Lecture Notes in Electrical Engineering 332, DOI 10.1007/978-3-662-45657-6_38

38.1 Introduction

Essential oil is a complex mixture of volatile aromatic substances, which is extracted from flowers, leaves, stems, roots, or fruits of natural aromatic plants. Many studies have shown that essential oil has a variety of biological activities such as antibacterial, anti-inflammatory, antioxidant, and insecticide [1–4]. There has been a wide spread interest in research of cinnamon essential oil. Cinnamon oil is a clear yellow to reddish brown oily liquid with sweet smell and spicy, which is extracted from the evergreen tree lauraceae *Cinnamomum cassia* bark, branches, or leaves. There are many chemotypes in cinnamon oil, including cinnamaldehyde type, linalool type, cinnamyl acetate type, and mixed type. Among them, trans-cinnamaldehyde has been proven to be the most abundant component in cinnamon oil with a variety of bioactivities compared with other components, such as anti-bacterial, antifungal, antioxidation, anti-inflammatory, antipathogenic, hypoglycemic, prevention of colorectal cancer, repairation of damaged pancreatic cells, insecticidal activities [5–13]. Due to the above advantages, cinnamon oil at present is widely used in flavors and fragrances, food, medicine, daily necessities, agro-chemicals, cosmetics, and many other industries. Before essential oil can be analyzed and used, it should be extracted from the matrix. Traditional methods used to extract essential oils are hydrodistillation, immersion, organic solvent extraction and soxhlet, which have many drawbacks, such as low extraction efficiency, toxic solvent residues, energy consumption. Moreover, losses of some heat-sensitive compounds maybe occur because of prolonged heating or hydrolytic effects. Applying of green technologies to extract essential oil from plants can overcome these shortcomings, such as supercritical carbon dioxide fluid extraction, microwave extraction, ultrasonic-assisted extraction [14–16]. Among them, ultrasonic-assisted extraction has been widely used in extraction of natural substances, which typically uses simple equipment, less solvent and low energy, easy to operate, obtains high extraction efficiency, and good quality products.

Ultrasonic is a high-frequency mechanical wave with a range of 20–60 kHz, and because of its cavitation, mechanical, thermal and some secondary effects, such as emulsifying and proliferation, it can increase speed and frequency of the movement of material molecules, accelerate effective ingredients into the extraction solvent and then obtain higher yield of products.

In this study, we applied ultrasonic-assisted extraction of essential oil from cinnamon bark. The effect of different parameters on the yield of cinnamon oil was investigated. In addition, we also conducted preliminary studies on inhibitory effect of cinnamon oil against *Escherichia coli* and *Bacillus subtilis* in order to provide a reference for extraction and application of cinnamon oil.

38.2 Materials and Methods

38.2.1 Plant Material

Fresh plant material was purchased from market in Tianjin, China. The samples were dried at 60 °C for 48 h in a electric air blowing drying oven and then were ground in different periods in order to obtain different powder granularity.

38.2.2 Microbial Strains and Bacteria Suspension Preparation

Escherichia coli (gram negative) and *Bacillus subtilis* (gram positive) were selected as representative bacteria to determine antibacterial activity of cinnamon oil. They were provided by Department of Biology, College of Agronomy & Resources and Environment, Tianjin Agricultural University. After activated, the bacteria were inoculated into liquid standard nutrient broth medium,and cultured at 37 °C with regularity shaking at 180 rpm for 18 h. The number of bacteria was counted using plate dilution method. The final concentration of colonies was 10^6 CFU/mL. Solid nutrient broth medium was supplemented with 2 % agar into liquid medium.

38.2.3 Preparation of Essential Oil

The extraction process was performed in an ultrasonic cleaning device (P180H model, Elma, Germany). Cinnamon powders were put into a stoppered flask, added extraction solvent, and then subjected to ultrasonic-assisted extraction under settled conditions. After extraction, the solution was filtered and dried under anhydrous sodium sulfate. After drying, the extract was again filtered and the filtrate was concentrated by vacuum rotary evaporator to recover solvent. The extraction rate of cinnamon oil was calculated using the following formula:

$$X(\%) = \frac{Q_1}{Q_2} \times 100\,\% \tag{38.1}$$

where Q_1 is the quality of extracted cinnamon oil; Q_2 is the quality of cinnamon powder in test.

38.2.4 Antimicrobial Activity Assay

The initial concentration of the obtained cinnamon oil was 1.00 mg/mL, and the oil was diluted twofold with *N,N*-dimethylaniline (DMF) as solvent in order to obtain a series of different concentrations of cinnamon oil samples.

In order to investigate the antibacterial activity of cinnamon oil, disk diffusion method was carried out according to the procedure described by Vijayakumar [17]. DMF was used as negative control. Three replicates were used for each treatment. After incubation, diameters of the inhibition zone were measured. Percentage of growth inhibition (%) was calculated according to the following formula:

$$I(\%) = \frac{M_a - M_b}{M_a} \times 100\,\% \tag{38.2}$$

where M_a is a diameter of the inhibition zone treated with cinnamon oil; M_b is a diameter of the control.

38.3 Results

38.3.1 Single Factor Test for the Extraction of Cinnamon Oil

In the present study, ultrasonic-assisted method was used for the extraction of essential oil from cinnamon bark. Many influenced factors on the oily yield were also studied by single factor test. The optimized extraction conditions were also obtained through orthogonal test.

38.3.1.1 Extraction Solvent Selection

Petroleum ether, ethyl ether, ethanol (95 %), cyclohexane, dichloromethane (AR) were applied as extracting solvents to examine their effects on the oily yield. Results were shown in Fig. 38.1.

As can be seen from Fig. 38.1, petroleum ether (b.p. 60–90 °C) was the best extraction solvent with the highest extraction yield, and the cinnamon oil obtained was a clear yellow oily liquid with significant aromatic odor, so petroleum ether (b.p. 60–90 °C) was selected as solvent to extract cinnamon oil in our study.

38.3.1.2 Effect of Extraction Time on the Yield of Cinnamon Oil

Extraction processes were carried out for 10, 20, 30, 40, 50 and 60 min, respectively, to survey effect of time on the yield. Results can be seen in Fig. 38.2.

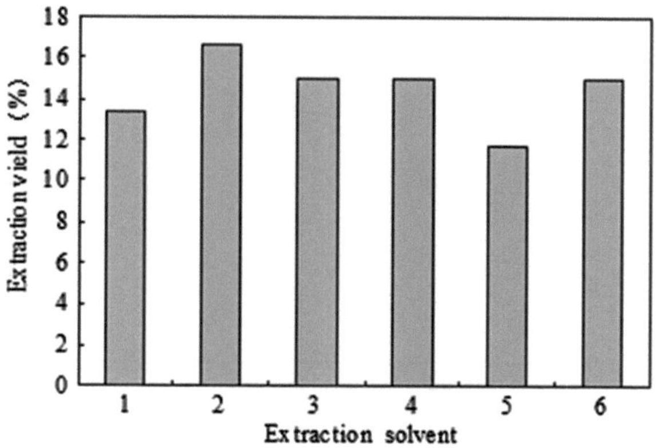

Fig. 38.1 Effect of different solvents on the extraction yield of cinnamon oil extraction conditions: 60–80 mesh powder; solid–liquid ratio was 6:60 (W/V); 330 W; 30 min; room temperature. *1* Petroleum ether (b.p. 30–60 °C); *2* Petroleum ether (b.p. 60–90 °C); *3* ethyl ether; *4* ethanol (95 %); *5* cyclohexane; *6* dichloromethane

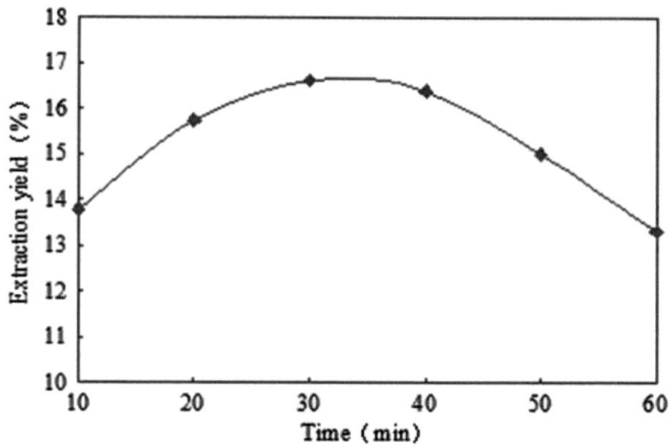

Fig. 38.2 Effect of extraction time on the extraction yield of cinnamon oil extraction conditions: 60–80 mesh powder; solid–liquid ratio was 6:60 (W/V); 330 W; room temperature

As it was expected, the yield of cinnamon oil enhanced with the increase of extraction time and reached maximum at 30 min and then declined. One possible explanation is that ultrasonic extraction is a diffusion and mass transfer process, and before the process reaching equilibrium, the extraction yield can increase. However, long extraction time maybe result in losses of volatile substances in essential oil, which cause the decline of yield.

38.3.1.3 Effect of Powder Granularity on the Yield of Cinnamon Oil

Cinnamon powders were sieved into five different sizes, i.e., 10–20, 20–40, 40–60 and 60–80, 80–100 mesh, then were subjected to ultrasonic extraction, respectively.

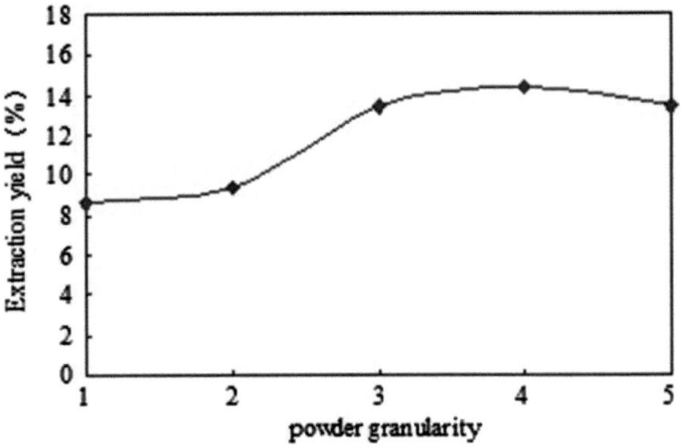

Fig. 38.3 Effect of powder granularity on the extraction yield of cinnamon oil extraction conditions: solid–liquid ratio was 6:60 (W/V); 330 W; 30 min; room temperature. *1* 10–20 mesh; *2* 20–40 mesh; *3* 40–60 mesh; *4* 60–80 mesh; *5* 80–100 mesh

As was shown in Fig. 38.3, the yield improved with the increasing of powder granularity, and under the same conditions, 60–80 mesh cinnamon powder was the best choice for maximum yield.

38.3.1.4 Effect of Ultrasonic Power on the Yield of Cinnamon Oil

In our work, eight different ultrasonic powers were employed to extract essential oil (Fig. 38.4). The results showed that maximum yield could be obtained under 198 W power, but the rate turned to decline with power increasing.

38.3.1.5 Effect of Solid–Liquid Ratio on the Yield of Cinnamon Oil

Some experiments were conducted to determine the effect of solid–liquid ratio on the yield of cinnamon oil (Fig. 38.5). As solid–liquid ratio is increasing, extraction rate raised, and reached maximum when the solid–liquid ratio was 6:60 (W/V), and then decreased. This indicated that excessive solvent could absorb a large amount of ultrasonic energy, which resulting in the decrease of temperature in extracted system and the yield reduction.

38.3.1.6 Effect of Temperature on the Yield of Cinnamon Oil

The highest rate of extraction was achieved at 40 °C (Fig. 38.6) and then reduced as the temperature is increasing. It is mainly attributed to the serious evaporation of extraction solvent and losses of some components of the essential oil under high

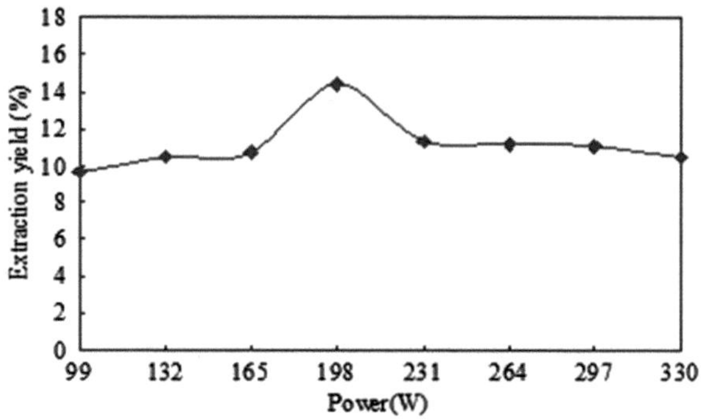

Fig. 38.4 Effect of ultrasonic power on the extraction yield of cinnamon oil extraction conditions: 60–80 mesh powder; solid–liquid ratio was 6:60 (W/V); 30 min; room temperature

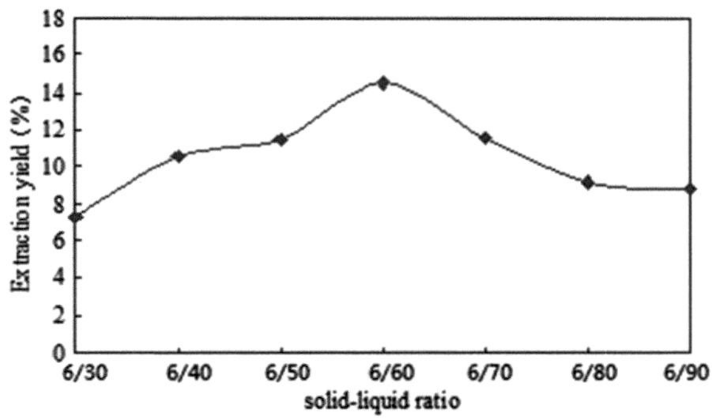

Fig. 38.5 Effect of ultrasonic power on the extraction yield of cinnamon oil extraction conditions: 60–80 mesh powder; solid–liquid ratio was 6:60 (W/V); 30 min; room temperature

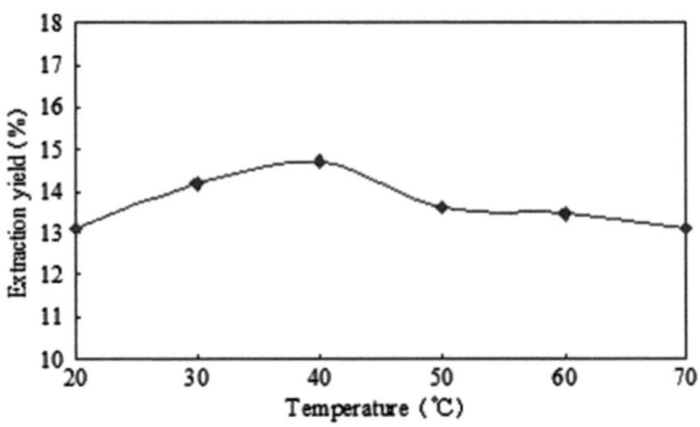

Fig. 38.6 Effect of temperature on the extraction yield of cinnamon oil extraction conditions: 60–80 mesh powder; solid–liquid ratio was 6:60 (W/V); 198 W; 30 min

temperature and high power. In addition, we also observed that the influence of temperature on the extraction yield was relatively small, so the temperature was not considered in the next orthogonal test.

38.3.2 Orthogonal Test for the Extraction of Cinnamon Oil

38.3.2.1 Orthogonal Test

Ultrasonic-assisted extraction of essential oil from cinnamon bark was optimized by using a three-level orthogonal test design (Table 38.1). The results obtained show that ultrasonic power has the largest effect on oily yield, followed by powder granularity and extraction time, yet solid–liquid ratio has the minimal impact. The best extraction conditions are as follows: 165 W, 80–100 mesh, 40 min, and 6:60 (W/V) solid–liquid ratio, respectively.

Table 38.1 Three-level orthogonal test and experimental results for ultrasonic-assisted extraction of cinnamon oil

Number	Factors				Yield(%)
	Power (W)	Solid–liquid ratio (W/V)	Powder granularity (mesh)	Time (min)	
1	165	6:50	40–60	20	11.4580
2	165	6:60	60–80	30	13.4660
3	165	6:70	80–100	40	14.3747
4	198	6:50	60–80	40	13.8900
5	198	6:60	80–100	20	13.6310
6	198	6:70	40–60	30	11.5254
7	231	6:50	80–100	30	10.7494
8	231	6:60	40–60	40	11.0217
9	231	6:70	60–80	20	9.9319
K_1	39.2987	36.0974	34.0051	35.0208	
K_2	39.0464	38.1187	37.2879	35.7409	
K_3	31.7030	35.8320	38.7551	39.2865	
k_1	13.0996	12.0325	11.3350	11.6736	
k_2	13.0155	12.7062	12.4293	11.9136	
k_3	10.5677	11.9440	12.9184	13.0955	
R	2.5319	0.7622	1.5833	1.4219	

K_1, K_2, K_3 are the sum of deviation under the three same level, respectively; R is the range value

Table 38.2 Verification test and results

Number	1	2	3
Yield (%)	14.7653	14.9051	14.7397
Average (%)		14.8034	
Relative standard deviation (%)		0.0060	

38.3.2.2 Verification Test

Three tests were conducted to verify the repeatability of optimum extraction conditions, and the results were listed in Table 38.2. Verification test demonstrated that the optimum extraction technology had good stability, and under this condition, the oily yield could reach 14.8034 %. In addition, content of trans-cinnamaldehyde in cinnamon oil was 82.62 % by HPLC analysis. All the results show that ultrasonic-assisted extraction of essential oil from cinnamon bark is feasible.

38.3.3 Antibacterial Activity of Cinnamon Oil Against E. coli and B. subtilis

Disk diffusion method was used in our study to preliminarily investigate antibacterial activity of cinnamon oil against *E. coli* and *B. subtilis*. The diameters of inhibition zone and the percentage of growth inhibition were listed in Tables 38.3 and 38.4.

The antibacterial results revealed that cinnamon oil had a strong antibacterial activity against *E. coli* and *B. subtilis*. However, with the decrease of concentration, the inhibitory effect of cinnamon oil also weakened, and this was probably ascribed to the dilution effect of essential oil, which resulted in the content of antibacterial active ingredients declining. In addition, it also could be seen that inhibitory effect of cinnamon oil on *E. coli* was better than on *B. subtilis*.

Table 38.3 Antibacterial activity of cinnamon oil against *Escherichia coli*

Concentration (mg/mL)	Diameter of the inhibition zone (mm)	Inhibition percentage (%)
1.00	25.33 ± 0.58	76.31 ± 0.53
0.50	12.67 ± 029	52.62 ± 1.07
0.25	9.00 ± 0.20	33.31 ± 1.48
0.17	8.17 ± 0.29	25.42 ± 0.72
control	6.0 ± 0	0

Table 38.4 Antibacterial activity of cinnamon oil against *Bacillus subtilis*

Concentration (mg/mL)	Diameter of the inhibition zone (mm)	Inhibition percentage (%)
1	17.67 ± 0.29	66.03 ± 0.55
0.50	11.67 ± 0.58	48.48 ± 2.62
0.25	8.03 ± 0.15	25.29 ± 1.41
0.17	7.0 ± 0	14.3 ± 0
Control	6.0 ± 0	0

38.4 Discussion

38.4.1 Ultrasonic-Assisted Extraction Conditions of Cinnamon Oil

It can be observed from the orthogonal test, ultrasonic power has the most influence on oily yield. However, over-high ultrasonic power leads to a reduction of the yield, and one possible explanation is that with the power increasing, temperature of the extraction system also enhances, which results in some losses of volatile components of essential oil and the decline of the yield. In addition, powder granularity also has a great impact on the yield, which indicates that fine materials is benefit for the extraction, yet too small powders might cause adhesion between materials, then hinder the dissolution of essential oil, and our result is in agreement with that of previous observation [18]. Besides, long extraction time and over large solid–liquid ratio are not conducive to the ultrasonic-assist extraction of cinnamon oil.

38.4.2 Antibacterial Activity of Cinnamon Oil

According to the result of disk diffusion assay, cinnamon oil shows a strong inhibitory effect against *E. coli* and *B. subtilis*. This may be due to the present of its main constituent in cinnamon oil, such as trans-cinnamaldehyde [19]. In our study, the content of trans-cinnamaldehyde was 82.62 % by HPLC analysis. However, as essential oil is a very complex mixture, it is not sure that trans-cinnamaldehyde is the only component reflecting antibacterial activity in cinnamon oil. In addition, many researchers have also proved that eugenol and linalool are also components in cinnamon oil with high antibacterial activities [20, 21]. Furthermore, some components in cinnamon oil may also have a synergistic antibacterial effect against microorganisms [22].

38.4.3 Inhibitory Mechanism

In recent years, many researchers have been extensively studied on the inhibitory mechanism of essential oils. Nanasombat and Wimuttigosol revealed that the hydrophobicity of essential oils may enable them to partition in the lipids of bacterial cell membrane and render them more permeable, leading to leakage of ions and cell contents, and extensive loss of critical molecules and ions will cause death of cells [23]. In addition, trans-cinnamaldehyde was also detected in dead cells by transmission electron microscopy (TEM) in Becerril's study [24]. However, since cinnamon oil is a complicated mixture, the use of more sophisticated equipments and means are further required to investigate whether other components even with small content in cinnamon oil also having inhibitory effects on microorganisms.

38.5 Conclusion

In conclusion, our research shows that the most important factor affecting ultrasonic-assisted extraction essential oil from cinnamon bark is ultrasonic power, followed by powder granularity, time, and solid–liquid ratio. The best extraction conditions are as follows: 165 W, 80–100 mesh, 40 min, and 6:60 (W/V) solid–liquid ratio, respectively. Under this condition, the yield of cinnamon oil extracted from cinnamon bark is 14.8034 % and the content of trans-cinnamaldehyde is 82.62 %. Ultrasonic-assisted extraction can provide shorter extraction time and save a lot of energy, which make it a good alternative for conventional extraction methods. In addition, antibacterial experiment demonstrates that cinnamon oil obtained by ultrasonic-assisted extraction has a good inhibitory effect against *E. coli* and *B. subtilis* and it can be applied as a natural antibacterial agent in many areas, such as food, flavors and fragrances, and pharmaceutical sectors.

Acknowledgments The research was supported by Tianjin Agricultural University Scientific Research and Development Fund program, project number: 2012N16.

References

1. Rasooli I (2003) Bacterial susceptibility to and chemical composition of essential oils from Thymus kotschyanus and Thymus persicus. J Agric Food Chem 51:2200–2205
2. Chao LK, Hua K-F, Hsu H-Y et al (2005) Study on the antiinflammatory activity of essential oil from leaves of Cinnamomum osmophloeum. J Agric Food Chem 53:7274–7278
3. Jirovetz L, Buchbauer G, Stoilova I et al (2006) Chemical composition and antioxidant properties of Clove leaf essential oil. J Agric Food Chem 54:6303–6307
4. Ntalli NG, Ferrari F, Giannakou I et al (2010) Phytochemistry and nematicidal activity of the essential oils from 8 Greek Lamiaceae aromatic plants and 13 terpene components. J Agric Food Chem 58:7856–7863

5. Das M, Rath CC (2012) Bacteriology of a most popular street food (Panipuri) and inhibitory effect of essential oils on bacterial growth. J Food Sci Technol 49:564–571
6. Cheng S-S, Lin J-Y, Hsui Y-R et al (2006) Chemical polymorphism and antifungal activity of essential oils from leaves of different provenances of indigenous cinnamon (Cinnamomum osmophloeum). Bioresour Technol 97:306–312
7. Wang CY, Wang SY, Chen C (2008) Increasing antioxidant activity and reducing decay of Blueberries by essential oils. J Agric Food Chem 56:3587–3592
8. Tung Y-T, Huang C-C, Ho S-T et al (2011) Bioactive phytochemicals of leaf essential oils of Cinnamomum osmophloeum prevent Lipopolysaccharide/D-galactosamine (LPS/D-GALN)-induced acute hepatitis in mice. J Agric Food Chem 59:8117–8123
9. Ali SM, Khan AA, Ahmed I et al (2005) Antimicrobial activities of eugenol and cinnamaldehyde against the human gastric pathogen Helicobacter pylori. Ann Clin Microbiol Antimicrob 4:1476–1482
10. Broadhurst CL, Polansky MM, Anderson RA (2000) Insulin-like biological activity of culinary and medicinal plant aqueous extracts in vitro. J Agric Food Chem 48:849–852
11. Wondrak GT, Villeneuve NF, Lamore SD et al (2010) The cinnamon-derived dietary factor cinnamic aldehyde activates the Nrf2-dependent antioxidant response in human epithelial colon cells. Molecules 15:3338–3355
12. Yuan HD, Huang B, Chung SH (2011) Protective effect of cinnamaldehyde on Streptozotocin-induced damage in rat pancreatic β-cells. Food Sci Biotechnol 20:1271–1276
13. Shen F, Xing M, Liu L et al (2012) Efficacy of trans-cinnamaldehyde against Psoroptes cuniculi in vitro. Parasitol Res 110:1321–1326
14. Guan W, Li S, Yan R et al (2007) Comparison of essential oils of clove buds extracted with supercritical carbon dioxide and other three traditional extraction methods. Food Chem 101:1558–1564
15. Lucchesi ME, Chemat F, Smadja J (2004) Solvent-free microwave extraction of essential oil from aromatic herbs: comparison with conventional hydro-distillation. J Chromatogr A 1043:323–327
16. Chao L, Yang RF, Fu XF (2008) Ultrasonic-assisted supercritical CO_2 extraction of volatile oil from compound recipe of clove and cinnamon. J S China Univ Technol 36:67–71
17. Vijayakumar A, Duraiandiyan V, Jeyaraj B et al (2012) Phytochemical analysis and in vitro antimicrobial activity of Illicium griffithii Hook. f. and Thoms extracts. Asian Pac J Trop Dis 2:190–199
18. Kuang X, Bin L, Rui K et al (2011) Granularity and antibacterial activities of ultra-fine cinnamon and clove powders. J Food Saf 31:291–296
19. Mall J-L, Chen C-P, Hsieh P-C (2001) Antimicrobial effect of extracts from Chinese chive, cinnamon, and corni fructus. J Agric Food Chem 49:183–188
20. Devi KP, Sakthivel R, Nisha SA (2013) Eugenol alters the integrity of cell membrane and acts against the nosocomial pathogen proteus mirabilis. Arch Pharm Res 36:282–292
21. Cheng B-H, Lin C-Y, Yeh T-F et al (2012) Potential source of S-(+)-Linalool from Cinnamomum osmophloeum ct. Linalool leaf: essential oil profile and enantiomeric purity. J Agric Food Chem 60:7623–7628
22. Shan B, Cai Y-Z, Brooks JD et al (2007) Antibacterial properties and major bioactive components of cinnamon stick (Cinnamomum burmannii): activity against foodborne pathogenic bacteria. J Agric Food Chem 55:5484–5490
23. Nanasombat S, Wimuttigosol P (2011) Antimicrobial and antioxidant activity of spice essential oils. Food Sci. Biotechnol 20:45–53
24. Becerril R, Gomez-Lus R, Goni P et al (2007) Combination of analytical and microbiological techniques to study the antimicrobial activity of a new active food E. coli and S. aureus. Anal Bioanal Chem 388:1003–1011

Chapter 39
Geranyl Butyrate Production by *Candida antarctica* Lipase B-Displaying *Pichia pastoris*

Zi Jin, Janvier Ntwali, Ying Lin, Huang Kui, Suiping Zheng and Shuangyan Han

Abstract Enzyme-displaying whole-cell yeast is becoming an alternative to immobilized enzyme. Here, a recombinant *Pichia pastoris* displaying Candida antarctica lipase B (CALB) was used to catalyze the synthesis of geranyl butyrate with geraniol and butanoic acid as substrates in n-heptane. Response surface methodology (RSM) with a four-variable five-level central composite design (CCD) was applied when biocatalyst concentration, geraniol concentration, butanoic acid/geraniol molar ratio, and water activity were taken as influential factors to improve the production of geranyl butyrate. The results indicated that both substrates molar ratio and water activity affected significantly the geranyl butyrate yield ($p < 0.05$). Biocatalyst concentration 20.9 g/L, geraniol concentration 0.57 mol/L, butanoic acid/geraniol molar ratio 1.21:1, and water activity 0.53 were the optimum of the ester synthesis. Under the above conditions, the geraniol conversion rate was 98.5 % in 10 mL reaction after 9 h reaction, which was equivalent to 126 g/L geranyl butyrate. When the esterification reaction was scaled up in batch stirred reactors, 97.87 % of geraniol conversion rate in 200 mL as well as 97.56 % in 500 mL could be obtained.

Keywords Geranyl butyrate · Yeast surface display · Candida antarctica lipase B · Response surface methodology · Batch stirred reactor

39.1 Introduction

Geranyl butyrate is one of the terpenoids. It has strong and pleasant odor which is used extensively as flavor ingredients in the food, cosmetics, and pharmaceutical industries [1]. In industrial scale, most geranyl esters are obtained by chemical

Z. Jin · J. Ntwali · Y. Lin · H. Kui · S. Zheng · S. Han (✉)
School of Bioscience & Bioengineering, South China University of Technology, Guangzhou, People's Republic of China
e-mail: syhan@scut.edu.cn

T.-C. Zhang and M. Nakajima (eds.), *Advances in Applied Biotechnology*, Lecture Notes in Electrical Engineering 332, DOI 10.1007/978-3-662-45657-6_39

synthesis, while the low yield, poor product quality as well as the additional separation and purification steps limiting a wide application [2]. Meanwhile, pushed by the market pressure which prefers "natural" rather than "artificial" or "synthetic," more and more efforts have been put in aromatic esters production using natural procedures [1, 3]. Enzymatic reaction delivers "natural product" in terms of its origin, and thus, it became valuable to produce flavor esters, especially in the food industry [1, 4].

As we know, immobilized enzyme offers the economic incentives of enhanced thermal and chemical stability, ease of handling, easy recovery, and reuse relative to non-immobilized forms [5]. Yeast surface display technique made it effective and convenient to construct a whole-cell biocatalyst by conventional genetic engineering as well as no extra work is required to either purify or immobilize the enzymes. Further, enzyme-displaying yeast whole-cell has been proved to own the good regioselectivity and stability as well as simple recycle course for ester synthesis [6–8]. In this case, the yeast cell not only serves as host for enzyme production, but also acts as immobilization medium for the enzyme, while about 104 enzyme molecules can be immobilized on the cell surface [9].

Recently, a CALB (Candida antarctica lipase B)-displaying *Pichia pastoris* was developed in our laboratory with a new endogenous CFEM protein Gcw12 from *Komagataella pastoris* GS115 as anchor, which functioned as cell surface receptors or signal transducers, or as adhesion molecules in host pathogen interactions (NCBI accession: XP_002490642) [10, 11]. Paul et al. who performed lipase catalyzed synthesis of geranyl and citronellyl esters using different commercial lipases, indicated that CALB are very efficient biocatalysts for the direct esterification reactions [12]. Although lipases catalyzed geranyl ester has been reported by different authors, low substrate concentration, small reaction scale, unsatisfying yield, and high cost enzyme draw the necessity to extend a new approach to this field of biocatalysis [1–3, 13]. Moreover, none of the work published so far has explored the possibility of whole-cell or enzyme-displaying yeast as biocatalyst to produce terpenoid esters.

In this study, the esterification of geranyl butyrate was carried out by CALB-displaying *P. pastoris* whole-cell biocatalyst in n-heptane. Response surface methodology (RSM) with a four-variable five-level central composite design (CCD) was adopted in this study to get a better understanding of the relationships between the important reaction parameters (biocatalyst concentration, geraniol concentration, butanoic acid/geraniol molar ratio, and initial water activity) and the yield of geranyl butyrate. Furthermore, scale-up batch stirred reaction systems (200 and 500 mL) were explored based on the obtained model.

39.2 Materials and Methods

39.2.1 Materials

Geranyl butyrate (used as standard) and heptane (HPLC grade) were purchased from Sigma-Aldrich Co. (St. Louis, MO, USA). Geraniol (97 % purity) was purchased from Aladdin Reagents Co. (Shanghai, China), and all other chemicals were also obtained commercially and of analytical grade. The recombinant strain of CALB-displaying *P. pastoris* GS115/CALB-GCW12 was constructed previously in our laboratory [11]. Yeast extract peptone dextrose medium (YPD), buffered glycerol-complex medium (BMGY), and buffered methanol-complex medium (BMMY) were used to culture *P. pastoris* [11].

39.2.2 Preparation of CALB-Displaying P. pastoris Whole-Cell Biocatalyst

A monocolony of CALB-displaying *P. pastoris* was inoculated into 10 mL of YPD lipid medium and cultured for 1 day at 30 °C, and the seed culture above (10 %) was transferred into 100 mL of BMGY medium and cultured for 1 day at 30 °C. Then the fermentation broth was centrifuged at 7,000 rpm for 10 min, and the sediment was resuspended in 100 mL of BMMY medium and cultured for 5 days at 30 °C. After that, these cells were harvested and washed for several times. The fresh CALB-displaying *P. pastoris* were resuspended with 50 mM Tris-HCl buffer (pH 8.0) in which trehalose as a freeze-drying protectant was added, followed by lyophilization with Christ Alpha 2-4 Freeze Dryer (Christ, Osterode, Germany) for 24 h. To obtain defined values of water activity, the enzyme and the reactants were equilibrated with saturated salt solutions at 25 °C for 48 h in fully sealed containers. The salts used were as follows: LiCl (aw = 0.11), $MgCl_2$ (aw = 0.33), $Mg(NO_3)_2$ (aw = 0.53), NaCl (aw = 0.75), and K_2SO_4 (aw = 0.97) [3].

39.2.3 Esterification Synthesis

The esterification reactions were carried out in screw-capped flasks containing a required amount of heptanes and mixture of butanoic acid and whole-cell biocatalyst powder which has been preheated at 55 °C for 10 min. Geraniol was then added and the reaction was incubated at 55 °C and shaken with 200 rpm for 9 h. During the reaction period, samples were taken for subsequent analysis. Biocatalyst content with specific aw, geraniol concentration and butanoic acid/geraniol molar ratio varied according to the experimental design (Table 39.1).

Table 39.1 Coded and actual values of variables for the central composite design

Variables	Levels				
	−2	−1	0	+1	+2
Biocatalyst concentration in g/l (*A*)	10	15	20	25	30
Geraniol concentration in mol/l (*B*)	0.2	0.3	0.4	0.5	0.6
Substrates molar ratio butanoic/geraniol (*C*)	3:1	2:1	1:1	1:2	1:3
Initial enzyme water activity (*D*)	0.11	0.33	0.53	0.75	0.97

39.2.4 Experiment Design

A four-variable five-level CCD was used for ester synthesis optimization. The coded and non-coded values of four independent variables including biocatalyst concentration (*A*), geraniol concentration (*B*), butanoic acid/geraniol molar ratio (*C*), and water activity (*D*) were described in Table 39.1. The response value was geraniol conversion rate and results from the design were analyzed in order to be fitted in the following second-order polynomial model: one after, as below.

$$Y = \beta_0 + \sum_{j=1}^{k} \beta_j X_j + \sum_{j=1}^{k} \beta_{jj} X_j^2 + \sum\sum_{i<j} \beta_{ij} X \qquad (39.1)$$

where Y is the predicted response, β_0, β_j, β_{jj}, and β_{ij} are the regression coefficients for intercept, linearity, square, and interaction terms, respectively; X_i and X_j represent the coded independent variables. Contour plots were provided by holding two variables constant and varying the two other variables. Optimum was estimated and the model was verified.

39.2.5 Scale-Up Reaction

The scale-up reaction was carried out in stirred reactor shown in Fig. 39.1. Reagents were poured in the three neck flasks and placed in a water bath with the temperature of 55 °C. The stirrer was set at 300 rpm, and a condenser filled with molecular sieves was set up at one of the flask's neck. As the reaction was going on the azeotropic mixture of water and organic solvent would evaporate through the condenser, the water in the mixture captured by the molecular sieves and the remaining organic compounds would re-condense and return to the reaction mixture.

Reaction parameters were the one optimized by RSM but on a large scale. The necessary amount of n-heptane was added to a 500-mL spherical bottom flask to make a reaction mixture of 200 mL, and substrates and biocatalyst were then added

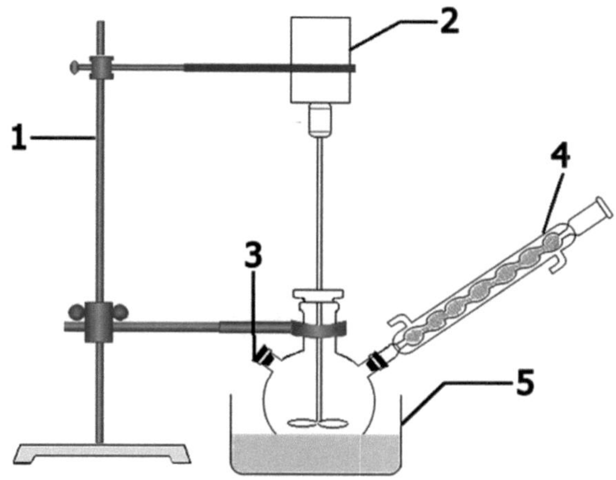

Fig. 39.1 Stirred batch reactor used for geranyl butyrate scale-up production. (*1*) Stand, (*2*) stirrer, (*3*) inlet, (*4*) condenser with molecular sieves, and (*5*) water bath

in concentrations as calculated by optimization. The mixture was left to react for 9 h. Then the 500-mL reaction system was set up by further scale-up appropriately.

39.2.6 Determination of Conversion Rate

Conversion rate was determined by gas chromatography. 20 μL of samples were dissolved in 980 μL of n-hexane and analyzed by gas chromatography using Agilent 7890A gas chromatograph equipped with flame ionization detector (FID) and a 0.25 μm × 0.320 mm × 30 m HP-INNOWAX capillary column. Nitrogen was the gas carrier at flow rate of 30 mL/min. The initial column oven temperature was kept at 100 °C for 0.5 min then upgraded until 150 °C at a rate of 100 °C/min and increased again to 158 °C at 2 °C/min. The injector temperature was kept at 270 °C and detector at 260 °C. Geranyl butyrate was used as internal standard.

39.3 Results and Discussion

39.3.1 Model Fitting

Preliminary experiments showed that esterification of geraniol with butanoic acid catalyzed by CALB-displaying whole-cell biocatalyst reached the maximal value when reaction for 9 h, so the time of reaction was set to be 9 h. In this work, 30 experiments were performed in order to estimate the coefficients of the model and the results are presented in Table 39.2. All reactions were conducted in triplicates.

Table 39.2 Central composite design and experimental data for 4-factor, 5-level response surface analysis

Run	*A*	*B*	*C*	*D*	Response
	Biocatalyst concentration	Geraniol concentration	Substrates molar ratio	Water activity	Geraniol conversion rate (%)
1	−1	−1	−1	−1	96.93
2	1	−1	−1	−1	97.40
3	−1	1	−1	−1	92.46
4	1	1	−1	−1	97.52
5	−1	−1	1	−1	43.65
6	1	−1	1	−1	46.02
7	−1	1	1	−1	44.73
8	1	1	1	−1	44.16
9	−1	−1	−1	1	92.19
10	1	−1	−1	1	96.19
11	−1	1	−1	1	86.24
12	1	1	−1	1	95.98
13	−1	−1	1	1	48.60
14	1	−1	1	1	49.00
15	−1	1	1	1	45.21
16	1	1	1	1	49.88
17	−2	0	0	0	85.32
18	2	0	0	0	93.97
19	0	−2	0	0	89.18
20	0	2	0	0	93.04
21	0	0	−2	0	60.02
22	0	0	2	0	33.33
23	0	0	0	−2	35.10
24	0	0	0	2	54.84
25	0	0	0	0	94.52
26	0	0	0	0	93.20
27	0	0	0	0	87.96
28	0	0	0	0	94.43
29	0	0	0	0	95.32
30	0	0	0	0	94.37

To determine whether the quadratic model was significant, an analysis of variance (ANOVA) was presented in Table 39.3. The *P*-values were used as tools to check the significance of each coefficient and showed the interaction strength of each parameter. When the *P*-value of the model was smaller than 0.0001, indicating that a probability value for the multiple regression was very small, the model was obviously suitable for use in this experiment. And the "lack of fit *F*-value" of 3.65 implied the "lack of fit" was not significant relative to the pure error. Using

Table 39.3 Analysis of variance for factors and their interaction on geraniol conversion

Source	Sum of square	df	Mean Square	F-Value	P-value[a]Prob.> F	
Model	16370.04	14	1169.29	57.94	< 0.0001	*significant*
A-biocatalyst concentration	2.89	1	2.89	0.14	0.7103	
B-Geraniol concentration	2.83	1	2.83	0.14	0.7134	
C-Substrates molar ratio	10369.54	1	10369.54	513.79	< 0.0001	
D-Water activity	138.21	1	138.21	6.85	0.0194	
AB	8.50	1	8.50	0.42	0.5262	
AC	7.92	1	7.92	0.39	0.5406	
AD	8.24	1	8.24	0.41	0.5326	
BC	6.61	1	6.61	0.33	0.5757	
BD	0.78	1	0.78	0.04	0.8465	
CD	80.59	1	80.59	3.99	0.0642	
A^2	12.69	1	12.69	0.63	0.4401	
B^2	2.69	1	2.69	0.13	0.7204	
C^2	7586.20	1	7586.20	375.88	< 0.0001	
D^2	3871.94	1	3871.94	191.85	< 0.0001	
Residual	302.74	15	20.18			
Lack of Fit	266.22	10	26.62	3.65	0.0831	*not significant*
Pure Error	36.52	5	7.30			
Cor Total	16672.78	29				

[a] Significant when *P*-value <0.5. R^2 = 0.9137, Adj. R^2 = 0.9649

designed experimental data from Table 39.2, polynomial equation was regressed and was given below: one after, as below.

$$y = 7.83 + 0.74A + 0.73B + 118.10C + 5.09D + 0.73AB + 0.92AC + 0.72AD - 0.84BC - 0.22BD - 2.94CD - 0.68A^2 - 0.31B^2 - 33.91C^2 - 11.85D^2 \quad (39.2)$$

where *A*, *B*, *C*, and *D* represented biocatalyst concentration, geraniol concentration, substrates molar ratio, and water activity, respectively. Among these parameters, it could be concluded that the linear effect of *C* and the quadric effect of C^2, D^2 were the primary determining factors of the responses on geraniol conversion rate as they had the largest coefficient. The linear effect of *D* and the interaction effect of *C*, *D* were the secondary determining factors.

39.3.2 Mutual Effect of Parameters

For better understanding of the influence of each factor and of their interaction on response, contour plots and three-dimensional graphs were usually used [14]. The contour plots that gave the variation of geraniol conversion rate with independent variables were shown in Fig. 39.2. Each contour plot represented an infinite number of combinations of two test variables with the other two maintained at their zero-level. The negligible interaction between the corresponding variables made a circular contour of response surfaces, while the significant interaction made an elliptical or saddle nature of the contour plots [14, 15].

Figure 39.2a, d, e shows the interaction between geraniol concentration and other three parameters (biocatalyst concentration, substrates molar ratio, and water activity).The mutual interactions between geraniol concentration and all the other three parameters were very poor. With the increase of geraniol concentration, the response geraniol conversion rate almost unchanged, which represented that geraniol had no inhibitory effect on the enzyme esterification activity, proving the high efficiency of the CALB-displaying yeast. This phenomenon had been reported by Trusek-Holownia et al. who worked with CALB on geranyl esters synthesis reaction kinetics [16], and some articles about other lipase [17, 18].

The interaction between the biocatalyst concentration and substrates molar ratio (Fig. 39.2b) as well as the interaction between the biocatalyst concentration and water activity (Fig. 39.2c) followed nearly the same pattern with geraniol concentration, which could also be predicted form a large *P*-value of biocatalyst concentration (Table 39.3). The geraniol conversion rate increased very slightly with increasing of biocatalyst concentration. Higher biocatalyst concentration usually resulted in more available active sites, but in the presence of high concentrations of enzyme-displaying yeasts, the cells aggregated, thus might bring about insufficient active sites as well as limit the diffusion of substrate [6, 19]. A previous study had shown that enzyme concentration had a little but positive effect on conversion rate [15].

A contour plot of geraniol conversion rate as a function of the water activity and substrates molar ratio was presented in Fig. 39.2f. These two parameters played a vital role in the geranyl butyrate synthesis. In this contour plot, the geraniol conversion rate initially increased until the critical value 2:1 then higher concentrations of butyric acid corresponded with in decreased conversion rate. Similar behavior was also observed for the esterification of geranyl propionate catalyzed by Novozym 435 from *Candida. antarctica* [4], as well as the esterification of citronellyl and geraniol esters of propionic, butyric, and caproic acids catalyzed by esterase

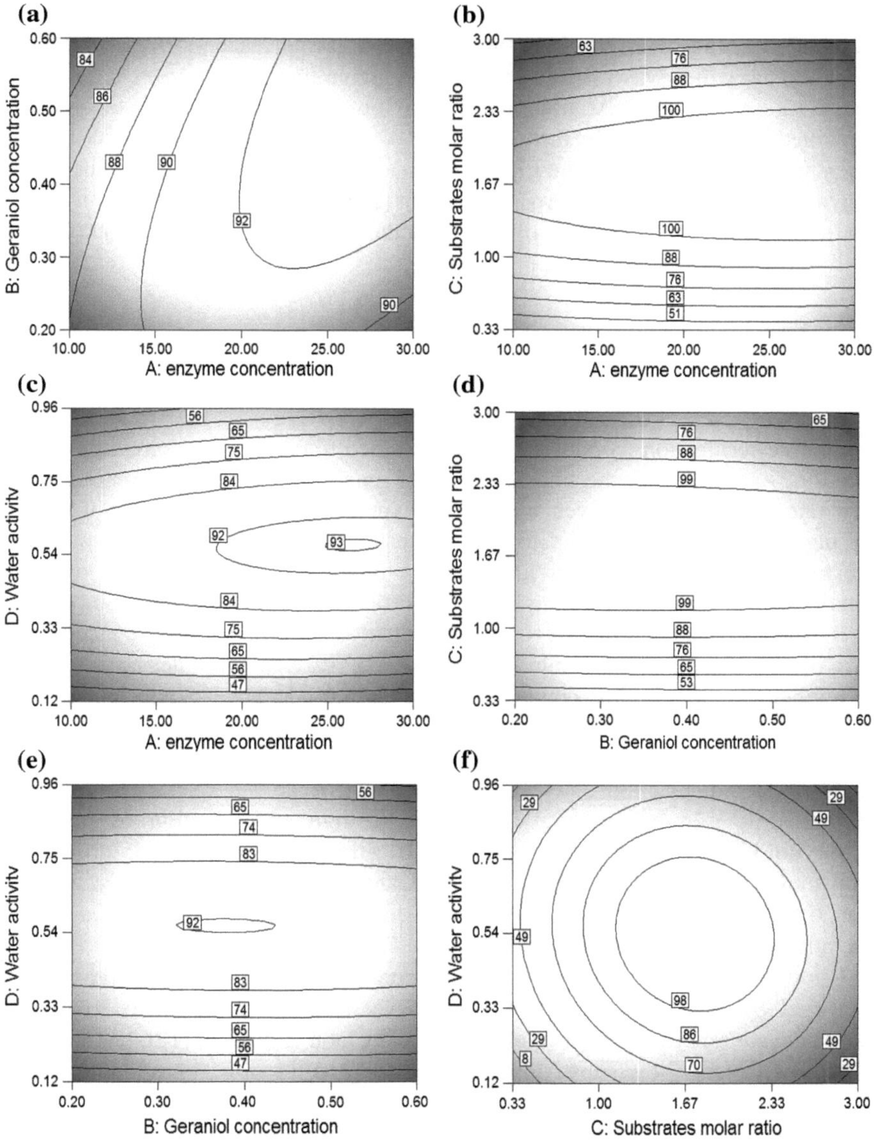

Fig. 39.2 Response surface plot showing the effect of parameters and their mutual interaction on geraniol conversion rate. **a** The interaction between biocatalyst concentration and geraniol concentration. **b** The interaction between biocatalyst concentration and substrates molar ratio. **c** The interaction between biocatalyst concentration and water activity. **d** The interaction between geraniol concentration and substrates molar ratio. **e** The interaction between geraniol concentration and water activity. **f** The interaction between substrates molar ratio and water activity

30,000 from *M. miehei* [17]. On the other hand, when the other three parameters maintained at their zero-level, geraniol conversion rate reached 95 % with water activity 0.54, which was 1.7-fold and 2.7-fold higher than that of water activity 0.96 and 0.12, respectively. At high water activities, low conversion rate could be explained as a result of the increasing rate of the hydrolysis reaction [20], while when too low water concentration, conversion rate also went down due to the dehydration of the enzyme microenvironment [21]. However, there are other reports that suggest that water activity of 0.25 was the most favorable for geraniol esters catalyzed by Novozym 435 [22].

39.3.3 Attaining Optimum Condition

Optimal esterification conditions were obtained through the regression model (Eq. 39.2) according to the limit criterion of maximum response geraniol conversion rate. The solutions given by Design-Expert Version 7.0.0 were not unique. Taking into account the need to produce ester with high conversion rate, high substrates concentrations, and low enzyme input, we determined the optimum conditions of synthesis of geranyl butyrate catalyzed by CALB-displaying *P. pastoris* whole-cell as follows: biocatalyst content 20.9 g/L, geraniol concentration 0.57 mol/L, butanoic acid/geraniol molar ratio of 1.21:1, and water activity 0.53.

Further experiment was conducted to verify the adequacy of the model using optimal conditions provided by the model. The predicted value was 100 % molar conversion and the results obtained after conducting these experiments were 95.33 % at 6 h reaction and could reach 98.84 % at 9 h reaction, which was equivalent to 126 g/L geranyl butyrate. The high concentration of substrates (0.69 mol/L butyric acid and 0.57 mol/L geraniol) in this work proves the robustness of the whole-cell biocatalyst. Meanwhile, the loss of activity was less than 5 % after the cells has been reused for 5 batches, which showed similar operational stability as the RML-displaying *P. pastoris* (data not shown) [23].

Up to now, there are numerous reports about the synthesis of geranyl butyrate catalyzed by lipases. Almost complete conversions of the reactants to geranyl butyrate were reported by Kanwar et al. [1] and Claon and Akoh [12], but they both worked with fairly low substrate concentration of 0.1 mol/L. Varma and Madras [13] investigated the transesterification of various esters to geranyl butyrate in supercritical carbon dioxide, but the conversions were less than 50 %. Laboret et al. achieved a yield of 106 g/L of geranyl butyrate in 10 h using commercial crude lipase powder esterase 30,000 from *M. miehei* [17]. In comparison, the high concentration of product and the low cost of catalyst are superior to the previous work, which would make CALB-displaying *P. pastoris* whole-cell biocatalyst promising for potential use for the commercial production.

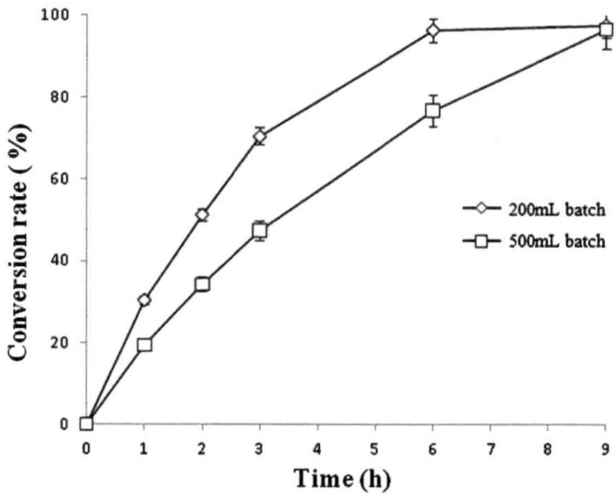

Fig. 39.3 Synthesis of geranyl butyrate in stirred batch reactors

39.3.4 Scale-Up Production of Geranyl Butyrate in an Agitated Reactor

Under the optimized conditions by RSM, the scale-up esterification of geraniol and butyric acid was carried out in a batch stirred reactor (reaction mixture 200 or 500 mL). In esterification process, water removal is important especially when you are dealing with large amount of the reactants because too much water shifts the equilibrium to hydrolysis [3, 20]. In the scale-up reactions here, water was removed either by molecular sieves or by azeotropic distillation (Fig. 39.1). The benefit of molecular sieves during the reaction on the ester yield for the control of the water content in the system has also been stated by Knežević-Jugović et al. [24]. As shown in Fig. 39.3, the reaction was faster in the case of 200 mL batch but slower in the 500 mL batch, suggesting that the increase in volume has an influence on the conversion rate. This influence had been encountered in previous work on esterification of (Z)-3-Hexen-1-yl acetate using Novozym 435 [25]. The point is under the optimized conditions by RSM, the conversion rate reached to 97.87 % in 200 mL as well as 97.56 % in 500 mL (Fig. 39.3), which are in good consistence with that in 10 mL reaction system. It may be explained by the fact that the higher volume requires more powerful stirring for the sheer force to be uniformly distributed. This result indicated that CALB-displaying yeast whole-cell biocatalyst is promising in geranyl butyrate synthesis on large scale.

39.4 Conclusion

Optimized esterification of geraniol with butyric acid catalyzed by CALB-displaying *P. pastoris* whole-cell biocatalyst was successfully achieved. It has been found that substrates molar ratio and water activity were the parameters most significantly affecting conversion rate. Under the optimized conditions, the geraniol conversion rate of 98.5 % after 9 h reaction with geraniol concentration 0.57 mol/L was equivalent to 126 g/L geranyl butyrate, which was also a high yield among the existing reports. Moreover, the model from CCD had been well reproduced in stirred reactors with 200 or 500 mL reaction medium which gave a high yield similar to that in shaking flasks. Thus, this work demonstrated that CALB-displaying *P. pastoris* whole-cell biocatalyst here is well suited for the synthesis of geranyl butyrate and promising in commercial application for further large-scale production.

Acknowledgments This study was financially supported by National High-tech R&D Program (863 Program) of China (2012AA022205) and Guangdong Province Key Technology R&D Program of Emerging Industries of Strategic Importance (2012A080800013).

References

1. Kanwar SS, Gehlot S, Verma ML, Gupta R, Kumar Y, Chauhan GS (2008) Synthesis of geranyl butyrate with the poly (acrylic acid-co-hydroxy propyl methacrylate-cl-ethylene glycol dimethacrylate) hydrogel immobilized lipase of *Pseudomonas aeruginosa* MTCC-4713. J Appl Polym Sci 110:2681–2692
2. Damnjanovic JJ, Zuza MG, Savanovic JK, Bezbradica DI, Mijin DZ, Boskovic-Vragolovic N, Knezevic-Jugovic ZD (2012) Covalently immobilized lipase catalyzing high-yielding optimized geranyl butyrate synthesis in a batch and fluidized bed reactor. J Mol Catal B Enzym 75:50–59
3. Karra-Chaabouni M, Pulvin S, Thomas D, Touraud D, Kunz W (2002) Role of water activity on the synthesis of geranyl butyrate by a Mucor miehei esterase in a solvent-free system. Biotechnol Lett 24:1951–1955
4. Paroul N, Grzegozeski LP, Chiaradia V, Treichel H, Cansian RL, Oliveira JV, de Oliveira D (2010) Production of geranyl propionate by enzymatic esterification of geraniol and propionic acid in solvent-free system. J Chem Technol Biotechnol 85:1636–1641
5. Mateo C, Palomo JM, Fernandez-Lorente G, Guisan JM, Fernandez-Lafuente R (2007) Improvement of enzyme activity, stability and selectivity via immobilization techniques. Enzym Microb Technol 40:1451–1463
6. Han S-Y, Pan Z-Y, Huang D-F, Ueda M, Wang X-N, Lin Y (2009) Highly efficient synthesis of ethyl hexanoate catalyzed by CALB-displaying *Saccharomyces cerevisiae* whole-cells in non-aqueous phase. J Mol Catal B: Enzym 59:168–172
7. Tanino T, Aoki T, Chung W-Y, Watanabe Y, Ogino C, Fukuda H, Kondo A (2009) Improvement of a Candida antarctica lipase B-displaying yeast whole-cell biocatalyst and its application to the polyester synthesis reaction. Appl Microbiol Biotechnol 82:59–66
8. Wang H, Lang Q, Li L, Liang B, Tang X, Mascini M, Liu A (2013) Yeast surface displaying glucose oxidase as whole-cell biocatalyst: construction, characterization and its electrochemical glucose sensing application. Anal Chem 85:6107–6112
9. Liang XX, Wang BB, Sun YF, Lin Y, Han SY, Zheng SP, Cui TB (2013) Quantitative evaluation of Candia antarctica lipase B displayed on the cell surface of a *Pichia pastoris* based on an FS anchor system. Biotechnol Lett 35:367–374
10. Bujalowski W (2003) Expanding the physiological role of the hexameric DnaB helicase. Trends Biochem Sci 28:116–118
11. Zhang L, Liang S, Zhou X, Jin Z, Jiang F, Han S, Zheng S, Lin Y (2013) Screening for glycosylphosphatidylinositol (GPI)-modified cell wall proteins in *Pichia pastoris* and their

recombinant expression on the cell surface. Appl Environ Microbiol. doi:10.1128/AEM.00824-13
12. Claon PA, Akoh CC (1993) Enzymatic synthesis of geraniol and citronellol esters by direct esterification in n-hexane. Biotechnol Lett 15:1211–1216
13. Varma MN, Madras G (2010) Kinetics of enzymatic synthesis of geranyl butyrate by transesterification in various supercritical fluids. Biochem Eng J 49:250–255
14. Güvenç A, Kapucu N, Kapucu H, Aydoğan Ö, Mehmetoğlu Ü (2007) Enzymatic esterification of isoamyl alcohol obtained from fusel oil: optimization by response surface methodology. Enzym Microb Technol 40:778–785
15. Rodriguez-Nogales JM, Roura E, Contreras E (2005) Biosynthesis of ethyl butyrate using immobilized lipase: a statistical approach. Process Biochem 40:63–68
16. Trusek-Holownia A, Noworyta A (2007) An integrated process: ester synthesis in an enzymatic membrane reactor and water sorption. J Biotechnol 130:47–56
17. Laboret F, Perraud R (1999) Lipase-catalyzed production of short-chain acids terpenyl esters of interest to the food industry. Appl Biochem Biotechnol 82:185–198
18. Stamatis H, Christakopoulos P, Kekos D, Macris B, Kolisis F (1998) Studies on the synthesis of short-chain geranyl esters catalysed by *Fusarium oxysporum* esterase in organic solvents. J Mol Catal B Enzym 4:229–236
19. Xu Y, Wang D, Mu XQ, Zhao GA, Zhang KC (2002) Biosynthesis of ethyl esters of short-chain fatty acids using whole-cell lipase from *Rhizopus chinesis* CCTCC M201021 in non-aqueous phase. J Mol Catal B Enzym 18:29–37
20. Barahona D, Pfromm PH, Rezac ME (2006) Effect of water activity on the lipase catalyzed esterification of geraniol in ionic liquid [bmim] PF6. Biotechnol Bioeng 93:318–324
21. Gandhi NN, Patil NS, Sawant SB, Joshi JB, Wangikar PP, Mukesh D (2000) Lipase-catalyzed esterification. Catal Rev 42:439–480
22. Peres C, Gomes da Silva MD, Barreiros S (2003) Water activity effects on geranyl acetate synthesis catalyzed by Novozym in supercritical ethane and in supercritical carbon dioxide. J Agric Food Chem 51:1884–1888
23. Huang D, Han S, Han Z, Lin Y (2012) Biodiesel production catalyzed by Rhizomucor miehei lipase-displaying *Pichia pastoris* whole cells in an isooctane system. Biochem Eng J 63:10–14
24. Knežević-Jugović ZD, Damnjanović JJ, Bezbradica DI, Mijin DŽ (2008) The immobilization of lipase on sepabeads: coupling, characterization and application in geranyl butyrate synthesis in a low aqueous system. Chem Ind Chem Eng Q 14:245–249
25. Shih I-L, Hung S-H, Chen F-Y, Ju H-Y, Shieh C-J (2007) Optimized synthesis of lipase-catalyzed l-menthyl butyrate by Candia rugosa lipase. Food Chem 100:1223–1228

Chapter 40
Metabolic Analysis of a *Corynebacterium glutamicum IdhA* Mutant During an Efficient Succinate Production Using pH-Control Under Oxygen Deprivation

Chen Wang, Heng Cai, Zhihui Zhou, Hong-gui Wan and Ping-kai Ouyang

Abstract *Corynebacterium glutamicum* produce succinate and lactate from glucose under oxygen deprivation without growth. By deleting the lactate dehydrogenase genes (*ldhA*), lactate formation could be completely abolished and yield of succinic acid increased up to 0.78 g g^{-1}. With increasing concentration of carbonate, which could regular the pH of culture, the glucose consumption rate has been improved, leading to increase the succinic acid production rates. Interestingly, depending on the pH conditions, there is a sharp increase in the presence of higher concentration carbonate. This phenomenon was investigated by the key metabolic enzymes analysis during different pH condition. A pyruvate carboxylase and malate dehydrogenase played a significant role in succinic production under oxygen deprivation. Succinic acid production by a *ldhA* mutant was investigated with free pH and constant pH 8.0. Succinic acid production and yield were higher under constant pH 8.0 conditions compared with culture with free pH. In an efficient succinic acid production process, *C. glutamicum* ATCC 13032 Δ*ldhA* accumulated 67.2 g l^{-1} succinic acid within 48 h with an average succinic production rate 0.14 g l^{-1} per cells (dry weight) (cdw) per hour.

Keyword *Corynebacterium glutamicum* · Succinic acid · pH control · Metabolic analysis

C. Wang · H. Cai (✉) · Z. Zhou · H.-g. Wan · P.-k. Ouyang
Nanjing University of Technology, Nanjing, China
e-mail: cheng@njtech.edu.cn

T.-C. Zhang and M. Nakajima (eds.), *Advances in Applied Biotechnology*,
Lecture Notes in Electrical Engineering 332, DOI 10.1007/978-3-662-45657-6_40

40.1 Introduction

Succinic acid is regarded as one of the 12 top value-added chemicals produced from biomass [29]. It can be used as a precursor of several commodity chemicals including tetrahydrofuran, 1,4-butanediol, γ-butyrolactone, and N-methyl pyrrolidone as well as biodegradable polymers such as polybutylene succinate [5, 32]. Besides its applications in surfactant, detergent extender, foaming agent, ion chelator, food additive [30]. Presently, it is primarily produced by catalytic hydrogenation of petrochemical. Bioconversion represents the chemical synthetic process production of succinic acid from renewable biomass by anaerobic bacteria has become more attractive economically [31].

Succinate is excreted as an end product of catabolism during anaerobic fermentation by several promising candidate bacteria including *Anaerobiospirillu succiniproducens,* anaerobic bacteria isolated from animal feces [9], which is one of well-known succinic acid wild-type producer that naturally produce succinic acid over 50 g l^{-1} under oxygen deprivation conditions; a Gram-negative bacterium *Actinobacillus succinogenes* FZ53 produce 105.8 g l^{-1} succinic acid in 78 h [25].

Metabolic engineering of *Escherichia coli* and *Corynebacterium glutamicum* for anaerobic has been shown previously to increase succinic acid yield [28]. An efficient *E. coli* AFP111-*pyc* produces succinic acid as a major product with by-products such as acetic acid and ethanol under anaerobic conditions [6, 8], a high titer and yield was obtained of 99.2 g l^{-1} and 1.1 g g^{-1}, respectively [28]. A high-cell density succinate production process of *C. glutamicum* shows the highest succinic acid titer of 146 g l^{-1} within 46 h [16]. Succinate titer and productivity goals are based on those obtained in industrial glutamic acid production: 150 g l^{-1} at 5 g l^{-1} h^{-1} [14]. In order to achieve this goal, genetical modifying of the succinic acid producing microorganisms and enhancing the bioprocesses efficiency is significant and inevitable.

Corynebacterium glutamicum is used for the industrial production of various amino acids [10]. Moreover, due to arrested cell growth under oxygen deprivation, it has great potential capacity for efficient production ethanol [11] or organic acid: D-lactate [17], succinate [15]. Lactate and succinate are the two major products during anaerobic fermentation with little ethanol, formate, acetate. To obtain an efficient succinic acid production of *C. glutamicum*, modifying in by-product formation pathways, such as lactate, acetate. In recent years, amounts of metabolic engineering *C. glutamicum* research have been reported [2, 11, 15]. However, present fermentative productions of succinate are still insufficient for industrial application, it is in great need of fermentative process research to improve production and yield of succinic acid.

In the present study, productivity of succinic acid by *C. glutamicum* under oxygen deprivation was investigated to demonstrate the prospect for practical application of the bioprocess. The metabolic analysis of *C. glutamicum ΔldhA* strain reveals that succinic acid production with constant pH should be an efficient culture process.

40.2 Materials and Methods

40.2.1 Bacterial Strains, Plasmids, and Culture Conditions

All bacterial strains and plasmids used or constructed in this study are listed in Table 40.1. For genetic manipulations, *E. coli* strains were grown at 37 °C in LB medium [21]. *Corynebacterium glutamicum* strains were routinely cultivated at 30 °C in nutrient-rich medium containing urea (2 g l^{-1}), yeast extract (2 g l^{-1}), casamino acid (7 g l^{-1}), $(NH_4)_2SO_4$ (7 g l^{-1}), KH_2PO_4 (0.5 g l^{-1}), K_2HPO_4 (0.5 g l^{-1}), $MgSO_4{\cdot}7H_2O$ (0.5 g l^{-1}), $FeSO_4{\cdot}7H_2O$ (6 mg l^{-1}), $MnSO_4{\cdot}H_2O$ (4.2 mg l^{-1}), biotin (100 μg l^{-1}), and thiamine(100 μg l^{-1}) with glucose (20 g l^{-1}). In anaerobic production, glucose (80 g l^{-1}) and sodium carbonate (300 mM) were added to the mineral salts medium, including KH_2PO_4 (0.5 g l^{-1}), K_2HPO_4 (0.5 g l^{-1}), $MgSO_4{\cdot}7H_2O$ (0.5 g l^{-1}), $FeSO_4{\cdot}7H_2O$ (6 mg l^{-1}), $MnSO_4{\cdot}H_2O$ (4.2 mg l^{-1}), biotin (100 μg l^{-1}), and thiamine (100 μg l^{-1}). If appropriate, the final antibiotic concentrations were as follows: for *E. coli* 50 μg ml^{-1} ampicillin and 50 μg ml^{-1} kanamycin, and for *Corynebacterium glutamicum*, 10 μg ml^{-1} of kanamycin.

Table 40.1 Strains, plasmids, and primers in this study

Strain, plasmid or primer	Relevant characteristics	Source or reference
E. coli		
JM109	*recA1 endA1 gyrA96 thi hsdR17 supE44 relA1* Δ(*lac-proAB*)/F′ [*traD36 proAB* + *lacI* ^{q} *lacZ*ΔM15]	Takara
C. glutamicum ATCC 13032	Wild type strain, biotin auxotroph	[1]
Δ*ldhA*	Markerless *ldhA* mutant	This work
Plasmids		
pUC18	Ap^r; a-lac/MCS M13 ori	Takara
pUC18-Δ*ldhA*	Ap^r; pUC18 with a 1.2 kb *Hin*dIII DNA fragment containing Δ*ldhA* gene	This work
pK18mobsacB	Kan^r; vector for allelic exchange in *C. glutamicum*	[23]
pK18mobsacB-Δ*ldhA*	Kan^r; pK18mobsacB with a 1.2 kb *Hin*dIII DNA fragment containing Δ*ldhA* gene	This work
Primer		
ldh1	5′-CCCAAGCTTGGACATAATCGGGCATAA-3′	This work
ldh2	5′-CCCAAGCTTGATCAACCGCTGCGTT-3′	This work

40.2.2 DNA Techniques

Plasmid DNA was isolated by using a Plasmid Miniprep Kit (Biomiga, USA) according to the manufacturer's instructions. When extracting DNA from *C. glutamicum* by using 4 mg/ml lysozyme at 37 °C for 30 min. Genomic DNA of *C. glutamicum* was isolated by using a TIANamp Bacteria DNA Kit (TIANGEN, China). Oligonucleotides were synthesized by Genscript Corporation. Routine methods like PCR, restriction, or ligation were carried out according to standard protocols [21]. PCR products were generated with Prime STAR HS DNA Polymerase (Takara, Japan). DNA fragment or agraose gel DNA fragment was purificated by Gel/PCR Extraction Kit (Biomiga, USA). Transformation of *E. coli* cells was performed by the $CaCl_2$ procedure [21]. Transformation of *C. glutamicum* was performed as described previously [26].

40.2.3 Construction of Deletion Mutants, Chromosomal Gene Replacements, and Plasmids

C. glutamicum mutants with in-frame deletions of *ldhA* (Δ*ldhA*) was constructed a two-step homologous recombination procedure as described previously [23]. Based on *C. glutamicum* ATCC 13032 whole genome sequence, two oligonucleotide primers were designed in order to amplify a portion of the *ldhA* gene. The oligonucleotide primers used in a PCR with *C. glutamicum* ATCC 13032 chromosomal DNA as a template. The 1.4-kb PCR product bearing *ldhA* gene was digested *Hind*III and inserted into the same site of pUC18, yielding pUC18-*ldhA* (Table 40.1). The 4.0-kb pUC18-*ldhA* delete 149 bp fragment digested with *Eco*RV and *Nae*I cohesive ends and self-ligated yielding pUC18-Δ*ldhA* (Table 40.1). The 1.2-kb *Hind*III Δ*ldhA* DNA fragment inserted into the *Hind*III-digested pK18mobsacB plasmid DNA yielding pK18mobsacB Δ*ldhA* (Table 40.1). *C. glutamicum* ATCC 13032 was transformed by electroporation with plasmids pK18mobsacB Δ*ldhA*. The transfer of the resulting deletion plasmids into *C. glutamicum* and selection for the first and second recombination events were performed as described previously [11].

40.2.4 Aerobic Culture for Cell Propagation

Corynebacterium glutamicum ATCC 13032 Δ*ldh* cells were precultured with constant agitation at 200 rpm at 30 °C 12 h in nutrient-rich medium containing glucose (20 g l^{-1}). 150 ml of preculture was inoculated in 3L of nutrient-rich medium containing glucose (20 g l^{-1}) in a 5L bioreactor (KF-5L, KoBioTech, Korea). Cultivation was performed at 30 °C for 10 h with constant agitation at 400 rpm.

40.2.5 Conditions for Succinic Acid Production

For succinic acid production, the cells from the 3L culture were harvested by centrifugation (5,000 rpm, 4 °C, 10 min). The cell precipitate was subsequently washed twice with mineral salts medium. An appropriate amount of washed cells was resuspended in 2L mineral salts medium containing glucose (80 g l^{-1}) and Na_2CO_3 (300 mM), and transferred into a 5L bioreactor (KF-5L, KoBioTech, Korea). The cell suspension was kept at 30 °C and stirred at 180 rpm.

40.2.6 Enzyme Assay

For measurement of intracellular enzyme activities, cells were harvested by centrifugation (8,000 rpm, 10 min, 4 °C) and washed twice with cold 100 mM Tris-HCl (pH 8.0). The washed cells were resuspended in the same buffer and sonicated on ice for 180 cycles (a working period of 3 s in a 2 s interval for each cycle) at a power output of 200 W by an ultrasonic disruptor (JY92-II, Scientz Biotechnology Co., Ningbo, China). The cell debris was removed by centrifugation (10,000 rpm, 15 min, 4 °C), and the supernatant was used for activity determination. Enzyme activities were measured using a spectrophotometer (UV-2800, Unico, USA) with the temperature controlled at 30 °C. Reactions were initiated by adding cell extract or substrate to give a final volume of 3 mL.

Assay for PC (pyruvate carboxylase) 100 mM Tris-HCl (pH 7.3), 25 mM $NaHCO_3$, 5 mM $MgCl_2$, 3 mM pyruvate, 4 mM ATP [7]. Assay for LDH [18], 30 mM sodium pyruvate (pH 7.5), 20 mM MOPS (pH 7.0), 0.2 mM NADH, MDH [33], 100 mM Tris-HCl (pH 8.0), 2 mM OAA, 0.2 mM NADH; ICL [27], 100 mM KH_2PO_4 (pH 7.0), 5 mM $MgCl_2$, 1 mM DTT, 0.2 mM NADH, 5 mM isocitrate, 12.8U LDH; PCK [27], 100 mM Tris-HCl (pH 6.6), 10 mM $MgCl_2$, 5 mM $MnCl_2$, 1 mM DTT, 10 mM ADP, 75 mM $NaHCO_3$, 0.2 mM NADH, 33 U MDH, 10 mM PEP; PK [19], 100 mM Tris-HCl (pH 7.5), 5 mM ADP, 1 mM DTT, 10 mM KCl, 15 mM $MgCl_2$, 10 mM PEP, 0.25 mM NADH, 12.8 U LDH; PPC [4], 100 mM Tris-HCl (pH 8.0), 5 mM PEP, 10 mM $MgCl_2$, 24 mM $NaHCO_3$, 0.15 mM NADH, 0.5 mM acetyl-CoA, 33 U MDH. The total protein concentration in crude cell extract was measured by Bradford's method [3] with bovine serum albumin as a standard.

40.2.7 Analytical Methods

The cell growth was measured in the optical density at 600 nm (OD_{600}) with a spectrophotometer (UV-2800, UNICO, USA). The biomass was calculated from OD_{600} values by a mentally determined correlation factor of 0.25 g dry cell weight

(CDW)/liter for an OD_{600} of 1 [13]. The concentrations of glucose were measured by a glucose analyzer containing glucose oxidase (SBA-40E, Institute of Biology, China). Organic acids in the culture supernatants were quantified by HPLC (Chromeleon server monitor, P680 pump, Dionex, USA) equipped with UV detector and conductivity meter and Grace PrevaiTM column operating at 215 nm with 25 mmol l^{-1} KH_2PO_4 (pH 2.5) mobile phase at a flow rate of 1.0 ml min^{-1}.

40.3 Results

40.3.1 Inactivation of Lactate Dehydrogenase for Anaerobic Succinate Production

During anaerobic metabolism, *C. glutamicum* produced organic acids such as acetate, lactate, which is the dominant product of anaerobic glucose catabolism. In order to accumulate succinate which is the end product of anaerobic metabolism, the lactate dehydrogenase encoded by the *ldhA* genes must be inactivated. For this purpose, a Δ*ldhA* mutant was constructed which is completely unable to oxidize pyruvate to lactate. The enzymatic activities and product formation were compared to the genetically modified strain with the wild type.

To confirm the deletion of *ldhA* in *C. glutamicum* ATCC 13032, its enzymatic activities were measured. The LDH activity of *C. glutamicum* ATCC 13032 showed up to 0.47 ± 0.002 μmol mg^{-1} min^{-1}. Whereas no LDH activity was detected in cell extracts of the *C. glutamicum* Δ*ldhA*. Independent batch cultivations by two stages—aerobic growth in nutrient-rich medium with glucose were performed in 500 ml flasks, anaerobic in mineral salts medium with glucose and carbonate were performed in 100 ml bottles. During the aerobic phase, the wild type is higher to the recombinant *C. glutamicum* in glucose consumption, growth.

Corynebacterium glutamicum Δ*ldhA* exhibited a lower growth rate and 29 % decreased biomass formation compared with wild type (Fig. 40.1). The wild type and engineered cells consumed glucose with an uptake rate of 1.06 and 0.94 g l^{-1} h^{-1}, respectively. Under oxygen deprivation, *C. glutamicum* cell growth is arrested, the intracellular enzyme retain the capability to convert glucose to organic acids. At the end of the cultivation after 16 h, the wild-type strain produced succinate in the supernatant reached 4.83 g l^{-1}. The succinate yield 0.19 g g^{-1} represented 24 % of the *C. glutamicum* Δ*ldhA* 0.78 g g^{-1}. The *C. glutamicum* Δ*ldhA* uptaked glucose with a rate of 1.06 g l^{-1} h^{-1} and produced succinate with a rate of 0.83 g l^{-1} h^{-1}. Simultaneously, 6.85 g l^{-1} acetate and 5.79 g l^{-1} pyruvate were formed. As expected strain *C. glutamicum* Δ*ldhA* was unable to produce lactate, while the wild-type accumulated lactate 35.29 g l^{-1}.

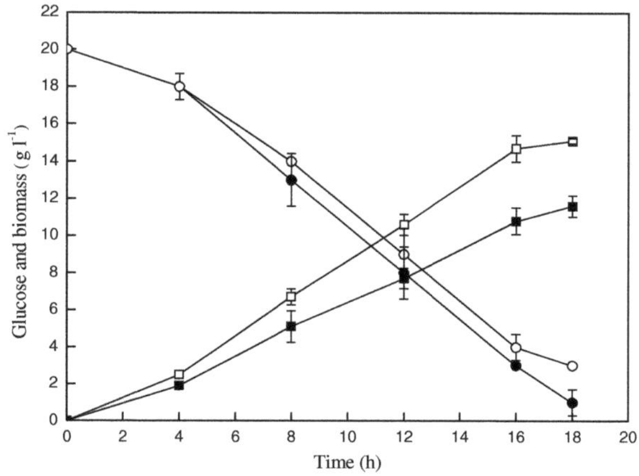

Fig. 40.1 The time course of glucose concentration and biomass by *C. glutamicum* during the aerobic phase. Symbols of *C. glutamicum* ATCC 13032 (*empty symbols*), *C. glutamicum* Δ*ldhA* (*filled symbols*), glucose (*circle symbols*), biomass (*square symbols*) are shown. Initial glucose was 20 g l^{-1}. Standard deviations are indicated by *bars* or are within each symbol. Data are averages based on duplicate experiments

40.3.2 Effects of Carbonate Concentration on Succinic Production

Carbonate was significant to succinic acid production in the previous study. Addition of carbonate to the reaction mixture during succinate production by *C. glutamicum* could lead to increasing substrate consumption rates [12]. Therefore, effects of carbonate on glucose consumption and succinic acid production rate of *C. glutamicum* during 16 h anaerobic cultivation were investigated in more detail.

With the addition of $NaHCO_3$, Na_2CO_3, K_2CO_3, respectively, glucose consumption and succinate yield increased with increasing carbonate concentration, up to 400 mM, where a max production was realized (Fig. 40.2). In the presence of Na_2CO_3, the glucose consumption and succinate production improved slightly from

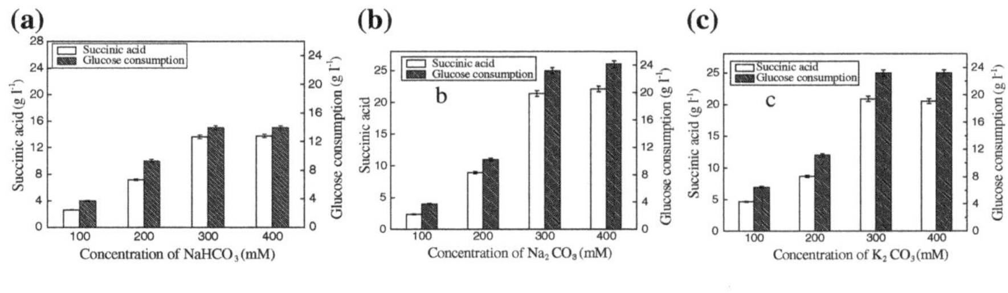

Fig. 40.2 Succinic acid production and glucose consumption by *C. glutamicum* Δ*ldhA* under various carbonate concentrations (**a** $NaHCO_3$, **b** Na_2CO_3, **c** K_2CO_3)

100 to 200 mM. Interestingly, there was a sharp increase with the addition of 300 mM of Na_2CO_3. A succinic acid production of 21.39 g L^{-1} in the presence of 300 mM was 2.38-fold higher than that in the presence of 200 mM. A similar phenomenon appeared by adding to K_2CO_3. However, above 300 mM, there was a slight increase or decrease in glucose consumption and succinic acid yield.

40.3.3 Succinic Acid Production by C. glutamicum ΔldhA Under Oxygen Deprivation with Free pH

The culture pH was critical factor affecting succinic acid production; *C. glutamicum* Δ*ldhA* produced succinate in mineral salts medium with 300 mM Na_2CO_3 (initial pH 9.4). With the succinic acid production, it displayed that the decrease of 2.25 pH unit after 18 h in anaerobic phase. It showed glucose consumption coupled accumulation of succinate from the beginning of the cultivation up to pH 7.21. An amount (24.57 g l^{-1}) of succinic acid was formed from the amount of glucose consumed 25 g l^{-1}. Below pH 7.2, the strain was unable to produce succinate, indicating that it is unsuitable for succinic acid production at lower pH. At 30 h cultivation, the pH was maintained at 8 by supplementing 10 N of sodium hydroxide solution. The succinic acid production capacity was recovered. The glucose consumption rate and succinate production rate were 0.83 and 0.85 g l^{-1} h^{-1}, respectively. When maintaining the pH at 9, the strain had a slightly decreased succinate production rate 0.76 g l^{-1} h^{-1}.

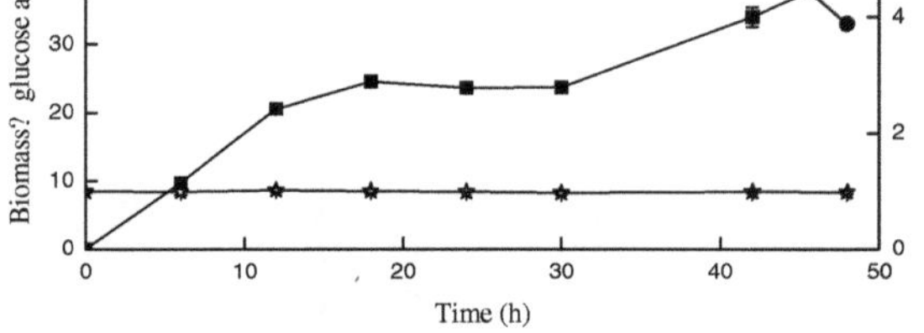

Fig. 40.3 Succinic acid production by *C. glutamicum* Δ*ldhA* in different pH conditions. Symbols of glucose (*filled circle*), succinic acid (*filled square*), biomass (*filled star*), pH (*filled triangle*) are shown. Initial glucose, Na_2CO_3, and biomass were 80 g l^{-1}, 300 mM, and 9 g dry cell l^{-1}, respectively. pH was maintained at the value of 8.0 and 9.0 by supplementing 10 N of sodium hydroxide solution at 30 and 42 h, respectively. Standard deviations are indicated by *bars* or are within each symbol. Data are averages based on duplicate experiments

Table 40.2 Organic acid production parameters of different *C.glutamicum* succinate during anaerobic cultivation

Corynebacterium gltamicum	Yield ($g\ g^{-1}$)	Glucose uptake rate ($g\ l^{-1}\ h^{-1}$)	Succinate production rate ($g\ l^{-1}\ h^{-1}$)	Succinate ($g\ l^{-1}$)	Lactate ($g\ l^{-1}$)	Acetate ($g\ l^{-1}$)
ATCC 13032	0.19	1.63	0.3	4.83	35.29	3.07
ΔldhA	0.78	1.06	0.83	13.32	0	6.85

40.3.4 Metabolic Analysis During Succinate Production in Different pH

In order to identify the influence of different pH for succinic acid production in *C. glutamicum ΔldhA*, it was essential to analyze several relevant enzymes which existed in producing succinate pathway. The specific activities of PC, MDH, ICL, PCK, PK, PPC were measured as described in the Enzyme assay section.

The recombinant strain showed no detectable ICL activity (Table 40.2) for all pH condition tested. When the pH value increased from 7.2 to 8.0, the PC activities increased by 2.8-fold compared to that of the pH 7.2. MDH were 5.1-fold higher than that of the pH 7.2. However, PCK had a little fluctuation from pH 7.2 to pH 9.0. Moreover, the activities of PPC had a slight increase. The strain lacked of producing succinate at pH 7.2, PC and MDH were lower compared with pH 8.0, suggesting that PC and MDH may be the predominant enzyme in producing succinate. Under optimal pH 8.0 conditions, the succinic acid production rate was 0.85 $g\ l^{-1}\ h^{-1}$ (Fig. 40.3).

40.3.5 Succinic Acid Production by C. glutamicum ΔldhA with Constant pH Under Oxygen Deprivation

Corynebacterium glutamicum ΔldhA did not produce succinic acid below pH 7.2. The strain had produced succinate well and showed a similar succinate yield above pH 8.0. To determine the efficiency of succinic acid production with constant pH, succinic acid production control pH at 8.0 by supplementing NaOH. Though the rate of succinic acid production and glucose consumption decreased as time went by, succinic acid concentration reached 67.17 $g\ l^{-1}$ at 0.14 $g\ l^{-1}\ h^{-1}\ g^{-1}$ (CDW) of whole volumetric productivity at 48 h. The yield of succinic acid was 0.85 $g\ g^{-1}$. These results showed that an appropriate pH condition was significant for producing succinic acid effectively (Fig. 40.4).

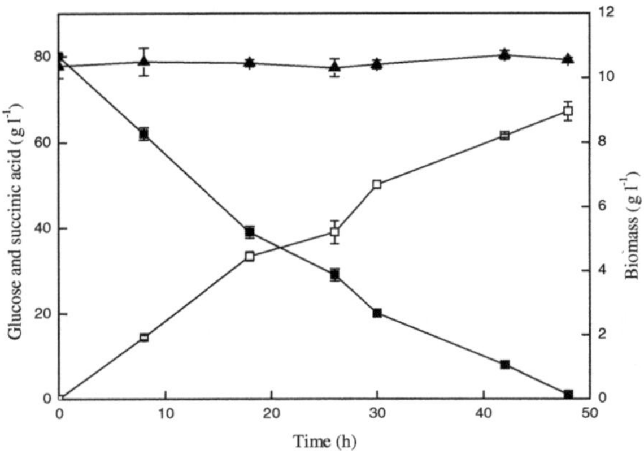

Fig. 40.4 Succinic acid production by *C. glutamicum ΔldhA* with constant pH 8.0. Symbols of glucose (*filled square*), succinic acid (*empty square*), biomass (*filled triangle*) are shown. Initial glucose, Na_2CO_3, and biomass were 80 g l^{-1}, 300 mM, and 10.5 g dry cell l^{-1}, respectively. pH was maintained the value of 8.0 by supplementing 10 N of sodium hydroxide solution. Standard deviations are indicated by *bars* or are within each symbol. Data are averages based on duplicate experiments

40.4 Discussion

Corynebacterium glutamicum under oxygen deprivation can produce organic acids such as lactate, succinate, acetate in mineral salts medium. The specific aims were to reduce by-product formation and to increase the succinate yield. In this work, a *ldhA* mutant engineering *C. glutamicum* was constructed. With addition of carbonate, which is a co-substrate for anaplerotic enzymes, led to an increase in the succinic production and glucose consumption rate. The metabolism of *C. glutamicum* was analyzed by measuring succinic acid production, glucose consumption, and various key metabolic enzymes in different pH conditions. In order to improve succinic production and yield, the pH was controlled. A succinic acid titer of 67.2 g l^{-1} and specific production rate 0.14 g l^{-1} h^{-1} g^{-1} be achieved.

The deletion of the *ldhA* genes encoding lactate dehydrogenase led to the biomass and glucose consumption decreased during aerobic metabolism. It is known that LDH mainly play a role during anaerobic metabolism. However, it affected sightly the cell growth in aerobic phase. For comparison with succinate producer *C. glutamicum ΔldhA*, *C. glutamicum* ATCC 13032 was created. The engineering strain could accumulate succinic acid is higher 2.76-fold than that of the wild type. Moreover, the major by-product lactate formation could be completely abolished. The succinate yield increased from 0.19 to 0.78 g g^{-1} as the recombinant strain lacked to produce lactate pathway. The metabolic flux flowed to succinic acid was improved. Whereas, LDH as a key enzyme coupled lactate production to reoxidation of NADH formed during glycolysis [24], it was deleted which led to limit the regeneration of NAD^+ from the NADH formed in glycolysis and accumulation of

NADH intracellular. The NAD^+/NADH ratio affected the glucose consumption and succinate production [12].

Carbonate plays a significant role in succinate production. On the one hand, it could increase flux from phosphoenlpyruvate (PEP) to succinic acid via the TCA reductive arm [22]. It would appear that increase of NAD^+/NADH ratio in the presence of carbonate, thus increasing glucose consumption [15]. On the other hand, carbonate could regulate the pH of culture. The succinic acid production and glucose consumption rate increase accompany with the carbonate concentration increase. Interestingly, it is worth to notice that there was a sharp increase from 200 to 300 mM by supplementing Na_2CO_3 or K_2CO_3. Furthermore, up to 400 mM, there is almost no improvement in producing succinic acid. These facts suggest that the cell maybe produced succinic acid with certain range; the succinate may be formed in alkaline condition.

To confirm above facts, the metabolism of *C. glutamicum ΔldhA* was analyzed by measuring succinic acid production rate and key enzyme activities during cultures with different pH. Results obtained at a lower pH condition that the modified strain lacks for succinic acid production ability. Once controlling the pH in certain range, the capacity of succinate production is recovery. We attribute these facts to metabolism key enzyme expression. Furthermore, the enzyme activities analysis corroborates our judgement. Specially, PC and MDH activities have a significant decline at lower pH. PC was reported to be the major bottleneck for derivatives of TCA cycle intermediates [20]. MDH is a NADH-dependent enzyme. The decrease of MDH led to limit the regeneration of NAD^+ from NADH formed in glycolysis. Furthermore, it reduces glucose consumption rate and succinic acid production. Our data suggest that an appropriate pH is important to succinate production.

As the approach to increase succinic acid production, fermentations automatically controlled were carried out. In the present study, succinic acid production by *C. glutamicum ΔldhA* studied was enhanced at a constant pH value of 8.0. *C. glutamicum ΔldhA* can produce 67.2 g l^{-1} within 48 h (Table 40.3), the volumetric succinic acid productivities (1.44 g l^{-1} h^{-1}) was slightly higher than the succinic acid producer *E. coli*AFP111-*pyc* [28] (1.3 g l^{-1} h^{-1}). Due to a *C. glutamicum* R *ΔldhA*-pCRA717 [16] produced at high cell density (50 g l^{-1} CDW), while in the present study, the biomass is only 10 g l^{-1}. The specific production rate is 2.3-fold higher than that of the *C. glutamicum* R *ΔldhA*-pCRA717. These results suggest that *C. glutamicum* ATCC13032 *ΔldhA* has a strong capacity of producing

Table 40.3 Enzymatic activities of *C. glutamicum ΔldhA* in different pH

	Succinate production rate (g l^{-1} h^{-1})	Activies (μmol mg^{-1} min^{-1})					
		PC	MDH	PCK	PK	PPC	ICL
pH7.2	0	0.12 ± 0.02	0.08 ± 0.02	0.1 ± 0.01	0.02 ± 0.01	0.06 ± 0.01	ND
pH8.0	0.85	0.34 ± 0.01	0.41 ± 0.02	0.08 ± 0.01	0.06 ± 0.01	0.06 ± 0.01	ND
pH9.0	0.76	0.31 ± 0.01	0.37 ± 0.02	0.1 ± 0.01	0.08 ± 0.01	0.06 ± 0.01	ND

succinic acid per cell. Further modification is needed to perform in order to improve the yield of succinate production.

In summary, our results show for the first study, whereby differences in the metabolic key enzyme activities regarding succinic acid production by engineering *C. glutamicum* under different culture pH conditions are reported. Adjustment of the fermentation process is still necessary to improve succinic acid production and yields by the *C. glutamicum* strains of the present study. It is no doubt that fermentation process control is supposed to be adapted for each particular strain, as they appeared different succinate production responses toward similar culture conditions. There were several improvements of the culture process, firstly, optimized cultivation conditions for biomass production during aerobic phase. Secondly, broadening its substrate utilization was in order to produce succinic acid from lignocellulosic biomass. Furthermore, the improvement of genetically modified strain may increase the overall efficiency of the process.

Acknowledgments This work was supported by the 973 Program of China (grant no. 2011CB707405).

References

1. Abe S, Takayama K, Kinoshita S (1967) Taxonomical studies on glutamic acid producing bacteria. J Gen Appl Microbiol 13:279–301
2. Boris L, Melanie B, Michael B (2012) Toward homosuccinate fermentation: metabolic engineering of *Corynebacterium glutamicum* for anaerobic production of succinate from glucose and formate. Appl Environ Microbiol 9:3325–3337
3. Bradford MM (1976) A rapid and sensitive method for the quantitation of microgram quantities of protein utilizing the principle of protein-dye binding. Anal Biochem 72:248–254
4. Canovas JL, Kornberg HL (1969) Phosphoenolpyruvate carboxylase from *Escherichia coli*. Methods Enzymol 13:288–292
5. Carole TM, Pellegrino J, Paster MD (2004) Opportunities in the industrial biobased products industry. Appl Biochem Biotechnol 113:871–885
6. Chattrjee R, Millard CS, Champion K, Clark DP, Donnelly MI (2001) Mutation of the ptsG gene results in increased production of succinate in fermentation of glucose by *Escherichia coli*. Appl Environ Microbiol 67:148–154
7. Uy Davin, Delaunay Stephane, Engasser Jean-Marc, Goergen Jean-Louis (1999) A method for the determination of pyruvate carboxylase activity during the glutamic acid fermentation with *Corynebacterium glutamicum*. J Microbiol Method 39:91–96
8. Donnelly MI, Millard CS, Stols L (1998) Mutant *E. coli* strain with increased succinic acid production. US Patent 5,770,435
9. Glassner DA, Datta R (1992) Process for the production and purification of succinic acid. US Patent 5,143,834
10. Hermann T (2003) Industrial production of amino acids by coryneform bacteria. J Biotechnol 104:155–172
11. Inui M, Kawaguchi H, Murakami S, Alain AV, Yukawa H (2004) Metabolic engineering of *Corynebacterium glutamicum* for fuel ethanol production under oxygen-deprivation conditions. J Mol Microbiol Biotechnol 8:243–254

12. Inui M, Murakami S, Okino S, Kawaguchi H, Vertes AA, Yukawa H (2004) Metabolic analysis of *Corynebacterium glutamicum* during lactate and succinate productions under oxygen deprivation conditions. J Mol Microbiol Biotechnol 7:182–196
13. Kabus A, Niebisch A, Bott M (2007) Role of cytochrome bd oxidase from *Corynebacterium glutamicum* in growth and lysine production. Appl Environ Microbiol 73:861–868
14. McKinlay JB, Vieille C, Zeikus JG (2007) Prospects for a bio-based succinate industry. Appl Microbiol Biotechnol 76:727–740
15. Okino S, Inui M, Yukawa H (2005) Production of organic acids by *Corynebacterium glutamicum* under oxygen deprivation. Appl Microbiol Biotechnol 68:475–480
16. Okino S, Noburyu R, Suda M, Jojima T, Inui M, Yukawa H (2008) An efficient succinic acid production process in a metabolically engineered *Corynebacterium glutamicum* strain. Appl Microbiol Biotechnol 81:459–464
17. Okino S, Suda M, Fujikura K, Inui M, Yukawa H (2008) Production of D-lactic acid by *Corynebacterium glutamicum* under oxygen deprivation. Appl Microbiol Biotechnol 78:449–454
18. Pamela KB, Fairoz MJ, Norizan L, David PC (1997) The ldhA gene encoding the fermentative lactate dehydrogenase of *Escherichia coli*. Microbiol 143:187–195
19. Peng L, Shimizu K (2003) Global metabolic regulation analysis for *Escherichia coli* K12 based on protein expression by 2-dimensional electrophoresis and enzyme activity measurement. Appl Microbiol Biotechnol 61:163–178
20. Peters-Wendisch PG, Schiel B, Wendisch VF, Katsoulidis E, Mockel B, Sahm H, Eikmanns BJ (2001) Pyruvate carboxylase is a major bottleneck for glutamate and lysine production by *Corynebacterium glutamicum*. J Mol Microbiol Biotechnol 2:295–300
21. Sambrook J, Fritsch EF, Maniatis T (1989) Molecular cloning: a laboratory manual. Cold Spring Harbor, New York
22. Samuelov NS, Lamed R, Lowe S, Zeikus JG (1991) Influence of CO_2–HCO_3—levels and pH on growth, succinate production, and enzyme activities of *Anaerobiospirillum succiniciproducens*. Appl Environ Microbiol 57:3013–3019
23. Schäfer A, Tauch A, Jager W (1994) Small mobilizable multi-purpose cloning vectors derived from the *Escherichia coli* plasmids pK18 and pK19: selection of defined deletions in the chromosome of *Corynebacterium glutamicum*. Gene 145:69–73
24. Toyoda K, Teramoto H, Inui M, Yukawa H (2009) The *ldhA* gene, encoding fermentative l-lactate dehydrogenase of *Corynebacterium glutamicum*, is under the control of positive feedback regulation mediated by LldR. J Bacteriol 13:4251–4258
25. Urbance SE, Pometto AL, DiSpirito AA, Demirci A (2003) Medium evaluation and plastic composite support ingredient selection for biofilm formation and succinic acid production by *Actinobacillus succinogenes*. Food Biotechnol 17:53–65
26. Van der Rest ME, Lange C, Molenaar D (1999) A heat shock following electroporation induces highly efficient transformation of *Corynebacterium glutamicum* with xenogeneic plasmid DNA. Appl Microbiol Biotechnol 52:541–545
27. Van der Werf MJ, Guettle MV, Jain MK, Zeikus JG (1997) Environmental and physiological factors affecting the succinic acid product ratio during carbohydrate fermentation by *Actinobacillus* sp. 130Z. Arch Microbiol 167:332–342
28. Vemuri GN, Eiteman MA, Altman E (2002) Succinate production in dual-phase *Escherichia coli* fementations depends on the time of transition from aerobic to anaerobic conditions. J Ind Microbiol Biotechnol 28:325–332
29. Werpy T, Petersen G, Aden A et al (2004) Top value added chemicals from biomass. Department of Energy, Washington, DC, pp 31–33
30. Wendisch VF, Bott M, Eikmanns BJ (2006) Metabolic engineering of *Escherichia coli* and *Corynebacterium glutamicum* for biotechnological production of organic acids and amino acids. Curr Opin Microbiol 9:268–274

31. Willke Th, Vorlop KD (2004) Industrial bioconversion of renewable resources as an alternative to conventional chemistry. Appl Microbiol Biotechnol 66:131–142
32. Zeikus JG, Jain MK, Elankovan P (1999) Biotechnology of succinic acid production and markets for derived industrial products. Appl Microbiol Biotechnol 51:545–552
33. Zeikus JG, Fuchs G, Kenealy W, Thauer RK (1977) Oxidoreductases involved in cell carbon synthesis of *Methanobacterium thermoautotrophicum*. J Bacteriol 132:604–613

Chapter 41
Effects of Stimulators on Lutein and Chlorophyll Biosyntheses in the Green Alga *Chlorella pyrenoidosa* Under Heterotrophic Conditions

Tao Li, Dongqing Bai, Lin Tian, Ping Li, Yihan Liu and Yue Jiang

Abstract In this study, the effects of some stimulators on lutein and chlorophyll biosyntheses in dark grown *Chlorella pyrenoidosa* were investigated. The aim of this study was to find out some stimulators that could effectively stimulate lutein and chlorophyll biosyntheses in *C. pyrenoidosa* in darkness. Vitamin A showed a profound effect in stimulating lutein biosynthesis, 70.9 % of increase of lutein content was observed when 4 mM vitamin A was supplemented in the culture. Similarly, chlorophyll biosynthesis was also improved by vitamin A. In addition, β-carotene and zeaxanthin also showed a positive effect on lutein accumulation in *C. pyrenoidosa* under heterotrophic conditions. The lutein content increased by 27.1 and 19.8 % when 0.1 mg/mL zeaxanthin and β-carotene were added into the culture, respectively. Besides, added zeaxanthin also significantly improved the lutein yield, by 45.4 %. In conclusion, lutein biosynthesis could be effectively stimulated by the selected stimulators in the green alga *C. pyrenoidosa* under heterotrophic conditions.

Keywords Lutein · *Chlorella pyrenoidosa* · Heterotrophic · Stimulators

T. Li (✉) · L. Tian · P. Li
Department of Basic Science, Tianjin Agricultural University, Tianjin 300384
People's Republic of China
e-mail: litao823@gmail.com

T. Li · D. Bai
Tianjin Key Laboratory of Aqua-ecology and Aquaculture, Tianjin, People's Republic of China

Y. Liu
The College of Biotechnology, Tianjin University of Science and Technology, Tianjin 300457, People's Republic of China

Y. Jiang (✉)
College of Food Science and Technology, Jiangnan University, Wuxi, People's Republic of China
e-mail: yjiang@hotmail.com.hk

T.-C. Zhang and M. Nakajima (eds.), *Advances in Applied Biotechnology*,
Lecture Notes in Electrical Engineering 332, DOI 10.1007/978-3-662-45657-6_41

41.1 Introduction

The carotenes, i.e., β-carotene, phytoene, lycopene, and their oxygenated derivatives, i.e., lutein, zeaxanthin and astaxanthin, can be produced by plants and microorganisms. These natural occurring carotenoids have important biological and physiological effects on human diseases [10, 11]. Now most of the carotenoids are produced from plants, such as lutein from marigold, but this production way is however limited by the planting area, weather conditions, and high manpower cost. On the contrary, the microbial carotenoids production is performed in an enclosed bioreactor by fermentation and could avoid those drawbacks, but the carotenoids contents are quite low. Therefore, it is quite necessary to find out an effective way to enhance the microbial carotenoids biosynthesis. Previous studies have shown that the carotenoids biosynthesis could be stimulated by certain compounds extracted from fungi, bacteria, yeast, etc. Other compounds, such as terpenes, ionones, amines alkaloids, antibiotics, vitamin A, abscisic acid as well as trisporic acid, also stimulated the β-carotene biosynthesis in *Blakeslea trispora* effectively [6, 8]. Some intermediates of TCA cycle and glycolysis which are reported to participate in the carbon skeletons formation of carotenoid were able to improve carotenoids biosynthesis greatly in many organisms [3].

Lutein and chlorophyll can be produced by the green alga *Chlorella pyrenoidosa* under heterotrophic conditions. The production of lutein could be significantly enhanced by employing high cell density culture techniques [9, 11]. However, previous studies did not focus on the improvement of lutein accumulation in a single algal cell and the cellular lutein content was still very low even though the lutein yield increased.

The objective of this study was to explore effective strategies to enhance the cellular lutein biosynthesis and production in the green alga *C. pyrenoidosa* under heterotrophic conditions. The outcome of this study will provide reference for the production of photosynthetic pigments by dark grown *C. pyrenoidosa* in large industrial scale in the future.

41.2 Materials and Methods

41.2.1 Microalga and Cultivation Conditions

The algal strain *C. pyrenoidosa* was purchased from Carolina Biological Supply Co., Burlington, USA. The medium was prepared according to a previous report [10]. The concentration of glucose and nitrate potassium in the medium was 40 and 8.25 g/L, respectively. The alga was inoculated into 250 mL Erlenmeyer flask containing 100 mL medium at inoculation volume of 10 % (v/v). The heterotrophic cultivation was performed by incubating the algal cells at 28 with orbital shaking at 180 rpm in darkness. Stimulators or intermediates dissolved in DMSO were added into the culture after 24 h of cultivation to investigate their effects on enhanced lutein biosynthesis. The algal cells were then harvested after 72 h of continuous cultivation for subsequent analyses. The culture supplemented with same volume of DMSO was set as control.

41.2.2 Biomass and Pigment Analyses

The biomass was determined by filtering 5 mL liquid culture through a pre-dried filter paper (Whatman) and the obtained wet algal pellets were rinsed twice with deionized water. Afterwards the pellets were dried at 105 °C to a constant weight

The lutein and chlorophyll contents were determined by following the report by Baroli [2]. The wet algal pellets obtained after centrifugation (3000×g, 5 min) were frozen at −70 °C for at least 1 h. Then the cell pellets were lyophilized for 36 h at a freeze dryer. Afterwards the dry cell pellets were grounded into a powder in a mortar. The pigments were extracted with acetone until the cell debrises became colorless. Then the supernatant containing extracted pigments was filtered through a 0.22 μm organic membrane (Millipore). Each sample (20 μL) was separated on a Waters Spherisorb 5 μm ODS 4.6 × 250 mm analytical column (Waters). Samples were eluted at a flow rate of 1.0 mL/min with a linear gradient from 100 % solvent A [acetonitrile: methanol: 0.1 M Tris-buffer, pH 8.0 (84:2:14, by vol.)] to 100 % solvent B [methanol: ethyl acetate (68:32, by vol.)] for 15 min followed by 10 min of solvent B. Individual pigments were identified by comparing their absorption spectra to those of standards (Sigma-Aldrich). The concentration of each pigment was calculated with corresponding standard curve.

41.2.3 Statistical Analysis

The data were expressed as average ± S.D. Statistical significance was evaluated by independent paired two sample *t*-test (OriginPro 7.0) when comparing two groups. $P < 0.05$ and 0.01 were recognized as significant and extremely significant, which were denoted with * and **, respectively.

41.3 Results and Discussion

41.3.1 Effects of Vitamin A on Lutein and Chlorophyll Biosyntheses

The final concentration of vitamin A in the culture was 0.2, 1.0, 2.0, 4.0 mM, respectively. Low concentration of vitamin A (0.2 mM) could significantly stimulate cell growth and the cell dry weight increased by 29.3 % compared with the control

($p < 0.05$). However, high concentration of vitamin A (2 and 4 mM) inhibited cell growth severely, decreased by 51.8 and 86.2 %, respectively (2 mM, $p < 0.01$; 4 mM, $p < 0.01$) (Fig. 41.1). Previous study showed that high concentration of vitamin A did not suppress the growth of *B. trispora* [5]. Previous study showed that high concentration of vitamin A did not suppress the growth of *B. trispora* [5].

It was found that high concentration of vitamin A induced a significant increase of cellular lutein content compared with the control (Fig. 41.2). The lutein content in the algal cells cultivated in the medium supplemented with 1 mM vitamin A reached 3.4 mg/g, increased by 27.6 % compared with the control. Moreover, the cellular lutein content even increased by 70.9 % ($p < 0.01$) when 4 mM vitamin A was supplemented, although the lutein yield was lower than the control due to the growth retardation caused by high concentration of vitamin A (Fig. 41.2). But low concentration of vitamin A caused a significant increase of lutein yield, increased by 37.5 % c when 0.2 mM vitamin A was added into the culture ($p < 0.01$) (Fig. 41.2). Similarly, vitamin A also has been reported to stimulate the carotenogenesis in other microorganism. Supplementation of 1,000 ppm vitamin A acetate in the medium caused more than twofold increase of β-carotene production in *B. trispora* [5].

The chlorophyll biosynthesis in *C. pyrenoidosa* under heterotrophic conditions was also stimulated by vitamin A (Fig. 41.3). The cellular chlorophyll a and b contents significantly increased by 52.9 and 65.2 %, respectively, when 4 mM vitamin A was supplemented ($p < 0.01$). The effects of vitamin A on

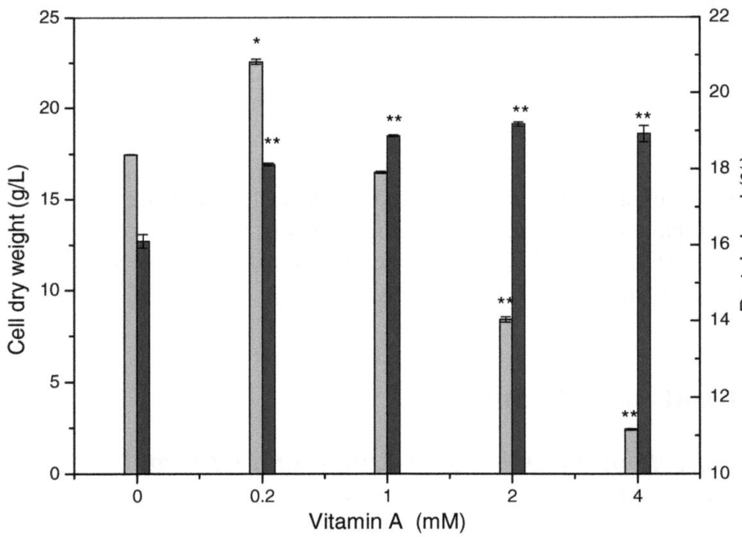

Fig. 41.1 Effects of supplemented vitamin A on the growth and protein synthesis of *C. pyrenoidosa* under heterotrophic condition. ▭, cell dry weight; ■, protein level. Data points were expressed as averages ±S.D. of three independent measurements

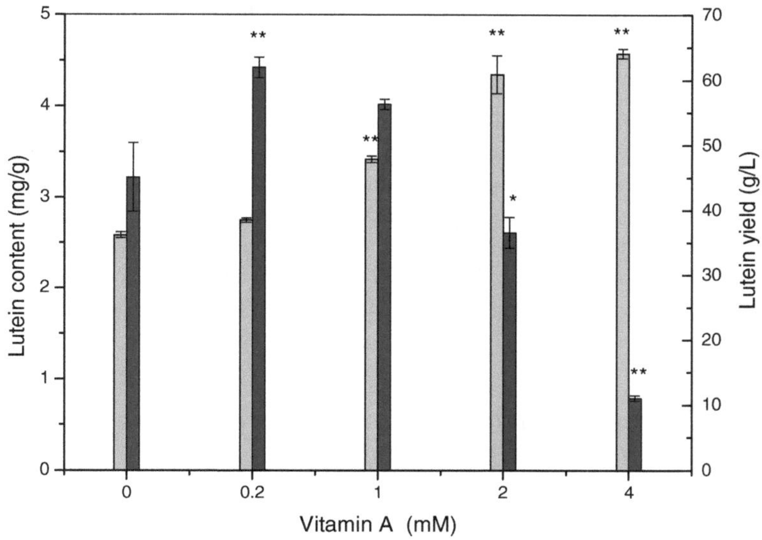

Fig. 41.2 Effects of supplemented vitamin A on lutein biosynthesis and production in *C. pyrenoidosa* under heterotrophic conditions. [light grey box], lutein content; [dark grey box], lutein yield. Data points were expressed as averages ±S.D. of three independent measurements

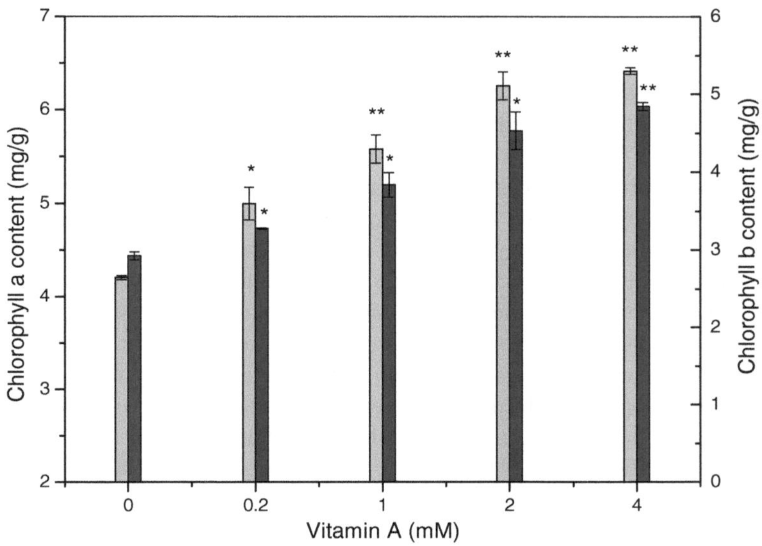

Fig. 41.3 Effects of supplemented vitamin A on chlorophyll biosynthesis in *C. pyrenoidosa* under heterotrophic conditions. [light grey box], chlorophyll a content; [dark grey box], chlorophyll b content. Data points were expressed as averages ±S.D. of three independent measurements

carotenogenesis has been reported in fungi and yeast et al. [6, 8]. No reports have ever been reported in plant and algae so far. In our study, we found that the change of lutein and chlorophyll contents in *C. pyrenoidosa* was synchronous when vitamin A was supplemented during the cultivation (Figs. 41.2 and 41.3). The protein level in *C. pyrenoidosa* also elevated after supplementing vitamin A in the culture (Fig. 41.1). Previous studies indicated that vitamin A might interact with many crucial proteins involved in signal transduction pathways [1, 4]. In our study, we speculated that this interaction would possibly activate the expression of nuclear-located photosynthetic genes, such as genes encoding light-harvesting complex II and key enzymes involved in chlorophyll and lutein biosynthetic pathways. As a result, both lutein and chlorophyll biosyntheses were enhanced simultaneously.

41.3.2 Effects of Carotenoid Intermediates on Lutein and Chlorophyll Biosyntheses

The carotenoid intermediates including zeaxanthin, lycopene and β-carotene were added into the culture of *C. pyrenoidosa* after 24 h of cultivation. Supplemented zeaxanthin and β-carotene (0.1 mg/mL) enhanced lutein biosynthesis significantly, by 27.1 and 19.8 %, respectively ($p < 0.05$) (Fig. 41.4). But no significant change of lutein content was observe if we improved the amount of added zeaxanthin and β-carotene (0.5 mg/mL) (Fig. 41.4). Besides, addition of zeaxanthin also significantly improved the lutein yield, by 45.4 % ($p < 0.05$) (Fig. 41.4), whereas β-carotene did not show significant effect on enhanced lutein yield. As for lycopene, no obvious influence on the lutein content and yield of *C. pyrenoidosa* under heterotrophic conditions was observed (Fig. 41.4).

Similarly, supplemented zeaxanthin also stimulated the chlorophyll biosynthesis in the green alga *C. pyrenoidosa* under heterotrophic conditions. The content of chlorophyll a and b contents increased by 12.1 and 16.3 %, respectively, when 0.1 mg/mL zeaxanthin was supplemented in the culture (Fig. 41.5). In plants and green algae, carotenoid cyclization is a key branching point for the biosynthesis of lutein, zeaxanthin, astaxanthin and so on. Lutein and zeaxanthin are synthesized in two diverse branch pathways using α-carotene and β-carotene as precursors, respectively [7].

In our study, the supplemented β-carotene or zeaxanthin might lead to feedback inhibition on β-carotene and zeanxanthin biosynthetic pathway (Fig. 41.6). As a result, more lycopene molecules were employed for the lutein biosynthesis in *C. pyrenoidosa* under heterotrophic conditions. Although from the viewpoint of industrial application, supplementation of zeaxanthin, and β-carotene in the lutein production by *C. pyrenoidosa* has no application value due to their high price. However, our study provides a hint that the lutein biosynthesis might be enhanced greatly if we block the zeaxanthin branch pathway or increase the expression level of lycopene ε-cyclase that is responsible for the lutein biosynthesis in the green alga *C. pyrenoidosa* under heterotrophic conditions.

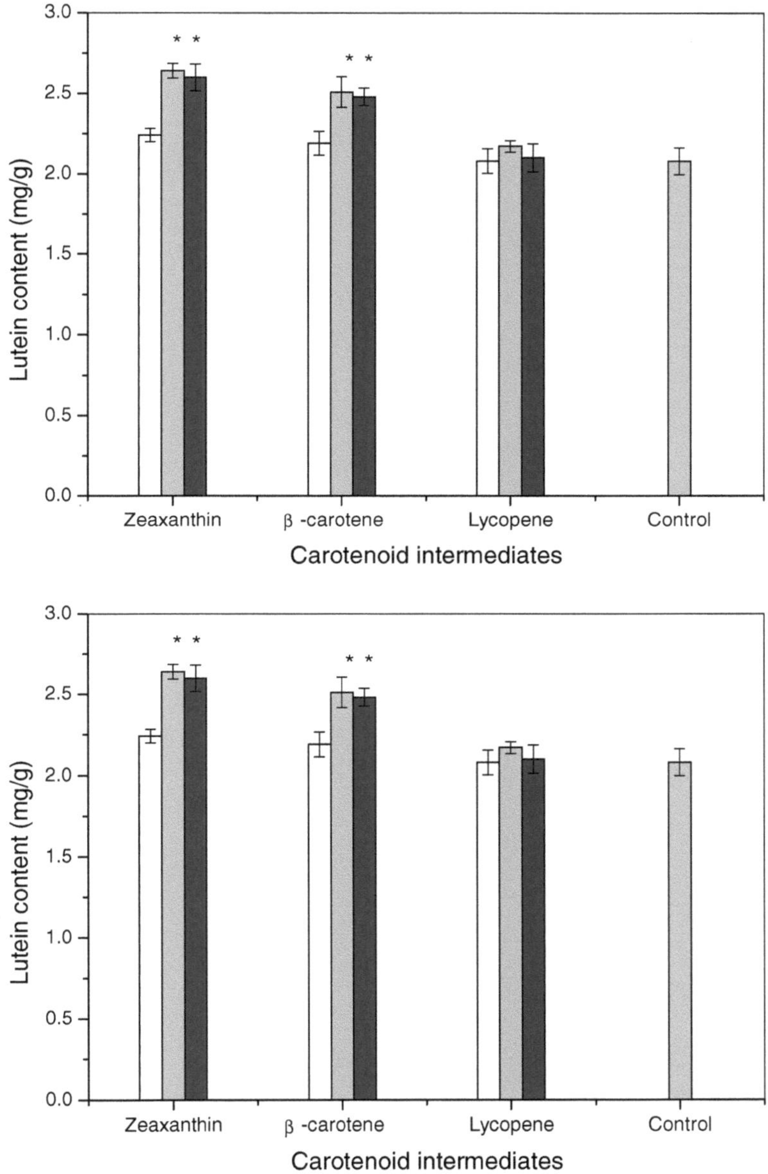

Fig. 41.4 Effects of supplemented carotenoid intermediates on lutein biosynthesis and production in *C. pyrenoidosa* under heterotrophic conditions. ☐, 0.02 mg/mL; ☐, 0.1 mg/mL; ☐, 0.5 mg/mL. Data were expressed as averages ±S.D. of three independent measurements

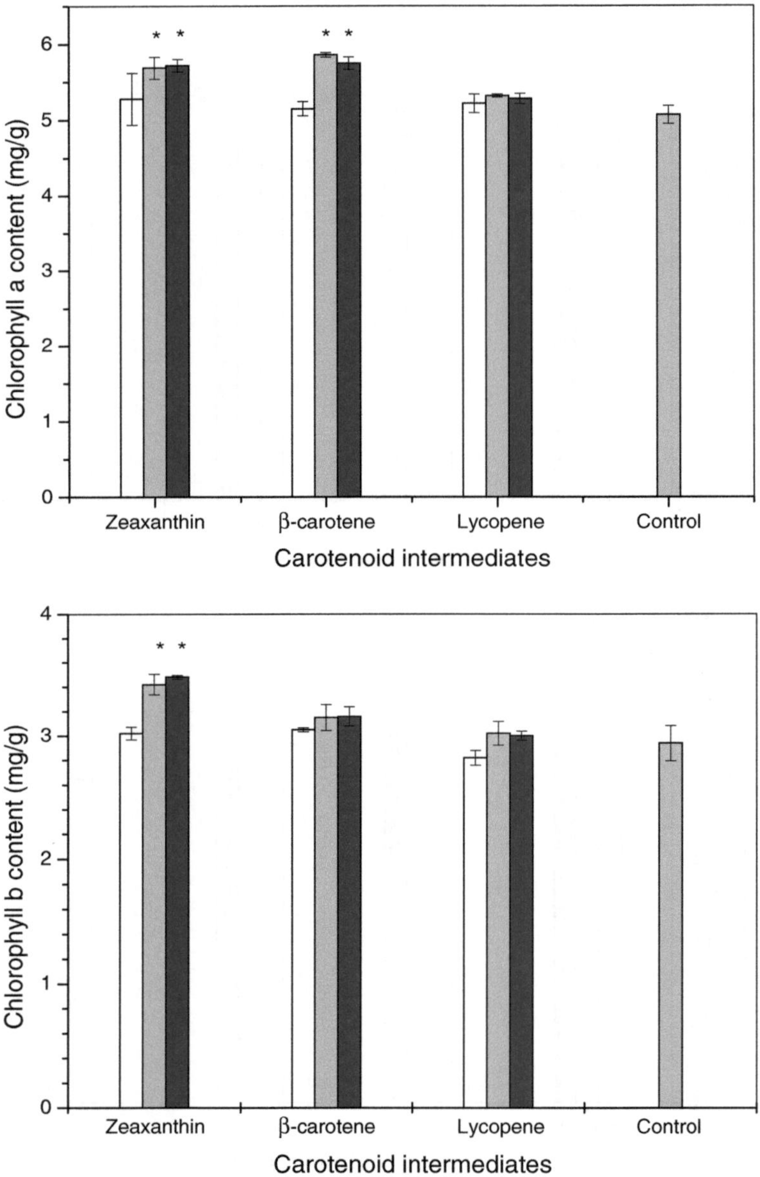

Fig. 41.5 Effects of supplemented carotenoid intermediates on chlorophyll biosynthesis in *C. pyrenoidosa* under heterotrophic conditions. ☐, 0.02 mg/mL; ▒, 0.1 mg/mL; ■ 0.5 mg/mL. Data were expressed as averages ±S.D. of three independent measurements

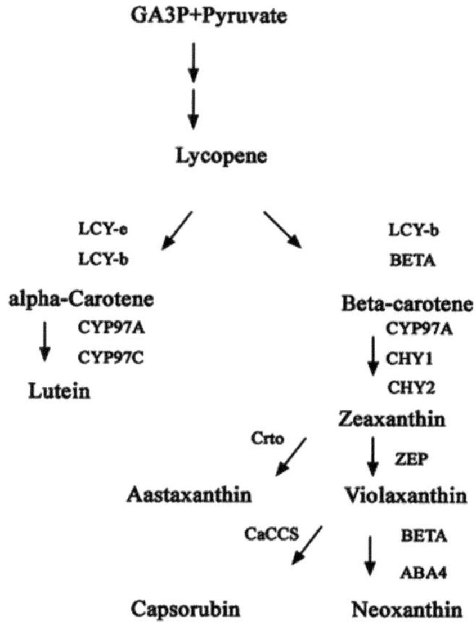

Fig. 41.6 The carotenoid biosynthetic pathway in plants and *green* alga. GA3P, glyceraldehyde-3-phosphate; Lcy-e, lycopene ε –cyclase; Lcy-b, lycopene β-cyclase; BETA, chromoplast-specific beta-cyclase; CYP97A, cytochrome P450 hydroxylase; CYP97C, cytochrome P450 hydroxylase; CHY1, CHY2, non-heme carotene hydroxylases; ZEP, zeaxanthin epoxidase; Crto, ketolase; CaCCS, capsanthin/capsorubin synthase. Adapted from Giuliano et al. [12]. Copyright 2008 by Elsevier Ltd

41.4 Conclusions

Our study showed that high concentration of vitamin A had a profound effect in stimulating lutein and chlorophyll biosyntheses, whereas, low concentration of vitamin A significantly enhanced lutein production. The carotenoid intermediates, such as β-carotene and zeaxanthin, also show significant effects on stimulating lutein and chlorophyll biosyntheses. The output of the research will be useful for the understanding of photosynthetic pigments biosynthesis in dark-grown *C. pyrenoidosa.*

Acknowledgments The research was supported by large-scale equipment sharing fundating of Tianjin Agricultural University, Funding project for Young and Middle-aged Backbone Teachers in University of Tianjin.

References

1. Apostolatos H, Apostolatos A, Vickers T, Watson JE, Song S, Vale F, Cooper DR, Sanchez-Ramos J, Patel NA (2010) Vitamin A metabolite, all-trans-retinoic acid, mediates alternative splicing of protein kinase C delta VIII (PKC delta VIII) isoform via splicing factor SC35. J Biol Chem 285(34):25987–25995. doi:10.1074/jbc.M110.100735

2. Baroli I (2003) Zeaxanthin accumulation in the absence of a functional xanthophyll cycle protects *Chlamydomonas reinhardtii* from photooxidative stress. Plant Cell Online 15 (4):992–1008. doi:10.1105/tpc.010405
3. Bhosale P (2004) Environmental and cultural stimulants in the production of carotenoids from microorganisms. Appl Microbiol Biotechnol 63(4):351–361
4. Bushue N, Wan Y-JY (2010) Retinoid pathway and cancer therapeutics. Adv Drug Deliv Rev 62(13):1285–1298
5. Choudhari SM, Ananthanarayan L, Singhal RS (2008) Use of metabolic stimulators and inhibitors for enhanced production of β-carotene and lycopene by *Blakeslea trispora* NRRL 2895 and 2896. Bioresour Technol 99(8):3166–3173
6. Dandekar S, Modi V, Jani U (1980) Chemical regulators of carotenogenesis by *Blakeslea trispora*. Phytochemistry 19(5):795–798
7. Fraser PD, Bramley PM (2004) The biosynthesis and nutritional uses of carotenoids. Prog Lipid Res 43(3):228–265
8. Govind N, Amin A, Modi V (1982) Stimulation of carotenogenesis in *Blakeslea trispora* by cupric ions. Phytochemistry 21(5):1043–1044
9. Liu L-M, Du G-C, Li Y, Li H-Z, Chen J (2005) Enhancement of pyruvate production by Torulopsis glabrata through supplement of oxaloacetate as carbon source. Biotechnol Bioprocess Eng 10(2):136–141
10. Shi X-M, Chen F, Yuan J-P, Chen H (1997) Heterotrophic production of lutein by selected Chlorella strains. J Appl Phycol 9(5):445–450
11. Wu Z, Shi X (2007) Optimization for high-density cultivation of heterotrophic chlorella based on a hybrid neural network model. Lett Appl Microbiol 44(1):13–18. doi:10.1111/j.1472-765X.2006.02038.x
12 Giuliano G et al (2008) Metabolic engineering of carotenoidbiosynthesis in plants. Trends Biotechnol 26:139–145

Chapter 42
Pharmacophore-Based Virtual Screening and Result Analysis of Histone Methyltransferase SMYD3 Inhibitors

Shaodan Liu, Ziyue Sun, Yonghui Zhong, Qingxin Cui, Xuegang Luo and Yujie Dai

Abstract Histone methyltransferase SMYD3 is a kind of protease relating to different kinds of tumors, which makes it a promising target for anticancer drugs. In order to find new small molecule inhibitors of SMYD3 for oncotherapy, a pharmacophore was created based on the crystal structure of SMYD3 with sinefungin (SFG) from Protein Data Bank (PDB ID: 3PDN) by using the software of Ligand Scout 3.1. The virtual screening of ZINC database was conducted with the pharmacophore using MOE software. Two hundred and four hits which meet at least five pharmacophore characteristics were obtained and four structural characteristics were summarized from these hits, which have laid a data foundation for the discovery of novel SMYD3 inhibitors.

Keywords Histone methyltransferase · SMYD3 · Virtual screening · Pharmacophore

42.1 Introduction

SMYD3 (SET and MYND domain-containing protein 3) is a kind of histone methyltransferase which belongs to the SMYD family. This kind of histone methyltransferases has two functional groups, including the SET domain which can methylate histone H3 and the MYND domain related to protein–protein

S. Liu · Z. Sun · Y. Zhong · X. Luo · Y. Dai (✉)
Key Laboratory of Industrial Fermentation Microbiology, Ministry of Education,
College of Bioengineering, Tianjin University of Science and Technology,
No. 29 of 13th St. TEDA, Tianjin 300457, People's Republic of China
e-mail: yjdai@126.com

Q. Cui
College of Pharmacy, Nankai University, Tianjin 300071, People's Republic of China

T.-C. Zhang and M. Nakajima (eds.), *Advances in Applied Biotechnology*,
Lecture Notes in Electrical Engineering 332, DOI 10.1007/978-3-662-45657-6_42

interactions [1]. The SET domain of SMYD3 can receive the methyl group from S-adenosyl methionine (SAM) and catalyze the di-/trimethylation of histone H3K4 specifically, then leads to changes in chromatin structure and affects transcription of the genes. What's more, SMYD3 can combine with downstream genes, cell cycle genes, and signal transduction related genes. The combination influences gene's expression, also promotes tumor cell proliferation and apoptosis. As for the MYND domain, it can be combined with sequences in gene promoter area, prompts SET domain's methylation function and affects gene transcription. By now, SMYD3 has been proved to have close relations with the occurrence and development of colon cancer, liver cancer, breast cancer, and prostate cancer, so it can be a new tumor target for anticancer drugs [2–5].

The development of a new drug often needs a large investment and high technologies. It is a long time to go and usually companied with high risks. With the rapid development of modern science and technology, Computer-Aided Drug Design (CADD) emerged. The application of CADD method not only accelerates the speed of drug discovery, but also improves the hit rate of successful research and development. In this research, CADD was employed for virtual screening and structure analysis of novel SMYD3 inhibitors.

42.2 Materials and Methods

42.2.1 Experimental Materials

The crystal structure of SMYD3 with the ligand sinefungin (SFG) (3PDN) was downloaded from RCSB Protein Data Bank (http://www.rcsb.org/). Pharmacophore model was created by using the software of Ligand Scout 3.1 [6]. Protein modification and virtual screening were done by using the software of MOE 2013.08 [7]. The small molecule database for the virtual screening of SMYD3 inhibitors was ZINC database [8].

42.2.2 Experimental Methods

42.2.2.1 Generation of Pharmacophore

The PDB file 3PDN.pdb downloaded from RCSB Protein Data Bank was modified by deleting all the water molecules and adding H atoms to the protein in MOE software. Then it was opened with Ligand Scout 3.1 and the pharmacophore was generated based on the interaction between the ligand Sinefungin (SFG, Fig. 42.1) and the receptor of SMYD3. It was saved as a PH4 format pharmacophore file for the virtual screening of SMYD3 inhibitors.

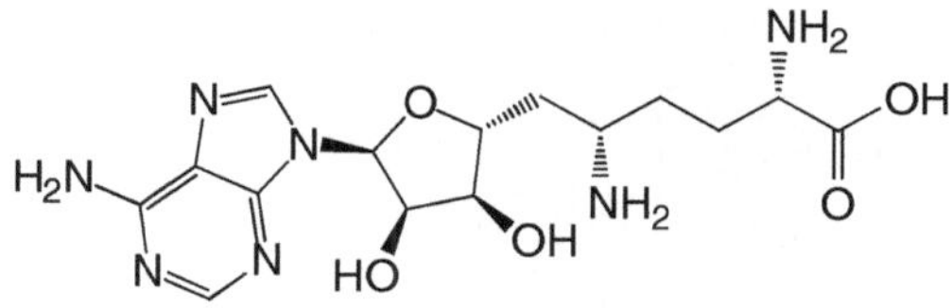

Fig. 42.1 The formula of SFG

42.2.2.2 Virtual Screening

The virtual screening of SMYD3 inhibitors using the pharmacophore generated as 2.2.1 was conducted using MOE 2013.08. Before virtual screening is being run, the 3d sdf format ZINC database files should be conversed to MDB format. Then the PH4 format pharmacophore file was opened and the virtual screening was performed with at least five pharmacophore queries match.

42.3 Result and Discussion

In this study, the pharmacophore model was generated using the ligand SFG as the template of SMYD3 inhibitor. It has a strong SMYD3 inhibiting activity. The interaction between SFG and the acceptor represents the common characteristics which SMYD3 inhibitors possess. The main characteristics of the pharmacophore generated are shown in Figs. 42.2 and 42.3.

From the pharmacophore generated, we can see that SFG forms 13 H-bonds with amino acid residues of SMYD3. It takes as electron pair donors when it forms hydrogen bonds with ASN16A, ASN205A, SER202A, CYS180A, ASN181A, TYR239A, and HIS206A (green lines in Figs. 42.2 and 42.3). However, it takes as electron pair acceptors when it forms hydrogen bonds with TYR124A, ASN16A, ASN132A, and HIS206A red lines in Figs. 42.2 and 42.3. Also SFG forms 2 π-π interactions with PHE259A (blue lines). These interactions give an explanation of the potent SMYD3 activity inhibition of SFG.

Based on at least five pharmacophore queries to match, 556 ZINC database files, which include more than 1.2 million compounds, were screened and 204 hits were obtained. In the screening result 204 molecules, some hits have very similar structure with that of SFG. Most of those molecules contain adenosine part or adenosine-like part. We are aiming at finding novel SMYD3 inhibitors through virtual screening, so this kind of molecules are excluded in the afterward analysis.

With the analysis and comparison, the result can be summarized into four categories. The structure descriptions and activity predictions for representative compounds of each category are shown in Table 42.1.

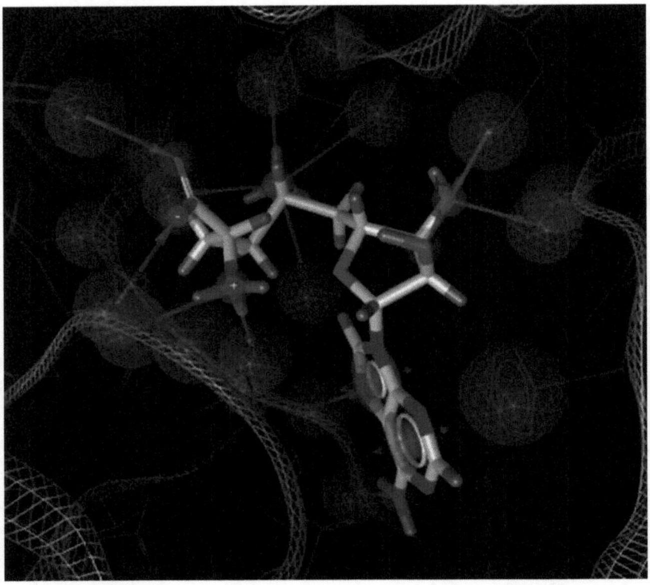

Fig. 42.2 The pharmacophore generated between SFG and SMYD3(3D)

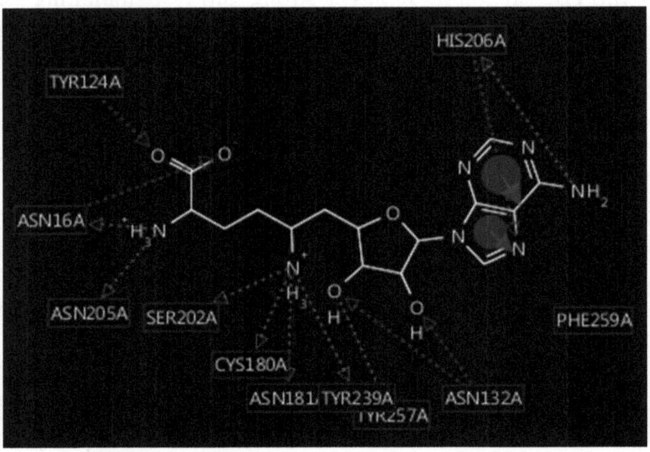

Fig. 42.3 The pharmacophore generated between SFG and SMYD3(2D)

Table 42.1 Representatives, structure descriptions, and activity predictions for four categories of 204 hits

Category	Representative	Structure description	Activity prediction
1	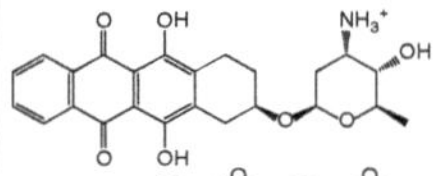 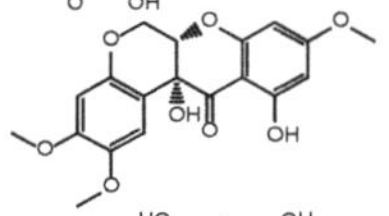 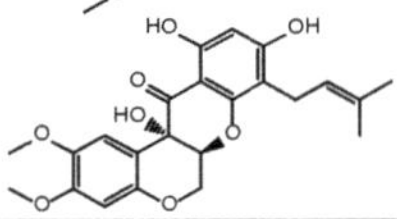	With tetracyclic or similar structure, the whole molecule is like a strip	The strip-like molecule is easy to enter the cavity of the protein active site. Some side chains may enhance the interaction and get ideal inhibitors
2	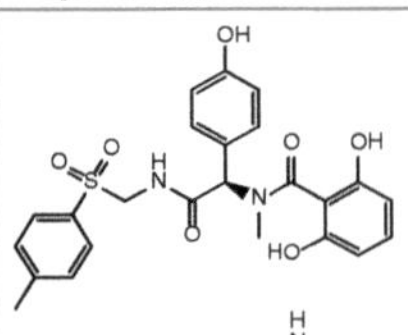 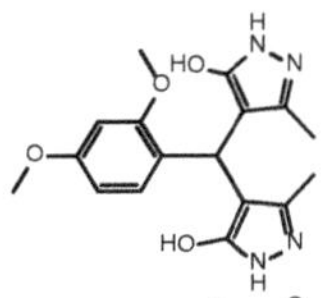 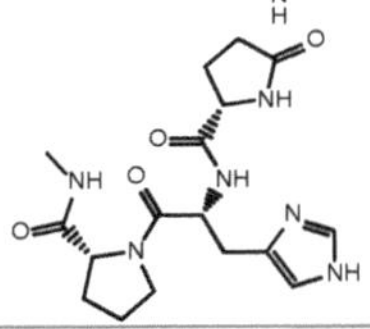	With a carbon atom as the center, three branches with aromatic or other rings on the end are stretching out in different directions	This kind of structure is very stereo, and once it combined with acceptor, it is hard to escape from the active site
3	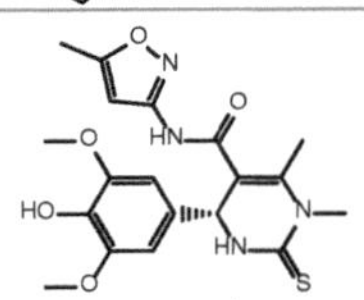 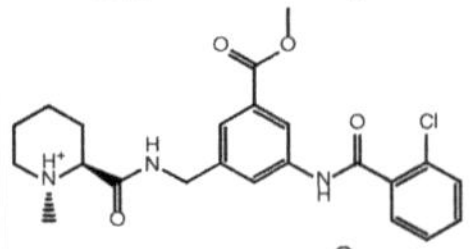 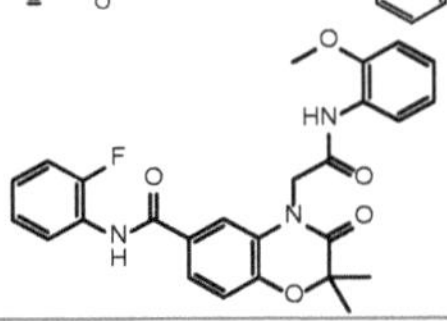	With a ring as the center, two branches with aromatic or other rings on the end are stretching out in different directions	This kind of molecules is not that stereo, but easier to get into the cavity. Aromatic ring on the branches can increase combination interaction

(continued)

Table 42.1 (continued)

Category	Representative	Structure description	Activity prediction
4	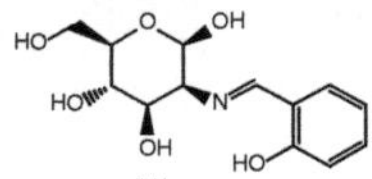 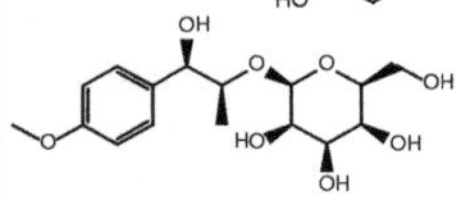 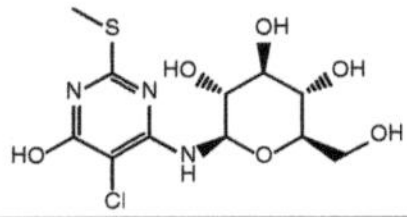	Usually has two parts of ring structure bridged together. One part is an ether ring modified with several hydroxyl groups	The hydroxyl groups help form H-Bonds, and enhance combination with acceptor

42.4 Conclusion

SMYD3 is a kind of protease relating to different kinds of tumors, which makes it a promising anticancer target. But up to now, there are few SMYD3 inhibitors reported. It is necessary to find new chemical entities with the help of CADD. The virtual screening result using the pharmacophore generated from the crystal structure of SMYD3 complexed with SFG will provide directions and technical support for the practical screening of SMYD3 inhibitors.

Acknowledgments This work was supported by the National Natural Science Foundation of China (No. 21272171).

References

1. Foreman KW, Brown M, Park F, Emtage S, Harriss J, Chhaya D, Zhu L, Crew A, Arnold L, Shaaban S, Tucker P (2011) Structural and functional profiling of the human histone methyltransferase SMYD3. PLoS One 6(7):e22290
2. Luo XG, Pan H, Liu ZP, Guo S (2009) Research progress of SMYD3 in the process of carcinogenesis. Chem Life 2:247–249
3. Chen LB, Xu JY, Yang Z, Wang GB (2007) Silencing SMYD3 in hepatoma demethylates RIZI promoter induces apoptosis and inhibits cell proliferation and migration. World J Gastroenterol 13(43):5718
4. Wang JJ, Zhang LM, Dai YP, Guo Y (2010) Effect and possible mechanism of microRNA silencing SMYD3 on invasiveness of breast cancer cell in vitro. Chin J Cancer Prev Treat 19:1532–1536
5. Chen J (2012) A preliminary study of histone methyltransferase SMYD3 in the role of prostate cancer cells PC3[D]. Shandong University, Shandong

6. LigandScout, version 1.0; Inte:Ligand GmbH, Clemesn-Maria-Hofbaurer-G. 6, 2344, Maria Enzersdorf, Austria. http://www.inteligand.com
7. Molecular Operating Environment (MOE) 2013.08 (2013) Chemical Computing Group Inc., 1010 Sherbooke St. West, Suite #910, Montreal, QC, Canada, H3A2R7
8. Irwin JJ, Shoichet BK (2005) ZINC—a free database of commercially available compounds for virtual screening. J Chem Inf Model 45:177–182

Chapter 43
Effects of K_2HPO_4 on the Growth of *Nostoc Flagelliforme* in Liquid Media with Different Carbon Sources

Hexin Lv, Feng Xia, Shiru Jia, Xianggan Cui and Nannan Yuan

Abstract Liquid fermentation is a promising strategy for sustainable production of *Nostoc flagelliforme* cells. In order to elucidate whether phosphate could promote the growth of *N. flagelliforme* in liquid fermentation, the effects of K_2HPO_4 on *N. flagelliforme* growth were investigated in liquid media with 15 mM $NaHCO_3$ and 20 % CO_2(v/v), respectively, in this study. Results showed that K_2HPO_4 could promote *N. flagelliforme* growth under culture conditions with both kinds of carbon sources, respectively. The optimal concentration of K_2HPO_4 for growth and protein contents of cells is 40–60 mg/L for the existence of each carbon source. The maximum dry cell weights of 0.75 g/L and 1.19 g/L were achieved in media with 15 mM $NaHCO_3$ and with 20 % CO_2, respectively, when the concentration of K_2HPO_4 is 60 mg/L. In media with 15 mM $NaHCO_3$, the photosynthetic and respiration rates' uptake of nutrients and protein contents increased significantly when K_2HPO_4 concentration increased to 60 mg/L and then reached a plateau, however, higher concentrations of K_2HPO_4 inhibited EPS accumulation. By contrast, in media with 20 % CO_2, all growth characteristics above of *N. flagelliforme* were promoted, and reached a plateau at higher K_2HPO_4 concentrations. These results provide valuable information for the cultivation of *N. flagelliforme* cells.

Keywords *Nostoc flagelliforme* · CO_2 mitigation · K_2HPO_4 · $NaHCO_3$ · Liquid fermentation

43.1 Introduction

Nostoc flagelliforme is a terrestrial filamentous cyanobacterium which is distributed in some arid and semiarid areas of China and has been used as food for more than 2000 years, and its herbal value was recognized more than 400 years ago [6]. It was

H. Lv · F. Xia · S. Jia (✉) · X. Cui · N. Yuan
Tianjin Key Laboratory of Industrial Microbiology, School of Bioengineering, TEDA, Tianjin University of Science & Technology, 13th street, 300457 Tianjin, China
e-mail: jiashiru@tust.edu.cn

T.-C. Zhang and M. Nakajima (eds.), *Advances in Applied Biotechnology*,
Lecture Notes in Electrical Engineering 332, DOI 10.1007/978-3-662-45657-6_43

reported that *N. flagelliforme* and its exopolysaccharides (EPS) in culture media contain antitumor and antiviral components such as acidic polysaccharide nostoflan [9]. *N. flagelliforme* grows slowly in their natural environment. It plays a major role in soil and water conservation in the arid areas of northern China. Growing market demand for *N. flagelliforme* has led to overharvesting and grassland degradation. Therefore, using artificial cultivation of *N. flagelliforme* to meet the market demand and to protect the environment is necessary [6]. Many attempts of culturing the natural colonial filaments of *N. flagelliforme* have been performed both in the lab and fields [7, 8, 15]. It was reported that the growth of *N. flagelliforme* could be significantly promoted by adding 15 mM $NaHCO_3$ and 20 % CO_2 [10, 11]. Phosphorus is an essential nutrient in algae cultivation, in order to further enhance the yield of *N. flagelliforme*, the effects of K_2HPO_4 on the growth photosynthesis, accumulation of EPS, photosynthetic pigments' contents, and uptake of nutrients in two kinds of carbon sources were investigated.

43.2 Materials and Methods

43.2.1 Algal Strains and Growth Conditions

Dissociated cells of *N. flagelliforme* were obtained according to the previous methods [16]. Culture medium modified $BG11_0$ with 0, 20, 40, 60, 80, and 100 mg/L K_2HPO_4 concentrations were employed [11]. The cultivation methods for $NaHCO_3$ and 20 % CO_2 were the same as that in our previous reports [11, 18].

43.2.2 Dry Cell Weight and EPS Determination

Dry cell weight (DCW) and the amount of EPS were determined according to previously described methods [1].

43.2.3 Chlorophyll a Contents and Oxygen Evolution Rate

Chlorophyll *a* contents, photosynthetic rates, and respiration rates were determined according to previously described methods [17]. The rates of photosynthesis and respiration were assayed by measuring the rate of O_2 evolution using an Oxy-lab O_2 electrode (Cambridge, UK) in a double jacket thermoregulated glass vessel [17].

43.2.4 Nitrate, Phosphate, Protein Contents Measurements

Nitrate contents were determined by the salicylic acid-sulfuric acid method [4]. Phosphate contents were determined by the molybdenum blue colorimetric method [13]. Protein content was measured by the Kjeldahl method [2]. Rates of carbon sequestration and photosynthetic efficiencies were determined according to previous reports [11].

43.2.5 Statistical Analyses

Statistical significance analyses were performed by SPSS v19.0. One-way ANOVA followed by the Least Significant Difference Test (LSD) in the post hoc analysis was used in this study.

43.3 Results

43.3.1 The DCW Changes in Media with Different K_2HPO_4 Concentrations

The DCW after 15 days cultivation in gradient K_2HPO_4 concentrations are presented in Fig. 43.1 (15 mM $NaHCO_3$) and Fig. 43.2 (20 % CO_2), respectively. The DCW was increased gradually and then reached a plateau in both kinds of carbon sources, respectively. The DCW reached 0.75 and 1.19 g/L at 60 mg/L K_2HPO_4 in

Fig. 43.1 Dry cell weight change at gradient K_2HPO_4 concentrations in media with 15 mM $NaHCO_3$

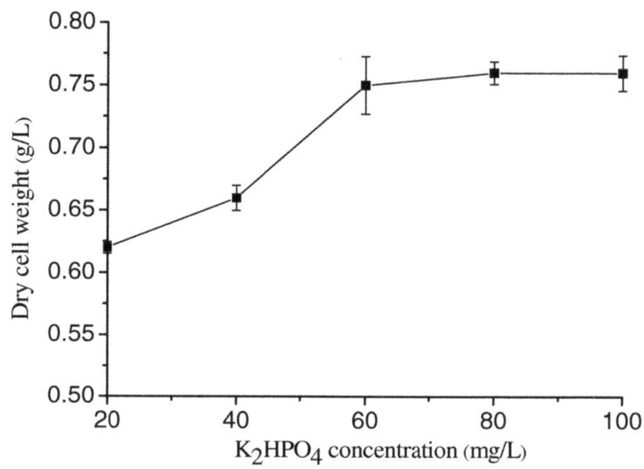

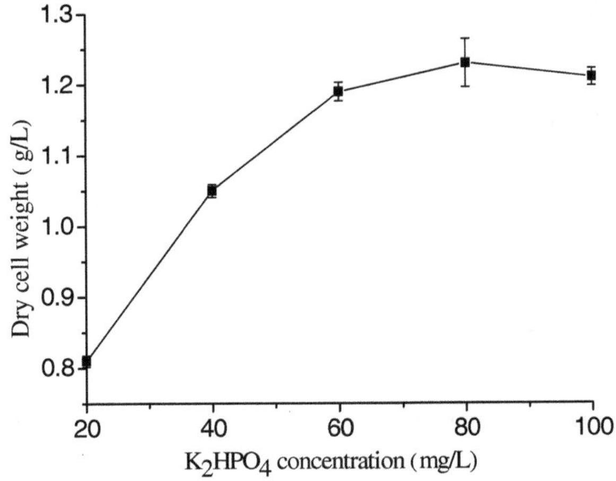

Fig. 43.2 Dry cell weight at gradient K_2HPO_4 concentrations in media with 20 % CO_2

media with 15 mM $NaHCO_3$ and 20 % CO_2, respectively. The DCW reached a plateau and showed no significant difference among 60, 80, and 100 mg/L K_2HPO_4 ($P > 0.05$).

43.3.2 The Photosynthesis, Cell Components, and Nutrients Uptake at Different K_2HPO_4 Concentrations in Media with 15 mM $NaHCO_3$

The photosynthesis, cell components, and nutrients uptake at different K_2HPO_4 concentrations in media with 15 mM $NaHCO_3$ were presented in Table 43.1. The photosynthetic and respiration rates increased gradually before 60 mg/L K_2HPO_4 in media with 15 mM $NaHCO_3$. There was no significant difference among 60, 80, and 100 mg/L K_2HPO_4 ($P > 0.05$). The chlorophyll *a* contents showed no significant difference among all concentrations but 40 mg/L K_2HPO_4, at which the chlorophyll *a* content is 18.84 mg/g. The EPS amounts were decreased along with the increase of K_2HPO_4 concentrations. The protein contents were increased along with the increase of K_2HPO_4 concentrations. In consistent with the increase of DCW, the uptakes of nitrate and phosphate increased along with the increase of K_2HPO_4 concentrations.

Table 43.1 Growth characteristics at gradient K_2HPO_4 concentrations in media with 15 mM $NaHCO_3$

K_2HPO_4 (mg/L)	O_2 evolution rates (μmol $\mathbf{O_2}$/(mg (Chl·*a*) ·h))			Cell components			Uptake of Nutrients	
	Respiration	Net photosynthesis	True photosynthesis	Chl *a* (mg/g)	EPS (mg/L)	Protein (%)	NO_3^- (g/L)	PO_4^{3-} (mg/L)
20	17.47 ± 0.87	60.45 ± 3.77	77.92 ± 4.22	13.79 ± 0.23	61.95 ± 2.11	21 ± 1.70	0.37 ± 0.07	3.41 ± 0.67
40	24.48 ± 1.22	66.21 ± 4.62	90.69 ± 5.89	18.84 ± 0.33	61.48 ± 2.33	28 ± 1.78	0.56 ± 0.11	6.84 ± 0.88
60	26.48 ± 1.35	78.94 ± 5.88	105.42 ± 6.98	13.91 ± 0.12	40.58 ± 1.89	45 ± 1.98	0.71 ± 0.12	10.25 ± 0.87
80	25.44 ± 2.09	80.47 ± 6.23	105.91 ± 7.67	12.86 ± 0.27	42.69 ± 1.49	50 ± 2.11	0.83 ± 0.09	13.72 ± 0.61
100	28.56 ± 2.86	79.45 ± 4.88	108.01 ± 8.44	12.94 ± 0.18	39.48 ± 2.06	47 ± 1.45	0.88 ± 0.11	17.12 ± 0.79

43.3.3 The Photosynthesis, Cell Components, and Nutrients Uptake at Different K_2HPO_4 Concentrations in Media with 20 % CO_2

The photosynthesis, cell components, and nutrients uptake at different K_2HPO_4 concentrations in media with 20 % CO_2 were presented in Table 43.2. The photosynthetic and respiration rates were increased along with the increase of K_2HPO_4 concentrations. The increases in respiration rates are not significant ($P > 0.05$), but an increase trend existed. Photosynthesis rates and chlorophyll *a* contents were increased significantly after reaching 60 mg/L K_2HPO_4 but there was no significant difference ($P > 0.05$) among the later three K_2HPO_4 concentrations. The optimal K_2HPO_4 concentration is less than 40 g/L for EPS productivity. The EPS amounts at 40, 60, 80, and 100 mg/L K_2HPO_4 showed no significant difference ($P > 0.05$). Protein contents in *N. flagelliforme* cells were increased along with the increase of K_2HPO_4 concentrations, and reached a maximum of 58 % at 80 mg/L K_2HPO_4. The uptakes of nitrate in the latter four concentrations showed no significant difference. The uptakes of phosphate were increased along with the increase of K_2HPO_4 concentrations.

43.3.4 The Rates of Carbon Sequestration and Photosynthetic Efficiencies at Different K_2HPO_4 Concentrations in Media with 20 % CO_2

Carbon content in *N. flagelliforme* was 0.4 (w/w) as previous report [11]. In order to assess the effects of K_2HPO_4 concentrations on the efficiencies of carbon sequestration and use of light from ambient environments, the rates of carbon sequestration and photosynthetic efficiencies at different K_2HPO_4 concentrations in media with 20 % CO_2 were determined. Results showed that the K_2HPO_4 concentrations had no effects on carbon sequestration of *N. flagelliforme*. Photosynthetic efficiencies were increased along with the increase of K_2HPO_4 concentrations. The photosynthetic efficiency reached a maximum of 7.06 % at 80 mg/L K_2HPO_4 (Table 43.3).

43.4 Discussion

Phosphorus is an essential macronutrient for growth and development of living organisms. It is a constituent of key molecules such as ATP, nucleic acids, or phospholipids, and as phosphate, pyrophosphate, ATP, ADP, or AMP, plays a crucial role in energy transfer, metabolic regulation, and protein activation [12].

Table 43.2 Growth characteristics at gradient K_2HPO_4 concentrations in media with 20 % CO_2

K_2HPO_4 (mg/L)	O_2 evolution rates (μmol O_2/(mg (Chl·*a*)·h))			Cell components			Uptake of Nutrients	
	Respiration	Net photosynthesis	True photosynthesis	Chl *a* (mg/g)	EPS (mg/L)	Protein (%)	NO_3^- (g/L)	PO_4^{3-} (mg/L)
20	20.77 ± 2.20	54.63 ± 3.44	75.40 ± 5.02	10.37 ± 0.18	39.32 ± 1.01	24 ± 1.77	0.65 ± 0.10	3.28 ± 0.89
40	24.72 ± 1.44	64.63 ± 2.99	89.35 ± 4.23	22.81 ± 0.22	45.82 ± 1.19	40 ± 1.33	0.81 ± 0.07	6.75 ± 0.61
60	26.07 ± 2.33	73.28 ± 5.19	99.35 ± 3.87	25.91 ± 0.19	46.12 ± 2/07	54 ± 1.28	0.90 ± 0.14	10.25 ± 0.92
80	28.37 ± 1.98	74.07 ± 3.77	102.44 ± 6.36	28.97 ± 0.25	46.23 ± 1.92	58 ± 2.14	0.93 ± 0.08	13.84 ± 0.59
100	30.71 ± 1.66	76.08 ± 4.15	106.79 ± 5.35	30.57 ± 0.30	48.37 ± 2.27	57 ± 2.29	0.84 ± 0.11	14.53 ± 0.88

Table 43.3 Rates of carbon sequestration and photosynthetic efficiency at gradient K_2HPO_4 concentrations in media with 20 % CO_2

K_2HPO_4 (mg/L)	Rate of carbon sequestration ($gCO_2/(L \cdot d)$)	photosynthetic efficiency (%)
20	0.09 ± 0.01	4.51 ± 0.01
40	0.12 ± 0.03	5.97 ± 0.03
60	0.13 ± 0.04	6.82 ± 0.04
80	0.14 ± 0.01	7.06 ± 0.01
100	0.12 ± 0.02	6.01 ± 0.02

Plants have evolved adaptive responses to cope with growth under conditions of limited phosphate availability and several genes responsive to Pi starvation have been isolated recently from vascular plants [14]. In the present study, we investigated the effects of K_2HPO_4 on the growth characteristics of *N. flagelliforme* under two kinds of carbon sources, respectively. 60 mg/L K_2HPO_4 is an optimal concentration for many growth characteristics such as DCW, photosynthesis, and protein content of *N. flagelliforme* cells in both kinds of carbon sources, respectively. The chlorophyll *a* contents were promoted by K_2HPO_4 concentrations in media with 20 % CO_2 but not in 15 mM $NaHCO_3$, which indicated that pigments synthesis was affected by both carbon source species and phosphorus availability. The promoting effect of K_2HPO_4 on photosynthesis of *N. flagelliforme* is consistent with the previous report that the availability of phosphate could induce adaptations of the rate of photosynthesis and photosynthate partitioning in Brassica nigra suspension cells [5]. Further works in the future will be needed to elucidate the underlying mechanisms of the promoting effect of K_2HPO_4 on photosynthesis in *N. flagelliforme*.

Previous report showed that the protein content was not influenced by phosphorus availability in dinoflagellate *Heterocapsa sp* [3]. More work at molecular level to the conflict results in these two algae species are needed in the future. Protein content is a major factor in the quality of algae on the market, so the results are meaningful to the industrial use of *N. flagelliforme*. It was anticipated that the EPS production would increase along with the increase of K_2HPO_4 concentrations because of the alkaline property of K_2HPO_4 [11], however, the EPS secreting are inhibited by higher concentrations of K_2HPO_4 in media with 15 mM $NaHCO_3$, therefore, the regulation of EPS production in *N. flagelliforme* are complicated and could be not simply attributed to pH adjustment in media.

Taken together, considering the DCW, protein contents, and EPS amounts in media with two kinds of carbon sources, we concluded that 60 mg/L K_2HPO_4 in media with 20 % CO_2 is the most favorable growth condition for *N. flagelliforme* cells production.

Acknowledgments This project is funded by China Postdoctoral Science Foundation (Grant No. 2014M561189) and the Program for Changjiang Scholars and Innovative Research Team in University (PCSIRT) (IRT1166).

References

1. Bai X, Su J, Zhao S, Jia S (2004) Study on the determination methods of extracellular polysaccharide inculture medium of *Nostoc flagelliforme* cells. Sci Tech Food Industry 25:146–148
2. Bellomonte G, Costantini A, Giammarioli S (1987) Comparison of modified automatic Dumas method and the traditional Kjeldahl method for nitrogen determination in infant food. J AOAC Int 70:227–229
3. Berdalet E, Latasa M, Estrada M (1993) Effects of nitrogen and phosphorus starvation on nucleic acid and protein content of *Heterocapsa sp.* J Plankton Res 16:303–316
4. Cataldo D, Maroon M, Schrader L (1975) Rapid colorimetric determination of nitrate in plant tissue by nitration of salicylic acid. Commun Soil Sci Plant Anal 6:71–80
5. Duff SMG, Gautam S, Plaxton WC (1994) The role of acid phosphatases in plant phosphorus metabolism. Physiol Plant 90:791–800
6. Gao K (1998) Chinese studies on the edible blue-green alga, *Nostoc flagelliforme*: a review. J Appl Phycol 10:37–49
7. Gao K, Yu A (2000) Influence of CO_2, light and watering on growth of *Nostoc flagelliforme* mats. J Appl Phycol 12:185–189
8. Gao K, Ye C (2003) Culture of the terrestrial cyanobacterium, *Nostoc flagelliforme* (cyanophyceae), under aquatic conditions. J Phycol 39:617–623
9. Kanekiyo K, Lee JB, Hayashi K et al (2005) Isolation of an antiviral polysaccharide, nostoflan, from a terrestrial cyanobacterium, *Nostoc flagelliforme*. J Nat Prod 68:1037–1041
10. Lv H, Jia S, Xiao Y et al (2013) Effect of $NaNO_3$ on the growth and photosynthesis of *Nostoc flagelliforme* cells. China Brewing 32:13–16
11. Lv H, Jia S, Xiao Y et al (2014) Growth characteristics of *Nostoc flagelliforme* at intermittent elevated CO_2 concentrations. Phycol Res 62:250–256
12. Marschner H (1995) Mineral nutrition of higher plants. Academic press, San Diego
13. Murphy J, Riley J (1962) A modified single solution method for the determination of phosphate in natural waters. Anal Chim Acta 27:31–36
14. Raghothama KG (1999) Phosphate acquisition. Annu Rev Plant Physiol 50:665–693
15. Su J, Jia S, Qiao C et al (2005) Culture of *Nostoc flagelliforme* on solid medium. Korean J Environ Biol 23:135–140
16. Su J, Jia S, Chen X et al (2008) Morphology, cell growth, and polysaccharide production of *Nostoc flagelliforme* in liquid suspension culture at different agitation rates. J Appl Phycol 20:213–217
17. Yu H, Jia S, Dai Y (2009) Growth characteristics of the cyanobacterium *Nostoc flagelliforme* in photoautotrophic, mixotrophic and heterotrophic cultivation. J Appl Phycol 21:127–133
18. Yuan N, Jia S, Dai Y et al (2012) Effect of $NaHCO_3$ on growth and photosynthesis of *Nostoc flagelliforme* cells. China Brewing 31:34–36

Chapter 44
Production of Alkyl Polyglucoside Using *Pichia pastoris* GS115 Displaying *Aspergillus aculeatus* β-Glucosidase I

Yajun Kang, Binru Wei, Dongheng Guo and Suiping Zheng

Abstract *Pichia pastoris* GS115 displaying *Aspergillus aculeatus* β-glucosidase I was used as whole-cell biocatalyst catalyzing the synthesis of 1-hexyl glucoside, 1-octyl glucoside, and 1-decyl glucoside through reverse hydrolysis in water–alcohol two-phase system, while glucose and primary alcohols were used as substrate. In this study, factors affecting the synthetic yield were optimized in our study, including water content, glucose concentration, whole-cell biocatalyst content, the pH value of sodium acetate buffer, and the reaction temperature. In 5 mL reaction volume, the optimal reaction conditions of HG were: 10 % pH 3.0 buffer, 0.2 mol/L glucose, 10 g/L whole-cell biocatalyst; the optimal reaction conditions of DG: 15 % pH 3.0 buffer, 0.6 mol/L glucose, 10 g/L whole-cell biocatalyst; the optimal reaction conditions of DG: 20 % pH 2.0 buffer, 0.2 mol/L glucose, 40 g/L whole-cell biocatalyst; three reactions were conducted under 55 °C. After 72 h the maximum yield of HG was 11.5 %, that of OG and DG were 5.3 and 3.2 % after 96 h, respectively. The reutilization experiments inducted that the whole-cell biocatalyst could be used at least for five times without obvious decrease in synthesis ability.

Keywords *P. pastoris* surface display · *A. aculeatus* β-glucosidase I · Alkyl polyglucoside · Reutilization

44.1 Introduction

β-glucosidases (EC: 3.2.1.21) [1] have many functions in living organisms such as: glycolipid and exogenous glycoside metabolism, cell wall lignifications, and biomass conversion. In industry, β-glucosidase was used as catalysts to synthesize

Y. Kang · B. Wei · D. Guo · S. Zheng (✉)
School of Bioscience and Bioengineering, South China University of Technology, Guangzhou 510006, China
e-mail: spzheng@scut.edu.cn

T.-C. Zhang and M. Nakajima (eds.), *Advances in Applied Biotechnology*, Lecture Notes in Electrical Engineering 332, DOI 10.1007/978-3-662-45657-6_44

alkyl/akyl polyglucosides [2, 3], food flavor enzymes to enhance the flavor and taste [4], and auxiliary additives to improve the efficiency of the cellulose degradation which is catalyzed by cellulases.

Alkyl polyglucoside (APG) is widely used as low-toxic and green surfactant. Traditionally, the chemical method is used to synthesize APG. However, the chemical method is very inefficient and inconvenient, such as it requires numerous steps of protection and deprotection, and it produces many by-products. Those disadvantages could be avoided when β-glucosidase was used as the catalyst because of its regioselectivity and stereoselectivity. β-glucosidase can catalyze transglycosylation and reverse hydrolysis, leading to the synthesis of APG [5]. The transglycosylation needs reactive sugar donors and an alcohol acceptors, while reverse hydrolysis needs glycoside donors and alcohol acceptors.

Pichia pastoris, a methylotrophic yeast specie, has increasingly been used for heterologous protein expression. As a higher eukaryotic expression system, *P. pastoris* expression system has many advantages, such as post-protein processing, protein folding, and posttranslational modification [6]. *P. pastoris* is easy to manipulate and has the advantage of 10–100-fold higher heterologous protein expression levels, since *P. pastoris* can be grown to higher cell densities. Heterologous protein expression is driven by the strong promoter (PAOX1) regulating the production of alcohol oxidases. These features made *P. pastoris* an established protein expression host, and reasonably useful as a displaying system. Zhang [7] used both bioinformatic and experimental experiments to find 13 GPI-modified cell wall proteins in *P. pastoris* using the CALB as the report gene. *Aspergillus aculeatus* β-glucosidase I has been reported that it can hydrolyze soluble and insoluble cello-oligosaccharides. Apiwatanapiwat et al. [8] modified a yeast strain coexpressing α-amylase, glucoamylase, endoglucanase, cellobiohydrase, and β-glucosidase that could directly produce ethanol from cassava pulp without addition of any hydrolytic enzymes during fermentation. Ito et al. [9] constructed a yeast strain for APG synthesis from genetically inducing the display of β-glucosidase I from *A. aculeatus* No. F-50 on the surface, the highest yield of APG was 27.3 % of the total sugar.

In this study, recombinant *P. pastoris* GS115 displaying *A. aculeatus* β-glucosidase I by *P. pastoris* cell wall protein Gcw21p found by Zhang [7] was constructed. The Gcw21p is a kind of GPI-modified cell wall protein which has C-terminal GPI attachment signal sequences and N-terminal secretion signal sequences, and it does not exist in the transmembrane domains. The target protein is at the N-terminal of the cell wall protein Gcw21p. The reaction conditions of APG synthesis have been optimized, and the reutilization of the whole-cell biocatalyst has been conducted.

44.2 Materials and Methods

44.2.1 Materials, Strains, and Media

Hexyl β-D-glucopyranoside, octyl β-D-glucopyranoside, and decyl β-D-glucopyranoside (used as standard) were purchased from Pure Chemical Analysis Co. (Bornew, Belgium). 4-Nitrophenyl-β-D-glucuronide (*p*NPG) was of High-performance liquid chromatography (HPLC) grade and purchased from Sigma–Aldrich Co. (St. Louis, MO, USA). Other analytical reagents of analytical grade were also obtained commercially.

Escherichia coli Top10 was used as the host strain for DNA amplification. *P. pastoris* GS115 was used for cell surface display. *E. coli* was cultured at LB medium (1 % tryptone, 0.5 % yeast extract, 1 % NaCl) containing 50 μg/mL kanamycin. Yeasts were grown either in MD plate (2 % glucose, 1.34 % YNB, 2 % agar) or BMGY/BMMY medium (1 % yeast extract, 2 % peptone, 100 mM potassium phosphate (pH 6.0), 1.34 % YNB, 400 μg/L biotin, 1 % glycerol or 2 % methanol). The *P. pastoris* GS115 was cultured under conditions: 30 °C, 250 rpm.

44.2.2 Construction of Recombinant Plasmid and Yeast Strain

The recombinant expression plasmid pK21GCW-ABGL used for the cell surface display of *A. aculeatus* BGL1, was constructed as follows: DNA sequences encoding mature BGL1 and FLAG peptide tag at the N-terminus, was amplified by PCR. The upstream and downstream primers 5′-AGCTTACGTAATGGATTACAAGGA TGACGACGATAAGGATGAACTGGCTTCTCTCC and 5′-CCGACGCGTTTG-CACCTTCGGGAGCGCCGCGTG-3′, respectively, for plasmid pICAS-abgl were used as the templates. The PCR products were digested with *Sna*B I/*Mlu* I, and then introduced into the corresponding site of the cell surface displaying plasmid pKCALB-GCW21 (constructed before in our lab) to replace the CALB gene. The recombinant plasmids were linearized and subsequently integrated into the host train *P. pastoris* GS115. The transformants were selected on MD plates by incubating at 30 °C for 72 h. The strains were cultivated in BMGY for 24 h. The cultures were centrifuged at 6,000 rpm for 5 min and then the cells were resuspended in BMMY medium containing 2 % (v/v) methanol while the OD600 was controlled at 1. At an OD600 of 1, 100 % methanol was added to the culture every 24 h to maintain the expression of the fusion proteins.

44.2.3 Measurement of the Hydrolysis Activity in Recombinant *P. pastoris* Cells

The method was modified to assay β-glucosidase hydrolysis activity. β-glucosidase activity was measured with 1 mM *p*NPG in 0.2 M sodium acetate, pH 5.0, in a 500 μL reaction at 40 °C for 30 min. The reaction was stopped by adding 500 μL 2 M sodium carbonate, pH 10, and the *p*-nitrophenol released was measured by its absorbance at 405 nm. One unit of β-glucosidase hydrolytic activity was defined as the amount of enzyme used to release 1 μmol of *p*-nitrophenol in 1 min.

44.2.4 Preparation of ABGL-Displaying Whole-Cell Biocatalyst

The cultivation condition was at 30 °C on an orbital shaker at 250 rpm for 168 h. Subsequently, the cells were collected, washed with 0.2 M sodium acetate buffer (pH 5.0) for three times, then the cells were resuspended in 2 mL 0.2 M sodium acetate buffer (pH 5.0) containing 1 % trehalose which was used as the freeze-drying protectant. After that, the cells were lyophilized in Christ Alpha 2–4 Freeze Dryer (Christ, Osterode, Germany) for 24 h. Finally, the dry powder was used as the whole-cell biocatalyst.

44.2.5 Assay of the Synthetic Activity of the Whole-Cell Biocatalyst

APG synthesis was performed in 5 mL reaction volume, including 0–2 mL sodium acetate buffer with the pH range from 2.0 to 6.0 and 3–5 mL primary alcohol with 0.1–1 mol/L glucose content and 5–60 g/L whole-cell biocatalyst. The reaction was constantly shaken at 200 rpm at 30–60 °C. Sample was taken every 1.5 h, the organic phase was separated by centrifugation for 5 min at 15,000×*g* and at room temperature, and then the amount of APG was measured.

44.2.6 High-Performance Liquid Chromatography Detection

The APGs were analyzed by HPLC using ELS Detector and C18 chromatographic column. Samples were filtered by 0.22 μm filter membrane before detected by HPLC. The mobile phase was methanol–water mixture (80:20, v/v). The HPLC detection was conducted under conditions: 1.0 ml/min flow rate, 30 °C for C18 column, 60 °C for drift tube, 30 psi for nitrogen pressure.

44.3 Results and Discussion

44.3.1 Display of ABGL on Yeast Cell Surface

The yeast cell surface displaying ABGL was used for the synthesis of alkyl polyglycosides. Yeast strains displaying BGL1 on their surface were constructed to synthesize the APG from primary alcohols and glucosides. The gene encoding ABGL was fused with the gene encoding the Gcw21p cell wall protein and it was expressed under the AOX1 promoter (Fig. 44.1a). The constructed plasmid was linearized and subsequently integrated into the host train *P. pastoris* GS115. The resulting transformants were selected by incubation at 30 °C for 72 h on MD plates. The transformants were cultured in BMGY/BMMY medium. After methanol induction, the enzyme activity and OD600 increased substantially up to a maximum hydrolysis activity of 3,738 U/g (Fig. 44.1b). The hydrolysis activity had slight decrease after being cultured for more than 120 h.

44.3.2 Optimization of the Synthetic Conditions

Generally, during the reverse hydrolysis reaction, the product yield would change along with the buffer content, alcohol, biocatalyst, the value of pH, and temperature. The optimal synthetic conditions had been ascertained for three kinds of products.

44.3.2.1 Influence of Water Content

No product could be detected when the reaction was conducted without water. Within certain limits, the APG yield increased with water content. As shown in

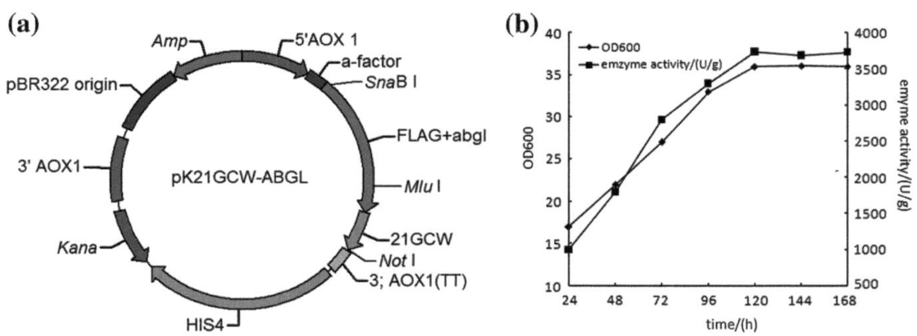

Fig. 44.1 Construction of plasmid pK21GCW-ABGL used for displaying ABGL on the yeast cell surface (**a**) and time course of *Pichia pastoris* GS115 displaying *Aspergillus aculeatus* β-glucosidase I by high cell-density fermentation in 250 mL shake flasks (**b**)

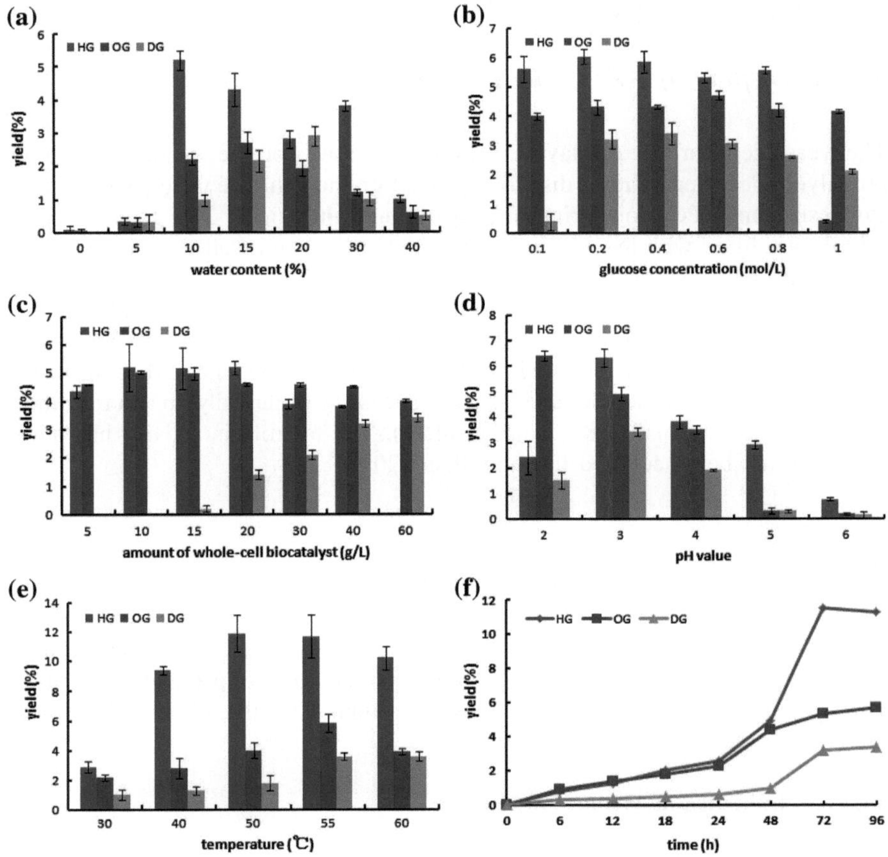

Fig. 44.2 **a** The influence of the kinetic parameters: influence of water content. **b** Influence of glucose concentration. **c** Influence of whole-cell biocatalyst content. **d** Influence of the pH value. **e**, **f** Influence of the temperature and final synthetic result under the optimal conditions. *Blue bar* stands for 1-hexyl glucoside, *red bar* stands for 1-octyl glucoside, and *green bar* stands for 1-decyl glucoside. The data points represent the averages of the results of three independent experiments

Fig. 44.2a, the optimal water content was increased along with the chain lengths of alcohol as glycosyl acceptor was increased. The optimal water content for the highest 1-hexyl glucoside, 1-octyl glucoside, and 1-decyl glucoside were 10, 15, and 20 % (v/v). Some studies had reported that the water content should be between 4 and 20 % (v/v) depending on the catalyst and the alcohol used to achieve a high yield of alkyl polyglucoside [5, 10]. The figure also showed that the APG yield decreased when the water content kept going on.

44.3.2.2 Influence of Glucose Concentration

Figure 44.2b illustrated the influence of glucose concentration on the synthesis of 1-hexyl, 1-octyl, and 1-decyl glucoside. The water content was fixed at their optimum value. Glucose concentration was varied from 0.1 to 1 mol/L, and the enzyme concentration was constant. At first, increasing glucose concentration in the aqueous phase led to a fast increase in 1-decyl glucose yield in the organic phase, but there were not obvious changes for the 1-hexyl and 1-octyl glucoside yield. A sharp decrease in HG yield was observed when the glucose concentration exceeded 0.8 M, which resulted from the substrate inhibition.

44.3.2.3 Influence of Whole-Cell Biocatalyst Content

The influence of whole-cell biocatalyst concentration was presented in Fig. 44.2c. It was clear that higher biocatalyst concentration promoted the reaction while the substrates were enough. Reactions with varying enzyme concentration from 5 to 40 g/L were performed to synthesis 1-hexyl and 1-octyl glucoside, but there were not significant changes. However, the yield dropped to undetectable when the whole-cell biocatalyst content went up to 60 g/L, it suggested that high catalyst concentration would inhibit the synthetic reaction. Situation was different when it came to 1-decyl glucoside. When the whole-cell biocatalyst content was less than 15 g/L, no products could be detected. In order to achieve a higher yield of 1-decyl glucoside, the whole-cell biocatalyst content should be more than 40 g/L.

44.3.2.4 Influence of the pH Value and the Temperature During the Synthetic Process

The influence of pH values on APG yields was shown in Fig. 44.2d. The pH varied from 2 to 6. For the maximum synthesis of 1-hexyl and 1-decyl glucoside, pH value should be adjusted to 3. At pH 2, 1-octyl glucoside yield reached to the highest. When the pH exceeded 4, APG production would be decreased. The result was consistent with the conclusion that the reaction system was conducted under acid circumstance in many chemical ways to synthesis APG.

Temperature played an important role in the thermodynamic equilibrium of the reaction and the activity of biocatalyst. As shown in Fig. 44.2e, the reaction temperatures range from 30 to 60 °C at 10 °C increments. Within certain limits, the APG yield increased along with the temperature. The temperature profile indicated that 55 °C was the optimal temperature for synthesizing APG. With higher temperature than 55 °C, the biocatalyst would be inactivated, and then the APG production dropped significantly.

44.3.2.5 Final Synthesis Under the Optimal Conditions

After the optimization experiments, synthetic experiment was conducted under the optimal conditions of each APG. Figure 44.2f showed the result: After 72 h the maximum HG could be 11.5 % that of OG and DG could be 5.3 and 3.2 % after 96 h. The results were slightly lower than the result of Xiayan [11], whose whole-cell catalyst was GS115/pKFS-ABGL. The difference between two strains was the different cell wall protein with different attachment means, which would influent the exposure of the enzyme's active site, which in turn could influence the enzyme activity.

44.3.3 Reutilization of the Whole-Cell Biocatalyst

In order to reuse the whole-cell biocatalyst, a series of experiments were conducted to study the efficiency of the whole-cell biocatalyst after several reactions. After each batch reaction (one batch for 72 h), the whole-cell biocatalyst was recovered by centrifugation. To remove glucose and alcohol absorbed by the whole-cell biocatalyst, it was washed with acetone, and then, it was used for the next batch reaction. The activity of the first batch was regarded as 100 %. As shown in Fig. 44.3, after continuous recover cycle, the activity of the whole-cell biocatalyst

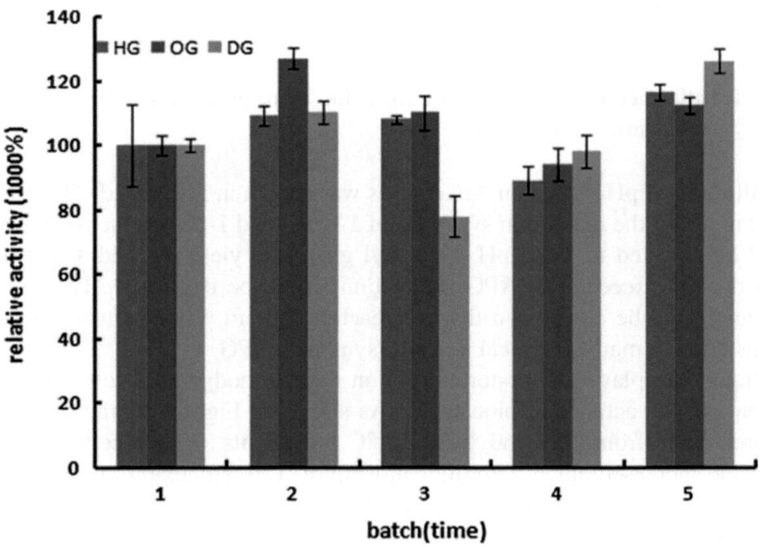

Fig. 44.3 Influence of operational stability of whole-cell biocatalyst on the alkyl glucoside production. *Blue bar* stands for 1-hexyl glucoside, *red bar* stands for 1-octyl glucoside, and *green bar* stands for 1-decyl glucoside. The data points represent the averages of the results of three independent experiments

did not significantly decrease. However, the relative yield of the third synthesis reaction of 1-decyl glucoside declined to 78 %, as the whole-cell biocatalyst became clot.

44.4 Conclusion

P. pastoris GS115 displaying *A. aculeatus* β-glucosidase I whole-cell biocatalyst could be an alternative to commercial immobilized β-glucosidase. The former is characterized by its low cost and simple recycling procedure. In this study, *P. pastoris* GS115 displaying *A. aculeatus* β-glucosidase I via Gcw21p cell wall protein was constructed. The optimal conditions for synthesizing 1-hexyl, 1-octyl, and 1-decyl glucosides were determined, and the recycling experiments of the whole-cell biocatalyst were conducted to ensure the reutilization efficiency. *P. pastoris* GS115 displaying *A. aculeatus* β-glucosidase I whole-cell biocatalyst showed excellent tolerance for high glucose concentrations and solvent. The biocatalyst was used for five batches of synthesis, and the activity loss was not very obvious except the mistake made by manual operations. Thus, whole-cell biocatalyst is promising for further large-scale production of alkyl polyglycoside, or application to cellulose hydrolysis.

References

1. Ketudat Cairns JRK, Esen A (2010) β-glucosidase. Cell Mol Life Sci 67:3389–3405
2. Vic, G, Thomas D (1992) Enzyme-catalysed synthesis of alkyl β-D-glucosides in organic media. Tetrahedron Lett 33(32):4567–4570
3. Lirdprapamongkol K, Svasti J (2000) Alkyl glucoside synthesis using *Thai rosewood* β-glucosidase. Biotechnol Lett 22:1889–1894
4. Gil JV, Lamuela-Raventos RM (2000) The use of transgenic yeasts expressing a gene encoding a glycosyl-hydrolase as a tool to increase reseratrol content in wine. Int J Food Microbiol 59(3):183–197
5. Van Rantwijk F, Oosterom MWV, Sheldon RA (1999) Glycosidase-catalysed synthesis of alkyl glycosides. Mol Catal B Enzym. doi:10.1016/S1381-1177(99)00042-9
6. Daly R, Hearn MTW (2005) Expression of heterologous proteins in *Pichiapastoris*: a useful experimental tool in protein engineering and production. J Mol Recognit 18:119–138
7. Zhang L, Liang S, Lin Y (2013) Screening for glycosyphosphatidylinositol-modified cell wall proteins in *Pichia pastoris* and their recombinant expression on the cell surface. Appl Environ Microbiol 79(18):5519
8. Apiwatanapiwat W, Murata Y, Kosugi A et al (2011) Direct ethanol production from cassava pulp using a surface-engineered yeast strain co-displaying two amylases, two cellulases, and β-glucosidase. Appl Microbiol Biotechnol 90(1):377–384
9. Ito J, Ebe T, Shibasaki S (2007) Production of alkyl glucoside from cellooligosaccharides using yeast strains displaying *Aspergillus aculeatus* β-glucosidase 1. J Mol Catal B Enzym. doi:10.1016/j.molcatb.2007.08.008

10. Rather MY, Mishra S, Chand S (2010) β-glucosidase catalyzed synthesis of octyl-β-D-glucopyranoside using whole cells of *Pichia etchellsii* in micro aqueous media. J Biotechnol 150(4):490–496
11. Xiayan H, Duanyu L, Suiping Z (2014) Optimization of the synthesis of alkyl glucoside catalyzed by *Aspergillus aculeatus* β-glucosidase-displaying *Pichia pastoris*. Biotechnol Bull 6:205-210

Chapter 45
Enhancement of Gellan Production in *Sphingomonas paucimobilis* JLJ by Heterogeneous Expression of *Vitreoscilla* Hemoglobin

Qinglong Ji, Dan Li, Xiangqian Li, Ting Li and Lin Yuan

Abstract A heterologous expression of *Vitreoscilla* hemoglobin (VHb) for improving cell growth and fermentation products has been successfully demonstrated in various hosts. To improve a commercially used strain for gellan, vgb gene was expressed in *Sphingomonas paucimobilis* JLJ under the control of the oxygen-dependent vgb gene promoter (P_{vgb}). The promoter was maximally induced under microaerobic conditions. Biochemical activity of expressed vgb was confirmed by PCR, SDS–PAGE, and the CO-difference spectra analysis which exhibited a characteristic absorption maximum near 420 nm. The results showed that the expression of vgb could enhance cell growth. The composition of fermentation medium and the physical conditions of cultivations were investigated. The optimum medium consists of (g/L): Sucrose 30, Soybean meal 5, $NaNO_3$ 1, KH_2PO_4 1, K_2HPO_4 1.5, $MgSO_4$ 0.6. The optimum fermentation conditions were confirmed: pH 6.0, seed age 16 h, inoculum volume 10 %, 200 rpm, 30 °C, and the maximum gellan yield of vgb^+ recombinant was 1.713 % ± 0.02 % after cultivated 72 h, 21 % higher than that of the parental strain. VHb exhibited positive effect on cell growth and optimize gellan yield. The results provided potential applications of vgb gene in other industrially important strains to solve the increasing demand of aerobic processes in polysaccharides production.

Keywords Gellan · *Sphingomonas paucimobilis* · Vgb · *Vitreoscilla* hemoglobin

Q. Ji · D. Li · X. Li · T. Li · L. Yuan (✉)
College of Life Science, Inner Mongolia University, Huhhot, China
e-mail: yuan0079@163.com

T.-C. Zhang and M. Nakajima (eds.), *Advances in Applied Biotechnology*,
Lecture Notes in Electrical Engineering 332, DOI 10.1007/978-3-662-45657-6_45

45.1 Introduction

The use of microbial polysaccharides in the food, pharmaceutical, and chemical industries has increased steadily during the past decade. The biopolymer gellan is a more recent addition to the family of microbial polysaccharides that is gaining much importance due to its novel property of forming thermoreversible gels when heated and cooled. In its native form, gellan is a linear anionic exopolysaccharides (EPS) based on a tetrasaccharide repeat unit composed of two molecules of D-glucose, one of L-rhamnose, and one of D-glucuronic acid [1]. The native gellan is partially esterified with acyl substituents (1 mol of glycerate and 0.5 mol of acetate) per repeat unit. It has applications in diverse fields in the food, pharmaceutical, and many other industries. Further research and development in biopolymer technology is expected to expand its use. To meet the growing demand of gellan gum, various traditional processes have been employed to improve the fermentative production, such as isolation of improved mutant strains and optimization of culture conditions [2, 3]. However, inadequate oxygenation, resulting in poor cell growth and reduced product yields, is one major limitation encountered in large-scale bioreactors. Gellan production was reported to almost completely cease when the dissolved oxygen (DO) concentration was adjusted to zero [4]. Furthermore, the DO also played a critical role in the viscosity and molecular weight of the polymer produced, which was closely related to the rheological properties and applications of gellan gum. Conventional treatment available for high oxygen supply usually depends on increasing the agitation speed or injecting pure oxygen, which cannot avoid undesirable high cost, particularly in industrial scale. Therefore, it will be valuable to identify new strategies to solve this difficult engineering problem.

A genetic strategy for improving productivity under oxygen starvation, complementary to bioprocess approaches for enhanced aeration, based on the cloned *Vitreoscilla* hemoglobin gene (vgb) has been described. *Vitreoscilla* hemoglobin (VHb) of the Gram-negative aerobe *Vitreoscilla* is the first described bacterial hemoglobin [5], and it enables the organism to survive in hypoxic environment. The gene vgb encoding the hemoglobin was first cloned from the chromosome of *Vitreoscilla* in 1988 and then expressed in *Escherichia coli* [6]. Subsequently, fundamental advances in understanding this oxygen-binding protein promoted its application in a variety of aerobic heterologous hosts to improve cell growth, oxidative metabolism, protein synthesis, and metabolite productivity, especially in oxygen-limited conditions [7–11]. Although the exact mechanism by which VHb causes these effects is unknown, it has been hypothesized that due to its unusual kinetic parameters for oxygen binding and release, VHb is able to scavenge oxygen molecules from solution and provide them for cellular activities in heterologous organisms. As indicated previously, the application of VHb is perhaps the most suitable inverse metabolic engineering approach to date to alleviate the defects of hypoxic conditions and thus to improve cellular activities under extreme oxygen limitation.

In this work, vgb gene was introduced into *Sphingomonas paucimobilis* JLJ with a broad-host-range plasmid under control of the native oxygen-dependent promoter from *Vitreoscilla stercoraria* ATCC15218. We improved the yield of gellan through enhancing the oxygen supply during high-density fermentation by the expression of vgb gene in JLJ.

45.2 Materials and Methods

45.2.1 Strains and Plasmids

Escherichia coli Trans1-T1 Phage Resistant [F^-φ80(lacZ)ΔM15ΔlacX74hsdR (r_K^- m_K^+)ΔrecA1398endA1tonA; TransGen, Beijing, China] was used as a host for cloning. *Sphingomonas paucimobilis* JLJ was obtained from the Fermentation Laboratory of Inner Mongolia University. A phylogenetic analysis is based on the morphological characteristics and 16S ribosomal RNA gene sequences of JLJ, JLJ is a member of *Sphingomonas* strains that secrete gellan polysaccharides. *Vitreoscilla stercoraria* ATCC15218 was obtained from the American Type Culture Collection. Plasmid pEASY-T1 (TransGen, Beijing, China) was used for the construction of pEASY-T1-P_{vgb}-VHb.

45.2.2 Media and Culture Conditions

Escherichia coli was grown in Luria–Bertani (LB) medium at 37 °C. *Vitreoscilla stercoraria* ATCC15218 was grown in ATCC medium 138: Beggiatoa medium at 30 °C for 48 h. Beggiatoa medium (g/L): Yeast extract 2, $CaCl_2$ 0.1, Sodium acetate 0.5, Agar 20. JLJ was maintained in preservation medium (g/L): Yeast extract 3, Beef extract 3, Tryptone 5, Sucrose 20, Agar 20, pH 7.2. The seed medium (g/L): Sucrose 25, Yeast extract 7, KH_2PO_4 0.5, K_2HPO_4 0.5, $MgSO_4$ 0.6, pH 6.0. For growth experiments, single colonies were inoculated into 50 ml of seed culture medium in 250 ml flasks and grown at 30 °C and 200 rpm for 16 h ($OD_{600} \approx 2.5$). Preculturers were used for inoculation at a volume ratio of 1:10 to the fermentation medium. Fermentation medium (g/L): Sucrose 30, Soybean meal 5, $NaNO_3$ 1, KH_2PO_4 1, K_2HPO_4 1.5, $MgSO_4$ 0.6, pH 6.0. The optimum fermentation conditions were confirmed: pH 6.0, seed age 16 h, inoculum volume 10 %, 200 rpm, 30 °C, 72 h.

45.2.3 Construction of Vgb Expression Plasmid

Isolation of *Vitreoscilla* genomic DNA was performed using TIANamp Bacteria DNA Kit (TIANGEN, Beijing, China). The vgb fragment and its native promoter sequences (853 bp) were amplified in polymerase chain reactions (PCR) using *Vitreoscilla* genomic DNA as template, oligonucleotides 5′-GCTCTAGAGAAGC

TTACAGGACGC TGG-3′ as forward primer containing a XbaI site(underlined) and 5′-CGAGCTCGCAAGGCACACCTGAAGAC-3′ as reverse primer containing a SacI site(underlined). After removal of the excess primers and enzymes by using a Universal DNA Purification Kit, the obtained vgb operon fragment was digested with XbaI and SacI. The resultant fragment was purified and cloned into the compatible sites of pEASY-T1 vector to generate pEASY-T1-P_{vgb}-VHb, which was then transformed into *E. coli* Trans1-T1 Phage Resistant.

45.2.4 Transformation of JLJ

Recombinant of pEASY-T1-Pvgb-VHb and the empty vector were transferred into JLJ by the electroporation method. Inoculate a single colony of JLJ in slant culture into 30 ml of seed culture medium in a 250 ml flask. Grow the cells at 30 °C shaking at 200 rpm to an OD_{600}(1:10 dilution) of approximately 0.5–0.7. The preparation of competent *Sphingomonas* cells was using 10 % (v/v) Glycerol with 0.05 % Tween 80. Place a 1.5 ml microfuge tube on ice, add 50 μl of competent cell and 10 μl vector (isolated from Trans1-T1 Phage Resistant). Mix well and incubate on ice for 10 min. Set the Bio-Rad Gene Pulser condition at 2.0 kV, 25 μF, 200 Ω. Transfer the mixture of cells and vector to a cold 0.1 cm electroporation cuvette and tap the suspension to the bottom. Place the cuvette in the ShockPod. Push the chamber lid down to close. After the shock, immediately add 1 ml of seed medium to the cuvette. Gently transfer the diluted cells to 5 ml seed medium. Grow the cells at 30 °C shaking at 200 rpm overnight. Cells were then resuspended into the sterile deionized water, and the appropriate dilutions were plated on selective medium with 50 μg/ml ampicillin (Amp) and 30 μg/ml kanamycin (Kan) to select recipient cells.

45.2.5 Confirmation of Transformants by PCR and Restriction Enzyme Digestion Analysis

Isolation of plasmid DNA of the transformants was performed using TIANprep Mini Plasmid kit (TIANGEN, Beijing, China). PCR was carried out, using the following primers designed to amplify a 660 bp internal portion of vgb: forward, 5′-TGCTGCTACACCATACTGA-3′; reverse, 5′-CAAGGCACACCTGAAGAC-3′. The recombinant plasmid was digested by restriction enzymes XbaI and SacI for double digestion.

45.2.6 Analysis of VHb Protein

Expression of vgb gene was confirmed by sodium dodecyl sulfate–polyacrylamide gel electrophoresis (SDS–PAGE) and the activity of VHb protein was detected by CO-difference spectra.

After being cultured for 24 h, cells were harvested by centrifugation (6,000 g, 10 min) at 4 °C, washed with 50 mmol/L potassium phosphate buffer (pH 7.0) and resuspended in the same phosphate buffer. Subsequently, the cell pellets were interrupted in an ice bath by an ultrasonic cell disruptor during one hundred cycles of sonification, 5 s each with a 3 s pause. The crude protein was extracted from cell lysate by centrifugation to remove cell debris.

The crude protein extracts were analyzed by SDS–PAGE (stacking gel 5 % and separating gel 15 %). The protein marker was blended with 9 recombinant proteins, 6.5–200 kDa range, obtained from TaKaRa. Samples mixed with SDS-loading buffer were boiled at 100 °C for 10 min before applying to the gel. After complete electrophoresis was reached, the gel was stained with Coomassie brilliant blue.

Carbon monoxide (CO) difference spectra assay was used to detect VHb biological activity. To avoid interference from the other proteins, the crude protein extracts were centrifuged at 12,000×*g* for 15 min. After centrifugation, sodium dithionite was added to the supernatant, until a saturated concentration was reached. Then the supernatant was divided into two after being reduced at room temperature for 20–30 min: one part was directly poured into a reference cell, and the other was exposed to CO for 3 min and placed without shaking in the dark for 5 min, before being poured into a detector cell. The spectral scan was performed between 400 and 470 nm.

45.2.7 Assays of Gellan Gun

The fermented broth (5.00 g) was diluted with redistilled water (30 mL) and centrifuged at 10,000×*g* for 30 min. After centrifugation, the supernatant was added to 120 mL concentration above 90 % ethanol extraction homogenized for 30 min. The fibrous precipitate was then separated by filtration, dried at 80 °C and weighed. Then the rate of the gellan gum was showed by $X(\%) = M_1/5 \times 100\ \%$. X showed gellan rate (%), M_1 showed the dried precipitate (g).

45.2.8 Stability of Recombinant Plasmid

Plasmid maintenance was investigated under nonselective conditions. The JLJ was inoculated in 30 ml of liquid preservation medium (not contain agar) and incubated on a rotary shaker (200 rpm) at 30 °C. The culture was transferred to the fresh preservation medium every 24 h for ten times, every time transfer is one generation. Every generation culture were diluted appropriately and then plated on both selective plates (preservation medium contain 100 μg/mL Amp and 100 μg/mL Kan) and nonselective plates (reservation medium not contain antibiotic) at 30 °C for 3 days, then grown colonies are counted, respectively. Stability was expressed as a fraction of colonies that retain the ampicillin and kanamycin resistance after growth under nonselective conditions.

45.2.9 Univariate Optimization of the Vgb$^+$

Recombinant strains vgb$^+$ fermentation conditions of the optimization mainly includes seed age, inoculum size, initial pH, and temperatures.

45.3 Results

45.3.1 Construction of Strains

Plasmid pEASY-T1-P_{vgb}-VHb was transformed into JLJ to construct vgb-bearing strain. The resistant colonies on the selection plates were picked randomly, and plasmid DNA was isolated for PCR using specific primers that amplified a 660 bp portion of vgb gene, and results of restriction enzyme analysis showed that a recombinant plasmid, pEASY-T1-Pvgb-VHb, contains a DNA fragment of about 850 bp was successfully constructed. In the PCR, restriction analysis and DNA sequencing analysis, the desired fragment of vgb gene was observed (Fig. 45.1a, b) and no false positives were detected. The empty vector pEASY-T1 was also transformed into JLJ to construct control strain. All the vgb-bearing transformants examined (ten randomly chosen strains) showed similar improved growth and gellan production compared to the empty vector control (also randomly chosen). And no significant difference between the empty vector control and parental strain was detected during cultivation process. Based on the premise that plasmid pEASY-T1 introduced had no effect on parental strain, one randomly chosen vgb-bearing transformant was selected and named vgb$^+$ for further analysis of the expressed vgb and its ability to produce gellan gum. And the corresponding gellan-producing wild-type parental strain was named WT and used as reference strain throughout this study.

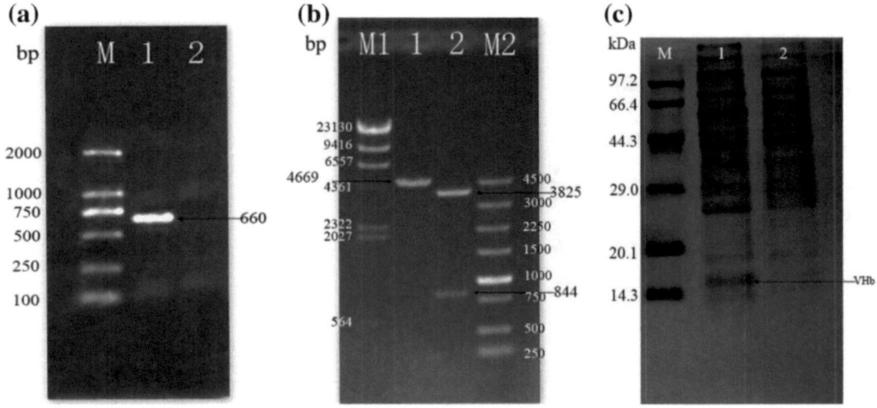

Fig. 45.1 Construction of the vgb-expressing strain and SDS–PAGE analysis. **a** PCR analysis of vgb^+ and WT strains. *Lane 1* vgb^+; *Lane 2* WT; M, DL2000. **b** Restriction analysis of the plasmid DNA isolated from vgb^+. *Lane 1* pEASY-T1-P_{vgb}-VHb vector digested with Xba I; *Lane 2* pEASY-T1-P_{vgb}-VHb vector digested with Xba I and Sac I; M1,λ-Hind III digest; M2, 250 bp Ladder. **c** SDS–PAGE analysis of the crude extract of vgb^+ and WT strains. *Lane 1* vgb^+; *Lane 2* WT; *M* Premixed Protein Marker (Low)

45.3.2 Expression of Vgb in JLJ

To further confirm the expression of vgb gene in recombinant strain, SDS–PAGE analysis of protein from both vgb^+ and WT cells were performed. As shown in Fig. 45.1c, the major protein bands observed for vgb^+ cells were comparable with the VHb size of approximately 16 kDa. As a negative control, such protein band was not shown in the sample of WT cells.

To verify that the synthesized VHb is biologically active, CO-binding assay was performed with the crude protein extracts. CO-difference spectrum (Fig. 45.2) clearly showed a pronounced absorption peak near 420 nm. On the contrary, the sample of WT cells showed no absorption from 400 to 470 nm.

45.3.3 Effect of Vgb Expression on Cell Growth and Gellan Production

To explore the VHb effects on cell growth and gellan production, fermentation experiments were carried out in bioreactors. The WT strain and the vgb^+ strain were grown in shake flask cultures, wherein the available oxygen concentration would be expected to decline monotonically with time (as the cell density increases) throughout the growth phase of the experiment. As shown in Fig. 45.3, compared with the WT strain, the vgb^+ strain grew more slowly during the initial phase of

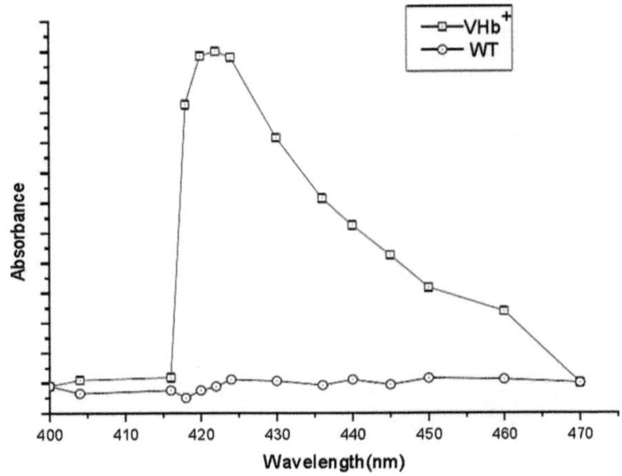

Fig. 45.2 CO-difference spectra of the recombinant strain vgb$^+$ (*squares*) and WT strain (*circles*). The baseline (*dotted line*) was for vgb$^+$ sample bubbled with air (the WT baseline was similar). VHb gives a characteristic peak near 420 nm

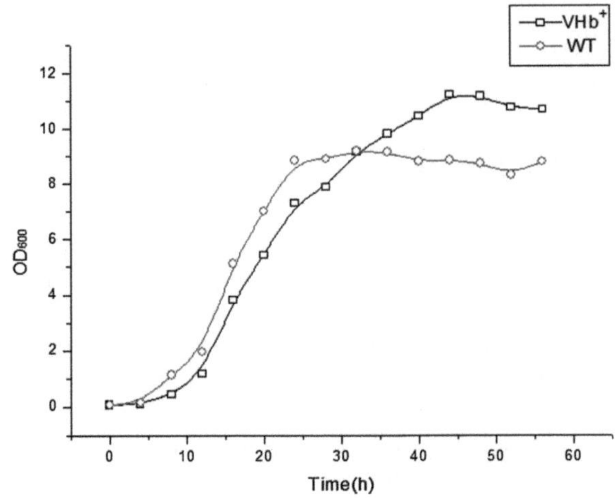

Fig. 45.3 Growth curve of the recombinant strain vgb$^+$ (*squares*) and WT (*circles*) strain

cultivation, but showed a higher biomass yield at the end. The WT strain displayed a lower cell density OD_{600} was about 9.3, while the vgb$^+$ strain achieved the higher cell density OD_{600} was about 11.2 at 44 h. The differences in cell growth occurred since 22 h after incubation when the WT strain reached stationary phase. The oxygen-dependent vgb gene promoter (P_{vgb}) was induced in response to oxygen limitation. The vgb$^+$ strain began to outgrow WT strain and reached a significantly higher cell density 10 h later and remained there until the end of cultivation.

The vgb$^+$ and WT strains were grown in Erlenmeyer flasks under different fermentation conditions to examine the optimization of fermentation conditions on recombinant strains vgb$^+$. As shown in Fig. 45.4, significant differences between vgb + and WT strains on secreted polysaccharide yields as well as broth viscosity were observed in all the fermentations. The culture conditions of WT were not

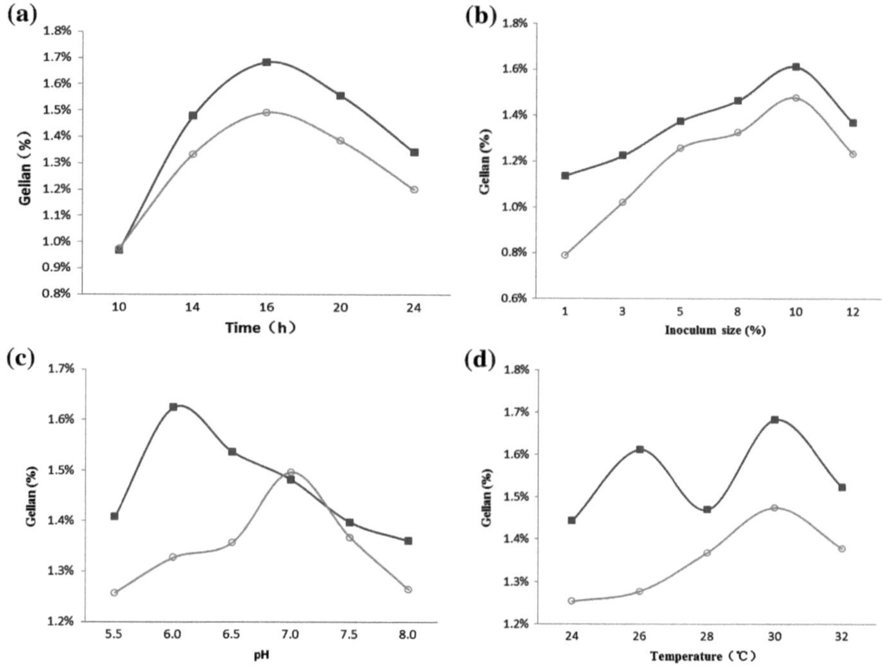

Fig. 45.4 The optimization of fermentation conditions of the recombinant strain vgb$^+$ (*solid squares*) and WT (*open circles*) strain. **a** Effects of seed age on gellan production; **b** effects of inoculum size on gellan production; **c** effects of initial pH on gellan production; **d** effects of temperature on gellan production

suitable for the strain vgb$^+$. The maximum gellan yield of vgb$^+$ recombinant was approx. 21 % higher than that for the WT strain under the optimum culture conditions of vgb$^+$.

45.3.4 Stability of Vgb-Expressing Plasmid

The segregational stability of the vgb-expressing plasmid pEASY-T1-P_{vgb}-VHb was investigated in Erlenmeyer flask culture. After a week of cultivation under nonselective condition, plasmids in the transformants were segregationally stable up to 80 %. Besides, the existence of plasmid was confirmed by PCR as certain vgb fragment was amplified and observed. These results indicated that the vgb-bearing vector produced little metabolic burden on the host strain and was able to keep basically stable at a reasonable level throughout the cultivation required for gellan production.

45.4 Discussion

Genetic engineering using vgb to express VHb in heterologous hosts has been used to increase the cell growth and production of products. In this work, we expressed vgb gene in *S. paucimobilis* JLJ under the control of the oxygen-dependent vgb gene promoter (P_{vgb}) to determine the effect of VHb on growth and gellan secretion.

As indicated by SDS–PAGE assay (Fig. 45.1c), the vgb gene was successfully expressed in vgb^+ strain. We further confirmed the biological activity of the synthesized VHb by CO-difference spectra analysis as specific absorption maximum near 420 nm was detected (Fig. 45.2). These results suggested that the P_{vgb} was capable of driving the constitutive and functional expression of vgb gene in JLJ.

Improvement of growth properties had been detected in various hosts with exogenous expression of VHb. To investigate whether the strain improvement in this study was because of the expression of the vgb gene, randomly chosen transformants were tested and compared with the two control groups (randomly selected strains bearing empty vectors and WT strain). Noticeable differences between vgb-expressing strains and those two control groups were observed, while no significant difference between the two control groups was detected (data not shown). These results suggested that the superior properties obtained in vgb-bearing strain were likely to be associated with the expression of exogenous vgb gene instead of a spontaneous mutation.

As indicated in Fig. 45.3, faster growth was also observed and higher final cell density was reached in the culture of the recombinant strain containing vgb gene than that of the WT strain after the VHb was induced. As previously described, VHb interacts with bacterial respiratory membranes, facilitating oxygen uptake and its transfer to the terminal oxidases. Here, such enhanced growth of vgb^+ cells was likely due to the increased utilization of nutrients provided by the advantage of energy metabolism for this obligate aerobe.

Consistent with the improved cell growth, gellan synthesis of vgb^+ strain was also enhanced (Fig. 45.4a–d), although most of the fermentation conditions were changed. Such as the optimum pH was changed because of active VHb. Previous research has proved that gellan is a closely growth-associated product, and thus any factor or process available stimulating growth, improved O_2 supply because of active VHb in this study, would enhance gellan production. In fact, oxygenation not only plays a key role in cell growth but also has a direct effect on gellan accumulation. The results showed that the higher production was achieved under the higher oxygen transfer condition. It was reported that lower temperature cultures can enlarge VHb effect on cell performance [12]. So the improvement of gellan yield at 26 °C was higher compared with that at 30 °C.

In many industrial fermentation processes, oxygen supply is always the most difficult engineering problem that leads to high cost of production and limitation for applications, particularly when a high cell density is needed to achieve economic efficiency. For instance, the improvement in oxygenation mainly depends on an increase in agitation or the volume of aeration, which means that more energy is consumed for the process (increasing production cost), and over high agitation has been shown to lead to mechanical damage to polysaccharides. Thus, the extension of this VHb technology of metabolic engineering to the genus *Sphingomonas* might be quite exciting as the DO level is the rate-limiting factor in gellans synthesis, especially in large-scale cultures.

Our research is the first application of *Vitreoscilla* hemoglobin as an effective metabolic engineering strategy under its native promoter in JLJ to regulate cell growth and optimize the yield of gellan gum. Our study constructed successfully the stable expression system with native promoter (P_{vgb}) for functional expression of vgb gene in *S. paucimobilis* JLJ. The native promoter P_{vgb} is regulated by dissolved oxygen (DO) concentration in *Vitreoscilla*. So the P_{vgb} promoter is induced by low DO levels (not the expensive inducers) also in *S. paucimobilis* JLJ to express soluble hemoglobin and enable cells to adapt to hypoxic environments, which is suitable for industrial fermentation requirements. The results provided potential applications of VHb in other industrially important strains to solve the increasing demand of aerobic processes in polysaccharides production.

References

1. Fialho A, Moreira L, Granja A (2008) Occurrence, production, and applications of gellan current state and perspectives. J Appl Microbiol Biotechnol 79:889–900
2. Nancy EH, Yamini NP, Russell JC (2004) Organization of genes required for gellan polysaccharide biosynthesis in *Sphingomonas elodea* ATCC31461. J Microbiol Biotechnol 31:70–82
3. Lobas D, Schumpe S (1992) The production of gellan exopolysaccharide with *Sphingomonas paucimobilis* E2 (DSM 6314). J Appl Microbiol Biotechnol 37:411–415
4. Wu XC, Chen YM, Li YD (2010) Constitutive expression of *Vitreoscilla* haemoglobin in *Sphingomonas elodea* to improve gellan gum production. J Appl Microbiol 110:422–430
5. Wakabayashi S, Matsubara H, Webster DA (1986) Primary sequence of a dimeric bacterial haemoglobin from *Vitreoscilla*. J Nat 322(31):481–483
6. Khosla C, Bailey JE (1988) The *Vitreoscilla*hemoglobin gene Molecular cloning, nucleotide sequence and genetic expression in *Escherichia coil*. J Mol Gen Genet 214:158–161
7. Geckil H, Barak Z (2004) Enhanced production of acetoin and butanediol in recombinant Enterobacter aerogenes carrying *Vitreoscilla* hemoglobin gene. J Bioprocess Biosyst Eng 26:325–330
8. Frey AD, Bailey JE, Kallio PT (2000) Expression of Alcaligenes eutrophus flavohemoprotein and engineered *Vitreoscilla* hemoglobin-reductase fusion protein for improved hypoxic growth of *Escherichia coli*. J Appl Environ Microbiol 66(1):98–104
9. Patel SM, Stark BC (2000) Cloning and expression of *Vitreoscilla* hemoglobin gene in *Burkholderia* sp. strain DNT for enhancement of 2,4-dinitrotoluene degradation. J Biotechnol Prog 16:26–30
10. Guan B, Ma H, Wang YP (2011) *Vitreoscilla* hemoglobin (VHb) overexpression increases hypoxia tolerance in zebrafish (Danio rerio). J Mar Biotechnol 13:336–344

11. Kallio PT, Bailey JE (1996) Intracellular expression of *Vitreoscilla* hemoglobin (VHb) enhances total protein secretion and improves the production of α–amylase and neutral protease in *Bacillus subtilis*. J Biotechnol Prog 12:31–39
12. Wu JM, Wang SY, Fu WC (2012) Lower temperature cultures enlarge the effects of *Vitreoscilla* hemoglobin expression on recombinant *Pichia pastoris*. Int J Mol Sci 13:13212–13226

Chapter 46
Enhanced Adenosine Production by *Bacillus subtilis* at Condition with Comprehensively Controlled Dissolved Oxygen and pH During Fermentation

Yue Liu, Juhua He, Qingyang Xu, Chenglin Zhang, Ning Chen and Xixian Xie

Abstract Adenosine has potent effects on cardiovascular diseases and has been widely used as an antiarrhythmic agent. To improve its production, we investigated the effects of dissolved oxygen (DO) and pH on production by *Bacillus subtilis*. Based on the kinetic parameters at different DO levels, we proposed a two-stage DO strategy to control DO level at 30–40 % before 20 h of fermentation and 10–20 % after 20 h of fermentation, and confirmed that using this strategy could increase adenosine yield to 19.2 g/L in 52 h, which is increased by 78.6, 66.7, 9.5, 18.6, and 32.2 %, compared to the conditions with DO uncontrolled or controlled at 0–10, 10–20, 20–30, and 30–40 %, respectively. On this basis, pH was adjusted to further boost adenosine production. The results showed that the two-stage DO plus pH-shift method was the optimal way for adenosine production, under which, adenosine yield was further improved by 21.4 %, reaching 23.3 g/L at 56 h. To our knowledge, this is the first report on adenosine production verifying that the two-stage DO plus pH-shift method is effective for enhancing adenosine yield by *B. subtilis*.

Keywords *Bacillus subtilis* · Adenosine · Optimization · Two-stage dissolved oxygen strategy · pH-shift method

Y. Liu · J. He · Q. Xu · C. Zhang · N. Chen (✉) · X. Xie (✉)
Metabolic Engineering Laboratory, College of Biotechnology, Tianjin University of Science and Technology, No. 29, 13 Main Street, Tianjin Economic and Technological Development Area, Tianjin 300457, China
e-mail: ningch66@gmail.com

X. Xie
e-mail: xixianxie@tust.edu.cn

T.-C. Zhang and M. Nakajima (eds.), *Advances in Applied Biotechnology*,
Lecture Notes in Electrical Engineering 332, DOI 10.1007/978-3-662-45657-6_46

46.1 Introduction

Adenosine is an endogenous nucleoside and a breakdown product of AMP, which plays a significant role in physiological regulation and function modulation of numerous cell types. Nowadays, adenosine has been proven to have potent effects on cardiovascular diseases and widely used as an antiarrhythmic agent to investigate and manage both narrow complex and, less often, broad complex tachycardias [1].

Adenosine can be produced through several approaches such as RNA degradation, chemical synthesis, and fermentation [2–4]. At present, microbial fermentation has attracted increasing attention because it is more sustainable and economical compared with complicated multistep chemical synthesis and others. Many methods have been adapted to improve adenosine production, including selecting bacteria with high adenosine yield [5, 6] and medium optimization [7, 8]. However, there are only a few reports on systematical control of fermentation process to improve adenosine production.

It is well known that oxygen is one of essential factors for growth, maintenance, and fermentation of aerobic bacteria [9, 10]. Oxygen availability strongly affects growth and biosynthesis of numerous microorganisms by influencing their metabolic pathways and changing their metabolic fluxes [11, 12] Adenosine fermentation process is a typical aerobic bioprocess. However, until recent, there is only few reports on the effects of oxygen on adenosine fermentation by *Bacillus subtilis*, showing that low oxygen condition may contribute to a higher adenosine yield (3.63 g/L) than high oxygen condition [13], and on the effects of pH on cell growth and metabolites accumulation in adenosine fermentation [14].

To our knowledge, the preferable pH for metabolic production may not be always consistent with that necessary for cell growth [15]. In general, adenosine fermentation always requires a slightly acidic condition [16, 17], but the optimum pH for *B. subtilis* growth is approximately 7.0. Therefore, optimization of adenosine production by varying pH conditions deserves further investigation.

In the light of these findings, the objectives of this study were to investigate the effects of various dissolved oxygen (DO) concentrations in combination with controlled pH condition on adenosine production. In ultimate, we determined the optimal DO and pH conditions for adenosine production. As a result, a two-stage DO plus pH-shift method was designed according to the kinetic analysis and confirmed by experiments.

46.2 Materials and Methods

46.2.1 Strain and Media

The adenosine-producing *B. subtilis* strain XGL used in this study was obtained from earlier work in our laboratory and stored at the Culture Collection of Tianjin University of Science and Technology. *B. subtilis* XGL, an adenosine-producing mutant with characteristics of xanthine and histidine deficiencies (His^-/Xan^-) and 8-azaguanine and sulfaguanidine resistance ($SG^r/8\text{-}AG^r$), was used for breeding.

The seed media contain 20 g/L glucose, 5 g/L yeast extract, 25 g/L corn steep liquor, 0.4 g/L $MgSO_4{\cdot}7H_2O$, 1 g/L KH_2PO_4, 5 g/L monosodium glutamate, 8 g/L tryptone, 5 g/L urea, 0.03 g/L xanthine, and 0.03 g/L histidine. The fed-batch fermentation media compose of 80 g/L glucose, 12 g/L monosodium glutamate, 15 g/L corn steep liquor, 1.5 g/L sodium gluconate, 18 g/L yeast extract, 2 g/L KH_2PO_4, 2 g/L $CaCl_2$, 5 g/L $MgSO_4{\cdot}7H_2O$, 0.006 g/L $FeSO_4{\cdot}7H_2O$, 0.006 g/L $MnSO_4$, 12 g/L $(NH_4)_2SO_4$, 0.03 g/L Xanthine, and 0.03 g/L Histidine. The pH of both seed and fermentation media was adjusted based on specific experimental requirements.

46.2.2 Culture Methods

46.2.2.1 Seed Culture

One loopful of fresh culture was transferred from a agar slant culture into a flask (1,000 mL) containing 100 mL seed medium and incubated at 33 °C at 200 rpm for 8 h. The pH was controlled as required.

46.2.2.2 Fermentation in the Stirred-Tank Bioreactor

Adenosine fermentation was performed in a 5 L jar bioreactor (Biotech-2002 Bioprocess controller, Baoxing, Shanghai, China) with a working volume of 3 L. 10 % (v/v) of seed culture was inoculated into the bioreactor and cultured at 34 °C at aeration rate of 1.17 vvm. In all cases, pH was automatically controlled at a set value with 25 % NH_4OH (80 %, w/v) and 2 M HCl solution. Glucose solution (80 %, w/v) was added to the bioreactor when the residual glucose concentration was below 25 g/L to maintain its concentration at about 25 g/L and DO level was controlled by adjusting agitation speeds.

46.2.2.3 Analytical Methods

The DO, pH, and temperature were monitored automatically with electrodes attached to the stirred-tank bioreactor. The 100 % DO is defined as the content of oxygen in the still air. Cell growth was determined by optical density of the fermentation broth with a spectrophotometer at 600 nm (UV TU-1900, Purkinje, Beijing, China). Dry cell weight (DCW) was monitored as previously reported [18]. Samples were centrifuged (10,000×*g*, 4 °C, 5 min) and the supernatants were analyzed for sugars and adenosine. The concentration of glucose was measured by a SBA-40E biosensor analyzer (Biology Institute of Shandong Academy of Sciences, China). Adenosine was measured by high performance liquid chromatography (HPLC, Series 1200, Agilent Technologies, Santa Clara, CA, USA).

46.3 Statistical Analysis

Each fed-batch fermentation experiment was independently repeated three times under the same conditions, and analyzed values were the mean of three corresponding samples. One-way analysis of variance (ANOVA) followed by Dunnett's multiple comparison test was used to determine significant difference, and the statistical significance was defined as $p \leq 0.05$.

46.4 Results and Discussion

46.4.1 Adenosine Production Without DO Control

During the process of fermentation, DO is one of the most important parameters affecting cell growth, target metabolites biosynthesis, and cell metabolism maintenance. Previous studies on fermentation process of L-phenylalanine, L-isoleucine, and transglutaminase have verified that too high or too low DO level is detrimental to cell growth and product synthesis [19–21].

In order to profoundly recognize the influence of DO on adenosine production, we investigated the changes of DO, DCW, adenosine, and glucose levels during adenosine fermentation without DO control.

As shown in Fig. 46.1, DO level could be clearly divided into two stages. At the initial fermentation stage (0–20 h), DO level declined linearly and then remained at low level. At the late fermentation stage, DO level increase slowly. The fermentation process indicated that the demand for oxygen increased rapidly at the initial fermentation stage and then decreased slowly at the late fermentation stage. By the

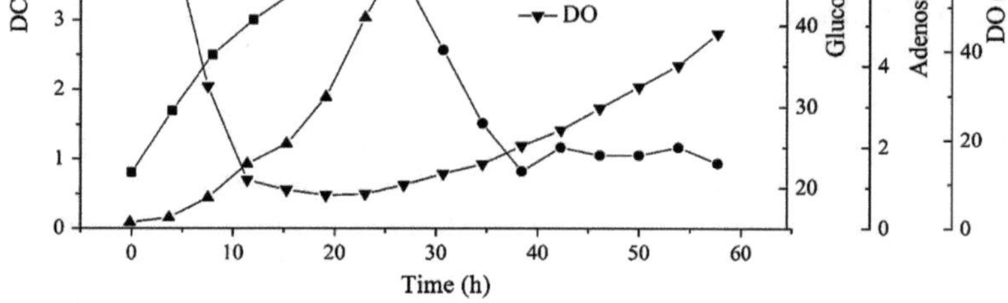

Fig. 46.1 Fermentation profiles of *Bacillus subtilis* with DO uncontrolled

end of fermentation, the DCW, adenosine, and glucose contents were 6.4, 11.3, and 80 g/L, respectively (Fig. 46.1).

It has been reported that low DO level could decrease metabolic flux to tricarboxylic acid cycle (TCA) [22], causing ineffective reoxidization of NAD(P)H or FADH and subsequently generating insufficient ATP for aerobic cell growth. As a result, DO should be maintained at a sufficient level for cell growth at the initial stage. At the late stage, it is desired to drive metabolic flux toward adenosine production rather than biomass formation. Therefore, it's unnecessary to constantly maintain DO at high level. Furthermore, Yu et al. [13] have shown that a relatively lower DO level is benefit to adenosine production. Therefore, we hypothesized that DO level in adenosine fermentation process should be controlled differently at the two stages. However, the most suitable DO levels at different stages in fermentation process need to be determined. Thus, fermentation tests with various DO levels were conducted to better understand the effects of different oxygen supply levels on adenosine production.

46.4.2 Effects of Various DO Levels on Adenosine Fermentation

In our study, the influences of DO concentrations (0–10, 10–20, 20–30, and 30–40 %) on cell growth, residual glucose concentration, and adenosine production were detected and shown in Fig. 46.2. Clearly, higher DO level improved cell growth. DCW reached 5.9, 6.9, 7.9, and 9.0 g/L under the DO levels of 0–10, 10–20, 20–30, and 30–40 %, respectively (Fig. 46.2a). It is also shown that more glucose was consumed with the increase of DO levels (Fig. 46.2b). Similar phenomena have been found in L-isoleucine and alginate fermentation by *Brevibacterium lactofermentum* [20] and *Azotobacter vinelandii*, respectively [23]. Besides,

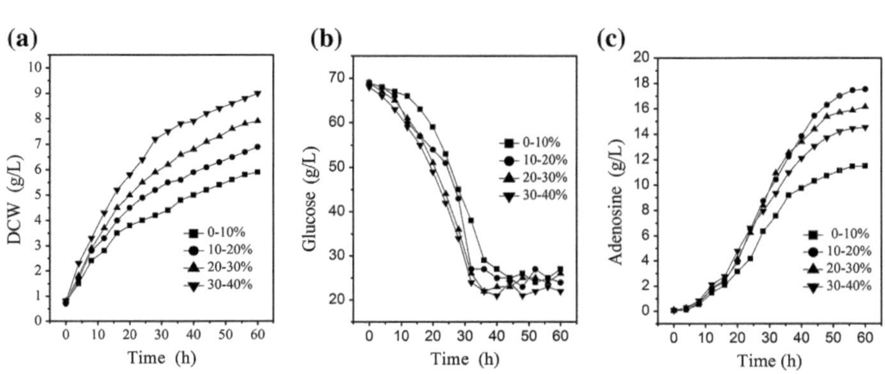

Fig. 46.2 Time profiles of DCW (**a**), glucose consumption (**b**), and adenosine (**c**) during fermentation processes under different DO levels of 0–10, 10–20, 20–30, and 30–40 %

it has been proved that adenosine production was obviously reduced if DO level was too high (30–40 %) or too low (0–10 %). The highest adenosine production (17.5 g/L) was achieved when DO level was 10–20 % (Fig. 46.2c).

46.4.3 Two-Stage DO Strategy for Adenosine Fermentation

According to Fig. 46.2, a relatively low DO level was favorable to adenosine production, while a comparatively high DO level could promote cell growth and glucose consumption. In order to find out the characteristics of the above processes at various DO levels, three kinetic parameters with the whole process, including specific cell growth rate, specific glucose consumption rate, and specific adenosine formation rate were calculated base on the data of Fig. 46.2.

As shown in Fig. 46.3a, the specific growth rate reached the highest (0.288 h^{-1}) when DO level was kept at 30–40 % during the initial 20 h, while the specific adenosine rate reached the highest when DO level was 10–20 % after 20 h (Fig. 46.3c). And Fig. 46.3b shows that higher DO level will result in an earlier reduction of glucose consumption rate. In Table 46.1, several key parameters were detected. Table 46.1 indicates that it is impossible to simultaneously achieve the goals of high cell productivity, high glucose consumption, and high adenosine production using single constant DO concentration during the fermentation process. Because cells grew poorly to reach the appropriate density for producing more adenosine at 10–20 % DO level, and grew too fast and consumed too many substrates at the initial stage at very high DO level, which was detrimental to adenosine synthesis at late stage, a two-stage DO strategy of controlling DO level at 30–40 % at the initial 0–20 h, and at 10–20 % thereafter until the end of fermentation was proposed to maximize adenosine production.

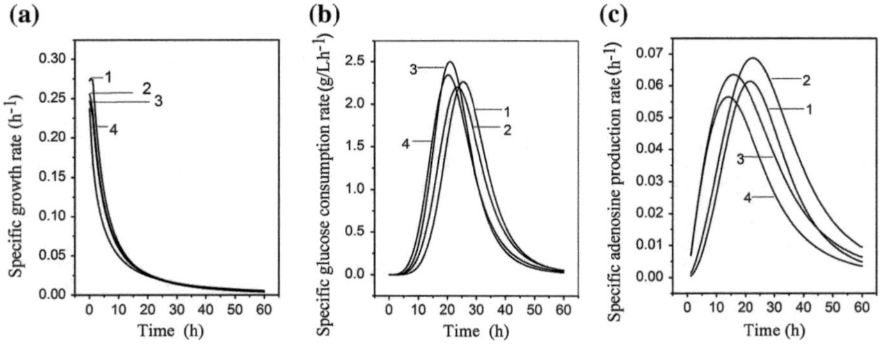

Fig. 46.3 Kinetic parameters of adenosine fermentation on specific growth rate (**a**), specific glucose consumption rate, (**b**) and specific adenosine production rate (**c**) under different DO levels of 0–10 % (curve 1), 10–20 % (curve 2), 20–30 % (curve 3), and 30–40 % (curve 4)

Table 46.1 Comparison of parameters in adenosine fermentation under different dissolved oxygen level

Parameters	DO level				
	0–10 %	10–20 %	20–30 %	30–40 %	Two-stage[a]
Initial glucose concentration (g/L)	±1	69 ± 1	69 ± 1	68 ± 1	69 ± 1
Culture time[b] (h)	60	60	60	60	52
Maximal dry cell weight (g/L)	5.9 ± 0.25	6.9 ± 0.4	7.9 ± 0.5	9.0 ± 0.4	7.4 ± 0.45
Maximal adenosine production (g/L)	11.5 ± 0.7	17.5 ± 0.6	16.2 ± 0.8	14.5 ± 0.4	19.2 ± 0.5
Maximal glucose consumption rate (1/h)	0.45 ± 0.03	0.43 ± 0.03	0.56 ± 0.02	0.63 ± 0.02	0.63 ± 0.03
Maximal specific growth rate (1/h)	0.234 ± 0.009	0.254 ± 0.007	0.264 ± 0.007	0.288 ± 0.006	0.283 ± 0.005
Adenosine yield on glucose (g/g)	0.138 ± 0.003	0.175 ± 0.003	0.160 ± 0.005	0.143 ± 0.003	0.225 ± 0.003
Average specific growth rate (1/h)	0.031 ± 0.001	0.035 ± 0.001	0.037 ± 0.002	0.041 ± 0.001	0.037 ± 0.001
Adenosine productivity (g/(Lh))	0.192 ± 0.012	0.292 ± 0.010	0.270 ± 0.013	0.242 ± 0.007	0.369 ± 0.008
Average specific glucose consumption rate (1/h)	0.22 ± 0.006	0.24 ± 0.004	0.27 ± 0.005	0.30 ± 0.005	0.29 ± 0.006
Average specific adenosine production rate after 20 h (1/h)	0.027 ± 0.001	0.036 ± 0.001	0.024 ± 0.001	0.017 ± 0.001	0.034 ± 0.002

[a] Change of DO level from 30–40 to 10–20 % after 20 h

[b] Fermentation time when the adenosine production reached maximal

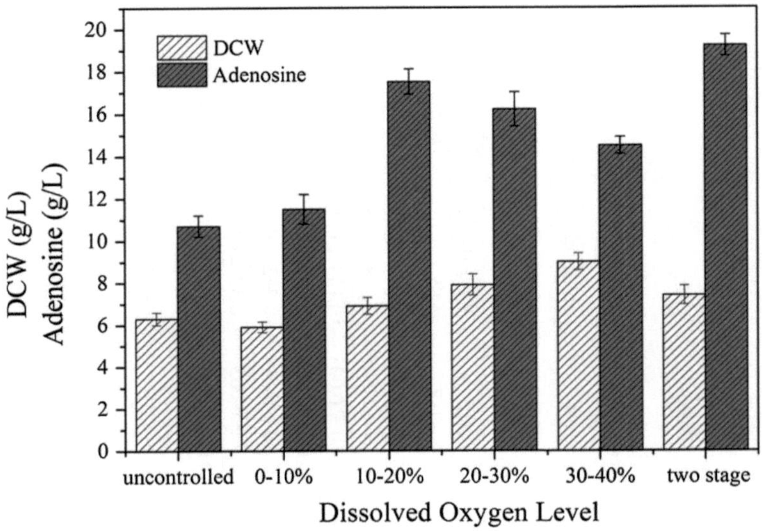

Fig. 46.4 Effect of DO on DCW and adenosine in the fermentation

Table 46.1 shows the fermentation process with the two-stage DO strategy. At this condition, adenosine yield reached as high as 19.2 g/L, representing an increase of 78.6, 66.7, 9.5, 18.6, and 32.2 % compared with that at uncontrolled DO condition and DO level controlled at 0–10, 10–20, 20–30, or 30–40 %, respectively (Fig. 46.4), and adenosine yield and glucose conversion rate reached the highest of 0.369 g/(Lh) and 0.225 g/g, respectively (Table 46.1). Overall, with this strategy, adenosine yield, DCW, and glucose conversion rate were all reached their peaks.

Li et al. have reported that DO level is related to three potential pathways: the Embden Meyerhof-Parnas Pathway (EMP), the hexose monophosphate (HMP) pathway and tricarboxylic acid (TCA) cycle [24]. A relatively lower DO could restrict the rate of TCA cycle, resulting in a relatively higher HMP/TCA flux ratio and subsequent redistribution of metabolic flux from glycolysis to the HMP pathway [25], which is in favor of adenosine production. However, oxygen limitation will probably restrict TCA fluxes to provide enough ATP for rapid cell growth, causing insufficient cell density for producing more adenosine. At the initial 20 h, high DO level (30–40 %) allowed higher specific growth rate and TCA cycle flux to provide enough ATP for rapid cell growth. However, sustained high DO level would cause substrate scantiness at the late stage, resulting in low adenosine yield. Therefore, a relative low DO level (10–20 %) was chosen at the late stage (from 20 h to the end), supposed to drive the metabolic flux toward HMP pathway, producing more precursor (PRPP) and accelerate the regeneration of NAD^+ from NADH for adenosine synthesis.

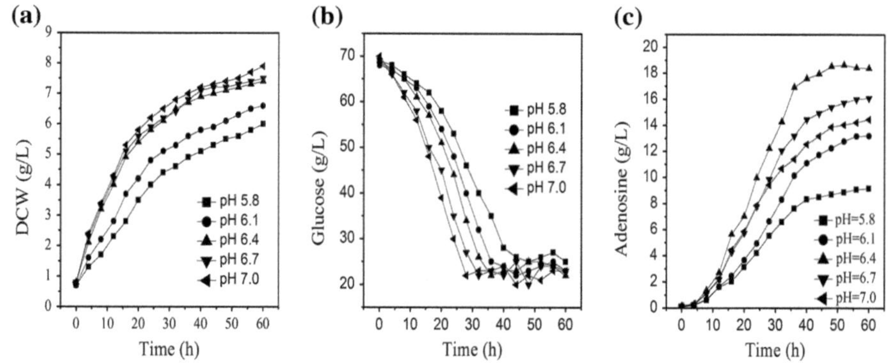

Fig. 46.5 Time profiles of DCW (**a**), glucose consumption (**b**), and adenosine (**c**) during fermentation processes under different pH of 5.8, 6.1, 6.4, 6.7, and 7.0

46.4.4 Effect of Different pH on the Cell Growth and Adenosine Production

Besides DO, pH has been proven to be another significant factor in the aerobic fermentations process [20, 26]. Therefore, on the basis of two-stage DO strategy, we further investigated the effects of various pH (5.8, 6.1, 6.4, 6.7, and 7.0) on adenosine fermentation.

Figure 46.5a shows that cell growth was obviously inhibited at pH 5.8 and 6.1 and the highest DCW (7.9 g/L) was attained at pH 7.0. Figure 46.5b shows that glucose concentration fells to a low levels more rapidly with pH ascended, and Fig. 46.5c shows that the adenosine yield reached the highest at pH 6.4. All these indicated that the optimal pH for cell growth is different from that for adenosine biosynthesis. In addition, slightly acidic pH can suppress cell growth, but enhance adenosine accumulation, and neutral environment is conducive to the growth of *B. subtilis*. A similar tendency was also observed in cAMP fermentation [18] and succinic acid fermentation [26].

46.4.5 Kinetic Analysis of Adenosine Fermentation at Different pH Conditions

According to the fermentation process at various pH, the kinetic parameters, including specific cell growth rate, specific glucose consumption rate, and specific adenosine formation rate were calculated.

Figure 46.6a shows that the maximum cell growth rate was achieved at 20 h at pH 7.0, and the duration of log phase shortened with pH increasing. Figure 46.6b shows that specific adenosine formation rate achieved the maximum at pH 6.4 at

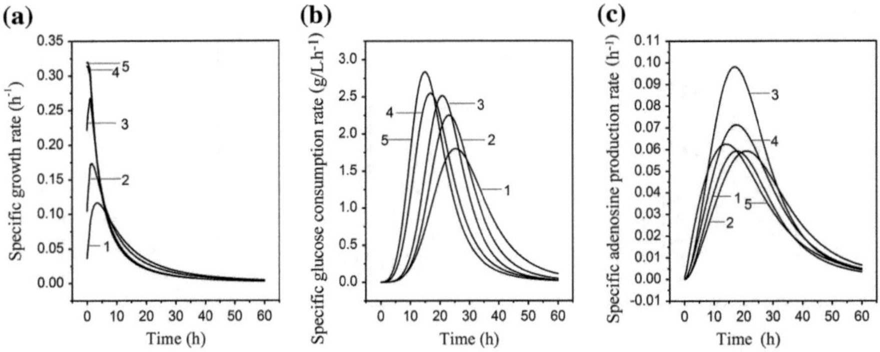

Fig. 46.6 Kinetic parameters of adenosine fermentation on specific growth rate (**a**), specific glucose consumption rate (**b**), and specific adenosine production rate (**c**) under different pH of 5.8 (curve 1), 6.1 (curve 2), 6.4 (curve 3), 6.7 (curve 4), and 7.0 (curve 5)

20 h. Thus, we concluded that the suitable pH for cell growth was different from that for adenosine biosynthesis. Obviously, at the initial adenosine fermentation process (about 20 h), the average specific growth rate reached the highest at pH 7.0 (Table 46.2), and the specific glucose consumption rate also reached at the highest point (Fig. 46.6b) at the late stage (after 20 h), the average specific adenosine production rate achieved the maximal value of 0.034 (1/h) at pH 6.4 (Fig. 46.6c). To obtain the highest adenosine yield, it was necessary to take into account both cell growth and adenosine biosynthesis. Therefore, a pH-shift method was adapted, that was, control pH at 7.0 before 20 h to improve cell growth, and at 6.4 after 20 h to facilitate adenosine biosynthesis.

46.4.6 pH-Shift Method for Adenosine Fermentation

On the basis of two-stage DO strategy, pH was controlled at 7.0 for the initial 20 h of fermentation process to obtain maximal specific cell growth rate, and then decreased to 6.4 thereafter till the end of the fermentation. Table 46.2 shows that at this condition, adenosine yield reached maximum of 23.3 g/L, which was 21.4 % higher than that at single two-stage DO strategy, and glucose conversion rate reached 0.203 g/g, which was 45.0, 38.1, 21.2 and 65.0 % higher than that at pH maintained at 5.8, 6.1, 6.7, and 7.0, respectively (Table 46.2). Herein, the combined DO strategy plus pH-shift method is an effective way for increasing adenosine yield.

Figure 46.7 summarizes adenosine yield, adenosine productivity, and glucose conversion rate at experiments with uncontrolled DO condition, two-stage DO strategy, and two-stage DO plus pH-shift strategy. Under two-stage DO plus pH-shift strategy, adenosine yield, adenosine productivity, and glucose conversion rate

Table 46.2 Comparison of parameters in adenosine fermentation under different pH condition

Parameters	pH					
	5.8	6.1	6.4	6.7	7.0	pH-shift[a]
Initial glucose concentration (g/L)	69 ± 1	68 ± 1	69 ± 1	69 ± 1	70 ± 1	69 ± 1
Culture time[b] (h)	60	56	52	60	60	56
Maximal dry cell weight (g/L)	6.0 ± 0.15	6.6 ± 0.18	7.4 ± 0.15	7.5 ± 0.20	7.9 ± 0.16	7.7 ± 0.16
Maximal adenosine production (g/L)	9.5 ± 0.5	13.7 ± 0.6	19.2 ± 0.5	16.8 ± 0.4	15.0 ± 0.4	23.3 ± 0.5
Maximal glucose consumption rate (g/L)	0.37 ± 0.04	0.40 ± 0.02	0.63 ± 0.02	0.93 ± 0.01	1.25 ± 0.02	0.75 ± 0.03
Maximal specific growth rate (1/h)	0.115 ± 0.003	0.139 ± 0.004	0.283 ± 0.003	0.331 ± 0.003	0.339 ± 0.005	0.359 ± 0.003
Adenosine yield on glucose (g/g)	0.140 ± 0.004	0.147 ± 0.006	0.225 ± 0.003	0.160 ± 0.005	0.123 ± 0.002	0.233 ± 0.003
Adenosine productivity (g/(Lh))	0.158 ± 0.008	0.245 ± 0.010	0.369 ± 0.009	0.280 ± 0.007	0.250 ± 0.007	0.416 ± 0.009
Average specific growth rate in first 20 h (1/h)	0.075 ± 0.001	0.085 ± 0.001	0.094 ± 0.001	0.096 ± 0.001	0.101 ± 0.001	0.098 ± 0.001
Average specific glucose consumption rate (1/h)	0.22 ± 0.005	0.26 ± 0.004	0.29 ± 0.005	0.37 ± 0.003	0.47 ± 0.004	0.35 ± 0.004
Average specific adenosine production rate after 20 h (1/h)	0.021 ± 0.001	0.029 ± 0.001	0.034 ± 0.001	0.027 ± 0.001	0.021 ± 0.001	0.039 ± 0.001

[a] In the first 20 h, DO level controlled 30–40 %, pH 7.0 then DO controlled 10–20 %, pH 6.4

[b] Fermentation time when the adenosine production reached maximal

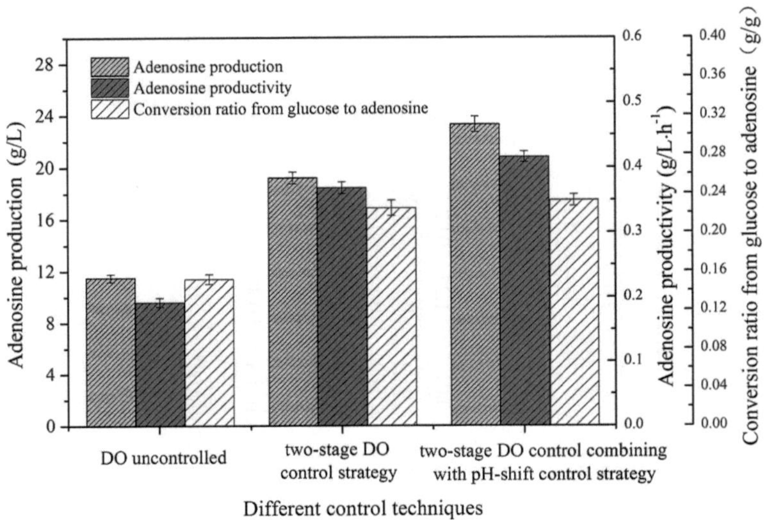

Fig. 46.7 Effect of different control techniques on adenosine production, adenosine productivity, and conversion ratio from glucose to adenosine in the fermentation

were enhanced by 102.6, 116.7, 53.3 %, respectively, compared to those in experiment with uncontrolled DO level, and 21.4, 12.7 and 3.6 %, respectively, compared to those under two-stage DO strategy.

To our best knowledge, this is the first report on the improvement of adenosine production by comprehensively optimizing DO and pH, and it has reached a comparable advanced level in the world within a shorter fermentation period. The results are very useful to large-scale adenosine as well as other bio-products fermentation.

46.5 Conclusion

In this study, we investigated the effects of different DO levels and pH on adenosine fermentation and found that relatively high DO level and pH were beneficial for cell growth, while relatively low DO level and pH can promote adenosine synthesis. Based on these findings, we developed a novel two-stage DO strategy plus pH-shift method (30–40 % DO and pH 7.0 at initial 20 h fermentation as well as 10–20 % DO and pH 6.4 thereafter). At this condition, adenosine yield, DCW and glucose conversion rate reached maximum of 23.3 g/L, 7.7 g and 0.233 g/g, respectively. This method is simple and reproducible, and has application potentials for highly efficient, large-scale adenosine production.

Acknowledgments We gratefully acknowledge the excellent support of Shanshan Du in their operation of the fermentation process control. This investigation is supported by the National Natural Science Foundation of China (31100054) and by Program for Changjiang Scholars and Innovative Research Team in University (IRT 1166).

References

1. Mallet M (2004) Proarrhythmic effects of adenosine: a review of the literature. Emerg Med J 21:408–410
2. Kunst F, Ogasawara N, Moszer I et al (1997) The complete genome sequence of the gram-positive bacterium *Bacillus subtilis*. Nature 390:249–256
3. Sauer U, Cameron DC, Baily JE (1998) Metabolic capacity of Bacillus subtilis for the production of purine nucleosides, riboflavin, and folic acid. Biotechnol Bioeng 59:227
4. Schallmey M, Singh A, Ward OP (2004) Developments in the use of *Bacillus* species for industrial production. Can J Microbiol 50:1–17
5. Karasawa M, Ichiumi T, Nakamatsu W (1992) Manufacture of adenosine with *Bacillus*. JP 04030797 A2
6. Nishiyama T, Karasawa M, Yamamoto K et al (1995) Production of adenosine by a growth-improved mutant of *Bacillus subtilis*. Industrial production of adenosine.III. Nippon Nogei Kagaku Kaishi 69:1341–1347
7. Nakamatsu W, Nishiyama T, Ichiumi T et al (1985) Adenosine by bacterial fermentation. JP 60176596 A2
8. Nishiyama T, Nakamatsu T, Shirota T (1994) Adenosine production by a mutant of *Bacillus subtilis* resistant to adenine analogs. Industrial production of adenosine.II. Nippon Nogei Kagaku Kaishi 68:809–814
9. Kaya-Çeliker H, Angardi V, Çalık P (2009) Regulatory effects of oxygen transfer on overexpression of recombinant benzaldehyde lyase production by *Escherichia coli* BL21 (DE3). Biotechnol J 4:1066–1076
10. Liu YS, Wu JY, Ho KP (2006) Characterization of oxygen transfer conditions and their effects on *Phaffia rhodozyma* growth and carotenoid production in shake-flask cultures. Biochem Eng J 27:331–335
11. Li H, Chen Y, Gu G (2008) The effect of propionic to acetic acid ratio on anaerobic–aerobic (low dissolved oxygen) biological phosphorus and nitrogen removal. Bioresour Technol 99:4400–4407
12. Yegneswaran P, Gray M, Thompson B (1991) Effect of dissolved oxygen control on growth and antibiotic production in *Streptomyces clavuligerus* fermentations. Biotechnol Prog 7:246–250
13. Yu WB, Gao SH, Yin CY, Zhou Y, Ye BC (2011) Comparative transcriptome analysis of *Bacillus subtilis* responding to dissolved oxygen in adenosine fermentation. PLoS ONE 6: e20092. doi:10.1371/journal.pone.0020092
14. Feng X, Xu H, Yao J, Li S, Zhu H, Ouyang P (2010) Kinetic analysis and pH-shift control strategy for propionic acid production with *Propionibacterium freudenreichii* CCTCC M207015. Appl Biochem Biotechnol 160:343–349
15. Çalik P, Bilir E, Çalik G, Özdamar TH (2003) Bioreactor operation parameters as tools for metabolic regulations in fermentation processes: influence of pH conditions. Chem Eng Sci 58:759–766
16. Haneda K, Hirano A, Kodaira R, Ohuchi S (1971) Accumulation of nucleic acid-related substances by microorganisms. Agric Biol Chem 35:1906–1912
17. Haneda K, Komatsu K, Kodaira R, Ohsawa H (1972) Stabilization of adenosine-producing mutants derived from *Bacillus* sp. No. 1043. Agric Biol Chem 36:1453–1460

18. Cao J, Chen X, Ren H, Zhang J, Li L, Chen Y, Xiong J, Bai J, Ying H (2012) Production of cyclic adenosine monophosphate by *Arthrobacter* sp. A302 using fed-batch fermentation with pH-shift control. World J Microbiol Biotechnol 28:121–127
19. Meiying Z, Guocheng D, Jian C (2002) pH control strategy of batch microbial transglutaminase production with *Streptoverticillium mobaraense*. Enzyme Microb Technol 31:477–481
20. Peng Z, Fang J, Li J, Liu L, Du G, Chen J, Wang X, Ning J, Cai L (2010) Combined dissolved oxygen and pH control strategy to improve the fermentative production of L-isoleucine by *Brevibacterium lactofermentum*. Bioprocess Biosyst Eng 33:339–345
21. Shu CH, Liao CC (2002) Optimization of L-phenylalanine production of *Corynebacterium glutamicum* under product feedback inhibition by elevated oxygen transfer rate. Biotechnol Bioeng 77:131–141
22. Huang WC, Chen SJ, Chen TL (2006) The role of dissolved oxygen and function of agitation in hyaluronic acid fermentation. Biochem Eng J 32:239–243
23. Peña C, Trujillo-Roldán MA, Galindo E (2000) Influence of dissolved oxygen tension and agitation speed on alginate production and its molecular weight in cultures of *Azotobacter vinelandii*. Enzyme Microb Technol 27:390–398
24. Li L, Chen X, Cheng J, Zhang C, Bai J, Chen Y, Niu H, Ying H (2012) Bi-stage control of dissolved oxygen to enhance cyclic adenosine monophosphate production by *Arthrobacter* A302. Bioprocess Biosyst Eng 35:1281–1286
25. Chen XC, Song H, Fang T, Cao JM, Ren HJ, Bai JX, Xiong J, Ouyang PK, Ying HJ (2010) Enhanced cyclic adenosine mono-phosphate production by *Arthrobacter* A302 through rational redistribution of metabolic flux. Bioresour Technol 101:3159–3163
26. Liu YP, Zheng P, Sun ZH, Ni Y, Dong JJ, Wei P (2008) Strategies of pH control and glucose-fed batch fermentation for production of succinic acid by *Actinobacillus succinogenes* CGMCC1593. J Chem Technol Biotechnol 83:722–772

Chapter 47
The Semi-continuous Cultivation of *Nostoc flagelliforme* Cells

Lifang Yue, Yupeng Xiao, Guojuan Sun, Shiru Jia, Yujie Dai and Xing Zheng

Abstract The semi-continuous cultivation is an important technique for large-scale liquid photoautotrophic cultivation of microalages, in which the excessive cells can be removed and the nutrient elements in the culture medium can be supplemented in time. In this paper, the semi-continuous cultivation of *Nostoc flagelliforme* cells was conducted and the cultivation conditions were optimised in order to achieve maximum biomass production. After single factor and Box-Behnken experiment with 15 days, the largest biomass amount (0.305 g/l) can be harvested at the best culture conditions (update rate 33 %, update cycle 1 day, and sodium nitrate concentration 1.6 g/l). This study offered important basic data for the implementation of large-scale photoautotrophic cultivation of *N. flagelliforme* cells.

Keywords *Nostoc flagelliforme* · Large-scale cultivation · Semi-continuous cultivation · Single factor experiment · Box-Behnken experime

47.1 Introduction

Currently, the solid culture and liquid suspension culture of *Nostoc flagelliforme* are the main ways reported for cultivation of *N. flagelliforme* [3, 4, 6, 11, 12]. The liquid suspension culture has been attracted more attention because it is easily

L. Yue · Y. Xiao · G. Sun · S. Jia
Key Laboratory of Industrial Fermentation Microbiology, Ministry of Education, College of Bioengineering, Tianjin University of Science and Technology, Tianjin 300457, People's Republic of China

X. Zheng
Fuqing Xin Da Ze Spirulina Cooperation Limited, Fuqing 350313, People's Republic of China

Y. Dai (✉)
Tianjin Key Laboratory of Industrial Microbiology, School of Bioengineering, Tianjin University of Science & Technology, TEDA, 13th Street, Tianjin 300457, China
e-mail: yjdai@126.com

T.-C. Zhang and M. Nakajima (eds.), *Advances in Applied Biotechnology*, Lecture Notes in Electrical Engineering 332, DOI 10.1007/978-3-662-45657-6_47

controlled and *N. flagelliforme* cells can get a faster growth rate, however, it is only in laboratory scale and batch style [12, 15]. The semi-continuous cultivation is an important technique that can be applied in large-scale liquid photoautotrophic cultivation of microalages, in which the excessive cells can be removed and the nutrient elements in the culture medium can be supplemented in time [9, 16]. And it is efficient for the cultivation of microorganisms and can be easily operated.

Light, nutrients and aeration are the main factors that influence the growth of *N. flagelliforme* cells in the liquid suspension cultivation [5, 7, 13, 14]. During the cultivation, the growth of *N. flagelliforme* may be hindered by many factors, such as, the shortage of light energy for individual cells in consequences of increased cell density, the accumulation of metabolites of *N. flagelliforme* and osmotic pressure increases in consequences of evaporation of water during the cultivation [2, 8, 10, 17]. Thus, the supplement and replacement of part culture medium intermittently not only eliminates the adverse effects mentioned above, but also to maintains the cell density and growth rate of *N. flagelliforme* cells.

The semi-continuous cultivation refers to the process of the microbial cultivation that, the culture medium was partly refreshed intermittently. It is efficient for the cultivation of microorganisms and can be easily operated. The photosynthesis of photosynthetic microorganisms was enhanced with the increase of the light transmittance in consequence of the decrease of cell density with some fresh culture medium was supplemented and some microorganisms were harvested during the semi-continuous cultivation.

Update rate (the volume ratio of fresh culture medium added to the original culture medium), update cycle (the time interval between the adjacent two times of culture medium update) and sodium nitrate are the key factors in the semi-continuous cultivation. Maximum biomass production can not be achieved, because the cell density is too low to make good use of the light energy under high update rate and short update cycle. On the other hand, the growth rate is also hindered by the shortage of nutrients and light energy with the low update rate and long update cycle. Thus, proper update rate and update cycle are necessary to obtain the high growth rate and biomass for the semi-continuous cultivation. Sodium nitrate is a kind of nitrogen source and is also very important to *N. flagelliforme* cultivation. Thus, in this paper, single factor and Box-Behnken experiments were conducted in order to examine the effect of update rate, update cycle and the concentration of sodium nitrate on harvested biomass in the semi-continuous cultivation of *N. flagelliforme* cells.

47.2 Materials and Methods

The *N. flagelliforme* cells were obtained from the Tianjin Key Lab of Industrial Microbiology (Tianjin, China). The cells were cultured in BG110 culture medium. Reagents were analytical grade. Biomass production was measured by weighting the dry *N. flagelliforme* cells [15].

47.2.1 Method of the Semi-continuous Cultivation

N. flagelliforme cells in logarithmic phase were added into a 10 l cylindrical glass bottle with 10 % of inoculum. The culture temperature was 25 °C. Light intensity was 2,000 lux. Light-dark cycle ratio was 24:0. The aeration rate was 0.8 vvm. The total volume of cultivation is 8 l. The semi-continuous cultivation was conducted from the fifth day. The entire culture period was 15 days.

47.2.2 Experimental Design

Each experiment was replicated at least three times and the optimal value of update rate, update cycle and concentration of sodium nitrate were optimised with single factor experiment.

The biomass of *N. flagelliforme* cells was taken as the responses of the experiments. A second order polynomial equation model was used to fit the biomass with interaction terms [1]. Once the regression equation model was developed, the response surface methodology was used to optimise the three variables, and identify the behaviour of the system within the optimised ranges. The experimental design, regression and graphical analysis were obtained using the Design Expert Version 8.0.6.1 (Stat-Ease Inc., Minneapolis, USA).

47.3 Results

47.3.1 Single Factor Experiment

The update rate was set 10, 15, 20, 30 and 40 %. Update cycle was set 0.5, 1, 1.5, 2, and 2.5 days. Sodium nitrate concentration was set 0, 0.5, 1, 1.5, 2 and 2.5 g/l. The results of single factor experiment are shown in Fig. 47.1. It can be seen that the largest biomass amount was harvested at update rate of 30 %, update cycle of 1 day and sodium nitrate concentration of 1.5 g/l.

47.3.2 Statistical Analysis and Model Fitting

In order to optimise the cultivation conditions and verify the influence of the synergy of variables, the Box-Behnken experimental design was carried out with three variables at each of three levels. The codings and levels for each factor are listed in Table 47.1, and the experimental results are listed in Table 47.2.

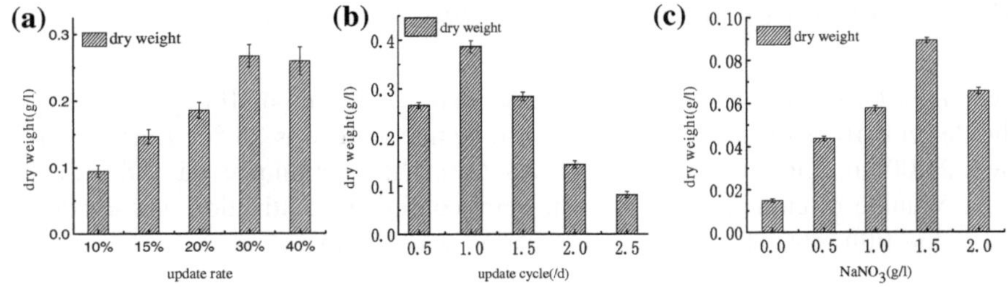

Fig. 47.1 Single factor experiment about the effects of update rate (**a**), update cycle (**b**), $NaNO_3$ concentration (**c**) on *Nostoc flagelliforme* biomass

Table 47.1 Coding and level of variables chosen for Box-Behnken Design

Factor	Level		
	−1	0	+1
X_1(update rate) (%)	20	30	40
X_2 (update cycle) (d)	0.5	1	1.5
X_3 (Concentration of $NaNO_3$) (g/l)	1.0	1.5	2.0

Table 47.2 Experimental design and results for Box-Behnken experiments

Run no	X_1	X_2	X_3	Response biomass (g/l)
1	0	−1	1	0.274
2	0	1	1	0.185
3	0	0	0	0.305
4	−1	1	0	0.164
5	0	−1	−1	0.264
6	−1	0	1	0.242
7	1	0	−1	0.246
8	−1	0	−1	0.215
9	1	0	1	0.258
10	1	−1	0	0.288
11	0	0	0	0.305
12	−1	−1	0	0.242
13	0	0	0	0.298
14	1	1	0	0.212
15	0	1	−1	0.166
16	0	0	0	0.29
17	0	0	0	0.296

From the experimental conditions and corresponding response values shown in Table 47.2, the model for the biomass (Y) was obtained. The regression equation in terms of actual factors is as follows:

Table 47.3 The analysis of variance for the response surface model Eq. 47.1

Source	Sum of squares	*df*	Mean square	*F*-value	*p*-value Prob > *F*
Model	0.035252	9	0.003917	46.66594	<0.0001
X_1	0.002485	1	0.002485	29.60748	0.0010
X_2	0.014535	1	0.014535	173.1697	<0.0001
X_3	0.000578	1	0.000578	6.886222	0.0342
$X_1\ X_2$	1×10^{-6}	1	1×10^{-6}	0.011914	0.9161
$X_1\ X_3$	5.63×10^{-5}	1	5.63×10^{-5}	0.670156	0.4400
$X_1^2X_3$	2.02×10^{-5}	1	2.02×10^{-5}	0.241256	0.6383
X_1^2	0.003104	1	0.003104	36.97679	0.0005
X_2^2	0.008583	1	0.008583	102.2598	<0.0001
X_3^2	0.004151	1	0.004151	49.45941	0.0002
Residual	0.000588	7	8.39×10^{-5}		
Lack of fit	0.000425	3	0.000142	3.478706	0.1299
Pure error	0.000163	4	4.07×10^{-5}		
Cor total	0.03584	16			
$R^2 = 0.9836$	$R^2_{\text{Adj}} = 0.9625$				

$$Y = -0.4191 + 0.0191X_1 + 0.2594X_2 + 0.4073X_3 + 1.0 \times 10^{-4}X_1X_2 - 7.5 \times 10^{-4}X_1X_3 + 9.0 \times 10^{-3}X_2X_3 - 2.7150 \times 10^{-4}X_1^2 - 0.1806X_2^2 - 0.1256X_3^2 \quad (47.1)$$

The analysis of variance for the response surface model Eq. 47.1 are shown in Table 47.3. The *F*-values (46.67) and the *P*-values <0.0001 suggested that the model was highly significant. Meanwhile, the lack of fit was not significant ($p > 0.05$). And there is well relationship between predicted and actual values (Fig. 47.2). All the results indicated that the factors in the model were well

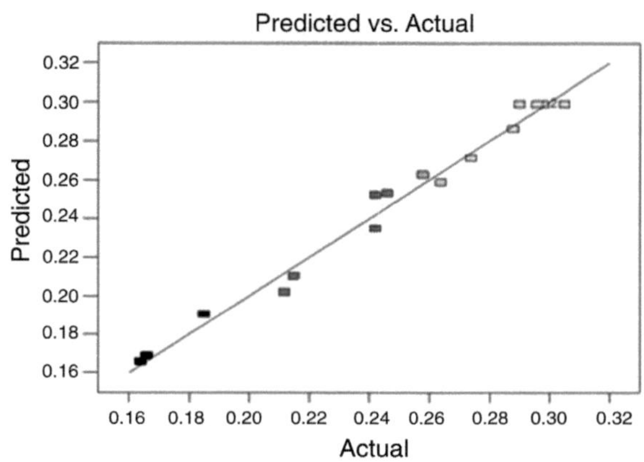

Fig. 47.2 Relationships between predicted and actual values for the response surface model equation

correlated. The high values of R^2 (0.9836) as well as R^2_{Adj} (0.9625) for the model fitting also indicated that the models fitted the experimental data very well. Therefore, the regression model has a high confidence and could accurately represent the variables chosen in the experiment.

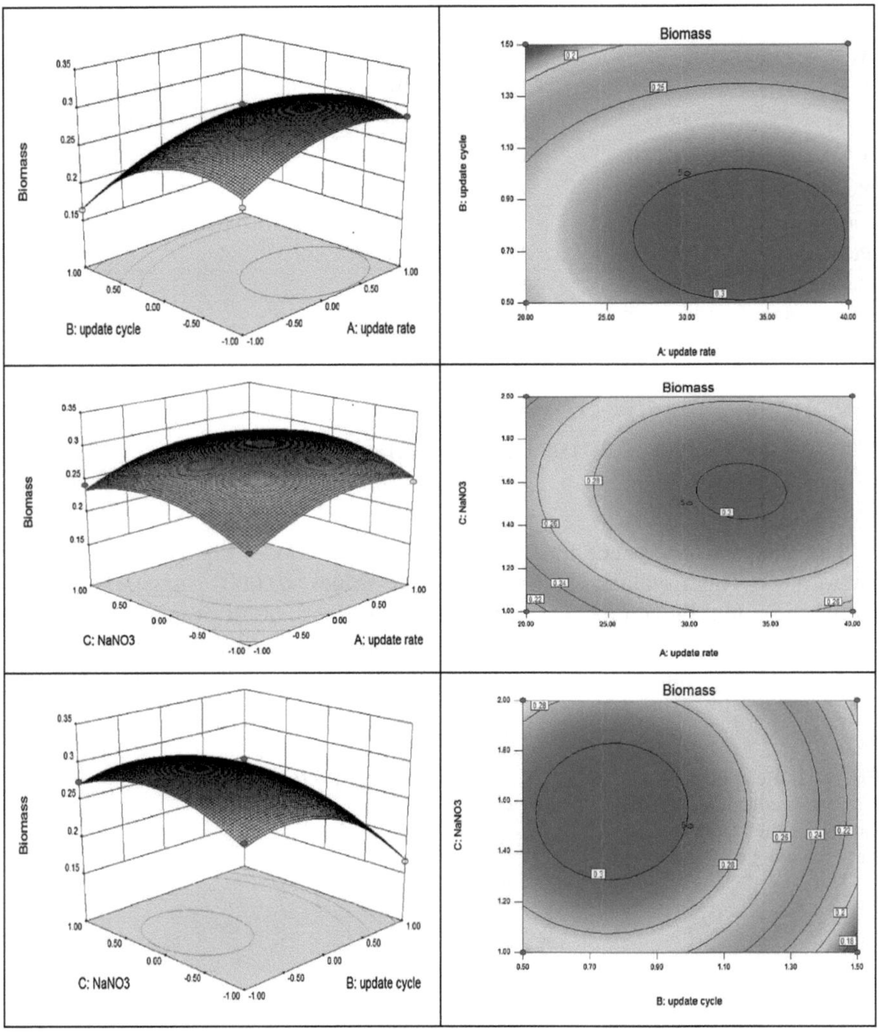

Fig. 47.3 Three-dimensional response surface curves and the contour shapes based on the regression Eq. 47.1

47.3.3 Effects of Factors on Response

Three-dimensional response surface curves and the contour shapes based on the regression Eq. 47.1 are shown in Fig. 47.3. The oval degrees of the contours reflect the strength of pairwise interaction effects between update rate, update cycle and the concentration of sodium nitrate. We can know from the contour shapes that the interaction effects between update rate and update cycle, update rate and sodium nitrate are stronger than the interaction effect between update cycle and sodium nitrate.

The slopes of three-dimensional response surface curves reflect the influence strength of the update rate, update cycle and sodium nitrate on biomass. It can be seen from Fig. 47.1 that the steepest slope for biomass vs update cycle indicates that the update cycle influences the biomass most. There is a maximum response for biomass in three-dimensional response surface curves. According to the RSM analysis, the largest biomass amount (0.305 g/l) can be harvested at the optimum cultivation conditions (update rate of 33 %, update cycle of 0.76 day, and concentration of sodium nitrate 1.6 g/l).

47.4 Discussion

Under the optimum cultivation condition, three parallel repetition tests were conducted to test the applicability of the model equations. The average harvested biomass was 0.301 ± 0.006 g/l. The error between the practical and the calculated values is smaller than 5 %. Thus, it can be concluded that the biomass can be predicted accurately using the multivariate quadratic regression model with the cultivation data of *N. flagelliforme* cells in different update rates, update cycles and the concentrations of sodium nitrate. Taking into account the convenience for implementation, the suitable conditions for semi-continuous cultivation of *N. flagelliforme* cells are update rate of 33 %, update cycle 1 day and sodium nitrate concentration 1.6 g/l.

Acknowledgments This work was supported by the National Natural Science Foundation of China (No. 31271809).

References

1. Bera MB, Panesar PS, Panesar R, Singh B (2008) Application of reverse micelle extraction process for amylase recovery using response surface methodology. Bioprocess Biosyst Eng 31:379–384
2. Changhai W, Yingying S, Ronglian X, Liqin S (2005) Effect of liquid circulation velocity and cell density on the growth of Parietochloris incisa in flat plate photobioreactors. Biotechnol Bioprocess Eng 10:103–108

3. Chen X, Jia S, Yue S, Wang N, Li C, Wang Y (2010) Effect of solid bed-materials on vegetative cells of Nostoc flagelliforme. J Appl Phycol 22:341–347
4. Gao K (1998) Chinese studies on the edible blue-green alga, Nostoc flagelliforme: a review. J Appl Phycol 10:37–49
5. Gao K, Qiu B, Xia J, Yu A (1998) Light dependency of the photosynthetic recovery of Nostoc flagelliforme. J Appl Phycol 10:51–53
6. Gao K, Ye C (2003) Culture of the terrestrial cyanobacterium, Nostoc flagelliforme (cyanophyceae), under aquatic conditions1. J Phycol 39:617–623
7. Gao K, Ye C (2007) Photosynthetic insensitivity of the terrestrial cyanobacterium Nostoc flagelliforme to solar UV radiation while rehydrated or desiccated1. J Phycol 43:628–635
8. Harada K-I, Ozaki K, Tsuzuki S, Kato H, Hasegawa M et al (2009) Blue color formation of cyanobacteria with β-cyclocitral. J Chem Ecol 35:1295–1301
9. Imamoglu E, Dalay MC, Sukan FV (2010) Semi-continuous cultivation of Haematococcus pluvialis for commercial production. Appl Biochem Biotechnol 160:764–772
10. Lee Y-K (2001) Microalgal mass culture systems and methods: their limitation and potential. J Appl Phycol 13:307–315
11. Liu X-J, Chen F (2003) Cell differentiation and colony alteration of an edible terrestrial cyanobacterium Nostoc flagelliforme, in liquid suspension cultures. Folia Microbiol 48:619–626
12. Liu X-J, Jiang Y, Chen F (2005) Fatty acid profile of the edible filamentous cyanobacterium *Nostoc flagelliforme* at different temperatures and developmental stages in liquid suspension culture. Process Biochem 40:371–377
13. Qiu B, Gao K (1999) Dried field populations of Nostoc flagelliforme (Cyanophyceae) require exogenous nutrients for their photosynthetic recovery. J Appl Phycol 11:535–541
14. Qiu B, Gao K (2002) Daily production and photosynthetic characteristics of Nostoc flagelliforme grown under ambient and elevated CO_2 conditions. J Appl Phycol 14:77–83
15. Su J, Jia S, Chen X, Yu H (2008) Morphology, cell growth, and polysaccharide production of Nostoc flagelliforme in liquid suspension culture at different agitation rates. J Appl Phycol 20:213–217
16. Wang L, Wang Y, Chen P, Ruan R (2010) Semi-continuous cultivation of Chlorella vulgaris for treating undigested and digested dairy manures. Appl Biochem Biotechnol 162:2324–2332
17. Zhao X, Bi Y, Chen L, Hu S, Hu Z (2008) Responses of photosynthetic activity in the drought-tolerant cyanobacterium, *Nostoc flagelliforme* to rehydration at different temperature. J Arid Environ 72:370–377

Chapter 48
Study on Ecological Diversity of Pectase and Its Producing Strains

Jing Xiao, Xiwang Zhou, Xiaolong Zhang and Ruiming Wang

Abstract Pectase is a heterogeneous group of related enzymes that hydrolyze the pectic substances. In this paper, seven strains producing high activity of alkaline pectate lyase were screened from alkaline soil and then the classification and activities of pectase from all these stains were detected. The results showed that seven strains had enzyme activity of pectinesterase and pectate lyase, but strains 410, 418, and 601 produced protopectinase, and its activity reached 0.341, 0.51, 0.28 u/mL separately. *Bacillus* sp. 410 and 418 showed pectin hydrolysis enzymes activity of 0.035 and 0.363 u/mL. When tested by pectin with 60 % esterification, only *Bacillus* sp. 520 had hydrolyzes activity. All these show that the pectase is abundant in different strains. Analysis of pectase system can provide fundamental basis for the study of application of wild strain, construction, and evaluation of engineering bacteria.

Keywords *Bacillus* · Pectate lyase · Screen · Enzyme system

48.1 Introduction

Pectase, a general term for a group of enzymes, is used to degrade pectin. It has extensive and important uses in industrial and environmental aspects, such as collecting pharmaceutical raw materials, curing stomach stones, food, fermentation, textiles, biotechnology and so on [1–4]. The production requirements in different application fields correspond to appropriate character of pectase. Even if acting on different objects in the same fields, such as different processing of linen, ramie, cotton fiber in textile industry, it also need corresponding pectase to match with all kinds of pectin substrates. It is important for us to research enzymes of pectase in

J. Xiao (✉) · X. Zhou · X. Zhang · R. Wang
Shandong Provincial Key Laboratory of Microbial Engineering, Qilu University of Technology, Jinan 250353, China
e-mail: xiaojing8168@163.com

T.-C. Zhang and M. Nakajima (eds.), *Advances in Applied Biotechnology*,
Lecture Notes in Electrical Engineering 332, DOI 10.1007/978-3-662-45657-6_48

various *Bacillus* [5]. Then we can determine the application direction of pectase by analyzing different resource of strains. Also it can acknowledge the value and range of pectase by researching mixed-fermentation of different strains and combination of enzymes. Finally, we can achieve the goal that we could promote the development of enzyme production. This treatise was based of alkaline pectase. During the experiment, we screened seven *Bacillus* from the strains in saline soil, and analyzed the *Bacillus* further, and determined the activity types and sizes [6].

48.2 Materials and Methods

48.2.1 Strains, Reagents, Medium, and Instrument

Bacillus sp. 410, *Bacillus* sp. 418, *Bacillus* sp. 427, *Bacillus* sp. 514, *Bacillus* sp. 520, *Bacillus* sp. 521 (the *Bacillus* which is used to product pectase were selected from saline soil).

Pectin, 30, 60, 90 % esterified potassium salt from citrus fruit (Sigma, USA); poly-D-galacturonic acid rom citrus peel (Sigma, USA); D-galacturonic acid (Sigma, USA); ammoniated ruthenium oxychloride (Sigma, USA); CTAB (Sangon); other reagents such as carbazole, sodium hexametaphosphate etc. (China).

Screening Medium (g/L): pectin 5, yeast extract 5, peptone 5, magnesium sulfate 2, dipotassium phosphate 1, agarophyte 20, pH 7.0–7.5. Seed medium (g/L): sucrose 20, peptone 5, corn steep liquor 30, magnesium sulfate 1, dipotassium phosphate 18.4, monopotassium phosphate 6, pH 7.0. Fermentation medium (g/L): lactose 20, peptone 3, calcium chloride 3, pectin 20, yeast extract 3, magnesium chloride 0.2, manganese sulfate 0.04, dipotassium phosphate 1, pH 7.0.

Analytical instrument: ZD-3A automatic potentiometric titrator; WFJ2000 ultraviolet and visible spectrophotometer (Unico, Shanghai, China); DW-3 lyophilizer (the corporation of genic, China).

48.3 Experimental Method

Screening of strains: The soil sample collected from saline-alkali soil was coated on the filter medium plate after a series of dilution. In order to get the single colony, the filter medium plate was cultivated upside down at 37 °C for 24 h. At the same time, the process of using CTAB reveal the image of colony which was selected by transparent zone before going on next operation. At last, selecting single colony that the diameter of single colony and transparent zone greater than three carry on fermentation after dropping ruthenium red developer. The strains that we got from preliminary screening were fermented and will be used as a kind of seed liquid after 12 h constant temperature cultivation under the environment of 37 °C, 100 r/min. Then, the fermentation will begin with 8 % inoculum concentration in shake flask under the environment of 37 °C and subsequent 4 h constant temperature cultivation with the speed of 200 r/min. In the next moment, we measured the activity of alkaline pectase which can also be called pectate lyase to achieve the goal of selection. The seed liquid contributed 10 % volume of flask.

Activity analysis of pectate lyase (acid): Definition of unit enzyme activity: Under the environment of 45 °C and pH 9.0, the number of enzyme we used that makes poly galacturonic acid translate into 1 μmol unsaturated poly galacturonic acid minute by minute. Enzyme activity determination: In test tube, 2 mL 0.2 % polygalacturonic acid which the solvent was 0.2 mol/L buffer solution of glycine—NaOH with 0.44 mmol/L $CaCl_2$ under the environment of pH 9.0 was taken as zynolyte, then the action will begin after dropping 20 enzyme-solution. After that, 3 mL 0.03 mol/L phosphoric acid was used to terminate the action after 15 min action under the environment of 45 °C; at last, measuring the absorbance values at 235 nm to determine the process of action. In this part, the inactive enzyme was used for black control. And the molar absorption coefficient of unsaturated poly galacturonic acid is 4,600 L mol^{-1} cm^{-1} at 235 nm. Also, the activity determination of pectate lyase acid is the same as that of pectate lyase; and in this part, substrate we chose was 30, 60, and 90 % of pectin that was determined.

Activity determination method of pectic acid hydrolysis enzyme: The drawing of standard curve and activity determination method can reference Zhang Fei's methods [7]. The activity determination of pectate lyase acid is the same as that of pectate lyase; and in this part, substrate we chose was 30, 60, and 90 % of pectin that was determined.

Protopectinase activity determination method: The preparation of protopectinase substrate: After smashing the substance that we got from the white layer of citrus peel, the 2 % Sodium hexametaphosphate solution under the environment of pH 4.0 was used as eluent to elute the broken matter until the eluent wouldn't color with carbazole sulfuric acid; then we can got the white precipitate under the environment of rotating speed with 5,000 r/min at 15 min; after that white spongy will be produced by the process of Vacuum freeze drying; finally, the protopectinase substrate can be obtained by shattering the white spongy. The protopectinase enzyme determination: In this part, the method of carbazole sulfuric acid colorimetry was used to measure the activity. On the condition of pH 5.0 and 37 °C that was ruled, the quantity of protopectinase catalyzed every hour, which was equal to 1 μmol galacturonic acid of pectic substance, was regarded as Unit of enzyme activity [8].

Pectinesterase activity determination methods (pH-stat): In this part, 0.05 mol/L NaOH was used to adjust the pH of 1 % pectin solution, the volume of which is 30 mL, until the pH reached 7.0. And then we use the 0.05 mol/L NaOH solution to keep the balance of pH 7.0 after adding 10 mL crude enzyme solution. The graph was drawn by the micro Moore and reaction time, and the slope of initial curve represents the pectinesterase activity on chart. At the same time, the active unit of PE is mol/s which means 1 mol NaOH was consumed every second.

48.4 Results

Among the production of alkaline pectase strains that were selected from the saline-alkali soil samples, *Bacillus* take up 75.8, including litter *Serratia marcescens*, hydrogen producing bacteria, etc. We can find that *Bacillus subtilis* and *Bacillus megaterium* etc. which could produce alkaline pectinase occupied vantage point in Saline-alkali soil of this zone.

Pectin lyase (acid) activity. When polygalacturonic acid sodium salt and different degree of esterification of pectin which are in the order of 30, 60, and 90 % was used for substance, all of the seven *Bacillus* can release the unsaturated galacturonic acid by trans-elimination to cut glycosidic bond of pectin under the environment of alkaline condition, thereinto, and the range of splitting activity respectively is 0.95–1.31, 0.35–0.85, 1.37–1.53, 0.30–0.67 u/mL. According to the result of experiment, it reflects the advantage of different strains that match with appropriate substance; for example, the degradation of vigor that the *B. subtilis* 521 acts on pectic acid and 30 % esterification of pectin is higher than other strains and the best of degree of esterification of pectin that the *B. subtilis* 418 and *B. subtilis* 427 react respectively is 60, 90 %. The result is shown in Fig. 48.1.

Under the environment of laboratory, although seven strains we selected by detection of activity, the which of the presents suitable activity of pectate lyases corresponding effect for different degree of esterification of pectin, the ability of depolymerization is higher than that we imaged.

Analysis results of pectic acid hydrolysis enzyme activity. In seven *Bacillus* we selected, the activity of *Bacillus* sp. 410 pectic acid hydrolysis and *Bacillus* sp. 418 pectic acid hydrolysis enzyme respectively is 0.035, 0.363 u/mL. When different degrees of methyl esterification of pectin which are in that order 30, 60, and 90 % was used for substance, the *Bacillus* sp. 520 acts on 60 % methyl esterification of pectin merely and the activity of it is 1.25 u/mL, but other strains could not measure relevant activity of pectic acid hydrolysis enzyme. The pectic-hydrolysis enzyme and pectate lyase are derived from pectate depolymerase; on the one hand, it can degrade the plant fiber of environment to promote recycling of carbon element; on

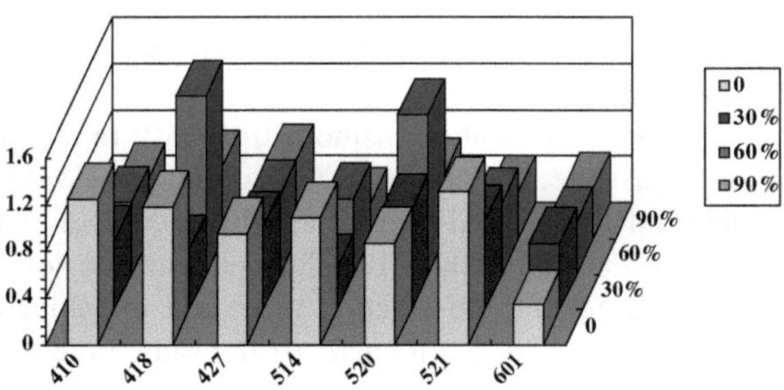

Fig. 48.1 Activity of cleaving glycosidic linkages of different strains on pectins with different degrees of esterification

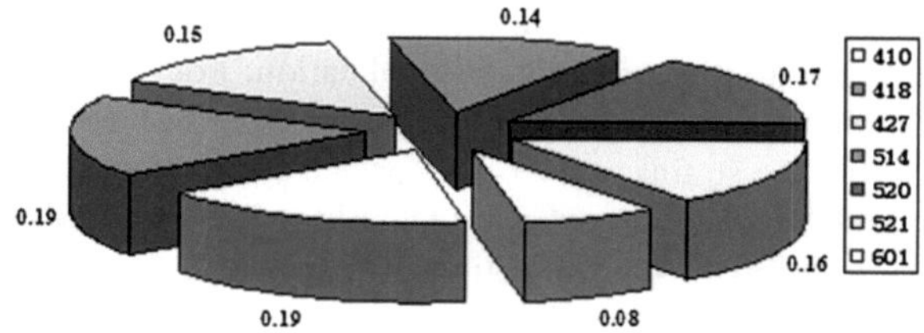

Fig. 48.2 Pectinesterases activity of different stains (mol/L S)

the other hand, microorganism can serve as demand of carbon source from products of pectin degradation. Because of the environment that *Bacillus* lived in, the performance of *Bacillus* present a series of characters of pectase and the effective range of alkaline pectinase was wide. But it will present its weaknesses on the condition of acidity.

Analysis results of protopectinase activity. After we got about 100 g (wet weight) orange peel from 2,500 kg citrus that we bought, we finally got approximate 10.4 g protopectin by eleven rinses to measure the activity of protopectinase. The results of experiment show that the strains of *Bacillus* sp. 410; *B. subtilis* 418; *Bacillus* sp. 601 can produce protopectinase and its activity successively is 0.341; 0.51; 0.28 u/mL.

Analysis results of pectinesterase activity: According to experiment, we can find that the pectinesterase acts on molecular of pectin and releases H^+ that made the pH of reaction system become lower and lower. On the condition of balance that uses 0.05 mol/L NaOH solution as titrant maintain the pH of system, the automatic potentiometric titrator was used to measure the activity of pectinesterase which was reflected by consume of alkali. Analysis results of pectinesterase activity shown in Fig. 48.2.

All of the strains we selected can produce pectinesterase, and the production of every strains is roughly similar except the *Bacillus* sp. 601, the activity of which is very low, reaching to 0.08 mol/L S. Particularly, the pectinesterase activity of *Bacillus* sp. 410 and *B. subtilis* 418 is higher than others, reaching up to 0.19 mol/L S.

48.5 Conclusion

These experimental results demonstrated that the *Bacillus* is an important source of production of industrial enzymes. When the pectin was regarded as substance and the alkaline pectinase (pectate lyase) was detected as detection target, all of seven *Bacillus* we selected from saline-alkali soil samples have well degradation activity of pectin, and the range of selection which acts on different pectin substance with methylation modification was relatively wide. Also, we can find that the pectin system of *Bacillus* sp. 410 and *Bacillus* sp. 418 had abundant types, including all kinds of pectin such as hydrolase, lyases, esterase and protopectinase; and other strains had their beneficial character in the aspect of pectin type. What's more, we can obtain better selection by direction of application. For example, we can select preponderant alkaline pectase just as the strains of 520, 521 for textile, paper etc. application direction; the strains of 410, 418, or 520 were selected for the process of sour fruit and vegetable juice processing technology. When *Bacillus* was used for the feed industry to improve the utilization of feed and keep the balance of crude fiber nutrition, the strain of 418 is the best choice.

In addition, the molecular biology of pectase is developing rapidly with the research of pectase, and the research of high expressions that single pectase gene translated into the host has been extensively spread. Research of the function of pectase and the enzyme system will be benefit for revealing the enzymatic properties and the application mechanism of enzyme.

References

1. Xue CH, Zhang YQ, Li ZJ (2005) The research progress of pectin and pectase. J Food Sci Biotechnol 6:94–99
2. Gerhaed D, Aagelika L (2000) Degradation of pectins with different degrees of esterification by Bacteroides thetaiotaomicron isolated from human gut flora. Appl Environ Microbiol 66:1321–1327
3. Petersen S (2001) Enzymes to upgrade plant nutrients. Feed Mix 9:12–15
4. Sakai T, Sakamoto T, Hallaert J, Vandamme EJ (1993) Pectin, pectinase and protopectinase: production, properties, and applications. Adv Appl Microbiol 39:213–229
5. Nevadita S, Madhu R, Mukesh S (2013) Microbial pectinase: sources, characterization and applications. Rev Environ Sci Bio/Technol 12:45–60
6. Hoondal GS, Tiwari RP, Tewari R (2002) Microbial alkaline pectinases and their industrial applications: a review. Appl Microbiol Biotechnol 59:409–418
7. Zhang F, Yue LL, Fei J et al (2004) The research of pectase activity determination method. Acta Agriculturae Boreali-Occidentalis Sinica 13:134–137
8. Sakamoto T, Hours R, Aandsakai T (1994) Purification, characterization, and production of two pectic transeliminases with protopectinase activity from Bacillus subtillis. Biosci Biotechnol Biochem 58:353–358

Part III
Biological Separation and Biological Purification

Chapter 49
The Extraction and Regeneration of Resin XAD-16 in the Purification of Epothilones

Can Li, Lin Zhao, Xiaona Wang, Qiang Ren and Xinli Liu

Abstract The XAD-16 resins, which can capture the epothilones in situ and remove them from the culture broth, are always used in isolation and purification of epothilones during the fermentation process. In this paper, the effects of five different organic solvents (methanol, alcohol, acetone, ethyl acetate, carbon tetrachloride) on the epothilone extraction and regeneration of macroporous adsorption resin XAD-16 were surveyed. Of five organic solvents, methanol showed a much better effect on epothilone yield than the others. Also, the differences in structure between the fresh-resin and the regenerative-resin with methanol by atomic force microscope were analyzed. As the results, pore size of methanol-resin was larger than fresh-resin's, which lead it more conducive to the adsorption of epothilones and regenerative-resin treated with methanol was sturdier. Result of the impact experiment showed the macroporous adsorption resin XAD-16 could be used for at least five times under methanol treatment.

Keywords Epothilones · *Sorangium cellulosum* · Regeneration · Macroporous adsorption resin

49.1 Introduction

Epothilones are a kind of 16-membered macrolides with strong stabilizing activities on polymerized microtubules which mimicked taxol® [1]. Ixabepilone, one derivative of epothilone, has been approved for clinical use by the U.S.

Although many organisms that can produce epothilones have been found, microbial fermentation of *Sorangium cellulosum* is still the main way to produce epothilones [2]. However, the yields of epothilones were so low and it cost too

C. Li · L. Zhao · X. Wang · Q. Ren · X. Liu (✉)
Shandong Provincial Key Laboratory of Microbial Engineering,
Qi Lu University of Technology, Jinan 250353, People's Republic of China
e-mail: vip.lxl@163.com

T.-C. Zhang and M. Nakajima (eds.), *Advances in Applied Biotechnology*,
Lecture Notes in Electrical Engineering 332, DOI 10.1007/978-3-662-45657-6_49

much in its extracting. In the fermentation process, various bioactive small molecules produced by microbial cultivation are degraded in the culture broth or may repress the formation of additional product. The inclusion of hydrophobic adsorber resin beads to capture these products in situ and remove them from the culture broth, which can reduce or prevent the degradation and repression during fermentation.

The XAD-16 resin, which can capture the epothilones in situ and remove them from the culture broth, could isolate the epothilones from other impurities and reduce the feedback inhibition. So, the epothilones yield of the cultivations containing the XAD-16 adsorption resin was approximately 10 times higher than that observed in the cultivations without the XAD-16 adsorption resin [3]. The XAD-16 resins play an important role in absorbing epothilones during microbial fermentation of *S. cellulosum*. However, high price and large demand of XAD-16 resin always raise production costs.

In this paper, the effects of five organic solvents on the regeneration of macroporous adsorption resin XAD-16 were surveyed.

49.2 Experimental Section

49.2.1 Strains and Materials

So2161 is a *S. cellulosum* strain that was isolated from an alkaline soil sample in our laboratory which has been morphologically and phylogenetically classified using previously reported methods [4]. It could be routinely cultivated on solid CNST agar plates and in liquid M26 medium at 30 °C (pH7.2) for 3 days. After cultivation in M26, seeds were then carried out in 300 mL glass flasks filled with 50 mL of liquid fermentation medium, cultured for 7 days at 30 °C [5].

All the organic solvents used in this paper (methanol, alcohol, acetone, ethyl acetate, carbon tetrachloride) were all analytical reagent supplied by TJSHIELD Ltd.

49.2.2 Regeneration of the Used XAD-16 Resin

The "used" XAD-16 resins were got from the fermentation of epothilones in 500 L fermenter, which were once extracted by methanol or other organic solvents for epothilones isolation. Here, we named it "the used resins" for comparison with the new resins. Regenerative resins means the used resins had been treated and regenerated.

5 g used XAD-16 resins were immersed and washed for thrice in 5 mL of different organic solvents until they were fully swelled and no obvious pigment could be seen in the eluate. Subsequently, they were washed with distilled water to

remove the organic solvents, then by 2 % (v/v) hydrochloric acid, followed by 1 % (mass fraction) sodium hydroxide to eliminate other impurities. Finally the pH value needed to be adjusted to neutrality and stored in distilled water.

49.2.3 Fermentation of Epothilones with Shake Flask

Strain So2161 was first inoculated on the CNST agar plates and cultured for 3–4 days at 30 °C. Then it was inoculated into liquid M26 medium and shook for 3–4 days. Cells were collected and inoculated into fermentation medium which contained 0.2 % potato starch, 0.2 % glucose, 0.2 % soybean powder, 0.15 % $MgSO_4$, 0.15 % $CaCl_2$, $FeCl_3$-EDTA 1 mL/L, trace element solution 1 mL/L, 0.2 % corn steep liquor, and 2 % regenerative XAD-16 resin [6]. 2 % fresh XAD-16 resin was added as control. All experiments were carried out in 300-mL Erlenmeyer flasks containing 50 mL fermentation medium supplemented as described for each condition. Then, the shake flasks were incubated at 30 °C for 7 days with shaking at 180 rpm. After fermentation, the resin beads were collected, washed with distilled water, air dried, and extracted with 5 mL methanol overnight. The extract was dried in vacuum at 40 °C, and then redissolved in 100 μL methanol for HPLC analysis (Shimadzu, Japan) [7], equipped with two LC-20AT pumps and an SPD-10A UV-V is detector.

Extractions were centrifugation at 10,000 r/min for 10 min, a 20 μL aliquot was injected into a C18 analytical column (250 × 4.6 mm, 5 μm, at 35 °C column temperature, Agilent). Epothilones were detected at 249 nm wave length, eluted with 70 % methanol and 30 % water at a flow rate of 1.0 mL/min [8]. Retention times of epothilone A and B were 7.8 and 8.7 min (compared with the standard epothilone A&B) (Fig. 49.1). The titers were quantitated and quantified based on standard sample and a standard curve generated from purified epothilones A and B (data not shown). And compared the production of epothilones under different grow conditions.

49.2.4 Observation of the Resins with Atomic Force Microscope

AFM images had been performed with a Mutimode 8 Nanoscope V system (Bruker, USA). Two scanners have been used: the J scanner with a maximum lateral (x, y) scan range of 125 μm and a maximum vertical (z) range of 5 μm for the biggest images; and the E scanner for the smallest areas analyzed (horizontal and vertical ranges of 0.4 μm). The scanned areas have gone from 18 × 18 nm to 15 × 15 μm. The smallest areas of 18 × 18 nm had been used to get the accurate optimal resolution to detect the pore size distribution.

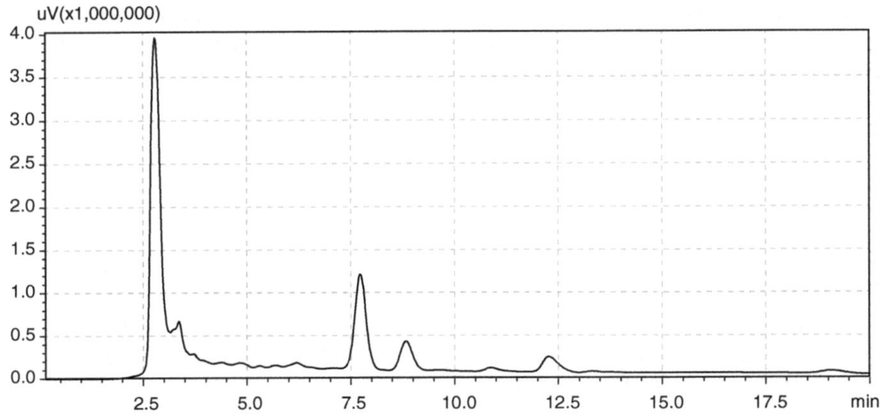

Fig. 49.1 The HPLC chromatogram of strain So2161 fermentation products (Retention times of epothilone A is 7.8 min and epothilone B is 8.7 min)

The block faces of resins were treated with pure ethanol for varying time prior to AFM examination. Then, the resins were mounted in special holders which at the same time fit the microtome and are suitable for the examination of the surface tissue by AFM [9]. Then fix the dissected resins to the 1 cm^2 slides with glue.

49.2.5 Study of the Durability of Regenerative Resins

Respectively, 2 g of regenerative resins with five different organic solvents were added to fermentation medium to a total volume of 100 mL in each simulated bioreactor (1 L shake flask) [3]. Control experiment was set for adding the fresh-resin XAD-16. The simulated bioreactor was unbaffled, with the rotor of magnetic stirring apparatus as the mixer (5 cm impeller length, and 1 cm width). The simulated bioreactor was set at 200 rpm by the magnetic stirring apparatus, and the temperature maintained at 30 °C. The agitation rate profile during the 10 days run was varied to mimic that of actual epothilone fermentation. After 10 days incubation, the resins were harvested from the culture, washed by ddH_2O, and observed with Dissecting Microscope [10].

49.2.6 Test of Regeneration Times

The used XAD-16 resins were collected after fermentation, repeated the above section to verifying the max regeneration times of the used XAD-16 resin.

49.3 Results and Discussion

49.3.1 Regeneration Ability Analysis

The eluants were detected by HPLC, and results were showed in Fig. 49.2. It indicated that the yield of epothilones in acetone, carbon tetrachloride, and ethyl acetate conditions were lower than the others. Methanol could significantly promote the adsorption ability of regenerative XAD-16 resins, which lead the yield of epothilone much higher than control.

49.3.2 Atomic Force Microscope Analysis

Comparing Fig. 49.3a with b, it could be seen that the size of pore was increased steeply with the treatment of methanol. Larger pore were more conducive to the adsorption of epothilones, which is consistent with the above conclusion.

Figure 49.4 was obtained by Origin Pro v8.0 (Information on http://www.crsky.com/soft/31361.html) [11, 12] which showed that the fresh-resin XAD-16 had pore sizes between 0.3 and 0.5 nm. However, the pore sizes of methanol-regenerative-resin XAD-16 were between 0.5 and 0.7 nm, which were clearly larger than the fresh-resin.

49.3.3 Durability Analysis

The ethyl acetate-regenerative-resin particles showed many cracks on the surface after 10 days impacting (Fig. 49.5d). While the methanol-resin and fresh-resin had

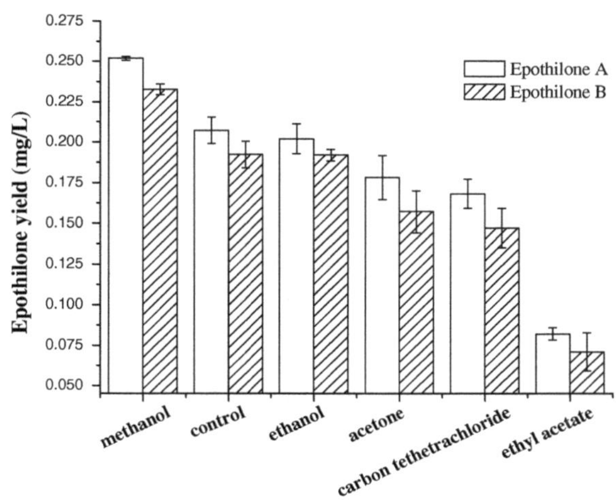

Fig. 49.2 The yield of epothilones of regenerative resins with different organic solvents

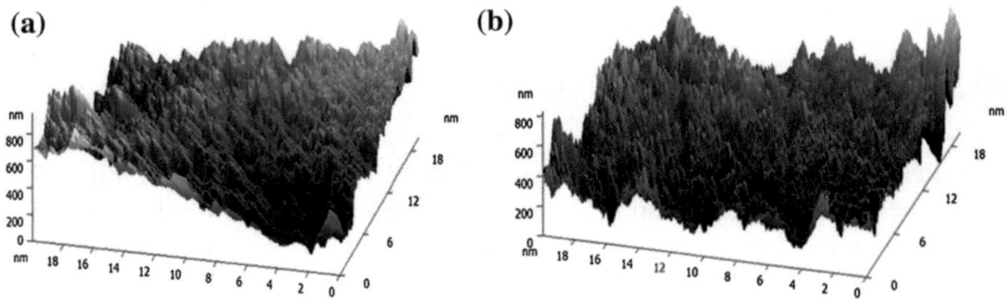

Fig. 49.3 The AFM images of the resins with scanned areas of 18 × 18 nm^2 used to obtain the pore size distribution(**a** the methanol-regenerative-resin XAD-16; **b** the fresh resin XAD-16)

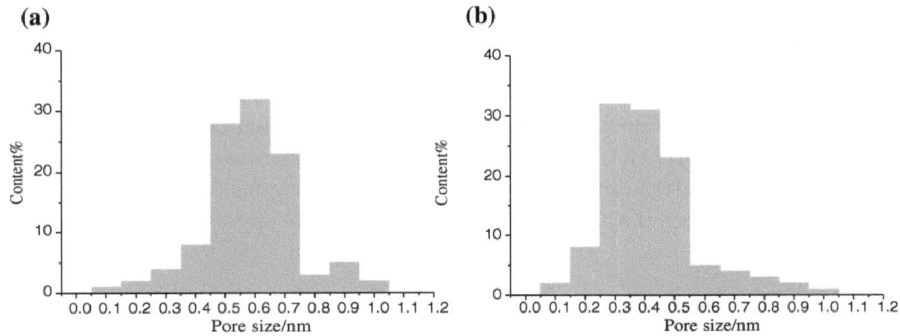

Fig. 49.4 Pore size distribution of the methanol-regenerative-resin XAD-16 (**a**) and the fresh-resin XAD-16 (**b**)

few chipping off of cup shaped pieces. Other solvents (ethanol, acetone, and carbon tetrachloride) were similar to the ethyl acetate and results were not shown in this paper.

Small fragments of resin beads were readily visible by day 10. Through the analysis of Nano Measurer 1.2 software (Information on http://emuch.net/html/201112/3712881.html) [13], we found that there were more fragments in regenerative resins with alcohol, acetone, carbon tetrachloride, and ethyl acetate (The resin particle size distributions of the samples could be determined, as there were no cellular aggregates present to interfere with the analysis). The relevant resin bead size distribution parameters of the various samples from this run were summarized in Table 49.1. The mean particle size (gained from Nano Measurer 1.2 software) represented the bead size in which 50 % of the resin beads in the sample are sizable

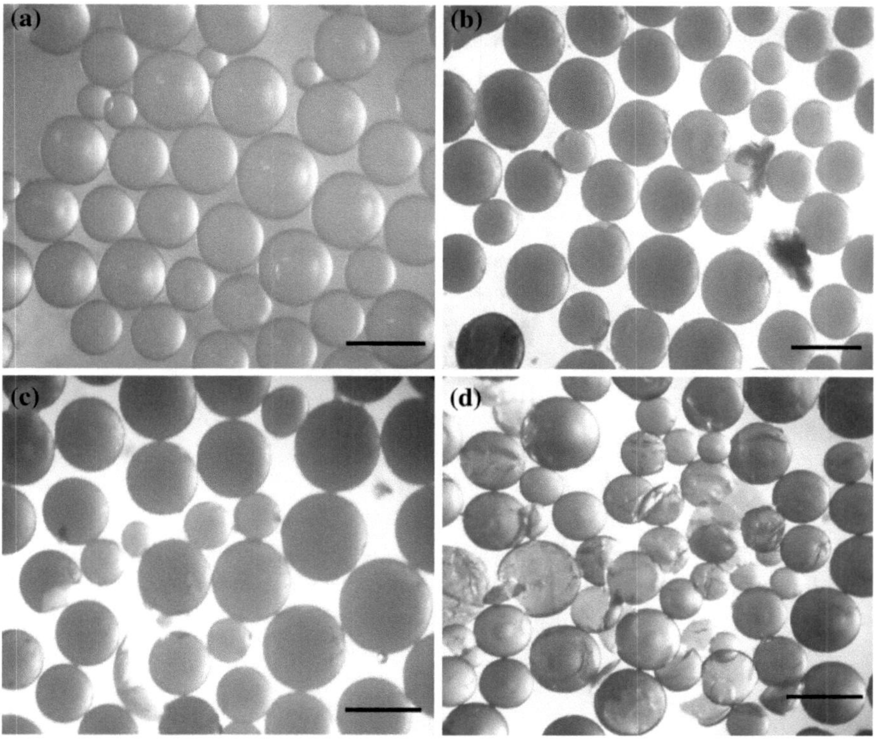

Fig. 49.5 Dissecting Microscopic images of XAD-16 resin before experiment (**a** fresh-resin) and after (**b** fresh-resin, **c** methanol-regenerative-resin, **d** ethyl acetate-regenerative-resin) the resin fracture simulation 10 days of agitation. The bar is 400 μm in the figure

to that particular size, the accumulation of small (<100 μm diameter) fragments present in the resin bead distribution (data not shown) and indicated the degree of broken resin.

Table 49.1 Resin particle distribution properties after the simulation cultivation

Resins	Mean particle size (μm)	Percentage of particles <100 μm (%)
Fresh resins	590	0
Regenerative resins with methanol	550	3.1
Regenerative resins with alcohol	500	14.1
Regenerative resins with acetone	498	15.2
Regenerative resins with carbon tetrachloride	493	16.5
Regenerative resins with ethyl acetate	490	18.7

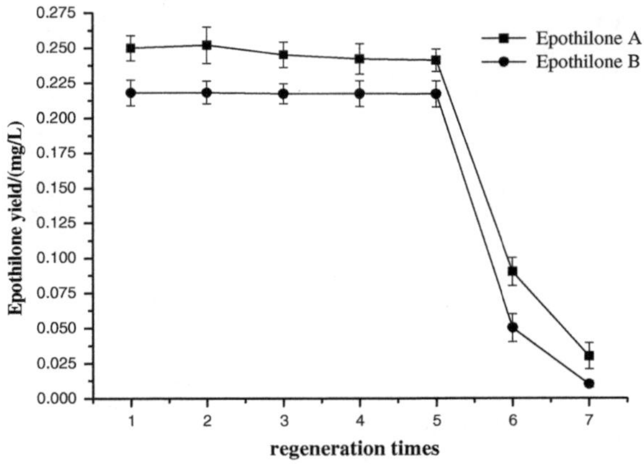

Fig. 49.6 Epothilones yield of different regeneration times

49.3.4 Analysis of the Regeneration Times

Using methanol as regenerant, macroporous adsorption resin XAD-16 could be used for at most five times (Fig. 49.6), for it become more and more fragile with the increase of regeneration times, and the resin beads gradually into significant fracture.

49.4 Conclusions

From the results, we know that organic solvent can be used as regenerant. Among the five organic solvents, methanol is the best regenerant which is more effective for the adsorption of epothilones during *S. cellulosum* So2161 cultivations.

Also, the size of pore was increased steeply with the addition of methanol, in which it was between 0.5 and 0.7 nm clearly larger than the fresh resins (between 0.3 and 0.5 nm). Larger pore was more conducive to the adsorption of epothilones, which lead a lower product concentration in the broth by removing feedback inhibition, might be one reason for higher production of epothilones in methanol-regenerative-resin.

Furthermore, resins under high agitation rates for long time resulted in significant fractures. Losing of fractions lead a low recovery, then we got a low yield of epothilones. It might be another reason for why the yield of epothilones was higher than others in the sturdier methanol-regenerative-resin.

Considering the yield of epothilones, we choose methanol as the best regenerant for used XAD-16 resin. Furthermore, Macroporous adsorption resin XAD-16 can be used for five times with methanol, which could significantly reduce the cost of industrial producing.

Acknowledgments This work received financial supports from Provincial Post-Doctoral Innovation Foundation (201203006) and Research projects of Science and Technology Division, Shandong (2012GSF12107).

References

1. Reichenbach H, Höfle G (2008) Discovery and development of the epothilones: a novel class of antineoplastic drugs. Drugs R.D. 9:1–10
2. Zhao L, Zhao BB, Lu CH, Li YZ, Shen YM (2010) Two epothilones from *Sorangium cellulosum* strain So0157-2. Chin J Nat Med 8:0298–0300
3. Scott F, Hiroko T, Jorge G, Peter L (2006) Characterization of product capture resin during microbial cultivations. J Ind Microbiol Biotechnol 33:445–453
4. Yan ZC, Wang B, Li YZ, Gong X, Zhang HQ, Gao PJ (2003) Morphologies and phylogenetic classification of cellulolytic myxobacteria. Syst Appl Microbiol 26:104–109
5. Li YZ, Hu W, Wu BH, Yan ZC, Chen Q (2001) Diversity of metabolites and their bio-activities in myxobacterium *Sorangium cellulosum*. Wei Sheng Wu Xue Bao 41:716–722
6. Gerth K, Bedorf N, Höfle G, Irschik H, Reichenbach H (1996) Epothilones A and B: antifungal and cytotoxic compounds from *Sorangium cellulosum* (Myxobacteria) production, physic-chemical and biological properties. J Antibiot 49:560–563
7. Gong GL, Sun X, Liu XL, Hu W, Cao WR, Liu H, Liu WF, Li YZ (2007) Mutation and a high-throughput screening method for improving the production of Epothilones of *Sorangium*. J Ind Microbiol Biotechnol 34:615–623
8. Benigni D, Chiang S, Comezoglu T, Davis L (2003) Methods for the preparation, isolation and purification of epothilone B and X-ray crystal structures of epothilone B. US Patent 17:56–62
9. Carvalho AL, Maugeri F, Silva V, Herna´ndez A, Palacio L, Pradanos P (2011) AFM analysis of the surface of nanoporous membranes: application to the nanofiltration of potassium clavulanate. J Mater Sci 46:3356–3369
10. Kakaboura A, Fragouli M, Rahiotis C, Silikas N (2007) Evaluation of surface characteristics of dental composites using profilometry, scanning electron, atomic force microscopy and gloss-meter. J Mater Sci Mater Med 18:155–163
11. Sui L, Yu LY, Zhang YH (2007) Surface and catalytic properties of K/V catalysts coating on porous Al_2O_3 substrate. J Dispersion Sci Technol 28:285–289
12. Li HQ, Sun LM, Shen GX, Qi L (2012) The nanoscale phase distinguishing of PCL-PB-PCL blended in epoxy resin by tapping mode atomic force microscopy. Nanoscale Res Lett 7:153–156
13. Breibach M, Bathen D, Schmid H, Ebener H (2002) Stability of adsorber resins under mechanical compression and ultrasonication. Polym Adv Techmol 13:391–400

Chapter 50
Conversion Process of High Color Value Gardenia Red Pigment

Shangling Fang, Wei Jiang, Jinghua Cao, Xu Xu, Yanyan Jing and Maobin Chen

Abstract Pretreatment conditions of geniposide and conversion process conditions of gardenia red pigment were optimized, gardenia red pigment was refined by acid precipitation or resin, and the effects of refining were compared, conversion process of high color value gardenia red pigment was studied, the results suggested that 5 g sodium hydroxide was added to the 50 mL concentrated geniposide at 40 °C and stirred for 3 h, 12.5 g citric acid was added with stirring until it was dissolved, then 30 mL pretreated geniposide and 6 g sodium glutamate were added to 50 mL fermentation broth of *Aspergillus niger*, it was kept at 50 °C in vacuum incubator for 25 h, then the fermentation broth was put into the boiling water for 2 h. Under the above conditions, the color value of gardenia red pigment which refined by acid precipitation or resin was 53 and 63, respectively. It was improved significantly when compared with color value was 4 under the initial conditions.

Keywords Gardenia red pigment · Geniposide · Refined · Color value

50.1 Introduction

Gardenia red pigment is widely used as a natural food colorant, it has advantages such as soluble in water, easy to use, stable under conditions of high temperature and pH [1]. The natural pigment with development value has great dyeing to protein, carbohydrates, and nontoxic [2]. Gardenia fruit is rich in gardenia yellow pigment, a plenty of geniposide exists in the waste after purification of gardenia yellow pigment, and geniposide could be converted into gardenia red pigment through biochemical methods in this article. The study mainly focused on the methods of preparation of the high color value gardenia red pigment.

S. Fang · W. Jiang · J. Cao · X. Xu · Y. Jing · M. Chen (✉)
Key Laboratory of Fermentation Engineering of Ministry of Education, Hubei Collaborative Innovation Center for Industrial Fermentation College of Bioengineering, Hubei University of Technology, Wuhan 430068, Hubei, China
e-mail: hgchenmaobin@163.com

T.-C. Zhang and M. Nakajima (eds.), *Advances in Applied Biotechnology*,
Lecture Notes in Electrical Engineering 332, DOI 10.1007/978-3-662-45657-6_50

50.2 Materials and Methods

50.2.1 Materials

Geniposide was prepared at the laboratory.

Strain: *Aspergillus niger* was isolated from Culture Collection Center of Hubei University of Technology.
Slant culture medium (PDA medium): potato 20 g, glucose 2 g, agar 2 g, H_2O 100 mL.
Seed culture medium: glucose 1 %, yeast extract 0.5 %, sodium nitrate 0.3 %, magnesium sulfate heptahydrate 0.2 %, potassium dihydrogen phosphate 0.2 %.
Fermentation medium: wheat bran (The bran was added to boiling water, the mixed solution was filtered through four layers of gauze) 2 %, ammonium sulfate 0.3 %, potassium dihydrogen phosphate 0.2 %, magnesium sulfate heptahydrate 0.2 %.

All the culture mediums should be sterilized at 115 °C for 30 min.

50.2.2 Preparation Process of Gardenia Red Pigment

50.2.2.1 Preparation of Geniposide

Geniposide was prepared according to the method of reference [3] reported.

50.2.2.2 Pretreatment of Geniposide

6 g sodium hydroxide was added to 50 mL concentrated geniposide at 50 °C and stirred for 5 h (reaction of alkaline hydrolysis), then 12.5 g citric acid was added with stirring until it was dissolved, the pretreated geniposide was stored at 4 °C for further use.

50.2.2.3 Preparation of Fermentation Broth of *A. niger*

After being inoculated onto PDA slant medium and cultured at 30 °C for 5 days, *A. niger* was inoculated into 50 mL seed culture medium at 30 °C for 3 days by shaking culture with 180 rpm, then the clump of mold was added to 50 mL fermentation medium at 30 °C for 5 days by shaking culture with 180 rpm.

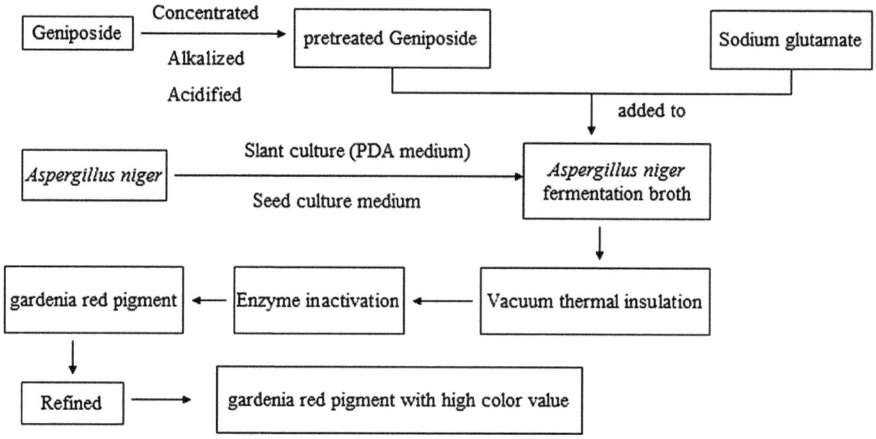

Fig. 50.1 Process flow diagram for preparation of gardenia red pigment

50.2.2.4 Conversion of Gardenia Red Pigment

30 mL pretreated geniposide and 5 g sodium glutamate were added to fermentation broth of *A. niger*, keeping it at 50 °C in vacuum incubator for 20 h. Then the fermentation broth was put into the boiling water for 1 h in order to deactivate enzyme, the clump of mold was removed by filtration, the filtrate was centrifuged at 8,000 rpm for 10 min, the supernatant liquid was obtained and saved, the absorbance of the supernatant liquid containing gardenia red pigment was detected at $A_{535\ \mathrm{nm}}$.

50.2.3 Optimization of Preparation Process of Gardenia Red Pigment

50.2.3.1 Optimization of Geniposide Pretreatment

Research on the optimal type of acid and alkali
Sodium hydroxide, potassium hydroxide, and sodium bicarbonate were chosen as the alkali to be added at the step of alkaline hydrolysis, respectively. Citric acid and 36 wt% concentrated hydrochloric acid were chosen as the acids to be added at the step of acidify, respectively. Rest of the operation methods were operated according to Sect. 50.2.2.

Research on the optimal dosage of acid and alkali
After the optimization of the optimal type of acid and alkali, different dosages of alkali were chosen, there were 4, 5, 6 and 7 g, respectively. Different dosages of acid were chosen, there were 8, 10, 12.5 and 15.0 g, respectively. Rest of the operation methods were operated according to Sect. 50.2.2.

Research on the optimal pretreatment temperature of geniposide
After the optimization of the optimal type and dosages of acid and alkali, the geniposide was pretreated at 40, 50, 60, and 70 °C, respectively. Rest of the operation methods were operated according to Sect. 50.2.2.

Research on the optimal alkalization time of geniposide
On the basis of the test results above, the geniposide was pretreated under the conditions of alkalization 3, 4, 5, 6, 7, and 8 h, respectively. Rest of the operation methods were operated according to Sect. 50.2.2.

50.2.3.2 Optimization of Transformation Conditions of Gardenia Red Pigment

Research on optimal dosage of geniposide
On the basis of optimum conditions of geniposide pretreatment, different dosages of geniposide were chosen, there were 10, 30, 50, and 70 mL, respectively. Sodium glutamate was added in geniposide and its dosages were 1.7, 5.0, 8.3 and 11.7 g, respectively. Rest of the operation methods were operated according to Sect. 50.2.2.4.

Research on optimal dosage of sodium glutamate
Different dosages of sodium glutamate were chosen, there were 4, 5, 6 and 7 g, respectively. Rest of the operation methods were operated according to Sect. 50.2.2.4.

Research on optimal temperature of vacuum thermal insulation
Different degrees of temperature of vacuum thermal insulation including were chosen, there were 40, 50, 60 and 70 °C, respectively. Rest of the operation methods were operated according to Sect. 50.2.2.4.

Research on optimal time of vacuum thermal insulation
Different times of vacuum thermal insulation were chosen, there were 10, 15, 20, 25 and 30 h, respectively. Rest of the operation methods were operated according to Sect. 50.2.2.4.

Research on optimal time of enzyme inactivation
Different times of enzyme inactivation were chosen, there were 0.5, 1, 1.5, 2, 2.5 and 3 h, respectively. Rest of the operation methods were operated according to Sect. 50.2.2.4.

50.2.4 Purification of Gardenia Red Pigment

Acid precipitation reaction: Gardenia red pigment was mixed with water at the ratio of 1:1, the pH of the mixture was adjusted to 2.8–3.0 with 18 wt% hydrochloric acid. After standing for 1 day, the mixture was centrifuged at 8,000 rpm for 20 min, the precipitate was dried in oven at 60 °C.

Resin method: After being added to GDX-103 macroporous resin column, the gardenia red pigment was eluted with deionized water until the eluent was colorless after the adsorption equilibrium, then eluted with 70 % (V/V) ethanol and 60 % (V/V) ethanol which containing 1‰ (mass concentration) sodium bicarbonate with the flow rate of 2 mL/min. These two types of eluent were collected, the gardenia red pigment powder was obtained by concentrating and drying the eluent.

50.2.5 Determination of Color Value of Gardenia Red Pigment

The color value of gardenia red pigment was determined according to the method of Ref. [4] reported.

50.3 Results and Discussion

50.3.1 Determination of Maximum Absorption Wavelength of Gardenia Red Pigment

Gardenia red pigment was appropriately diluted with deionized water; deionized water was also used as blank control. Both of the sample and blank control were scanned by UV—visible scanner. As shown in Fig. 50.2, it was only one absorption peak in the vicinity of 535 nm between the band of 280 and 800 nm. The result is consistent with the Ref. [5] which described the maximum absorption wavelength of gardenia red pigment was at the range of 532 ± 3 nm. So the maximum absorption wavelength of gardenia red pigment was 535 nm.

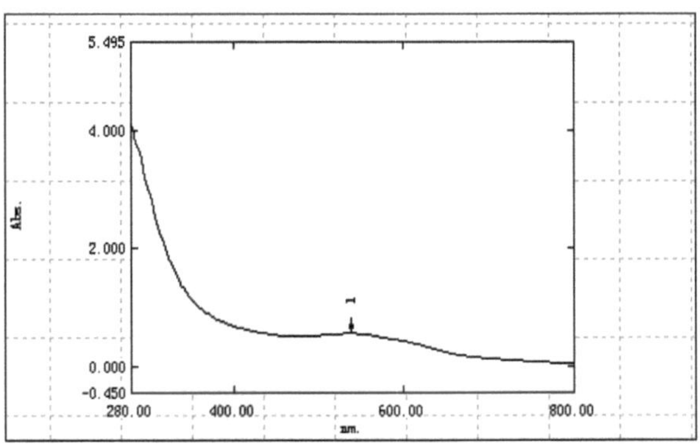

Fig. 50.2 Absorption spectrum of gardenia red pigment

50.3.2 Optimization of Pretreatment Methods of Geniposide

50.3.2.1 Type of Acid and Alkali

As shown in Table 50.1, after alkaline hydrolysis by using sodium hydroxide, citric acid was added in geniposide, the gardenia red pigment which was obtained finally had bright color and high absorbance value.

50.3.2.2 Dosage of Acid and Alkali

As shown in Table 50.2, the color of gardenia red pigment was bright and high when 5 g sodium hydroxide and 12.5 g citric acid were added.

50.3.2.3 Pretreatment Temperature of Geniposide

It was shown in Table 50.3 that the optimal pretreatment temperature of geniposide was 40 °C. Under the condition of 40 °C, the absorbance value of gardenia red pigment was 0.361, it was higher than the others.

Table 50.1 Results of the optimal type of acid and alkali

Type of alkali	Type of acid	Color and absorbance
Sodium hydroxide	Citric acid	Red 0.600
Sodium hydroxide	Concentrated hydrochloric acid	Yellow —
Potassium hydroxide	Citric acid	Red 0.576
Potassium hydroxide	Concentrated hydrochloric acid	Yellow —
Sodium bicarbonate	Concentrated hydrochloric acid	Yellow —
Sodium bicarbonate	Citric acid	Red 0.562

Table 50.2 Results of the optimal dosage of acid and alkali

Dosage of sodium hydroxide (g)	Dosage of citric acid (g)	Color and Absorbance
6	10	Orange —
6	12.5	Red 0.366
6	15	Red 0.540
5	12.5	Red 0.546
7	12.5	Red 0.358
4	8	Dark red 0.518

Table 50.3 Results of optimal pretreatment temperature of geniposide

Pretreatment temperature (°C)	Color and absorbance
40	Red 0.361
50	Red 0.299
60	Red 0.223
70	Red 0.279

50.3.2.4 Alkalization Time of Geniposide

As shown in Table 50.4, after alkalization for 7 h, the gardenia red pigment which conversed obviously was obtained and its absorbance value was 0.446. When alkalization time was kept for 3 h, the absorbance value was 0.435, the result was similar with the alkalization time when it was 7 h. In order to save time, 3 h of alkalization time was selected. Comparing with 7 h, efficiency could be improved, so 3 h was chosen as the optimal alkalization time.

50.3.3 Optimization of Conversion Conditions of Gardenia Red Pigment

50.3.3.1 Dosage of Geniposide

As shown in Table 50.5, the effect of conversion of gardenia red pigment was the best when 50 mL geniposide was added. Similarly, 30 mL geniposide could also be used comparing with the dosage of 50 mL. Considering that much more raw materials can be saved out, the dosage of 30 mL was selected finally.

50.3.3.2 Dosage of Sodium Glutamate

As shown in Table 50.6, both of color and absorbance were the best under the condition of 6 g sodium glutamate was added while the dosage of geniposide was 30 mL.

Table 50.4 Results of the optimal alkalization time of geniposide

Alkalization time (h)	Color and absorbance
3	Red 0.435
4	Red 0.332
5	Red 0.367
6	Red 0.370
7	Red 0.446
8	Red 0.353

Table 50.5 Results of the optimal dosage of geniposide

Dosage of geniposide (mL)	Dosage of sodium glutamate (g)	Color and absorbance
10	1.7	Red 0.423
30	5.0	Red 0.592
50	8.3	Red 0.602
70	11.7	Red 0.381

Table 50.6 Results of the optimal dosage of sodium glutamate

Dosage of geniposide (mL)	Dosage of sodium glutamate (g)	Color and absorbance
30	4	Red 0.710
30	5	Red 0.714
30	6	Red 0.728
30	7	Red 0.723

50.3.3.3 Temperature of Vacuum Thermal Insulation

Table 50.7 showed that the best temperature of vacuum thermal insulation for conversing of geniposide red pigment was 50 °C.

50.3.3.4 Time of Vacuum Thermal Insulation

As shown in Table 50.8, the absorbance was up to 0.716 when the time of vacuum thermal insulation reached to 25 h, the quality of gardenia red pigment was better than the others.

50.3.3.5 Time of Enzyme Inactivation

Table 50.9 showed that the best enzyme inactivation time was 2 h for conversing of geniposide red pigment. Under the condition above, the absorbance was up to 0.767.

Table 50.7 Results of the optimal temperature of vacuum thermal insulation

The temperature of vacuum thermal insulation (°C)	Color and absorbance
40	Red 0.689
50	Red 0.713
60	Red 0.702
70	Red 0.681

Table 50.8 Results of the optimal time of vacuum thermal insulation

Time of vacuum thermal insulation (h)	Color and absorbance
10	Red 0.566
15	Red 0.634
20	Red 0.656
25	Red 0.716
30	Red 0.710

Table 50.9 Results of the optimal time of enzyme inactivation

Time of enzyme inactivation (h)	Color and absorbance
0.5	Red 0.631
1.0	Red 0.653
1.5	Red 0.664
2.0	Red 0.767
2.5	Red 0.754
3.0	Red 0.744

50.3.4 Refined Results of Gardenia Red Pigment

Gardenia red pigment which obtained in the initial conditions was concentrated and dried. Its color value $E_{1\,\mathrm{cm}}^{1\,\%}$ (535 nm) was 4. Under the optimal conditions of the pretreatment of gardenia and conversion processes of gardenia red pigment, the transformation liquid was prepared. After acid precipitation or resin method, two types of gardenia red pigment were obtained from the transformation liquid, the color value $E_{1\,\mathrm{cm}}^{1\,\%}$ (535 nm) of gardenia red pigment were 53 and 63, respectively. The color value was improved significantly comparing with the initial conditions.

50.4 Conclusion

The maximum absorption wavelength of gardenia red pigment was 535 nm. The optimal process of geniposide pretreatment were shown as follow: 5 g sodium hydroxide was added to the 50 mL concentrated geniposide at 40 °C and stirred for 3 h (reaction of alkaline hydrolysis), then 12.5 g citric acid was added to acidify with stirring until it was dissolved, the pretreated geniposide was stored at 4 °C for further use. The optimal process of transformation conditions of gardenia red pigment were shown as follow: 30 mL pretreated geniposide and 6 g sodium glutamate were added to 50 mL fermentation broth of *A. niger*, keeping it at 50 °C in vacuum incubator for 25 h, then the fermentation broth was put into the boiling water for 2 h in order to deactivate enzyme. Gardenia red pigment which obtained in the initial conditions was concentrated and dried, its color value $E_{1\,\mathrm{cm}}^{1\,\%}$ (535 nm) was 4. Under the optimal conditions of the pretreatment of gardenia and conversion processes of gardenia red pigment, the transformation liquid was prepared. After acid precipitation and resin method, two types of gardenia red pigment were obtained from the transformation liquid, the color value $E_{1\,\mathrm{cm}}^{1\,\%}$ (535 nm) of gardenia red pigment was 53 and 63, respectively, the color value was improved significantly when compared with the initial conditions.

References

1. Zhang M, Li SJ (2004) The study on stability of gardenia red pigment. J China Food Addit 1:39–42
2. Zhang QH, Zhu XD, Zhang B, Wu XY, Li SJ (2009) Purification of gardenia red pigment by macroporous adsorption resin. J Sci Technol Food Ind 30:229–231
3. Jing YY, Li SJ, Chen MB, Jiang W, Fang SL (2012) Purification of gardenia yellow pigment and conversion of gardenia blue pigment. J Chem Bioeng 29:68–71
4. Jing YY, Jiang W, Liu J, Chen MB, Fang SL (2013) Preparation of gardenia yellow pigment and selected strains to convert gardenia red pigment. J China Brew 8:112–116
5. Lin Y, Li SJ (2009) Gardenia red produced by biotransformation. J Mod Agric Sci 16:18–23

Chapter 51
Efficient Purification and Active Configuration Investigation of R-phycocyanin from *Polysiphonia urceolata*

Li-ping Zhu, Shi-gan Yan and Ai-jie Lv

Abstract An efficient method for purification of R-phycocyanin (R-PC) from *Polysiphonia urceolata* using anion exchange chromatography with pH elution combined with ammonium sulfate precipitation was proposed. R-PC was conveniently purified on a DEAE-Sepharose Fast Flow chromatography column with elution of pH 4.8 eluants after ammonium sulfate precipitation. The R-PC obtained was pure and with a recovery of 38.83 %. The absorption maxima of the purified RPC were at 550 and 618 nm, and the fluorescence emission maximum was at 632 nm with an exciting wavelength of 580 nm. The aggregation state of R-PC was $(\alpha\beta)_3$ and the molecular weights of the α and β subunits were 16.7 and 18.9 kDa, respectively. The efficient method has the potential to be applied in the R-PC commercial production. The active configuration investigation showed that RPC presented a good structural stability in the pH range of 4–9.

Keywords *Polysiphonia urceolata* · R-phycocyanin · Purification · Anion exchange chromatography · Active configuration

51.1 Introduction

R-phycocyanin (R-PC) is a minor phycobiliproteins found in some red alga such as *Porphyridium cruentum, Polysiphonia urceolata, Porphyra haitanensis, Griffithsia tenuis*, and so on. Because of solely possessing phycoerythrobilin and phycocyanobilin prosthetic groups, R-PC is considered as a surviving intermediate species in an

L.-p. Zhu · S.-g. Yan (✉)
College of Food and Bioengineering, Shandong Provincial Key Laboratory of Microbial Engineering, Qilu University of Technology, Jinan 250353, Shandong, China
e-mail: yanshigan@126.com

A.-j. Lv
School of Chemistry and Pharmaceutical Engineering, Qilu University of Technology, Jinan 250353, Shandong, China

T.-C. Zhang and M. Nakajima (eds.), *Advances in Applied Biotechnology*,
Lecture Notes in Electrical Engineering 332, DOI 10.1007/978-3-662-45657-6_51

evolutionary pathway leading from phycoerythrin to phycocyanin [2]. R-PC is widely applied as natural dyes, fluorescence markers, therapeutics, and immunological regulator [11]. Its β subunit was taken as a model to study the refolding/unfolding conformational changes [5]. So it is a great necessity to purify R-PC in view of the great commercial perspective as well as scientific interest.

There are some purification methods to obtain R-PC. Multi-step hydroxyapatite chromatography combined with size exclusion chromatography are usually developed to purify R-PC [2, 11, 12]. However, these methods are time-consuming, complicated, and low-purity, which finally increases the production cost and limits the extensive applications. Therefore, it is essential to develop simplified and efficient methods for the purification of R-PC. In this paper, an efficient method for rapid purification of R-PC from *P. urceolata* using one-step anion exchange chromatography was described.

51.2 Materials and Methods

51.2.1 Organisms

Marine red alga *P. urceolata*, widely distributing near the Chinese coast and enriching in large amounts of R-PE [6, 10, 13] and small amounts of R-PC [3], is an inexpensive and good organism for extraction of phycobiliproteins [7].

Polysiphonia urceolata Grev. was kindly collected on the intertidal zone of Qingdao, east of China from March to April, rinsed with seawater, removed contaminating materials, then stored at −20 °C.

51.2.2 Pretreatments

Frozen algae (100 g of wet weight) were suspended in 500 ml 20 mM sodium acetate buffer (pH 5.8) and thawed at 4 °C for 24 h. After filtered by gauze, the filtrate was saturated to 60 % (w/v) with ammonium sulfate, stood at 4 °C for 24 h, and then centrifuged at 15,000 g at 4 °C for 20 min. The pellet was suspended in a minimum volume of 20 mM sodium acetate buffer (pH 5.8). The suspension was dialyzed overnight against the same buffer, and then centrifuged at 15,000 g at 4 °C for 20 min. The supernatant was collected for further purification.

51.2.3 Anion Exchange Chromatography

The supernatant was applied on a DEAE-Sepharose Fast Flow column (4 cm × 20 cm), which had been pre-equilibrated with 20 mM sodium acetate buffer (pH 5.8) containing 0.05 M NaCl. After washed with 80 ml of the same buffer, the column was eluted with 20 mM sodium acetate buffer containing 0.05 M NaCl with a straight elution of pH 4.8 to recover RPC with volume of 80 ml at a rate of 1 ml min^{-1}. The blue eluants was collected.

51.2.4 Spectral Measurements

The absorption spectrum was determined using UV/VIS-550 spectrophotometer (Jasco, Japan). The scan wavelength range was 250–700 nm. The purity was evaluated according to the absorbance ratio of $A_{\lambda max}/A_{280}$ [1]. The fluorescence emission spectrum was determined using FP-6500 spectrofluorometer (Jasco, Japan) with an excitation wavelength of 580 nm for RPC. All spectra were recorded at room temperature.

51.2.5 Gel Electrophoresis and Molecular Weight Determination

Native-PAGE and SDS-PAGE were performed using a vertical slab discontinuous gel electrophoresis apparatus (Bio-Rad, U.S.A.) with the same stacking gel of 5 % and separating gel of 7.5 and 15 %, respectively. The gels were stained with Commassie Brilliant Blue R250.

The molecular weight of RPC was determined by HPLC on a Shimadzu instrument SPD-M10AVP (Japan) with a gel filtration column (TSK G3000sw, 7.5 mm × 60 mm). A Shimadzu LC-10AS pump and a Shimadzu SPD-M10Avp diode array detector were used for HPLC analysis, the elution was monitored at 280 nm at 25 °C. The column was pre-equilibrated with 0.05 M sodium phosphate buffer (pH 7.0) at 0.5 ml min^{-1} and was calibrated using gel filtration molecular weight marker MW-GF-1000.

51.2.6 RPC Stability to pH Variations

Concentrated RPC was reserved in solutions with a protein concentration of 0.5 mg/ml at varied pH of 3.0, 4.0, 5.0, 6.0, 7.0, 8.0, 9.0, 10.0 for 12 h. Absorption spectrum of RPC in different solution was monitored from 400 to 700 nm using UV/VIS-550 spectrophotometer (Jasco, Japan). The pH was strictly determined using Thermo Orion 868 pH meter (Thermo Orion, USA).

51.3 Results

51.3.1 Pretreatments of RPC

The cell-free crude extract was got after cell disruption and filtration, which contained 59.95 mg R-PC and the ratio A_{618}/A_{280} was as low as 0.0802. After ammonium sulfate precipitation, 46.18 mg R-PC was recovered and the ratio A_{618}/A_{280} remarkably increased to 0.1973. The yields of R-PC after this step were 91.59 % (Table 51.1). Ammonium sulfate precipitation of 60 % saturation was directly used to remove the impurities from algae crude extraction.

51.3.2 Anion Exchange Chromatography for Purification of R-PC

Different proteins can be eluted with different pH eluants according to their individual pIs (R-PE at 3.7, R-PC at 4.6) [11]. R-PC can be eluted before R-phycoerythrin from *P. urceolata* using anion exchange chromatography with continuous pH elution [4], which will avoid R-PC from being contaminated by R-phycoerythrin. On a DEAE-Sepharose Fast Flow column, 17.93 mg R-PC was obtained after anion-exchange chromatography, representing 38.83 % of the R-PC was recovered from the algal crude extract. The purity of R-PC (A_{618}/A_{280}) reached 3.53 (Table 51.1). The efficiency and the recovery of the anion-exchange chromatography described were both higher than previously developed multi-step chromatography methods [2, 11, 12].

51.3.3 Spectral Measurements

The obtained R-PC presented two-peak typical absorption maxima at 550 and 618 nm (Fig. 51.1), which was corresponded to R-PC as the earlier published [3]. The ratio A_{615}/A_{280} of the R-PC was 3.53. The fluorescence emission maximum was at 632 nm with an exciting wavelength of 580 nm (Fig. 51.1), which was consistent with the published [2].

Table 51.1 Variation of recoveries and purities of RPC in each step

Purification step	Yield (mg)	Recovery (%)	Purity (A_{618}/A_{280})
Crude algal extract	59.95	100	0.0802
Ammonium sulfate extract	46.18	91.59	0.1973
Anion-exchange chromatography	17.93	38.83	3.53

100 g frozen alga was used in the preparation of RPC

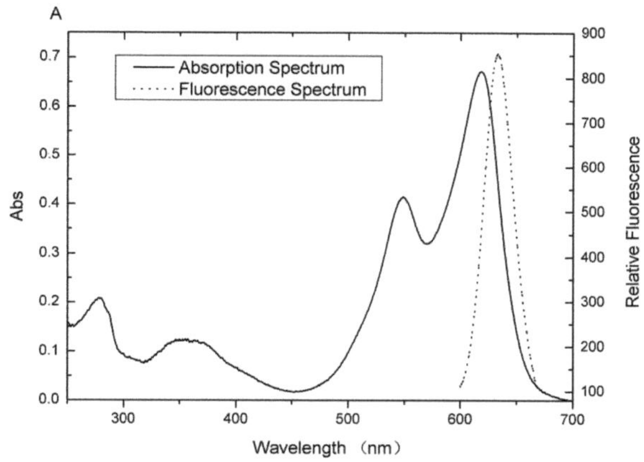

Fig. 51.1 Absorption and fluorescence spectra of RPC from *Polysiphonia urceolata* after anion exchange chromatography. Absorption and fluorescence spectra of R-PC. The absorbance maxima were at 550 and 618 nm in the absorption spectrum (*solid line*). The fluorescence emission maximum was at 632 nm in the fluorescence spectrum (*dashed line*) when excited by 580 nm

51.3.4 Gel Electrophoresis and Molecular Weight Determination

Native-PAGE only yielded one band. SDS-PAGE presented two bands corresponding to the α and β subunits of R-PC with the molecular weights of 16.7 and 18.9 kDa (Fig. 51.2).

According to the standard curve using the marker MW-GF-1000 on HPLC, the molecular weight of R-PC obtained was 104.9 kDa (Figs. 51.3 and 51.4). So the aggregation states of R-PC were presumed to be $(\alpha\beta)_3$. The aggregation state and molecular weights of R-PC were both similar to those of *G. pacifica* and *P. cruentum* [2, 9, 11].

51.3.5 Active Configuration of RPC to pH Variations

RPC displayed a good stability to pH variations between pH 4 and 9, With pH 4–9 the absorption spectra of RPC at 550 and 618 nm are relatively stable, indicating the changes of energy absorbing ability and aggregation state are insignificant. But at pH 3, both peaks fell dramatically and the 618-nm-peak below the 550-nm-peak, suggesting that RPC trimer has completely dissociated. When pH differed from neutrality to pH 10.0 in 12 h, both peaks were less than those at pH 3, furthermore, the absorption peaks from 550 to 618 nm left shift to 520 and 580 nm, respectively, and the absorbance at 580 nm was less than that at 520 nm as shown in Fig. 51.5. This result showed there might be further structural changes in RPC monomers, indicating dramatic changes in RPC structure.

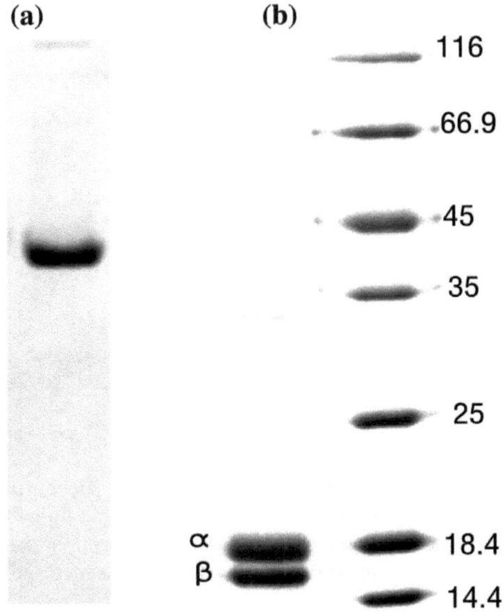

Fig. 51.2 Native- and SDS-PAGE of R-PC from *Polysiphonia urceolata*. **a** Native-PAGE. Electrophoresis was run in a 7.5 % separating gel. **b** SDS-PAGE. Electrophoresis was run in a 15 % separating gel. The molecular weights of the α and β subunits corresponding to the two bands with 16.7, 18.9 kDa, respectively

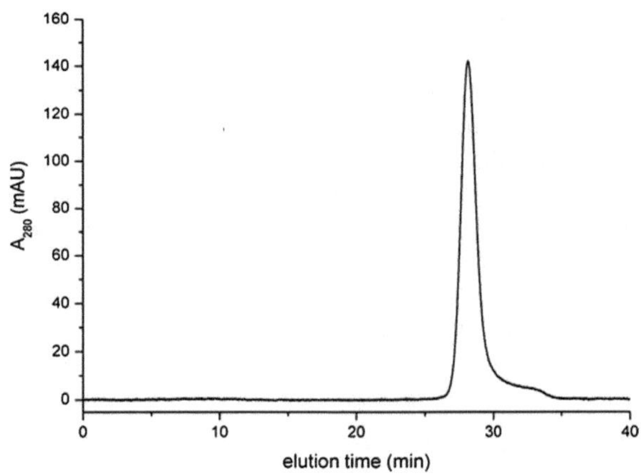

Fig. 51.3 HPLC profile of R-PC from *Polysiphonia urceolata* recording the absorbance at 280 nm

51.4 Discussion

Simplified and scalable preparation of pure PBPs is required for commercial application and scientific research. The market price of pure PBPs ($A_{\lambda max}/A_{280} > 4.0$) for pharmaceutical or fluorescent uses has been reported as high as \$50/mg (Market corporation 2005). This high value makes attractive the attempt to develop efficient methods for the recovery and purification of PBPs. It is noted that 50–90 % of the phycobiliprotein production cost lies in the purification steps [8].

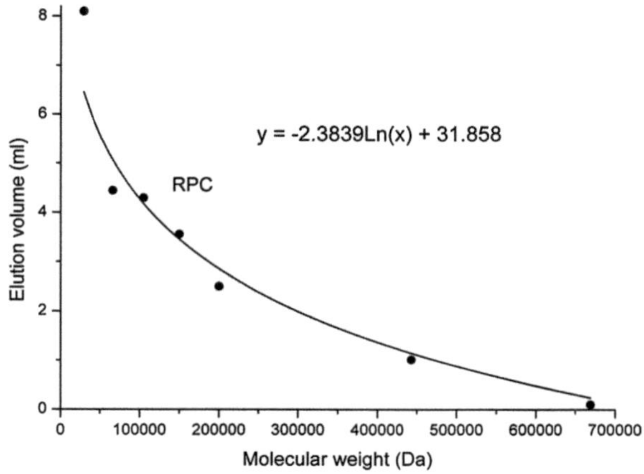

Fig. 51.4 Calibration curve of protein molecular weight size standard. According to the standard curve, the molecular weights of the R-PC was about 104.9 kDa

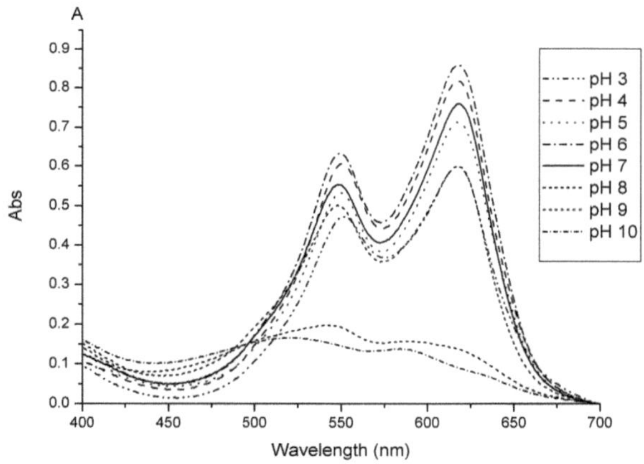

Fig. 51.5 Effects of solvent pH on the absorption spectra of R-PC

The traditional methods for separation of PBPs in the previous published, involving ammonium sulfate precipitation, hydroxylapatite chromatography and gel filtration chromatography, etc. [6, 13]. There is numerous research work on high-purity or large-scale purification of CPC because it has potential anti-oxidation property and it is a major constituent of cyanobacteria. However, the lower content of RPC in some red macroalgae such as *P. cruentum, P. urceolata, P. haitanensis* and *G. tenuis*, makes R-PC more difficult to be extracted. Among the limited reported methods for purification of R-PC, multi-step hydroxylapatite chromatography combined with size exclusion chromatography were usually developed [2, 11, 12], but the efficiency and the recovery are both less than anion-exchange chromatography combined with ammonium sulfate precipitation as described in this paper. Crude and concentrated R-PC was extracted through ammonium sulfate precipitation, and then pure RPC could be obtained by a further anion-exchange chromatography.

Comparing with ion-strength gradient elution, anion-exchange chromatography method with optimal pH stepwise elution not only simplified the desalting process, but also shows high efficiency. The purity and the recovery of the purified RPC in this paper are 3.53 and 38.83 %.

In summary, a simple and efficient purification of RPC from *P. urceolata* using DEAE—Sepharose Fast Flow chromatography with optimal pH straight elution combined with ammonium sulfate precipitation was proposed in this paper. Rather than pH gradient elution, optimal pH elution reduced the separation time and made the purification simpler without gradient mixer. The efficient method simplified the R-PC purification process, reduced the processing time and consequently decreased the production cost, so it has the potential to be applied in the R-PC commercial production.

Acknowledgments We wish to gratefully acknowledge the support of the National High Technology R&D Program (2002AA302213) for this work.

References

1. Bennett A, Bogorad L (1973) Complementary chromatic adaption in a filamentous blue-green alga. J Cell Biol 58:419–435
2. Glazer AN, Hixson CS (1975) Characterization of R-phycocyanin. Chromophore content of R-phycocyanin and C-phycoerythrin. J Biol Chem 250:5487–5495
3. Jiang T, Zhang JP, Chang WR, Liang DC (2001) Crystal structure of R-phycocyanin and possible energy transfer pathways in the phycobilisome. Biophys J 81:1171–1179
4. Liu LN, Chen XL, Zhang XY, Zhang YZ, Zhou BC (2005) One-step chromatography method for efficient separation and purification of R-phycoerythrin from *Polysiphonia urceolata*. Biotechnology 116:91–100
5. Ma YH, Xie J, Zhang CX, Zhao JQ (2007) Three-stage refolding/unfolding of the dual-color beta-subunit in R-phycocyanin from *Polysiphonia urceolata*. Biochem Biophys Res Commun 352:787–793
6. MacColl R, Eisele LE (1996) R-phycoerythrins having two conformations for the same aggregate. Biophys Chem 61:161–167
7. Pan ZZ, Zhou BC, Tseng CK (1986) Comparative studies on spectral properties of R-phycoerythrin from the red seaweeds from Qingdao. Chin J Oceanol Limnol 4:353–359
8. Patil G, Chethana S, Sridevi AS, Raghavarao KS (2006) Method to obtain C-phycocyanin of high purity. J Chromatogr A 1127:76–81
9. Watson BA, Waaland SD, Waaland JR (1986) Phycocyanin from the red alga *Anotrichium tenue*: modification of properties by a colorless polypeptide (Mr 30000). Biochemistry 25:4583–4587
10. Yu LH, Zeng FJ, Zhou BC (1991) Subunit composition and chromophore content of R-phycoerythrin from the red alga *Polysiphonia urceolata* Grev. Chin J Biochem Biophys 23:127–133
11. Zeng FJ, Lin QS, Jiang LJ (1992) Isolation and characterization of R-phycocyanin from the red alga *Porphyra haitanensis*. Acta Biochimica et Biophysica Sinica 24:545–551
12. Zhang JP, Zhang JM, Zhao JQ, Jiang LJ (1997) Study on the structure and function of R-phycocyanins from *Polysiphonia urceolata* 1. Seperation and characteristic of the structure of R-phycocyanins. Acta Biophysica Sinica 13:173–178
13. Zhang YZ, Chen XL, Wang LS, Zhou BC, He JA, Shi DX, Pang SJ (2002) In vitro assembly of R-phycoerythrin from marine red alga *Polysiphonia urceolata*. Sheng Wu Hua Xue Yu Sheng Wu Wu Li Xue Bao (Shanghai) 34:99–103

Chapter 52
Concentration of Sinigrin from Indian Mustard (*Brassica juncea* L.) Seeds Using Nanofiltration Membrane

Tianxin Wang, Hao Liang and Qipeng Yuan

Abstract Sinigrin has drawn more attention of pathologists and immunologists, which exhibits a wide range of biological activities. In the present study, the feasibility of concentration of sinigrin using nanofiltration membranes was studied using one-factor-at-a-time method with the permeate flux and sinigrin retention as responses. After seven commercial membranes were compared, including NF, NF-270, SPESI, STARMEN 120, STARMEN 128, STARMEN 220, STARMEN 240, we investigated the effects of temperature, pressure, and feed concentration on the concentration process of sinigrin from crude aqueous extracts of Indian mustard (*Brassica juncea* L.) seeds followed by primary ultrafiltration. The results indicated that NF-270 membrane showed the best concentration capability to sinigrin. Their ranges of permeate flux and sinigrin retention were 7.44–23.47 L/(m^2 h) and 90.77–97.13 %, respectively. Furthermore, the high flux and high retention were explained by swelling ratio, SEM, and ATR-FTIR, which were applied to determine the differences of nanofiltration membrane before and after soaking in aqueous solution. Finally, it could be concluded that the concentration process using nanofiltration provided a novel, rapid, and economical method for concentration of sinigrin.

Keywords Sinigrin · *Brassica juncea* L. · Indian mustard · Nanofiltration · Concentration

T. Wang · H. Liang · Q. Yuan (✉)
State Key Laboratory of Chemical Resource Engineering, Beijing University of Chemical Technology, Beijing 100029, China
e-mail: yuanqp@mail.buct.edu.cn

T. Wang
Key Laboratory of Food Nutrition and Safety, Ministry of Education, Tianjin University of Science and Technology, Tianjin 300457, China

T.-C. Zhang and M. Nakajima (eds.), *Advances in Applied Biotechnology*,
Lecture Notes in Electrical Engineering 332, DOI 10.1007/978-3-662-45657-6_52

52.1 Introduction

Indian mustard (*Brassica juncea* L.) seed is one of the oldest and most popular herbal medicines belonging to the family *Brassicaceae*, which distributed originally in Central Asia and presently cultivated in other areas including China, Japan, India, Southeast Asia, and Africa. Owing to its good therapeutic performance, Indian mustard seed was widely applied as a single or combined herbal to eliminate cavity effusion, relieve joint pain and numbness, alleviate cough and reduce swelling in Traditional Chinese Medicine.

Sinigrin (Fig. 52.1) is a major sulfur-containing glucosinolate compound commonly found in Indian mustard seed. In the last decades, more and more interest was paid in the field of physiological effects of sinigrin, since it played an important role in the pest- and disease-resistant mechanism [1]. Its major breakdown product allyl isothiocyanate is well known as a powerful antibacterial and antifungal compound [2]. Sinigrin also may be involved in anticarcinogenesis effects in vivo. Cooking will inactivate myrosinase and after ingestion, sinigrin could reach the large intestine. Biologically active isothiocyanate would be mediated to form by resident microflora before adsorption and delivery to tissues [3], where it acts as an anticancer agent and an inducer of anticarcinogenic phase II enzymes, such as glutathione transferase, UDP-glucuronosyltransferase and quinine reductase [4].

In the previous studies, sinigrin can be obtained primarily by water extraction [5], because sinigrin is water-solute due to its ionized sulfate and hydrophilic thioglucose moieties. And other units of separation and purification were also always undertaken in an aqueous solution [6, 7]. In order to concentrate the sinigrin, evaporation process was followed. However, the evaporation process consumed a lot of steam and energy, and it may result environmental pollution and other serious problems. Thus, an alterative process for concentration is necessary.

In recent years, membrane separation techniques have been widely used in many fields because of its gentleness, cleanness, environmental protection, and low cost. And purification and concentration are the two important application of nanofiltration. As a relatively new pressure-driven separation process, nanofiltration has also made rapid progress and extended its application fields. It has been widely and successfully used in food, pharmaceutical industry, and water treatment [8, 9]. Based on these works, concentration by nanofiltration is one of the possible and promising techniques replacing evaporation technology. Typically, the molecular weight cut-off (MWCO) of nanofiltration membranes ranges from 200 to 1,000 Da. Thus, the high rejection of sinigrin (MW = 359.4 Da) can be highly expected.

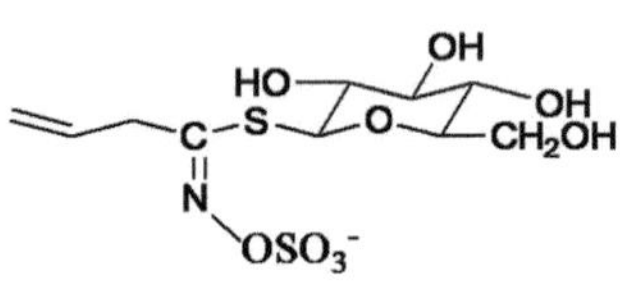

Fig. 52.1 Chemical structure of sinigrin

In the present study, we studied the feasibility of sinigrin concentration with nanofiltration, and the effects of process variables, such as temperature, pressure, and feed concentration, on the process using one-factor-at-a-time methodology. Furthermore, the explanation of the difference of nanofiltration membrane before and after soaking in solvent, which resulted in the corresponding high flux and high retention, has been illuminated by swelling ratio, scanning electron microscope (SEM), and attenuated total reflectance-Fourier transform infrared (ATR-FTIR).

52.2 Experimental

52.2.1 Reagents and Materials

Indian mustard (*Brassica juncea* L.) seeds were purchased from Anhui Fangmin Medicine Co., Ltd (AnHui, China). Sinigrin standard was purchased from Appli-Chem Company (Darmstadt, Germany). Methanol and trifluoroacetic acid (TFA) from Dikma Technologies Inc. (California, USA) were HPLC grade. Petroleum ether from Beijing Chemical Works (Beijing, China) was analytical grade. Deionized water was purified by a Milli-Q water purification system (Bedford, Massachusetts, USA). All solutions prepared for HPLC were filtered through 0.45 μm nylon membranes before used.

52.2.2 Apparatus

All nanofiltration experiments were performed using a C-40B (Nitto Denko, Osaka, Japan) Dead-end test cell, which comprises a cylindrical stainless steel chamber with removable end plates. A membrane was placed on top of a porous stainless steel support disc. Buna-N (nitrile rubber) O-rings were used as sealing parts, giving an active membrane area of 32 cm^2. The cell, with a volume of 380 mL, was pressurized with compressed high-purity nitrogen gas and the pressure can be applied up to 4,000 kPa. A thermostated water bath and a vortex were used to control the temperature and provide stirring, respectively.

52.2.3 Sinigrin Solution Sample Preparation

Ground myrosinase-inactivated seed meals were defatted with petroleum ether, and then were stirred for 5 min in a tenfold excess (w/v) of boiling water three times [5]. After the supernatant was filtered, the extracted solution was filtered though an ultrafiltration apparatus with average pore size 0.02–0.2 μm (Tianjin, China) in order to primarily remove the impurities, such as solute-protein and polysaccharide with the molecular weight over 1,000 Da. The solutions before and after nanofiltration membrane are prepared for quantitatively analysis by HPLC.

52.2.4 Membranes and Membrane Pretreatment

NF-270, NF, and SPESI membranes were purchased from Amei Ande Membrane Technology Co., Ltd. (USA) in a "dry" form, and STARMEN 120, STARMEN 128, STARMEN 220, and STARMEN 240 were purchased from KOCH Membrane Company (California, USA) in a wet form. Table 52.1 shows the membranes properties in our experiments, according to the information offered by the manufacturers.

Prior to every flux experiment, circular discs were cut and then pretreated by soaking them into the deionized water for at least 5 days so as to saturate the membranes with the carrier solvent.

52.2.5 Experimental Procedure

Solvent permeation experiment was carried out before every filtration experiment until the final stable flux was obtained. Afterward, the filtration cell was loaded with 100 mL feed solution. Filtration experiment was conducted and stopped when 50 mL feed solution permeated across the membrane. All experiments were stopped at VRF = 2

$$\mathrm{VRF} = \frac{V_I}{V_R} \tag{52.1}$$

where VRF is the volume reduction factor, V_I is the volume of initial feed, and V_R is the volume of retentate. The volume of permeate was measured with a measuring cylinder, and high stirring rate at 300 rpm was applied.

Table 52.1 Membrane properties (MWCO and hydrophilicity/hydrophobicity) provided by the manufacturers

Membrane type	MWCO (Da)	Nature
AMFOR NF-270	300	Hydrophilic
AMFOR NF	400	Hydrophilic
AMFOR SPESI	300	Hydrophilic
STARMEM122	220	Hydrophobic
STARMEM228	280	Hydrophobic
STARMEM240	400	Hydrophobic
STARMEM120	200	Hydrophobic

The permeate flux (*J*) was obtained by

$$J = \frac{V}{At} \tag{52.2}$$

where *V* is the volume of permeate, *A* is membrane area, and *t* is time.

The retention was calculated by

$$R = \left(1 - \frac{C_P}{C_R}\right) \times 100\ \% \tag{52.3}$$

where C_P and C_R are the final concentrations in the permeate and retentate, respectively.

For the selection of the membrane process, feed filtration was performed in triplicate at 25 °C and 15 bar, and the initial total solids concentration is 0.5 mg/mL. The basic criteria of membrane selection were solute retention and permeate flux. Therefore, a membrane with both good sinigrin retention and high permeate flux would be desirable. According to these principles, the desired membrane was selected and was used for further experiments.

52.2.6 HPLC Analysis of Sinigrin

Quantification of sinigrin concentration was carried out by a Shimadzu HPLC apparatus equipped with Shimadzu model LC-20AT pumps, an LC-20A UV detector, a CTO-10ASVP column oven (Shimadzu, Kyoto, Japan), and a reversed-phase C18 column (250 mm × 4.6 mm, 5 μm, Diamonsil™, USA). The column temperature was maintained at 30 °C. The elution started of 1.0 % v/v methanol/water (0.1 % v/v TFA), then methanol was raised to 70 % during 20 min and maintained for 2 min to purge the column. The flow rate employed was 1.0 mL/min throughout the run. 5 μL samples were injected into the column. The UV detector was set at 235 nm.

52.3 Results and Discussion

52.3.1 Membrane Selection

The results of the performances of chosen membranes, measured in terms of sinigrin retention and permeate flux are presented in Table 52.2 (25 °C, 15 bar, 0.5 mg/mL). All hydrophilic membranes produced acceptable solvent flux. While, solvent flux for all hydrophobic membranes was so low that retention experiments were not performed. Thus, the hydrophilic membranes, including NF, NF-270, and SPESI, were

Table 52.2 Experimental results for membrane selection

Membrane type	MWCO (Da)	Flux of pure solvent ($L \cdot m^{-2}\,h^{-1}$)	Flux of extraction solution ($L \cdot m^{-2}\,h^{-1}$)	Sinigrin retention (%)
NF-270	300	67.53 ± 3.37	9.50 ± 0.47	96.83 ± 0.87
NF	400	60.35 ± 3.02	8.48 ± 0.42	93.04 ± 1.04
SPESI	300	30.92 ± 4.54	7.47 ± 1.37	94.84 ± 0.85

chosen to determine their retentions. Both the permeate fluxes and sinigrin retentions for hydrophilic membranes are all desirable. Moreover, the concentration performance of NF-270 is better than that of the other two significantly at $p < 0.05$. Taken all these factors into account, we adopted NF-270 to operate the following experiments to investigate the effects of operation temperature, pressure, and feed concentration on the performance of nanofiltration.

52.3.2 *Effects of Independent Variables on Permeate Flux and Retention*

The effects of independent variables on permeate flux are shown in Table 52.3. It is obvious that temperature has a positive effect on flux. It is probable for the following reasons. First, the viscosity of the feed solution reduces with increasing temperature, which enhances the mass transfer of each component. Second, the diffusion coefficient increases with the increasing operating temperature. High operating pressure results in high permeate flux generally. This phenomenon can be explained on the basis of the solution-diffusion mechanism. In our work, permeate flux increased with pressure at low level. However, the flux remains approximately constant ($p > 0.05$). Similar phenomenon was also observed by other researchers [10]. This is probably due to the membrane blocking or fouling resulting from the concentration polarization. Because of concentration polarization, the total solids concentration on the membrane surface will be higher than the feed concentration. Further, if the surface concentration is higher than the solubility, surface precipitation may form and cause the blocking of the membrane. As expected, feed concentration had a negative effect on permeate flux. It is probably because high feed concentration will induce concentration polarization phenomenon. Concentration polarization problems often exist in many nanofiltration processes when solute concentrations are high [11].

The effect of independent variables on sinigrin retention is also shown in Table 52.2. It can be seen that feed concentration has positive effect on sinigrin retention. This increasing retention may be resulted from the concentration polarization, which decreased the average pore size of the membrane. The retention increased with concentration increasing [11]. The retention is insignificantly changed with the pressure, probably because concentration polarization already occurred and fouling did not increase with pressure. Further, with the temperature

Table 52.3 Effect of independent variables on flux of pure solvent, flux of extraction solution, and sinigrin retention

	Flux of pure solvent (L·m^{-2} h^{-1})	Flux of extraction solution (L·m^{-2} h^{-1})	Sinigrin retention ratio (%)
Temperature (15 bar, 0.5 mg/mL)			
20 °C	61.54 ± 1.89	7.69 ± 0.47	96.35 ± 0.64
25 °C	67.69 ± 2.17	8.68 ± 0.51	94.94 ± 0.75
30 °C	68.18 ± 2.20	9.15 ± 0.46	93.45 ± 0.44
35 °C	74.21 ± 3.48	10.12 ± 0.50	91.88 ± 0.62
40 °C	95.99 ± 5.51	11.96 ± 0.87	90.20 ± 1.01
Pressure (25 °C, 0.5 mg/mL)			
10 bar	37.20 ± 6.74	7.44 ± 0.36	96.14 ± 0.76
15 bar	67.69 ± 2.17	8.68 ± 0.51	94.94 ± 0.75
20 bar	72.67 ± 2.41	9.89 ± 0.61	97.13 ± 0.67
25 bar	87.89 ± 5.13	10.12 ± 0.52	96.17 ± 0.56
Feed concentration (25 °C, 15 bar)			
0.1 mg/mL	-	23.47 ± 3.45	90.77 ± 0.60
0.2 mg/mL	-	16.29 ± 1.88	90.83 ± 0.71
0.3 mg/mL	-	13.11 ± 0.81	91.25 ± 0.63
0.4 mg/mL	-	9.88 ± 1.71	92.74 ± 0.82
0.5 mg/mL	-	8.68 ± 0.51	94.94 ± 0.75

increasing, the concentration polarization is not opting to form, due to the high solubility under the higher temperatures, resulting in the decreased retention.

Table 52.3 shows that the retention is always higher than 90 % suggesting that the NF-270 membrane is appreciated to the concentration process of sinigrin. And the permeation flux can be increased by raising operation temperature, increased pressure, or declined feed concentration, without obvious decrease of rejection.

52.3.3 Explanation of the Nanofiltration with the High Flux and High Sinigrin Retention Using Hydrophilic Membranes

52.3.3.1 Effect of Solute's Molecule Charge

It has been reported that besides the molecular size, the molecular charge is another important parameter affecting separation in aqueous nanofiltration processes. When the molecular size is much smaller than the membrane pores, the molecular charge can be the decisive factor in determining retention of the molecule (charge effect); when the molecules have similar size as the membrane pores, retention of the molecules is mainly a result of size exclusion (sieve effect) [12].

As shown in Table 52.2, when water is used as solvent, the retention of negatively charged sinigrin is obviously high for all hydrophilic membranes (NF-270, NF, and SPESI). This phenomenon can be well explained by the charge effect, in particular for the NF membrane, whose MWCO (400 Da) is higher than the solute's molecular weight.

52.3.3.2 Effect of the Hydrophilic Membranes

To gain further insight into the effect of the hydrophilic membranes on sinigrin concentration, the microstructure changes of the hydrophilic membranes were analyzed with swelling ratio, SEM and ATR-FTIR. As shown in Fig. 52.2, NF-270 showed the highest swelling ratio, which is consistent with its highest flux. Among the three hydrophilic membranes, the penetrate flux increased with the swelling ratio increasing, which exhibited all hydrophilic membranes become swollen significantly. And the results are also supported by the SEM images. In Fig. 52.3, the surfaces of the membranes after soaking in the aqueous solution were wrinkled and became rougher than those of dry membranes. And the degrees of different membranes were significantly relative to the respective swelling ratio. NF-270, which is the most wrinkled and roughest, exhibited the highest swelling ratio. While, the SPESI membrane, whose swelling ratio is the lowest among the three, showed few differences between before and after soaking. The change strongly suggests that the skin had become more hydrophilic, resulted in an increase of the number of hydrophilic sites in an order of NF-270 > NF > SPESI.

Despite clear indications given by SEM images that the active layer undergoes changes, resulting in increased swelling of top active layer, it was still not clearly known the reason why this change was caused predominantly, such as an increase in the amount of hydrophilic groups in the polymer, a decrease in the degree of

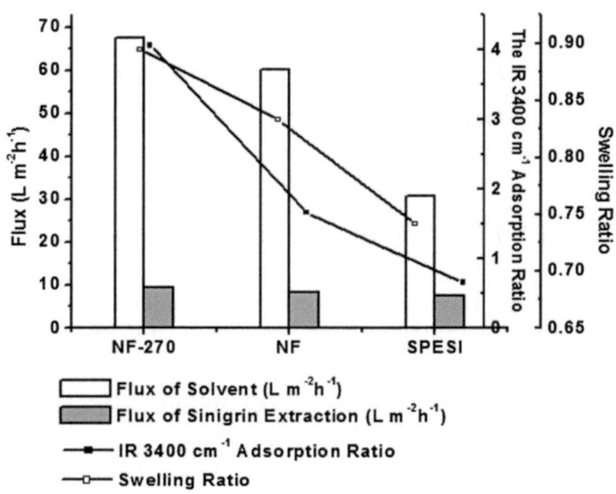

Fig. 52.2 The comparison of flux, swelling ratio, and IR (3,400 cm^{-1}) absorption ratio of different hydrophilic membranes

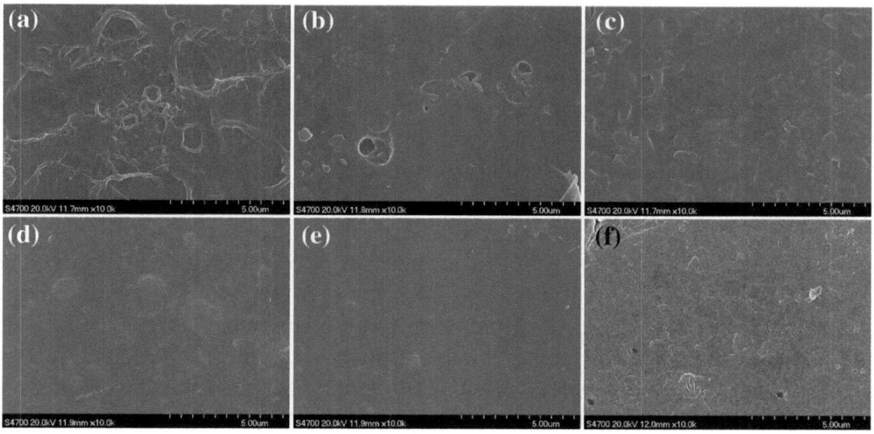

Fig. 52.3 SEM images of a top active layer of NF-270 (**a**), NF (**b**), SPESI (**c**) membrane after soaking and NF-270 (**d**), NF (**e**), SPESI (**f**) before soaking

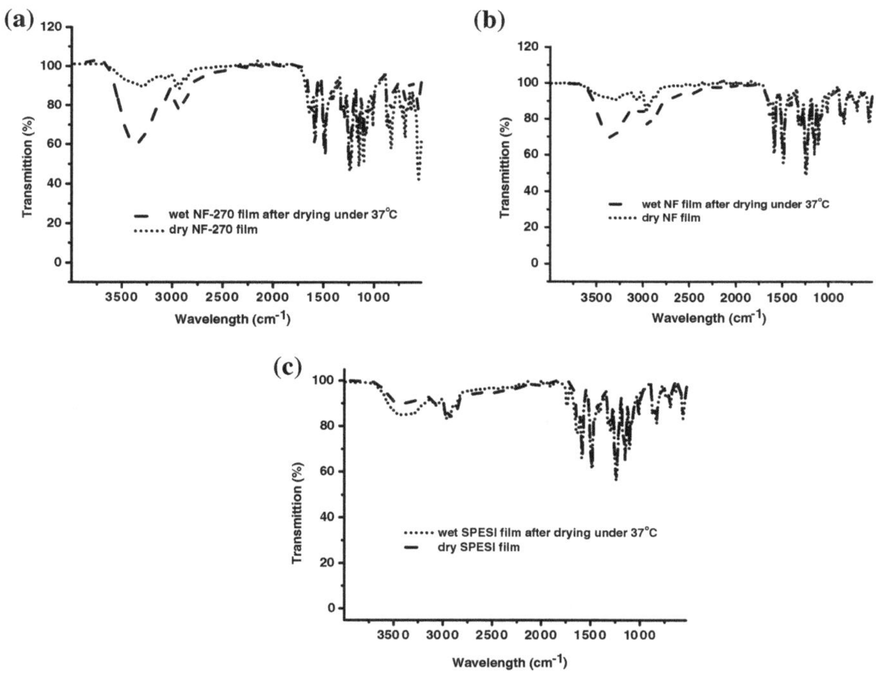

Fig. 52.4 A typical series of ATR-FTIR spectra of a sample of isolated active layer of different hydrophilic membranes in the course of hydration by exposure to nitrogen of 50 % relative humidity. **a** NF-270; **b** NF; **c** SPESI

cross linking or for both. Thus, the analysis of sorption of water vapor in the low relative humidity range determined by ATR-FTIR help us to insight the changes. IR spectroscopy has been widely used to study hydrated polymers [13], since the absorption bands of water are rather intense and are sensitive to interactions with environment. The ATR mode offers the possibility of concentrating on a micron-thin surface layer, which makes it a highly suitable tool for studying the active layer of composite membranes.

Figure 52.4 shows a typical series of spectra of isolated active layers before and after soaking in the aqueous solution, to nitrogen of 50 % relative humidity. The systematic increase in the intensity of the -OH stretching vibration band (3,400 cm^{-1}) after exposure to solvent is easily observed, which means the number of hydrophilic sites was increased, maybe resulted in the increase of flux. And we can see in Fig. 52.2 that there is a positive correlation between flux, swelling ratio, and IR 3,400 cm^{-1} absorption ratio. NF-270 exhibited the highest IR 3,400 cm^{-1} absorption ratio, resulting in a high degree of swelling ratio, leading to a high flux. The ATR-FTIR data, therefore, indicate that the increase in swelling is due, at least in part, to an increase of the number of hydrophilic sites.

52.4 Conclusion

The feasibility of concentration of sinigrin by nanofiltration was systemically studied. After seven commercial membranes were compared, the effects of temperature, pressure, and feed concentration, for the concentration of sinigrin, were investigated. The results indicated that NF-270 membrane showed the highest permeate flux and the best retention capability to sinigrin. Permeate flux increased with increasing temperature and pressure, while decreased with increasing feed concentration. The retention increased with increasing concentration and declined temperature. Furthermore, swelling ratios and SEM images of nanofiltration membranes before and after soaking declared that the membranes were swollen. This result was also supported by ATR-FTIR. The spectra showed that the peak of -OH stretching vibration band (3,400 cm^{-1}) was increased significantly. The content of hydroxyl in membrane was increased, leading to the improvement permeability of solvent. Based on the results mentioned above, it could be concluded that the concentration process using nanofiltration provided a novel, rapid, and economical method for the concentration of sinigrin.

Acknowledgments The authors acknowledge the support of the Natural Science Foundation of China (20776009, 20976009) and (2008066005), and the Young Scholars Funds of Beijing University of Chemical Technology (QN. 0809).

References

1. Sarwar M, Kirkegaard JA (1998) Biofumigation potential of brassicas. Plant Soil 201:103–112
2. Rosa EAS, Rodrigues PMF (1999) Towards a more sustainable agriculture system: the effect of glucosinolates on the control of soil-borne diseases. J Hort Sci Biotechnol 74:667–674
3. Krul C, Humblot C, Philippe C et al (2002) Metabolism of sinigrin (2-propenyl glucosinolate) by the human colonic microflora in a dynamic in vitro large-intestinal model. Carcinogenesis 23:1009–1016
4. Wallig MA, Kingston S, Staack R et al (1998) Induction of rat pancreatic glutathione S-transferase and quinone reductase activities by a mixture of glucosinolate breakdown derivatives found in brussels sprouts. Food Chem Toxicol 36:365–373
5. Szmigielska AM, Schoenau JJ, Levers V (2000) Determination of glucosinolates in canola seeds using anion exchange membrane extraction combined with the high-pressure liquid chromatography detection. J Agric Food Chem 48:4487–4491
6. Charpentier N, Bostyn S, Coïc JP (1998) Isolation of a rich glucosinolate fraction by liquid chromatography from an aqueous extract obtained by leaching dehulled rapeseed meal (*Brassica napus L.*). Ind Crops Prod 8:151–158
7. Fahey JW, Wade KL, Stephenson KK et al (2003) Separation and purification of glucosinolates from crude plant homogenates by high-speed counter-current chromatography. J Chromatogr A 996:85–93
8. Warczok J, Ferrando M, Güell C (2004) Concentration of apple and pear juices by nanofiltration at low pressures. J Food Eng 63:63–71
9. Wu LH (1997) Nanofiltration membrane-a new separating material and its application in pharmaceutical industry. Membr Sci Tech 17(5):11–14
10. Su B, Wang Z, Wang J et al (2005) Concentration of clindamycin phosphate aqueous ethanol solution by nanofiltration. J Membr Sci 251:189–200
11. Ahn KH, Cha HY, Yeom IT et al (1998) Application of nanofiltration for recycling of paper regeneration wastewater and characterization of filtration resistance. Desalination 119:169–176
12. Robinson JR, Tarleton ES, Millington CR et al (2004) Solvent flux through dense polymeric nanofiltration membranes. J Membr Sci 230:29–37
13. Yarwood J, Sammon C, Mura C et al (1999) Vibrational spectroscopic studies of the diffusion and perturbation of water in polymeric membranes. J Mol Liq 80:93–115

Chapter 53
Optimization of Crude Polysaccharides Extraction from *Dioscorea esculenta* by Response Surface Methodology

Kaihua Zhang, Liming Zhang, Na Liu, Jianheng Song and Shuang Zhang

Abstract In this study, response surface methodology (RSM) was employed to optimize the extraction process of crude polysaccharides from *Dioscorea esculenta* (lesser yam) performed using water decoction. Response surface methodology, based on a three-level, three-variable central Box–Behnken design (BBD), was employed to obtain the best possible combination of extraction time (X_1: 2.0–3.0 h), extraction pH (X_2: 9–11), and water to the raw material ratio (X_3: 20–30 mL/g) for maximum polysaccharide extraction. The experimental data obtained were fitted to a second-order polynomial equation using multiple regression analysis and also analyzed by appropriate statistical methods (ANOVA). The optimum extraction conditions were as follows: extraction time of 2.66 h, extraction pH of 10.07, and the ratio of water to raw material of 26.36 mL/g. Under these conditions, the experimental yield was 4.92 ± 0.037 %, which is well in close agreement with the value predicted by the model 4.96 %.

Keywords Optimization · Lesser yam · Polysaccharide · Extraction · Response surface methodology

53.1 Introduction

Yams, the tubers of *Dioscorea* spp., are important staple foods in many tropical countries [1, 2]. Even more interestingly, yams have also been used as health food and herbal medicinal ingredients in traditional Chinese medicine [3].

The yam polysaccharides were isolated and purified from the roots of *Dioscorea opposita Thunb* [4] and *D. nipponica Makino* [5], and biological activity tests also showed that yam polysaccharides have a strong effect in antitumor activity in cancer

K. Zhang · L. Zhang (✉) · N. Liu · J. Song · S. Zhang
Key Laboratory of Industrial Fermentation Microbiology, Ministry of Education, Tianjin University of Science and Technology, Tianjin 300457, People's Republic of China
e-mail: zhlm@tust.edu.cn

T.-C. Zhang and M. Nakajima (eds.), *Advances in Applied Biotechnology*,
Lecture Notes in Electrical Engineering 332, DOI 10.1007/978-3-662-45657-6_53

bearing murine models, and antioxidants, immune regulation, antiaging, hypoglycemic effect, and so on [6, 7]. However, little attention was devoted to the extraction and purification of the polysaccharides of *D. esculenta*, which is grown in southwest of China. It is well known that genetic variations and environmental conditions influence polysaccharide properties. Therefore, the optimization of extracting parameters for polysaccharide of this yam plant is developed.

Response surface methodology (RSM) is an effective statistical technique for optimizing complex processes. The main advantage of RSM is the reduced number of experimental trials needed to evaluate multiple parameters and their interactions. Box–Behnken design (BBD) is a type of response surface design. It is an independent quadratic design in that it does not contain an embedded factorial or fractional factorial design. In this design, the treatment combinations are at the midpoints of edges of the process space and at the center. These designs are rotatable (or near rotatable) and require three levels of each factor. It is more efficient and easier to arrange and interpret experiment in comparison with others. It is widely used in many researches [8, 9]. The objective of this study was to optimize the extraction time, extraction pH, and ratio of liquid to solid by RSM for maximum yield of lesser yam polysaccharides.

53.2 Materials and Methods

53.2.1 Materials

Dioscorea esculenta (lesser yam) of 1-year-old was collected from Mengwang County of Jinghong city, Yunnan province, China, and was identified by Professor Wenbin Hou, Tianjin Institute of Pharmaceutical Research, China. The yam tubers were washed with tap water, rinsed with deionized water, and peeled, fibrous roots were removed and trimmed to get rid of defective parts. Then the tubers were sliced into pieces with the thickness between 3 and 5 mm and dried at 60 °C in a drying oven until weight constancy. Samples were ground using a grinder and passed through a 40-mesh sieve.

53.2.2 Extraction of Crude Polysaccharides

Extractions were carried out according to the method of Han et al. [10]. Each pretreated sample was extracted by hot distilled water in designed extraction pH, extraction temperatures, extraction times, and ratios of water to raw material. The extract was processed in room temperature, filtered, and then precipitated using 100 % ethanol, with ratio of ethanol to extracting solution at 4:1. After being left overnight, the precipitates were collected by centrifugation at 4,000 rpm for 15 min, redissolved in deionized water, deproteinated by the method of Sevag et al. [11], and then freeze-dried to obtain yam polysaccharides. The percentage polysaccharides extraction yield (%) is calculated as follows:

$$\text{Polysaccharides extraction yield } (\%,\ \text{w/w}) = \frac{\text{Dried crude polysaccharides weight (g)}}{\text{Powder weight (g)}} \times 100\,\% \quad (53.1)$$

53.2.3 Experimental Design

After determining the preliminary range of extraction variables through single-factor test, a three-level, three-factor Box–Behnken factorial design (BBD) was performed [12]. Extraction time (X_1), extraction temperature (X_2), and ratio of water to raw material (X_3) were the independent variables selected to be optimized for the extraction of yam crude polysaccharides. The range of independent variables and their levels were presented in Table 53.1. Polysaccharides extraction yield was taken as the response for the combination of the independent variables given in Table 53.2. All the experiments were carried out at random in order to minimize the effect of unexplained variability in the observed responses due to systematic errors. The variables were coded according to the equation:

$$X_i = \frac{X_i - X_0}{\Delta X_i} \quad (53.2)$$

where X_i is the (dimensionless) coded value of the variable X_i, X_0 is the value of X_i at the center point, and ΔX is the step change. The behavior of the system was explained by the following quadratic equation:

$$Y = A_0 + \sum_{i=1}^{3} A_i X_i + \sum_{i=1}^{3} A_{ii} X_i^2 + \sum_{i=1}^{2} \sum_{j=i+1}^{3} A_{ij} X_{ij} \quad (53.3)$$

where Y is the dependent variable, A_0 is constant, and A_i, A_{ii}, and A_{ij} are coefficients estimated by the model. X_i and X_j are levels of the independent variables. They respectively delegate the linear, quadratic, and cross-product effects of the X_1, X_2, and X_3 factors on the response. The model evaluated the effect of each independent

Table 53.1 Independent variables and their levels used in the response surface design

Independent variables	Levels		
	−1	0	+1
Extraction time (X_1) (h)	2	2.5	3
Extraction pH (X_2)	9	10	11
Ratio of water to raw material (X_3) (mL/g)	20	25	30

Table 53.2 Box–Behnken experimental design and results for yield of yam polysaccharides

No.	X_1/extraction time (h)	X_2/extraction pH	X_3/ratio of water to raw material (mL/g)	Prediction (%)	Yield (%)
1	0	−1	−1	4.67	4.71
2	0	0	0	4.95	4.95
3	−1	0	+1	4.83	4.88
4	+1	+1	0	4.84	4.86
5	0	0	0	4.95	4.99
6	−1	0	−1	4.71	4.71
7	+1	−1	0	4.78	4.80
8	−1	+1	0	4.72	4.70
9	0	−1	+1	4.77	4.75
10	+1	0	+1	4.88	4.88
11	−1	−1	0	4.75	4.72
12	0	+1	−1	4.69	4.71
13	0	0	0	4.95	4.94
14	0	0	0	4.95	4.93
15	0	0	0	4.95	4.92
16	+1	0	−1	4.81	4.76
17	0	+1	+1	4.80	4.78

variable to a response. Analysis of the experimental design and calculation of predicted data were carried out by using Design-Expert 8.0.6.1 software to evaluate the response of the independent variables.

53.3 Results and Discussion

53.3.1 Effect of Extraction Time on Yield of Yam Polysaccharides

Extraction time is an important factor to associate with final concentration of polysaccharides, extraction efficiency, and energy cost [13, 14]. In this study, the effect of extraction time on extraction yield of polysaccharides from lesser yam was investigated and the result was shown in Fig. 53.1. Firstly, the extraction time was set at 1.0, 1.5, 2.0, 2.5, and 3.0 h when other factors (extraction pH, ratio of water to raw material) were fixed at 10, 25:1. It could be found that the extraction yield increased as extraction time ascended from 1.0 to 2.5 h, and then increased slowly when the extraction time exceeded 2.5 h (Fig. 53.1a). This might be due to the time requirement of the exposure of the yam polysaccharides to the release medium where the liquid penetrated into the raw materials, dissolved the yam polysaccharides, and subsequently diffused out from the raw materials. It is reported that a long

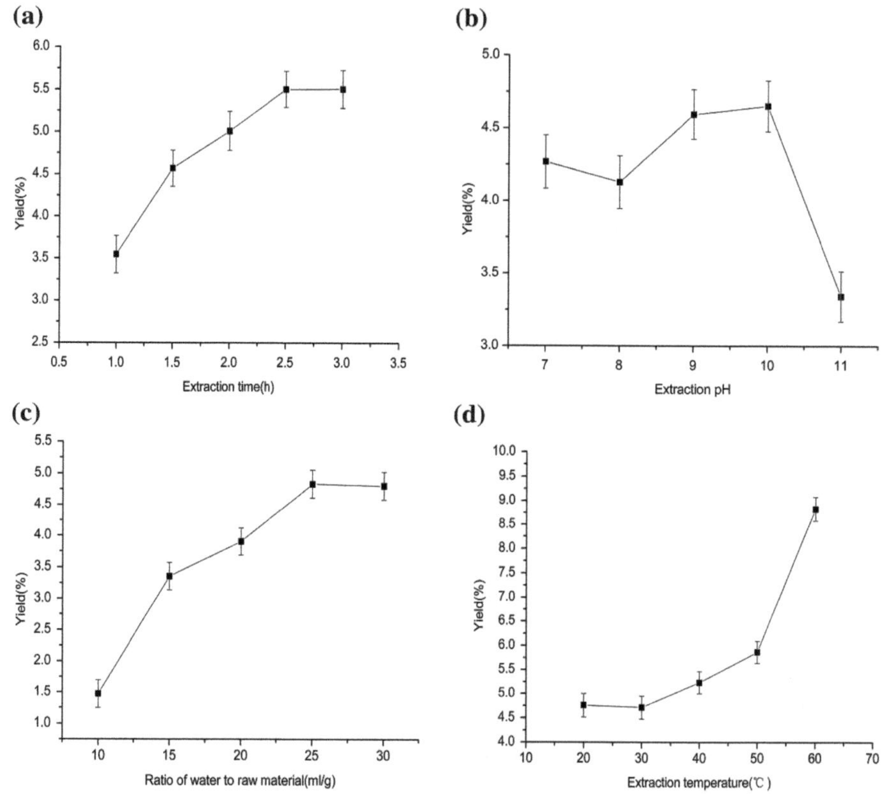

Fig. 53.1 Effect of extraction parameters on yield of yam polysaccharides (**a** extraction time, h; **b** extraction pH; **c** ratio of water to raw material, mL/g; **d** extraction temperature, °C)

extraction time benefits on the production of polysaccharides [15]. This result indicates that extraction time of 2.5 h is enough to the polysaccharides in the present work.

53.3.2 Effect of Extraction pH on Yield of Yam Polysaccharides

As is known to all, polysaccharides can be divided into acidic and alkaline polysaccharides in nature. And the amount of acidic polysaccharides is far more than the alkaline and neutral polysaccharides. So the selection of pH for the experiment is 7–11. It could be found that the extraction yield increased as extraction pH ascended from 8 to 10, and then declined when the extraction pH exceeded 11 (Fig. 53.1b). This result may be due to high pH that could damage certain structures of yam polysaccharides.

53.3.3 *Effect of the Ratio of Water to Raw Material on Yield of Yam Polysaccharides*

Water to raw material ratio will significantly affect extract yield [16]. Different ratio of water to raw material will significantly affect extract yield. If ratio of water to raw material is too small, yam polysaccharides in raw material cannot be completely extracted up. If ratio of water to raw material is too big, this will cause high process cost [17]. Therefore, it is necessary to select a suitable ratio of water to raw material for extraction of polysaccharides [18]. In this study, ratio of water to raw material was set at 10, 15, 20, 25, and 30 to investigate the effect of different ratio of water to raw material on the yield of yam polysaccharides. The extraction time was fixed at 2.5 h, the extraction pH is 10. We found that the yield of yam polysaccharides continued to increase evidently with the increasing ratio of water to raw material. A possible explanation is that increase in ratio of water to raw material may increase diffusivity of the solvent into cells and enhance desorption of the yam polysaccharides from the cells. But the yield of yam polysaccharides started to increase slowly after the ratio of water to raw material exceeded 25 (Fig. 53.1c).

53.3.4 *Effect of Extraction Temperature on Yield of Yam Polysaccharides*

When the temperature is higher than 60 °C, the starch will become gelatinized and part of them change into polysaccharide. So the temperature was set at 20, 30, 40, 50, and 60 °C, while the extraction time and ratio of water to raw material were set at 2.5 h and 25:1, respectively. As shown in Fig. 53.1d, extraction yield of polysaccharides always increased as extraction temperature ascended from 20 to 60 °C, that have an adverse effect on the response surface optimization experiment. So, the finally selected three factors are extraction time, extraction pH, and the ratio of water to raw material, respectively.

53.3.5 *Optimization of the Extraction Parameters of Yam Polysaccharides*

53.3.5.1 Statistical Analysis and the Model Fitting

Response surface optimization is more advantageous than the traditional optimization method, it is less laborious and time-consuming than other approaches required to optimize a process. There were 17 runs for optimizing the four individual parameters in BBD. This design was applied to the yam polysaccharides by heat water extraction. The data were analyzed by multiple regression analysis using the Design-Expert 8.0.6.1 and the following polynomial equation was derived to represent yam polysaccharides yield as a function of the independent variables tested. And Y is the predicted yam polysaccharides yield and X_1, X_2, and X_3 are the coded values for extraction time, extraction pH, and ratio of water to raw material, respectively. Table 53.2 shows the process variables and experimental data. The results of the analysis of variance, goodness of fit, and the adequacy of the models were summarized. The percentage yield range from 4.70 to 4.99 %. The maximum yield of yam polysaccharides (4.99 %) was recorded with extraction time 2.5 h, extraction pH = 10, ratio of water to raw material 25 mL/g. The application of RSM suggested, based on parameter estimates, an empirical relationship between the response variable (extraction yield of yam polysaccharides) and the test variable under consideration. By applying multiple regression analysis on the experimental data, the response variable and the test variables are related by the following second-order polynomial equation:

$$\begin{aligned} Y = {} & 4.95 + 0.036X_1 + 8.750 \times 10^{-3}X_2 + 0.050X_3 + 0.020X_1X_2 - 0.012X_1X_3 \\ & - 7.500 \times 10^{-3}X_2X_3 - 0.053X_1^2 - 0.120X_2^2 - 0.086X_3^2 \end{aligned} \tag{53.4}$$

The fit statistics of extraction yield (Y) for the selected quadratic predictive model is shown in Table 53.3. The model P-value (Prob > F) was very low (<0.0029), which implied that the model was significant. For the model fitted, the coefficient of determination (R^2) for model (0.9291) was close to 1.0, which indicated that only 7.09 % of the total variations were not explained by the model. The F-value of 3.92 and P-value of 0.1102 for extraction yield implied that the lack of fit was insignificant relative to the pure error due to noise. The value of the adjusted determination coefficient (adjusted $R^2 = 0.8380$) also represented the satisfactory correlation between actual and predicted values. At the same time, a very low value 0.84 of coefficient of the variation (CV) clearly indicated a very high degree of precision and a good deal of reliability of the experimental values.

The model was found to be adequate for prediction within the range of experimental variables. The regression coefficient values of Eq. (53.4) were listed in Table 53.4.

Table 53.3 Analysis of variance for the fitted quadratic polynomial model of extraction of yam polysaccharides

Source	SS	DF	MS	F−value	Prob > F
Model	0.15	9	0.017	10.19	0.0029
Residual	0.011	7	1.642×10^{-3}		
Lack of fit	8.575×10^{-3}	3	2.858×10^{-3}	3.92	0.1102
Pure error	2.920×10^{-3}	4	7.300×10^{-4}		
Cor total	0.16	16			
	$R^2 = 0.9291$	$R^2_{adj} = 0.8380$	CV = 0.84		

Table 53.4 Estimated regression model of relationship between response variable (yield of yam polysaccharides) and independent variables (X_1, X_2 and X_3)

Variables	SS	DF	MS	*F*-value	*P*-value
X_1	0.011	1	0.011	6.40	0.0392
X_2	6.125×10^{-4}	1	6.125×10^{-4}	0.37	0.5607
X_3	0.020	1	0.020	12.18	0.0101
X_1X_2	1.600×10^{-3}	1	1.600×10^{-3}	0.97	0.3565
X_1X_3	6.250×10^{-4}	1	6.250×10^{-4}	0.38	0.5568
X_2X_3	2.250×10^{-4}	1	2.250×10^{-4}	0.14	0.7222
X_1^2	0.012	1	0.012	7.20	0.0314
X_2^2	0.064	1	0.064	38.79	0.0004
X_3^2	0.031	1	0.031	18.74	0.0034

The *P*-values were used to check the significance of each coefficient. The smaller value of P indicated the more significant corresponding coefficient [19]. It can be seen from this table that the linear coefficients (X_1), quadratic term coefficients (X_1^2, X_2^2, and X_3^2), and cross-product coefficients (X_1X_2) were significant, with very small *P*-values ($P < 0.05$). The other term coefficients were not significant ($P > 0.05$). The full model filled Eq. (53.3) was made three-dimensional and contour plots to predict the relationships between the independent variables and the dependent variables.

53.3.5.2 Optimization of Extraction Conditions

The 3D response surface and 2D contour plots are the graphical representations of regression equation. They provide a method to visualize the relationship between responses and experimental levels of each variable and the type of interactions between two test variables. The shapes of the contour plots, circular or elliptical, indicate the significance between the variables. Circular contour plot indicates that the interactions between the corresponding variables are negligible, while elliptical contour plot indicates that the interactions between the corresponding variables are significant. The graphical representations of the regression Eq. (53.4), called the response surfaces and the contour plots were obtained using Design-Expert, and the results of extraction yield of yam polysaccharides affected by extraction time, extraction pH, and ratio of water to raw material are presented in Fig. 53.2.

The 3-D plot and the contour plot in Fig. 53.2a, which gives the ratio of water to raw material (0 level), show that extraction yield of yam polysaccharides increased evidently with increase of extraction pH from 9 to 10, but beyond 10.5, the extraction yield of yam polysaccharides declined slowly as the pH ascended. The extraction yield of yam polysaccharides increased very slowly with the extraction time. Figure 53.2b shows the 3-D plot and the contour plot at varying extraction time and ratio of water to raw material at fixed extraction pH (0 level). From Fig. 53.2, it can be seen that the extraction yield of yam polysaccharides increased evidently with increase in ratio of water to raw material from 20 to 25 mL/g, but

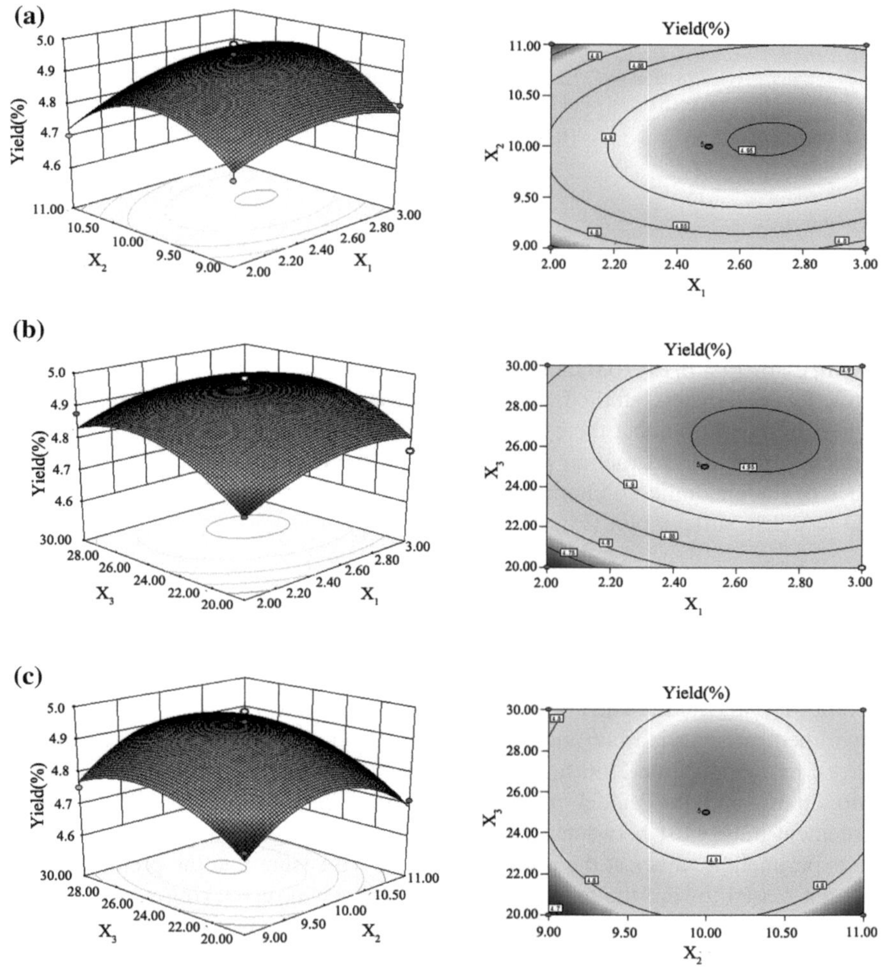

Fig. 53.2 Response surface plots and contour plots showing the effect of the extraction time, extraction pH, and ratio of water to raw material on the response yield

beyond 25 mL/g, extraction yield of yam polysaccharides declined slowly as the ratio of water to raw material ascended. The 3-D plot and the contour plot based on independent variables' ratio of water to raw material and extraction were shown in Fig. 53.2c, while the extraction time was kept at a zero level. An increase in the extraction yield of yam polysaccharides could be significantly achieved with the increase in ratio of water to raw material. It was obvious that the extraction yield of yam polysaccharides was increased rapidly with the increasing extraction pH from 9 to 10, but beyond 10.5, the extraction yield of yam polysaccharides declined slowly as the temperature ascended.

Table 53.5 Predicted and experimental values of the responses at optimum and modified conditions

	Extraction time (h)	Extraction pH	Ratio of water to material (mL/g)	Yield of yam polysaccharides (%)
Optimum conditions	2.66	10.07	26.36	4.96
Modified conditions	2.75	10	26	4.92 ± 0.037

According to Fig. 53.2, and above single parameter study, it can be concluded that optimal extraction condition of yam polysaccharides were extraction time 2.66 h, extraction pH = 10.07, ratio of water to raw material 26.36. Among the three extraction parameters studied, the ratio of water to raw material was the most significant factor to affect the extraction yield of yam polysaccharides, followed by extraction time and extraction pH according to the regression coefficients' significance of the quadratic polynomial model (Table 53.4) and gradient of slope in the 3-D response surface plot (Fig. 53.2).

53.3.6 Verification of the Models

In order to validate the adequacy of the model equations, Eq. (53.4), a verification experiment was carried out under the optimal conditions (within the experimental range): extraction time 2.66 h, extraction pH = 10.07, and water to raw material ratio 26.36, and the model predicted a maximum response of 4.96 %. Good agreement must exist between the values predicted using the model equations and the experimental values at the points of interest. To ensure that the predicted result was not biased toward the practical value, experimental rechecking was performed using this deduced optimal condition. This set of conditions was determined to be optimal by the RSM optimization approach and was also used to validate experimentally and predict the values of the responses using the model equation. A mean value of 4.92 ± 0.037 % (n = 3) was gained, obtained from real experiments, demonstrated the validation of the RSM model. The results of analysis confirmed that the response model was adequate for reflecting the expected optimization (Table 53.5), and the model of Eq. (53.4) was satisfactory and accurate.

53.4 Conclusion

The response surface method proved to be useful for optimization of technology of yam polysaccharides extraction. Statistical analysis proved to be a useful and powerful tool in developing optimum extraction conditions. In the current work, the extraction temperature and number of extraction exhibited the most and least significant ($P < 0.05$) impact on the extraction yield of yam polysaccharides, respectively. The present work revealed that the desirable yam polysaccharides was extracted by using extraction time 2.66 h, extraction pH = 10.07, and water to raw material ratio of 26.36 (mL/g). Under the most suitable conditions, the experimental yield of yam polysaccharides was 4.92 ± 0.037 %, which was close to the predicted yield value 4.96 %.

References

1. Akanbi CT, Gureje PO, Adeyemi IA (1996) Effect of heat moisture pre-treatment on physical characteristics of dehydrated yam. J Food Eng 28:45–54
2. Omonigho SE, Ikenebomeh MJ (2000) Effect of temperature treatment on the chemical composition of pounded white yam during storage. Food Chem 71:215–220
3. Liu SY, Wang JU, Shyu YT et al (1995) Studies of yams (Dioscorea spp.) in Taiwan. J Chin Med 6(2):111–126
4. Luo DH (2014) Structural investigation of a polysaccharide (DMB) purified from Dioscorea nipponica Makino. Carbohydr Polym 103:261–266
5. Zhang N, Kang TG, Yin HB (2011) Optimal extraction of polysaccharide from Dioscorea nipponica by central composite design and response surface methodology. J Chin Med Mater 34(1):123–126
6. Zhao GH, Chen ZD, Li ZX et al (2003) Effects of Chinese yam polysaccharide on immune function of cancer bearing mice. Acta Nutrimenta Sinica 25:110–112
7. Li JJ, Miao J, Zhang L et al (2010) Research progress of the bioactivity and analytical method of polysaccharide in Dioscorea batatas Decene. J Anhui Agric 38(35):20008–20009
8. Khajeh M (2011) Response surface modeling of lead pre-concentration from food samples by miniaturised homogenous liquid–liquid solvent extraction: Box-Behnken design. Food Chem 129:1832–1838
9. Sun Y, Liu J, Kennedy JF (2010) Application of response surface methodology for optimization of polysaccharides production parameters from the roots of Codonopsis pilosula by a central composite design. Carbohydr Polym 80:949–953
10. Han J, Jiang X, Zhang L (2011) Optimisation of extraction conditions for polysaccharides from the roots of Isatis tinctoria L. by response surface methodology and their in vitro free radicals scavenging activities and effects on IL-4 and IFN-γmRNA expression in chicken lymphocytes. Carbohydr Polym 86:1320–1326
11. Sevag MG, Lackman DB, Smolens J (1938) The isolation of components of streptococcal nucleoproteins in serologically active form. J Biol Chem 124:425
12. Box GEP, Behnken DW (1960) Some new three level designs for the study of quantitative variables. Technometrics 2:455–475
13. Hou XJ, Chen W (2008) Optimization of extraction process of crude polysaccharides from wild edible BaChu mushroom by response surface methodology. Carbohydr Polym 72:67–74
14. Ye CL, Jiang CJ (2011) Optimization of extraction process of crude polysaccharides from Plantago asiatica L. by response surface methodology. Carbohydr Polym 84:495–502
15. Liu ZD, Wei GH, Guo YC (2006) Image study of pectin extraction from orange skin assisted by microwave. Carbohydr Polym 64:548–552
16. Govender S, Pillay V, Chetty DJ, Essack SY et al (2005) Optimisation and characterisation of bioadhesive controlled release tetracycline microspheres. Int Pharm J 306:24–40
17. Zhu CP, Gao GT, Li JK (2010) Response surface analysis for optimizing microwave-assisted extraction of Pleurotus ostreatus polysaccharide. Food Sci 31:68–72

18. Yin G, Dang Y (2008) Optimization of extraction technology of the Lycium barbarum polysaccharides by Box-Behnken statistical design. Carbohydr Polym 74:603–610
19. Guo X, Zou X, Sun M (2010) Optimization of extraction process by response surface methodology and preliminary characterization of polysaccharides from Phellinus igniarius. Carbohydr Polym 80:344–349

Chapter 54
Nanofiltration Extraction and Purification Method for Cyclic Adenosine Monophosphate (cAMP) from Chinese Date Fruit

Chunxia Wang, Yihan Liu, Hongbin Wang, Lianxiang Du and Fuping Lu

Abstract Cyclic adenosine monophosphate (cAMP) is a kind of important bioactive substances. The cAMP content of Chinese date fruit is particularly abundant and stable, more than any of the other plants researched so far. In this chapter, we report the development of nanofiltration extraction method for cAMP from date fruit (jujube) paste. The paste was hydrolyzed at 50 °C with 2.5 g/L pectinases (pectin lyase, cellulose, and hemicellulose) for 3 h. The yield efficiency of date juice achieved 70 %, and most importantly, the extracted cAMP content increased from 10.8 to 15.7 mg/100 g. The results prove that pectic substances have a protective effect to ring adenosine phosphate After resin adsorption, elution with formic acid, and filtration through a 150–500 Da Nanofiltration membrane, the crude cAMP content was increased to 36 %. The purification percentage of cAMP using silica gel column chromatography was as much as 97 %. The date fruits were soaked and pureed, and then subjected to enzyme hydrolysis, filtration, resin adsorption, formic acid elution, and nanofiltration. Pure cAMP was obtained after purification via a silica gel chromatography column. The byproduct of cAMP extraction could be further developed, which might enable a nonpolluting industrial chain of products.

54.1 Introduction

Chinese date fruit, also known as Jujube or red date, is a shrub or small tree that is popular in many parts of the word for its edible fruit and as a Chinese herbal medicine [1]. The Chinese date fruits are a potential source of cyclic adenosine

C. Wang (✉) · Y. Liu · H. Wang · L. Du · F. Lu
College of Bioengineering, Tianjin University of Science and Technology, Tianjin 300457, China
e-mail: cxwang@tust.edu.cn

T.-C. Zhang and M. Nakajima (eds.), *Advances in Applied Biotechnology*,
Lecture Notes in Electrical Engineering 332, DOI 10.1007/978-3-662-45657-6_54

monophosphate (cAMP) and cyclic guanosine monophosphate (cGMP). cAMP is important to the regulation of intracellular metabolism, fulfilling many biological functions as second messengers [2, 3]. Its regulatory role in vivo is quite broad, for example, activating cAMP-dependent protein kinase in myocardial intracellular and substrate proteins [4, 5]. It has proved effective in the treatment of hypertension, diabetes, cancers, cardiac shock, and cardiovascular diseases [6, 7].

The cAMP content in particular is 100–600 nmol/g, which is the highest among 180 researched natural plants [8, 9].

In the 1980s, Hanabusa et al. [10] were the first to extract cAMP from Chinese date fruits imported from China. The jujubes were soaked in water before extraction, homogenized by machine, and centrifuged at 12000 × *g* for 15 min. The supernatant was isolated, and the residue was again homogenized with water and centrifuged. Finally, the supernatant was mixed and cAMP was harvested. The cAMP in jujubes could be extracted with hot water or by using an organic solvent such as methanol or ethanol, as reported by researchers [11]. Zhang et al. [12] used an ultrasound and water bath extraction method to extract cAMP from jujubes.

In the present study, a new process using a mixture of enzymes to extract jujube cAMP was developed. Compared with the nonenzyme or single-enzyme methods, using a combination of enzymes greatly improved the jujube juice yield. To extract cAMP, nanofiltration was employed.

The jujube fruits were soaked and pureed, and then subjected to enzyme hydrolysis, filtration, resin adsorption, and formic acid elution. Pure cAMP was obtained after purification via a silica gel chromatography column. Our method could maximize the extraction volume of jujube cAMP, and has great practical value for industrialized production.

54.2 Materials and Methods

54.2.1 Materials and Chemical Reagents

Equipment used in the study included a membrane separation machine (S500) and stainless steel tubular membrane (150/200; Hyflux, Shanghai), centrifugal spray-drying equipment (ZPG-25) (Changzhou Ernuo Drying Equipment, Changzhou, China), a freeze dryer (DW3; Eyela) and a 10-L stainless steel extraction irrigator (DC-TQ-10; Shanghai Dacheng Lab Equipment, Shanghai). The chromatography analysis was performed on an Agilent 1,100 high performance liquid chromatography (HPLC) instrument with a G1367D autosampler, G1312B binary gradient pump, G1316B column oven, and TC-C18 column (4.6 mm × 250 mm, 10 μm; Agilent Technologies, Palo Alto, CA).

Mildew- and pest-free dried Chinese date fruits were purchased in the market (Tianjin-jinghai). Cyclic AMP (purity 98 %) was from Merck & Co. (Germany). Pectinlyase (pectolyase), cellulose, and hemicellulose were purchased from Tianjin

Nuoao Technology Development (Tianjin, China). Strong-base anion exchange resin (205 × 7, 80–100 mesh) was purchased from the Chemical Plant of Nankai University (Tianjin, China). Silicone (200–300 mesh) was provided by Qingdao Haiyang (Qingdao, China).

54.2.2 Nanofiltration Obtained of Chinese Date cAMP Samples

Chinese date fruit were cleaned and soaked in water at room temperature overnight (the ratio of fruit to water was 1:6–8, w/w) and homogenized into a slurry. The slurry was transferred to the extraction tank with added pectinases and hydrolyzed at 50–60 °C for 2–3 h. After filtering, the date juice was retrieved. The juice was transferred to the mixing tank to be mixed with resin.

After specific adsorption and filtration, the resin (i.e., containing cAMP) was washed with water, with a volume ~two-fold larger than that of the resin, to remove impurities. This was further purified through the resin column, and eluted with 0.5 M formic acid at the flow rate 1.8–2.2 mL/min. The eluent (formic acid) was detected with a UV scanner. When the absorption peak of the eluent was at 256–258 nm and the area of the absorption peaks <0.1, this indicated that the targeted product had been eluted cleanly with the formic acid. The formic acid eluent was filtered through the nanofiltration system at the flow rate of 30–35 L/h at 0.5–1 Mpa pressure. The intercepted components contained the targeted products and other compounds with a relative molecular weight of 100–500. The cAMP was trapped by nanofiltration and the obtained products were concentrated using a rotary evaporator. When the volume of the concentrate was 10 % of the formic acid elution, the operation was stopped.

After centrifugation and spray drying, the crude cAMP was harvested. The crude powder was dissolved in methanol and separated in the silica gel chromatography column, using a mixture of methanol, ethyl acetate, and benzene as the elution solvent. The colorless elution was concentrated to dryness, and dissolved with an appropriate amount of water. After freeze-drying, the pure cAMP was obtained.

54.2.3 Test Method

The protocol for HPLC was in accordance with a previously published method 17. The chromatography column used octadecylsilane-bonded silica as filler. The mobile phase was 0.05 M potassium dihydrogen phosphate (containing 0.01 M tetrabutylammonium bromide, pH 4.3 adjusted with phosphoric acid)–acetonitrile, 85:15 (v/v). The detection wavelength was 258 nm.

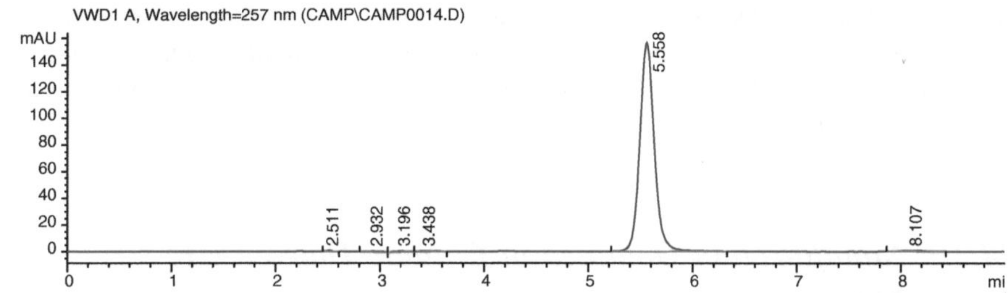

Fig. 54.1 A representative chromatogram of cAMP in standard solvent

To prepare the standard solution, 10 mg cAMP standard was dissolved in 5 mL water in a 10-mL volumetric flask and 1 mL of 1 M hydrochloric acid solution was added. This was placed in a 42 °C water bath for 30 min, then cooled to room temperature and the pH was adjusted to 7.0 with sodium hydroxide. The result was diluted with water to a final concentration of 0.2 mg/mL. A 20-μL sample was injected into the chromatography column and the main peak and the adjacent peaks of impurities were monitored to meet the requirements (Fig. 54.1).

To prepare the test sample, the date extract (above) was weighed and diluted with water to a final concentration of 0.1 mg/mL. A 20-μL sample was injected into the chromatography column at a flow rate of 1.0 mL/min. The chromatogram was recorded and the amount of cAMP in the sample was calculated using external standard calibration and the peak area.

54.2.4 Optimization of Composite Enzyme Ratio and Conditions of Enzymatic Hydrolysis Reaction

A composite of pectinase, cellulase, and hemicellulose enzymes were added to date pastes and the proportion of complex enzyme activity, enzymatic hydrolysis reaction temperature, solid–liquid ratio, enzyme dosage, and pH value were optimized via an orthogonal array test.

54.3 Result

54.3.1 Extraction of Chinese Date Fruit cAMP and the Influences of Enzyme Types on the Extraction Efficiency of cAMP

Under the standardized conditions, solid–liquid ratio 1:10, extraction temperature 50 °C, extraction time 3 h, pH 6.0, and 5 g/L enzymes from dates juice, the cAMP extraction efficiency of different enzymes (pectinlyase, cellulase, hemicellulase,

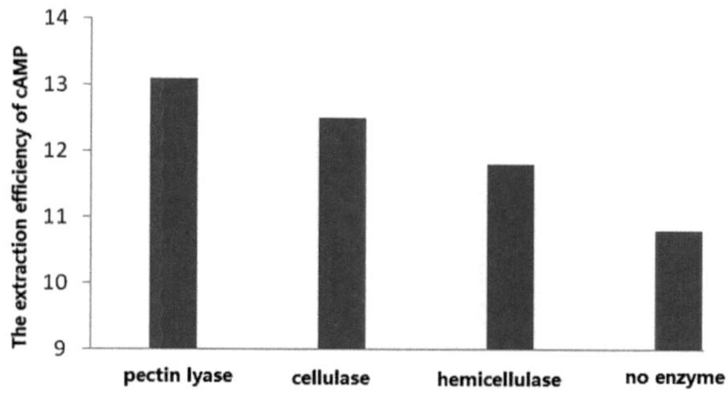

Fig. 54.2 Comparison of pectin lyase, cellulase, hemicellulase, or no enzyme on cAMP extraction efficiency

or none) was compared (Fig. 54.2). The results showed that under identical conditions pectinase, cellulase, and hemicellulase, in that order, improved the cAMP extraction rate.

54.3.2 Optimal Enzyme Ratio

Although cAMP extraction efficiencies of pectinase, cellulase, and hemicellulase were tested separately, theoretically the three enzymes might perform collaboratively because of their related molecular structures. To determine the best ratio of combinations of these enzymes, an $L_9(3^3)$ orthogonal array was used (Table 54.1). The optimal ratio was determined to be pectin lyase:cellulase:hemicellulase = 4:2:1 (Table 54.2). The influence of pectin lyase was greater than that of cellulase, and the influence of cellulase was greater than that of hemicellulase.

Table 54.1 The factor level table of the orthogonal array test

	Spiked amounts (μg)		
Level	A: Pectin lyase	B: Cellulase	C: Hemicellulase
1	100	50	25
2	200	100	50
3	300	150	75

Table 54.2 Orthogonal array test results

Test[1] no.	A: Pectin lyase	B: Cellulase	C: Hemicellulase	cAMP (mg/100 g)
1	1	1	1	14.3
2	1	2	2	14.8
3	1	3	3	14.7
4	2	1	2	15.4
5	2	2	3	15.6
6	2	3	1	15.5
7	3	1	3	14.9
8	3	2	1	15.1
9	3	3	2	15.0
K1	43.8	44.6	44.9	–
K2	46.4	45.5	45.2	–
K3	45.1	45.2	45.2	–
k1	14.6	14.9	15.0	–
k2	15.5	15.2	15.1	–
k3	15.0	15.1	15.1	–
Range	5.26	1.76	0.38	–
Optimum	A2	B2	C2	–
Major factor	1	2	3	

[1] *Experimental conditions* temperature, 50 °C; *time* 3 h; *solid–liquid ratio* 1:10; *pH value* 6.0

54.3.3 Optimal Experimental Conditions for Enzymatic Hydrolysis

The main factors affecting enzymatic hydrolysis were pH, temperature, solid–liquid ratio, and enzyme dose. Based on the results of the experiments, an $L_9(3^4)$ orthogonal table was used to conduct an orthogonal array test for selecting the optimal reaction temperature, solid–liquid ratio, enzyme dose, and pH value. We found that the influence of these factors on cAMP extraction efficiency, ranked high to low, was: pH > temperature > enzyme dose > solid–liquid ratio (Table 54.3). Determined by horizontal trend analysis, the optimal temperature was 50 °C, solid–liquid ratio 1:8, enzyme dose 0.5 %, and pH 5.5. By comprehensive analysis of the orthogonal

Table 54.3 Factor level table of orthogonal array test

Level	A: Temperature (°C)	B: Solid–liquid ratio (w/v)	C: Enzyme dose (%)	D: pH
1	45	1:6	0.25	5.0
2	50	1:8	0.5	5.5
3	55	1:10	0.75	6.0

Table 54.4 Results of the orthogonal array test

Test no.	A: Temperature	B: Solid–liquid ratio	C: Enzyme dose	D: pH	cAMP (mg/100 g)
1	1	1	1	1	14.7
2	1	2	2	2	15.4
3	1	3	3	3	15.2
4	2	1	2	3	15.6
5	2	2	3	1	15.1
6	2	3	1	2	15.6
7	3	1	3	2	15.3
8	3	2	1	3	15.4
9	3	3	2	1	15.0
K1	45.3	45.6	45.7	44.8	–
K2	46.2	45.9	45.9	46.2	–
K3	45.7	45.8	45.6	46.2	–
k1	15.1	15.2	15.2	14.9	–
k2	15.4	15.3	15.3	15.4	–
k3	15.2	15.3	15.2	15.4	–
Range	1.92	0.55	0.67	2.86	–
Optimum	A2	B2	C2	D2	–
Major factors	2	4	3	1	–

array test results, the ultimate optimal conditions were A2B2C2D2. Namely, the temperature of 50 °C, the solid–liquid ratio of 1:8, enzyme dose of 0.5 % and initial pH of 5.5 were fixed (Table 54.4).

54.3.4 Verification Experiment and Determination of Enzymatic Hydrolysis Process

Under the optimal conditions determined by theoretical analysis (50 °C, solid–liquid ratio 1:8, enzyme dose 0.5 %, and pH 5.5), experiments to verify were conducted. The proportion of pectinase, cellulase, and hemicellulase with the best activity was 4:2:1 and the reaction time was 3 h. Under these conditions, the cAMP content in the extraction was 15.7 mg/100 g.

Thus the date juice extraction process was finalized as: clean date fruits and keep in water at room temperature overnight (ratio of fruit to water 1:8), blend in a beating, extract with the multifunction extractor, add 2.5 g/L of enzyme mixture (pectinase:cellulase:hemicellulose = 4:2:1), hydrolyze for 3 h at 50 °C, filter, and harvest the juice.

54.3.5 The Determination of Crude cAMP

Based on the enzymatic hydrolysis process described above, 30 % crude cAMP was harvested from Chinese date juice in the three experimental batches after adsorption, elution, nanofiltration, and concentration. The cAMP contents were determined by HPLC (Table 54.5 and Fig. 54.3). The retention time of the absorption peak of the sample at 5–5.1 min was the same as that of the standard solvent (Figs. 54.3 and 54.1, respectively). The mean cAMP content was 36 % using the enzyme mixture (Table 54.5) and the cAMP content was 30 % after nanofiltration and concentration when only pectinase was applied, as reported by others. The extraction efficiency increased to 20 % in our study.

54.3.6 Purification Using Silica Gel Column Chromatography

Silica gel with the mixed solvent (the ratio of methanol: ethyl acetate: benzene was 3:1:1) was stirred into a semi-fluid, and then loaded into the column (5 cm × 30 cm). Quantitative crude powder was dissolved in methanol and injected at the top of the silica gel column, which was eluted with the mixed solvent. The colorless elution was collected, and concentrated to dryness. The harvested sample was dissolved with an appropriate amount of water and freeze-dried to obtain the purified sample powder.

Table 54.5 The results of 30 % crude cAMP

Batch	Reference peak area	Sample peak area	Reference content (mg/mL)	Sample content[1] (%)
1	5316430	3618637	0.1	34
2	5342834	3717280	0.1	35
3	5548091	4470707	0.1	40

[1] Mean, 36 %

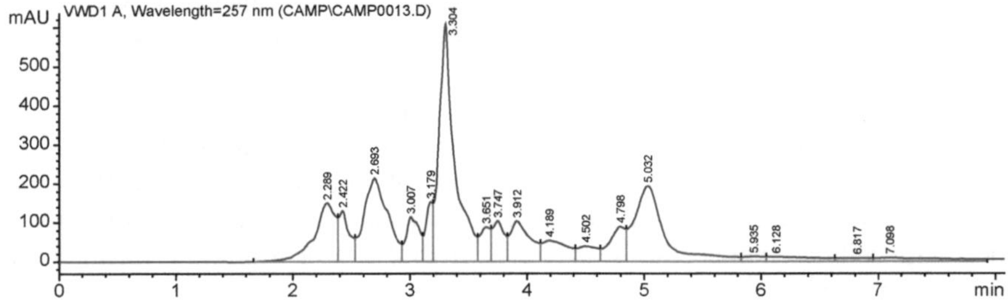

Fig. 54.3 A representative chromatogram of the cAMP in a single sample

Table 54.6 The results of three batch samples after purification

Batch	Reference peak area	Sample peak area	Reference content (mg/mL)	Sample content[1] (%)
1	5484567	5382273	0.1	98
2	5348280	5144848	0.1	96
3	5348015	5186725	0.1	97

[1] Mean, 97 %

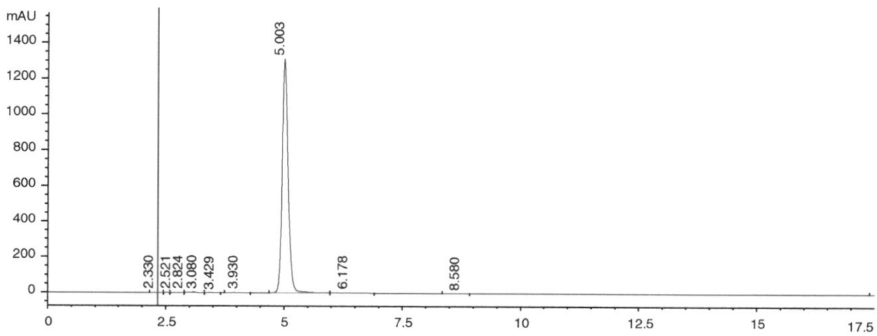

Fig. 54.4 Chromatogram of the crude cAMP after purification via silica gel column chromatography

The cAMP content of the samples was determined using HPLC (Table 54.6 and Fig. 54.4). The average cAMP content of three batches reached 97 % after purification via silica gel column chromatography, proving that purification was excellent.

54.4 Discussion

Because of the instability of cAMP in ethanol, the alcohol precipitation method is not suitable for extraction of cAMP [10]. Published methods for extracting cAMP have included microwave-assisted extraction [13], water extraction [11], and ultrasonic extraction [12]. All these methods require tremendous energy consumption and time. In the present study, strongly basic anion exchange resin was mixed with date juice to make the functions of specific exchange and adsorption, after. Cyclic AMP combined with resin was eluted with formic acid and the resultant effluent was filtered by membrane separation. Finally, cAMP was harvested. Using an optimized enzymatic hydrolysis process, the juice yield increased to 70 % and the cAMP content increased from 10.8 mg/100 g (reported for the original method) [14] to 15.7 mg/100 g.

According to the method published previously [15], components with the relative molecular weight of 100–500 Da were present in the concentrate after filtration through a 150 Da nanofiltration membrane. Water, formic acid, and other small molecular components were satisfactorily separated from the filtrate. The target component and filtrate was analyzed using HPLC. The results showed that the product of the nanofiltration was the same substance as the test standard. Cyclic AMP and other substances with the relative molecular weight of 100–500 Da were included in the final solution. Although many miscellaneous peaks were present, the crude cAMP (>30 %) was obtained after spray-drying.

In the present study, the application of modern enzyme engineering technology and membrane separation technology retained the activity of the active ingredient and evaluated product yield. The crude cAMP had the same absorption peak as its test standard, and the average crude cAMP content was 36 %. Compared with treatment with a single enzyme (pectinase) and nanofiltration membrane concentration, the extraction efficiency was evaluated at 20 %. After purification via silica gel column chromatography, the average cAMP content per sample reached 97 %. This is slightly lower than the 98.43 % reported by Mi et al. [16] who used ion exchange chromatography to extract cAMP from *Ziziphus* fruits and purified by silica column chromatography.

Acknowledgments This work was supported by the National High-Tech Research and Development Plan ('863' Plan) (no. 2013AA102106), the National Natural Science Fund (31101219), and the programme for Changjiang Scholars and Innovative Research Team in University (IRT1166).

References

1. Fan J, Lu L, Shang HW (2003) Progress on research and development of Zizyphus Jujuba. Food Sci 24:161–163
2. Sutherland EW, Alan RG, Butcher RW (1968) Some aspects of the biological role of adenosine 3',5'-monophosphate (cyclic AMP). Circulation 37:279–306
3. Botsford JL, Harman JG (1992) Cyclic AMP in prokaryotes. Microbiol Mol Biol Rev 56:100–122
4. Daniel PB, Walker WH, Habener JF (1998) Cyclic AMP signaling and gene regulation. Annu Rev Nutr 8:353–383
5. Park D, Kim C, Cheon M et al (1997) cAMP and protein kinase C elevate LH beta mRNA levels by activating transcription rather than stabilizing mRNA in rat pituitary cells. Mol Cells 7:98–103
6. Withers DJ (1997) Signalling pathways involved in the mitogenic effects of cAMP. Clin Sci (Lond) 92:445–451
7. Dun SX (1994) The nutritional and medicinal value of jujube and Chinese wolfberry. Food Research and Development, 3, pp 34–37. http://epub.cnki.net/KNS/brief/result.aspx?dbPrefix=CJFQ
8. Hanabusa K, Cyong J, Takahashi M (1981) High-level of cyclic AMP in the jujube plum. Plants Med 42:380–384
9. Liu MJ, Wang YH (1991) Study on AMP contents of *Zizyphus jujube* Mill, *Zizyphus spinosus* Hu and other twelve horticultural plants. J Hebei Agric Univ 14:20–22
10. Cyong JC, Kiyomichi H (1980) Cyclic adenosine monophosphate in fruits of Zizyphus jujuba. Photochemistry 19:2747–2748

11. Wang HH, Lin YL, Liao JF et al (2003) The anti-amnesic effects of the water extract of Jujubae fructus. Chin Pharm J (Taipei) 55(5):361–369
12. Cui ZQ, Meng XJ (2006) Study on ultrasonic extraction of cAMP from winter-dates [J]. Food Sci Tech 12:46–49
13. Cui ZQ, Meng XJ (2007) Study on microwave-assisted extraction of cAMP from winter-dates. Food Sci Tech 28:163–166
14. Zhang MJ, Li W, Pang XM et al (2012) Study on extraction cAMP from dates and its quantitative determination. Food Ferment Ind 38:228–231
15. Wang CX, Lu FP, Liu YH et al (2011) The application of nanofiltration in the extraction of date cyclic adenosinemonophosphate (cAMP). Food Ferment Ind 2:190–195
16. Mi D, Wang Y, Li MR (2007) Research on extraction of cAMP from fruits of Ziziphus. J Shanghai Normal Univ (Nat Sci) 36:77–79

Chapter 55
Effect of Manchurian Walnut Extracts on Cancer Cells Proliferation

Changcai Zhao, Xing Niu, Rui Huang, Jiali Dong, Yuyin Li and Aipo Diao

Abstract Extracts from the Manchurian walnut have the antibacterial, anti-inflammatory, and anticancer properties. In this study, different extracts were obtained from Manchurian walnut peels, flowers, and leaves by the Soxhlet extraction method, and used to study the antitumor activity. MTT assay shows that the ethyl acetate phase of the peels inhibits cell proliferation of colon cancer cell SW480 with an IC_{50} at 56 μg/mL. Hoechst 33258 staining illustrates that the ethyl acetate phase of the peels induces cell apoptosis of SW480 cells. These results indicate that the ethyl acetate of Manchurian walnut peels inhibits cell proliferation and promotes cell apoptosis of cancer cells.

Keywords Manchurian walnut · SW480 cells · Cell proliferation · Apoptosis

55.1 Introduction

Colorectal cancer is one of the most common malignancies in the world with approximately one million new cases diagnosed every year [1]. The high incidence of the disease and its associated morbidity and mortality has brought colorectal cancer to the center of cancer research, namely its etiology, diagnosis, and treatment [2]. The current treatment of colorectal cancer mainly depends on surgery, chemotherapy, radiotherapy, and targeted therapy. However, the curative effect of these treatments are less than satisfactory [3]. Clearly, development of novel approach for colorectal cancer treatment is highly warranted. In China, Traditional Chinese

C. Zhao · X. Niu · R. Huang · J. Dong · Y. Li (✉) · A. Diao (✉)
Key Laboratory of Industrial Fermentation Microbiology of the Ministry of Education, College of Bioengineering, Tianjin University of Science and Technology, Tianjin 300457, China
e-mail: liyuyin@tust.edu.cn

A. Diao
e-mail: diaoaipo@tust.edu.cn

T.-C. Zhang and M. Nakajima (eds.), *Advances in Applied Biotechnology*, Lecture Notes in Electrical Engineering 332, DOI 10.1007/978-3-662-45657-6_55

Medicine (TCM) has played a positive role in colorectal cancer treatment. There is a great need to establish effective herbal formula for colorectal cancer treatment [4].

Manchurian walnut (Juglans regia L.) belonging to the family Juglandaceae has a long history of use in traditional folk medicine for the treatment of various ailments [5]. Different parts of the tree are reported to possess antimicrobial, antiinflammatory, immunomodulatory, diuretic, laxative, and anticancer properties [6–9].

In this study, we evaluated the anticancer potential of the extracts from Manchurian walnut flowers, leaves, and peels, performing different methods of extraction and concentration, and explore which parts contain the effective anticancer ingredients.

55.2 Materials and Methods

55.2.1 Cells and Reagents

The human colon cancer cells SW480 were cultured in Dulbecco's modified Eagle's medium (DMEM) containing 10 % fetal bovine serum (FBS), 2 mM glutamine, 100 U/mL penicillin, and 100 mg/mL streptomycin (GIBCO-BRL, Gaithersburg, MD) at 37 °C in a humidified incubator with 5 % CO_2. The powder of Manchurian walnut flowers, leaves, and peels was dissolved in DMSO (dimethysulfoxide), then kept at −20 °C. 3-(4,5-dimethylthiazol-2-yl) -2,5-diphenyltetrazolium, 2-(4-Amidinoph-enyl)-6-indolecarbamidine dihydrochloride DMSO were purchased from Solarbio. Organic solvents were the petroleum ether, anhydrous ethanol, and ethyl acetate.

55.2.2 Extraction and Separation of Different Parts of the Manchurian Walnut

45 g powder of the manchurian walnut peels, flowers, or leaves was extracted using Soxhlet extraction method in different organic solvents (petroleum ether, anhydrous ethanol, and ethyl acetate). The extracts were then concentrated and three organic phase with different parts of the manchurian walnut were obtained, respectively. All concentrated phase were dissolved in DMSO for further test.

55.2.3 Cell Proliferation Assay

Cell proliferation was determined by MTT assay. The mitochondrial dehydrogenase in living cells reduced 3-(4, 5-dimethylthiazol-2-yl) -2,5-diphenyltetrazolium bromide, MTT (yellow) to formazan dyes (purple). SW480 cells were seeded onto

96-well plate, and treated with the manchurian walnut extracts in different concentration for 48 h. Each concentration set three vice-hole. After treatment with the extracts, 20 μL MTT solution (0.5 mg/mL in PBS) per well was added to 96-well plate and the plate was incubated at 37 °C for 4 h. Finally, 200 μL DMSO was added to the plate per well to dissolve the formazan dyes and the absorbance was measured at 570 nm by Microplate Reader.

55.2.4 Cell Apoptosis Assay

Cell apoptosis was determined by Hoechst 33258 staining. When the cells go into apoptosis, chromatin will condense, after Hoechst 33258 staining, the nucleus of normal cells showed normal blue under the fluorescence microscope, while the nucleus of apoptotic cells form dense stain or chunky dense stain, and some even show white. SW480 cells were seeded onto the cover slips, and then cells were treated with the extracts for 24 or 48 h. Then the cover slips were taken out and washed twice with PBS before fixed in 4 % paraformaldehyde for 5 min. After washing twice with PBS the cover slips were treated with 0.2 % Triton-100 for 10 min. The cover slips were soaked in blocking solution (3 % BSA) at room temperature for 30 min before washing twice with PBS. The cells on the cover slips were stained with Hoechst 33258 for 30 min in the dark at room temperature, then mounted with mounting medium after washing 3 times with PBS.

55.3 Results

55.3.1 Extraction of the Different Parts of Manchurian Walnut

The peels, flowers, and leaves of Manchurian walnut were extracted by Soxhlet extraction method, nine kinds of concentrated phases were obtained through different extraction processes, including peels P (I), P (II), P (III); flowers F (I), F (II), F (III); leaves L (I), L (II), L (III) (Fig. 55.1).

55.3.2 Effects of the Manchurian Walnut Extracts on the SW480 Cell Proliferation

The same amount of SW480 cells were seeded onto each well of 96-well plate and incubated at 37 °C for 48 h, cell proliferation was detected by MTT assay after the treatment with the extracts from the Manchurian walnut peels, flowers, and leaves, with adding the DMSO as the negative control.

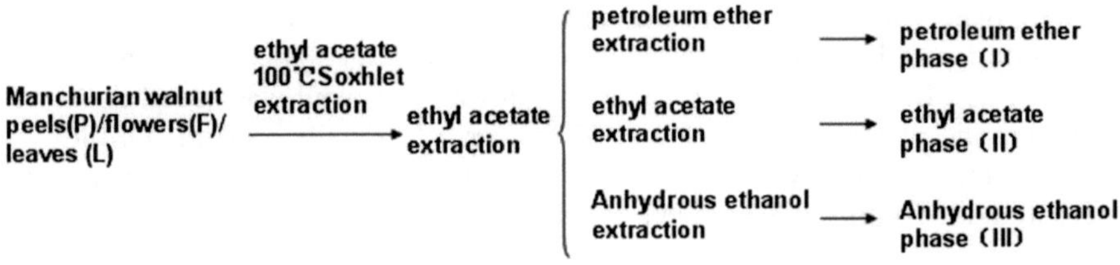

Fig. 55.1 The extraction process of Manchurian walnut

For the three kinds of concentrated phases of the Manchurian walnut peels, each of them had certain inhibition to the growth of SW480 cells, with the ethyl acetate phase P (II) inhibition was stronger than the other two phases, and the inhibitory remained at low concentration of 10 μg/mL (Fig. 55.2a). For the extracts of the Manchurian walnut flowers, the petroleum ether phase F (I) and ethyl acetate phase F (II) had the inhibition on the growth of SW480 cells at the concentration of 1,000 μg/mL, and the inhibitory effect of F (II) was stronger (Fig. 55.2b). For the extracts of the Manchurian walnut leaves, each phase had certain inhibitory effect on the growth of SW480 cells at the concentration of 1,000 μg/mL, with more inhibitory effect for L (II) (Fig. 55.2c). These results indicate that the ethyl acetate phase of the extracts from Manchurian walnut has stronger inhibitory effect on the growth of SW480 cells.

We further studied the inhibitory effect of the ethyl acetate phase of the extracts from the Manchurian walnut peels, flowers, and leaves on the growth of SW480 cells, and found that the ethyl acetate phase P (II) of peels had obvious inhibitory

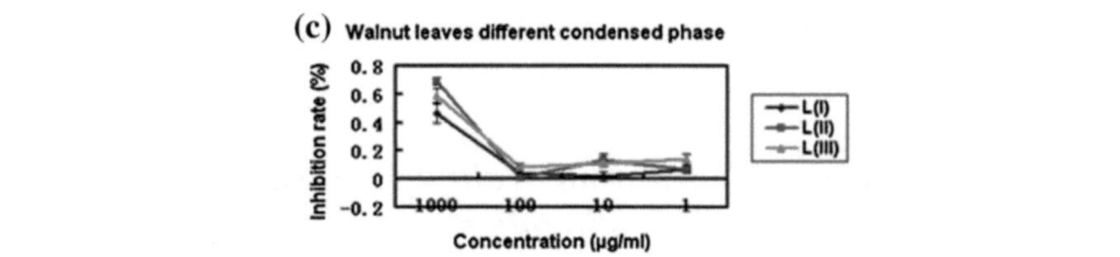

Fig. 55.2 Effects of the manchurian walnut extracts on the SW480 cells proliferation. The SW480 cells were treated for 48 h with Walunt peels different extracts (**a**), Walunt flowers different extracts (**b**), or with Walunt leaves different extracts (**c**) in different concentration

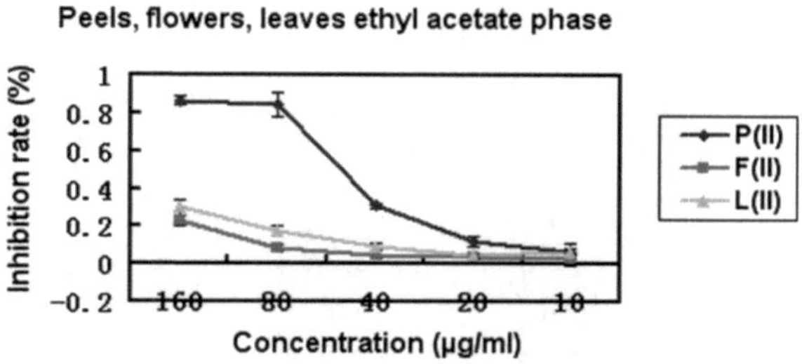

Fig. 55.3 Inhibitory effect of the ethyl acetate phase of the extracts from the Manchurian walnut peels, flowers, and leaves on the growth of SW480 cells

effect, and the ethyl acetate phase of leaves and flowers had low inhibition, particularly at low concentration. Moreover, based on the inhibition rate of 84 % for the peels P (II) at the concentrations of 80 μg/mL, and the inhibition rate of 31 % at 40 μg/mL, we calculated that the IC_{50} of the peels P (II) extract for the inhibition of SW480 cells growth is accurately 56 μg/mL (Fig. 55.3).

55.3.3 The Manchurian Walnut Extracts Induces Apoptosis of SW480 Cells

To test whether the ethyl acetate phase of the Manchurian walnut peel extracts induces cell apoptosis, SW480 cells were treated with different concentrations of the peels P (II) for 24 or 48 h, then stained with Hoechst 33258 to detect the apoptotic cells. With increasing concentration of P (II), the number of apoptotic cells with dense stain or chunky dense stain of nucleus increased significantly compared with the control, and the treatment of 80 μg/mL P (II) induces cell apoptosis obviously. Total cells are reducted in the view, clearly. These results indicate that the ethyl acetate phase of Manchurian walnut peel extracts could promote SW480 cells apoptosis (Fig. 55.4).

55.4 Discussion

This research got different condensed phase through the extract of the Manchurian walnut parts of the flowers, leaves, and peels (Fig. 55.1). Determined by MTT experiments showed that compared with flowers and leaves, after the ethyl acetate phase of the peels P (II) treaded SW480 cell, cell proliferation capacity was decreased obviously. In other words, the inhibition rate of P (II) was very high to SW480 cells. It was illustrated that P (II) could inhibit the proliferation of colon cancer cells (Figs. 55.2 and 55.3). Hoechst 33258 staining results showed that the

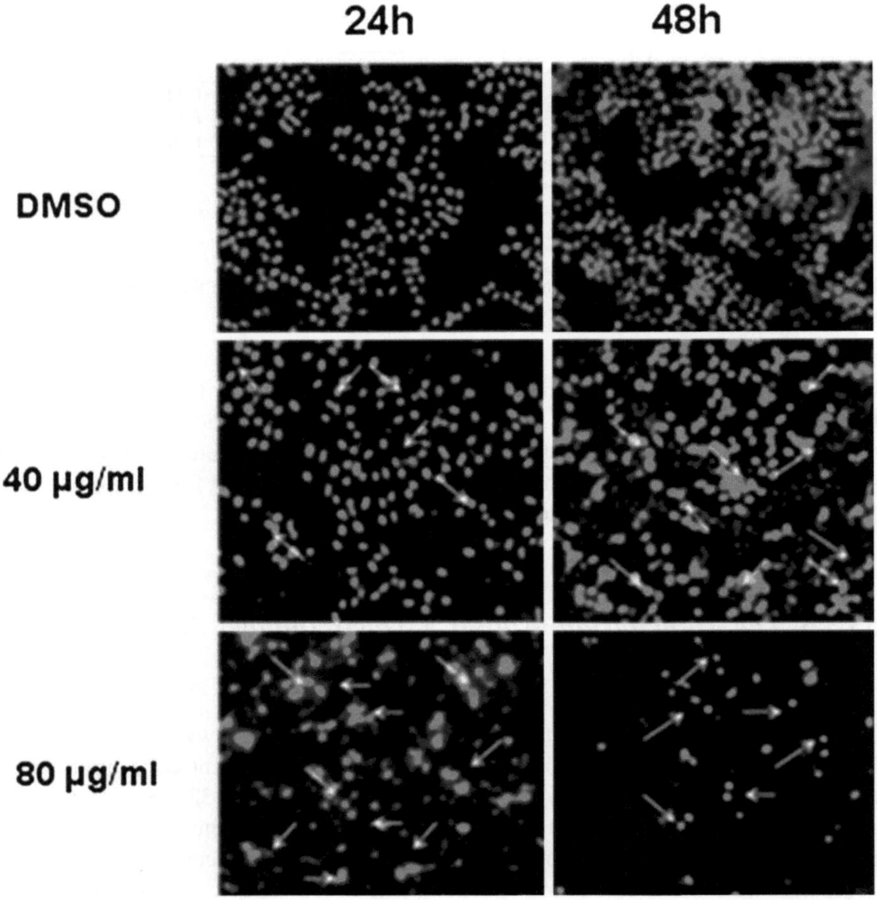

Fig. 55.4 The ethyl acetate phase of the Manchurian walnut peel extracts induces SW480 cell apoptosis

ethyl acetate phase of peel P (II) acting to cells, the amount and degree of colon cancer cells apoptosis significantly enhanced. The result showed that the ethyl acetate phase of peel P (II) could promote cell apoptosis (Fig. 55.4).

In traditional folk medicine, walnuts are commonly used in antibacterial, anti-inflammatory, treatment of cancer, contain biologically active ingredients which are mostly flavonoids. Juglone has cytotoxic properties when administered to cell cultures and it possesses antiviral, antibacterial, and antifungal properties [10]. Quinones include a wide range of phenolic compounds classification, which contains an array of pharmacological properties and is widely distributed in nature. Quinone-rich plants have been widely applied in traditional folk medicine to treat various diseases, including cancer [11]. The main components of the peel acetate phase P (II) are flavonoids, quinones, or others, that need further analysis to extract.

Ki67 is a nuclear protein associated with cell proliferation [12]. Its expression levels are positively correlated with some clinical–pathological variables in patients [13]. Ki67 serving as a mediator of malignant behavior in cancer suggests that inhibition of Ki67 can be incorporated into a novel cancer therapy [14]. Considering that the ethyl acetate of Manchurian walnut peels P (II) inhibits cell proliferation and promotes cell apoptosis of cancer cells, whether it inhibits Ki67 protein expression needs further study.

Acknowledgments This research is supported by the program for National college students' innovative entrepreneurial training plan (Item no. 201310057073) and Changjiang Scholars and Innovative Research Team in University (IRT1166).

References

1. Cantero-Muñoz P, Urién MA, Ruano-Ravina A (2011) Efficacy and safety ofintraoperative radiotherapy in colorectal cancer: a systematic review. Cancer Lett 306(2):121–133
2. Khuloud B, Jan S, Stefanie D et al (2012) The anticancer effect of saffron in two p53 isogenic colorectal cancer cell lines. BMC Complement Altern Med. doi:10.1186/1472-6882-12-69
3. Storli KE, Søndenaa K, Bukholm IR et al (2011) Overall survival after resection for colon cancer in a national cohort study was adversely affected by TNM stage, lymph node ratio, gender, and old age. Int J Colorectal Dis 26(10):1299–1307
4. Shan D, Bing H, Hongmei A et al (2013) Teng-Long-Bu-Zhong-Tang, a Chinese herbal formula, enhances anticancer effects of 5–fluor ouracil in CT26 colon carcinoma. BMC Complement Altern Med. doi:10.1186/1472-6882-13-128
5. Aithal KB, Rao SB, Sunil K et al (2012) Tumor growth inhibitory effect of juglone and its radiation sensitizing potential. In vivo and in vitro studies. Integr Cancer Ther 11(1):68–80
6. Bhatia K, Rahman S, Ali M et al (2006) In vitro antioxidant activity of Juglans regiaL bark extract and its protective effect on cyclophosphamide induced urotoxicity in mice. Redox Rep 11(6):273–279
7. Blumenthal M (1998) Walnut Hull. The complete german commission E monographs. American Botanical Council, Austin
8. Duke JA, Ayensu ES (1985) Medicinal plants of China. Ann Soc Belg Med Trop 65 (2):117–120
9. Bhargava UC, Westfall BA (1968) Antitumor activity of Juglansniga (black walnut) extractives. J Pharm Sci 57(10):1674–1677
10. Paulsen MT, Ljungman M (2005) The natural toxin juglone causes degradation of p53 and inducesrapid H2AX phosphorylation and cell death in human fibroblasts. Toxicol Appl Pharmacol 209(1):1–9
11. Babula P, Adam V, Havel L et al (2007) Naphthoquinones and their pharmacological properties. Ceska Slov Farm 56(3):114–120
12. Iatropoulos MJ, Williams GM (1996) Proliferation markers. Exp Toxicol Pathol 48 (2–3):175–181
13. Balleine RL, Webster LR, Davis S et al (2008) Molecular grading of ductal carcinoma in situ of the breast. Clin Cancer Res 14(24):8244–8252
14. Fang L, Cheng Q, Li W et al (2014) Antitumor activities of an oncolytic adenovirus equipped with a double siRNA targeting Ki67 and hTERT in renal cancer cells. Virus Res 181:61–71

Chapter 56
Extraction and Purification of Lumbrokinase from the "Ohira the 2nd" Earthworm

Tianjun Li, Jian Ren, Tao Li and Yingchao Wang

Abstract Lumbrokinase is an ideal drug hypertension to cardiovascular diseases. This essay discussed the purification process of lumbrokinase and decided its molecular weight with "Ohira the 2nd" earthworms as raw material. After ammonium sulfate precipitation, the crude enzyme was prepared, which was carried through chromatography of Sephadex G-75 and then DEAE-52. Certain high-purity lumbrokinase components were obtained, whose protein molecular weight was 33KD and specific activity was 1,784 U/mg, after SDS-PAGE test. These lumbrokinase components showed a direct fibrinolytic and plasminogen activation dual role, and a strong thermal stability and pH stability, which was good to thrombolysis.

Keywords "Ohira the 2nd" earthworms lumbrokinase · Purification · Detection

56.1 Introduction

Cardiovascular diseases including hypertension, cerebral hemorrhage, coronary heart disease, and cerebral infarction are frequently occurring in modern society and very harmful to human health. In China, every year people who died from cerebrovascular are more than 50 % of total deaths. Now the drugs for the clinic treatment of cardiovascular diseases have some drawbacks, such as expensive palliatives, easy to lose activity, and easily leading to bleeding. So looking for various sources of thrombolytic drugs is considered necessary. PPA which can activate the anticoagulant system, fibrinolysis, platelet aggregation, and improve microcirculation, is an ideal drug for the cardiovascular diseases treatment.

In recent years, earthworms production had greatly increased due to the improvement of earthworm breeding techniques, but most of the studies only

T. Li (✉) · J. Ren · T. Li · Y. Wang
Department of Basic Science, Tianjin Agriculture University, Tianjin 300384, China
e-mail: litianjun65@163.com

T.-C. Zhang and M. Nakajima (eds.), *Advances in Applied Biotechnology*,
Lecture Notes in Electrical Engineering 332, DOI 10.1007/978-3-662-45657-6_56

focused on the earthworms feeding, and studying its efficacy is rarely worth. It has been demonstrated that the development of lumbrokinase as a drug would be more valuable and the market value is much higher than the production of feed earthworms. However, few studies focused on the separation and purification of lumbrokinase from the earthworms have been reported up to now. Therefore our study focusing on the development of purification process of lumbrokinase has high academic, economic, and practical values.

56.2 Materials and Methods

56.2.1 Materials

"Ohira the 2nd" Earthworms are obtained from animal health, Ltd. of Tianjin.

pH 7.75 PBS solution; 10 % ammonium persulfate; 1.30 % gel stock solution; 1.5 M Tris–HCl separation gel buffer pH 8.8 (4×); 1.0 M Tris–HCl stacking gel buffer pH 6.8 (8×); electrode buffer pH 8.3 (1×).

56.2.2 Experimental Methods

56.2.2.1 Preparation of Crude Enzyme

Put fresh earthworms into the water for a day to spit out the dirt of the belly and then put them into the refrigerator. 1 kg frozen earthworms was suspended in 1 L saline and homogenized at room temperature. Afterward the mixture was stirred for 10 h and then filtered with gauze. The supernatant was obtained after centrifugation at a speed of 3,000 r/min for 30 min. Saturation of ammonium sulfate was prepared according to the table, and a certain amount of ammonium sulfate was added to make the saturation 20S, and then allowed to stand overnight at 4 °C, supernatant was centrifuged at 3,000 r/min for 30 min, and save the precipitate. Supernatant was added in an amount ammonium sulfate, respectively the mixture was kept at room temperature for 10 h and then centrifuged at 3,000 r/min for 30 min. The crude protein pellets was resolubilized with buffer and dialyzed with dialysis membrane over night. Then the crude enzyme solution was kept in the refrigerator for the subsequent analyses.

Purification

56.2.2.2 Lumbrokinase

The crude enzyme was dissolved in 0.02 M, pH 7.8 Tris–HCl buffer solution, and centrifuged. The supernatant was collected by Sephadex G-75 gel filtration chromatography. The sample volume was 3 ml, flow rate 20 ml/h, 8 min/tube, each tube were measured protein absorption A280 value, the value of the collection to the A280 < 0.003. The active moiety was collected.

The solution of enzyme is activated in the previous step for ion exchange chromatography DEAE-52. Then the active enzyme solution was eluted with 0.1 and 0.2 mol/L, NaCl solution. The constant pump flow rate is set as 20 mL/h. Per tube was 4 ml of the collection of fluid. The protein in the eluant was detected by measuring the optical density (OD) at 280 nm. Determination of protein peaks was collected and some activity of the purified solution is Lumbrokinase.

56.2.2.3 Lumbrokinase Purity Testing of the Finished Product: The Use of SDS-PAGE.

.

56.3 Results and Discussions

56.3.1 Separation and Purification of Lumbrokinase

56.3.1.1 Ammonium Sulfate Fractionation and Membrane Separation

After extracting earthworm tissue by ammonium sulfate precipitation, desalt by dialysis with a dialysis membrane, and then freeze-dry crude enzyme. Relationship between the amounts of precipitation of ammonium sulfate saturation is shown in Fig. 56.1.

Figure 56.1 shows that with increasing saturation of ammonium sulfate, the protein precipitate also increases. However after the ammonium sulfate saturation reaches a peak of about 70S, although the amount of saturated ammonium sulfate increases the amount of precipitated protein decreases. Thus fractional ammonium sulfate precipitation occurs at 20–70S ammonium sulfate saturation.

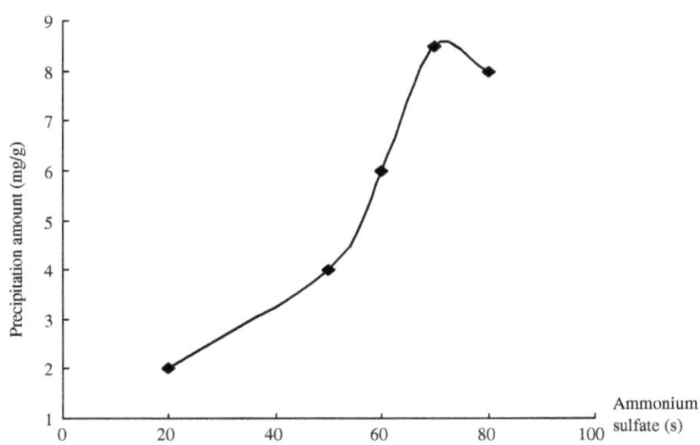

Fig. 56.1 Ammonium sulfate precipitation *curve*

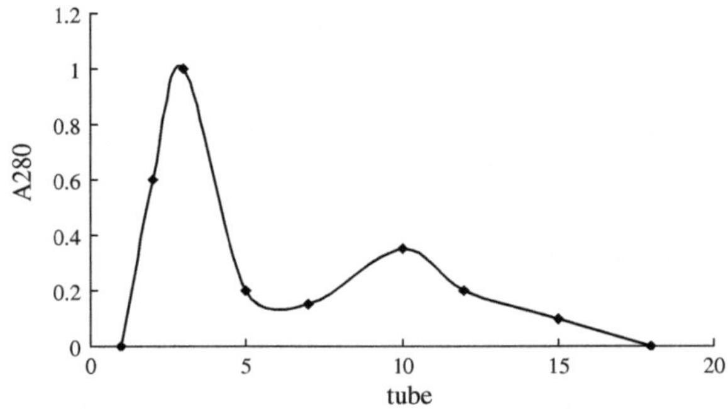

Fig. 56.2 Lumbrokinase sephadex G-75 gel filtration chromatography

56.3.1.2 Sephadex G-75 Gel Filtration Chromatography

The crude enzyme was dissolved in 0.02 M, pH 7.8 Tris–HCl buffer solution, and centrifuged. The supernatant was collected by Sephadex G-75 gel filtration chromatography with flow rate of 20 mL/h. Twenty tubes were collected at a rate of 8 min/tube. Each tube was measured for protein absorption A280 values and plotted in figure and protein absorption peaks were determined.

The crude enzyme solution in PPA after Sephadex G-75 gel filtration chromatography, protein absorption was measured for its values at 280 nm, the two proteins showed an absorption peak appears, the enzyme activity occurs (Fig. 56.2). At the peak of the 2, 6 to 20 collect all the enzyme. The yield was 75 %, and the resulting eluant was freeze-dried to obtain the semi PPA.

56.3.1.3 DEAE-52 Fiber Resin Ion Exchange Chromatography

The active enzyme solution in the previous step for ion exchange chromatography DEAE-52. Then the active enzyme solution was eluted with 0.1 and 0.2 mol/L, NaCl solution. Flow rate of 20 ml/h (6 s/d), the distribution of the collection, each tube 5 ml, protein absorption detector A280 value and select a certain number of tubes was measured to determine the dynamic range of peak activity.

Shown in Fig. 56.3, the crude enzyme solution in PPA after DEAE-52 ion exchange chromatography resin fiber, its absorption of protein was measured at

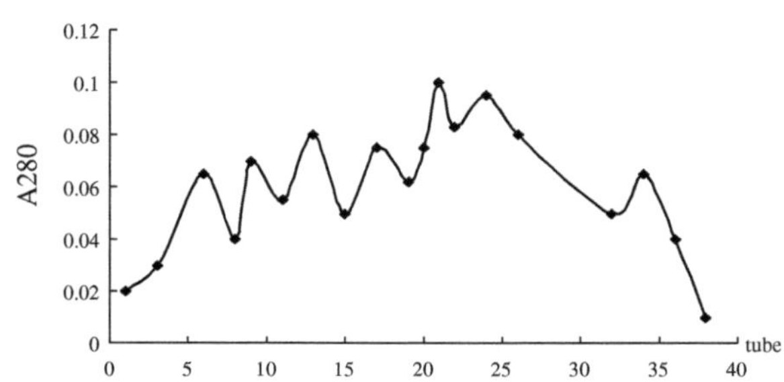

Fig. 56.3 Lumbrokinase DEAE-52 fiber resin ion exchange chromatography

280 nm. The results show peak seven protein, enzyme activity in two protein absorption peak to peak of the first six, 10–34 to collect all of the enzyme solution obtained was 60 %. The resulting enzyme solution was freeze-dried to give pure PPA.

56.3.2 Lumbrokinase Activity Assay

The purity of the different purification steps each PPA for detection of enzyme activity obtained as shown in Table 56.1.

From Table 56.1, the PPA obtained after ammonium sulfate precipitation of specific activity of 1,216 U/mg lumbrokinase, purification factor of 3.3 times. After Sephadex G-75 gel filtration, we can get live as 1,623 U/mg than the PPA, purification factor of 4.47 times. After DEAE-fiber resin ion exchange chromatography, we can obtain a specific activity of 1,784 U/mg the PPA, and less than half of the total energy loss. Separation and purification of such route is feasible, but also for the separation and purification after lumbrokinase further study provides a theoretical basis.

56.3.3 Determination of Protein Lumbrokinase Purity and Molecular Detection

After SDS-PAGE, an electrophoretic separation obtained with Fig. 56.4: The protein bands with a standard one, the strip two by ion exchange with PPA, i.e., pure belt strip over molecular sieves three chromatography lumbrokinase band, band 4 with crude enzyme by ammonium sulfate precipitation. Standards-based protein obtained the separation and purification of the protein molecular weight lumbrokinase 33KD.

Table 56.1 The activity of different purity lumbrokinase

Step	Total protein/mg	Total activity/U	U ratio live (U/mg)	Purification factor	Yield (%)
Homogenates	8,800	3,200,000	363	1	100
Ammonium sulfate	2,500	3,040,000	1,216	3.3	95.1
Precipitation					
Sephadex G-75	1,500	2,435,000	1,623	4.47	76.1
DEAE ion exchange	900	1,606,500	1,784	4.91	50.2

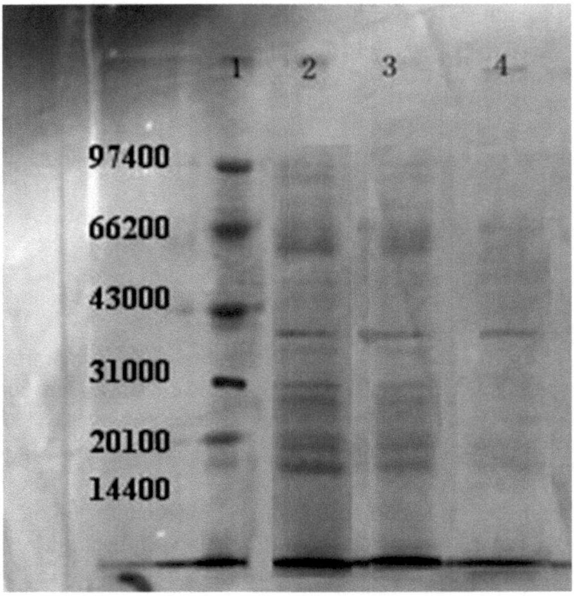

Fig. 56.4 Lumbrokinase electrophoresis

56.4 Conclusions

Among ammonium sulfate precipitation, membrane separation, Sephadex G-75 gel filtration chromatography, and DEAE-fiber resin ion exchange chromatography, lumbrokinase thrombolytic drug component was obtained in high purity by SDS-PAGE electrophoresis in a single band of protein molecular weight 32 ± 2 KD and a specific activity of 1,784 U/mg. It has a dual role of direct fibrinolytic and plasminogen activation, and it has strong thermal stability and good pH stability.

Chapter 57
Study on Green Crystallization Process for L-glutamic Acid Production

Zhi-hua Li, Cheng-lin Zhang and Qing-yang Xu

Abstract To investigate the production of L-glutamic acid with high yield as well as improved purity, the optimization of crystallization process was conducted. During this study, various physicochemical parameters (e.g., initial temperature, cooling rate, acid adding rate, ultrasonic time, and stirring speed) of concentrated isoelectric crystallization method were evaluated to optimize the yield and purity of L-glutamic acid. The optimum crystallization parameters are as follows: acid adding rate 0.5 mL/min, ultrasonic time 10 min, and stirring speed 200 rpm. The yield of L-glutamic acid at optimal condition was 95.4 %, attaining a 6.5 % growth. The purity of crystallized product exceeded 99 %, giving a rise of 4 %. The optimal crystallization process with higher yield and improved purity reduces the energy consumption and thus promotes sustainable development.

Keywords L-glutamic acid · Metastable phase · Crystallization · Optimization · Green

57.1 Introduction

L-glutamic acid (GA) is one of the most important amino acid products, with a wide range of industrial applications including food, medical, and chemical industries [1]. L-glutamic acid has the greatest commercial demand of single amino acid, exceeding

Z.-h. Li · C.-l. Zhang · Q.-y. Xu
College of Biotechnology, Tianjin University of Science and Technology, Tianjin 300457, China

Z.-h. Li · C.-l. Zhang · Q.-y. Xu
National and Local United Engineering Lab of Metabolic Control Fermentation Technology, Tianjin 300457, China

Z.-h. Li · C.-l. Zhang · Q.-y. Xu (✉)
Tianjin Engineering Lab of Efficient and Green Amino Acid Manufacture, Tianjin 300457, China
e-mail: xuqingyang@tust.edu.cn

T.-C. Zhang and M. Nakajima (eds.), *Advances in Applied Biotechnology*, Lecture Notes in Electrical Engineering 332, DOI 10.1007/978-3-662-45657-6_57

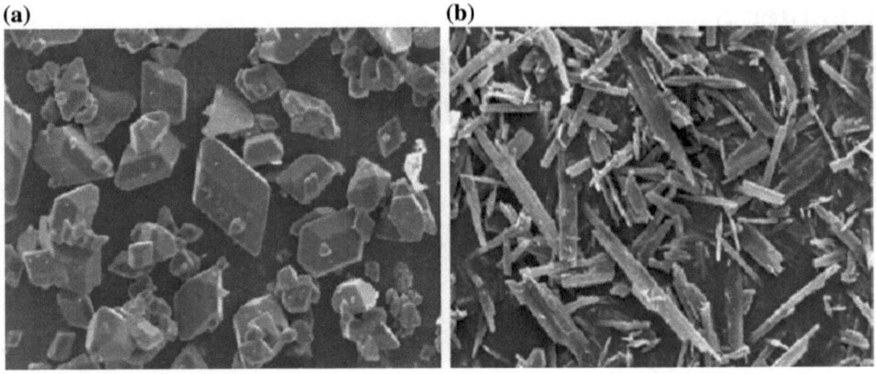

Fig. 57.1 Images of L-glutamic acid crystals: **a** alpha form, **b** beta form

1.5 million tons per year on a global basis. Therefore, many efforts have been devoted to the production of L-glutamic acid especially with cheaper available raw materials. Industrially L-glutamic acid is generally produced by fermentation. L-glutamic acid crystals have two known polymorphs with contrasting morphologies, the metastable alpha form has a prismatic morphology and the stable beta form shows needlelike platelets [2]. Alpha form crystals are preferred because these are easier to filter when compared to the beta crystals. Apart from the chemical property of material, the polymorph and size distribution of crystals are also affected by other external factors such as temperature, supersaturation ratio, stirring speed, impurity, etc. Currently, two methods namely isoelectric crystallization method and ion-exchange method are normally used for L-glutamic acid recovery from fermentation broth [3]. Isoelectric crystallization method was first developed by the Ajinomoto Company. Compared to ion-exchange method, the yield of L-glutamic acid manufactured by isoelectric crystallization method is low (generally 88 %), due to crystal transformation. However, isoelectric crystallization can reduce material and energy consumption and cut down on water pollution in the environment (Fig. 57.1).

In this paper, the effects of temperature (initial temperature and cooling rate), surface tension of fermentation broth, stirring speed and dropping speed of acid on isoelectric crystallization method were investigated. In additional, L-glutamic acid crystallization under optimal process was performed.

57.2 Materials and Methods

57.2.1 Materials

The fermentation broth was supplied by Ningxia Shenghua Milai Bioengineering Co. Ltd. Sodium hydroxide (AR grade) was purchased from Tianjin Northern Tianyi Chemical Reagent Factory. Hydrochloric acid (36–38 %) was purchased from Tianjin Yingda Rare Chemical Reagents Factory. Concentrated sulfuric acid was purchased from Tianjin No. 3 Chemical Reagent Factory. Para aminoazobenzene 4-sulfonic acid (AR grade) was purchased from Beijing Chemical Plant.

57.2.2 Characterization

57.2.2.1 L-glutamic Acid Concentration Determination

The pH of the fermentation broth was adjusted to 6.5–7.5 by adding diluted acid or alkali. Then, the L-glutamic acid content was determined on SBA-40 biosensor.

57.2.2.2 Measurement of Crystal Size Distribution

Crystal size distribution was performed on Malvern Mastersizer 2,000 laser diffraction particle size analyzer. D_{50} represents a particle diameter meaning that the number of particles having this particle diameter or less is 50 % of the total number of particles.

57.2.2.3 Measurement of the Metastable Zone Width

Metastable zone width (MSZW, the difference between the saturation temperature and the temperature at which crystals are detected) for L-glutamic acid was detected by laser.

57.2.3 Methods

The effect of initial temperature (35, 40, 45, 50, 55, 60 °C) and cooling rate (0.2, 0.5, 0.8, 1.0, 1.5, 2.0 °C/min) on metastable zone width was investigated.

The effect of ultrasonic time (from 1 to 9 h at an interval of 1 h) on L-glutamic acid crystallization was studied. The effects of stirring speed (120, 160, 180, 200, and 220 rpm) and acid adding rate (0.1, 0.5, 1.0, and 1.5 mL/min) on L-glutamic acid crystallization were determined. The temperature was 20 °C during the whole experiments.

57.3 Results and Discussions

57.3.1 The Effect of Temperature on Metastable Zone Width

The crystallization of L-glutamic acid is mainly composed of two separate steps, namely nucleation and crystal growth. Nucleation process involves phase transition. The change of Gibbs free energy in this process can be expressed as:

$$\Delta G = \Delta G_v + \Delta G_s \quad (57.1)$$

(1) where ($\Delta G_v < 0$), is the change of Gibbs free energy in the process of new phase formation, ($\Delta G_s > 0$) is the surface energy of the new phase interface. In case that the nucleus is supposed as spherical crystal, the Eq. 57.1 can be denoted as:

$$\Delta G = \frac{4}{3}\pi r^3 \Delta G_v^0 + 4\pi r^2 \Delta G_s^0 \quad (57.2)$$

(2) where ΔG_v^0 and ΔG_s^0 are the decrease of Gibbs free energy per unit volume of the new phase formation and the increase of Gibbs free energy per unit area between the new and old phase interface.

Supersaturation ratio affects not only the formation of nucleus, but also the growth rate which has a great influence on the particle size distribution of crystals. As shown in Fig. 57.2, AB is the solubility curve and CD is the super solubility curve which is roughly in parallel to curve AB. The two curves divide Fig. 57.2 into three parts: stable zone, metastable zone, and unstable zone. There is no possibility for the L-glutamic acid crystallization in stable zone due to the unsaturated solution. Nucleation process will happen in metastable zone and rapidly crystallization arises in unstable zone.

The nucleation rate is greatly affected by temperature, supersaturation ratio, and agitation. According to the classical nucleation theory, the homogeneous nucleation (HON) rate [4] could be expressed as Eq. 57.3:

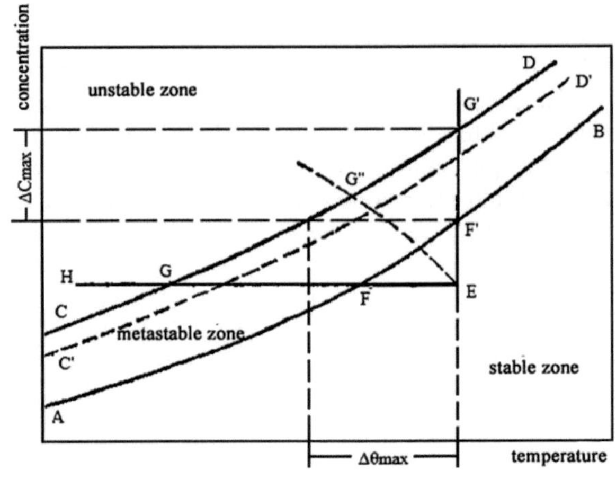

Fig. 57.2 Schematic diagram of state of solutions

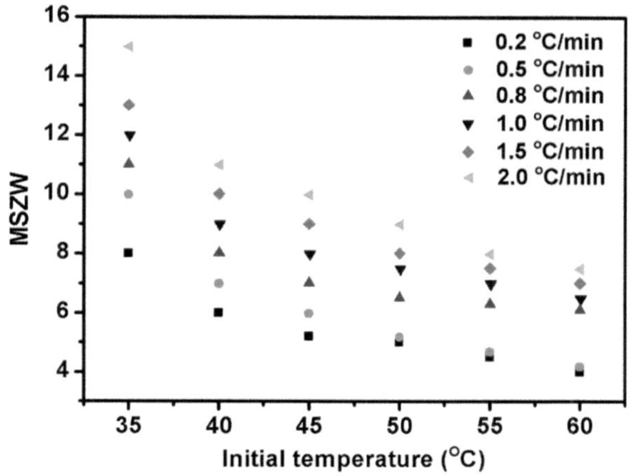

Fig. 57.3 Effect of the temperature on MSZW

$$J = A\exp\left[-\frac{fr_{SL}^{3}M^{2}N_{a}}{\rho^{2}v^{2}R^{3}T^{3}(\ln S)^{2}}\right] \tag{57.3}$$

where f is shape factor, M is molecular weight $\left(\frac{kg}{mol}\right)$, ρ is crystalline density $\left(\frac{kg}{m^3}\right)$, R is gas constant (J mol^{-1} K^{-1}), S is supersaturation ratio, r_{SL} is surface tension $\left(\frac{J}{m^2}\right)$, N_a is Avogadro constant (mol), v is the number of ion per mole, T is Kelvin temperature (F).

The initial temperature and cooling rate have enormous influence on metastable zone according to Eq. 57.3. Thus, the effects of different temperature and cooling rate were investigated. The results are shown in Fig. 57.3. At the same initial temperature, metastable zone becomes wider with increasing cooling rate. Furthermore, at the same cooling rate, metastable zone becomes wider as lowering the initial temperature. The result indicates that spontaneous nucleation is hardly achieved because of the larger nucleation barrier at lower temperature. In addition, the concentration difference emerged in low-temperature region is smaller than that in high-temperature region with the decrease of identical temperature, leading to smaller nucleation impetus.

57.3.2 The Effect of Ultrasonic Wave on Crystallization

The effect of ultrasonic wave on L-glutamic acid crystallization is presented in Fig. 57.4. The result reveals that the ultrasonic irradiation enhances the nucleation rate which ultimately reduces the crystallization time. In order to understand how ultrasound promotes crystallization, the surface tension of L-glutamic acid solution under ultrasound was measured. Figure 57.5 shows that surface tension decreases

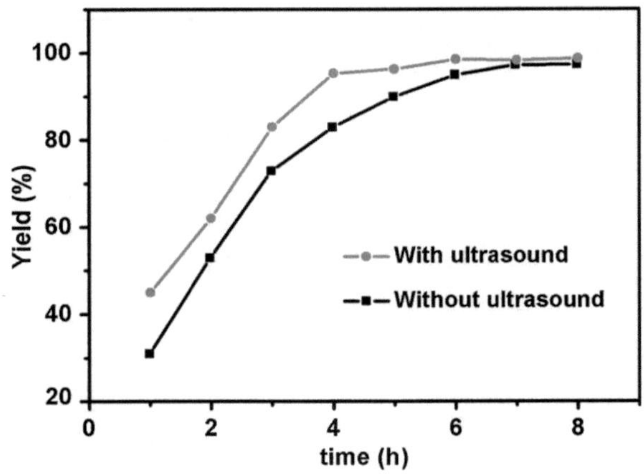

Fig. 57.4 Effect of the ultrasonic time on the yield of L-glutamic acid

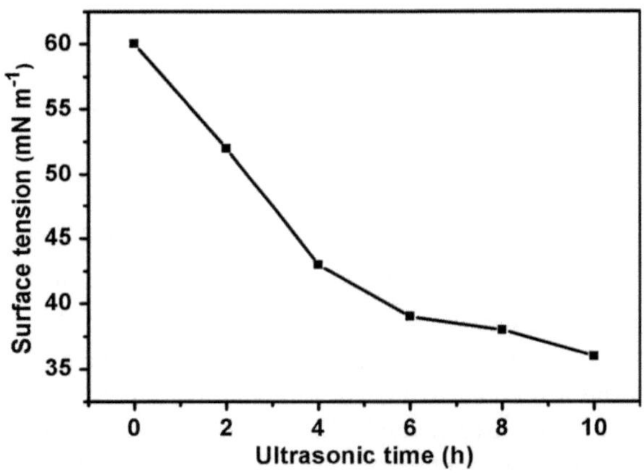

Fig. 57.5 Effect of the ultrasonic time on the surface tension of L-glutamic acid solution

with increasing ultrasonic time. Apparently, the rate of nucleation increases with the decrease in surface tension according to the Eq. 57.3.

57.3.3 The Effect of Stirring Speed

The effect of stirring speed on the yield in the crystallization was investigated. The results are shown in Fig. 57.6. The stirring speed has scarcely impact on the yield when under 200 rpm. However, the yield drops rapidly when the stirring speed increases to 220 rpm. This was described to the formation of beta form L-glutamic acid confirmed by optical images. At a high mechanical agitation speed, large crystal is easily shredded which leads to increased tendency of beta form formation. And the beta form L-glutamic acid causes the sharp decline in L-glutamic acid yield.

Fig. 57.6 Effect of stirring speed on the yield

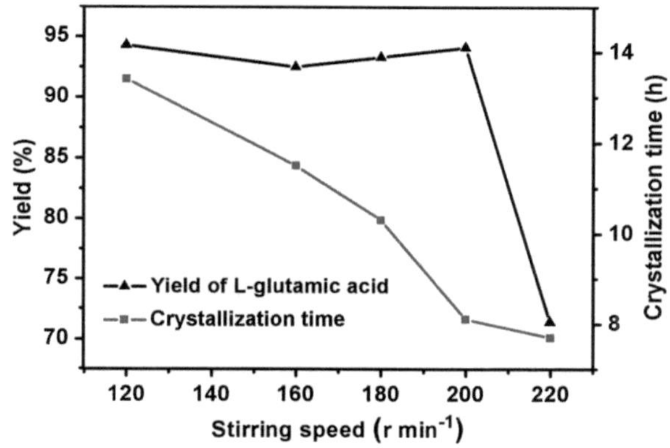

In addition, the crystallization time decreases with increasing stirring speed. Thus, 200 rpm is the most suitable stirring speed under comprehensive consideration.

57.3.4 The Effect of Acid Adding Rate

As shown in Fig. 57.7, the acid adding rate has a great effect on the yield of L-glutamic acid. It was found that the polymerization degree of crystal increased higher with faster acid adding rate (i.e., the polymerization degree obtained at acid adding rate 1.5 mL/min is two times higher than that obtained at acid adding rate 0.1 mL/min). As the increasing of the acid adding rate, the D_{50} is falling down. The ratio of alpha form crystal can reach above 97 % with acid adding rate being 0.5 mL/min, while the ratio of alpha crystal decreases extremely with acid adding rate being 1 mL/min. The decrease of the ratio of alpha form crystal is a result of the formation of beta crystal at excessively fast pH decreasing generated by fast acid adding rate. Thus, the formation of alpha crystal with contented particle size and

Fig. 57.7 Effect of acid adding rate on crystallization

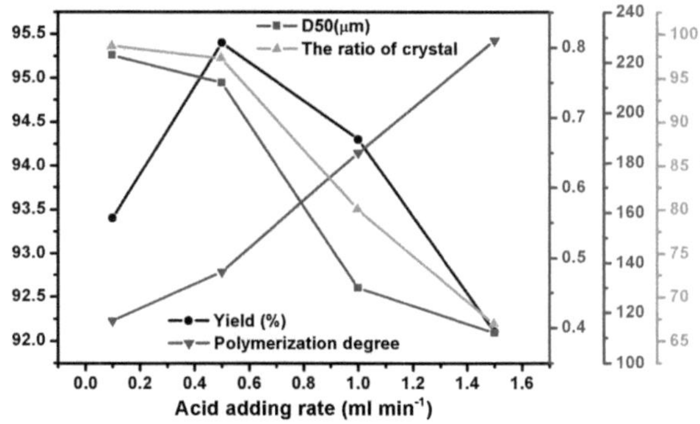

size distribution is difficult to attain with fast acid adding rate. In terms of the polymerization degree, slower acid adding rate is better. However, too long operating time will affect the production efficiency. Considering all affects, 0.5 mL/min is the best choice.

57.3.5 Performance Enhancement of L-glutamic Acid Crystallization Under Optimal Process

Concentrated isoelectric crystallization methods were utilized to promote L-glutamic acid crystallization from fermentation broth treated with ceramic membrane and ultrafiltration membrane. Various factors (i.e., ultrasound and stirring speed) influencing the performance of crystallization were evaluated. The optimum conditions obtained are as follows: acid adding rate 0.5 mL/min, ultrasonic time 10 min, and stirring speed 200 rpm.

57.3.5.1 Increasing Yield

As can be seen from Table 57.1, the yield of L-glutamic acid crystallization at optimal condition reached 95.4 and 6.5 % higher than the yield obtained at initial condition.

57.3.5.2 Increasing Purity

The purity of crystallized production of 3 parallel experiments was 99.3 %, 99.6 % and 99.5 %, respectively. A rise of 4.0 % was achieved compared to the purity before optimization. Such a good result is ascribed to the Optimization of process. In addition, the energy consumption reduction in the Optimization of process is successfully reached, which ultimately makes L-glutamic acid crystallization process environment-friendly (Table 57.2).

Table 57.1 Comparison of the results obtained at optimal condition and initial condition

	Output (g/L)	Yield (%)
Optimal condition	145	95.4
Initial condition	136	89.5
Increased value	6	6.5

Table 57.2 The experimental results of glutamic acid extracted by optimal condition

Number	L-glutamic acid fermentation broth			L-glutamic acid product			Polymorph
	Volume (L)	Concentration (g/L)	Mass (g)	Dry weight (g)	Purity (%)	Mass (g)	
1	41	105	4,305	4127.3	99.3	4098.4	alpha
2	50	112	5,600	5375.1	99.6	5353.6	alpha
3	46	106	4,876	4709.3	99.5	4685.8	alpha

57.4 Conclusion

In conclusion, the effect of external factors such as temperature, stirring speed, and ultrasound on the L-glutamic acid crystallization with concentrated isoelectric crystallization method was investigated. Spontaneous nucleation is more likely to occur in the high-temperature zone due to larger impetus generated by concentration gradient. The yield of L-glutamic acid increases under ultrasound because of the decrease of surface tension (i.e., nucleation rate increases with decreasing surface tension according to Eq. 57.3). The most suitable stirring speed is 200 rpm considering the yield and crystallization time. Furthermore, the optimal acid adding rate is 0.5 mL/min. The performance of crystallization improves significantly under the optimum operating conditions above. The yield of L-glutamic acid reached to 95.4 % and the purity exceeded 99 %, thus the subsequent crystal transition was omitted. The modified method in our paper makes L-glutamic acid crystallization process energy-saving, economically and environment-friendly.

Acknowledgments This work was supported by National High Technology Research and Development Program (2013AA102106), Tianjin Research Program of science and technology (14JCYBJC23500), and Program for Changjiang Scholars and Innovative Research Team in University (IRT 1166).

References

1. Sarvamangala DD, Vijayalakshmi P (2013) Production and crystallization of L-glutamic acid. Int J Pharm Sci Rev Res 18:58–60
2. Roelands CPM, Ter Horst JH, Kramer HJM, Jansens PJ (2007) Precipitation mechanism of stable and metastable polymorphs of L-glutamic acid. AICHE J 53:354–362
3. Kumar R, Vikramachakravarthi D, Pal P (2014) Production and purification of glutamic acid: a critical review towards process intensification. Chem Eng Process 81:59–71
4. Roelands CPM, Ter Horst JH, Kramer HJM, Jansens PJ (2005) The unexpected formation of the stable beta phase of L-glutamic acid during pH-shift precipitation. J Cryst Growth 275: e1389–e1395

Chapter 58
Synthesis of Poly (β-L-malic Acid) by the Optimization of Inorganic Nitrogen Complexing with Growth Factors Using *Aureobasidium pullulans* CGMCC3337

Changsheng Qiao, Yumin Song, Zhida Zheng, Xujia Fan and Shiyun Yu

Abstract Poly (β-L-malic acid) (PMLA) can be used as a prodrug or used for drug delivery system, and it has attracted industrial interest. In previous studies, we could obtain 38 g/L PMLA with 10 g/L yeast extract as a nitrogen source, but the complex broth component made it difficult to the downstream process. To decrease the pressure in the extraction process, we used inorganic nitrogen resources and growth factors to replace yeast extract. Response surface methodology (RSM) was applied to optimize the fermentation medium formulation to improve the production of poly (β-L-malic acid) (PMLA) with *Aureobasidium pullulans CGMCC3337*. It was confirmed that ammonium nitrate 2.06 g/L, adenine 0.02 g/L, cytosine 0.012 g/L aspartate (Asp) 0.42 g/L, histidine(His) 0.6 g/L, leucine(Leu) 0.4 g/L, and threonine(Thr) 0.28 g/L could replace the yeast extract in PMLA production. The model developed based on RSM successfully predicted PMLA productivity (R^2 = 0.9697). With the optimum medium established, 37.62 g/L PMLA was accumulated in the fermentation broth, and the impurities in the fermentation broth was reduced apparently.

Keywords *Aureobasidium pullulans* · Poly (β-malic acid) · Inorganic nitrogen · Growth factors · Response surface methodology (RSM)

58.1 Introduction

PMLA is a well-known biodegradable, bioabsorbable, and water-soluble polymer having pendant carboxylic groups, which allow further conjugation of functional groups or molecules, including drugs via chemical modification. These advantages

C. Qiao (✉) · Y. Song · Z. Zheng · X. Fan · S. Yu
Key Laboratory Of Industrial Microbiology, Ministry of Education, Tianjin University of Science and Technology, No.29, 13th Avenue, TEDA, Tianjin 300457, China
e-mail: qiaochangsheng@tust.edu.cn

T.-C. Zhang and M. Nakajima (eds.), *Advances in Applied Biotechnology*, Lecture Notes in Electrical Engineering 332, DOI 10.1007/978-3-662-45657-6_58

make it important to investigate high purity and high-yield production of PMLA. For example, it is reported that a PMLA-based drug delivery system and biodegradable nanoparticles could be used for tumor targeting [1, 2]. So PMLA can be used as a prodrug or for drug delivery systems, which has attracted a great deal of interest in industries [3–5]. PMLA can be obtained by chemical synthesis or microbial fermentation. Recently, biosynthesis has been extensively investigated. Leathers [6] did significant work on PMLA production by strains of *Aureobasidium pullulans*. 9–11 g/L PMLA could be gained by adding yeast extract and peptone to the medium. Under optimized conditions, using sodium nitrate as the nitrogen source, 9.8 g/L PMLA was produced after nine days of fermentation in the stirred-tank reactors with an overall yield of 0.11 g PMLA/g glucose [7]. In preliminary studies, we had successfully accumulated 38.9 g/L β-PMLA with the addition of 10 g/L yeast extract, which was bad for the downstream process of β-PMLA production. The application of PMLA was vastly restricted in medicine and other industries.

Therefore, the objective of the present investigations is to optimize the cultural medium formulation with the addition of inorganic nitrogen source and amino acids, to acquire the highest possible production of PMLA, to reduce impurities in the fermentation broth, to easily investigate the metabolic flux analysis and metabonomics analysis, and to provide a kind of cheap medium components for PMLA production by the strain *A. pullulans* CGMCC3337.

58.2 Materials and Methods

58.2.1 Microorganism

Aureobasidium pullulans CGMCC3337 was isolated in the Key Laboratory of Industrial Microbiology, Ministry of Education, Tianjin University of Science and Technology, and stored in China General Microbiological Culture Collection Center.

58.2.2 Medium and Culture Method

The compositions of the inoculum medium were as follows (g/L): glucose 80.28, yeast extract 3, $(NH_4)_2SO_4$ 3.06, succinic acid 2, K_2CO_3 0.4, KH_2PO_4 0.1, $CaCO_3$ 27.7, $ZnSO_4 \cdot 7H_2O$ 5×10^{-3}, $MgSO_4 \cdot 7H_2O$ 0.1. The medium was autoclaved for 20 min at 121 °C after the pH of the medium was adjusted to 4.5.

For the inoculums preparation, cells cultivated on potato-glucose agar inoculated to 50 ml seed medium in 500 ml baffle flasks and incubated at 25 °C, 200 rpm for 2 days on a rotary shaker. Fermentation was carried out in 500 ml baffled flasks, and each flask contains 50 ml sterile fermentation medium, which was inoculated with 5 ml of the prepared inoculum. Then the flasks were incubated for 7 days on a rotary incubator at 25 °C, 200 rpm.

58.2.3 Preliminary Experiments

We had developed a semisynthetic medium using yeast extract as the nitrogen source. To replace the yeast extract, we study the effect of nitrogen on PMLA production. Various inorganic salts were tested individually. Fifteen amino acids and a number of growth factors as trace elements were also evaluated for their suitability for the PMLA production. Carbon source and metal irons were kept at the initial levels. Since optimizing amino acids and growth factors involved in many parameters, only one parameter varied in the optimization process.

58.2.4 Experimental Design

Plackett–Burman design (PBD) was used to identify those variables that significantly influenced PMLA production. Then, experiments were performed along the steepest ascent path until the response showed no further increase. Based on the results of our preliminary experiments, the major factors were optimized by response surface methodology (RSM) design (Design-Expert 6).

58.2.5 Determination of Poly Malic Acid Yield

The PMLA content was determined by an HPLC method on a column of Prevail C^{18} after the PMLA was hydrolyzed to L-malic and was expressed as g/L [8].

58.3 Results and Discussion

58.3.1 Effects of Inorganic Nitrogen Sources to the Synthesis of PMLA

In order to replace yeast extract, inorganic nitrogen sources (NH_4Cl, $NaNO_3$, NH_4NO_3, Urea, NH_4HSO_4, $(NH_4)_2CO_3$, and $(NH_4)_2SO_4$) were individually evaluated for their performances in PMLA production by the strain *A. pullulans* CGMCC3337 (Table 58.1). To provide the optimal C/N, the addition of inorganic nitrogen source was consistent with nitrogen content in the yeast extract.

Table 58.1 Effect of inorganic nitrogen sources on PMLA production

Nitrogen source (g/L)	NH_4Cl	$NaNO_3$	NH_4NO_3	Urea	NH4HSO_4	$(NH_4)_2CO_3$	$(NH_4)_2SO_4$	Yeast extract
Concentration	2.75	4.37	2.06	1.54	5.92	2.67	3.38	5.92
Biomass	10.3	0.28	19.86	16.29	0.24	10.87	0.31	38.80
PMLA	4.5	0.18	17.92	5.63	0.43	2.37	0.57	20.67

Among the various inorganic nitrogen sources tested, NH_4NO_3 exhibited a prominent effect on the production ability of 19.86 g/L and maximum biomass is 17.92 g/L, followed by urea 16.29 g/L, 5.63 g/L, NH_4Cl 10.3 g/L, 4.5 g/L, $(NH_4)_2CO_3$ 10.87 g/L, 2.3 g/L, respectively. NH_4HSO_4, $(NH_4)_2SO_4$, and $NaNO_3$ exhibited no effect on the growth ability (data not shown). The yield and biomass of inorganic nitrogen were obviously lower than those of yeast extract. This might be a result of growth factors existence in yeast extract. Therefore, the effects of growth factors on synthesis of PMLA were investigated subsequently.

58.3.2 Effects of Growth Factors on PMLA Production

In the investigation process, adding relevant growth factors according to the content in the yeast extract, while the other nutrient contents were as follows: sugar 65, KH_2PO_4 0.15, $MgSO_4 \cdot 7H_2O$ 0.4, $CaCO_3$ 12 g/L. The concentration of PMLA increased 18.98, 7.22, 11.61, and 11.01 % compared to the original value of 19.8 g/L, with the addition of Asp (0.28), His (0.4), Leu (0.6), and Thr (0.38 g/L), as shown in Fig. 58.1a. When the fermentation medium added Ala, Glu, Ile, Met, Phe, and Ser, it had little effect on PMLA production. However, negative effects on PMLA production had been observed with Arg, Gly, Lys, Tyr, and Val. The effects of purines (Adenine, Guanine) and pyrimidines (Cytosine, Thymine) were investigated for the synthesis of PMLA. The results were shown in Fig. 58.1b. When various vitamins were added to the basic medium, the production of PMLA had no difference (data not shown). The addition of trace of growth factors almost had no effects on the

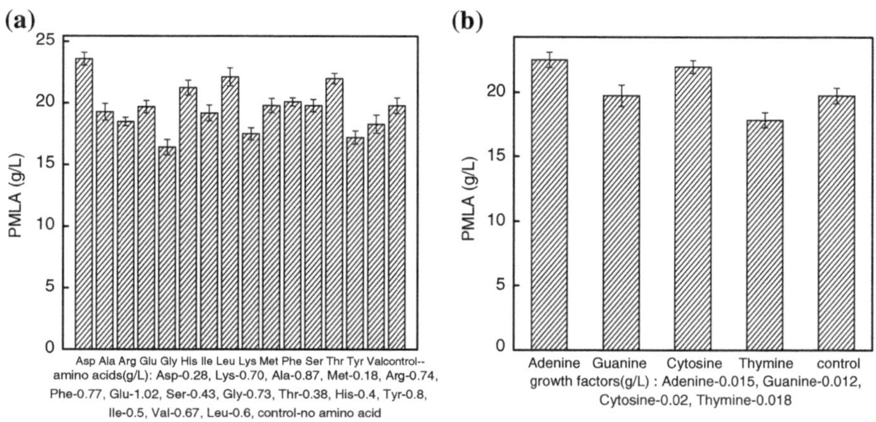

Fig. 58.1 Effects of growth factors on PMLA (g/L) production. **a** 15 kinds of amino acids (g/L); **b** purines (Adenine, Guanine) and pyrimidines (Cytosine, Thymine) (g/L)

Table 58.2 Variables and test levels for Plackett–Burman experiment

No.	Variables	Coefficient	*F*-value	*P*-value
X_1	sugar	−0.9	3.48	0.0993
X_2	KH_2PO_4	0.68	1.99	0.1961
X_3	$MgSO_4 \cdot 7H_2O$	−1.25	6.66	0.0326*
X_4	NH_4NO_3	−0.97	4.03	0.0796
X_5	$CaCO_3$	0.86	3.18	0.1124
X_6	Adenine	−1.9	15.46	0.0043*
X_7	Cytosine	−0.3	0.39	0.5503
X_8	Asp	2.35	23.69	0.0012*
X_9	His	0.46	0.91	0.3686
X_{10}	Leu	−0.63	1.7	0.2289
X_{11}	Thr	−1.01	4.36	0.0703

* Statistically significant at 95 % of probability level

biomass, respectively (data not shown). Finally, adding all the valuable factors to medium, we gained 29.8 g/L PMLA and 18.98 g/L biomass. All the results we investigated are helpful to the biomass accumulation. In order to further improve the production, RSM was used to optimize the addition of components in the medium (Table 58.2).

58.3.3 Screening of Significant Nutrients Using PBD

Twenty independent experiments were conducted according to the experimental design shown in Table 58.3. On the basis of the calculated coefficients, *F*-values and *P*-values (Table 58.2) X_3, X_6, X_8 were found to be the most significant variables affecting PMLA and chosen for further optimization in the next path of steepest ascent and BBD.

58.3.4 Method of Steepest Ascent

The method of steepest ascent was based on the zero level of the PBD and moved sequentially along the direction in which X_3, X_6 decreased and X_8 increased. The experimental design and results were shown in Table 58.4. The highest response was 34.76 with X_3 15, X_6 1.2, X_8 0.42 g/L.

Table 58.3 Plackett–Burman design of variables (in coded levels) with PMLA yield as response

Run	X_1	X_2	X_3	X_4	X_5	X_6	X_7	X_8	X_9	X_{10}	X_{11}	PMLA (g/L)
1	80	0.2	0.3	2.02	14	0.03	0.01	0.42	0.36	0.62	0.24	29.51
2	50	0.2	0.5	2.02	10	0.03	0.01	0.42	0.44	0.58	0.32	27.91
3	80	0.1	0.5	2.1	10	0.02	0.01	0.42	0.44	0.62	0.24	26.8
4	80	0.2	0.3	2.1	14	0.02	0.01	0.42	0.44	0.62	0.32	32.57
5	50	0.2	0.5	2.02	14	0.03	0.01	0.34	0.44	0.62	0.32	25.01
6	50	0.1	0.5	2.1	10	0.03	0.01	0.34	0.36	0.62	0.32	18.72
7	50	0.1	0.3	2.1	14	0.02	0.01	0.42	0.36	0.58	0.32	31.79
8	50	0.1	0.3	2.02	14	0.03	0.01	0.42	0.44	0.58	0.24	32.6
9	80	0.1	0.3	2.02	10	0.03	0.01	0.34	0.44	0.62	0.24	22.26
0	50	0.2	0.3	2.02	10	0.02	0.01	0.42	0.36	0.62	0.32	31.01
11	80	0.1	0.5	2.02	10	0.02	0.01	0.42	0.44	0.58	0.32	28.48
12	50	0.2	0.3	2.1	10	0.02	0.01	0.34	0.44	0.62	0.24	29.79
13	80	0.1	0.5	2.02	14	0.02	0.01	0.34	0.36	0.62	0.32	22.35
14	80	0.2	0.3	2.1	10	0.03	0.01	0.34	0.36	0.58	0.32	18.53
15	80	0.2	0.5	2.02	14	0.02	0.01	0.34	0.36	0.58	0.24	29.81
16	80	0.2	0.5	2.1	10	0.03	0.01	0.42	0.36	0.58	0.24	27.14
17	50	0.2	0.5	2.1	14	0.02	0.01	0.34	0.44	0.58	0.24	25.63
18	50	0.1	0.5	2.1	14	0.03	0.01	0.42	0.36	0.62	0.24	25.79
19	80	0.1	0.3	2.1	14	0.03	0.01	0.34	0.44	0.58	0.32	23.65
20	50	0.1	0.3	2.02	10	0.02	0.01	0.34	0.36	0.58	0.24	30.85

Table 58.4 Experimental design of steepest ascent and corresponding response

Step change value	$MgSO_4 \cdot 7H_2O$ (g/L)	Adenine (g/L)	Asp (g/L)	PMLA (g/L)
0	0.4	0.025	0.38	29.76
0 + 1Δ	0.35	0.023	0.4	32.08
0 + 2Δ	0.3	0.021	0.42	34.76
0 + 3Δ	0.25	0.019	0.44	28.48
0 + 4Δ	0.2	0.017	0.46	30.59
0 + 5Δ	0.15	0.015	0.48	31.57

58.3.5 Further Optimization of the Nutrients Using BBD

The design matrix of tested variables and the experimental results were represented in Table 58.5. The statistical analysis results were shown in Table 58.5, and the regression model was given as:

$$\begin{aligned} Y = & \; 37.61 + 0.52X_3 + 0.034X_6 + 0.26X_8 - 0.33X_3X_6 + 0.87X_3X_8 + 1.51X_6X_8 \\ & - 1.87X_3^2 - 2.82X_6^2 - 3.48X_8^2 \end{aligned} \tag{58.1}$$

Table 58.5 BBD matrix for optimization of PMLA

Run	Coded level			PMLA(g/L)	
	X_3	X_6	X_8	Experimental value	Predicted value
1	0.25	0.019	0.42	31.69	32.04
2	0.35	0.019	0.42	34	33.74
3	0.25	0.023	0.42	32.5	32.76
4	0.35	0.023	0.42	33.5	33.15
5	0.25	0.02	0.4	33.18	32.35
6	0.35	0.02	0.4	31.87	31.65
7	0.25	0.02	0.44	30.92	31.14
8	0.35	0.02	0.44	33.09	33.92
9	0.3	0.019	0.4	32.05	32.53
10	0.3	0.023	0.4	29.01	29.58
11	0.3	0.019	0.44	30.6	30.03
12	0.3	0.023	0.44	33.6	33.12
13	0.3	0.02	0.42	37.03	37.61
14	0.3	0.02	0.42	37.98	37.61
15	0.3	0.02	0.42	37.94	37.61
16	0.3	0.02	0.42	37.12	37.61
17	0.3	0.02	0.42	37.98	37.61

where Y is the predicted β-PMLA yield, X_3 is $MgSO_4 \cdot 7H_2O$, X_6 is Adenine, and X_8 is initial Asp. The quadratic regression model was evaluated by ANOVA. As shown in Table 58.6, the model F-value was 24.89, and the value of "Prob > F" was 0.0002, suggesting that the model was highly significant. The "Lack of Fit F-value" of 4.10 implied the lack of fit that was not significant relative to the pure error. There was a 10.32 % chance that a "Lack of Fit F-value" of this large could occur due to the noise. The model was found to be adequate for prediction within the range of variables employed. The determination of coefficient (R^2) was 0.9697.

Table 58.6 Analysis of variance (ANOVA) for the experiment results of the BBD

Source	Sum of squares	Degrees of freedom	Mean square	F-value	P-value
Model	125.08	9	13.9	24.89	0.0002*
Residual	3.91	7	0.56		
Lack of fit	2.95	3	0.98	4.1	0.1032
Pure error	0.96	4	0.24		
Cor total	128.99	16			

$R^2 = 0.9697$, $R^2_{Adj} = 0.9308$

* Statistically significant at 95 % of probability level

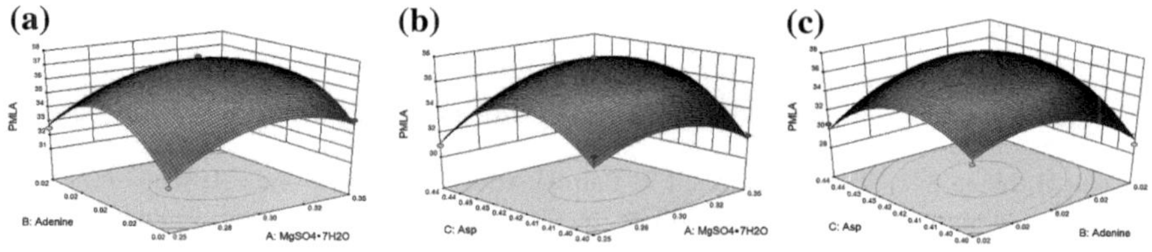

Fig. 58.2 Three-dimensional response surface plots for PMLA (g/L) production by *A. pullulans* CGMCC3337 showing variable interactions of: **a** MgSO4·7H2O (g/L) and Adenine (g/L); **b** MgSO4·7H2O (g/L) and Asp (g/L); **c** Adenine (g/L) and Asp (g/L)

The high R^2 values in this study indicated that the model equation could adequately predict the responses. The values of the adjusted determination coefficients (adjusted R^2) 0.9308 also indicated high significance for the model.

Three-dimensional response surface Fig. 58.2a, b, c represented regression equations to depict the relationships among the variables and to confirm these results. PMLA fermentation by *A. pullulans* was carried out under the optimum medium, which was improved by the RSM optimization approach. The mean value of the PMLA yield was 37.62 ± 0.21 g/L in triplicate tests, which was in agreement with the predicted value. The value also was closed to that of yeast extract (38 g/L). Yeast extract could be replaced by the complex of NH_4NO_3 complexing with growth factors to produce PMLA.

58.4 Conclusion

The present studies demonstrated that a synthetic fermentation medium would be optimized with poly (β-L-malic acid) fermentation. When the yeast extract was totally replaced by inorganic ammonium nitrate, the PMLA production was significantly increase with the addition of adenine (0.02 g/L), cytosine (0.012), Asp (0.42), His (0.6), Leu (0.4) and Thr (0.28 g/L) by response surface methodology. Finally, 37.62 ± 0.21 g/L PMLA was gained, which was closed to the value of yeast extract (10 g/L) addition. In this way, the impurities in fermentation broth were reduced apparently and the difficulty in the downstream process of PMLA production was reduced significantly. Meanwhile, PMLA produced under the condition ultimately reduced the cost of medium components. And the unclear flux of metabolic of PMLA would be much easier to be investigated in the further studies.

Acknowledgments This work was supported by Tianjin Higher Education Science and Technology Development Program of Tianjin Education Committee, Tianjin China (No. ZD200703) and Tianjin University of Science and Technology, scientific research funds (No. 20120118).

References

1. Braud C, Bunel C, Vert M (1985) Poly (β-malic acid): a new polymeric drug-carrier. Pol Bulle 13:293
2. Ding H, Inoue S, Ljubimov AV, Patil R, Portilla-Arias J, Hu JW, Konda B, Wawrowsky KA (2010) Inhibition of brain tumor growth by intravenous poly (β-L-malic acid) nanobioconjugate with pH-dependent drug release. Natl Acad Sci 107:18143
3. Ding H, Portilla-Arias J, Patil R, Black KL, Ljubimova JY (2011) The optimization of polymalic acid peptide copolymers for endosomolytic drug deliever. Biomat 32:5269
4. Fujita M, Lee B, Khazenzon N, Penichet M, Wawrowskyc KA, Patil R, Ding H, Eggehard Holler E, Black K (2007) Brain tumor tandem targeting using a combination of monoclonal antibodies attached to biopoly (β-l-malic acid). Control Relea 122:356
5. Ljubimova J, Fujita javascript:popRef('a1')Mjavascript:popRef('a1'), Ljubimov A, Torchilin V, Black K.L, Holler E (2008) Poly(malic acid) nanoconjugates containing various antibodies and oligonucleotides for multitargeting drug delivery. Nanomedic 3:247
6. Manitchotpisit P, Skory CD, Peterson SW, Price NPJ, Vermillion KE (2012) Poly(-L-malic acid) production by diverse phylogenetic clades of *Aureobasidium pullulans*. Ind Microbio Biotechn 39:125
7. Yuan LL, Li YQ, Wang Y, Zhang XH, Xu YQ (2008) Optimization of critical medium components using response surface methodology for Phenazine-1-Carboxylic acid production by *Pseudomonas sp* M-18Q. Biosci and Bioen 105:232
8. Zheng YF, Zhou H, Wang LY, Wei P (2010) Determination of poly (malic acid) in fermentation broth by reversed phase high performance liquid chromatography. Food Sci Tech 35:2

Chapter 59
Primary Characterization and In Vitro Antioxidant Activities of Polysaccharides from Yam Peel

Na Liu, Liming Zhang, Kaihua Zhang, Aiying Tian and Ruichao Li

Abstract The research purpose is to make full use of yam peel resources. In this present study, yam peel polysaccharide (YPP) was isolated from the *Dioscorea esculenta* peels. The primary characterization of YPP was explored, and we carried on the in vitro antioxidant experiment research. The FT-IR analysis was performed. And the absorptions at around 1,148 cm^{-1} indicated a pyranose form of sugars. In vitro, YPP exhibited a potent scavenging activity on the hydroxyl radical, DPPH radical and superoxide radical, especially scavenging activity for DPPH radicals. The IC_{50} value of YPP for eliminating hydroxyl radicals, DPPH radical, and superoxide anion radicals was respectively 9.5, 2.1, and 10.2 mg/ml. And the highest inhibition rates were, respectively, 60.2, 79.3, and 54.0 %. In this experiment, antioxidation test in vitro indicated that YPP was similar to other yam tuber polysaccharides. Furthermore, the further works on purification and isolation studies were done and the primary characterization of YPP were shown by chromatography. In general, these results suggested that the YPP may be explored as a novel and effective natural antioxidant to alleviate oxidative stress, and they may find promising applications as potent antioxidant agents and antiaging medicines.

Keywords Yam peel polysaccharides · Extraction · Antioxidant activity · Characterization

N. Liu · L. Zhang (✉)
Key Laboratory of Industrial Fermentation Microbiology, Ministry of Education,
Tianjin University of Science and Technology, Tianjin 300457, People's Republic of China
e-mail: zhanglmd@126.com

K. Zhang · A. Tian · R. Li
College of Bioengineering, Tianjin University of Science and Technology, Tianjin 300457,
People's Republic of China

T.-C. Zhang and M. Nakajima (eds.), *Advances in Applied Biotechnology*,
Lecture Notes in Electrical Engineering 332, DOI 10.1007/978-3-662-45657-6_59

59.1 Introduction

Yams (*Dioscorea* sp.) are widely distributed. They are economically important tuber and bulbil crops in tropical and subtropical regions worldwide [1, 2]. Yams are not only a common food in China regarded as a tonic, but also has been also used as traditional Chinese medicine (TCM) more than 2000 years for the treatment of diabetes, diarrhea, asthma, and other ailments [3]. The fresh yam is difficult to store and easy to deteriorate during storage, so it is desired to develop a stable form of dried yam products [4]. The traditional process consists of peeling, blanching, and drying the fresh tubers. However, the main residue is the yam peels during processing, which accounts for 10–20 % of the total yam weight. There are few industrial uses for yam peels, and they are a major agricultural waste in different regions of the planet.

Polysaccharides are composed of monosaccharides condensation polymers. It widely exists in animals, plants, and microorganisms. In recent years, many literatures focus on the extraction, chemical structure, and biological activities of nonstarchy polysaccharide from yams [5–7]. What is more, plant derived nonstarchy polysaccharides have emerged as an important class of bioactive natural products. A wide range of polysaccharides have been demonstrated to play an important role as free radical scavengers in the prevention of oxidative damage in living organism and can be explored as novel potent antioxidant [8–10].

In recent years, studies have been shown that yam polysaccharide is one of the important active components in yam and also has been proved that is efficacious for antitumor, decreasing blood glucose, and modulating immune function and so on [11, 12]. And in the past decade, the search for natural antioxidant compounds has gained considerable attention and the number of publications on antioxidants and oxidative stress has nearly quadrupled [13, 14]. It has been reported that many plant polysaccharides have strong antioxidant abilities and can be explored as novel potential antioxidants [15, 16]. However, little attention was devoted to yam peel, and the YPP as well as the antioxidant activity of the YPP in vitro. Therefore, the purpose of this study was to be concerned with the crude polysaccharide extracted from yam peel and to evaluate their antioxidant activities in vitro. In this study, the antioxidant activities of YPP were evaluated, including activities of hydroxyl free radical, DPPH radical, and superoxide anion scavenging in vitro. In addition, the works on purification and isolation studies were done and the primary characterization of polysaccharide from yam peel were shown.

59.2 Material and Methods

59.2.1 Material

The fresh *Dioscorea esculenta* was purchased from Jinghong City (Yunnan Province, China). The yams were taken out of the cold room, washed, and peeled. The peel of yam was washed by immersing in a water tank for 15 min, and then dried using a wash dryer. Blanching was carried out by immersing the peel in an 85 °C water bath for 30 s. The peel was freeze-dried in a local food processing plant. The final moisture content was about 6 % and the thickness was about 0.5 mm. The freeze-dried yam peel was stored in plastic bags at −20 °C until used. Before extraction, the peel was ground into powder and passed through a screener with mesh number of 80. All chemicals used in the experiment were analytical grade.

59.2.2 Methods

59.2.2.1 Extraction of Crude Polysaccharide

The samples were defatted in a Soxhlet apparatus with petroleum ether at 30–60 °C, and soaked with 80 % ethanol twice to remove some colored materials, oligosaccharides, and some small molecules [17]. The organic solvent was volatilized and pretreated powder was obtained.

The pretreated powder (10.0 g) was extracted with weak base solution in a designed extraction time, pH of the aqueous medium, and ratio of water to a raw material. While extraction temperatures were kept steady for a given temperature (50 °C, within ± 1.0 °C) in order to avoid the gelatinization of yam starch. The extract was left to cool at room temperature, and then centrifuged at 2,000 × g for 15 min to collect the supernatant.

The supernatant was concentrated to one-fifth of the initial volume using a rotary evaporator at 65 °C under vacuum. The resulting solution was mixed with 95 % (v/v) ethanol to a final concentration of 75 % (v/v) and incubated overnight. The precipitates were then collected by centrifugation (2,000 × g, 15 min), washed six times with dehydrated ethanol, and air-dried at 50 °C for 10 h to get the crude polysaccharides.

The percentage total polysaccharide yield (%, w/w) is calculated as the total polysaccharide content of extraction divided by dried sample weight.

59.2.2.2 Analysis of Polysaccharide

YPP was obtained by the finally optimum extraction conditions. Protein was removed by Sevag method [18]. Total sugar content of YPP was measured by phenol—sulfuric acid methods [19]. The protein content was determined by the method of Bradford [20]. All measurements were conducted in triplicate.

FT-IR spectrum was obtained using a Vector 22 FT-IR spectrophotometer (Bruker, Germany). The dried YPP was grinded with potassium bromide power and pressed into pellet for spectrometric measurement in the range of 4,000–450 cm^{-1}. Sixteen scans with the resolution of 4 cm^{-1} were collected for spectra, and each spectrum was normalized after the baseline correction.

59.2.2.3 Assay of Hydroxyl Radical Scavenging Activity

The scavenging activity of YPP on $HO^{\bullet}$ was measured according to the method of Zhang et al. [21], with some modifications. Reaction mixture contained 2.0 mL of 9 mmol/L $FeSO_4$, 2.0 mL of sample solutions in distilled water at different concentrations (0–16 mg/mL), 2.0 mL of 8.8 mmol/L H_2O_2 and 2.0 mL 9 mmol/L salicylic acid (dissolved by alcohol). 2.0 mL of 8.8 mmol/L H_2O_2 was added finally, and then the reaction mixture was incubated at ambient temperature for 30 min. Distilled water and ascorbic acid served as blank and positive control, respectively. The absorbance was recorded at 510 nm, and the scavenging activity of YPP was calculated according to the equation:

$$\text{Scavengin effect}\,(\%) = (1 - (A_s - A_b)/A_c) \times 100 \tag{59.1}$$

where A_c is the absorbance of control without sample, A_s is the absorbance of the test sample, A_b is the absorbance of the bank.

59.2.2.4 Assay of DPPH Radicals Scavenging Activity

The DPPH radical scavenging activity was carried out using the method described by Li et al. [22] with minor modification. Briefly, 1 mL DPPH solution (dehydrated alcohol solvent, 0.1 mmol/L) was added to 2.0 mL of various concentrations of YPP dissolved in distilled water to be tested and the reference compound (ascorbic acid, Vc). The mixture was shaken and left to stand for 30 min at room temperature in the dark. The absorbance was measured at 517 nm with a UV–vis spectrophotometer. The DPPH radical scavenging effect was calculated as Eq. (59.1).

59.2.2.5 Superoxide Anion Scavenging Activity

The superoxide anion scavenging activity of various concentrations polysaccharide samples was investigated based on the method of Marklund and Marklund [23] with some modifications. 2 mL of samples solution and 6 mL of 50 mmol/L Tris–HCl buffer (pH 8.2) were kept in water bath at 37 °C for 10 min. Then 1 mL of 7 mmol/L 1, 2, 3-Trihydroxybenzene (which was incubated at 37 °C for 10 min) was added quickly to terminate the reaction. The changes absorbance of the mixture were determined at 320 nm within 4 min. Ascorbic acid was used as a reference material. The superoxide anion scavenging activity was calculated according to the equation:

$$\text{Scavenging effect}\,(\%) = \left[\Delta\,\text{Abs. of sample} - \Delta\,\text{Abs.of blank}/_{\Delta\text{Abs. of control}}\right] \times 100 \tag{59.2}$$

59.2.2.6 Purification of Crude Polysaccharide

The freeze-dried crude polysaccharides were redissolved in deionized water and forced through a filter (0.45 nm), then applied to a column (300 × 26 mm) of DEAE-cellulose and eluted successively with deionized water, 0.2 M NaCl solutions for 200 min at a flow rate of 1.0 mL/min, respectively. Then, two peaks were collected and further purified on a Sephadex G-75 gel filtration column (600 × 16 mm), respectively. Then, the fractions were eluted with deionized water at a flow rate of 0.3 mL/min. The main polysaccharide fraction (YPPn and YPPa) were collected and lyophilized [24]. YPPn and YPPa were used for further study.

59.2.2.7 Analysis of Monosaccharide Composition

The monosaccharide of YPPn and YPPa were analyzed by HPLC. The samples were hydrolyzed with 2 M TFA (2 mL) at 120 °C for 2 h. After removing TFA with methanol, the hydrolyzed products were reduced with $NaBH_4$, followed by neutralization with dilute acetic acid and evaporated at 45 °C. The reduced products (alditols) were added with 1 mL pyridine and 1 mL acetic anhydride in a boiling water bath for 1 h. The acetylated products were analyzed by HPLC [24, 25].

59.2.2.8 Statistical Analysis

All experiments were performed in triplicate and the results were expressed as the mean ± standard deviation (SD). Comparison of means was performed by one-way analysis of variance (ANOVA), followed by Duncan's test. Statistical analysis was carried out by using the software Design-Expert 8.0.6 (Trial Version, Stat-Ease Inc. Minneapolis, MN, USA).

59.3 Results and Discussion

59.3.1 Analysis of YPP

The total sugar content and protein content of YPP obtained by the finally optimum extraction conditions was 93.2 ± 1.91 %, 2.5 ± 0.26 %, respectively.

In order to primarily characterize YPP, FT-IR analysis was performed. The infrared spectrum of this polysaccharide is given in Fig. 59.1. As shown in Fig. 59.1, the FT-IR spectrum of YPP displayed a broad and intense band at around 3,454 cm^{-1}, which was assigned to the hydroxyl groups stretching vibration. The weak peak toward 2,928 cm^{-1} was attributed to the C–H antisymmetrical stretching vibration. In addition, an asymmetrical stretching peak at 1,638 cm^{-1} indicated that there were carboxyl groups in YPP. The bands between 1,200 and 1,000 cm^{-1} in the spectrum were assigned to the valent vibrations of the C–O–C bond and glycosidic bridge [26]. The absorptions at around 1,148 cm^{-1} indicated a pyranose form of sugars.

59.3.2 Antioxidant Activity Analysis

The results of hydroxyl radical scavenging activity of YPP and ascorbic acid were given in Fig. 59.2a. A positive correlation between the concentration of YPP and the hydroxyl radical scavenging activity was observed from 0 to 14 mg/mL. The IC_{50} value of YPP and ascorbic acid for eliminating hydroxyl radicals was 9.5 and

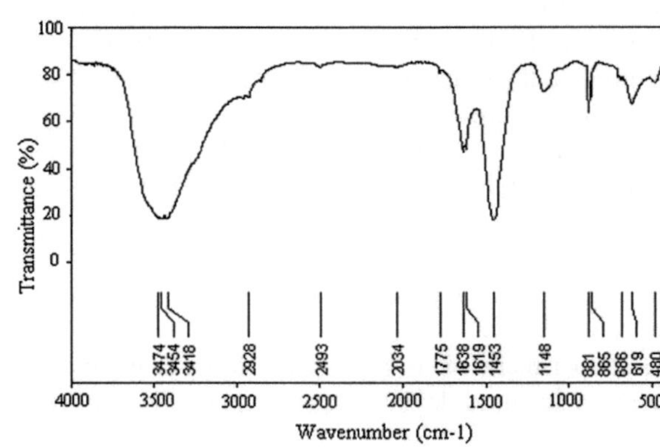

Fig. 59.1 Fourier transform infrared spectrometry (FT-IR) spectrum of YPP between 450 and 4,000 cm^{-1}. YPP, yam peel polysaccharide

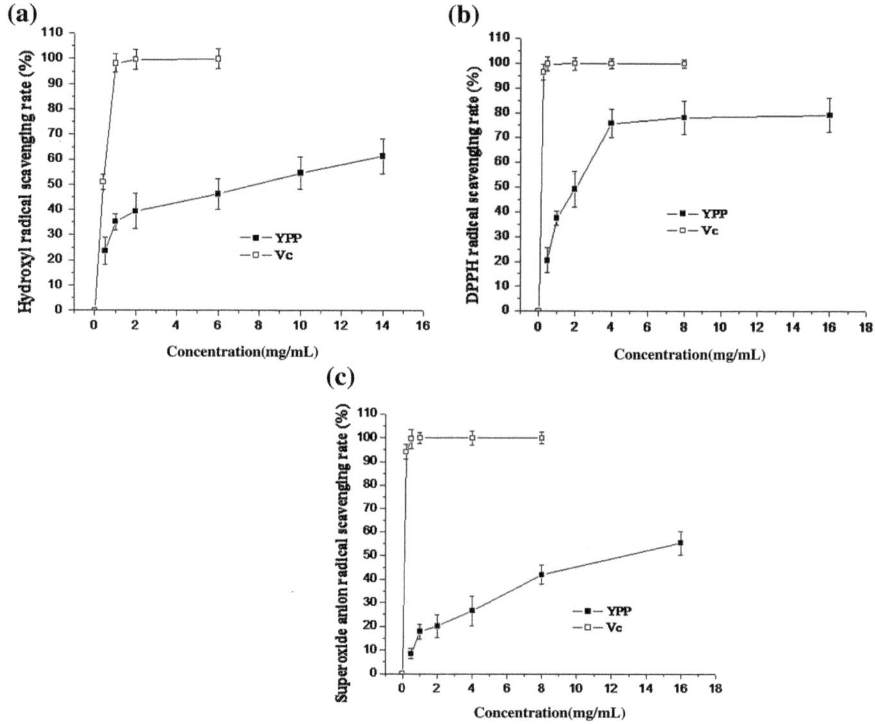

Fig. 59.2 Scavenging effects of different concentration of YPP and ascorbic acid on hydroxyl (**a**), DPPH (**b**), and superoxide anion radicals (**c**). Values are means ±SD (n = 3). YPP, yam peel polysaccharide; Vc, ascorbic acid

0.38 mg/mL, which indicated that the scavenging activity of YPP against hydroxyl radical was much lower than that of ascorbic acid.

The DPPH radical scavenging activity of YPP was investigated at different concentrations (0–16 mg/mL) and the result was presented in Fig. 59.2b. As can be seen from Fig. 59.2b, the superoxide anion radicals scavenging rate increased from 20.5 to 79.3 % when the concentration of YPP increased from 0.5 to 16 mg/mL, and exhibited an interesting radical scavenging activity with an IC_{50} value of 2.1 mg/mL. The ability to scavenge DPPH hydroxyl radical was lower than that of ascorbic acid (IC_{50} = 0.05 mg/mL), which was similar to other plant polysaccharides [27, 28].

The result of superoxide anion radicals scavenging ability of YPP was shown in Fig. 59.2c and compared with ascorbic acid as control of standards. The scavenging effects of YPP increased in a dose-dependent manner. The IC_{50} value of YPP and ascorbic acid for eliminating superoxide anion radicals was 10.2 and 0.12 mg/mL, which indicated that the scavenging activity of YPP against superoxide anion radical was much lower than that of ascorbic acid.

In this experiment, antioxidation test in vitro indicated that the YPP had scavenging activities of hydroxyl radical, DPPH radical, and superoxide anion radical scavenging, which was similar to other yam tuber polysaccharides [29, 30]. Researchers have shown that yam polysaccharides contain antioxidative activity [29, 31]; however, there are few reports regarding the antioxidant activities of yam peel polysaccharides. Yam peels, which are produced during processing and discarded as agricultural by-products, are rich in plant polysaccharide. Published data indicate that polysaccharides isolated from plants have high antioxidant activity on free radicals and can be explored as novel potential antioxidants [32, 33]. Accordingly, this search for plant-derived biomaterials will therefore stimulate research interest in producing antioxidant polysaccharides from yam tuber and its underutilized waste.

59.3.3 Isolation and Purification of YPP

Ion-exchange chromatography on a DEAE-cellulose column obtained two main peaks from deionized water and NaCl elution, and were named as YPP1 (eluted with deionized water), YPP2 (eluted with 0.05 M NaCl) (Fig. 59.3). The polysaccharides were further separated and sequentially purified through Sephadex G-750 column. Fractions (3 mL) were collected. Two major polysaccharide peaks, YPP1-1 and YPP2-1. YPP1-1 were collected and named neutral polysaccharides (YPPn). YPP2-1 were collected and named acidity polysaccharides (YPPa).

YPPn and YPPa were hydrolyzed by TFA into individual monosaccharides that were further reduced and acetylated for HPLC analysis. As can be seen in Fig. 59.4, YPPn was composed of mannose, glucose, fructose, galactose, xylose, and arabinose with a molar ratio: 1:0.15:0.08:0.27:0.14:0.06. YPPa was composed of mannose and glucose with a molar ratio: 1:0.83.

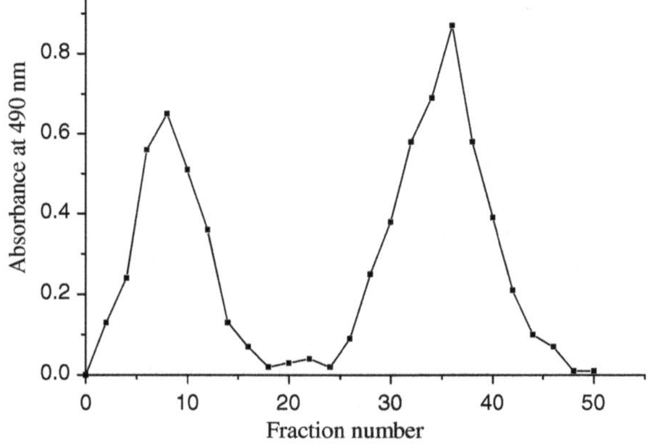

Fig. 59.3 Chromatography of eluted crude polysaccharide (YPP) on DEAE-cellulose column. YPP1 eluted with deionized water; YPP2 eluted with 0.2 M NaCl

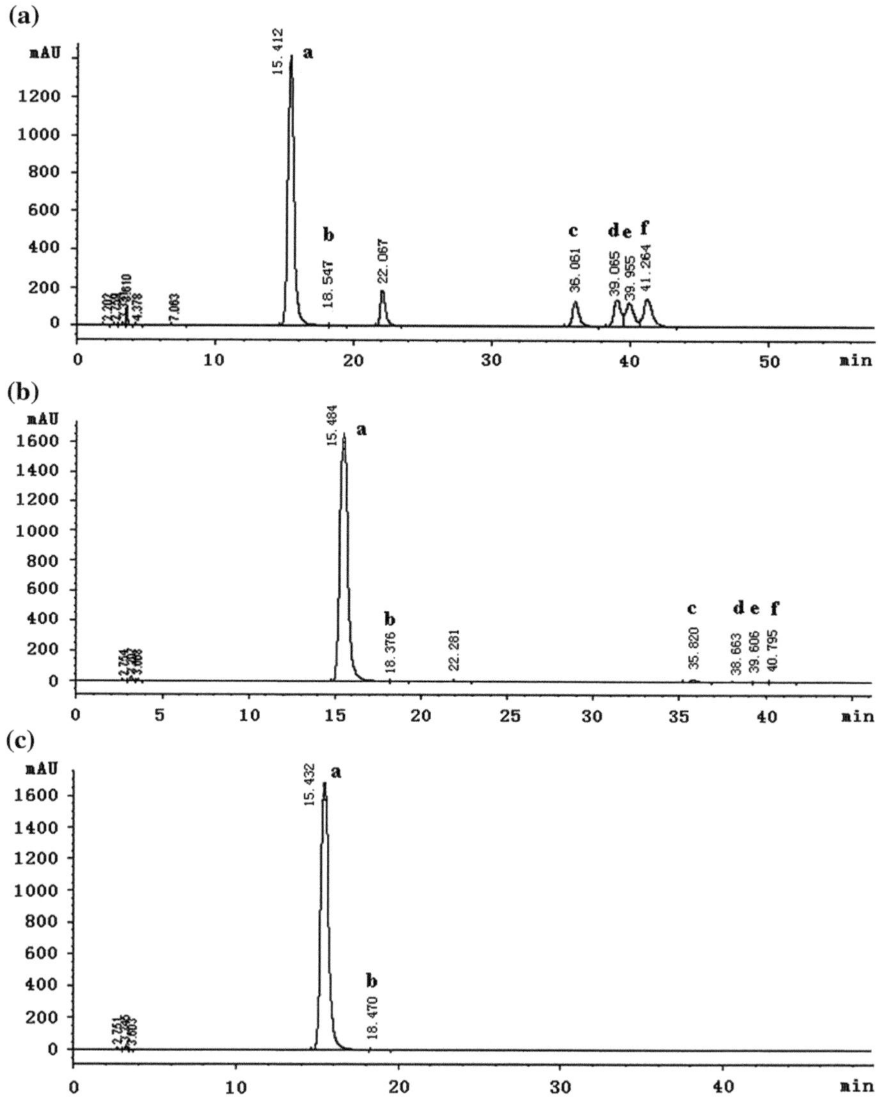

Fig. 59.4 HPLC profile of YPPn and YPPa (**a** standard monosaccharides, peaks from *left* to *right*: a. Man, b. Glu, c. Fru, d. Gal, e. Xyl, f.Ara; **b** monosaccharide composition of YPPn, peaks from *left* to *right*: a. Man, b. Glu, c. Fru, d. Gal, e. Xyl, f. Ara; **c** monosaccharide composition of YPPa, peaks from *left* to *right*: a. Man, b. Glu)

59.4 Conclusion

In this study, the total sugar content and protein content of YPP obtained by the finally optimum extraction conditions were 93.2 ± 1.91 %, 2.5 ± 0.26 %, respectively. The FT-IR analysis was performed. And the absorptions at around 1,148 cm^{-1} indicated a pyranose form of sugars. In vitro, YPP exhibited a potent scavenging activity on the hydroxyl radical, DPPH radical, and superoxide radical. The IC_{50} value of YPP for eliminating hydroxyl radicals, DPPH radical, and superoxide anion radicals was, respectively, 9.5, 2.1, and 10.2 mg/ml. And the highest inhibition rates were, respectively, 60.2, 79.3, and 54.0 %. And especially the antioxidant activity of DPPH radical was shown apparently. Furthermore, the further works on purification and isolation studies were done and the primary characterization of YPP were shown. The purification experiment obtained the YPPn and YPPa, and the HPLC analysis was performed on them. YPPn was composed of mannose, glucose, fructose, galactose, xylose, and arabinose. YPPa was composed of mannose and glucose. These results indicate that YPP has enormous potential for use as a novel natural antioxidant in functional food or feed additives. Thus, further works on structures, functions evaluation, and application studies are in progress to make it an effective product.

Acknowledgments This research was financially supported by the National Natural Science Foundation of China (Project No. 31271809).

References

1. Akanbi CT, Gureje PO, Adeyemi IA (1996) Effect of heat moisture pretreatment on physical characteristics of dehydrated yam. J Food Eng 28:45–54
2. Omonigho SE, Ikenebomeh MJ (2000) Effect of temperature treatment on the chemical composition of pounded white yam during storage. Food Chem 71:215–220
3. Zhao YQ, Wang AF (2000) Antoxidant Chinese yam polysaccharides and its pro-proliferative effect on endometrial epithelial cells. Tradit Chin Med Res 13:49–51
4. Akissoe N, Hounhouigan J, Mestres C, Nago M (2004) Effect of tuber storage and pre- and post-blanching treatments on the physicochemical and pasting properties of dry yam flour. Food Chem 85:141–149
5. Zhao GH, Chen ZD, Li ZX, Wang Y (2003) Effects of Chinese yam polysaccharide on immune function of cancer bearing mice. Acta Nutr Sinica 25:110–112
6. Luo D (2008) Identification of structure and antioxidant activity of a fraction of polysaccharide purified from *Dioscorea nipponica* Makino. Carbohydr Polym 71:544–549
7. Luo D (2012) Optimization of total polysaccharide extraction from *Dioscorea nipponica* Makino using response surface methodology and uniform design. Carbohydr Polym 90:284–288
8. Wang YF, Yang ZW, Wei XL (2012) Antioxidant activities potential of tea polysaccharide fractions obtained by ultra filtration. Int J Biol Macromol 50:558–564
9. He L, Ji PF, Gong XG, Li WQ, Cheng JW, Qian H, Song XL (2011) Physicochemical characterization, antioxidant and anticancer activities in vitro of a novel polysaccharide from Melia toosendanSieb.Et Zucc fruit. Int J Biol Macromol 49:422–427

10. Ye CL, Hu WL, Dai DH (2011) Extraction of polysaccharides and the antioxidant activity from the seeds of Plantago asiatica L. Int J Biol Macromol 49:466–470
11. Hsu CL, Chen W, Weng YM, Tseng CY (2003) Chemical composition, physical properties, and antioxidant activities of yam flours as affected by different drying methods. Food Chem 83:85
12. Zhao GH, Kan JQ, Li ZX, Chen ZD (2005) Structural features and immunological activity of a polysaccharide from *Dioscorea opposita* Thunb roots. Carbohydr Polym 61:125–131
13. Balavigneswaran CK, Kumar TSJ, Packiaraj RM, Veeraraj A, Prakash S (2013) Anti-oxidant activity of polysaccharides extracted from Isocrysis galbana using RSM optimized conditions. Int J Biol Macromol 60:100–108
14. Yang Q, Chen HG, Zhou X, Zhang JZ (2013) Optimum extraction of polysaccharides from Opuntia dillenii and evaluation of its antioxidant activities. Carbohydr Polym 97:736–742
15. Samavati V, Manoochehrizade A (2013) Polysaccharide extraction from Malva sylvestris and its anti-oxidant activity. Int J Biol Macromol 60:427–436
16. Chaiklahan R, Chirasuwan N, Triratana P, Loha V, Tia S, Bunnag B (2013) Polysaccharide extraction from Spirulina sp. and its antioxidant capacity. Int J Biol Macromol 58:73–78
17. Zhu T, Heo HJ, Row KH (2010) Optimization of crude polysaccharides extraction from Hizikia fusiformis using response surface methodology. Carbohydr Polym 82:106–110
18. Vilkas E, Radjabi NF (1986) The glucomannan system from Aloe vahombe (liliaceae). III. Comparative studies on the glucomannan components isolated from the leaves. Biochimie 68 (9):1123–1127
19. Dubois M, Gilles KA, Hamilton JK, Rebers PA, Smith F (1956) Colorimetric method for determination of sugars and related substances. Anal Chem 28:350–356
20. Bradford MM (1976) A rapid and sensitive method for the quantitation of microgram quantities of protein utilizing the principle of protein-dye binding. Anal Biochem 72:248–254
21. Zhang ZS, Wang XM, Yu SC, Lin Y (2011) Synthesized oversulfated and acetylated derivatives of polysaccharide extracted from Enteromorpha linza and their potential antioxidant activity. Int J Biol Macromol 49:1012–1015
22. Li XL, Zhou AG, Han Y (2006) Anti-oxidation and anti-micro organism activities of purification polysaccharide from Lygodium japonicum in vitro. Carbohydr Polym 66:34–42
23. Marklund S, Marklund G (1974) Involvement of the superoxide anion radical in the autoxidation of pyrogallol and a convenient assay for superoxide dismutase. Eur J Biochem 47:469–471
24. Dreher TW, Hawthorne DB, Grant BR (1979) Comparison of open-column and high performance gel permeation chromatography in the separation and molecular weight estimation of polysaccharides. J Chromatogr A 174:443–446
25. Erbing B, Jansson PE, Widmalm G, Nimmich W (1995) Structure of the capsular polysaccharide from the Klebsiella K8 reference strain 1015. Carbohydr Res 273:197–205
26. Zhao M, Yang N, Yang B, Jiang Y, Zhang G (2007) Structural characterization of water-soluble polysaccharides from Opuntia monacantha cladodes in relation to their anti-glycated activities. Food Chem 105:1480–1486
27. Ye CL, Huang Q (2012) Extraction of polysaccharides from herbal Scutellaria barbata D. Don (Ban-Zhi-Lian) and their antioxidant activity. Carbohydr Polym 89:1131–1137
28. Wang SP, Dong XF, Tong JM (2013) Optimization of enzyme-assisted extraction of polysaccharides from alfalfa and its antioxidant activity. Int J Biol Macromol 62:387–396
29. Shen S, Chen ZR, Huan X (2012) Study on the extraction technology and antioxidant activity of yam polysaccharide. J Anhui Agric Sci 40(20):10581–10584
30. Xu XQ, Liu ZF, Huo NR, Zhao Y, Tian X, Lei TT (2012) In vitro ontioxidation capacity and immunomodulatory efficacy in mice of yam polysaccharide. J Chin Cereals Oil Assoc 27 (7):42–46
31. Shang XY, Ren J, Cao G, Xu CL, Niu WN, Qin CG (2010) Preparation and antioxidant properties of polysaccharides from *Dioscorea Opposite* Thunb roots. Chem Res 21(2):72–76

32. Han J, Jiang X, Zhang L (2011) Optimisation of extraction conditions for polysaccharides from the roots of Isatis tinctoria L. by response surface methodology and their in vitro free radicals scavenging activities and effects on IL-4 and IFN-γ mRNA expression in chicken lymphocytes. Carbohydr Polym 86:1320–1326
33. Zhang Z, Wang F, Wang X, Liu X, Hou Y, Zhang Q (2010) Extraction of the polysaccharides from five algae and their potential antioxidant activity in vitro. Carbohydr Polym 82:118–121

Chapter 60
Optimization of Sample Preparation for the Metabolomics of *Bacillus licheniformis* by GC-MS

Hongbin Wang, Zhixin Chen, Jihan Yang, Yihan Liu and Fuping Lu

Abstract Metabolomic has become an important method in microbiology study. Sample preparation affects the quality of final metabolomics analysis strongly. There is no universal preparation method that can suit the metabolomics study of all kinds of microorganism, because of their various cell structure. In this study, a suitable sample preparation method for the metabolomics study of *Bacillus licheniformis* was explored and optimized. The main steps in metabolite sample preparation include quenching, estimation of leakage, and metabolite extraction. The result indicated that 60 % methanol, 0.9 % ammonium carbonate buffer was an appropriate quenching solution for *Bacillus licheniformis* by measuring intracellular metabolites, energy charge, and intracellular metabolites. Among the four different extraction methods (cold pure methanol, PM; cold methanol/water (70:30 v/v), MW; acetonitrile/methanol/water, (2:2:1 v/v/v), AMW; or acetonitrile/water (1:1 v/v), AW), MW was superior to others on the intracellular metabolites, which could effectively extract more intracellular metabolites. The results imply that the optimized preparation method for *Bacillus licheniformis* is critical for a reliable and accurate analysis of metabolome.

Keywords Metabolomic · *Bacillus licheniformis* · Sample preparation · Quenching · Extraction

60.1 Introduction

Bacillus licheniformis, a gram-positive forming-spore bacterium, is known for its high protein secretion capacity and is being applied extensively as a host for the industrial production of enzymes such as proteases and amylases [1]. In the

H. Wang · Z. Chen · J. Yang · Y. Liu · F. Lu (✉)
Key Lab of Industrial Fermentation Microbiology, Ministry of Education,
College of Bioengineering, Tianjin University of Science and Technology,
Tianjin 300457, China
e-mail: lfp@tust.edu.cn

T.-C. Zhang and M. Nakajima (eds.), *Advances in Applied Biotechnology*,
Lecture Notes in Electrical Engineering 332, DOI 10.1007/978-3-662-45657-6_60

microbiology study, metabolomics is currently a popular subject as a high-throughput analytical technology. It is of vital importance for quantitative analysis of the whole metabolome in a biological system [2]. The efficiency and reliability of sample preparation are of key importance for the comprehensive analysis of intracellular metabolite. Many studies have applied metabolomics techniques for microbial analysis [3–5]. But an optimized sample preparation method, generally including sampling, quenching and metabolite extraction, has received little attention in investigating *B. licheniformis*.

Quenching is the first step in metabolomic sample preparation. The perchloric acid quenching has been used to arrest the cellular metabolism in poly-gamma-glutamate-producing *B. licheniformis* [2], but the cold methanol quenching is still the most commonly used quenching method for bacteria [6, 7]. Although the cold methanol quenching can terminate metabolic process among cells instantaneously, serious losses of intracellular metabolites due to cell leakage were observed in bacteria [7, 8]. Ideally, all intracellular metabolites should be extracted completely, nonselectively, and reproducibly by an extraction solvent [9]. Cold methanol, cold chloroform, and cold perchloric acid have also been frequently used. Recently, a mixed solvent system containing acetonitrile with a medium polarity has been applied to bacteria.

In this paper, the quality of quenching and extraction conditions of *B. licheniformis* were investigated by detecting intracellular metabolites with GC-MS. We optimized and established efficient quenching and extraction procedures that minimized metabolite leakage. These results can be used as a customized protocol or an experimental strategy for *B. licheniformis*.

60.2 Experimental Section

60.2.1 Strain and Cultivation

All pure standards were purchased from Sigma-Aldrich. *B. licheniformis* ATCC14580 (China Center for Type Culture Collection) was grown at 37 °C on LB medium (per liter: tryptone 10 g, yeast extract 5 g, NaCl 10 g, pH 7.2). A single colony of *B. licheniformis* was inoculated into a 250 mL flask containing 50 mL LB broth, cultured at 37 °C and 180 rpm for 12 h until the mid-exponential phase.

60.2.2 Sampling

Cell suspension (10 mL) was sampled rapidly into five different cold (<−40 °C) methanol quenching solutions (30 mL): (A) pure methanol; (B) 60 % methanol; (C) 60 % methanol, 0.2 % ammonium carbonate buffer; (D) 60 % methanol, 0.9 %

ammonium carbonate buffer; (E) 60 % methanol, 2 % ammonium carbonate buffer. The solutions were mixed thoroughly and centrifuged for 10 min at 5,000 g and 4 °C. The supernatant was separated from the cell pellet. The cells were washed twice with 10 mL of 0.9 % ammonium carbonate buffer at −4 °C and centrifuged for 10 min at 5,000 g. Cell pellets were quickly frozen in liquid nitrogen and stored at −80 °C.

60.2.3 Extraction of Metabolites

Samples were thawed on ice (≤6 °C). The glass bead cell disruption method was carried out as described by Meyer et al. [10]. Four different extraction solution: pure methanol (MeOH), acetonitrile/water mixture (AW; 1:1, v/v), acetonitrile/methanol/water (AMW; 2:2:1, v/v/v), or methanol/water (MW; 1:1:, v/v) containing the cell sample was added to screw cap micro tube filled with 0.5 mL glass beads (Sartorius AG, diameter 0.10–0.11 mm). Next the cells were disrupted in a homogenizer (Precellys 24) using 3 cycles for 3 s at 5,000 rpm. The glass beads and the cell debris were separated from the supernatant by centrifugation for 5 min at 4 °C and 13,000 rpm, the aliquots were combined, and the glass beads were washed once with double-distilled water. The washing solutions were added to the metabolite extracts. The metabolite containing samples were restocked with double-distilled water to a final organic solution concentration of 10 %.The supernatant was concentrated to dryness and kept at −80 °C prior to derivatization and analysis by GC-MS.

60.2.4 Energy Charge Determination

For energy charge determination AMP, ADP, and ATP have been quantified. Intracellular ATP, ADP, and AMP were detected by HPLC. HPLC analysis was performed using a Waters 1525 Binary Pump with a 2489 UV/Visible Detector, a 2707 Autosampler, and an Agilent TC-C18 column (5 μm, 250 × 4.6 mm, Agilent). The 1.0 L mobile phase contained 13.6 g KH_2PO4, 0.608 g NaOH, 0.767 mL formic acid, 50 mL methanol, and 950 mL water. The methanol was chromatographic pure, and others were of analytical grade. The flow rate was 1.0 mL/min and the detection wavelength was 210 nm. Injecting volume was 20 μL and temperature of the column was 30 °C. The EC was calculated by the formula:

$$\text{Energy charge} = \frac{[\text{ATP}] + 0.5 \times [\text{ADP}]}{[\text{ATP}] + [\text{ADP}] + [\text{AMP}]} \tag{60.1}$$

60.2.5 GC-MS Analysis

Samples were derivatised for GC-MS by a two-step methox-imation/silylation derivatization procedure [11]. We added ribitol(1 mg/mL) to the samples as derivatization standards. The dried samples were first methoximated with a solution of 40 mg/mL methoxyamine hydrochloride (80 uL) in anhydrous pyridine at 30 °C for 90 min. Samples were then silylated with MSTFA (80 uL) at 37 °C for 120 min. The derived extracts were analyzed with an Agilent 7890 gas chromatograph coupled with an Agilent 5975C-GC/MSD. A 1 μL aliquot of the derivatives was injected into a HP-5MS capillary column (30 m × 250 μm, 0.25 μm film thickness) with a split less injection mode. The initial GC oven temperature was 70 °C, 4 min after injection the GC oven temperature was increased with 5 °C/min to 280 °C and kept at 280 °C for 5 min. Helium was used as a carrier gas and the helium flow was kept constantly at a flow rate of 1.0 mL/min. Detection was achieved using an MS detector in an electron impact mode and a full scan monitoring mode (m/z 30–600). The temperature of the ion source and the quadrupole were set at 230 °C and at 150 °C respectively.

60.3 Results and Discussion

60.3.1 Optimization of Quenching Solution

The efficiency of current quenching procedures with *B. licheniformis* was analyzed by GC-MS. Cell suspension was sampled rapidly into five different quenching solution to halt metabolism. As a result, differences in metabolite levels between these samples are due to the specific quenching method applied. A clustered heatmap in Fig. 60.1 detailed the levels of the identified metabolites quenched by different quenching solutions, in which 60 % methanol, 0.9 % ammonium carbonate was closely related to 60 % methanol, 2 % ammonium carbonate and these two groups showed almost equally superior performance and shared more common features. 60 % methanol, 0.9 % ammonium carbonate and pure methanol yielded the highest and lowest metabolite abundances, respectively.

Figure 60.2 shows the metabolite numbers of four classes on which we focused. In these classes, most metabolites are amino acids and other organic acids. Notably, 60 % methanol, 0.9 % ammonium carbonate quenching method produced the greatest number of amino acids, organic acids, fatty acids and sugars.

Most of the amino acids listed in Fig. 60.3a were more highly abundant with the 60 % methanol, 0.9 % ammonium carbonate quenching method than with others. Specifically, alanine, valine, leucine, and proline were significantly predominant in samples obtained using 60 % methanol, 0.9 % ammonium carbonate pretreatment. Other organic acids in Fig. 60.3b, like lactic acid, Fumaric acid, and D-Gluconic acid shared similar results as amino acids, indicating that *B. licheniformis* cells were

Fig. 60.1 Hierarchical clustering analysis of metabolites from *B. licheniformis* using five quenching solutions. Each column and each row represent a quenching solution and an individual metabolite, respectively

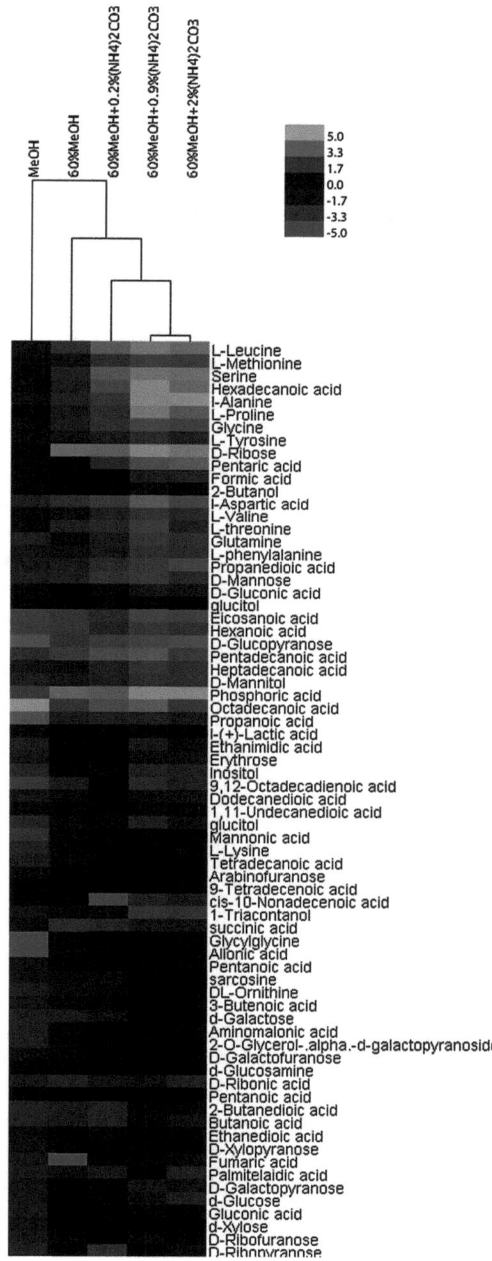

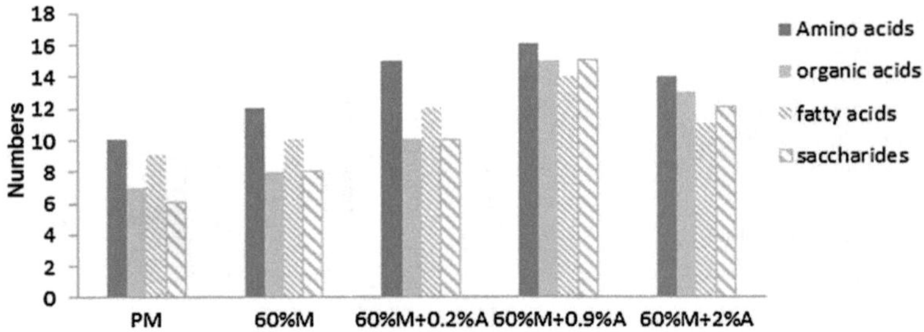

Fig. 60.2 Numbers of four classes of metabolites obtained through five quenching: solution: **a** pure methanol; **b** 60 % methanol; **c** 60 % methanol, 0.2 % ammonium carbonate buffer; **d** 60 % methanol, 0.9 % ammonium carbonate buffer; **e** 60 % methanol, 2 % ammonium carbonate buffer

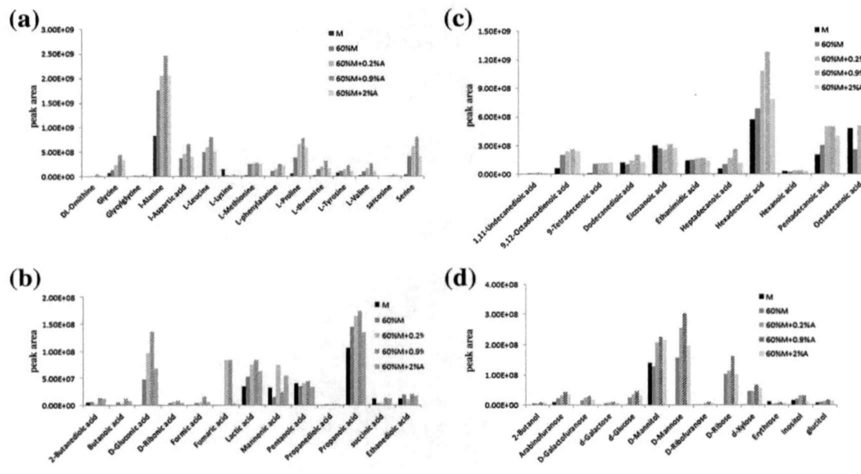

Fig. 60.3 Comparison of five quenching: methods in intracellular metabolite abundances **a** amino acids; **b** organic acids; **c** fatty acids; **d** sugars

also sensitive to cold methanol. Fatty acids show little change by different quenching methods (Fig. 60.3c). The levels of 9,12-Octadecadienoic Acid (linoleic acid) and Hexadecanoic acid in the four quenching methods expect pure methanol quenching method are similar. It is possible that partly the huge size of these molecules prevents their leakage from cell membrane. The same trend was also found in other research [8, 12]. Most sugars were detected at higher levels in the samples quenched by 60 % methanol, 0.9 % ammonium carbonate quenching method, especially for glucose, mannose acid, and ribose (Fig. 60.3d). So, 60 % methanol, 0.9 % ammonium carbonate quenching method showed better performance on *B. licheniformis*.

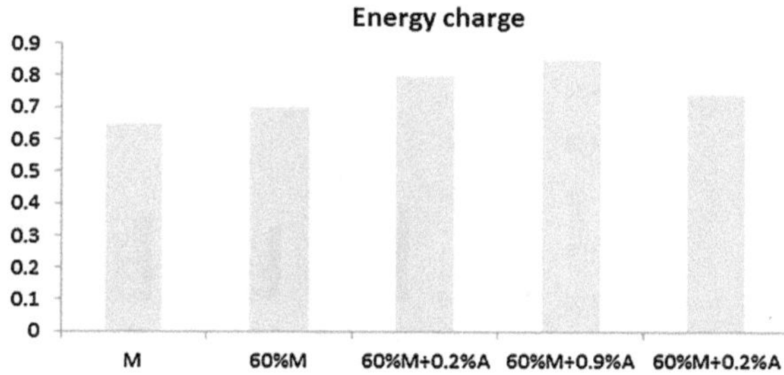

Fig. 60.4 The energy charge obtained by the different quenching: methods

60.3.2 Cellular Charge and Leakage

The EC has defined as an important physiological indictor which shows cellular metabolic activity [13]. As shown in Fig. 60.4, the EC values of all quenched cells are more than 0.60. The result showed that the 60 % methanol, 0.9 % ammonium carbonate quenching method that the EC value of all quenched cells is more high than others could effectively inhibit intracellular enzymes and stop the cellular metabolism immediately. Most literatures show that cellular leakage existed during quenching and it was unavoidable. In *Escherichia coli*, the percentage of ATP leakage could rise to about 70 % when methanol was used as a quenching solution [7]. Leakage was also evaluated in *B. licheniformis* by comparing the metabolites in the quenching supernatant fluid (Fig. 60.5) Metabolites levels in supernatant of quenched samples can straightly reflect the intracellular metabolite leakage. The 60 % methanol, 0.9 % ammonium carbonate and pure methanol quenching method yielded the lowest and highest metabolite abundances respectively. The possible effects of fluidity increase or thickness decrease in cell membranes and an increased membrane permeability were considered to be the main causes of the low metabolite recovery of the high concentration methanol quenching method. The intracellular metabolite leakage decrease as methanol concentrations increase. It is important that a right concentration is applied in quenching solution. Comparison of three kinds of ammonium carbonate concentrations, we find 0.9 % ammonium carbonate methanol quenching method is more suitable than others. It is possible that the osmotic pressure of the 0.9 % ammonium carbonate is near to cell osmotic pressure. The content of metabolites in the 60 % methanol, 0.9 % ammonium carbonate was then chosen for further use.

60.3.3 Optimization of the Extraction Solution

Extraction methods were evaluated and optimised for the efficient extraction metabolites [6, 14, 15]. During the extraction procedures, intracellular metabolites

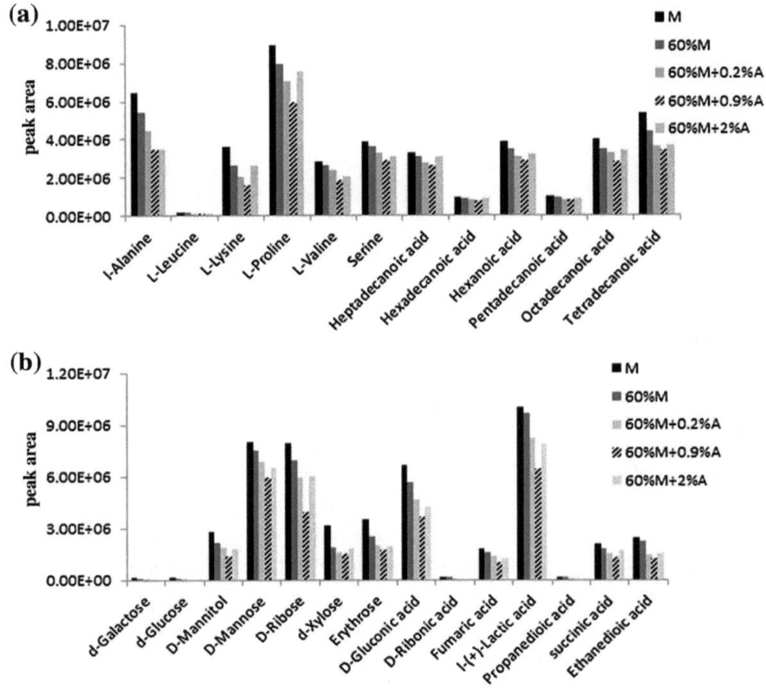

Fig. 60.5 Comparison of the leaked metabolites in the quenching supernatant fluid of five quenching methods

are exposed to various organic solvents. For an optimal metabolome analysis, this solvent should neither chemically nor physically modify or degrade the metabolites. Furthermore, the extraction solution must extract as many metabolites as possible [16]. For the comparison of the extraction potential, four different extraction solution are chosen: pure methanol (PM), methanol/water mixture (MW; 1:1, v/v), acetonitrile/methanol/water (AMW; 2:2:1, v/v/v), or acetonitrile/water (AW; 1:1, v/v). The different extraction solutions, tested in this study, result in highly varying relative metabolite concentrations. A clustered heatmap in Fig. 60.6 details the levels of the identified metabolites by different extraction solutions, in which methanol/water mixture (MW; 1:1, v/v) was closely related to acetonitrile/methanol/water (AMW; 2:2:1, v/v/v) and these two groups show almost equally superior performance and shared more common features. Thus, PW showed the highest peak intensity, whereas PM showed the lowest peak intensity. Extraction using PM, in particular, produced high metabolite recovery yields for amino acids, organic acids, phosphates, sugars, and fatty acids. These results strongly suggest that metabolite sample preparation methods need to be specifically developed, evaluated, and optimized for microorganisms.

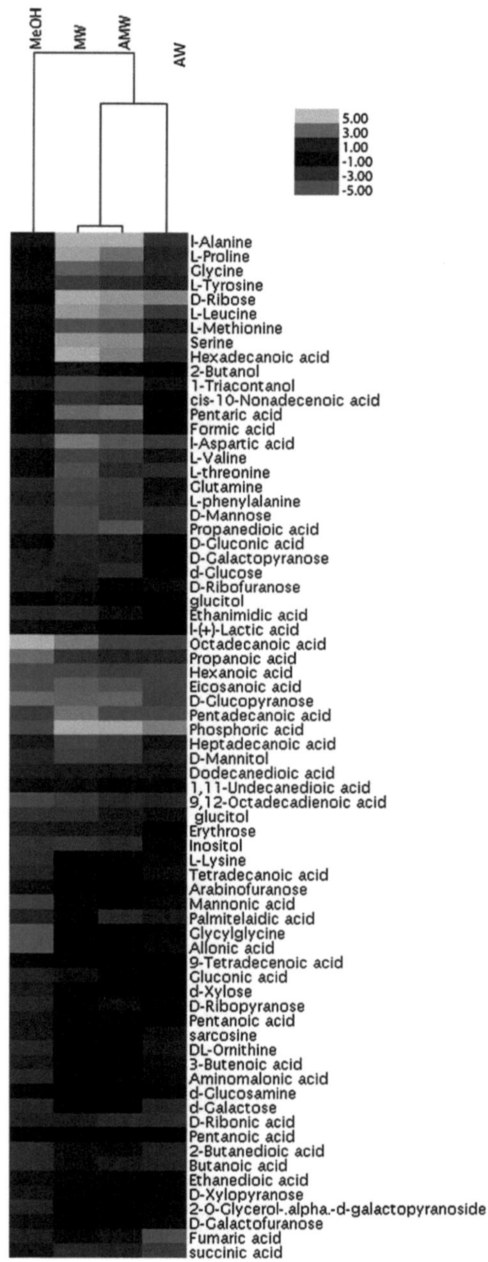

Fig. 60.6 Hierarchical clustering analysis of metabolites from *B. licheniformis* using pure methanol (PM), acetonitrile/water (AW; 1:1, v/v), acetonitrile/methanol/water (AMW; 2:2:1, v/v/v), or methanol/water (MW 1:1, v/v) as the extraction. Each column and each row represent an extraction solvent and an individual metabolite, respectively

Acknowledgments This research is supported by the National High-Tech Research and Development Plan ('863' Plan: 2011AA100905-4),the National Natural Science Foundation (31370076), the Tianjin Research Program of Application Foundation and Advanced Tech-nology (14JCYBJC23800), the Tianjin University of Science and Technology Research Science Foundation.

References

1. Tran PL, Cha H-J, Lee J-S, Park S-H, Woo E-J, Park K-H (2014) Introducing transglycosylation activity in *B. licheniformis* α-amylase by replacement of His235 with Glu. Biochem Biophys Res Commun 451(4):541–547
2. Li X, Long D, Ji J, Yang W, Zeng Z, Guo S, Ji Z, Qi G, Chen S (2013) Sample preparation for the metabolomics investigation of poly-gamma-glutamate-producing *B. licheniformis* by GC-MS. J Microbiol Meth 94(1):61–67
3. Shin MH, Lee DY, Liu KH et al (2009) Evaluation of sampling and extraction methodologies for the global metabolic profiling of Saccharophagus degradans. Anal Chem 82(15):6660–6666
4. Tian J, Sang P, Gao P, Fu R, Yang D, Zhang L, Zhou J, Wu S, Lu X, Li Y, Xu G (2009) Optimization of a GC-MS metabolic fingerprint method and its application in characterizing engineered bacterial metabolic shift. J Sep Sci 32(13):2281–2288
5. van der Werf MJ, Overkamp KM, Muilwijk B, Coulier L, Hankemeier T (2007) Microbial metabolomics: toward a platform with full metabolome coverage. Anal Biochem 370(1):17–25
6. Chen MM, Li AL, Sun MC, Feng Z, Meng XC, Wang Y (2014) Optimization of the quenching method for metabolomics analysis of Lactobacillus bulgaricus. J Zhejiang Univ Sci B 15(4):333–342
7. Link H, Anselment B, Weuster-Botz D (2008) Leakage of adenylates during cold methanol/glycerol quenching of Escherichia coli. Metabolomics 4(3):240–247
8. Canelas AB, Ras C, Pierick A, Dam JC, Heijnen JJ, Gulik WM (2008) Leakage-free rapid quenching technique for yeast metabolomics. Metabolomics 4(3):226–239
9. Canelas AB, ten Pierick A, Ras C et al (2009) Quantitative evaluation of intracellular metabolite extraction techniques for yeast metabolomics. Anal Chem 81(17):7379–7389
10. Meyer H, Liebeke M, Lalk M (2010) A protocol for the investigation of the intracellular Staphylococcus aureus metabolome. Anal Biochem 401(2):250–259
11. Halket JM, Waterman D, Przyborowska AM et al (2005) Chemical derivatization and mass spectral libraries in metabolic profiling by GC/MS and LC/MS/MS. J Exp Bot 56(410):219–243
12. Zhao C, Nambou K, Wei L et al (2014) Evaluation of metabolome sample preparation methods regarding leakage reduction for the oleaginous yeast Yarrowia lipolytica. Biochem Eng J 82:63–70
13. Sellick CA, Hansen R, Maqsood AR et al (2013) Effective quenching processes for physiologically valid metabolite profiling of suspension cultured mammalian cells. Anal Chem 81(1):174–183
14. Dettmer K, Nurnberger N, Kaspar H, Gruber MA, Almstetter MF, Oefner PJ (2011) Metabolite extraction from adherently growing mammalian cells for metabolomics studies: optimization of harvesting and extraction protocols. Anal Bioanal Chem 399(3):1127–1139
15. Mozzi F, Ortiz ME, Bleckwedel J, De Vuyst L, Pescuma M (2013) Metabolomics as a tool for the comprehensive understanding of fermented and functional foods with lactic acid bacteria. Food Res Int 54(1):1152–1161
16. Tredwell GD, Edwards-Jones B, Leak DJ, Bundy JG (2011) The development of metabolomic sampling procedures for Pichia pastoris, and baseline metabolome data. PLoS One 6(1): e16286

Chapter 61
Characterization of Recombinant L-Amino Acid Deaminase of *Proteus mirabilis*

Chenglin Zhang, Jia Feng, Xixian Xie, Qingyang Xu and Ning Chen

Abstract L-amino acid deaminases catalyze the deamination of L-amino acids. Up until now, two types of L-amino acid deaminase have been identified in *Proteus* species. To investigate enzymatic characteristics of L-amino acid deaminase from *Proteus mirabilis*, L-amino acid deaminase encoding gene (*pmta*) was cloned from *P. mirabilis* T-1. Prokaryotic expression system was established to express recombinant Pmta. Enzymatic characteristics of the enzymes were analyzed. Results showed that recombinant Pmta exhibited function of second type of L-amino acid deaminase. The Km and Vmax value of Pmta for histidine was 10.57 mmol/L and 202.06 μmol/min/mg, respectively. The optimal temperature and pH of recombinant Pmta was 40 °C and 7.0. The enzymatic characteristics of Pmta were different from those of Pm1 discovered in *P. mirabilis* KCTC, which was probably due to different amino acid sequences. The Pmta deaminase will be very useful in the preparation of commercially valuable materials including urocanic acid and 3-mercaptopyruvic acid.

Keywords *Proteus mirabilis* · L-amino acid deaminase · Enzymatic characteristics

An erratum of this chapter can be found under DOI 10.1007/978-3-662-45657-6_66

C. Zhang · J. Feng · X. Xie · Q. Xu · N. Chen (✉)
College of Biotechnology, Tianjin University of Science and Technology, Tianjin 300457, China
e-mail: ningch@tust.edu.cn

C. Zhang · X. Xie · Q. Xu · N. Chen
National and Local United Engineering, Lab of Metabolic Control Fermentation Technology, Tianjin 300457, China

T.-C. Zhang and M. Nakajima (eds.), *Advances in Applied Biotechnology*,
Lecture Notes in Electrical Engineering 332, DOI 10.1007/978-3-662-45657-6_61

61.1 Introduction

As a motile Gram-negative bacterium within the *Enterobacteriaceae*, *Proteus mirabilis* undergoes dramatic morphological changes in response to growth on solid surfaces [1, 2]. It is commonly found in water, soil, intestinal tract, and on the skin of human and animals. *P. mirabilis* is a frequent etiological agent especially in catheterized patients or individuals with structural abnormalities of the urinary tract and it can cause serious complications including acute pyelonephritis, bladder and kidney stones, and bacteremia [1, 2]. In addition, *P. mirabilis* has been reported as one of the causative agents of human pneumonia and other lung infection conditions, endotoxin-induced sepsis, and central nervous system infection [3–5].

L-Amino acid deaminases catalyze deamination of natural L-amino acids to generate the respective α-keto acids. It is known that *P. mirabilis* contain two types of amino acid deaminases [6–8]. One type deaminated a wide range of aliphatic and aromatic amino acids, whereas the other deaminated only a small range of basic amino acids, such as L-arginine and L-histidine [6–8]. Massad et al. cloned *pma* from *P. mirabilis* HI4320 and the Pma could catalyze such aromatic amino acids as L-phenylalanine, L-leucine, L-aspartic acid, L- methionine to corresponding α-keto acids, so it belonged to the first type of amino acid deaminase [8]. Then, Baek et al. discovered *pm1* from *P. mirabilis* KCTC the same strain and the Pm1 could deaminate only a narrow range of L-amino acids, especially basic molecules including L-arginine and L-histidine, indicating that it belong to the second type of amino acid deaminase [9, 10].

Iron acquisition by siderophores is required for bacterial survival [11], however, *P. mirabilis* does not produce any of such traditional siderophores as enterobactin, aerobactin, or the ferroxamine-type siderophores) [12, 13]. Drechsel et al. [14] first demonstrated novel roles as siderophores for α-keto acids produced by deaminases in the genera *Proteus*. Amino acid deaminases are commonly employed to produce organic acids. A typical example is the production of phenylpyruvic acid from L-phenylalanine by an amino acid deaminase from *P. mirabilis* [15–17].

In this study, a second type of amino acid deaminase, Pmta, encoding genes were cloned from *P. mirabilis* T-1 isolated from acute pyelonephritis patients. And the enzymatic characteristics were studied.

61.2 Materials and Methods

61.2.1 Strains and Plasmids

Escherichia coli DH5α, *E. coli* BL21(DE3), *P. mirabilis* T-1 was stored in our lab. pMD19-T simple vector [TAKARA Biotechnology(Dalian) Co., Ltd.] and pET-42a (laboratory stock) were used as cloning and expression vector, respectively.

61.2.2 Media and Culture Condition

E. coli and *P. mirabilis* T-1 were grown at 37 °C in Luria-Bertani medium [LB, 1 % (w/v) Tryptone, 0.5 %(w/v) yeast extract, and 1 %(w/v) NaCl]. Kanamycin (50 μg/ml) or ampicillin (100 μg/ml) was added for strain selection.

61.2.3 Primers

Primers used were listed in Table 61.1.

61.2.4 Cloning of Pmta

Pmta was amplified cloned from *P. mirabilis* T-1 using the primers PM-1 and PM-2. The PCR product was cloned into pMD19-T simple vector and was sequenced.

61.2.5 Amplification and Expression of Pmta Genes

As reported, the second type of amino acid deaminase is a membrane-bound protein, Pmta were predicted contained the same region. To achieve its soluble and high-level expression, the two genes with deletion of N-terminal transmembrane region (from 21 to 87th nucleotide) was first amplified by PCR using primes PM-3 and PM-5 (88-1416 bp) from *P. mirabilis T-1* and then using PM-5 and PM-4 from the above generated fragments, named *pmta*.

Fragments of *pmta* was digested by *Nde* I and *Xho* I and was cloned into the corresponding restriction sited of pET-42a, resulting in pET-pmta. Then, the recombinant plasmids were transformed to *E. coli* BL21(DE3), generating BL-pmta.

BL-pmta was cultured in LB medium at 37 °C to the midexponential stage of cell growth and expression of Pmta was induced by addition of IPTG (to 0.1 mmol/L), followed by incubation for further 4 h. Induced cells were harvested and disrupted by

Table 61.1 Primers

Primers	Sequence (5′-3′)
PM-1	ATGGCAATAAGTAGAAGAAAATTTATTCTT
PM-2	TTAGAAACGATACAGACTAAATGGTTTG
PM-3	CGAGAAGGGCGTTTTGTTCC
PM-4	GGCCATATGGCAATAAGTAGAAGACGAGAAGGGCGTTTTGTTCC (*Nde* I)
PM-5	GGCCTCGAGGAAACGATACAGACTAAATGGTTTGG (*Xho* I)

sonication. Cell lysates were separated into soluble and insoluble fractions by centrifugation at 10,000×*g* for 30 min. To confirm the expression of Pmta, SDS-PAGE was performed. Protein bands were visualized in gels by staining with coomassie brilliant blue R 250.

The 6×His tagged Pmta was purified using an Ni^{2+}-NTA affinity column. And the enzyme concentration was determined by the BCA assay kit (Bio-Rad, USA).

61.2.6 Amino Acid Deaminase Assay

Amino acid deaminase activity was determined by the method of Massad et al. [8]. The optimum temperature and pH was determined by measuring the amino acid deaminase activity at different temperature and pH with L-histidine as substrate as described above.

61.3 Results and Discussions

61.3.1 Cloning and Sequence Analysis of Pmta

Pmta was successfully cloned from *P. mirabilis* T-1. Sequencing analysis showed that the open reading frame of *pmta* was 1,416 bp in length and encoded a 472-amino acid protein. The gene exhibited 98.31 % (1,392/1,416) similarity with nucleotides of *pm1* belonging to the second amino acid deaminase of *P. mirabilis* KCTC 2566 [9]. Amino acid sequence of Pmta differed in 6 amino acid residues, Asp-42, Gly-55, His-97, Ser-150, Ala-318 and Ser-355, with respect to the corresponding Gly, Asp, Tyr, Ser, Val, and Ala residues of Pmta from *P. mirabilis* KCTC 2566. The above results indicated that Pmta from *P. mirabilis* might exhibit different characteristics from Pmta of *P. mirabilis* KCTC 2566.

61.3.2 Expression of Pmta in E. Coli BL21(DE3)

It is reported that unlike eukaryotic and other prokaryotic L-amino acid deaminases, L-amino acid deaminases from *Proteus* species are membrane-bound [10]. Transmembrane helices of Pmta located between residues 7 and 29, with 96.58 % probability were predicted by the program of TMHMM Server v.2.0 (http://www.cbs.dtu.dk/services/TMHMM-2.0/, data not shown), which is similar to Pm1, and thus it is difficult to obtain the purified protein in a soluble form. Up to now, even though a few methods have been routinely used to increase the chances of producing a soluble and active membrane-bound protein, appropriated conditions for soluble expression of a given membrane-bound protein are often failed to be

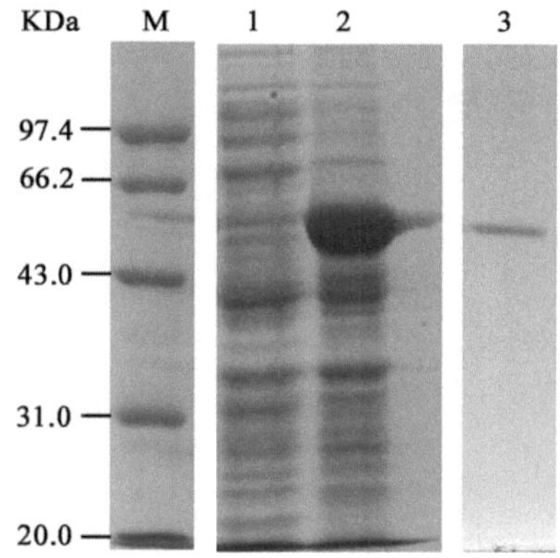

Fig. 61.1 SDS-PAGE analysis of recombinant Pmta expressed in *E. coli* BL21 (DE3). *M* marker; *Lane 1 E. coli* BL21(DE3) harboring pET42a; *Lane 2 E. coli* BL21 (DE3) harboring pET-pmta; *Lane 3* purified recombinant Pmta

established. To study the function of Pmta and to achieve its soluble and high-level expression, the gene without transmembrane region was amplified by PCR and was cloned into pET42a after excised with *Nde* I and *Xho* I. Expression of *pmta* was analyzed by SDS-PAGE. An additional band was detected at approximately 50 kDa, corresponding to the expected size of the 452-amino acid protein in lysate of BL-pmta, indicating that *pmta* was successfully expressed. Recombinant Pmta was purified (Fig. 61.1) and the concentration was 3.6 mg/L.

61.4 Enzymatic Characterization of Pmta

It was reported that the two types of amino acid deaminases differ in substrate. One type exhibits high activity to a wind range of L-amino acids including aliphatics and aromatics, whereas the other type deaminates a narrow range of L-amino acids, especially basic molecules. As described above, Pmta exhibited high similarity with Pm1 of *P. mirabilis* KCTC 2566. In the present study, it is found that Pmta

Table 61.2 Conversion rate of 20 L-amino acids by recombinant enzymes

L-amino acids	Conversion rate (%)	L-amino acids	Conversion rate (%)
L-Ala	15.9	L-Leu	2.1
L-Arg	58.1	L-Lys	9.6
L-Asn	6.7	L-Met	4.5
L-Asp	3.7	L-Phe	52.1
L-Cys	31.2	L-Pro	10.8
L-Gln	2.3	L-Ser	18.9
L-Glu	9.7	L-Thr	11.7
L-Gly	5.3	L-Trp	8.9
L-His	90.1	L-Tyr	5.7
L-Ile	2.1	L-Val	4.2

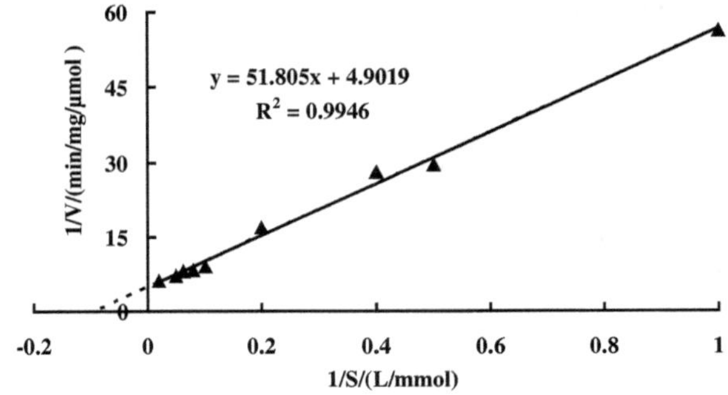

Fig. 61.2 Hyperbolic kinetics of recombinant enzymes

effectively deaminated basic amino acids such as L-histidine and L-arginine (Table 61.2), thus it belonged to the second type of L-amino acid deaminase. As is known, urocanate formed by histidinase or histidine ammonia-lyase after elimination of ammonium is a natural photoprotectant against UV radiation-induced DNA damage [18], so Pmta might be suitable to be explored in industrial production of urocanate.

Notably, Pmta also exhibited significant activity of deaminating L-cysteine to produce 3-mercaptopyruvic acid that was studied as a potential treatment for cyanide poisoning [19]. As it was reported, Pm1 efficiently deaminates L-glutamate to produce α-oxoglutarate and ammonia. However, Pmta had a relatively lower activity toward L-gluminate.

Purified Pmta was used to examine the kinetics of activity. The Km and Vmax value for histidine were 10.57 mmol/L and 202.06 μmol/min/mg, respectively (Fig. 61.2). We previously detected the kinetics of Pm1 activity, and the Km and Vmax value for histidine was 15.34 mmol/L and 186.97 μmol/min/mg, respectively (data not shown). The result indicated that Pmta had a higher deamination activity and substrate specificities when histidine was used as substrate.

The optimal temperature of recombinant Pmta was determined at the range of 10–70 °C. The results showed that the enzyme activity increased with the increasing temperature from 10–40 °C and decreased from 40–70 °C. The optimal

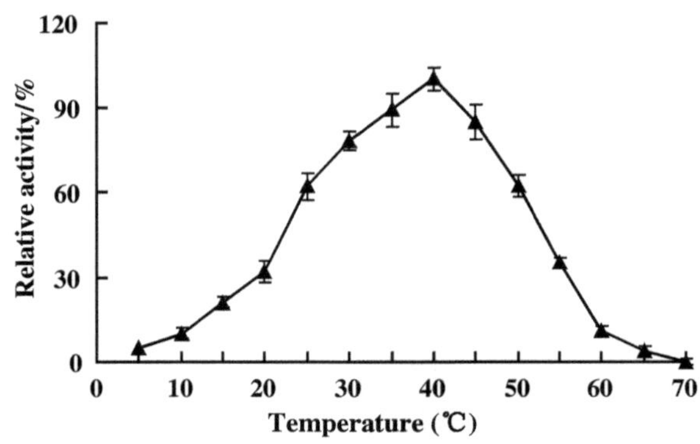

Fig. 61.3 Effect of temperature on recombinant enzyme activity

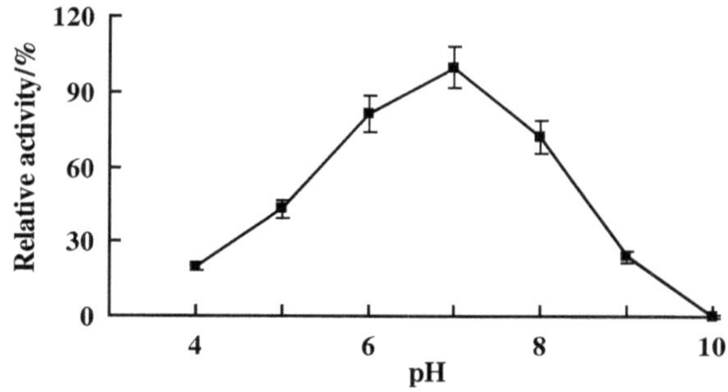

Fig. 61.4 Effect of pH on recombinant enzyme activity

temperature for recombinant Pmta was 40 °C (Fig. 61.3). The optimal pH of recombinant Pmta was determined at the range of 3.0–10.0. Results showed that the optimal pH for recombinant Pmta was 7.0 (Fig. 61.4). The differences in optimal temperature and pH of Pmta and Pm1 might be due to the divergence in amino acid sequence of the two enzymes.

61.5 Conclusion

This study was to determine the characteristics of Pmta. Homology alignment of Pmta and Pml showed that Pmta had relatively higher similarity to Pm1. Enzyme kinetic analysis indicated that the Km value of Pmta for histidine was 10.57 mmol/L and Vmax value was 202.06 μmol/min/mg and that Pmta had a higher activity and substrate specificities for histidine.

In the present study, we found that the enzymatic features of Pmta were different from Pm1, which might be due to the difference in primary structure of the two enzymes. Additionally, the Pmta deaminase will be very useful in the preparation of commercially valuable materials including urocanic acid and 3-mercaptopyruvic acid.

Acknowledgments This work was supported by the National High Technology Research and Development Program (No. 2013AA102106), by the Program for Changjiang Scholars and Innovative Research Team in University (No. IRT1166) and by the Tianjin Municipal Education Commission (No. 20120630).

References

1. Holt JG, Krieg NR, Sneath PHA et al (1994) Bergey's manual of determinative bacteriology, 9th edn. Williams and Wilkins, Baltimore
2. Hoeniger JFM (1964) Cellular changes accompanying the swarming of *Proteus mirabilis*.I. Observations on living cultures. Can J Microbiol 10:1–9

3. Warren JW, Tenney JH, Hoopes JM et al (1982) A prospective microbiologic study of bacteriuria in patients with chronic indwelling catheters. J Infect Dis 146(6):719–723
4. Mobley HLT, Belas R (1995) Swarming and pathogenicity of *Proteus mirabilis* in the urinary tract. Trends Microbiol 3(7):280–284
5. Mohr C, Brenner W, Miller JM (2000) Classification, identification, and clinical significance of *Proteus*, *Providencia*, and *morganella*. Clin Microbiol Rev 13(4):534–546
6. Chow AW, Taylor PR, Yoshikawa TT et al (1979) A nosocomial outbreak of infections due to a multiple resistant *Proteus mirabilis*: role of intestinal colonization as a major reservoir. J Infect Dis 139(6):621–627
7. Roberts JA, Fussell EN, Kaack MB (1990) Bacterial adherence to urethral catheters. J Urol 144(2 Pt 1):264–269
8. Massad G, Bahrani FK, Mobley HLT (1994) *Proteus mirabilis* fimbriae: identification, isolation, and characterization of a new ambient-temperature fimbria. Infect Immun 62 (5):1989–1994
9. Baek JO, Seo JW, Kwon O et al (2011) Expression and characterization of a second L-amino acid deaminase isolated from *Proteus mirabilis* in *Escherichia coli*. J Basic Microbiol 51 (2):129–135
10. Baek JO, Seo JW, Kwon O et al (2008) Heterologous expression and characterization of L-amino acid deaminase from *Proteus mirabilis* in *Escherichia coli*. Chi J Biotechnol 24 (12):2129
11. Dumanski AJ, Hedelin H, Edin-Liljegren A, Beauchemin D et al (1994) Unique ability of the *Proteus mirabilis* capsule to enhance mineral growth in infectious urinary calculi. Infect Immun 62(7):2998–3003
12. Lukomski S, Serwecinska L, Rozalski A et al (1991) Cell-free and cell-bound hemolytic activities of *Proteus penneri* determination by different Hly determinants. Can J Microbiol 37 (6):419–424
13. Rabsch W, Winkelmann G (1991) The specificity of bacterial siderophore receptors probed by bioassays. Biol Met 4(4):244–250
14. Drechsel H, Thieken A, Reissbrodt R et al (1993) Alpha-keto acids are novel siderophores in the genera *Proteus*, *Providencia*, and *Morganella* and are produced by amino acid deaminases. J Bacteriol 175(9):2727–2733
15. Coker C, Poore CA, Li X et al (2000) Pathogenesis of *Proteus mirabilis* urinary tract infection. Microbes Infect 2(12):1497–1505
16. Duerre J, Chakrabarty S (1975) L-amino acid oxidase of *Proteus rettgeri*. J Bacteriol 121 (2):656–663
17. Takahashi E, Ito K, Yochimoto T (1999) Cloning of L-amino acid deaminase gene from *Proteus vulgaris*. Biosci Biotechnol Biochem 63(12):2244–2247
18. Kolenbrander HM, Berg CP (1967) Role of urocanic acid in the metabolism of L-histidine. Arch Biochem Biophys 119(1):110–118
19. Nagahara N, Li Q, Sawada N (2003) Do antidotes for acute cyanide poisoning act on mercaptopyruvate sulfurtransferase to facilitate detoxification? Curr Drug Targets Immune Endocr Metabol Disord 3(3):198–204

Chapter 62
Screening for Strains Capable of 13β-ethyl-4-gonene-3, 17-dione Biotransformation and Identification of Product

Linlin Huang, Xiaoguang Liu, Yulan He, Pingping Wei, Shuhong Mao and Fuping Lu

Abstract Screening of strains for improvement in the yields of the desired steroid as well as production of novel steroid is highly demanded. In this study, a filamentous fungus *Aspergillus oryzae* capable of transforming steroid efficiently was screened from the collection of fresh bark of south China. A novel transformation product was purified, crystallized, and determined as [13, 17β-]furan-17-hydroxyl-4-gonene-3one by single-crystal X-ray diffraction. At the end of the transformation, the yields of the product reached to 95 %, which indicated *A. oryzae* can be efficiently applied in the field of steroid biotransformation with the novel biocatalytic ability.

Keywords Biotransformation · *Aspergillus oryzae* · 13β-ethyl-4-gonene-3, 17-dione

62.1 Introduction

Steroids are widely used as antiinflammatory, diuretic, anabolic, contraceptive, antiandrogenic, and anticancer agents for prevention and therapy of many severe diseases [1–3]. At present steroid production has become an important branch of healthcare industry and the second large category next to antibiotics. Microbial transformations are an efficient and often key means for the production of new

L. Huang (✉) · X. Liu · Y. He · P. Wei · S. Mao (✉) · F. Lu (✉)
Key Laboratory of Industrial Fermentation Microbiology, Ministry of Education, National Engineering Laboratory for Industrial Enzymes, The College of Biotechnology, Tianjin University of Science & Technology, Tianjin 300457, People's Republic of China
e-mail: shuhongmao@tust.edu.cn

F. Lu
e-mail: lfp@tust.edu.cn

T.-C. Zhang and M. Nakajima (eds.), *Advances in Applied Biotechnology*,
Lecture Notes in Electrical Engineering 332, DOI 10.1007/978-3-662-45657-6_62

biologically active steroid compounds due to its high chemo-, regio-, and stereoselectivity, which additionally fulfills the principles of "green chemistry" [4, 5].

Microorganisms enable efficient syntheses of known and novel steroids that have attracted considerable attention in the steroid manufacture [6]. About 1,500 species, containing *Streptomyces*, *Aspergillus,* and *Penicillium,* of fungi have been extensively studied about conversion of steroid [7–12]. For example, the 11α, 11β, and 16α-hydroxylation with high yield via biotransformation has been applied extensively for commercial production of functionalized steroid compounds such as adrenal cortex hormones, corticosteroids.

However, commercialized microbial process in the production of valuable steroid is still limited. Therefore, screening of strains for improvement in the yields of the desired steroid as well as production of novel steroid is highly demanded.

13β-ethyl-4-gonene-3, 17-dione is a key intermediate in the synthesis of Gestodene, which is widely used as contraceptive medicine [13]. Many strains have been discovered to mediate hydroxylation of 13β-ethyl-4-gonene-3, 17-dione in 11α and 15α position [14–17]. In our recent study, *A. oryzae* was observed to have the ability to transform 13β-ethyl-4-gonene-3, 17- dione efficiently, a new steroid derivative was obtained and identified as [13,17β-]furan-17-hydroxyl-4- gonene-3one by single-crystal X-ray diffraction. To the best of our knowledge, this conversion is a novel reaction in the field of microbial transformation of steroids.

62.2 Materials and Methods

62.2.1 Materials

Steroids (13β-ethyl-4-gonene-3, 17-dione) of the analytical grade were obtained from Beijing Zizhu Pharmaceutical Company Ltd. (Beijing, China). Silicagel GF254 thin layer chromatographic plates and silica gel (200–300 mesh) were purchased from Qingdao Marine Chemical Factory, China. All chemical solvents were of analytical grade or higher. HPLC grade acetonitrile (Fisher, Loughborough, UK) and ultrapure water were used for all HPLC analyses.

62.2.2 Sampling

Samples were collected from the fresh bark of trees of Qichun, Hubei province.

62.2.3 Mediums

The plate and agar-slant medium contains: 2 % (w/v) glucose, 20 % (v/v) potato extract, and 2 % (w/v) agar power. The fermentation medium contains: 2 % (w/v) glucose, 2 % (w/v) peptone, and 2 % (w/v) yeast extract, pH natural. Mediums were sterilized at 121 °C for 20 min.

62.2.4 Screening Procedure

62.2.4.1 Separation and Purification

The bark samples were cut into little pieces, and cultured on the plate medium. After being incubated for 1–2 days at 28 °C, some fungal colonies were chosen based on differences in colony morphology. Then, the pure colonies of each strain were obtained by repeated transferring and further subcultured on slant medium. After 3 days incubation, the purified culture was stored at 4 °C.

62.2.4.2 Screening Microorganism and Microbial Transformation

The strains was grown on an agar-slant at 28 °C for 3 days and then the spores were washed down with sterile water, then they were inoculated into a 250 mL flask with 50 mL fermentation medium with a final concentration of 2×10^6 spores/mL. After cultivated on a rotary shaker at 180 rpm at 28 °C for about 24 h, the steroid substrate 13β-ethyl-4-gonene-3, 17-dione (0.5 g) dissolved in 1 mL ethanol was added in the each fermentation medium. The steroid biotransformation was carried out at the same condition. After the biotransformation, the reaction mixture was extracted with ethyl acetate, and then analyzed by TLC and HPLC. All the experiments were repeated three times.

62.2.5 Extraction, Isolation, and Identification of the Transformed Products

At the end of transformation, the fermentation broth was extracted three times with ethyl acetate. The extract was evaporated under reduced pressure on a rotary evaporator. The concentrated residue was separated by column chromatography using silica gel and petroleum ether/ethyl acetate (1:1, v/v) as the eluent. The purified product was crystallized in dichloromethane and then determined using single-crystal X-ray diffraction.

62.2.6 Identification of the Strain

Purified colonies of the strain were grown in the fermentation medium for 48 h at 28 °C under agitation. Then, the mycelium was collected by centrifuge and washed twice with water. The extraction of DNA was performed using a DNA extraction Mini kit according to the manufacturer's instructions (ZOMNBIO). The primer pair of ITS-1 (5′-TCCGTAGGTGAACCTGCGG-3′) and ITS-4 (5′-TCCTCCGCT TATTGATATGC-3′) was used to amplify the internal transcribed spacer regions (ITS-1 and ITS-2) of the ribosomal DNA (rDNA). The PCR products were directly sequenced after purification. The sequence similarity searching for this strain was performed using the NCBI BLAST. A species name was assigned to the isolated strain in the blast search.

62.2.7 Analytical Methods

62.2.7.1 TLC

The course of biotransformation was controlled by means of TLC with petroleum ether/ethyl acetate (1:1, v/v) as a mobile phase, and then the plates were analyzed under the UV 254 nm.

62.2.7.2 HPLC

The sample of 10 μL was first loaded onto a SDB C18 column (4.6 mm × 150 mm, Agilent Technologies, USA). The condition for HPLC was: mobile phase water–acetonitrile (70:30, v/v); flow rate (0.8 mL min^{-1}); and wavelength (240 nm). The concentration of substrate and product was determined from calibration curves generated from eluent solutions of standard substrate and product.

62.3 Results and Discussion

In this study, 36 strains were isolated from the samples, and 8 strains were found to have the steroid biotransformation ability. A strain named 5# was observed to have the ability to transform 13β-ethyl-4-gonene-3, 17-dione efficiently and specifically, and as shown in Fig. 62.1, a single product was observed after 48 h incubation.

Fig. 62.1 TLC analysis of 13β-ethyl-4-gonene-3, 17-dione transformation by strain 5# (*Lane 1* standard of 13β-ethyl-4-gonene-3, 17-dione; *Lane 2* transformation products)

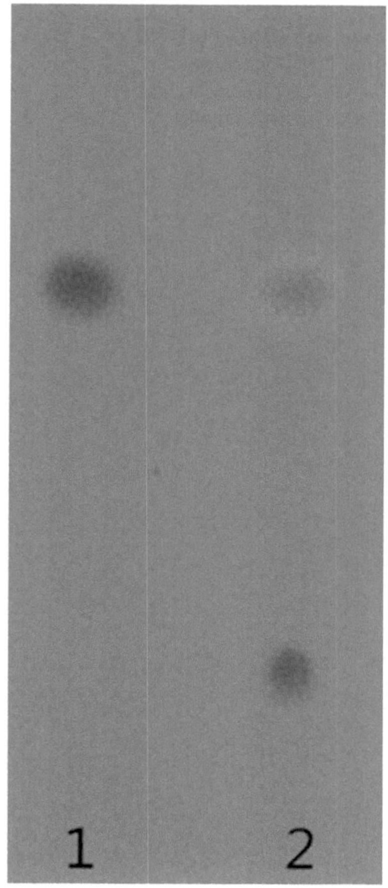

62.3.1 Identification of the Strain

As it can be seen from Fig. 62.2, a 576 bp of the ITS fragment was amplified from the genomic DNA by PCR. The strain isolated in this study was identified as *A. oryzae* which gave the highest bit score and showed 99 % similarity in the blast search. In addition, the morphology of this strain after 48 and 72 h of incubation at 28 °C was examined directly with a microscope. As shown in Fig. 62.3, morphological characteristics of asexual spores and associated spore producing structures of this strain further confirm the fungal species identity [18–21].

A. oryzae is widely used to produce commercial enzymes applied in food pharmaceutical fields [21]. However, few application of *A. oryzae* in steroid biotransformation has been reported.

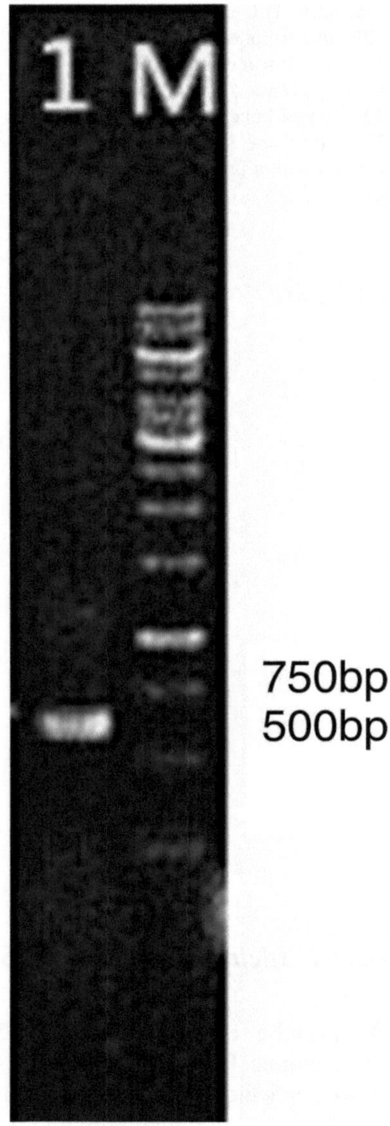

Fig. 62.2 A 576 bp DNA fragment amplified by PCR from ITS of strain 5# (*M* DNA molecular weight marker, *1* PCR product of ITS)

62.3.2 Identification of Transformation Products

After biotransformation, the reaction mixture was extracted three times with ethyl acetate. The transformation product was separated, recrystallized, and then identified as [13,17β-]furan-17-hydroxyl-4-gonene-3one by single-crystal X-ray diffraction. The CCDC 102 3284 was obtained after the crystallographic data was

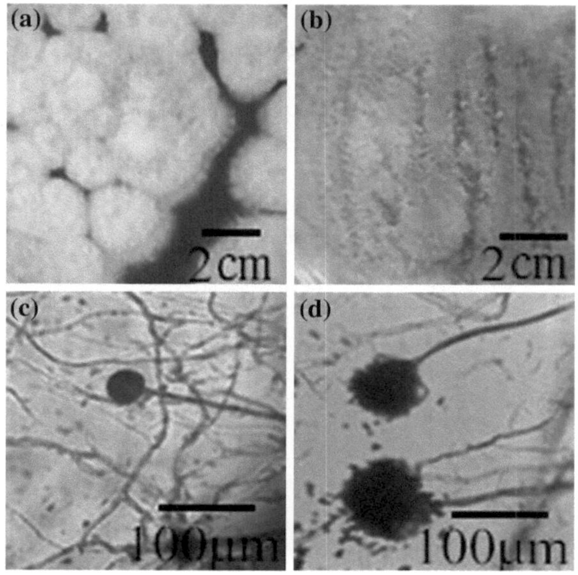

Fig. 62.3 Colonial morphology (**a** and **b**) and Microscopic characteristic (**c** and **d**) of strain 5#

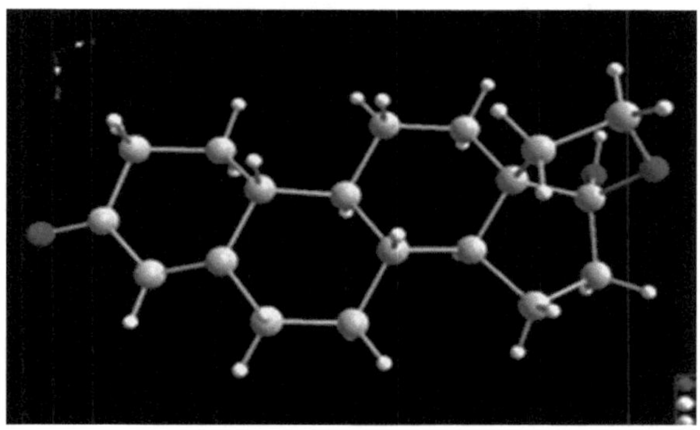

Fig. 62.4 X-ray crystal structure of [13, 17β-] furan-17-hydroxyl-4-gonene-3one

deposited with the Cambridge Crystallographic Data Centre. The data of the structure can be obtained on application to CCDC, 12 Union Road, Cambridge CB21EZ, UK (fax: +441223336033 or e-mail: deposit@ccdc.cam.ac.uk). The crystal structure of the novel metabolic product, as shown in Fig. 62.4, was reported for the first time. The biotransformation of 13β-ethyl-4-gonene-3, 17-dione was described in Scheme 62.1, which indicates that the strain of *A. oryzae* can be used in the novel steroid bioconversion.

Crystal data: C19H26O3, *Mr* = 302.4, crystal size 0.18 × 0.17 × 0.15 mm, Crystal system, space group: Monoclinic, $P_2(1)$, $a = 7.1505$ (7) Å, $b = 19.628(2)$ Å, $c = 11.4618(12)$ Å, $\alpha = 90°$, $\beta = 103.574°$, $\gamma = 90°$, $V = 1563.7(3)$ Å, $Z = 4$, Dcalcd = 1.284 mg/m^3, absorption coefficient = 0.085 mm^{-1}, measured temperature 173(2) K, reflection collected 9112, independent reflections 5460, final *R* indices

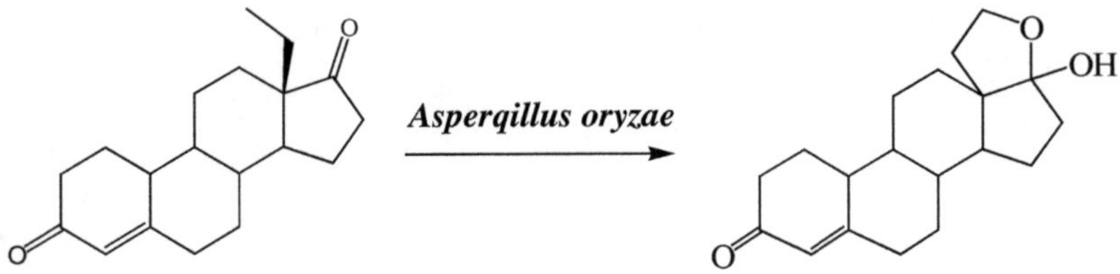

Scheme 62.1 Microbial transformation of 13β-ethyl-4-gonene-3, 17-dione by *A. oryzae*

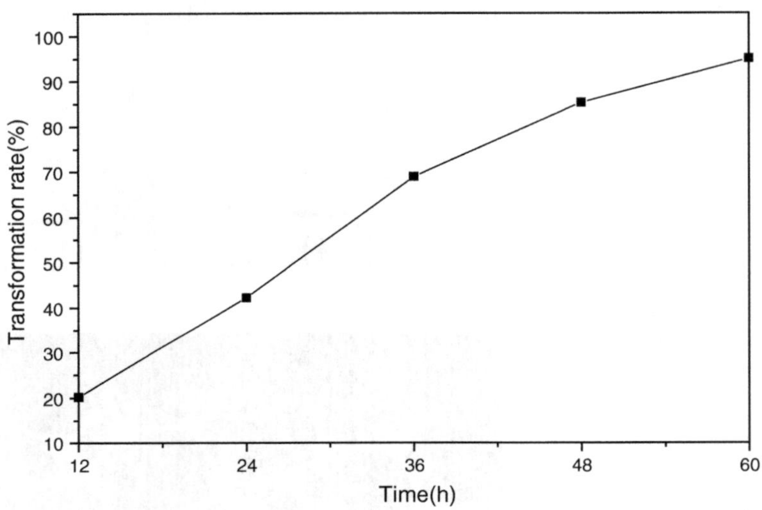

Fig. 62.5 Time course profile of the biotransformation of 13β-ethyl-4-gonene-3, 17-dione by *A. oryzae*

R_1 = 0.0418, wR_2 = 0.0975, *R* indices (all data) R_1 = 0.0461, wR_2 = 0.1004, GOF = 1.051.

The biotransformation of 13β-ethyl-4-gonene-3, 17-dione (1) in the culture of *A. oryzae* every 12 h was detected by HPLC for a time course study. As shown in Fig. 62.5, the yield of the product increased to about 42.2 % after 24 h incubation, followed by a gradual increase to about 68.9 % after additional 24 h. At the end of the biotransformation, its yield increases to 95 %.

62.4 Conclusion

A. oryzae was isolated in this study and observed to have the ability to transform 13β-ethyl-4-gonene-3, 17-dione efficiently. The novel metabolic product [13,17β-] furan-17-hydroxyl-4-gonene-3one was identified and reported for the first time.

This result indicated that the transformation of 13β-ethyl-4-gonene-3, 17-dione by *A. oryzae* is a novel microbial reaction in the field of microbial transformation of steroids.

Acknowledgments This work was supported by National High-tech Research and Development Program of China (863 Program) (No. 2011AA02A211) and Natural Science Foundation of China (No. 21206127).

References

1. Bhatti HN, Khera RA (2012) Biological transformations of steroidal compounds: a review. Steroids 77(12):1267–1290
2. Donova MV, Egorova OV (2012) Microbial steroid transformations: current state and prospects. Appl Microbiol Biotechnol 94(6):1423–1447
3. Hu XY, Cui L, Feng K (2012) Optimization of fermentation medium for 11α-Hydroxylation of steroid catalyzed by *Rhizopus nigricans* RN-M246. Chem Bioeng 6:029
4. Furuya T, Shibata D, Kino K (2009) Phylogenetic analysis of *Bacillus* P450 monooxygenases and evaluation of their activity towards steroids. Steroids 74(12):906–912
5. Kolet SP, Niloferjahan S, Haldar S et al (2013) Biocatalyst mediated production of 6β, 11α-dihydroxy derivatives of 4-ene-3-one steroids. Steroids 78(11):1152–1158
6. Fernandes P, Cruz A, Angelova B et al (2003) Microbial conversion of steroid compounds: recent developments. Enzyme Microb Technol 32(6):688–705
7. Zhou H, Lu W, Wen J et al (2009) Kinetic analysis of 11α-hydroxylation of steroids by *Rhizopus nigricans*. J Mol Catal B Enzym 56(2):136–141
8. Martin GD (2010) Biotransformation reactions by *Rhizopus spp*. Curr Org Chem 14(1):1–14
9. Mao S, Hua B, Wang N, Hu X et al (2013) 11α hydroxylation of 16α, 17-epoxyprogesterone in biphasic ionic liquid/water system by *Aspergillus ochraceus*. J Chem Technol Biotechnol 88(2):287–292
10. Hunter AC, Carragher NE (2003) Flexibility of the endogenous progesterone lactonisation pathway in *Aspergillus tamarii* KITA: transformation of a series of cortical steroid analogues. J Steroid Biochem Mol Bio 87(4):301–308
11. Kołek T, Szpineter A, Świzdor A (2008) Baeyer-Villiger oxidation of DHEA, pregnenolone, and androstenedione by *Penicillium lilacinu* AM111. Steroids 73(14):1441–1445
12. Gao JM, Shen JW, Wang JY et al (2011) Microbial transformation of 3β-acetoxypregna-5, 16-diene-20-one by *Penicillium citrinum*. Steroids 76(1):43–47
13. Stanczyk FZ, Archer DF (2014) Gestodene: a review of its pharmacology, potency and tolerability in combined contraceptive preparations. Contraception 89(4):242–252
14. Yang J, Yang S, Yang YL et al (2007) Microbial hydroxylation of 16α, 17α-dimethyl-17β-(l-oxopropyl) androsta-l, 4-dien-3-one to rimexolone by *Curvularia lunata* AS 3.4381. J Mol Catal B Enzym 47(3):155–158
15. Donova MV, Egorova OV, Nikolayeva VM (2005) Steroid 17β-reduction by microorganisms —a review. Process Biochem 40(7):2253–2262
16. Romano A, Romano D, Ragg E et al (2006) Steroid hydroxylations with *Botryodiplodia malorum* and *Colletotrichum lini*. Steroids 71(6):429–434
17. Andrushina VA, Druzhinina AV, Yaderets VV et al (2011) Hydroxylation of steroids by *Curvularia lunata* mycelium in the presence of methyl-β-cyclodextrine. Appl Biochem Biotechnol 47(1):42–48
18. Nelson PE, Toussoun TA, Marasas WFO (1983) *Fusarium* species: an illustrated manual for identification. Pennsylvania State University, Park

19. Klich MA (2002) Identification of common *Aspergillus* species. Centraalbureau voor Schimmelcultures (CBS), Utrecht
20. Samson RA, Hoekstra ES, Frisvad JC (2004) Introduction to food-and airborne fungi, 7th edn. Centraalbureau voor Schimmelcultures (CBS), Utrecht
21. Barbesgaard P, Heldt-Hansen HP, Diderichsen B (1992) On the safety of *Aspergillus oryzae*: a review. Appl Microbial Biotechnol 36(5):569–572

Chapter 63
Extent and Pattern of DNA Cytosine Methylation Changes Between Non-pollinated and Pollinated Ovaries from *Cymbidium hybridium*

Xiaoqiang Chen, Xiulan Li, Ning Sun and Wenqin Song

Abstract In this study, MSAP (methylation sensitive amplified polymorphism) technique was carried out to analyze differences in the methylation status between ovaries before and after pollination from *Cymbidium hybridium*. 72 selective primer combinations were used to check the status of DNA cytosine methylation and a total of 5,892 fragments were obtained. The results demonstrated DNA methylation events occurred in ovaries from *C. hybridium*. Both total and full methylation levels (14, 9.5 %) in the pollinated ovaries were lower than those in non-pollinated ovaries (11.4, 7.8 %), which suggested some demethylation had occurred. Furthermore, methylation patterns also varied between the two ovaries. Distinct patterns of DNA methylation arising through demethylation or de novo methylation might have specialized functions. This suggested the significance of epigenetic function in the development of orchid ovaries. The hypothesis that DNA methylation played a role in the ovary development of *C. hybridium* will help to illuminate the methylation status of given genes, to clone the fragments with different methylation patterns, and further shed novel insights into the molecular mechanisms of the ovary development of orchids from the point of view of epigenetics.

Keywords DNA methylation · MSAP · *Cymbidium hybridium* · Ovary development

---

X. Chen (✉) · N. Sun
College of Agronomy and Resources and Environment, Tianjin Agricultural University, Tianjin 300384, People's Republic of China
e-mail: chenxiaoqiang@tjau.edu.cn

X. Li · W. Song
College of Life Sciences, Nankai University, Tianjin 300071, People's Republic of China

T.-C. Zhang and M. Nakajima (eds.), *Advances in Applied Biotechnology*,
Lecture Notes in Electrical Engineering 332, DOI 10.1007/978-3-662-45657-6_63

63.1 Introduction

DNA methylation has been associated with numerous biological processes including differential gene expression, cell differentiation, genomic imprinting, gene silencing, etc. [1–3]. DNA methylation is a fundamental epigenetic mechanism for regulating the expression of genes related to plant growth and development [4]. Although DNA methylation likely has a conservative role in regulating the expression of genes, the levels and patterns of DNA methylation appear to vary drastically among different organisms [5]. For example, the level of 5mC is from 6 % of cytosines in Arabidopsis to 25 % in maize.

There are a number of methods to study DNA methylation by studying global levels of methylated cytosines or by examining specific gene sequences [6], such as direct sequencing, high-performance liquid chromatography, liquid chromatography/mass spectroscopy, methylation-sensitive restriction digestion, nucleotide primer extension assays, primer-specific PCR, etc.

MSAP is a modification of AFLP method and has been developed to investigate DNA methylation profiles in plants [7] using the isoschizomers *Hpa*II and *Msp*I, which share the same recognition sites but display differential sensitivity to cytosine methylation [8]. MSAP could detect the genetic diversity of the genome thoroughly without getting knowledge of any information of the nucleotide sequence [9] and there are some methylation studies of plants to be reported [5]. For instance, MSAP was used in the analysis of extent and pattern of cytosine methylation in the rice genome and found many sites were differentially methylated in seedlings and aged leaves of rice [10]. Cervera et al. [5] applied this approach to Arabidopsis ecotypes and found the percentage of methylation-sensitive polymorphisms to be lower in early flowering ecotypes than in late flowering ones. The analysis of DNA methylation in the process of the budding seeds from *Brassica napus* using MSAP revealed that distinct patterns of DNA methylation occurred through demethylation or de novo methylation might have specialized functions [11]. In all, according to recognition of tissue-specific MSAP bands, we would identify some functional genes differential expression in certain tissues.

Cymbidum hybridium used in this study is a kind of orchid belonging to the family of Orchidaceae, which is one of the largest families of flowering plants [12]. Different from other flowering plants whose oocytes are not fertilized with pollens until ovaries mature, the ovary development of orchids only starts after successful pollination.

However, little is known in the molecular mechanisms of ovary development after pollination. In this study, MSAP (methylation sensitive amplified polymorphism) analysis, a more global method, was carried out to analyze differences in the methylation status between ovaries of *C. hybridium* before and after pollination from a genome-wide perspective and further to shed novel insights into the molecular mechanisms of the ovary development of orchids from the point of view of epigenetics.

63.2 Materials and Methods

63.2.1 Plant Material

Cymbidium hybridium was used as plant material in this study. *C. Hybridium* used in this study was grown in a greenhouse in natural light and controlled temperature ranging from 23 to 27 ℃. After 15–20 days of their blossoming, artificial pollination was carried out. When the color of the columns obviously changed at 2 days after pollination, the pollinated and non-pollinated ovaries from the same orchid were separately frozen in liquid nitrogen immediately after collection and stored at −80 ℃.

63.2.2 DNA Extraction and Purification

Frozen materials of the pollinated and non-pollinated ovaries were ground in liquid nitrogen and simultaneously extracted by CTAB methods. The possible contaminated RNA was digested by *DNase* free *RNase* (Takara, Shiga, Japan) for 30 min at 37 ℃. The DNA was checked for quality and quantity by agarose gel electrophoresis (0.8 %) and fluorimetry (ND-1000, NanoDrop, Wilmington, USA). Total DNA was stored at −20 ℃.

63.2.3 MSAP Analysis

Two combinations of restriction enzymes, each at a concentration of 5 U/μL, were prepared by mixing *EcoR*I with one of the isoschizomers, *Hpa*II and *Msp*I. Two sets of digestion reactions were carried out simultaneously. In the first reaction, approximately 400 ng of DNA from pollinated ovaries was digested at 37 ℃ overnight with 3 U *EcoR*I /*Hpa* II each in 10 μL reaction volume. In the second reaction, the same amount of non-pollinated genomic DNA was digested with *EcoR* I/*Msp* I under the same reaction conditions.

Subsequently, the digested DNA fragments from the two reactions were ligated separately with an equal volume of the ligation solution containing 10 U DNA ligase, 5 pmol/L *EcoR*I adapter, 50 pmol/L *Hpa*II-*Msp*I adapter, and 1 mmol/L ATP at 16 ℃ overnight.

The ligation mixture was performed in a 1:10 dilution with TE and used as template DNA in the pre-selective amplification. The PCR reaction was performed in 25 μL of a reaction mix with 1 μL of diluted ligation mixture DNA, 0.5 μmol/L of each forward and reverse primers, 200 μmol/L dNTPs, 1x polymerase buffer, and 1 U *Taq* DNA polymerase (*Takara*) for 21 cycles with 30 s denaturation at 94 ℃, 1 min annealing at 56 ℃, and 1 min extension at 72 ℃. The PCR product from the above pre-amplification was diluted 1–20 (v:v) with TE buffer, and further used as template

Table 63.1 Sequence of adapters and primers used for MSAP analysis

	EcoR I (E) 5′–3′	*Hpa*II/*Msp* I (H/M) 5′–3′
Adapter-1	GACGATGAGTCTAGAA	CTCGTAGACTGCGTACC
Adapter-2	CGTTCTAGACTCATC	AATTGGTACGCAGTC
Pre-amplication primers	GACTGCGTACCAATTC (E00)	GATGAGTCTAGAACGG(H/M00)
Selective amplication primers	GACTGCGTACCAATTCAA (E11)	GATGAGTCTAGAACGGTA (H/M23)
	GACTGCGTACCAATTCAG (E13)	GATGAGTCTAGAACGGTC (H/M24)
	GACTGCGTACCAATTCAAC (E32)	GATGAGTCTAGAACGGTAA (H/M79)
	GACTGCGTACCAATTCACA (E35)	GATGAGTCTAGAACGGTAG (H/M81)
	GACTGCGTACCAATTCACC (E36)	GATGAGTCTAGAACGGTAT (H/M82)
	GACTGCGTACCAATTCACG (E37)	GATGAGTCTAGAACGGTCA (H/M83)
	GACTGCGTACCAATTCACT (E38)	GATGAGTCTAGAACGGTGT (H/M90)
	GACTGCGTACCAATTCAGC (E40)	GATGAGTCTAGAACGGTTC (H/M92)
	GACTGCGTACCAATTCAGG (E41)	

DNA for the next selective amplification. The ingredients of selective amplification were the same as described above with 1 μL of diluted pre-amplification mixture DNA. The selective amplification was performed by touchdown program using amplification primers. The conditions were as follows: 13 cycles at 94 ℃ for 30 s, 0.7 ℃/cycle from 65–56 ℃ for 30 s, and 72 °C for 1 min; another 23 cycles of PCR amplification were used after touchdown program. The denaturing step was 94 ℃ for 30 s, the annealing step was 56 ℃ for 30 s, and the extension step at 72 ℃ for 1 min; a final extension was at 72 ℃ for 10 min. The sequences of adapters, pre-amplification primers, and amplification primers are shown in Table 63.1.

The final selective amplification products were denatured, separated on a 6 % polyacrylamide sequencing gel, and visualized by silver staining. A 50-bp ladder was used to estimate the molecular size of fragments. All reactions were performed in triplicate to avoid false-positive results. If its results were reproducible, one sample was used for further analysis. Finally, only clear and reproducible bands that appeared in four independent PCR amplifications were scored and counted.

63.3 Results

63.3.1 MSAP Analysis

In order to detect the methylation status in the genomic DNAs of the pollinated and non-pollinated ones from *C. hybridium*, MSAP analysis was conducted (Fig. 63.1) with 72 pairs of MSAP selective amplification primers (Table 63.1). With each primer combination, there were four lanes corresponding to two different sources of genomic DNA, each restricted with one of the two pairs of enzymes *EcoR* I/*Hpa* II and *EcoR*I/*Msp*I. Thus, each of the fragments represented a recognition site cleaved by one or both of the isoschizomers. These MSAP fragments observed (Fig. 63.1) had good stability. A total of 5,892 fragments were amplified and the two different DNAs showed essentially the same degree of methylation as shown in Table 63.2.

H1, fragments obtained after digestion with EcoR I/Hpa II from the non-pollinated ovaries; M1, fragments obtained after digestion with EcoR I/Msp I from the non-pollinated ovaries. H2, fragments obtained after digestion with EcoR I/HpaII from the pollinated ovaries; M2, fragments obtained after digestion with EcoRI/Msp I from the pollinated ovaries. Different methylation patterns among fragments

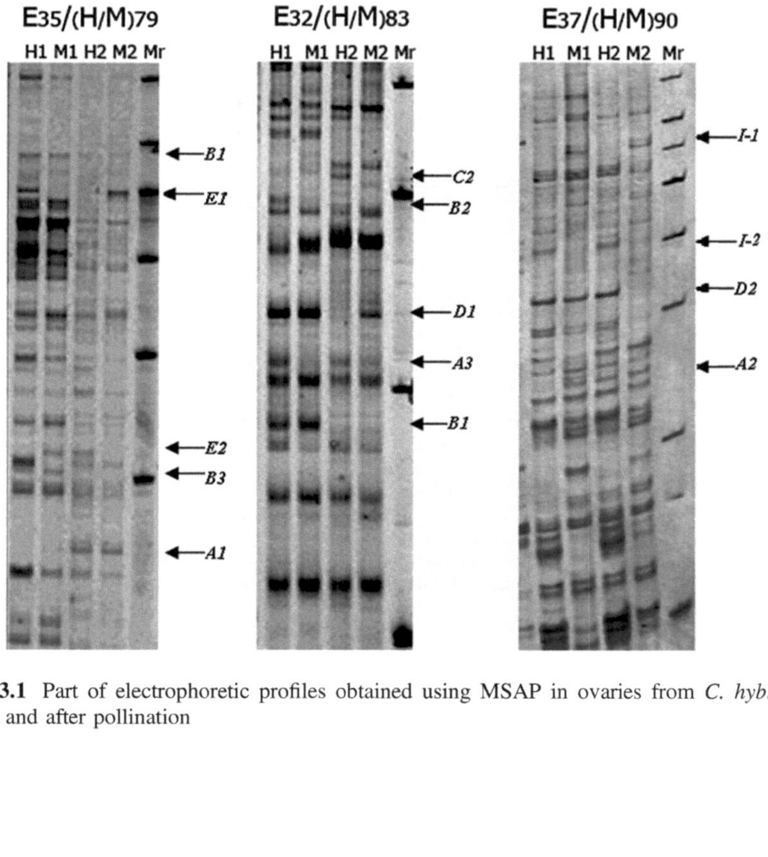

Fig. 63.1 Part of electrophoretic profiles obtained using MSAP in ovaries from *C. hybridium* before and after pollination

Table 63.2 Methylation levels of non-pollinated and pollinated ovaries from *C. hybridium*

	Non-pollinated ovaries	Pollinated ovaries
Total amplified fragments	2,942	2,950
Total methylated fragments	412	336
Total methylation ratio	14 %	11.4 %
Full methylated fragments[a]	280	231
Full methylation ratio[b]	9.5 %	7.8 %

[a] Full methylation, 5′C^{m} CGG in double strands
[b] Full methylated fragments/total amplified fragment

Table 63.3 Different patterns of methylation in non-pollinated and pollinated ovaries from *C. hybridium*, number of sites, and ratios of different patterns

Type	Non-pollinated ovaries		Pollinated ovaries		Sites no.	Ratios (%)
	H	M	H	M		
Type I					321	42.9
I-1	0	1	0	1	113	
I-2	1	0	1	0	34	
I-3	1	1	1	1	174	
Type II					427	57.1
A					169	39.6
A1	0	0	1	1	70	
A2	0	1	1	1	44	
A3	1	0	1	1	55	
B					121	28.3
B1	1	1	0	0	52	
B2	1	0	0	0	30	
B3	0	1	0	0	39	
C					32	7.5
C1	1	1	0	1	17	
C2	1	1	1	0	15	
D					78	18.3
D1	0	0	0	1	36	
D2	0	0	1	0	42	
E					27	6.3
E1	1	0	0	1	13	
E2	0	1	1	0	14	

Notes "0" stands for the absence of a fragment, and "1" stands for the present of a fragment. H and M are the combinations of enzymes of EcoR I/Hpa II and EcoR I/Msp I, respectively

of the ovaries before and after pollination are shown by *arrowheads* (Table 63.3). Primer combinations are indicated on the *top* of the figure. Mr indicates the molecular weight marker of 50-bp ladder.

63.3.2 DNA Methylation Levels in Pollinated and Non-pollinated Ovaries from *C. hybridium*

Fragments ranging from 100 to 500 bp were counted. For each lane, varying from 10 to 30 MSAP fragments were observed. Of the 5,892 fragments detected, there are 2,942 total amplified fragments for non-pollinated ovaries and 2,950 for pollinated ones respectively (Table 63.2). Of these, the total methylated fragments for two different DNAs are 412 and 336 respectively. Thus, approximately 14 % (412/2942) of 5′-CCGG sites were cytosine-methylated in the non-pollinated ovaries, while 11.4 % (336/2,950) in the pollinated ones. Of the non-pollinated ovaries, 280 were due to full methylation of the internal cytosine, which allowed cleavage by *Msp*I but not by *Hpa*II, whereas the remaining 132 were due to hemi-methylation of the external cytosine, which permitted cleavage by *Hpa*II but not by *Msp*I. Same as detected above, there were 231 full methylation fragments and 125 hemi-methylation fragments of non-pollinated ovaries.

63.3.3 DNA Methylation Patterns in Pollinated and Non-pollinated Ovaries from *C. hybridium*

Typically, the MSAP bands (748) generating 15 kinds of different distribution patterns could be grouped into two major types (Fig. 63.1, Table 63.3).

The same MSAP bands present in the two *EcoR* I/*Msp* I lanes or *EcoRI*/*Hpa*II lanes were classified into Type I, accounting for about 42.7 %(321/648)of the total number of bands. These sites are referred to as monomorphic, i.e., either present or absent in both genotype, representing no methylation differences between pollinated and non-pollinated ovaries. Type I which included three classes, I-1, I-2, and I-3. I-1, is full of methylation sites that permit cleavage by *Msp*I but not by *Hpa*II; the patterns of the bands display are as shown in Table 63.3 (0,1,0,1). I-2 belong to hemi-methylation sites, the patterns are like (1,0,1,0), and I-3 are (1,1,1,1). Those varying between the two different tissues were Type II, meaning approximately 57.1 % (427/574) of the total number of MSAP bands were polymorphic sites. These sites fell into a number of patterns that could be grouped into five major classes (classes A, B, C, D, and E) (Table 63.3).

As shown in Table 63.3, of these 427 polymorphic sites, Class A (39.6 %, 169/427) and Class B (28.3 %, 121/427) are most frequently observed. In class A, there are 169 amplification fragments in total consisting of three patterns—A1(0,0,1,1), A2(0,1,1,1), and A3(1,0,1,1). A common feature of the patterns in class A was that the bands could be detected both in the H and M lanes of pollinated ovaries but either not in H and M lanes of non-pollinated ovaries or only displayed in one of two lanes. This indicated that de-methylation occurred at those sites to a certain degree in pollinated ovaries compared to non-pollinated ovaries. Class B consists of three patterns— B1(1,1,0,0), B2(1,0,0,0), B3(0,1,0,0) and the fragments in Class B were detected in one or both lanes of non-pollinated ovaries but not in any lanes of pollinated ovaries, indicating both the external cytosine and the internal cytosine methylation of the DNA occurred in the pollinated ovaries. Classes C and D accounted for 18.3 % (78/427) and 7.5 % (32/427), respectively. Class C indicated increased methylation and Class D represented a decrease in the level of methylation in the pollinated ovaries compared to the non-pollinated ones. E1(1,0,0,1) and E2(0,1,1,0) consisted of Class E accounting for 6.3 % (27/427), which indicated some methylation differences existed after ovary pollination.

63.4 Discussion

In this study, the technique was carried out to analyze differences in the methylation status between ovaries before and after pollination from *C. hybridium*. 72 pairs of selective primer combinations were used to check the status of cytosine methylation of DNA samples and a total of 5,892 fragments were obtained. The reaction was performed in triplicate with each pair of primer combinations and all of the bands displayed good stability and consistency, which indicated that MSAP technique could be applied to methylation analysis of genomic DNA *C. hybridium* efficiently.

According to the reports, the ratios of the cytosine methylation are different in different plant species in higher plants [13]. 16.3 % of the methylation occurred in the rice genome [10], 5–7 % in Arabidopsis with different ecological type [14], 33 % in wheat, 15.7 % in the budding seeds from *Brassica napus* [11]. We showed that the total methylation levels of the DNA from *C. hybridium* range from 9.6 to 11.6 %. In fact, the methylation level we detected only reflect the extent of methylation in the CGs or parts of the CCGs, whereas methylation also occurred in the CAGs and CTGs in some plants [9]. Therefore, the methylation ratios of the cytosines in the overall genome from *C. hybridium* are likely significantly higher than the results detected in this paper. According to the result in this paper, there are many extensive cytosine methylation alterations occurring in the *C. hybridium* genome. In the total 626 methylation sites amplified from the non-pollinated and pollinated ovaries of *C. hybridium* (Table 63.2), 389 full methylation sites are detected which is near to twice those of the hemi-methylated sites (237). This showed that, in the ovaries from *C. hybridium,* methylation frequency of the ^{5m}CCGG and ^{5m}C^{5m}CGG in one DNA strand is lower than that of the C^{5m}CGG in both strands, which are different from the results obtained from Navel Oranges [15]. On the whole, different plants have different DNA methylation levels and patterns. In the DNA methylation levels in different tissues or different development stages of the same species also exist obvious differences [11]. Rossi et al. [4] analyzed the methylation pattern of maize and found the specialized genes expressed in endosperm were highly methylated in other tissues.

DNA methylation might have played an important role in the process of cell differentiation and development, thus plant gene expression could be controlled by DNA methylation and demethylation. Both the overall methylation levels and full methylation ratios of the non-pollinated and pollinated ovaries from *C. hybridium* detected in the study have differences. Both the overall methylation ratios (11.6 and 9.6 %) and full methylation ratios (7.1 and 6.1 %) showed that the methylation levels of the pollinated ovaries are slightly lower than those of the non-pollinated ones. In the development processes of the pollinated ovaries, a quantity of de-methylation events was detected, which could explain why the ovaries from *C. hybridium* did not begin to develop until they were pollinated. Pollination could urge some genes to be de-methylated and activated, thus cell division began and the ovaries developed subsequently.

According to the band types (Fig. 63.1) obtained in the analysis, the methylation patterns of the ovaries from *C. hybridium* after pollination also changed except at the methylation levels. There are large quantities of variation in the methylation bands types between the pollinated and the non-pollinated ovaries, and DNA methylation patterns are rich. Based on the analysis of the DNA methylation patterns by statistics, monomorphic and polymorphic amplification sites account for 42.9 and 57.1 % of the total methylation amplification sites separately, which show the wide occurrence of DNA methylation alterations, including the levels and patterns occurring in the pollinated and non-pollinated ovaries. The distribution of polymorphic bands suggests that methylation variation occurred in the pollinated and the non-pollinated ovaries and was mainly caused by de-methylation and de novo methylation. According to the different band types of occurrence in the pollinated and the non-pollinated ovaries, 427 polymorphic amplification sites obtained in this study could be divided into five distinct groups, designated as A, B, C, D, and E. Among these sites, most of the band types belong to A and B (Table 63.3) which accounts for about 67.9 % ((121+169)/427) significantly higher than the other three types. The methylation band types of group A were characterized as a phenomenon mainly involving de-methylation, which may be related to gene expression during development of the pollinated ovaries. Group B were the results of certain sites of DNA from the pollinated ovaries occurred during de novo methylation. All of the sites of groups C, D, and E were 32.1 % of the total MSAP sites, the methylation band types of which reflected that the methylation levels decreased or increased in the pollinated ovaries, a process involving incomplete demethylation and methylation. Maybe, these incomplete demethylation and methylation sites could serve as targets of methylation and demethylation and then become regulators inside the cell [1]. The underlying mechanism is unknown, nonetheless, it underscores the emerging view that different methylation patterns are specified and maintained by divergent genetic machineries in plants [16].

Acknowledgments This work was supported by the Natural Science Foundation of Tianjin, China (10JCYBJC09000) and Tianjin science and technology support plan key projects (09ZCKFNC01600).

References

1. Finnegan EJ, Peacock WJ, Dennis ES (2000) DNA methylation, a key regulation of plant development and other processes. Curr Opin Genet Dev 10:217–223
2. Martienssen RA, Colot V (2001) DNA methylation and epigenetic inheritance in plants and filamentous fungi. Science 293:1070–1074
3. Tariq M, Paszkowski J (2004) DNA and histone methylation in plants. Trends Genet 20:244–251
4. Rossi V, Motto M, Pellegrini L (1997) Analysis of the methylation pattern of the maize Opaque-2 (02) promoter and in vitro binding studies indicate that the 02 B-Zip protein and other endosperm factors can bind to methylated target sequences. J Biol Chem 272:13758–13771
5. Cervera MT, Ruiz-Garcia L, Martinez-Zapater JM (2002) Analysis of DNA methylation in *Arabidopsis thaliana* based on methylation-sensitive AFLP markers. Mol Genet Genomics 268:543–552
6. Jacobsen SE, Sakai H, Finnegan EJ (2000) Ectopic hypermethylation of flower-specific genes in Arabidopsis. Curr Biol 10:179–186
7. Liu B, Brubaker CL, Mergeai G et al (2001) Polyploid formation in cotton is not accompanied by rapid genomic changes. Genome 44:321–330
8. ReynaLopez GE, Simpson J, RuizHerresa J (1997) Differences in DNA methylation patterns are detectable during the dimorphic transition of fungi by amplification of restriction polymorphisms. Mol Gen Genet 253:703–710
9. Vos P, Hogers R, Bleeker M et al (1995) AFLP: a new technique for DNA fingerprinting. Nucleic Acids Res 23:4407–4414
10. Xiong LZ, Xu CG, Maroof MAS (1999) Patterns of cytosine methylation in an elite rice hybrid and its parental lines, detected by a methylation sensitive amplification polymorphism technique. Mol Gener Genet 261:439–446
11. Lu GY, Wu XM, Chen BY (2005) MSAP analysis of DNA methylation during and following *Brassica rapa* seed germination. Chin Sci Bull 24:2750–2756
12. Zhu GF, Guo ZF (2004) Progress on molecular biology of main ornamental Orchidaceae. Chin Bull Bot 21:471–477
13. Twyman RM (2001) Advanced molecular biology: a concise reference. Science Press, Beijing, pp 185–194
14. Lister R, O'Malley RC, Tonti-Filippini J et al (2008) Highly integrated single-base resolution maps of the epigenome in Arabidopsis. Cell 133:1–14
15. Hong L, Deng XX (2005) Analysis of DNA methylation in navel oranges based on MSAP marker. Scientia Agricultura Sinica 38:2301–2307
16. Chan SW, Henderson IR, Jacobsen SE (2005) Gardening the genome: DNA methylation in *Arabidopsis thaliana*. Nat Rev Genet 6:351–360

Part IV
Progress of Biotechnology

Chapter 64
The Application Status of Microbes in Salted Fish Processing

Yan Yan Wu, Gang You and Lai Hao Li

Abstract Salted fish is a traditional Chinese aquatic product, because of its rich nutrition and the salty, sweet, unique taste and flavor, it is loved by consumers. At present, empirical traditional processing methods are adopted in enterprises that processing salted aquatic products, which have many disadvantages not only in the production scale, but also its efficiency. In order to innovate the traditional salted fish processing technology, as well as to meet the demand of modern consumers for salted fish with lower salt, richer nutrition, higher safety, better quality and flavor, this paper reviews researches on the isolation of dominant microorganisms in traditional salted fish and the application of lactic acid bacteria and other microorganisms in the salted fish processing, further, puts forward the tendency of applying modern microbial technologies to the processing of salted fish products.

Keywords Salted fish · Microbe · Lactic acid bacteria · Application status

64.1 Introduction

Salted fishes are one of traditional seafood in China; ease of processing, preservation, and unique taste increase popularity of this product among consumers. Salted fish products such as smoked fish, pickled fish with wine in Southern China are fermented

---

Y.Y. Wu (✉) · G. You · L.H. Li
South China Sea Fisheries Research Institute, Chinese Academy of Fishery Sciences, Guangzhou 510300, People's Republic of China
e-mail: wuyygd@163.com

Y.Y. Wu · G. You · L.H. Li
Key Lab of Aquatic Product Processing, Ministry of Agriculture, Guangzhou 510300, People's Republic of China

G. You
Collage of Food Science and Engineering, Ocean University of China, Qingdao 266003, People's Republic of China

T.-C. Zhang and M. Nakajima (eds.), *Advances in Applied Biotechnology*, Lecture Notes in Electrical Engineering 332, DOI 10.1007/978-3-662-45657-6_64

aquatic products, whose fermentation process is mainly spontaneous which makes the products vulnerable to environment changes and quality hard to be controlled, in addition to the fact that they are mostly produced in workshops, they are very difficult to be industrialized. Awareness toward safe food product resulted in people's concerns about safety issues related to salted seafood. For example, the salted fish has crude salts nitrate or nitrite, which may come into being carcinogenic nitrosamines during curing.

The traditional salted fish was based on natural fermentation by the microflora of fish meats, the product quality relied on the microbial fermentation. The use of safe lactic acid bacteria could be a promising approach to develop new fish products. Such as fish sauce and silage fermentation, which have no undesirable odor, prolonged shelf life with nutritional advantages. GRAS (generally recognize as safe) strains, such as *Lactobacillus plantarum, Lactobacillus acidophilus, Leuconostoc mesenteroides*, and *Pediococcus pentosaceus* are occasionally used outside the dairy and pickle industries. Use of GRAS bacteria in salted fish products could expand lactic acid bacteria application. Purification of beneficial microorganism from traditional fermented fish products, and application of these beneficial microorganisms for fermentation process of salted fish could improve the efficiency and safety of products.

This paper reviews researches on the isolation of dominant microorganisms in traditional salted fish and the application of lactic acid bacteria and other microorganisms in the salted fish processing, further, puts forward the tendency of applying modern microbial technologies to the processing of salted fish products.

64.2 Isolation and Screening of Dominant Microorganisms in Traditional Salted Fish

In recent years, the halophilic bacteria from traditional fermented fish products has been analyzed and screened in many researches. Previous studies showed that lactic acid bacteria, micrococcus, yeast, and other microorganisms were the dominant bacteria in the pickled fish products processing.

Tian et al. [1] and Wood [2] had isolated and purified dominant lactic acid bacteria from traditional pickled fish, which mainly include *Lactobacillus casei*, *Lactobacillus*, and *Lactococcus* bacteria of sausages. Paludan-Müller et al. [3] had studied the fermentation process and phase change of bacteria in a Thailand fermented fish named plaa-som under different salt concentrations (6, 7, 9, 11 %). The advantage microbe of lactic acid bacteria and yeast were separated in the fermentation process. Five kinds of homofermentative lactic acid bacterium strains were isolated and screened by Nimnoi (2011) [4] from naturally fermented fish. There was no mucus and gas in final fermented product by these natural isolates. All of them can grow within 20–45 °C, of which the most acid-producing strains were the 3rd strain (identified as *P. pentosaceus*) and the 46th strain (identified as *L. plantarum*). The fermentation at 6 % salt was excellent in both strains as compared with other isolates. Three types of lactic acid bacterium

Table 64.1 Phase and prevailing microbes during the natural fermentation of salted fish

Fermentation phase	Prevailing microorganisms
Initiation phase	Various bacterial on the surface
Primary fermentation phase	Lactic acid bacteria, yeasts, micrococcus
Secondary fermentation phase	Yeasts
Post-fermentation phase	Yeasts, molds, spoilage bacteria

are *L. plantarum, L. mesenteroides*, and *Pediococcus pentosaceus* that had been isolated and identified, respectively, from salted fish, they have a good salt resistance and a significantly degradation of nitrite [5, 6].

Growth prevalence of microbes in different fermentation phases during the natural fermentation of salted fish is shown in Table 64.1 [7]. Xie et al. [8] found that *Lactobacillus plantanen, Lactobacillus curvatus, Lactobacillus alimentarius, P. pentosaceus, Pediococcus acidiactica* exist in cured fish and the distribution of lactic acid bacterium was different in different stages. Eleven strains of homofermentative, rod-shaped lactic acid bacteria and five strains of heterofermentative, sphere-shaped lactic acid bacteria were isolated from fermented fish (pla-ra and pla-chom) in Thailand, including two new species *L. acidophilus* sp. nov. and *Weissella thailandensis* sp. nov. [9]. Kanno et al. [10] analyzed the lactic acid bacteria in Narezushi (salted and fermented fish with rice of Japan) and identified four *L. plantarum* and one *L. mesenteroides* that were able to ferment lactose, grow in MRS containing 3 g/L bile, grow in broth adjusted to pH 3.6, and scavenge $DPPH^{-}$ and/or O^{2-} radicals.

64.3 Application of Lactic Acid Bacterium in Salted Fish

The intentional inoculation of purified beneficial starter cultures to fish is a contemporary approach in the processing of fermented fish, which leads to a high degree of control over the fermentation process.

Lactic acid bacterium, as a food preservative with beneficial dietary function, antimicrobial and antioxidant properties have been studied for many years; although, industrial fermentation through lactic acid bacterium was delayed in China. Moreover, fermentation was focused on dairy products, vegetable and fruits pickles and was less common in the processing of meat and aquatic products. The researches on kimchi, a kind of fermented food from the solanaceous vegetable, leaf and tuber vegetables, were relatively developed. Vegetable pickles are preserved vegetable made of a variety of fresh vegetables by fermentation through Lactic acid bacterium in low-salt content [11]. Anshu et al. [12] studied collective properties of the *L. plantarum* and *L. casei* as the starter culture inoculated in pickled vegetables. The results showed that the best fermentation matching model of lactic acid bacteria was either independent or mixed at 1:1 ratio; other abiotic factors of 3 % salt concentration, 5 % inoculation size fermented at 30 °C for 72 h were optimal for high quality and excellent taste. The average rate of pure inoculation fermentation was at least 3.26 times higher than the natural fermentation.

According to the application technology of microorganism in pickled vegetables, the beneficial microbes could separate and purify from naturally salted fish. Purified inoculum may apply to salted fish and fermented under controlled conditions. Eventually, fermented products would have better organoleptic quality with shorten curing time. The few studies on fermentation and preservation of fish through bio-leaving or curing agents, particularly lactic acid bacteria, were reported. Although, lactic acid bacteria have been widely used in vegetables and dairy products. Application of lactic acid bacteria fermentation technology to salted fish product, on one hand, could improve the fish products quality, flavor, and nutrition; on the other hand, ensure the safety, shorten the processing time, cut the costs, and provide the technical basis for industrial fermentation. Different microorganisms and materials commonly used in the fermented fish products are summarized in Table 64.2 [13–15]. The group of lactic acid bacteria occupies a central role in the processes of fermented products and beverages are tabulated in Table 64.3 [1, 16–18].

Controlled fermentation has effectively eliminated the defects of the traditional curing of salted fish, and improved the product safety; flavor and sensory quality with shorten fermentation time, especially low levels of nitrosamines. After microbial decomposition, the nitrite content of product was remarkably reduced, which effectively guarantees the safety of fermented meat products. Wu et al. [19] isolated five kinds of lactic acid bacteria from salted fish products, used them in the salted fish processing after the strains proportional distribution, it cannot only shorten the curing time, but also enhance the flavor, prevent the amine production, improve product quality and safety. Corbiere Morot-Bizot et al. [20] reported that lactic acid bacteria, micrococcus and *Staphylococcus aureus* can secrete protease, lipase, nitrate reductase, amino acid deaminase, and amino acid decarboxylase; which play an important role in products quality, color, and flavor. The results of these studies illustrated that microbial fermentation in salted fish is feasible, which can improve the organoleptic quality of cured fish products and shorten the processing time.

The products quality becomes better and the texture becomes tender after fermentation due to microbial enzymatic activities. The fermented smoked fish have closer texture, lower salinity, and the traditional fish aroma in the appropriate conditions, the total volatile basic nitrogen (TVB-N) and peroxide values were

Table 64.2 Microorganisms used in fermented fish products

Product name	Major ingredients	Microorganisms
Sikhae	Salt, seawater fish, millet	*L. mesenteroides, L. plantarum*
Narezushi	Salt, seawater fish, millet	*L. mesenteroides, L. plantarum*
Burong-isda	Salt, freshwater fish, rice	*L. brevis, Streptococcus* sp.
Pla-ra	Salt, freshwater fish, rice	*Pediococcus* sp.
BalaO-bala0	Salt, shrimp, rice	*L. mesenteroides, P. cerevisiae*
Kungchao	Salt, shrimp, rice	*P. cerevisiae*

Table 64.3 Fermented products and associated lactic acid bacteria subsp

Type of fermented product	Lactic acid bacteria
Cheeses	*L. lactis* subsp. *lactis, L. lactis* subsp. *lactis* var. *diacetylactis, L. lactis* subsp. *cremoris, Leuc. mesenteroides* subsp. *cremoris*
Butter and buttermilk	*L. lactis* subsp. *lactis, L. lactis* subsp. *lactis* var. *diacetylactis, L. lactis* subsp. *cremoris, Leuc. mesenteroides* subsp. *cremoris*
Yogurt	*Lb. delbrueckii* subsp. *bulgaricus, S. thermophilus*
Fermented sausage (Europe)	*Lb. sakei, Lb. curvatus*
Fermented sausage (USA)	*P. acidilactici, P. pentosaceus*
Fermented fish products	*Leuc. mesenteroides, Lb. plantarum, P. acidilactici*
Fermented vegetables	*Leuc. mesenteroides, P. cerevisiae, Lb. brevis, Lb. plantarum*
Fermented cereals	*Lb. sanfransiscensis, Lb. farciminis, Lb. fermentum, Lb. brevis,Lb. plantarum, Lb. amylovorus, Lb. reuteri, Lb. pontis, Lb. panis, Lb. alimentarius, W. cibaria*
Alcoholic beverages	*O. oeni Lb. sakei*

L. = *Lactococcus, Lb.* = *Lactobacillus, Leuc.* = *Leuconostoc, O.* = *Oenococcus, P.* = *Pediococcus S.* = *Streptococcus, W.* = *Weissella*

18.72 mg/100 g and 0.18 g/kg, respectively [21]. Tilapia as the raw material, in the condition of *L. acidophilus* (La) and *L. plantarum* (Lp) with mass ratio of 1:1, was inoculated with 2 % of the fish, just fermentation time 24 h, the cured fish protein content is as high as 42.50 %, fleshy, soft, bright color, soft texture [22]. The above results show that the microbial fermentation is the main flavor characteristics, which significantly reduces the fermentation time. The L8 strains of lactic acid bacteria isolated from kimchi inoculated into low-salted fish, in different salt concentrations and inoculation sizes, were improved the pickled fish quality and shortened the processing time [23]. Riebroy et al. [13] studied the inoculation of lactic acid bacteria on fermentation process and product quality of a fermented fish mince in Thailand named som-fug. The results show that the *P. acidilactici* fermentation rate was faster than the *L. plantarum* and *P. pentosaceus* rate, whose pH value fell to 4.5 in 36 h, while the texture value improved significantly. Therefore, inoculated lactic acid bacteria can shorten the fermentation time and improve the quality of the products.

The growth of lactic acid bacteria and pH reduction effectively inhibit the growth of pathogenic and spoilage bacteria in fermented fish products. Additionally, some harmful parasites were also killed during fermentation. So it can improve food preservation and safety. For example, 5 % NaCl and 4 % glucose added to the Moroccan sardines, artificial inoculation *Lactobacillus delbrueckii* subsp. *Delbrueckii*, under 30 °C fermentation after two weeks, *Escherichia coli, S. aureus, Salmonella* were not detected and the sulfite-reducing *Clostridia*, the total number of bacteria and yeast were also inhibited [14]. So it proved that lactic acid bacterium could prolong the preservation time of sardines. Oberman and Libudzise [24] isolated novel bacterial strain from naturally fermented fish and studied its physiological and metabolic characteristics; the fermented products have unique organoleptic quality, good esthetic quality, safety, shorten processing time, and extend shelf life.

Lactic acid bacteria effectively removed smell of the fish. Such as *Tetragenococcus halophilus* plays an important role in the successful fermentation of silver carp fish sauce to remove the smell of carp [25].

Fermentation could quicken to hydrolyze protein, and then increase water soluble protein content, water soluble solids content and total acidity, produce high amount of free amino acid, improve quality and flavor. As in silver carp in vaccinated *Plant lactobacillus* and *P. pentosaceus*, fermentation production of Yuzha fish products [26].

Lactic acid bacteria fermentation may help to slow down the fat oxidation of products. *Lactobacillus sakei* CECT 4808 and *Lb. curvatus* CECT 904^T were inoculated individually or in combination into rainbow trout fillets, the samples were vacuum-packed and stored at 4 ± 0.5 °C. Katikou, P. discovered that the antioxidant activity of strain *L. sakei* CECT 4808 is better, not only can prolong the shelf life, and to a certain extent, inhibit fat oxidation in the process of storage [27].

From traditional acid fish breeding strains of lactic acid bacteria as in the production of fermented fish good starter cultures, effectively improving the quality of sour fish, developed both keep the traditional acid flavor characteristics of fish, and can realize industrial production of high-quality fermented fish.

64.4 Research and Application Prospects

The traditional salted fish processing technology is relatively backward, mill processing, it is difficult to guarantee the quality of product, especially the traditional salted fish is high salt preserved for a long time, nitrite and nitroso compound is formed in products, with the risk of cancer, at the same time, due to the traditional method of processing salted fish, low moisture content, packaging pallet, so the products not only taste hard, nutrition, and easily polluted by microorganism in circulation process, prone to fat oxidation and deterioration. Traditional high-salt fish products becomes gradually unacceptable as the rise of modern people's living standard, thus the transformation to low-salt ones has come into being.

China is one of the first countries to use lactic acid bacteria fermented food. People has rich experience in the fermentation and fermentation strains resources, for the different fish products processing screening of fermentation microorganisms provided good conditions. In order to innovate the traditional salted fish processing technology, as well as to meet the demand of modern consumers for salted fish with lower salt, richer nutrition, higher safety, better quality and flavor, we should apply modern microorganism technology to salted fish processing. The intentional inoculation of purified beneficial starter cultures to fish is a contemporary approach in the processing of fermented fish, which leads to a high degree of control over the fermentation process. This technology could transform the backward traditional mode into the advanced industrial processing of cured fish products, which can satisfy the consumer's need for safer and higher-quality products.

In short, along with the people understanding of lactic acid bacteria function and application effects, the rapid development of biotechnology, consumer demand for pickled fish with low-salted, nutrition, safe, high quality, high flavor, new production of fermented salted fish will be an important development direction in the future of pickled industry.

Acknowledgments This research project is financially supported by The National Natural Science Foundation of China (31371800), the Basic Scientific Research Business Expenses Aid of Chinese Academy of Fishery Sciences (2013A1001, 2014C05XK01), and Special Promotion of Guangdong Marine Fishery Science and Technology (A201301C01).

References

1. Tian GJ, Shang YY, Huang ZY et al (2011) The dominant lactic acid bacteria fish separation, purification and characterization. Food Ferment Ind (Chin) 37(6):78–81
2. Wood BJB (1997) Microbiology of fermented foods, vol 1, 2. Springer, New York
3. Paludan-Müller C, Madsen M, Sophanodora P et al (2002) Fermentation and microflora of plaa-som, a Thai fermented fish product prepared with different salt concentrations. Int J Food Microbiol 73(1):61–70
4. Nimnoi P, Lumyong S (2011) Improving solid-state fermentation of monascus purpureus on agricultural products for pigment production. Food Bioprocess Tech 4(8):1384–1390
5. Liu FJ, Wu YY, Li LH et al (2012) Nitrite degradation characteristics of lactic acid bacterium from salted fish. Guangdong Agr Sci (Chin) 39(1):94–97
6. Wu YY, Liu FJ, Li LH et al (2012) Isolation and identification of nitrite-degrading lactic acid bacteria from salted fish. Adv Mater Res 393–395:828–834
7. Daeschel MA, Andersson RE, Fleming HP (1987) Microbial ecology of fermenting plant materials. FEMS Microbiol Lett 46(3):357–367
8. Xie J, Xiong SB, Zeng LB et al (2009) Lactic acid bacteria in cured fish and its growth properties. Food Ferment Ind (Chin) 35(6):32–36
9. Tanasupawat S, Shida O, Okada S et al (2000) Lactobacillus acidophilus sp. nov. and Weissella thailandensis sp. nov. isolated from fermented fish in Thailand. Int J Syst Evol Microbiol 50(4):1479–1485
10. Kanno T, Kuda T, An C et al (2012) Radical scavenging capacities of saba-narezushi, Japanese fermented chub mackerel, and its lactic acid Bacteria. Food Sci Technol 47(1):25–30
11. Hansen EB (2002) Commercial bacterial starter cultures for fermented foods of the future. Int J Food Microbiol 78(1–2):119–131
12. Anshu S, Arindam K, Sunita A et al (2011) Use of fermentation technology on vegetable residues for value added product development: A concept of zero waste utilization. Int J Food Ferment Technol 1(2):173–184
13. Riebroy S, Benjakul S, Visessanguan W (2008) Properties and acceptability of Som-fug, a Thai fermented fish mince, inoculated with lactic acid bacteria starters. LWT Food Sci and Technol 41(4):569–580
14. Ndaw A, Zinedine A, Faid M et al (2008) Effect of controlled lactic acid bacterial fermentation on the microbiological and chemical qualities of Moroccan sardines (Sardina pilchardus). Acta Microbiol Immunol Hung 55(3):295–310
15. Ly MH, Covarrubias-Cervantes M, Dury-Brun C et al (2008) Retention of aroma compounds by lactic acid bacteria in model food media. Food Hydrocolloid 22(2):211–217

16. Holzapfel WH, Geisen R, Schillinger U (1995) Biological preservation of foods with reference to protective cultures, bacteriocins and food-grade enzymes. Int J Food Microbiol 24(3):343–362
17. Ray B, Daeschel M (1992) Food biopreservatives of microbial origin. CRC Press, Boca Raton
18. Alvarado C, Garcia Almendarez BE, Martin SE, Regalado C (2006) Food-associated lactic acid bacteria with antimicrobial potential from traditional Mexican foods. Rev Latinoam Microbiol 48(3–4):260–268
19. Wu YY, You G, Li LH et al (2014) Comparison of flavor components between low-salt lactic acid fermented fish and traditional salted fish. J Fish China (Chin) 38(4):600–611
20. Corbiere Morot-Bizot S, Leroy S, Talon R (2006) Staphylococcal community of a small unit manufacturing traditional dry fermented sausages. Int J Food Microbiol 108(2):210–217. doi:10.1016/j.ijfoodmicro.2005.12.006
21. Zhou CY, Huang ZY, Tian GJ (2012) Application of lactobacilli in cured fish. Food Sci (Chin) 33(1):215–218
22. Zhou WJ, Wu YY, Li LH et al (2009) Utilization of compound lactobacillus in cured tilapia preparation. Food Sci (Chin) 30(23):242–245
23. Liu L, Wu ZF, Tu SF (2008) Study on effects of lactic acid bacteria on the quality of low-salt pickled fish. Beijing: China's scientific and technical papers online. http://www.paper.edu.cn/releasepaper/content/200807-256
24. Oberman H, Libudzisz Z (1997) Fermented milks. In: Wood BJB (ed) Microbiology of fermented foods. Blackie Academic and Professional, London, pp 308–350
25. Uchida M, Ou J, Chen BW et al (2005) Effects of soy sauce koji and lactic acid bacteria on the fermentation of fish sauce from freshwater silver carp Hypophthalmichthys molitrix. Fish Sci 71(2):422–430
26. Tan RC, Ouyang JM, Lu XL et al (2007) Fermentation conditions of yuzha by inoculated Lactobacillus plantarum and Pediococcus pertosaceus. Food Sci (Chin) 28(12):268–272
27. Katikou P, Ambrosiadis I, Georgantelis D et al (2007) Effect of Lactobacillus cultures on microbiological, chemical and odour changes during storage of rainbow trout fillets. J Sci Food Agric 87(3):477–484

Chapter 65
Construction and Functional Analysis of Luciferase Reporter Plasmids Containing ATM and ATR Gene Promoters

Li Zheng, Xing-Hua Liao, Nan Wang, Hao Zhou, Wen-Jian Ma and Tong-Cun Zhang

Abstract In eukaryotic cells, maintenance of genomic stability relies on the coordinated action of a network of cellular processes, including DNA replication, DNA repair, cell-cycle progression, and others. The DNA damage response (DDR) signaling pathway mediated by the ataxia telangiectasia-mutated (ATM) and ATM and Rad3-related (ATR) kinases is the central regulator of this network in response to DNA damage. The serine/threonine kinases ATM and ATR are the main kinases activated following various assaults on DNA. In this study, human ATM and ATR promoter luciferase reporter constructs were generated by PCR amplification. Then both PCR fragments respectively were digested and cloned into pGL3 vector. Finally, these promoter sequences were verified by sequencing. These results showed that luciferase reporter with ATM and ATR promoters were successfully constructed. Then the activation of the ATM promoter and ATR promoter following UV light treatments were detected in A431 cells by luciferase reporter assays. The results showed that UV damage could enhance transcriptional activity of ATM/ATR. Our research will provide useful tools for further deciphering ATM/ATR signaling and the pathways mediating the DNA damage response.

Keywords ATM · ATR · DNA damage · UV · Luciferase activity assay

65.1 Introduction

DNA damage is a key factor both in the evolution and treatment of cancer. DNA damage induced by ionizing (IR) and ultraviolet (UV) irradiation or caused by abnormal structures, such as stalled replication forks, triggers a complex cascade of

L. Zheng · X.-H. Liao · N. Wang · H. Zhou · W.-J. Ma · T.-C. Zhang (✉)
Key Laboratory of Industrial Fermentation Microbiology, Ministry of Education, College of Biotechnology, Tianjin University of Science and Technology, Tianjin 300457, China
e-mail: tony@tust.edu.cn

T.-C. Zhang and M. Nakajima (eds.), *Advances in Applied Biotechnology*,
Lecture Notes in Electrical Engineering 332, DOI 10.1007/978-3-662-45657-6_65

phosphorylation events that ultimately serve to influence or affect DNA repair, cell cycle delay, and apoptosis with the overall purpose of maintaining genome stability.

Naturally occurring UV radiation is the environmental mutagen responsible for the largest percentage of environmentally induced skin pathologies, including erythema and inflammation, degenerative aging changes, and cancer [1]. Humans are exposed to UV radiation primarily as a consequence of unprotected exposure to sunlight [2]. UV radiation has many deleterious effects on cells [3–5]. UV radiation produces both direct and indirect DNA damage, and each can result in mutagenesis in skin cells.

Two members of the phosphatidylinositol 3-kinase-related kinase (PIKK) family, the ataxia telangiectasia-mutated (ATM) and ATM and Rad3-related (ATR), play a central role in the damage recognition and initial phosphorylation events [6–9]. These kinases are activated by different forms of DNA damage. ATM responds to DNA double-strand breaks (DSBs), whereas ATR functions following exposure to other forms of DNA damage such as bulky lesions or stalled replication forks [8, 10, 11]. Recently, a growing body of evidence indicates that the roles of these PIKK kinases overlap. For example, overlapping activities of ATR and ATM kinases are required for proper maintenance of these phosphorylation events in response to UV irradiation [12–14].

In this study, ATM promoter and ATR promoter were amplified from human genome by PCR, respectively, and inserted into pGL-3 (luciferase vector) basic vector successfully. Furthermore, luciferase assays were performed in A431 cells to test the transcriptional activity of ATM and ATR. These results showed that DNA damage could activate the transcription of ATM promoter and ATR promoter. Our research will provide useful tools for further deciphering ATM/ATR signaling and the pathways mediating the DNA damage response.

65.2 Materials and Methods

65.2.1 Cell Culture

Human epidermoid carcinoma cell line A431 was obtained from American Type Culture Collection. A431 cells were grown in Dulbecco's modified Eagle's medium (DMEM) (GIBCO) supplemented with 10 % fetal bovine serum (FBS) at 37 °C in a 5 % CO_2 incubator.

65.2.2 Plasmid Construction

The human genome was extracted from A431 cells. ATM promoter fragment from −860 to +53 bp is amplified by PCR using the primers (F: 5′-TATGGTACC-CGTATTGCGTGGAGGATGGAG-3′; R: 5′-TTACTCGAGCAGCGACTTAGC-

GTTTGCGG-3′), and ATR promoter fragment from −504 to +115 bp is amplified by PCR using the primers (F: 5′-TTAGGTACCGTCCTCAACGAAACCTAA-CAGT -3′; R: 5′-TTACTCGAGACTAGTCAACCACGCCAACG-3′). PCR condition is as follows: pre-degeneration for 5 min at 94 °C, denaturation for 1 min at 94 °C, annealing for 1 min at 55 °C, and extension for 2 min at 72 °C. PCR reaction was carried out for 35 cycles and PCR products were visualized in 2 % agarose gels stained with ethidium bromide under UV transillumination.

The PCR product of ATM, the PCR product of ATR, and pGL3-Basic vehicle plasmid were digested with restriction enzyme *Kpn*I and *Xho*I at 37 °C for 1 h. These fragments of PCR product and pGL3-Basic vehicle plasmid were mixed with 2 μL T4 ligase buffer and 1 μL T4 DNA ligase, incubated at 16 °C for 24 h, and then transformed into competent *E. coli*. A single colony was picked and cultured in LB which contains ampicillin. These plasmids were extracted and sequenced.

65.2.3 Transfection and UV Treatment

A431 cells were seeded onto 24-well plates at the density of 1.5×10^4 cells/cm^2. A431 cells were transfected with 1 μg of ATM promoter luciferase reporter plasmid or ATR promoter luciferase reporter plasmid for 24 h using TurboFect reagent (Fermentas), respectively. Mainly, 1 μg total plasmids were added into 2 μL Turbo reagents, incubated for 20 min, and then added to the medium without serum. After transfection for 6 h, the medium without serum were replaced with the medium containing 10 % fetal calf serum. 24 h after transfection, cells were treated with 1, 2, 3, 4 and 5 J/m^2 UV stimulus. Untreatment by UV can be used as a negative control.

65.2.4 Luciferase Reporter Assays

After UV treatment 2 h, 50 μL of protein extracts (100 μL/well) were prepared for luciferase assays, which were measured by using a luciferase reporter assay system (Promega) on a Synergy™ 4 (Bioteck). All experiments were performed at least thrice with different preparations of plasmids and primary cells, producing qualitatively similar results. Values were normalized as the relative luciferase activity (fold). Columns represent the means of three independent experiments expressed relative to the luciferase activities obtained for untreated cells, which were arbitrarily defined as 1. The error bars represent the standard errors of the mean.

65.3 Result

65.3.1 The ATM Promoter and ATR Promoter Contained Key Sites

Based on the results of fragment competition test and homology, the ATM promoter contained some key sites which located at base pairs −739 to −730 (5′-ggggaactcc-3′), −601 to −590 (5′-ttccttccgaa-3′), −556 to −547 (5′-gggcttcccc-3′), and the ATR promoter contained some key sites which located at base pairs −470 to −460 (5′-ttcaagttgaa-3′), −433 to −423 (5′-ttcaagttgaa-3′). These regions of ATM and ATR promoter were similar to some sequences recognized by the proteins of NF-κB family and STAT3. Thus the NF-κB-like factors and STAT3 were involved in the protein-DNA interaction. It suggested that there might be various kinds of NF-κB-like factors and STAT3 involved in the regulation of ATM and ATR gene transcription at these key sites (Fig. 65.1).

65.3.2 Construction of ATM and ATR Promoter Luciferase Reporter Plasmids

To estimate the PCR amplification of ATM promoter and ATR promoter, agarose gel electrophoresis was performed. As shown in Fig. 65.2a, two bands emerged at the site of 914 and 620 bp, which represent PCR products of ATM promoter (from upstream 860 bp to downstream 53 bp) and ATR promoter (from upstream 504 bp to downstream 115 bp), respectively.

Then, these PCR products were digested by double restriction enzyme *Kpn*I and *Xho*I and cloned to the pGL3-Basic vector. Recombinant plasmids were extracted and purified, and agarose gel electrophoretic analysis was performed. Figure 65.2b represented agarose gel electrophoretic analysis of recombinant plasmids.

To confirm these recombinant plasmids, we digest these plasmids with two respective cloning restriction enzymes and then electrophoresed through agarose gel. As shown in Fig. 65.2c, these recombinant plasmids were cut into two bands,

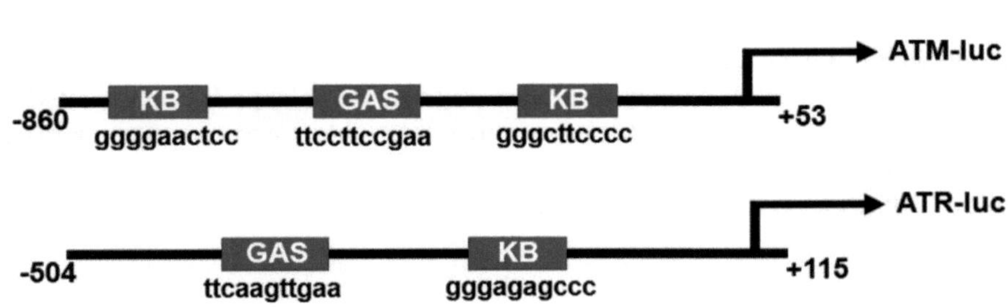

Fig. 65.1 The ATM promoter and ATR promoter contained key sites

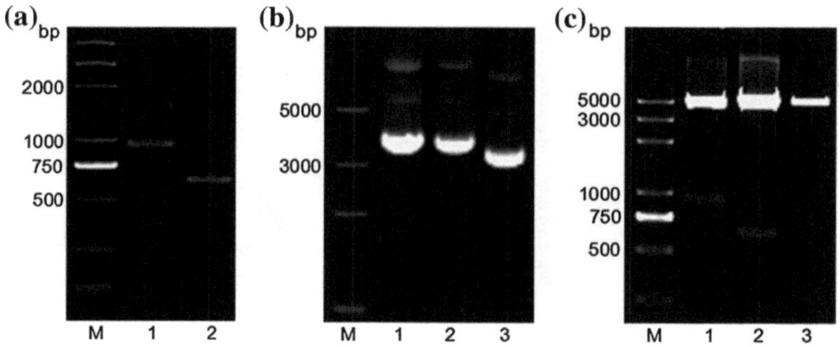

Fig. 65.2 Agarose gel electrophoretic analysis of ATM and ATR promoter luciferase reporter plasmids. **a** Agarose gel electrophoretic analysis of PCR product. *M* 1 kb DNA marker; *1* ATM gene promoter; *2* ATR gene promoter. **b** Agarose gel electrophoretic analysis of recombinant plasmids. *M* 1 kb DNA marker; *1* ATM recombinant plasmid; *2* ATR recombinant plasmid; *3* pGL3-Basic vehicle plasmid. **c** Agarose gel electrophoretic analysis of recombinant plasmids by digestion. *M* 1 kb DNA marker; *1* ATM recombinant plasmid; *2* ATR recombinant plasmid; *3* pGL3-Basic vehicle plasmid

respectively. The first lane contained two bands, about 3,000 and 914 bp, which represented pGL3-Basic vehicle plasmid and ATM promoter, respectively. The second lane contained two bands, about 3,000 and 620 bp, which represented pGL3-Basic vehicle plasmid and ATR promoter, respectively. These luciferase reporter plasmids of both ATM gene promoter and ATR gene promoter were further confirmed by sequencing. These results of DNA sequence alignments showed that luciferase reporter plasmid containing ATM promoter and ATR promoter were constructed successfully.

65.3.3 Luciferase Assay

65.3.3.1 The Efficiency of Transfection

The transfection efficiency was demonstrated using an EGFP (enhanced green fluorescent protein) expression plasmid. The transfection efficiency was approximately 60 %, demonstrating that DNA was transfected efficiently into A431 cells.

65.3.3.2 UV Stimulus Can Obviously Enhance the Transcriptional Activity of ATM Promoter

Luciferase Reporter Assay was performed to test the role of UV damage in regulating ATM transcription. Contrasting with control group untreated with UV stimulus, UV damage showed a significant effect to enhance the transcription activity of ATM promoter in a dose-dependent manner (Fig. 65.3).

Fig. 65.3 UV stimulus enhanced the transcriptional activity of ATM promoter

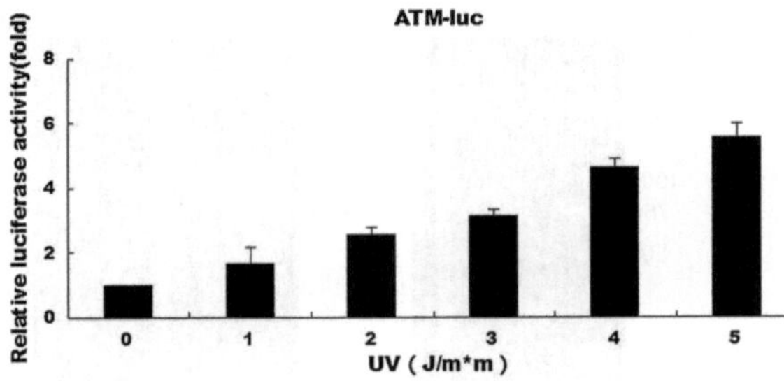

Fig. 65.4 UV stimulus enhanced the transcriptional activity of ATR promoter

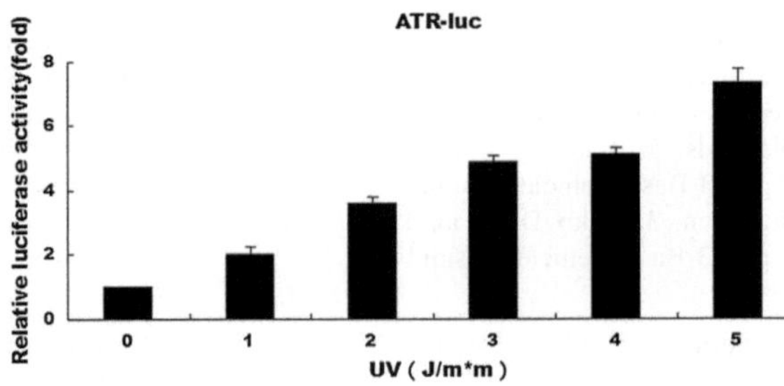

65.3.3.3 UV Stimulus Can Obviously Increase the Transcriptional Activity of ATR Promoter

Luciferase Reporter Assay was also performed to test the role of UV damage in regulating ATR transcription. Contrasting with control group untreated with UV stimulus, treatment of UV also increased the transcriptional activity of ATR promoter in a dose-dependent manner (Fig. 65.4).

65.4 Discussion

Various endogenous and exogenous agents, such as solar UV irradiation, continuously damage cellular DNA. If not repaired properly, this can result in mutations or chromosomal aberrations and may eventually trigger cancer. DNA damage responses are orchestrated by multiple signal transduction processes, key among which are the ATM/ATR pathways [15–17]. Activation of these pathways is crucial for the proper coordination of checkpoint and DNA repair processes. In recent years, it has become evident that DNA damage responses are central both for the evolution and therapy of cancer.

ATM and ATR are the master transducers of DNA signals, and they orchestrate a large network of cellular processes to maintain genomic integrity. These kinases are activated in response to DNA damage and subsequently phosphorylate targets responsible for such diverse activities as blocking cell cycle progression, coordinating DNA repair activities, and affecting transcription of DNA damage response genes. In response to DNA damage, hundreds of proteins are phosphorylated at Ser/Thr-Glu motifs and additional sites in an ATM- or ATR-dependent manner [18–21]. Therefore, the study of ATM/ATR is necessary, which will prompt us to study further.

These regions of ATM and ATR promoter are similar to some sequences recognized by the proteins of NF-κB family and STAT3 (the signal transducers and activators of transcription). Members of the NF-κB transcription factor family orchestrate a wide range of stress-like inflammatory responses, regulate developmental programs, and control the growth and survival of normal and malignant cells. STATs are identified as latent cytoplasmic transcription factors that are activated by cytokines and growth factors to mediate essential cellular processes such as cell growth, proliferation, and immune responses. There may be various kinds of NF-κB-like factors and STAT3 involved in the regulation of ATM and ATR gene transcription at these sites.

In this study, we successfully cloned ATM and ATR promoter luciferase reporter plasmids and found that UV damage could enhance the transcriptional activity of both ATM promoter and ATR promoter. Our studies will help to screen some novel transcription factors in regulating DNA damage and repair via ATM/ATR signal pathway.

Acknowledgments This work was financially supported by National Natural Science Foundation of China (No. 30970615, 31071126, 31000343, 31171303, 31171297, 31270837), Program for Changjiang Scholars and Innovative Research Team in University of Ministry of Education of China (IRT1166) and Research Fund for the Doctoral Program of Higher Education of China (20111208110001).

References

1. Elwood JM, Jopson J (1997) Melanoma and sun exposure: an overview of published studies. Int J Cancer 73:198–203
2. Godar DE (2005) UV doses worldwide. Photochem Photobiol 81:736–749
3. Wei Q, Lee JE, Gershenwald JE, Ross MI, Mansfield PF, Strom SS, Wang LE, Guo Z, Qiao Y, Amos CI et al (2003) Repair of UV light-induced DNA damage and risk of cutaneous malignant melanoma. J Natl Cancer Inst 95:308–315
4. Cleaver JE, Crowley E (2002) UV damage, DNA repair and skin carcinogenesis. Front Biosci 7:d1024–d1043
5. Krutmann J, Morita A, Chung JH (2012) Sun exposure: what molecular photodermatology tells us about its good and bad sides. J Invest Dermatol 132:976–984
6. Tibbetts RS, Cortez D, Brumbaugh KM, Scully R, Livingston D, Elledge SJ, Abraham RT (2000) Functional interactions between BRCA1 and the checkpoint kinase ATR during genotoxic stress. Genes Dev 14:2989–3002
7. Zhou BB, Elledge SJ (2000) The DNA damage response: putting checkpoints in perspective. Nature 408:433–439
8. Khanna KK, Lavin MF, Jackson SP, Mulhern TD (2001) ATM, a central controller of cellular responses to DNA damage. Cell Death Differ 8:1052–1065

9. Rouse J, Jackson SP (2002) Interfaces between the detection, signaling and repair of DNA damage. Science 297:547–551
10. O'Connell MJ, Walworth NC, Carr AM (2000) The G2-phase DNA-damage checkpoint. Trends Cell Biol 10:296–303
11. Shiloh Y (2001) ATM and ATR: networking cellular responses to DNA damage. Curr Opin Genet Dev 11:71–77
12. Jazayeri A, Falck J, Lukas C, Bartek J, Smith GC, Lukas J, Jackson SP (2006) ATM- and cell cycle-dependent regulation of ATR in response to DNA double-strand breaks. Nat Cell Biol 8:37–45
13. Muller C, Calsou P, Frit P, Cayrol C, Carter T, Salles B (1998) UV sensitivity and impaired nucleotide excision repair in DNA-dependent protein kinase mutant cells. Nucleic Acids Res 26:1382–1389
14. Hannan MA, Hellani A, Al-Khodairy FM, Kunhi M, Siddiqui Y, Al-Yussef N et al (2002) Deficiency in the repair of UV-induced DNA damage in human skin fibroblasts compromised for the ATM gene. Carcinogenesis 23:1617–1624
15. Cimprich KA, Cortez D (2008) ATR: an essential regulator of genome integrity. Nat Rev Mol Cell Biol 9:616–627
16. Guo Z et al (2000) Requirement for Atr in phosphorylation of Chk1 and cell cycle regulation in response to DNA replication blocks and UV-damaged DNA in Xenopus egg extracts. Genes Dev 14:2745–2756
17. Lavin MF, Kozlov S (2007) ATM activation and DNA damage response. Cell Cycle 6:931–942
18. Matsuoka S, Ballif BA, Smogorzewska A, McDonald ER III, Hurov KE, Luo J, Bakalarski CE, Zhao Z, Solimini N, Lerenthal Y et al (2007) ATM and ATR substrate analysis reveals extensive protein networks responsive to DNA damage. Science 316:1160–1166
19. Smolka MB, Albuquerque CP, Chen SH, Zhou H (2007) Proteome-wide identification of in vivo targets of DNA damage checkpoint kinases. Proc Natl Acad Sci 104:10364–10369
20. Stokes MP, Rush J, Macneill J, Ren JM, Sprott K, Nardone J, Yang V, Beausoleil SA, Gygi SP, Livingstone M et al (2007) Profiling of UV-induced ATM/ATR signaling pathways. Proc Natl Acad Sci 104:19855–19860
21. Bensimon A, Schmidt A, Ziv Y, Elkon R, Wang SY, Chen DJ, Aebersold R, Shiloh Y (2010) ATM-dependent and -independent dynamics of the nuclear phosphoproteome after DNA damage. Sci Signal 3:rs3

Erratum to: Characterization of Recombinant L-Amino Acid Deaminase of *Proteus mirabilis*

Chenglin Zhang, Jia Feng, Xixian Xie, Qingyang Xu and Ning Chen

Errataum to: Chapter 61 in: T.-C. Zhang and M. Nakajima (eds.), *Advances in Applied Biotechnology*, Lecture Notes in Electrical Engineering, DOI 10.1007/978-3-662-45657-6_61

Due to an unfortunate error, Table 61.2 in Chap. 61 Characterization of Recombinant L-Amino Acid Deaminase of *Proteus mirabilis* (Lecture Notes in Electrical Engineering, vol. 332, pp. 589–596) was mistaken.

The online version of the original chapter can be found under
DOI 10.1007/978-3-662-45657-6_61

C. Zhang · J. Feng · X. Xie · Q. Xu · N. Chen (✉)
College of Biotechnology, Tianjin University of Science and Technology,
Tianjin 300457, China
e-mail: ningch@tust.edu.cn

C. Zhang · X. Xie · Q. Xu · N. Chen
National and Local United Engineering, Lab of Metabolic Control Fermentation Technology,
Tianjin 300457, China

T.-C. Zhang and M. Nakajima (eds.), *Advances in Applied Biotechnology*,
Lecture Notes in Electrical Engineering 332, DOI 10.1007/978-3-662-45657-6_66

The correct table should be:

Table 61.2 Conversion rate of 20 L-amino acids by recombinant enzymes

L-amino acids	Conversion rate (%)	L-amino acids	Conversion rate (%)
L-Ala	14.37	L-Leu	2.54
L-Arg	55.25	L-Lys	10.19
L-Asn	8.63	L-Met	4.36
L-Asp	2.59	L-Phe	51.43
L-Cys	33.48	L-Pro	9.65
L-Glu	9.24	L-Ser	19.71
L-Gln	3.52	L-Thr	15.26
L-Gly	5.36	L-Trp	7.93
L-His	89.81	L-Tyr	4.96
L-Ile	2.47	L-Val	4.58

MIX
Papier aus verantwortungsvollen Quellen
Paper from responsible sources
FSC® C105338

If you have any concerns about our products, you can contact us on
ProductSafety@springernature.com

In case Publisher is established outside the EU, the EU authorized representative is:
Springer Nature Customer Service Center GmbH
Europaplatz 3, 69115 Heidelberg, Germany

Printed by Libri Plureos GmbH
in Hamburg, Germany